RECIPROCALS OF BASIC FUNCTIONS

18. $\displaystyle\int \frac{1}{1 \pm \sin u}\,du = \tan u \mp \sec u + C$

19. $\displaystyle\int \frac{1}{1 \pm \cos u}\,du = -\cot u \pm \csc u + C$

20. $\displaystyle\int \frac{1}{1 \pm \tan u}\,du = \tfrac{1}{2}(u \pm \ln|\cos u \pm \sin u|) + C$

21. $\displaystyle\int \frac{1}{\sin u \cos u}\,du = \ln|\tan u| + C$

22. $\displaystyle\int \frac{1}{1 \pm \cot u}\,du = \tfrac{1}{2}(u \mp \ln|\sin u \pm \cos u|) + C$

23. $\displaystyle\int \frac{1}{1 \pm \sec u}\,du = u + \cot u \mp \csc u + C$

24. $\displaystyle\int \frac{1}{1 \pm \csc u}\,du = u - \tan u \pm \sec u + C$

25. $\displaystyle\int \frac{1}{1 \pm e^u}\,du = u - \ln(1 \pm e^u) + C$

POWERS OF TRIGONOMETRIC FUNCTIONS

26. $\displaystyle\int \sin^2 u\,du = \tfrac{1}{2}u - \tfrac{1}{4}\sin 2u + C$

27. $\displaystyle\int \cos^2 u\,du = \tfrac{1}{2}u + \tfrac{1}{4}\sin 2u + C$

28. $\displaystyle\int \tan^2 u\,du = \tan u - u + C$

29. $\displaystyle\int \sin^n u\,du = -\frac{1}{n}\sin^{n-1} u \cos u + \frac{n-1}{n}\int \sin^{n-2} u\,du$

30. $\displaystyle\int \cos^n u\,du = \frac{1}{n}\cos^{n-1} u \sin u + \frac{n-1}{n}\int \cos^{n-2} u\,du$

31. $\displaystyle\int \tan^n u\,du = \frac{1}{n-1}\tan^{n-1} u - \int \tan^{n-2} u\,du$

32. $\displaystyle\int \cot^2 u\,du = -\cot u - u + C$

33. $\displaystyle\int \sec^2 u\,du = \tan u + C$

34. $\displaystyle\int \csc^2 u\,du = -\cot u + C$

35. $\displaystyle\int \cot^n u\,du = -\frac{1}{n-1}\cot^{n-1} u - \int \cot^{n-2} u\,du$

36. $\displaystyle\int \sec^n u\,du = \frac{1}{n-1}\sec^{n-2} u \tan u + \frac{n-2}{n-1}\int \sec^{n-2} u\,du$

37. $\displaystyle\int \csc^n u\,du = -\frac{1}{n-1}\csc^{n-2} u \cot u + \frac{n-2}{n-1}\int \csc^{n-2} u\,du$

PRODUCTS OF TRIGONOMETRIC FUNCTIONS

38. $\displaystyle\int \sin mu \sin nu\,du = -\frac{\sin(m+n)u}{2(m+n)} + \frac{\sin(m-n)u}{2(m-n)} + C$

39. $\displaystyle\int \cos mu \cos nu\,du = \frac{\sin(m+n)u}{2(m+n)} + \frac{\sin(m-n)u}{2(m-n)} + C$

40. $\displaystyle\int \sin mu \cos nu\,du = -\frac{\cos(m+n)u}{2(m+n)} - \frac{\cos(m-n)u}{2(m-n)} + C$

41. $\displaystyle\int \sin^m u \cos^n u\,du = -\frac{\sin^{m-1} u \cos^{n+1} u}{m+n} + \frac{m-1}{m+n}\int \sin^{m-2} u \cos^n u\,du$

$$= \frac{\sin^{m+1} u \cos^{n-1} u}{m+n} + \frac{n-1}{m+n}\int \sin^m u \cos^{n-2} u\,du$$

PRODUCTS OF TRIGONOMETRIC AND EXPONENTIAL FUNCTIONS

42. $\displaystyle\int e^{au}\sin bu\,du = \frac{e^{au}}{a^2+b^2}(a\sin bu - b\cos bu) + C$

43. $\displaystyle\int e^{au}\cos bu\,du = \frac{e^{au}}{a^2+b^2}(a\cos bu + b\sin bu) + C$

POWERS OF u MULTIPLYING OR DIVIDING BASIC FUNCTIONS

44. $\displaystyle\int u\sin u\,du = \sin u - u\cos u + C$

45. $\displaystyle\int u\cos u\,du = \cos u + u\sin u + C$

46. $\displaystyle\int u^2\sin u\,du = 2u\sin u + (2-u^2)\cos u + C$

47. $\displaystyle\int u^2\cos u\,du = 2u\cos u + (u^2-2)\sin u + C$

48. $\displaystyle\int u^n\sin u\,du = -u^n\cos u + n\int u^{n-1}\cos u\,du$

49. $\displaystyle\int u^n\cos u\,du = u^n\sin u - n\int u^{n-1}\sin u\,du$

50. $\displaystyle\int u^n\ln u\,du = \frac{u^{n+1}}{(n+1)^2}[(n+1)\ln u - 1] + C$

51. $\displaystyle\int ue^u\,du = e^u(u-1) + C$

52. $\displaystyle\int u^n e^u\,du = u^n e^u - n\int u^{n-1}e^u\,du$

53. $\displaystyle\int u^n a^u\,du = \frac{u^n a^u}{\ln a} - \frac{n}{\ln a}\int u^{n-1}a^u\,du + C$

54. $\displaystyle\int \frac{e^u\,du}{u^n} = -\frac{e^u}{(n-1)u^{n-1}} + \frac{1}{n-1}\int \frac{e^u\,du}{u^{n-1}}$

55. $\displaystyle\int \frac{a^u\,du}{u^n} = -\frac{a^u}{(n-1)u^{n-1}} + \frac{\ln a}{n-1}\int \frac{a^u\,du}{u^{n-1}}$

56. $\displaystyle\int \frac{du}{u\ln u} = \ln|\ln u| + C$

POLYNOMIALS MULTIPLYING BASIC FUNCTIONS

57. $\displaystyle\int p(u)e^{au}\,du = \frac{1}{a}p(u)e^{au} - \frac{1}{a^2}p'(u)e^{au} + \frac{1}{a^3}p''(u)e^{au} - \cdots$ [signs alternate: $+ - + - \cdots$]

58. $\displaystyle\int p(u)\sin au\,du = -\frac{1}{a}p(u)\cos au + \frac{1}{a^2}p'(u)\sin au + \frac{1}{a^3}p''(u)\cos au - \cdots$ [signs alternate in pairs after first term: $+ + - - + + - - \cdots$]

59. $\displaystyle\int p(u)\cos au\,du = \frac{1}{a}p(u)\sin au + \frac{1}{a^2}p'(u)\cos au - \frac{1}{a^3}p''(u)\sin au - \cdots$ [signs alternate in pairs: $+ + - - +$

eGrade Plus

www.wiley.com/college/anton

Based on the Activities You Do Every Day

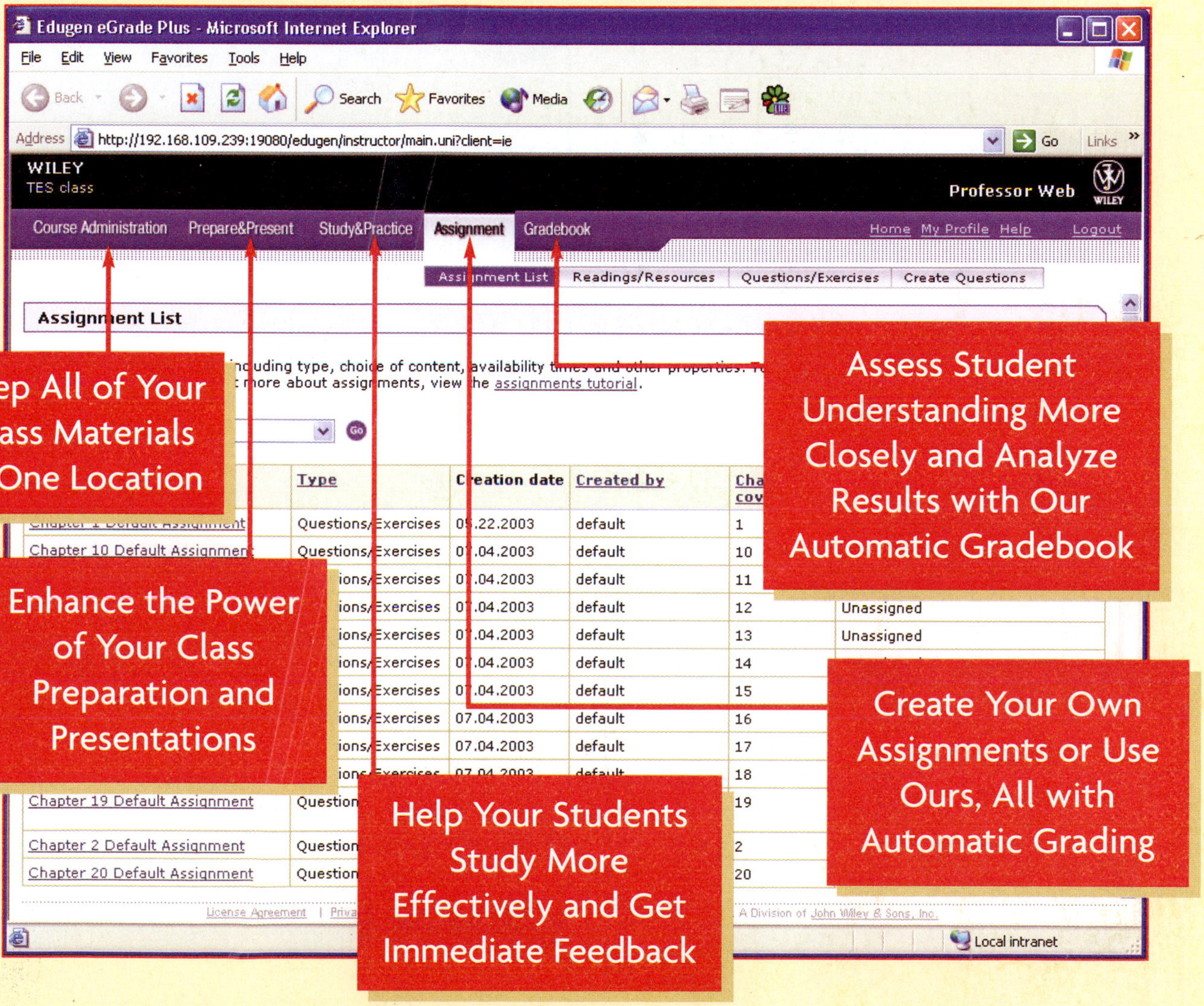

Edugen eGrade Plus - Microsoft Internet Explorer

File Edit View Favorites Tools Help

Back Search Favorites Media

Address http://192.168.109.239:19080/edugen/instructor/main.uni?client=ie Go Links

WILEY
TES class

Professor Web

Course Administration | Prepare&Present | Study&Practice | Assignment | Gradebook

Home My Profile Help Logout

Assignment List | Readings/Resources | Questions/Exercises | Create Questions

Assignment List

...ncluding type, choice of content, availability times and other properties. ...
... more about assignments, view the assignments tutorial.

	Type	Creation date	Created by	Cha cov	
Chapter 1 Default Assignment	Questions/Exercises	0 .22.2003	default	1	
Chapter 10 Default Assignment	Questions/Exercises	0 .04.2003	default	10	
	ions/Exercises	0 .04.2003	default	11	
	ions/Exercises	0 .04.2003	default	12	Unassigned
	ions/Exercises	0 .04.2003	default	13	Unassigned
	ions/Exercises	0 .04.2003	default	14	
	ions/Exercises	0 .04.2003	default	15	
	ions/Exercises	07.04.2003	default	16	
	ions/Exercises	07.04.2003	default	17	
	ions/Exercises	07.04.2003	default	18	
Chapter 19 Default Assignment	Question			19	
Chapter 2 Default Assignment	Question			2	
Chapter 20 Default Assignment	Question			20	

License Agreement | Priva... A Division of John Wiley & Sons, Inc.

Local intranet

Keep All of Your Class Materials in One Location

Enhance the Power of Your Class Preparation and Presentations

Help Your Students Study More Effectively and Get Immediate Feedback

Assess Student Understanding More Closely and Analyze Results with Our Automatic Gradebook

Create Your Own Assignments or Use Ours, All with Automatic Grading

All the content and tools you need, all in one location, in an easy-to-use browser format.

Choose the resources you need, or rely on the arrangement supplied by us.

Now, many of Wiley's textbooks are available with eGrade Plus, a powerful online tool that provides a completely integrated suite of teaching and learning resources in one easy-to-use website. eGrade Plus integrates Wiley's world-renowned content with media, including a multimedia version of the text. Upon adoption of eGrade Plus, you can begin to customize your course with the resources shown here.

See for yourself! Go to www.wiley.com/college/egradeplus for an online demonstration of this powerful new software.

Students,
eGrade Plus Allows You to:

Study More Effectively

Get Immediate Feedback When You Practice on Your Own

eGrade Plus problems link directly to relevant sections of the **electronic book content,** so that you can review the text while you study and complete homework online. Additional resources include **hyperlinks to the Student Study Guide, the Student Solutions Manual, calculus explorations,** and **Calculus Solutions, powered by JustAsk!**

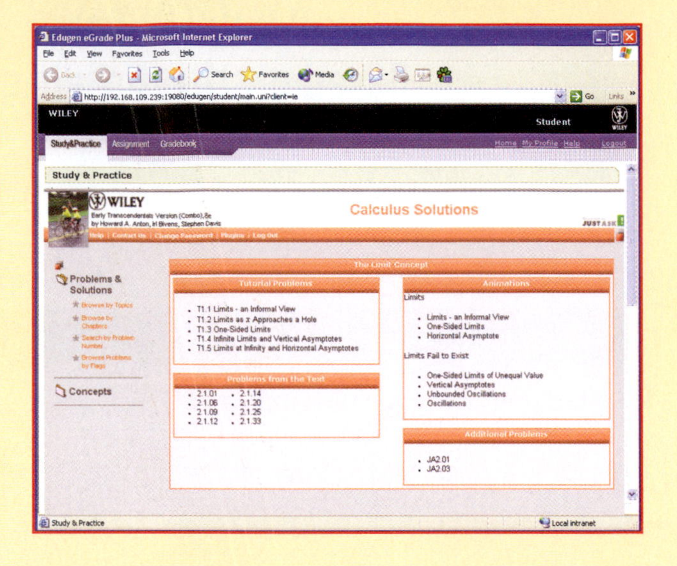

Complete Assignments / Get Help with Problem Solving

An **Assignment** area keeps all your assigned work in one location, making it easy for you to stay "on task." In addition, many homework problems contain a **link** to the relevant section of the **multimedia book,** providing you with a text explanation to help you conquer problem-solving obstacles as they arise. You will have access to a variety of resources for building your confidence and understanding.

Assignment List

Create or edit Assignments, including type, choice of content, availability times and other properties. To edit settings for an Assignment, select the Assignment name. To find out more about assignments, view the assignments tutorial.

Assignment name	Type	Creation date	Created by	Chapters covered	Status
Chapter 1 Default Assignment	Questions/Exercises	05.22.2003	default	1	Unassigned
Chapter 10 Default Assignment	Questions/Exercises	07.04.2003	default	10	Unassigned
Chapter 11 Default Assignment	Questions/Exercises	07.04.2003	default	11	Unassigned
Chapter 12 Default Assignment	Questions/Exercises	07.04.2003	default	12	Unassigned
Chapter 13 Default Assignment	Questions/Exercises	07.04.2003	default	13	Unassigned
Chapter 14 Default Assignment	Questions/Exercises	07.04.2003	default	14	Unassigned
Chapter 15 Default Assignment	Questions/Exercises	07.04.2003	default	15	Unassigned
Chapter 16 Default Assignment	Questions/Exercises	07.04.2003	default	16	Unassigned
Chapter 17 Default Assignment	Questions/Exercises	07.04.2003	default	17	Unassigned
Chapter 18 Default Assignment	Questions/Exercises	07.04.2003	default	18	Unassigned
Chapter 19 Default Assignment	Questions/Exercises	07.04.2003	default	19	Assigned by Professor Kimmel, past due
Chapter 2 Default Assignment	Questions/Exercises	07.04.2003	default	2	Unassigned
Chapter 20 Default Assignment	Questions/Exercises	07.04.2003	default	20	Unassigned

Keep Track of How You're Doing

A **Personal Gradebook** allows you to view your results from past assignments at any time.

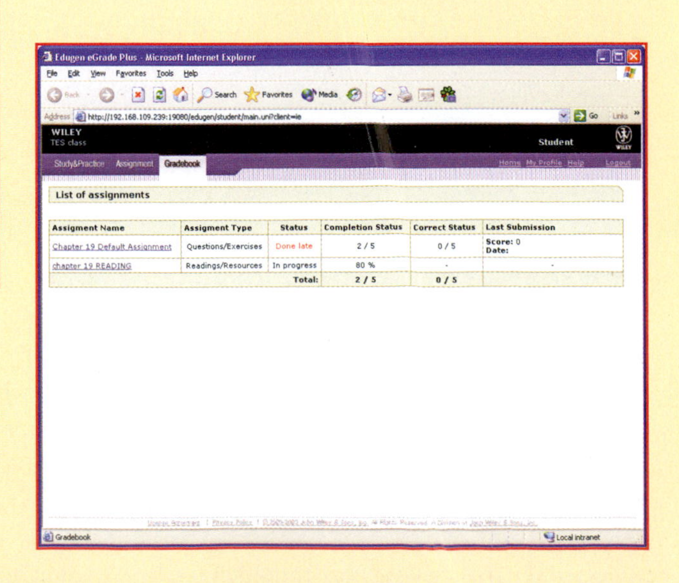

Calculus

Multivariable

eighth edition

Calculus

MULTIVARIABLE

HOWARD ANTON

Drexel University

IRL BIVENS

Davidson College

STEPHEN DAVIS

Davidson College

WILEY

JOHN WILEY & SONS, INC.

Associate Publisher: Laurie Rosatone
Freelance Developmental Editor: Anne Scanlan-Rohrer
Senior Marketing Manager: Angela Battle
Associate Editor: Jennifer Battista
Editorial Assistants: Danielle Amico/Kelly Boyle
Senior Production Editor: Ken Santor
Senior Designer: Karin Kincheloe
Cover Design: David Levy
Cover Photo: © Arthur Tilley/Taxi Getty Images
Text Design: Nancy Field
Photo Editor: Hilary Newman/Ellinor Wagoner
Illustration Editor: Sigmund Malinowski
Illustration Studio: Techsetters, Inc.

This book was set in Times Roman by Techsetters, Inc., and printed and bound by Von Hoffmann Press. The cover was printed by Von Hoffmann Press.

This book is printed on acid-free paper. ♾

The paper in this book was manufactured by a mill whose forest management programs include sustained yield harvesting of its timberlands. Sustained yield harvesting principles ensure that the numbers of trees cut each year does not exceed the amount of new growth.

ISBN 0-471-48237-4

Printed in the United States of America

10 9 8 7 6 5 4 3 2 1

ABOUT HOWARD ANTON

Howard Anton obtained his B.A. from Lehigh University, his M.A. from the University of Illinois, and his Ph.D. from the Polytechnic University of Brooklyn, all in mathematics. In the early 1960s he worked for Burroughs Corporation and Avco Corporation at Cape Canaveral, Florida, where he was involved with the manned space program. In 1968 he joined the Mathematics Department at Drexel University, where he taught full time until 1983. Since that time he has been an adjunct professor at Drexel and has devoted the majority of his time to textbook writing and activities for mathematical associations. Dr. Anton was president of the EPADEL Section of the Mathematical Association of America (MAA), served on the board of Governors of that organization, and guided the creation of the Student Chapters of the MAA. He has published numerous research papers in functional analysis, approximation theory, and topology, as well as pedagogical papers. He is best known for his textbooks in mathematics, which are among the most widely used in the world. There are currently more than one hundred versions of his books, including translations into Spanish, Arabic, Portuguese, Italian, Indonesian, French, Japanese, Chinese, Hebrew, and German. For relaxation, Dr. Anton enjoys traveling and photography.

ABOUT IRL BIVENS

Irl C. Bivens, recipient of the George Polya Award and the Merten M. Hasse Prize for Expository Writing in Mathematics, received his A.B. from Pfeiffer College and his Ph.D. from the University of North Carolina at Chapel Hill, both in mathematics. Since 1982, he has taught at Davidson College, where he currently holds the position of professor of mathematics. A typical academic year sees him teaching courses in calculus, topology, and geometry. Dr. Bivens also enjoys mathematical history, and his annual History of Mathematics seminar is a perennial favorite with Davidson mathematics majors. He has published numerous articles on undergraduate mathematics, as well as research papers in his specialty, differential geometry. He is currently a member of the editorial board for the MAA Problem Book series and is a reviewer for *Mathematical Reviews*. When he is not pursuing mathematics, Professor Bivens enjoys juggling, swimming, walking, and spending time with his son Robert.

ABOUT STEPHEN DAVIS

Stephen L. Davis received his B.A. from Lindenwood College and his Ph.D. from Rutgers University in mathematics. Having previously taught at Rutgers University and Ohio State University, Dr. Davis came to Davidson College in 1981, where he is currently a professor of mathematics. He regularly teaches calculus, linear algebra, abstract algebra, and computer science. A sabbatical in 1995–1996 took him to Swarthmore College as a visiting associate professor. Professor Davis has published numerous articles on calculus reform and testing, as well as research papers on finite group theory, his specialty. Professor Davis has held several offices in the Southeastern section of the MAA, including chair and secretary-treasurer. He is currently a faculty consultant for the Educational Testing Service Advanced Placement Calculus Test, a board member of the North Carolina Association of Advanced Placement Mathematics Teachers, and is actively involved in nurturing mathematically talented high school students through leadership in the Charlotte Mathematics Club. He was formerly North Carolina state director for the MAA. For relaxation, he plays basketball, juggles, and travels. Professor Davis and his wife Elisabeth have three children, Laura, Anne, and James, all former calculus students.

To
My Wife Pat
My Children: Brian, David, and Lauren

In Memory of
My Mother Shirley
My Father Benjamin
My Esteemed Colleague Albert Herr
My Benefactor Stephen Girard (1750–1831)

—HA

To
My Son Robert

—IB

To
My Wife Elisabeth
My Children: Laura, Anne, and James

—SD

PREFACE

ABOUT THIS EDITION

A major focus of this edition was to ***increase student comprehension*** through judicious streamlining of the exposition; the creation of new problem types, particularly the Quick Check and Focus on Concepts exercises; and revision of many examples to add more steps and reformat them for clarity.

Technology This edition provides many examples and exercises for instructors who want to use graphing calculators, computer algebra systems, or other programs. However, these are implemented in a way that allows the text to be used in courses where technology is used extensively, moderately, or not at all. New ***Technology Mastery*** comments direct students to useful "just in time" applications of technology. Exercises that require technology are marked with icons for easy identification.

Internet This text is supplemented by a Web site:

www.wiley.com/college/anton

Streamlined Exposition Every page, every explanation, and every example in the text was examined critically and the exposition was streamlined, where needed, to get students right to the heart of concepts. Many examples have been revised to make them clearer and more inviting. In addition, all appendices from the seventh edition were moved to the companion Web site. *Expanding the Calculus Horizon* modules are now posted on the text's Web site, and students are still directed to the modules by a preview paragraph and Web link at the end of appropriate chapters in the text.

NEW FEATURES IN THE EIGHTH EDITION

New and Updated Exercises

- New ***Quick Check Exercises*** starting each section's exercise set contain a basic set of 4–8 exercises designed to cover key skills and concepts in the section. Students can use these as a concise way of testing their knowledge of each section. Answers to the Quick Check exercises appear at the end of each section.
- New ***Focus on Concepts*** throughout each exercise set highlight exercises of a conceptual nature.

- ***Review Exercises*** replace the Supplementary Exercises at the end of each chapter. A selection of these exercises can be used to review important concepts within the chapter or to construct a chapter test. In addition, exercise sets were revised and expanded to include more variety and better pairings between odd and even exercises.

Margin Notes General margin comments call attention to ideas in the text or provide further insights. These general comments and the ***Technology Mastery*** comments replace the *For the Reader* comments from previous editions.

Derivative Notation The notation for the definition of the derivative has been brought into alignment with that used in standard calculus texts.

OTHER FEATURES

Flexibility This edition has a built-in flexibility that is designed to serve a broad spectrum of calculus philosophies—from traditional to reform. Technology can be emphasized or not, and the order of many topics can be permuted freely to accommodate the instructor's specific needs.

Trigonometry Review Deficiencies in trigonometry plague many students, so we have included a substantial trigonometry review in Appendix A.

Historical Notes The biographies and historical notes have been a hallmark of this text from its first edition and have been maintained. All of the biographical materials have been distilled from standard sources with the goal of capturing the personalities of the great mathematicians and bringing them to life for the students.

Graded Exercise Sets Section Exercise Sets are "graded" to begin with routine problems and progress gradually toward problems of greater difficulty.

Rigor The challenge of writing a good calculus book is to strike the right balance between rigor and clarity. Our goal is to present precise mathematics to the fullest extent possible in an introductory treatment. Where clarity and rigor conflict, we choose clarity; however, we believe it to be important that the student understand the difference between a careful proof and an informal argument, so we have tried to make it clear to the reader when the arguments being presented are informal or motivational. Theory involving ϵ-δ arguments appear in separate sections so that they can be covered or not, as preferred by the instructor.

Mathematical Level This text is written at a mathematical level that will prepare students for a wide variety of careers that require a sound mathematics background, including engineering, the various sciences, and business.

Computer Graphics This edition makes extensive use of modern computer graphics to clarify concepts and to develop the student's ability to visualize mathematical objects, particularly those in 3-space. For those students who are working with graphing technology, there are many exercises that are designed to develop the student's ability to generate and analyze mathematical curves and surfaces.

Applicability of Calculus One of the primary goals of this edition is to link calculus to the real world and the student's own experience. This theme is carried through in the examples, exercises, and modules. Applications given in the exercises have been chosen to provide the student a sense of how calculus can be applied.

Principles of Integral Evaluation The traditional Techniques of Integration is entitled "Principles of Integral Evaluation" to reflect its more modern approach to the material. The chapter emphasizes general methods and the role of technology rather than specific tricks for evaluating complicated or obscure integrals.

Appendix on Polynomial Equations Because many calculus students are weak in solving polynomial equations, we have included an appendix (Appendix B) that reviews the Factor Theorem, the Remainder Theorem, and procedures for finding rational roots.

Rule of Four The "rule of four" refers to presenting concepts from the verbal, algebraic, visual, and numerical points of view. In keeping with current pedagogical philosophy, we used this approach whenever appropriate.

SUPPLEMENTS

SUPPLEMENTS FOR THE STUDENT

Student Solutions Manual, Neil Wigley
The Student Solutions Manual provides students with detailed solutions to odd-numbered exercises from the text.
Multivariable ISBN: 0-471-67212-2

Student Study Guide, Brian Camp
The Student Study Guide contains key ideas and study suggestions, as well as sample tests for each section and chapter of the text.
Multivariable ISBN: 0-471-67213-0

SUPPLEMENTS FOR THE INSTRUCTOR

SUPPLEMENTS FOR THE INSTRUCTOR CAN BE OBTAINED BY SENDING A REQUEST ON YOUR INSTITUTIONAL LETTERHEAD TO MATHEMATICS MARKETING MANAGER, JOHN WILEY & SONS, INC., 111 RIVER STREET, HOBOKEN, NJ 07030, OR BY CONTACTING YOUR LOCAL WILEY REPRESENTATIVE.

Instructor's Manual, Irl Bivens and Stephen Davis
The Instructor's Manual provides suggested time allocations and teaching plans for each section in the text. Most of the teaching plans contain a bulleted list of key points to emphasize. The discussion of each section concludes with a sample homework assignment.
ISBN: 0-471-67207-6

Instructor's Solutions Manual, Neil Wigley
The Instructor's Solutions Manual contains detailed solutions to all exercises in the text.
Multivariable ISBN: 0-471-72429-7

Test Bank, Henry Smith
The Test Bank contains a variety of questions and answers for every section in the text.
ISBN: 0-471-67209-2

FOR THE STUDENT AND THE INSTRUCTOR

Web Horizon Modules

Selected chapters end with references to Web modules called *Expanding the Calculus Horizon*. As the name implies, these modules are intended to take the student a step beyond the traditional calculus text. The modules, all of which are optional, can be assigned either as individual or group projects and can be used by instructors to tailor the calculus course to meet their specific needs and teaching philosophies. For example, there are modules that touch on iteration and dynamical systems, equations of motion, application of integration to railroad design, collision of comets with Earth, and hurricane modeling. These can be found on the Web site,

www.wiley.com/college/anton

OTHER RESOURCES

eGrade Plus is a powerful online tool that provides instructors and students with an integrated suite of teaching and learning resources in one easy-to-use Web site. eGrade Plus is organized around the essential activities you and your students perform in class:

For Instructors

- **Prepare & Present:** Create class presentations using a wealth of Wiley-provided resources—such as an online version of the textbook, PowerPoint slides, and interactive simulations—making your preparation time more efficient. You may easily adapt, customize, and add to this content to meet the needs of your course.
- **Create Assignments:** Automate the assigning and grading of homework or quizzes by using Wiley-provided question banks, or by writing your own. Student results will be automatically graded and recorded in your gradebook. eGrade Plus can link homework problems to the relevant section of the online text, providing students with context-sensitive help.
- **Track Student Progress:** Keep track of your students' progress via an instructor's gradebook, which allows you to analyze individual and overall class results to determine their progress and level of understanding.
- **Administer Your Course:** eGrade Plus can easily be integrated with another course management system, gradebook, or other resources you are using in your class, providing you with the flexibility to build your course, your way.

For Students

Wiley's eGrade Plus provides immediate feedback on student assignments and a wealth of support materials. This powerful study tool will help your students develop their conceptual understanding of the class material and increase their ability to solve problems.

- **A "Study and Practice"** area links directly to text content, allowing students to review the text while they study and complete homework assignments. This package includes the following:
  - **Calculus Solutions powered by JustAsk!(TM)** include problems that correlate to chapter materials, interactive tutorials, detailed solutions and answers, and solution guidelines.

- **Calculus Explorations** comprise a series of interactive Java applets that allow students to explore the geometric significance of many major concepts of Calculus 1.
- **Algebra & Trigonometry Refresher** is a self-paced, guided review of key algebra and trigonometry topics that are essential for mastering calculus.
- **Student Solutions Manual** contains detailed solutions to selected problems in the text.
- **Student Study Guide** offers study hints and tips, key ideas and concepts, and sample quizzes and tests.
- **Calculus WebQuizzes** provide opportunity for student self-assessment.

- **An "Assignment"** area keeps all the work you want your students to complete in one location, making it easy for them to stay "on task." Students will have access to a variety of interactive problem-solving tools, as well as other resources for building their confidence and understanding. In addition, many homework problems contain a link to the relevant section of the multimedia book, providing students with context-sensitive help that allows them to conquer problem-solving obstacles as they arise.
- **A Personal Gradebook** for each student will allow students to view their results from past assignments at any time.

Please visit **www.wiley.com/college/anton**, or view our online demo at **www.wiley.com/college/egradeplus**. Here you will find additional information about the features and benefits of eGrade Plus, how to request a "test drive" of eGrade Plus for this title, and how to adopt it for class use.

The Faculty Resource Network The *Faculty Resource Network* is a peer-to-peer network of academic faculty dedicated to the effective use of technology in the classroom. This group can help you apply innovative classroom techniques, implement specific software packages, and tailor the technology experience to the specific needs of each individual class. Ask your Wiley representative for more details.

ACKNOWLEDGMENTS

It has been our good fortune to have the advice and guidance of many talented people whose knowledge and skills have enhanced this book in many ways. For their valuable help we thank:

REVIEWERS AND CONTRIBUTORS TO THE EIGHTH EDITION

Gregory Adams, *Bucknell University*
Bill Allen, *Reedley College–Clovis Center*
Jerry Allison, *Black Hawk College*
Stella Ashford, *Southern University and A&M College*
Christopher Barker, *San Joaquin Delta College*
David Bradley, *University of Maine*
Paul Britt, *Louisiana State University*
Andrew Bulleri, *Howard Community College*
Miriam Castroconde, *Irvine Valley College*
Neena Chopra, *The Pennsylvania State University*
Gaemus Collins, *University of California, San Diego*
Danielle Cross, *Northern Essex Community College*
Stephan DeLong, *Tidewater Community College–Virginia Beach Campus*
Ryness Doherty, *Community College of Denver*
T. J. Duda, *Columbus State Community College*
Peter Embalabala, *Lincoln Land Community College*
Laurene Fausett, *Georgia Southern University*
Richard Hall, *Cochise College*
Noal Harbertson, *California State University, Fresno*
Donald Hartig, *California Polytechnic State University*
Konrad Heuvers, *Michigan Technological University*
John Johnson, *George Fox University*
Grant Karamyan, *University of California, Los Angeles*
Cecilia Knoll, *Florida Institute of Technology*
Carole King Krueger, *The University of Texas at Arlington*
Richard Lane, *University of Montana*
James Martin, *Wake Technical Community College*
Vania Mascioni, *Ball State University*
Tamra Mason, *Albuquerque TVI Community College*
Roy Mathias, *The College of William & Mary*
John Michaels, *SUNY Brockport*
Darrell Minor, *Columbus State Community College*
Darren Narayan, *Rochester Institute of Technology*
Efton Park, *Texas Christian University*
Joanne Peeples, *El Paso Community College*
Richard Ponticelli, *North Shore Community College*
Holly Puterbaugh, *University of Vermont*
Robert Rock, *Daniel Webster College*
John Saccoman, *Seton Hall University*
Paul Seeburger, *Monroe Community College*
Charlotte Simmons, *University of Central Oklahoma*
Bryan Stewart, *Tarrant County College–Southeast Campus*
Bradley Stoll, *The Harker School*
Eleanor Storey, *Front Range Community College*
Richard Swanson, *Montana State University*
Helen Tyler, *Manhattan College*
Paramanathan Varatharajah, *North Carolina A&T State University*
David Voss, *Western Illinois University*
Jim Voss, *Front Range Community College*
Richard Watkins, *Tidewater Community College*
Jane West, *Trident Technical College*
Janine Wittwer, *Williams College*
Richard Zang, *University of New Hampshire*
Diane Zych, *Erie Community College–North Campus*

REVIEWERS AND CONTRIBUTORS TO EARLIER EDITIONS

Edith Ainsworth, *University of Alabama*
Loren Argabright, *Drexel University*
David Armacost, *Amherst College*
Dan Arndt, *University of Texas at Dallas*
Ajay Arora, *McMaster University*
Mary Lane Baggett, *University of Mississippi*
John Bailey, *Clark State Community College*
Robert C. Banash, *St. Ambrose University*
William H. Barker, *Bowdoin College*
George R. Barnes, *University of Louisville*
Scott E. Barnett, *Wayne State University*
Larry Bates, *University of Calgary*
John P. Beckwith, *Michigan Technological University*
Joan E. Bell, *Northeastern Oklahoma State University*
Harry N. Bixler, *Baruch College, CUNY*
Kbenesh Blayneh, *Florida A&M University*
Marilyn Blockus, *San Jose State University*
Ray Boersma, *Front Range Community College*
Barbara Bohannon, *Hofstra University*
David Bolen, *Virginia Military Institute*
Daniel Bonar, *Denison University*
George W. Booth, *Brooklyn College*

Phyllis Boutilier, *Michigan Technological University*
Linda Bridge, *Long Beach City College*
Mark Bridger, *Northeastern University*
Judith Broadwin, *Jericho High School*
John Brothers, *Indiana University*
Stephen L. Brown, *Olivet Nazarene University*
Virginia Buchanan, *Hiram College*
Robert C. Bucker, *Western Kentucky University*
Robert Bumcrot, *Hofstra University*
Christopher Butler, *Case Western Reserve University*
Carlos E. Caballero, *Winthrop University*
Cheryl Cantwell, *Seminole Community College*
James Caristi, *Valparaiso University*
Judith Carter, *North Shore Community College*
Stan R. Chadick, *Northwestern State University*
Hongwei Chen, *Christopher Newport University*
Chris Christensen, *Northern Kentucky University*
Robert D. Cismowski, *San Bernardino Valley College*
Patricia Clark, *Rochester Institute of Technology*
Hannah Clavner, *Drexel University*
Ted Clinkenbeard, *Des Moines Area Community College*
David Clydesdale, *Sauk Valley Community College*
David Cohen, *University of California, Los Angeles*
Michael Cohen, *Hofstra University*
Pasquale Condo, *University of Lowell*
Robert Conley, *Precision Visuals*
Mary Ann Connors, *U.S. Military Academy at West Point*
Cecil J. Coone, *State Technical Institute at Memphis*
Norman Cornish, *University of Detroit*
Fielden Cox, *Centennial College*
Terrance Cremeans, *Oakland Community College*
Gary Crown, *Wichita State University*
Lawrence Cusick, *California State University–Fresno*
Michael Dagg, *Numerical Solutions, Inc.*
Art Davis, *San Jose State University*
A. L. Deal, *Virginia Military Institute*
Charles Denlinger, *Millersville University*
William H. Dent, *Maryville College*
Blaise DeSesa, *Allentown College of St. Francis de Sales*
Blaise DeSesa, *Drexel University*
Debbie A. Desrochers, *Napa Valley College*
Dennis DeTurck, *University of Pennsylvania*
Jacqueline Dewar, *Loyola Marymount University*
Preston Dinkins, *Southern University*
Gloria S. Dion, *Educational Testing Service*
Irving Drooyan, *Los Angeles Pierce College*
Tom Drouet, *East Los Angeles College*
Clyde Dubbs, *New Mexico Institute of Mining and Technology*
Della Duncan, *California State University–Fresno*
Ken Dunn, *Dalhousie University*
Sheldon Dyck, *Waterloo Maple Software*
Hugh B. Easler, *College of William and Mary*
Scott Eckert, *Cuyamaca College*
Joseph M. Egar, *Cleveland State University*
Judith Elkins, *Sweet Briar College*
Brett Elliott, *Southeastern Oklahoma State University*
William D. Emerson, *Metropolitan State College*
Garret J. Etgen, *University of Houston*
Benny Evans, *Oklahoma State University*
Philip Farmer, *Diablo Valley College*
Victor Feser, *University of Maryland*
Iris Brann Fetta, *Clemson University*
James H. Fife, *Educational Testing Service*
Sally E. Fischbeck, *Rochester Institute of Technology*
Dorothy M. Fitzgerald, *Golden West College*
Barbara Flajnik, *Virginia Military Institute*
Daniel Flath, *University of South Alabama*
Ernesto Franco, *California State University–Fresno*
Nicholas E. Frangos, *Hofstra University*
Katherine Franklin, *Los Angeles Pierce College*
Marc Frantz, *Indiana University–Purdue University at Indianapolis*
Michael Frantz, *University of La Verne*
Susan L. Friedman, *Bernard M. Baruch College, CUNY*
William R. Fuller, *Purdue University*
Beverly Fusfield
Daniel B. Gallup, *Pasadena City College*
Bradley E. Garner, *Boise State University*
Carrie Garner
Susan Gerstein
Mahmood Ghamsary, *Long Beach City College*
Rob Gilchrist, *U.S. Air Force Academy*
G. S. Gill, *Brigham Young University*
Michael Gilpin, *Michigan Technological University*
Kaplana Godbole, *Michigan Technological Institute*
S. B. Gokhale, *Western Illinois University*
Morton Goldberg, *Broome Community College*
Mardechai Goodman, *Rosary College*
Sid Graham, *Michigan Technological University*
Bob Grant, *Mesa Community College*
Raymond Greenwell, *Hofstra University*
Dixie Griffin, Jr., *Louisiana Tech University*
Gary Grimes, *Mt. Hood Community College*
David Gross, *University of Connecticut*
Jane Grossman, *University of Lowell*
Michael Grossman, *University of Lowell*
Dennis Hadah, *Saddleback Community College*
Diane Hagglund, *Waterloo Maple Software*
Douglas W. Hall, *Michigan State University*
Nancy A. Harrington, *University of Lowell*
Kent Harris, *Western Illinois University*
Karl Havlak, *Angelo State University*
J. Derrick Head, *University of Minnesota–Morris*
Jim Hefferson, *St. Michael College*
Albert Herr, *Drexel University*
Peter Herron, *Suffolk County Community College*
Warland R. Hersey, *North Shore Community College*
Konrad J. Heuvers, *Michigan Technological University*
Dean Hickerson
Robert Higgins, *Quantics Corporation*
Rebecca Hill, *Rochester Institute of Technology*
Tommie Ann Hill-Natter, *Prairie View A&M University*
Holly Hirst, *Appalachian State University*
Edwin Hoefer, *Rochester Institute of Technology*
Louis F. Hoelzle, *Bucks County Community College*
Robert Homolka, *Kansas State University–Salina*
Henry Horton, *University of West Florida*
Joe Howe, *St. Charles County Community College*
Shirley Huffman, *Southwest Missouri State University*
Hugh E. Huntley, *University of Michigan*
Fatenah Issa, *Loyola University of Chicago*
Gary S. Itzkowitz, *Rowan University*
Emmett Johnson, *Grambling State University*
Jerry Johnson, *University of Nevada–Reno*
John M. Johnson, *George Fox College*
Wells R. Johnson, *Bowdoin College*
Kenneth Kalmanson, *Montclair State University*
Herbert Kasube, *Bradley University*
Phil Kavanagh, *Mesa State College*
David Keller, *Kirkwood Community College*
Maureen Kelley, *Northern Essex Community College*
Dan Kemp, *South Dakota State University*
Harvey B. Keynes, *University of Minnesota*
Lynn Kiaer, *Rose-Hulman Institute of Technology*
Vesna Kilibarda, *Indiana University Northwest*
Cecilia Knoll, *Florida Institute of Technology*
Holly A. Kresch, *Diablo Valley College*
Richard Krikorian, *Westchester Community College*
John Kubicek, *Southwest Missouri State University*
Paul Kumpel, *SUNY, Stony Brook*
Theodore Lai, *Hudson County Community College*
Fat C. Lam, *Gallaudet University*
Leo Lampone, *Quantics Corporation*
James F. Lanahan, *University of Detroit–Mercy*
Bruce Landman, *University of North Carolina at Greensboro*
Jeuel LaTorre, *Clemson University*
Kuen Hung Lee, *Los Angeles Trade–Technology College*
Marshall J. Leitman, *Case Western Reserve University*
Benjamin Levy, *Lexington H.S., Lexington, Mass.*
Darryl A. Linde, *Northeastern Oklahoma State University*
Phil Locke, *University of Maine, Orono*
Leland E. Long, *Muscatine Community College*
John Lucas, *University of Wisconsin–Oshkosh*
Stanley M. Lukawecki, *Clemson University*
Phoebe Lutz, *Delta College*
Nicholas Macri, *Temple University*
Michael Magill, *Purdue University*

Ernest Manfred, *U.S. Coast Guard Academy*
Melvin J. Maron, *University of Louisville*
Mauricio Marroquin, *Los Angeles Valley College*
Thomas W. Mason, *Florida A&M University*
Majid Masso, *Brookdale Community College*
Larry Matthews, *Concordia College*
Thomas McElligott, *University of Lowell*
Phillip McGill, *Illinois Central College*
Judith McKinney, *California State Polytechnic University, Pomona*
Joseph Meier, *Millersville University*
Robert Meitz, *Arizona State University*
Laurie Haskell Messina, *University of Oklahoma*
Aileen Michaels, *Hofstra University*
Janet S. Milton, *Radford University*
Robert Mitchell, *Rowan College of New Jersey*
Marilyn Molloy, *Our Lady of the Lake University*
Ron Moore, *Ryerson Polytechnical Institute*
Barbara Moses, *Bowling Green State University*
Eric Murphy, *U.S. Air Force Academy*
David Nash, *VP Research, Autofacts, Inc.*
Doug Nelson, *Central Oregon Community College*
Lawrence J. Newberry, *Glendale College*
Kylene Norman, *Clark State Community College*
Roxie Novak, *Radford University*
Richard Nowakowski, *Dalhousie University*
Stanley Ocken, *City College–CUNY*
Ralph Okojie, *Elizabeth City State University*
Ann Ostberg
Judith Palagallo, *The University of Akron*
Donald Passman, *University of Wisconsin*
David Patterson, *West Texas A&M*
Walter M. Patterson, *Lander University*
Steven E. Pav, *Alfred University*
Edward Peifer, *Ulster County Community College*
Gary L. Peterson, *James Madison University*
Lefkios Petevis, *Kirkwood Community College*
Robert Phillips, *University of South Carolina at Aiken*
Mark A. Pinsky, *Northeastern University*
Catherine H. Pirri, *Northern Essex Community College*
Thomas W. Polaski, *Winthrop University*
Father Bernard Portz, *Creighton University*
Irwin Pressman, *Carleton University*
Douglas Quinney, *University of Keele*
David Randall, *Oakland Community College*
B. David Redman, Jr., *Delta College*
Irmgard Redman, *Delta College*
Richard Remzowski, *Broome Community College*
Guanshen Ren, *College of Saint Scholastica*
William H. Richardson, *Wichita State University*
John Rickert, *Rose-Hulman Institute of Technology*
David Robbins, *Trinity College*
Lila F. Roberts, *Georgia Southern University*
David Rollins, *University of Central Florida*
Naomi Rose, *Mercer County Community College*
Sharon Ross, *DeKalb College*
David Ryeburn, *Simon Fraser University*
David Sandell, *U.S. Coast Guard Academy*
Avinash Sathaye, *University of Kentucky*
Ned W. Schillow, *Lehigh County Community College*
Dennis Schneider, *Knox College*
George W. Schultz, *St. Petersburg Junior College*
Dan Seth, *Morehead State University*
Richard B. Shad, *Florida Community College–Jacksonville*
George Shapiro, *Brooklyn College*
Parashu R. Sharma, *Grambling State University*
Michael D. Shaw, *Florida Institute of Technology*
Donald R. Sherbert, *University of Illinois*
Howard Sherwood, *University of Central Florida*
Mary Margaret Shoaf-Grubbs, *College of New Rochelle*
Bhagat Singh, *University of Wisconsin Centers*
Ann Sitomer, *Portland Community College*
Martha Sklar, *Los Angeles City College*
Henry Smith, *Southeastern Louisiana University*
Jeanne Smith, *Saddleback Community College*
John L. Smith, *Rancho Santiago Community College*
Wolfe Snow, *Brooklyn College*
Ian Spatz, *Brooklyn College*
Jean Springer, *Mount Royal College*
Rajalakshmi Sriram, *Okaloosa-Walton Community College*
Norton Starr, *Amherst College*
Mark Stevenson, *Oakland Community College*
Gary S. Stoudt, *University of Indiana of Pennsylvania*
John A. Suvak, *Memorial University of Newfoundland*
P. Narayana Swamy, *Southern Illinois University*
Richard B. Thompson, *The University of Arizona*
Skip Thompson, *Radford University*
Josef S. Torok, *Rochester Institute of Technology*
William F. Trench, *Trinity University*
Walter W. Turner, *Western Michigan University*
Thomas Vanden Eynden, *Thomas More College*
Paul Vesce, *University of Missouri–Kansas City*
Richard C. Vile, *Eastern Michigan University*
David Voss, *Western Illinois University*
Ronald Wagoner, *California State University–Fresno*
Shirley Wakin, *University of New Haven*
James E. Ward, *Bowdoin College*
James Warner, *Precision Visuals*
Peter Waterman, *Northern Illinois University*
Evelyn Weinstock, *Glassboro State College*
Bruce R. Wenner, *University of Missouri–Kansas City*
Candice A. Weston, *University of Lowell*
Bruce F. White, *Lander University*
Neil Wigley, *University of Windsor*
Ted Wilcox, *Rochester Institute of Technology*
Gary L. Wood, *Azusa Pacific University*
Yihren Wu, *Hofstra University*
Richard Yuskaitis, *Precision Visuals*
Michael Zeidler, *Milwaukee Area Technical College*
Michael L. Zwilling, *Mount Union College*

The following people read the eighth edition at various stages for mathematical and pedagogical accuracy and/or assisted with the critically important job of preparing answers to exercises:

Elka Block, *Twin Prime Editorial*
Dean Hickerson, *University of California, Davis*
Thomas Polaski, *Winthrop University*
Frank Purcell, *Twin Prime Editorial*
David Ryeburn, *Simon Fraser University*

CONTENTS

chapter fourteen PARTIAL DERIVATIVES 924

chapter fifteen MULTIPLE INTEGRALS 1018

chapter sixteen TOPICS IN VECTOR CALCULUS 1102

A Note to Readers

Cross-references in this special Multivariable version of *Calculus, Eighth Edition* refer to the Early Transcendentals single variable version (ISBN 0-471-48238-2). Readers using this Multivariable version in conjunction with the Late Transcendentals single variable version (ISBN 0-471-48274-9) instead of the Early Transcendentals single variable version can easily find the cross-references they are seeking by consulting this table:

cross-reference on page number	refers to Early Transcendentals	refers to Late Transcendentals
809	Section 7.7	Section 6.7
871	Definition 6.5.1	Definition 5.5.1
873	Chapter 6	Chapter 5
877	Theorem 7.4.3	Theorem 6.4.3
883	Theorem 6.6.3	Theorem 5.6.3
907	Section 6.7	Section 5.7
914	page 776	page 780
959	Section 3.8	Section 3.9
971	Section 3.7	Section 3.8
978	Exercise 24, Section 6.9	Exercise 24, Section 7.6
997	Theorem 5.4.2	Theorem 4.4.2
998	Figure 5.2.6	Figure 4.2.6
999	Theorem 5.2.4	Theorem 4.2.4
1001	Theorem 5.4.3	Theorem 4.4.3
1001	Section 5.4	Section 4.4
1005	Theorem 5.4.4	Theorem 4.4.4
1006	Section 1.7	Section 1.6
1022	Section 7.2	Section 6.2
1025	Definition 7.6.1	Definition 6.6.1
1031	Section 7.2	Section 6.2
1043	Section 7.5	Section 6.5
1049	Section 7.5	Section 6.5
1072	Section 7.3	Section 6.3
1089	page 1044	page 1048
1113	Definition 6.5.1	Definition 5.5.1
1114	Theorem 6.6.2	Theorem 5.6.2
1114	Definition 6.5.1	Definition 5.5.1
1115	Definition 6.5.1	Definition 5.5.1
1123	Definition 7.7.1	Definition 6.7.1
1123	Definition 7.7.3	Definition 6.7.3
1123	Section 7.7	Section 6.7
1130	Theorem 6.6.1	Theorem 5.6.1
1133	Theorem 6.6.3	Theorem 5.6.3
1137	Equation 5, Section 7.7	Equation 5, Section 6.7
1145	Exercise 29, Section 7.4	Exercise 25, Section 6.4

For *Multivariable* readers who do not have access to either the ET or LT single variable version of *Calculus, Eighth Edition*, references in this multivariable version to Section 1.8 (Parametric Equations), Chapter 9 (Mathematical Modeling with Differential Equations), and Chapter 11 (Analytic Geometry in Calculus) are available at no cost on the World Wide Web at www.wiley.com/college/anton for download or viewing as Adobe Acrobat© PDF files.

c h a p t e r   t w e l v e

THREE-DIMENSIONAL SPACE; VECTORS

What if angry vectors veer
Round your sleeping head,
and form. There's never
need to fear Violence of the
poor world's abstract storm.
—Robert Penn Warren
Poet

In this chapter we will discuss rectangular coordinate systems in three dimensions, and we will study the analytic geometry of lines, planes, and other basic surfaces. The second theme of this chapter is the study of vectors. These are the mathematical objects that physicists and engineers use to study forces, displacements, and velocities of objects moving on curved paths. More generally, vectors are used to represent all physical entities that involve both a magnitude and a direction for their complete description. We will introduce various algebraic operations on vectors, and we will apply these operations to problems involving force, work, and rotational tendencies in two and three dimensions. Finally, we will discuss cylindrical and spherical coordinate systems, which are appropriate in problems that involve various kinds of symmetries and also have specific applications in navigation and celestial mechanics.

Photo: *To fully describe the motion of a boat one must specify its speed and direction of motion at each instant. Speed and direction together describe a "vector" quantity. We will study vectors in this chapter.*

12.1 RECTANGULAR COORDINATES IN 3-SPACE; SPHERES; CYLINDRICAL SURFACES

In this section we will discuss coordinate systems in three-dimensional space and some basic facts about surfaces in three dimensions.

RECTANGULAR COORDINATE SYSTEMS

In the remainder of this text we will call three-dimensional space ***3-space***, two-dimensional space (a plane) ***2-space***, and one-dimensional space (a line) ***1-space***. Just as points in 2-space can be placed in one-to-one correspondence with pairs of real numbers using two perpendicular coordinate lines, so points in 3-space can be placed in one-to-one correspondence with triples of real numbers by using three mutually perpendicular coordinate lines, called the ***x-axis***, the ***y-axis***, and the ***z-axis***, positioned so that their origins coincide (Figure 12.1.1). The three coordinate axes form a three-dimensional ***rectangular coordinate system*** (or ***Cartesian coordinate system***). The point of intersection of the coordinate axes is called the ***origin*** of the coordinate system.

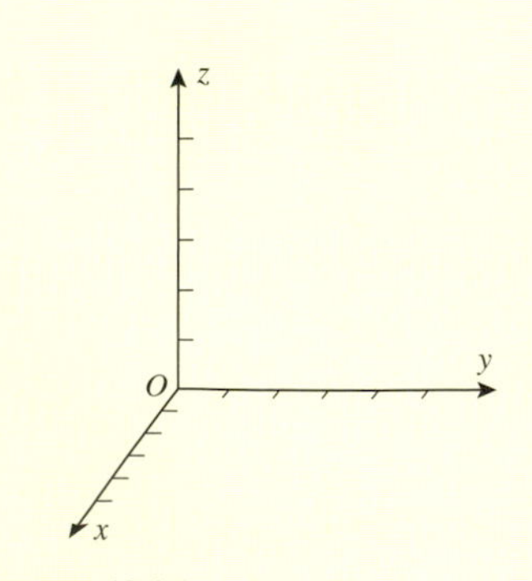

Figure 12.1.1

Rectangular coordinate systems in 3-space fall into two categories: ***left-handed*** and ***right-handed***. A right-handed system has the property that when the fingers of the right hand are cupped so that they curve from the positive x-axis toward the positive y-axis, the thumb points (roughly) in the direction of the positive z-axis (Figure 12.1.2). Similarly for

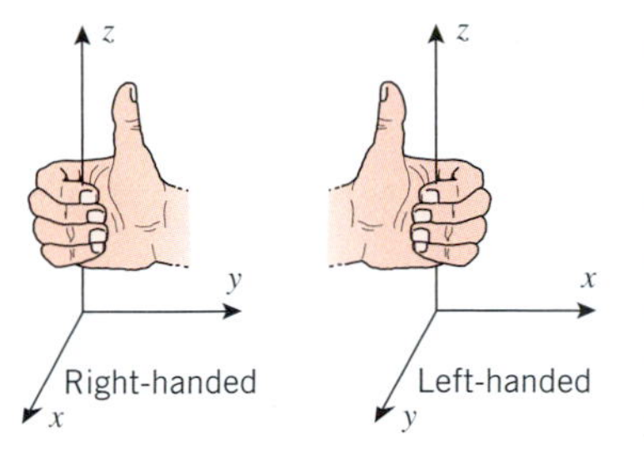

Figure 12.1.2

a left-handed coordinate system (Figure 12.1.2). We will use only right-handed coordinate systems in this text.

The coordinate axes, taken in pairs, determine three ***coordinate planes***: the ***xy-plane***, the ***xz-plane***, and the ***yz-plane*** (Figure 12.1.3). To each point P in 3-space we can assign a triple of real numbers by passing three planes through P parallel to the coordinate planes and letting a, b, and c be the coordinates of the intersections of those planes with the x-axis, y-axis, and z-axis, respectively (Figure 12.1.4). We call a, b, and c the ***x-coordinate***, ***y-coordinate***, and ***z-coordinate*** of P, respectively, and we denote the point P by (a, b, c) or by $P(a, b, c)$. Figure 12.1.5 shows the points $(4, 5, 6)$ and $(-3, 2, -4)$.

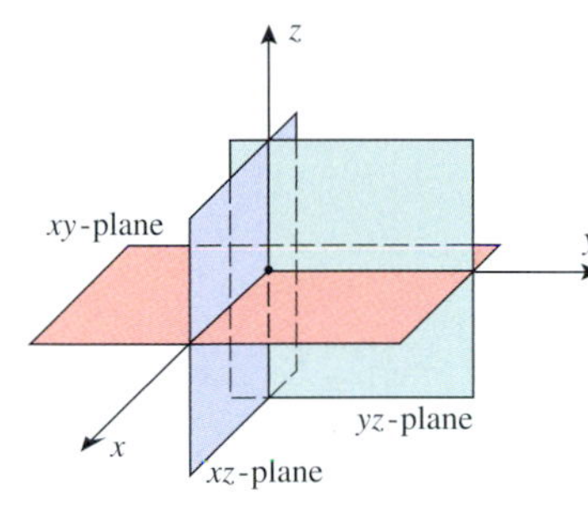

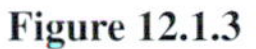

Figure 12.1.3

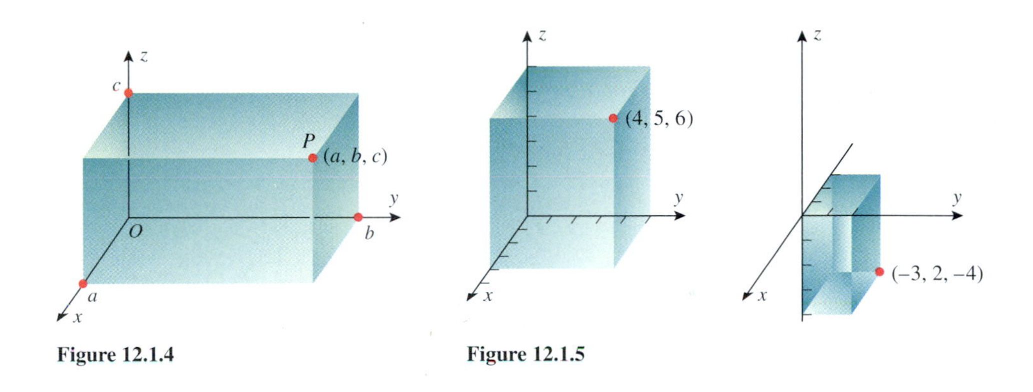

Figure 12.1.4

Figure 12.1.5

Just as the coordinate axes in a two-dimensional coordinate system divide 2-space into four quadrants, so the coordinate planes of a three-dimensional coordinate system divide 3-space into eight parts, called ***octants***. The set of points with three positive coordinates forms the ***first octant***; the remaining octants have no standard numbering.

You should be able to visualize the following facts about three-dimensional rectangular coordinate systems:

REGION	DESCRIPTION
xy-plane	Consists of all points of the form $(x, y, 0)$
xz-plane	Consists of all points of the form $(x, 0, z)$
yz-plane	Consists of all points of the form $(0, y, z)$
x-axis	Consists of all points of the form $(x, 0, 0)$
y-axis	Consists of all points of the form $(0, y, 0)$
z-axis	Consists of all points of the form $(0, 0, z)$

DISTANCE IN 3-SPACE

To derive a formula for the distance between two points in 3-space, we start by considering a box whose sides have lengths a, b, and c (Figure 12.1.6). The length d of a diagonal of the box can be obtained by applying the Theorem of Pythagoras twice: first to show that a diagonal of the base has length $\sqrt{a^2+b^2}$, then again to show that a diagonal of the box has length

$$d = \sqrt{\left(\sqrt{a^2+b^2}\right)^2 + c^2} = \sqrt{a^2+b^2+c^2} \tag{1}$$

Figure 12.1.6

We can now obtain a formula for the distance d between two points $P_1(x_1, y_1, z_1)$ and $P_2(x_2, y_2, z_2)$ in 3-space by finding the length of the diagonal of a box that has these points as diagonal corners (Figure 12.1.7). The sides of such a box have lengths

$$|x_2 - x_1|, \quad |y_2 - y_1|, \quad \text{and} \quad |z_2 - z_1|$$

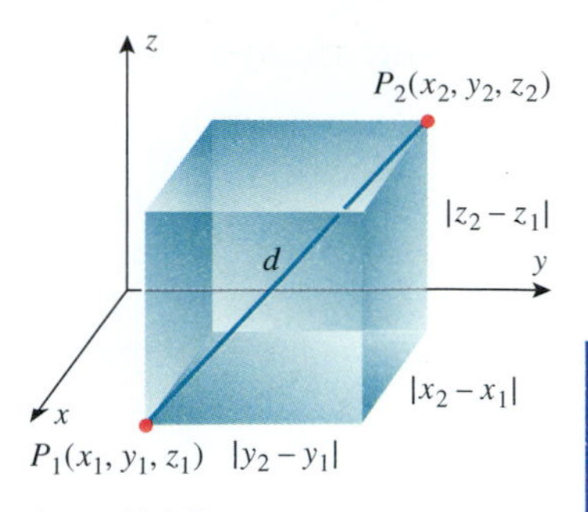

Figure 12.1.7

and hence from (1) the distance d between the points P_1 and P_2 is

$$d = \sqrt{(x_2 - x_1)^2 + (y_2 - y_1)^2 + (z_2 - z_1)^2} \tag{2}$$

(where we have omitted the unnecessary absolute value signs).

> Recall that in 2-space the distance d between points $P_1(x_1, y_1)$ and $P_2(x_2, y_2)$ is
>
> $$d = \sqrt{(x_2 - x_1)^2 + (y_2 - y_1)^2}$$
>
> Thus, the distance formula in 3-space has the same form as the formula in 2-space, but it has a third term to account for the additional dimension. We will see that this is a common occurrence in extending formulas from 2-space to 3-space.

▶ Example 1 Find the distance d between the points $(2, 3, -1)$ and $(4, -1, 3)$.

Solution. From Formula (2)

$$d = \sqrt{(4-2)^2 + (-1-3)^2 + (3+1)^2} = \sqrt{36} = 6 \blacktriangleleft$$

SPHERES

Recall that in an xy-coordinate system, the set of points (x, y) whose coordinates satisfy an equation in x and y is called the *graph* of the equation. Analogously, in an xyz-coordinate system, the set of points (x, y, z) whose coordinates satisfy an equation in x, y, and z is called the ***graph*** of the equation. For example, consider the equation

$$x^2 + y^2 + z^2 = 25 \tag{3}$$

This equation can be rewritten as

$$\sqrt{x^2 + y^2 + z^2} = 5$$

so the graph of (3) consists of all points that are at a distance of 5 units from the origin. Thus, the graph is a sphere of radius 5 centered at the origin (Figure 12.1.8).

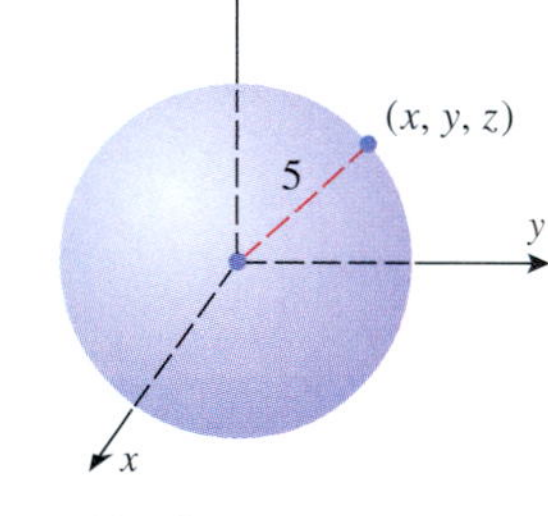

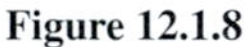
Figure 12.1.8

In general, the sphere with center (x_0, y_0, z_0) and radius r consists of those points (x, y, z) whose coordinates satisfy

$$\sqrt{(x - x_0)^2 + (y - y_0)^2 + (z - z_0)^2} = r$$

or, equivalently,

$$(x - x_0)^2 + (y - y_0)^2 + (z - z_0)^2 = r^2 \tag{4}$$

This is called the ***standard equation of the sphere*** with center (x_0, y_0, z_0) and radius r. Some examples are given in the following table:

> Recall that in 2-space the standard equation of the circle with center (x_0, y_0) and radius r is
>
> $$(x - x_0)^2 + (y - y_0)^2 = r^2$$
>
> Comparing this to (4) we see that the standard equation for the sphere in 3-space has the same form as the standard equation for the circle in 2-space, but with an additional term to account for the third coordinate.

EQUATION	GRAPH
$(x-3)^2 + (y-2)^2 + (z-1)^2 = 9$	Sphere with center (3, 2, 1) and radius 3
$(x+1)^2 + y^2 + (z+4)^2 = 5$	Sphere with center (–1, 0, –4) and radius $\sqrt{5}$
$x^2 + y^2 + z^2 = 1$	Sphere with center (0, 0, 0) and radius 1

If the terms in (4) are expanded and like terms are then collected, then the resulting equation has the form

$$x^2 + y^2 + z^2 + Gx + Hy + Iz + J = 0 \tag{5}$$

The following example shows how the center and radius of a sphere that is expressed in this form can be obtained by completing the squares.

▶ Example 2 Find the center and radius of the sphere

$$x^2 + y^2 + z^2 - 2x - 4y + 8z + 17 = 0$$

Solution. We can put the equation in the form of (4) by completing the squares:

$$(x^2 - 2x) + (y^2 - 4y) + (z^2 + 8z) = -17$$
$$(x^2 - 2x + 1) + (y^2 - 4y + 4) + (z^2 + 8z + 16) = -17 + 21$$
$$(x - 1)^2 + (y - 2)^2 + (z + 4)^2 = 4$$

which is the equation of the sphere with center $(1, 2, -4)$ and radius 2. ◀

In general, completing the squares in (5) produces an equation of the form

$$(x - x_0)^2 + (y - y_0)^2 + (z - z_0)^2 = k$$

If $k > 0$, then the graph of this equation is a sphere with center (x_0, y_0, z_0) and radius $\sqrt{k}$. If $k = 0$, then the sphere has radius zero, so the graph is the single point (x_0, y_0, z_0). If $k < 0$, the equation is not satisfied by any values of x, y, and z (why?), so it has no graph.

12.1.1 THEOREM. *An equation of the form*

$$x^2 + y^2 + z^2 + Gx + Hy + Iz + J = 0$$

represents a sphere, a point, or has no graph.

CYLINDRICAL SURFACES

Although it is natural to graph equations in two variables in 2-space and equations in three variables in 3-space, it is also possible to graph equations in two variables in 3-space. For example, the graph of the equation $y = x^2$ in an xy-coordinate system is a parabola; however, there is nothing to prevent us from writing this equation as $y = x^2 + 0z$ and inquiring about its graph in an xyz-coordinate system. To obtain this graph we need only observe that the equation $y = x^2$ does not impose any restrictions on z. Thus, if we find values of x and y that satisfy this equation, then the coordinates of the point (x, y, z) will also satisfy the equation for *arbitrary* values of z. Geometrically, the point (x, y, z) lies on the vertical line through the point $(x, y, 0)$ in the xy-plane, which means that we can obtain the graph of $y = x^2$ in an xyz-coordinate system by first graphing the equation in the xy-plane and then translating that graph parallel to the z-axis to generate the entire graph (Figure 12.1.9).

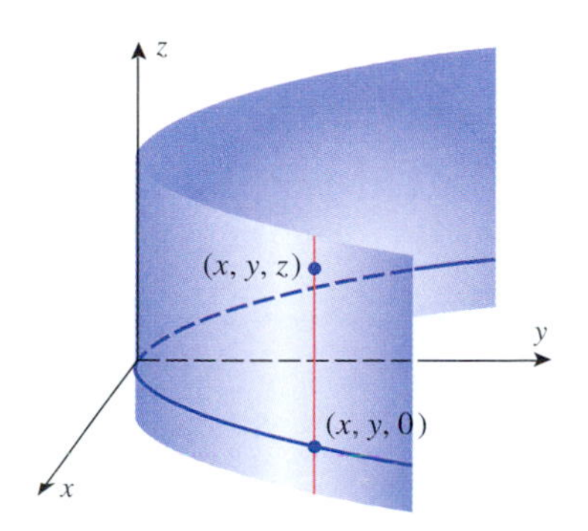

Figure 12.1.9

The process of generating a surface by translating a plane curve parallel to some line is called ***extrusion***, and surfaces that are generated by extrusion are called ***cylindrical surfaces***. A familiar example is the surface of a right circular cylinder, which can be generated by translating a circle parallel to the axis of the cylinder. The following theorem provides basic information about graphing equations in two variables in 3-space:

12.1.2 THEOREM. *An equation that contains only two of the variables x, y, and z represents a cylindrical surface in an xyz-coordinate system. The surface can be obtained by graphing the equation in the coordinate plane of the two variables that appear in the equation and then translating that graph parallel to the axis of the missing variable.*

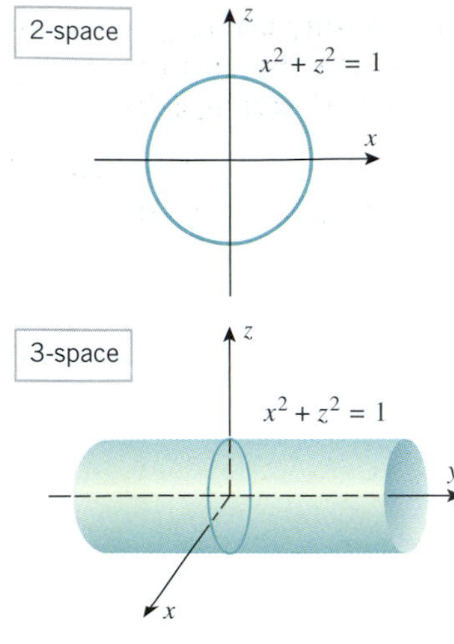

Figure 12.1.10

▶ **Example 3** Sketch the graph of $x^2 + z^2 = 1$ in 3-space.

Solution. Since y does not appear in this equation, the graph is a cylindrical surface generated by extrusion parallel to the y-axis. In the xz-plane the graph of the equation $x^2 + z^2 = 1$ is a circle (Figure 12.1.10). Thus, in 3-space the graph is a right circular cylinder along the y-axis. ◀

▶ **Example 4** Sketch the graph of $z = \sin y$ in 3-space.

Solution. (See Figure 12.1.11.) ◀

In an xy-coordinate system, the graph of the equation $x = 1$ is a line parallel to the y-axis. What is the graph of this equation in an xyz-coordinate system?

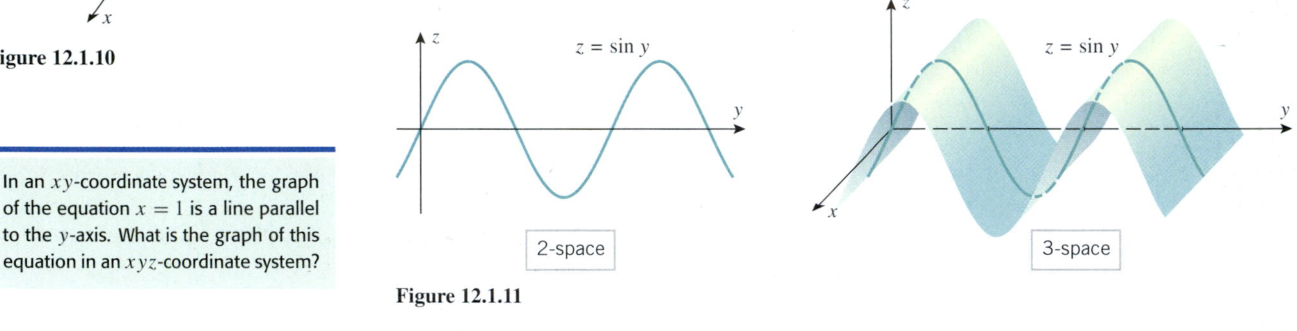

Figure 12.1.11

✔ QUICK CHECK EXERCISES 12.1 *(See page 792 for answers.)*

1. The distance between the points $(1, -2, 0)$ and $(4, 0, 5)$ is ________.

2. The graph of $(x-3)^2 + (y-2)^2 + (z+1)^2 = 16$ is a ________ of radius ________ centered at ________.

3. The shortest distance from the point $(4, 0, 5)$ to the sphere $(x-1)^2 + (y+2)^2 + z^2 = 36$ is ________.

4. Let S be the graph of $x^2 + z^2 + 6z = 16$ in 3-space.
   (a) The intersection of S with the xz-plane is a circle with center ________ and radius ________.
   (b) The intersection of S with the xy-plane is two lines, $x =$ ________ and $x =$ ________.
   (c) The intersection of S with the yz-plane is two lines, $z =$ ________ and $z =$ ________.

EXERCISE SET 12.1 Graphing Utility

1. In each part, find the coordinates of the eight corners of the box.

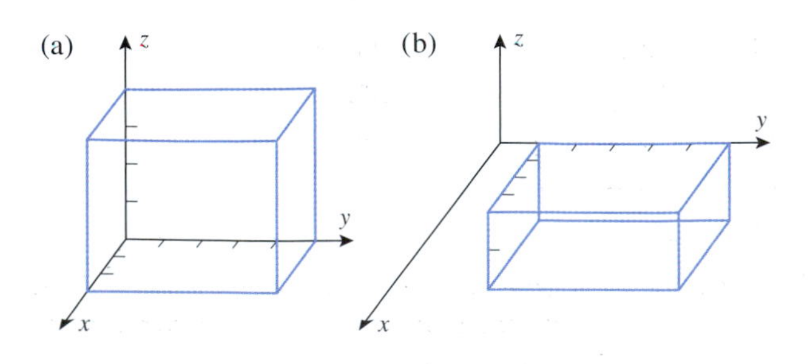

2. A cube of side 4 has its geometric center at the origin and its faces parallel to the coordinate planes. Sketch the cube and give the coordinates of the corners.

FOCUS ON CONCEPTS

3. Suppose that a box has its faces parallel to the coordinate planes and the points $(4, 2, -2)$ and $(-6, 1, 1)$ are endpoints of a diagonal. Sketch the box and give the coordinates of the remaining six corners.

4. Suppose that a box has its faces parallel to the coordinate planes and the points (x_1, y_1, z_1) and (x_2, y_2, z_2) are endpoints of a diagonal.
   (a) Find the coordinates of the remaining six corners.
   (b) Show that the midpoint of the line segment joining (x_1, y_1, z_1) and (x_2, y_2, z_2) is
$$\left(\tfrac{1}{2}(x_1 + x_2), \tfrac{1}{2}(y_1 + y_2), \tfrac{1}{2}(z_1 + z_2)\right)$$
   [*Suggestion:* Apply Theorem G.2 in Web Appendix G to three appropriate edges of the box.]

5. Interpret the graph of $x = 1$ in the contexts of
   (a) a number line (b) 2-space (c) 3-space.

6. Consider the points $P(3, 1, 0)$ and $Q(1, 4, 4)$.
   (a) Sketch the triangle with vertices P, Q, and $(1, 4, 0)$. Without computing distances, explain why this triangle is a right triangle, and then apply the Theorem of Pythagoras twice to find the distance from P to Q.
   (b) Repeat part (a) using the points P, Q, and $(3, 4, 0)$.
   (c) Repeat part (a) using the points P, Q, and $(1, 1, 4)$.

7. Find the center and radius of the sphere that has $(1, -2, 4)$ and $(3, 4, -12)$ as endpoints of a diameter. [See Exercise 4.]

8. Show that $(4, 5, 2)$, $(1, 7, 3)$, and $(2, 4, 5)$ are vertices of an equilateral triangle.

9. (a) Show that $(2, 1, 6)$, $(4, 7, 9)$, and $(8, 5, -6)$ are the vertices of a right triangle.
   (b) Which vertex is at the 90° angle?
   (c) Find the area of the triangle.

10. Find the distance from the point $(-5, 2, -3)$ to the
   (a) xy-plane (b) xz-plane (c) yz-plane
   (d) x-axis (e) y-axis (f) z-axis.

11. In each part, find the standard equation of the sphere that satisfies the stated conditions.
   (a) Center $(1, 0, -1)$; diameter $= 8$.
   (b) Center $(-1, 3, 2)$ and passing through the origin.
   (c) A diameter has endpoints $(-1, 2, 1)$ and $(0, 2, 3)$.

12. Find equations of two spheres that are centered at the origin and are tangent to the sphere of radius 1 centered at $(3, -2, 4)$.

13. In each part, find an equation of the sphere with center $(2, -1, -3)$ and satisfying the given condition.
   (a) Tangent to the xy-plane
   (b) Tangent to the xz-plane
   (c) Tangent to the yz-plane

14. (a) Find an equation of the sphere that is inscribed in the cube that is centered at the point $(-2, 1, 3)$ and has sides of length 1 that are parallel to the coordinate planes.
   (b) Find an equation of the sphere that is circumscribed about the cube in part (a).

15. A sphere has center in the first octant and is tangent to each of the three coordinate planes. Show that the center of the sphere is at a point of the form (r, r, r), where r is the radius of the sphere.

16. A sphere has center in the first octant and is tangent to each of the three coordinate planes. The distance from the origin to the sphere is $3 - \sqrt{3}$ units. Find an equation for the sphere.

17–22 Describe the surface whose equation is given.

17. $x^2 + y^2 + z^2 + 10x + 4y + 2z - 19 = 0$
18. $x^2 + y^2 + z^2 - y = 0$
19. $2x^2 + 2y^2 + 2z^2 - 2x - 3y + 5z - 2 = 0$
20. $x^2 + y^2 + z^2 + 2x - 2y + 2z + 3 = 0$
21. $x^2 + y^2 + z^2 - 3x + 4y - 8z + 25 = 0$
22. $x^2 + y^2 + z^2 - 2x - 6y - 8z + 1 = 0$

23. In each part, sketch the portion of the surface that lies in the first octant.
   (a) $y = x$ (b) $y = z$ (c) $x = z$

24. In each part, sketch the graph of the equation in 3-space.
   (a) $x = 1$ (b) $y = 1$ (c) $z = 1$

25. In each part, sketch the graph of the equation in 3-space.
   (a) $x^2 + y^2 = 25$ (b) $y^2 + z^2 = 25$ (c) $x^2 + z^2 = 25$

26. In each part, sketch the graph of the equation in 3-space.
   (a) $x = y^2$ (b) $z = x^2$ (c) $y = z^2$

27. In each part, write an equation for the surface.
   (a) The plane that contains the x-axis and the point $(0, 1, 2)$.
   (b) The plane that contains the y-axis and the point $(1, 0, 2)$.
   (c) The right circular cylinder that has radius 1 and is centered on the line parallel to the z-axis that passes through the point $(1, 1, 0)$.
   (d) The right circular cylinder that has radius 1 and is centered on the line parallel to the y-axis that passes through the point $(1, 0, 1)$.

28. Find equations for the following right circular cylinders. Each cylinder has radius a and is "tangent" to two coordinate planes.

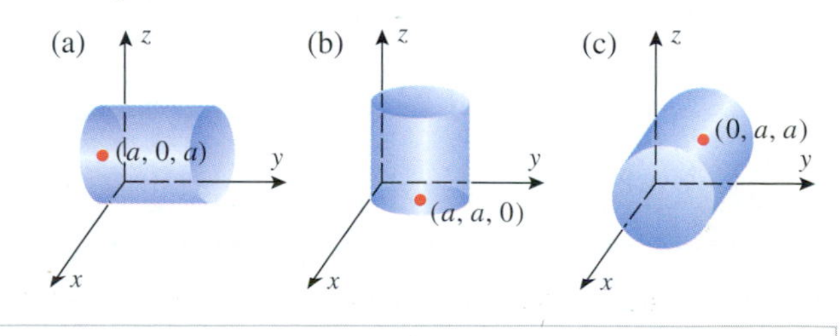

29–38 Sketch the surface in 3-space.

29. $y = \sin x$
30. $y = e^x$
31. $z = 1 - y^2$
32. $z = \cos x$
33. $2x + z = 3$
34. $2x + 3y = 6$
35. $4x^2 + 9z^2 = 36$
36. $z = \sqrt{3 - x}$
37. $y^2 - 4z^2 = 4$
38. $yz = 1$

39. Use a graphing utility to generate the curve $y = x^3/(1 + x^2)$ in the xy-plane, and then use the graph to help sketch the surface $z = y^3/(1 + y^2)$ in 3-space.

40. Use a graphing utility to generate the curve $y = x/(1 + x^4)$ in the xy-plane, and then use the graph to help sketch the surface $z = y/(1 + y^4)$ in 3-space.

41. If a bug walks on the sphere

$$x^2 + y^2 + z^2 + 2x - 2y - 4z - 3 = 0$$

how close and how far can it get from the origin?

42. Describe the set of all points in 3-space whose coordinates satisfy the inequality $x^2 + y^2 + z^2 - 2x + 8z \leq 8$.

43. Describe the set of all points in 3-space whose coordinates satisfy the inequality $y^2 + z^2 + 6y - 4z > 3$.

44. The distance between a point $P(x, y, z)$ and the point $A(1, -2, 0)$ is twice the distance between P and the point $B(0, 1, 1)$. Show that the set of all such points is a sphere, and find the center and radius of the sphere.

45. As shown in the accompanying figure, a bowling ball of radius R is placed inside a box just large enough to hold it, and it is secured for shipping by packing a Styrofoam sphere into each corner of the box. Find the radius of the largest Styrofoam sphere that can be used. [*Hint:* Take the origin of a Cartesian coordinate system at a corner of the box with the coordinate axes along the edges.]

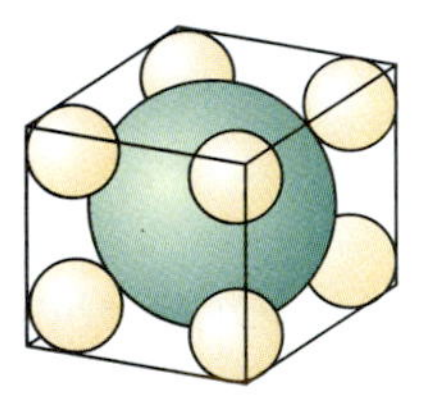

Figure Ex-45

46. Consider the equation

$$x^2 + y^2 + z^2 + Gx + Hy + Iz + J = 0$$

and let $K = G^2 + H^2 + I^2 - 4J$.

(a) Prove that the equation represents a sphere if $K > 0$, a point if $K = 0$, and has no graph if $K < 0$.

(b) In the case where $K > 0$, find the center and radius of the sphere.

47. (a) The accompanying figure shows a surface of revolution that is generated by revolving the curve $y = f(x)$ in the xy-plane about the x-axis. Show that the equation of this surface is $y^2 + z^2 = [f(x)]^2$. [*Hint:* Each point on the curve traces a circle as it revolves about the x-axis.]

(b) Find an equation of the surface of revolution that is generated by revolving the curve $y = e^x$ in the xy-plane about the x-axis.

(c) Show that the ellipsoid $3x^2 + 4y^2 + 4z^2 = 16$ is a surface of revolution about the x-axis by finding a curve $y = f(x)$ in the xy-plane that generates it.

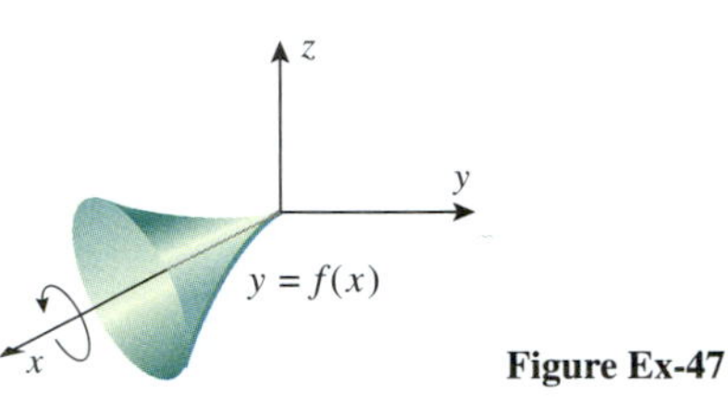

Figure Ex-47

48. In each part, use the idea in Exercise 47(a) to derive a formula for the stated surface of revolution.

(a) The surface generated by revolving the curve $x = f(y)$ in the xy-plane about the y-axis.

(b) The surface generated by revolving the curve $y = f(z)$ in the yz-plane about the z-axis.

(c) The surface generated by revolving the curve $z = f(x)$ in the xz-plane about the x-axis.

49. Show that for all values of θ and ϕ, the point

$$(a \sin\phi \cos\theta, a \sin\phi \sin\theta, a \cos\phi)$$

lies on the sphere $x^2 + y^2 + z^2 = a^2$.

✔ QUICK CHECK ANSWERS 12.1

1. $\sqrt{38}$ **2.** sphere; 4; $(3, 2, -1)$ **3.** $\sqrt{38} - 6$ **4.** (a) $(0, 0, -3)$; 5 (b) 4; -4 (c) 2; -8

12.2 VECTORS

Many physical quantities such as area, length, mass, and temperature are completely described once the magnitude of the quantity is given. Such quantities are called "scalars." Other physical quantities, called "vectors," are not completely determined until both a magnitude and a direction are specified. For example, winds are usually described by giving their speed and direction, say 20 mi/h northeast. The wind speed and wind direction together form a vector quantity called the wind velocity. Other examples of vectors are force and displacement. In this section we will develop the basic mathematical properties of vectors.

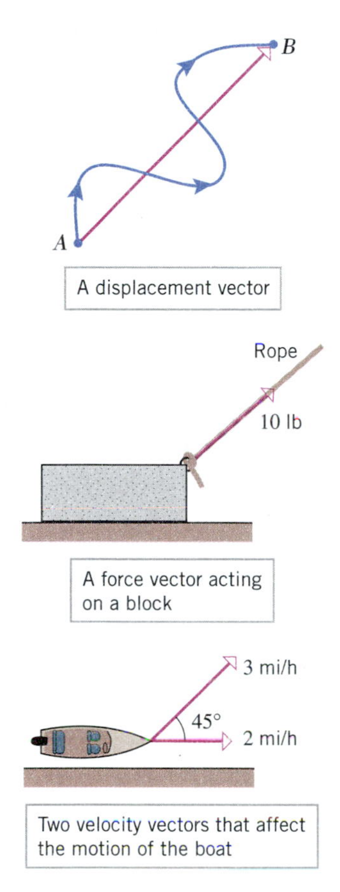

Figure 12.2.1

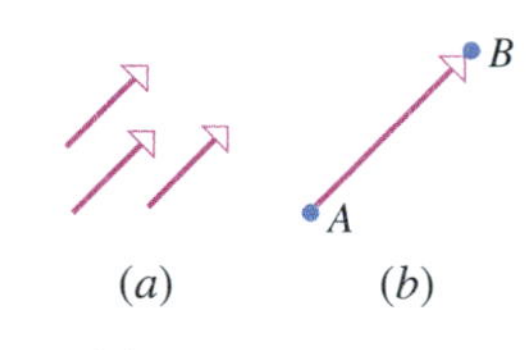

Figure 12.2.2

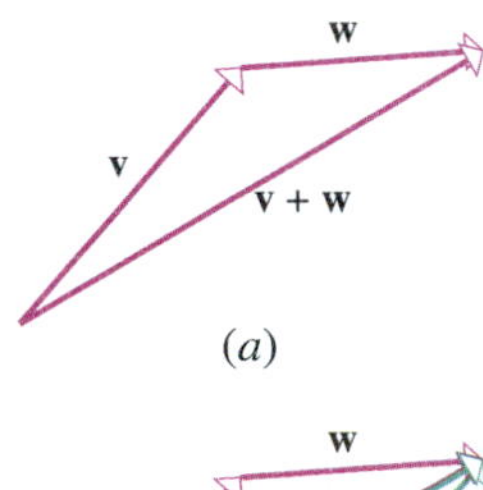

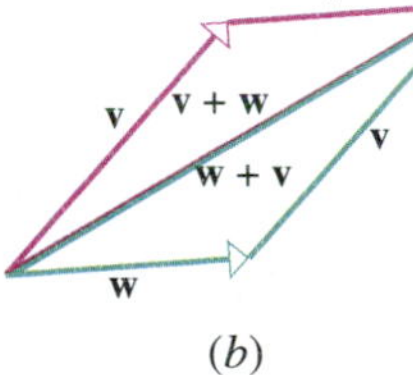

Figure 12.2.3

VECTORS IN PHYSICS AND ENGINEERING

A particle that moves along a line can move in only two directions, so its direction of motion can be described by taking one direction to be positive and the other negative. Thus, the *displacement* or *change in position* of the point can be described by a signed real number. For example, a displacement of 3 (= +3) describes a position change of 3 units in the positive direction, and a displacement of −3 describes a position change of 3 units in the negative direction. However, for a particle that moves in two dimensions or three dimensions, a plus or minus sign is no longer sufficient to specify the direction of motion—other methods are required. One method is to use an arrow, called a ***vector***, that points in the direction of motion and whose length represents the distance from the starting point to the ending point; this is called the ***displacement vector*** for the motion. For example, the first part of Figure 12.2.1 shows the displacement vector of a particle that moves from point A to point B along a circuitous path. Note that the length of the arrow describes the distance between the starting and ending points and not the actual distance traveled by the particle.

Arrows are not limited to describing displacements—they can be used to describe any physical quantity that involves both a magnitude and a direction. Two important examples are forces and velocities. For example, the arrow in the second part of Figure 12.2.1 represents a force vector of 10 lb acting in a specific direction on a block, and the arrows in the third part of that figure show the velocity vector of a boat whose motor propels it parallel to the shore at 2 mi/h and the velocity vector of a 3 mi/h wind acting at an angle of 45° with the shoreline. Intuition suggests that the two velocity vectors will combine to produce some net velocity for the boat at an angle to the shoreline. Thus, our first objective in this section is to define mathematical operations on vectors that can be used to determine the combined effect of vectors.

VECTORS VIEWED GEOMETRICALLY

Vectors can be represented geometrically by arrows in 2-space or 3-space; the direction of the arrow specifies the direction of the vector and the length of the arrow describes its magnitude. The tail of the arrow is called the ***initial point*** of the vector, and the tip of the arrow the ***terminal point***. We will denote vectors with lowercase boldface type such as **a**, **k**, **v**, **w**, and **x**. When discussing vectors, we will refer to real numbers as ***scalars***. Scalars will be denoted by lowercase italic type such as a, k, v, w, and x. Two vectors, **v** and **w**, are considered to be ***equal*** (also called ***equivalent***) if they have the same length and same direction, in which case we write $\mathbf{v} = \mathbf{w}$. Geometrically, two vectors are equal if they are translations of one another; thus, the three vectors in Figure 12.2.2a are equal, even though they are in different positions.

Because vectors are not affected by translation, the initial point of a vector **v** can be moved to any convenient point A by making an appropriate translation. If the initial point of **v** is A and the terminal point is B, then we write $\mathbf{v} = \overrightarrow{AB}$ when we want to emphasize the initial and terminal points (Figure 12.2.2b). If the initial and terminal points of a vector coincide, then the vector has length zero; we call this the ***zero vector*** and denote it by **0**. The zero vector does not have a specific direction, so we will agree that it can be assigned any convenient direction in a specific problem.

There are various algebraic operations that are performed on vectors, all of whose definitions originated in physics. We begin with vector addition.

> **12.2.1 DEFINITION.** If **v** and **w** are vectors, then the ***sum*** $\mathbf{v} + \mathbf{w}$ is the vector from the initial point of **v** to the terminal point of **w** when the vectors are positioned so the initial point of **w** is at the terminal point of **v** (Figure 12.2.3a).

In Figure 12.2.3*b* we have constructed two sums, $\mathbf{v}+\mathbf{w}$ (purple arrows) and $\mathbf{w}+\mathbf{v}$ (green arrows). It is evident that

$$\mathbf{v}+\mathbf{w}=\mathbf{w}+\mathbf{v}$$

and that the sum coincides with the diagonal of the parallelogram determined by $\mathbf{v}$ and $\mathbf{w}$ when these vectors are positioned so they have the same initial point.

Since the initial and terminal points of $\mathbf{0}$ coincide, it follows that

$$\mathbf{0}+\mathbf{v}=\mathbf{v}+\mathbf{0}=\mathbf{v}$$

> **12.2.2 DEFINITION.** If $\mathbf{v}$ is a nonzero vector and k is a nonzero real number (a scalar), then the ***scalar multiple*** $k\mathbf{v}$ is defined to be the vector whose length is $|k|$ times the length of $\mathbf{v}$ and whose direction is the same as that of $\mathbf{v}$ if $k>0$ and opposite to that of $\mathbf{v}$ if $k<0$. We define $k\mathbf{v}=\mathbf{0}$ if $k=0$ or $\mathbf{v}=\mathbf{0}$.

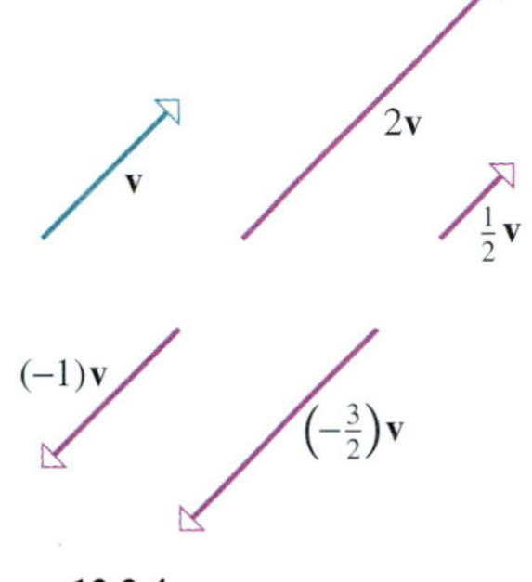

Figure 12.2.4

Figure 12.2.4 shows the geometric relationship between a vector $\mathbf{v}$ and various scalar multiples of it. Observe that if k and $\mathbf{v}$ are nonzero, then the vectors $\mathbf{v}$ and $k\mathbf{v}$ lie on the same line if their initial points coincide and lie on parallel or coincident lines if they do not. Thus, we say that $\mathbf{v}$ and $k\mathbf{v}$ are ***parallel vectors***. Observe also that the vector $(-1)\mathbf{v}$ has the same length as $\mathbf{v}$ but is oppositely directed. We call $(-1)\mathbf{v}$ the ***negative*** of $\mathbf{v}$ and denote it by $-\mathbf{v}$ (Figure 12.2.5). In particular, $-\mathbf{0}=(-1)\mathbf{0}=\mathbf{0}$.

Vector subtraction is defined in terms of addition and scalar multiplication by

$$\mathbf{v}-\mathbf{w}=\mathbf{v}+(-\mathbf{w})$$

The difference $\mathbf{v}-\mathbf{w}$ can be obtained geometrically by first constructing the vector $-\mathbf{w}$ and then adding $\mathbf{v}$ and $-\mathbf{w}$, say by the parallelogram method (Figure 12.2.6*a*). However, if $\mathbf{v}$ and $\mathbf{w}$ are positioned so their initial points coincide, then $\mathbf{v}-\mathbf{w}$ can be formed more directly, as shown in Figure 12.2.6*b*, by drawing the vector from the terminal point of $\mathbf{w}$ (the second term) to the terminal point of $\mathbf{v}$ (the first term). In the special case where $\mathbf{v}=\mathbf{w}$ the terminal points of the vectors coincide, so their difference is $\mathbf{0}$; that is,

$$\mathbf{v}+(-\mathbf{v})=\mathbf{v}-\mathbf{v}=\mathbf{0}$$

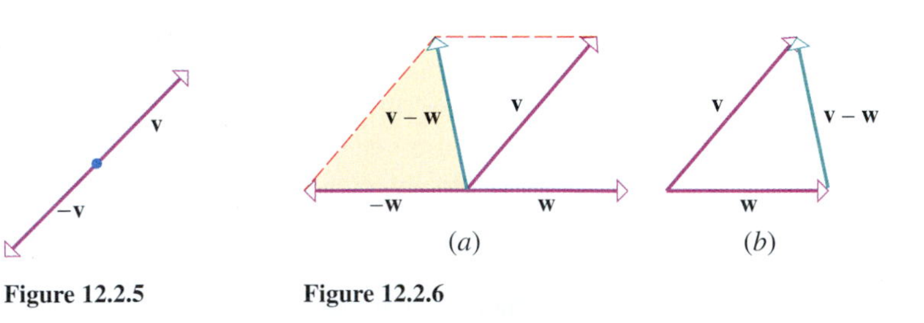

Figure 12.2.5

Figure 12.2.6

VECTORS IN COORDINATE SYSTEMS

Problems involving vectors are often best solved by introducing a rectangular coordinate system. If a vector $\mathbf{v}$ is positioned with its initial point at the origin of a rectangular coordinate system, then its terminal point will have coordinates of the form (v_1, v_2) or (v_1, v_2, v_3), depending on whether the vector is in 2-space or 3-space (Figure 12.2.7). We call these coordinates the ***components*** of $\mathbf{v}$, and we write $\mathbf{v}$ in *component form* as

$$\underbrace{\mathbf{v}=\langle v_1, v_2\rangle}_{\text{2-space}} \quad \text{or} \quad \underbrace{\mathbf{v}=\langle v_1, v_2, v_3\rangle}_{\text{3-space}}$$

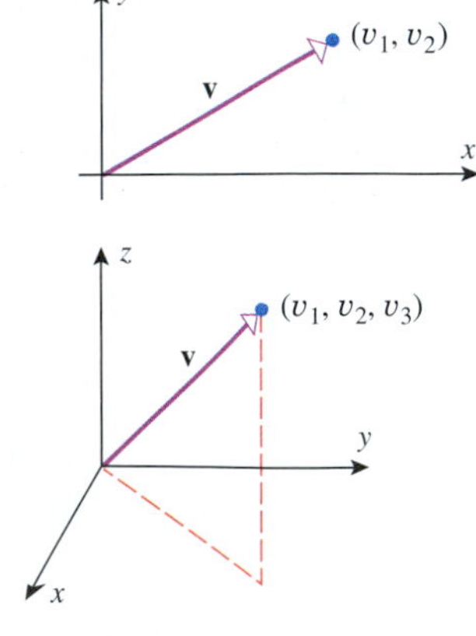

Figure 12.2.7

In particular, the zero vectors in 2-space and 3-space are

$$\mathbf{0} = \langle 0, 0 \rangle \quad \text{and} \quad \mathbf{0} = \langle 0, 0, 0 \rangle$$

respectively.

Components provide a simple way of identifying equivalent vectors. For example, consider the vectors $\mathbf{v} = \langle v_1, v_2 \rangle$ and $\mathbf{w} = \langle w_1, w_2 \rangle$ in 2-space. If $\mathbf{v} = \mathbf{w}$, then the vectors have the same length and same direction, and this means that their terminal points coincide when their initial points are placed at the origin. It follows that $v_1 = w_1$ and $v_2 = w_2$, so we have shown that equivalent vectors have the same components. Conversely, if $v_1 = w_1$ and $v_2 = w_2$, then the terminal points of the vectors coincide when their initial points are placed at the origin. It follows that the vectors have the same length and same direction, so we have shown that vectors with the same components are equivalent. A similar argument holds for vectors in 3-space, so we have the following result.

12.2.3 THEOREM. *Two vectors are equivalent if and only if their corresponding components are equal.*

For example,

$$\langle a, b, c \rangle = \langle 1, -4, 2 \rangle$$

if and only if $a = 1$, $b = -4$, and $c = 2$.

ARITHMETIC OPERATIONS ON VECTORS

The next theorem shows how to perform arithmetic operations on vectors using components.

12.2.4 THEOREM. *If $\mathbf{v} = \langle v_1, v_2 \rangle$ and $\mathbf{w} = \langle w_1, w_2 \rangle$ are vectors in 2-space and k is any scalar, then*

$$\mathbf{v} + \mathbf{w} = \langle v_1 + w_1, v_2 + w_2 \rangle \tag{1}$$
$$\mathbf{v} - \mathbf{w} = \langle v_1 - w_1, v_2 - w_2 \rangle \tag{2}$$
$$k\mathbf{v} = \langle kv_1, kv_2 \rangle \tag{3}$$

Similarly, if $\mathbf{v} = \langle v_1, v_2, v_3 \rangle$ and $\mathbf{w} = \langle w_1, w_2, w_3 \rangle$ are vectors in 3-space and k is any scalar, then

$$\mathbf{v} + \mathbf{w} = \langle v_1 + w_1, v_2 + w_2, v_3 + w_3 \rangle \tag{4}$$
$$\mathbf{v} - \mathbf{w} = \langle v_1 - w_1, v_2 - w_2, v_3 - w_3 \rangle \tag{5}$$
$$k\mathbf{v} = \langle kv_1, kv_2, kv_3 \rangle \tag{6}$$

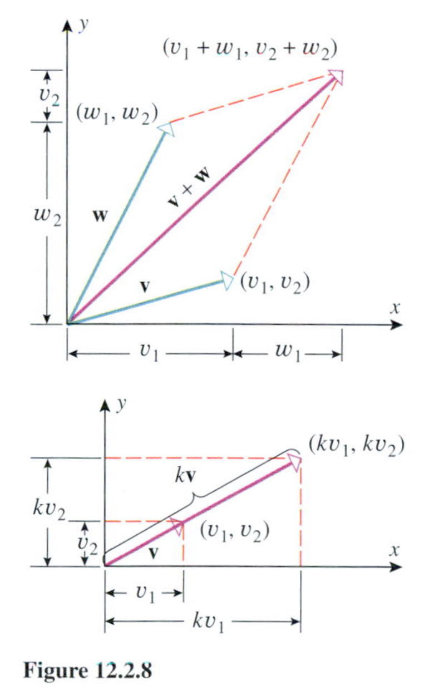

Figure 12.2.8

We will not prove this theorem. However, results (1) and (3) should be evident from Figure 12.2.8. Similar figures in 3-space can be used to motivate (4) and (6). Formulas (2) and (5) can be obtained by writing $\mathbf{v} + \mathbf{w} = \mathbf{v} + (-1)\mathbf{w}$.

Example 1 If $\mathbf{v} = \langle -2, 0, 1 \rangle$ and $\mathbf{w} = \langle 3, 5, -4 \rangle$, then

$$\mathbf{v} + \mathbf{w} = \langle -2, 0, 1 \rangle + \langle 3, 5, -4 \rangle = \langle 1, 5, -3 \rangle$$
$$3\mathbf{v} = \langle -6, 0, 3 \rangle$$
$$-\mathbf{w} = \langle -3, -5, 4 \rangle$$
$$\mathbf{w} - 2\mathbf{v} = \langle 3, 5, -4 \rangle - \langle -4, 0, 2 \rangle = \langle 7, 5, -6 \rangle$$ ◀

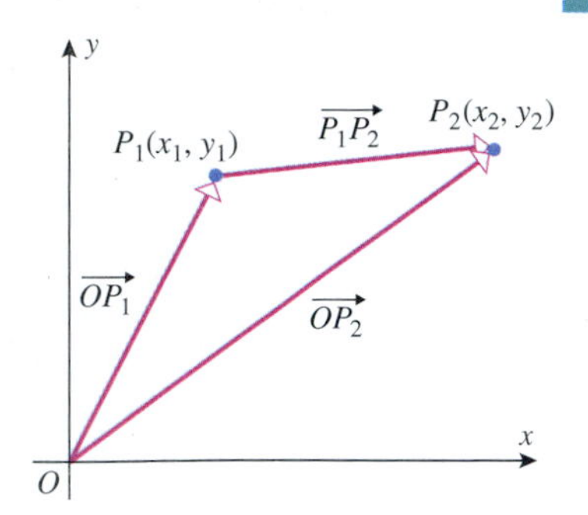

Figure 12.2.9

VECTORS WITH INITIAL POINT NOT AT THE ORIGIN

Recall that we defined the components of a vector to be the coordinates of its terminal point when its initial point is at the origin. We will now consider the problem of finding the components of a vector whose initial point is not at the origin. To be specific, suppose that $P_1(x_1, y_1)$ and $P_2(x_2, y_2)$ are points in 2-space and we are interested in finding the components of the vector $\overrightarrow{P_1P_2}$. As illustrated in Figure 12.2.9, we can write this vector as

$$\overrightarrow{P_1P_2} = \overrightarrow{OP_2} - \overrightarrow{OP_1} = \langle x_2, y_2 \rangle - \langle x_1, y_1 \rangle = \langle x_2 - x_1, y_2 - y_1 \rangle$$

Thus, we have shown that the components of the vector $\overrightarrow{P_1P_2}$ can be obtained by subtracting the coordinates of its initial point from the coordinates of its terminal point. Similar computations hold in 3-space, so we have established the following result.

12.2.5 THEOREM. *If $\overrightarrow{P_1P_2}$ is a vector in 2-space with initial point $P_1(x_1, y_1)$ and terminal point $P_2(x_2, y_2)$, then*

$$\overrightarrow{P_1P_2} = \langle x_2 - x_1, y_2 - y_1 \rangle \tag{7}$$

Similarly, if $\overrightarrow{P_1P_2}$ is a vector in 3-space with initial point $P_1(x_1, y_1, z_1)$ and terminal point $P_2(x_2, y_2, z_2)$, then

$$\overrightarrow{P_1P_2} = \langle x_2 - x_1, y_2 - y_1, z_2 - z_1 \rangle \tag{8}$$

Example 2 In 2-space the vector from $P_1(1, 3)$ to $P_2(4, -2)$ is

$$\overrightarrow{P_1P_2} = \langle 4 - 1, -2 - 3 \rangle = \langle 3, -5 \rangle$$

and in 3-space the vector from $A(0, -2, 5)$ to $B(3, 4, -1)$ is

$$\overrightarrow{AB} = \langle 3 - 0, 4 - (-2), -1 - 5 \rangle = \langle 3, 6, -6 \rangle$$ ◀

RULES OF VECTOR ARITHMETIC

The following theorem shows that many of the familiar rules of ordinary arithmetic also hold for vector arithmetic.

It follows from part (*b*) of Theorem 12.2.6 that the expression

$$\mathbf{u} + \mathbf{v} + \mathbf{w}$$

is unambiguous since the same vector results no matter how the terms are grouped.

12.2.6 THEOREM. *For any vectors* **u**, **v**, *and* **w** *and any scalars k and l, the following relationships hold:*

(*a*) $\mathbf{u} + \mathbf{v} = \mathbf{v} + \mathbf{u}$
(*b*) $(\mathbf{u} + \mathbf{v}) + \mathbf{w} = \mathbf{u} + (\mathbf{v} + \mathbf{w})$
(*c*) $\mathbf{u} + \mathbf{0} = \mathbf{0} + \mathbf{u} = \mathbf{u}$
(*d*) $\mathbf{u} + (-\mathbf{u}) = \mathbf{0}$
(*e*) $k(l\mathbf{u}) = (kl)\mathbf{u}$
(*f*) $k(\mathbf{u} + \mathbf{v}) = k\mathbf{u} + k\mathbf{v}$
(*g*) $(k + l)\mathbf{u} = k\mathbf{u} + l\mathbf{u}$
(*h*) $1\mathbf{u} = \mathbf{u}$

The results in this theorem can be proved either algebraically by using components or geometrically by treating the vectors as arrows. We will prove part (*b*) both ways and leave some of the remaining proofs as exercises.

PROOF (*b*) (ALGEBRAIC IN 2-SPACE). Let $\mathbf{u} = \langle u_1, u_2 \rangle$, $\mathbf{v} = \langle v_1, v_2 \rangle$, and $\mathbf{w} = \langle w_1, w_2 \rangle$. Then

$$\begin{aligned}(\mathbf{u}+\mathbf{v})+\mathbf{w} &= (\langle u_1, u_2\rangle + \langle v_1, v_2\rangle) + \langle w_1, w_2\rangle \\ &= \langle u_1+v_1, u_2+v_2\rangle + \langle w_1, w_2\rangle \\ &= \langle (u_1+v_1)+w_1, (u_2+v_2)+w_2\rangle \\ &= \langle u_1+(v_1+w_1), u_2+(v_2+w_2)\rangle \\ &= \langle u_1, u_2\rangle + \langle v_1+w_1, v_2+w_2\rangle \\ &= \mathbf{u}+(\mathbf{v}+\mathbf{w})\end{aligned}$$

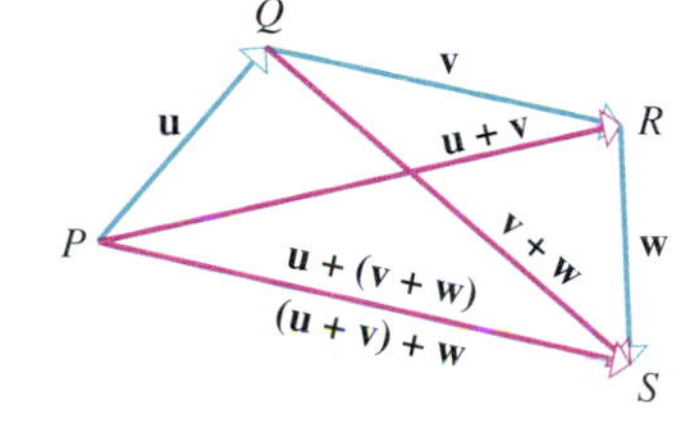

Figure 12.2.10

PROOF (*b*) (GEOMETRIC). Let $\mathbf{u}$, $\mathbf{v}$, and $\mathbf{w}$ be represented by $\overrightarrow{PQ}$, $\overrightarrow{QR}$, and $\overrightarrow{RS}$ as shown in Figure 12.2.10. Then

$$\mathbf{v}+\mathbf{w} = \overrightarrow{QS} \quad \text{and} \quad \mathbf{u}+(\mathbf{v}+\mathbf{w}) = \overrightarrow{PS}$$
$$\mathbf{u}+\mathbf{v} = \overrightarrow{PR} \quad \text{and} \quad (\mathbf{u}+\mathbf{v})+\mathbf{w} = \overrightarrow{PS}$$

Therefore,

$$(\mathbf{u}+\mathbf{v})+\mathbf{w} = \mathbf{u}+(\mathbf{v}+\mathbf{w}) \qquad ■$$

Observe that in Figure 12.2.10 the vectors **u**, **v**, and **w** are positioned "tip to tail" and that

$$\mathbf{u}+\mathbf{v}+\mathbf{w}$$

is the vector from the initial point of **u** (the first term in the sum) to the terminal point of **w** (the last term in the sum). This "tip to tail" method of vector addition also works for four or more vectors (Figure 12.2.11).

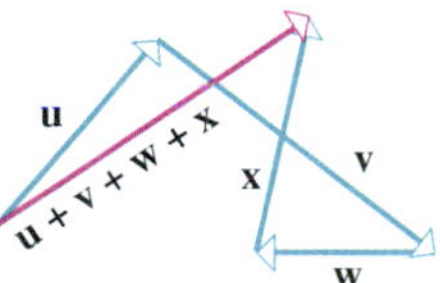

Figure 12.2.11

NORM OF A VECTOR

The distance between the initial and terminal points of a vector **v** is called the ***length***, the ***norm***, or the ***magnitude*** of **v** and is denoted by $\|\mathbf{v}\|$. This distance does not change if the vector is translated, so for purposes of calculating the norm we can assume that the vector is positioned with its initial point at the origin (Figure 12.2.12). This makes it evident that the norm of a vector $\mathbf{v} = \langle v_1, v_2 \rangle$ in 2-space is given by

$$\|\mathbf{v}\| = \sqrt{v_1^2 + v_2^2} \tag{9}$$

and the norm of a vector $\mathbf{v} = \langle v_1, v_2, v_3 \rangle$ in 3-space is given by

$$\|\mathbf{v}\| = \sqrt{v_1^2 + v_2^2 + v_3^2} \tag{10}$$

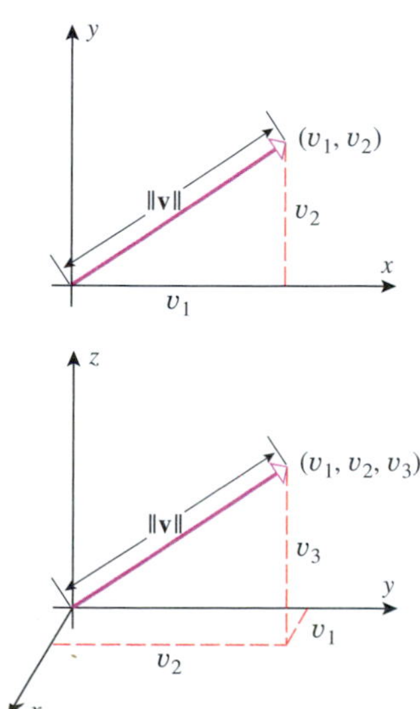

Figure 12.2.12

► **Example 3** Find the norms of $\mathbf{v} = \langle -2, 3 \rangle$, $10\mathbf{v} = \langle -20, 30 \rangle$, and $\mathbf{w} = \langle 2, 3, 6 \rangle$.

Solution. From (9) and (10)

$$\begin{aligned}\|\mathbf{v}\| &= \sqrt{(-2)^2 + 3^2} = \sqrt{13} \\ \|10\mathbf{v}\| &= \sqrt{(-20)^2 + 30^2} = \sqrt{1300} = 10\sqrt{13} \\ \|\mathbf{w}\| &= \sqrt{2^2 + 3^2 + 6^2} = \sqrt{49} = 7 \; ◄\end{aligned}$$

Note that $\|10\mathbf{v}\| = 10\|\mathbf{v}\|$ in Example 3. This is consistent with Definition 12.2.2, which stipulated that for any vector **v** and scalar k, the length of $k\mathbf{v}$ must be $|k|$ times the length of **v**; that is,

$$\|k\mathbf{v}\| = |k|\|\mathbf{v}\| \tag{11}$$

Thus, for example,

$$\begin{aligned}\|3\mathbf{v}\| &= |3|\|\mathbf{v}\| = 3\|\mathbf{v}\| \\ \|-2\mathbf{v}\| &= |-2|\|\mathbf{v}\| = 2\|\mathbf{v}\| \\ \|-1\mathbf{v}\| &= |-1|\|\mathbf{v}\| = \|\mathbf{v}\|\end{aligned}$$

This applies to vectors in 2-space and 3-space.

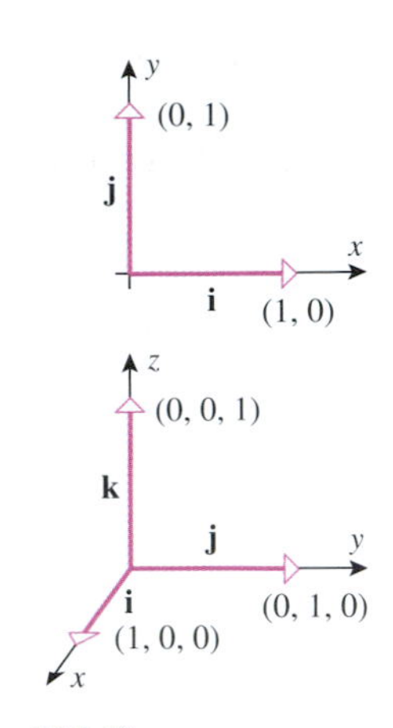

Figure 12.2.13

UNIT VECTORS

A vector of length 1 is called a ***unit vector***. In an xy-coordinate system the unit vectors along the x- and y-axes are denoted by $\mathbf{i}$ and $\mathbf{j}$, respectively; and in an xyz-coordinate system the unit vectors along the x-, y-, and z-axes are denoted by $\mathbf{i}$, $\mathbf{j}$, and $\mathbf{k}$, respectively (Figure 12.2.13). Thus,

$$\mathbf{i} = \langle 1, 0\rangle, \qquad \mathbf{j} = \langle 0, 1\rangle \qquad \text{In 2-space}$$

$$\mathbf{i} = \langle 1, 0, 0\rangle, \qquad \mathbf{j} = \langle 0, 1, 0\rangle, \qquad \mathbf{k} = \langle 0, 0, 1\rangle \qquad \text{In 3-space}$$

Every vector in 2-space is expressible uniquely in terms of $\mathbf{i}$ and $\mathbf{j}$, and every vector in 3-space is expressible uniquely in terms of $\mathbf{i}$, $\mathbf{j}$, and $\mathbf{k}$ as follows:

$$\mathbf{v} = \langle v_1, v_2\rangle = \langle v_1, 0\rangle + \langle 0, v_2\rangle = v_1\langle 1, 0\rangle + v_2\langle 0, 1\rangle = v_1\mathbf{i} + v_2\mathbf{j}$$

$$\mathbf{v} = \langle v_1, v_2, v_3\rangle = v_1\langle 1, 0, 0\rangle + v_2\langle 0, 1, 0\rangle + v_3\langle 0, 0, 1\rangle = v_1\mathbf{i} + v_2\mathbf{j} + v_3\mathbf{k}$$

▶ Example 4

2-SPACE	3-SPACE
$\langle 2, 3\rangle = 2\mathbf{i} + 3\mathbf{j}$	$\langle 2, -3, 4\rangle = 2\mathbf{i} - 3\mathbf{j} + 4\mathbf{k}$
$\langle -4, 0\rangle = -4\mathbf{i} + 0\mathbf{j} = -4\mathbf{i}$	$\langle 0, 3, 0\rangle = 3\mathbf{j}$
$\langle 0, 0\rangle = 0\mathbf{i} + 0\mathbf{j} = \mathbf{0}$	$\langle 0, 0, 0\rangle = 0\mathbf{i} + 0\mathbf{j} + 0\mathbf{k} = \mathbf{0}$
$(3\mathbf{i} + 2\mathbf{j}) + (4\mathbf{i} + \mathbf{j}) = 7\mathbf{i} + 3\mathbf{j}$	$(3\mathbf{i} + 2\mathbf{j} - \mathbf{k}) - (4\mathbf{i} - \mathbf{j} + 2\mathbf{k}) = -\mathbf{i} + 3\mathbf{j} - 3\mathbf{k}$
$5(6\mathbf{i} - 2\mathbf{j}) = 30\mathbf{i} - 10\mathbf{j}$	$2(\mathbf{i} + \mathbf{j} - \mathbf{k}) + 4(\mathbf{i} - \mathbf{j}) = 6\mathbf{i} - 2\mathbf{j} - 2\mathbf{k}$
$\|2\mathbf{i} - 3\mathbf{j}\| = \sqrt{2^2 + (-3)^2} = \sqrt{13}$	$\|\mathbf{i} + 2\mathbf{j} - 3\mathbf{k}\| = \sqrt{1^2 + 2^2 + (-3)^2} = \sqrt{14}$
$\|v_1\mathbf{i} + v_2\mathbf{j}\| = \sqrt{v_1^2 + v_2^2}$	$\|\langle v_1, v_2, v_3\rangle\| = \sqrt{v_1^2 + v_2^2 + v_3^2}$ ◀

The two notations for vectors illustrated in Example 4 are completely interchangeable, the choice being a matter of convenience or personal preference.

NORMALIZING A VECTOR

A common problem in applications is to find a unit vector $\mathbf{u}$ that has the same direction as some given nonzero vector $\mathbf{v}$. This can be done by multiplying $\mathbf{v}$ by the reciprocal of its length; that is,

$$\mathbf{u} = \frac{1}{\|\mathbf{v}\|}\mathbf{v} = \frac{\mathbf{v}}{\|\mathbf{v}\|}$$

is a unit vector with the same direction as $\mathbf{v}$—the direction is the same because $k = 1/\|\mathbf{v}\|$ is a positive scalar, and the length is 1 because

$$\|\mathbf{u}\| = \|k\mathbf{v}\| = |k|\|\mathbf{v}\| = k\|\mathbf{v}\| = \frac{1}{\|\mathbf{v}\|}\|\mathbf{v}\| = 1$$

The process of multiplying a vector $\mathbf{v}$ by the reciprocal of its length to obtain a unit vector with the same direction is called ***normalizing*** $\mathbf{v}$.

▶ Example 5 Find the unit vector that has the same direction as $\mathbf{v} = 2\mathbf{i} + 2\mathbf{j} - \mathbf{k}$.

Solution. The vector $\mathbf{v}$ has length

$$\|\mathbf{v}\| = \sqrt{2^2 + 2^2 + (-1)^2} = 3$$

so the unit vector $\mathbf{u}$ in the same direction as $\mathbf{v}$ is

$$\mathbf{u} = \tfrac{1}{3}\mathbf{v} = \tfrac{2}{3}\mathbf{i} + \tfrac{2}{3}\mathbf{j} - \tfrac{1}{3}\mathbf{k} \text{ ◀}$$

TECHNOLOGY MASTERY

Many calculating utilities can perform vector operations, and some have built-in norm and normalization operations. If your calculator has these capabilities, use it to check the computations in Examples 1, 3, and 5.

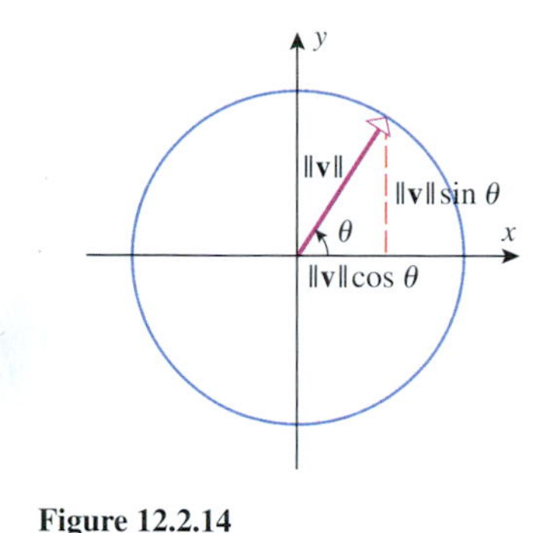

Figure 12.2.14

VECTORS DETERMINED BY LENGTH AND ANGLE

If $\mathbf{v}$ is a nonzero vector with its initial point at the origin of an xy-coordinate system, and if θ is the angle from the positive x-axis to the radial line through $\mathbf{v}$, then the x-component of $\mathbf{v}$ can be written as $\|\mathbf{v}\|\cos\theta$ and the y-component as $\|\mathbf{v}\|\sin\theta$ (Figure 12.2.14); and hence $\mathbf{v}$ can be expressed in trigonometric form as

$$\mathbf{v} = \|\mathbf{v}\|\langle\cos\theta, \sin\theta\rangle \quad \text{or} \quad \mathbf{v} = \|\mathbf{v}\|\cos\theta\,\mathbf{i} + \|\mathbf{v}\|\sin\theta\,\mathbf{j} \tag{12}$$

In the special case of a unit vector $\mathbf{u}$ this simplifies to

$$\mathbf{u} = \langle\cos\theta, \sin\theta\rangle \quad \text{or} \quad \mathbf{u} = \cos\theta\,\mathbf{i} + \sin\theta\,\mathbf{j} \tag{13}$$

▶ Example 6

(a) Find the vector of length 2 that makes an angle of $\pi/4$ with the positive x-axis.

(b) Find the angle that the vector $\mathbf{v} = -\sqrt{3}\,\mathbf{i} + \mathbf{j}$ makes with the positive x-axis.

Solution (a). From (12)

$$\mathbf{v} = 2\cos\frac{\pi}{4}\mathbf{i} + 2\sin\frac{\pi}{4}\mathbf{j} = \sqrt{2}\,\mathbf{i} + \sqrt{2}\,\mathbf{j}$$

Solution (b). We will normalize $\mathbf{v}$, then use (13) to find $\sin\theta$ and $\cos\theta$, and then use these values to find θ. Normalizing $\mathbf{v}$ yields

$$\frac{\mathbf{v}}{\|\mathbf{v}\|} = \frac{-\sqrt{3}\,\mathbf{i} + \mathbf{j}}{\sqrt{(-\sqrt{3}\,)^2 + 1^2}} = -\frac{\sqrt{3}}{2}\mathbf{i} + \frac{1}{2}\mathbf{j}$$

Thus, $\cos\theta = -\sqrt{3}/2$ and $\sin\theta = \frac{1}{2}$, from which we conclude that $\theta = 5\pi/6$. ◀

VECTORS DETERMINED BY LENGTH AND A VECTOR IN THE SAME DIRECTION

It is a common problem in many applications that a direction in 2-space or 3-space is determined by some known unit vector $\mathbf{u}$, and it is of interest to find the components of a vector $\mathbf{v}$ that has the same direction as $\mathbf{u}$ and some specified length $\|\mathbf{v}\|$. This can be done by expressing $\mathbf{v}$ as

$$\mathbf{v} = \|\mathbf{v}\|\mathbf{u}$$

v is equal to its length times a unit vector in the same direction.

and then reading off the components of $\|\mathbf{v}\|\mathbf{u}$.

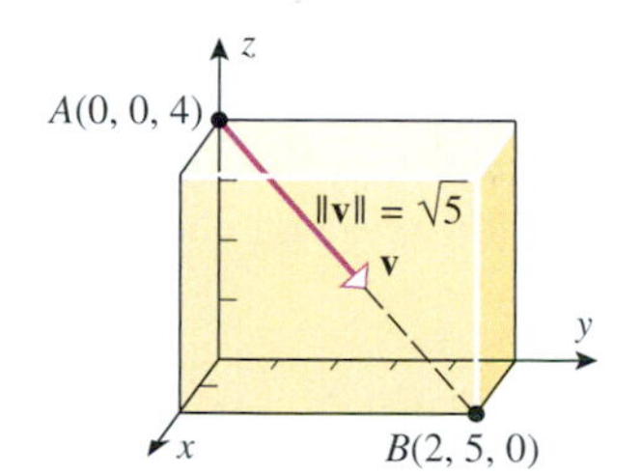

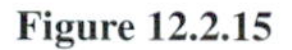
Figure 12.2.15

▶ Example 7

Figure 12.2.15 shows a vector $\mathbf{v}$ of length $\sqrt{5}$ that extends along the line through A and B. Find the components of $\mathbf{v}$.

Solution. First we will find the components of the vector $\overrightarrow{AB}$, then we will normalize this vector to obtain a unit vector in the direction of $\mathbf{v}$, and then we will multiply this unit vector by $\|\mathbf{v}\|$ to obtain the vector $\mathbf{v}$. The computations are as follows:

$$\overrightarrow{AB} = \langle 2, 5, 0\rangle - \langle 0, 0, 4\rangle = \langle 2, 5, -4\rangle$$

$$\|\overrightarrow{AB}\| = \sqrt{2^2 + 5^2 + (-4)^2} = \sqrt{45} = 3\sqrt{5}$$

$$\frac{\overrightarrow{AB}}{\|\overrightarrow{AB}\|} = \left\langle \frac{2}{3\sqrt{5}}, \frac{5}{3\sqrt{5}}, -\frac{4}{3\sqrt{5}} \right\rangle$$

$$\mathbf{v} = \|\mathbf{v}\|\left(\frac{\overrightarrow{AB}}{\|\overrightarrow{AB}\|}\right) = \sqrt{5}\left\langle \frac{2}{3\sqrt{5}}, \frac{5}{3\sqrt{5}}, -\frac{4}{3\sqrt{5}} \right\rangle = \left\langle \frac{2}{3}, \frac{5}{3}, -\frac{4}{3} \right\rangle$$ ◀

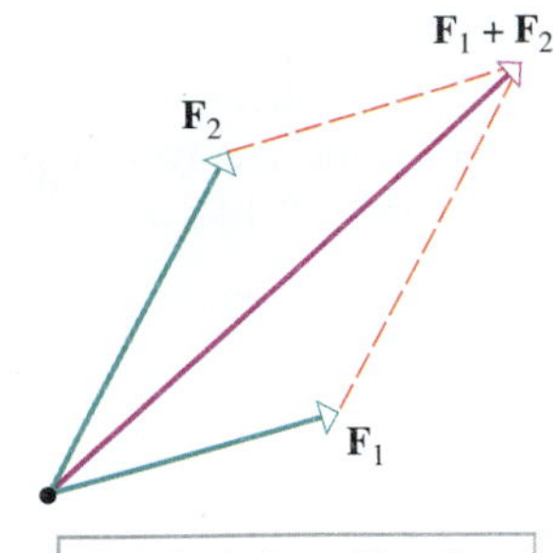

The single force $\mathbf{F}_1 + \mathbf{F}_2$ has the same effect as the two forces $\mathbf{F}_1$ and $\mathbf{F}_2$.

Figure 12.2.16

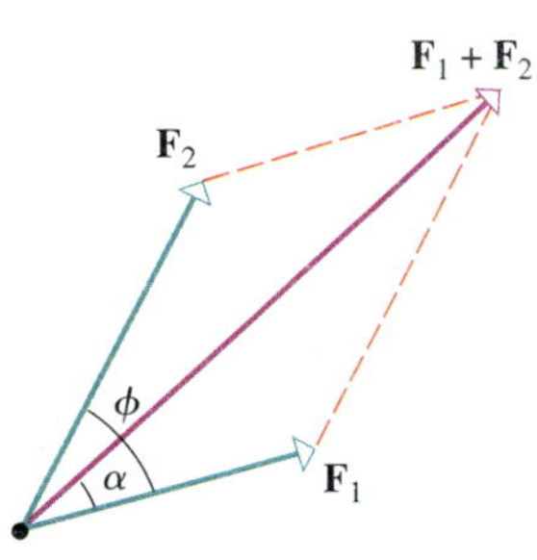

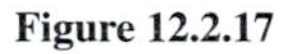
Figure 12.2.17

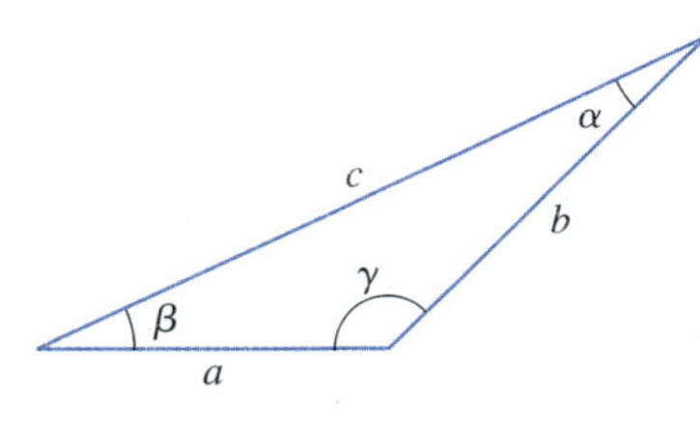

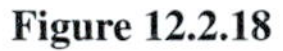
Figure 12.2.18

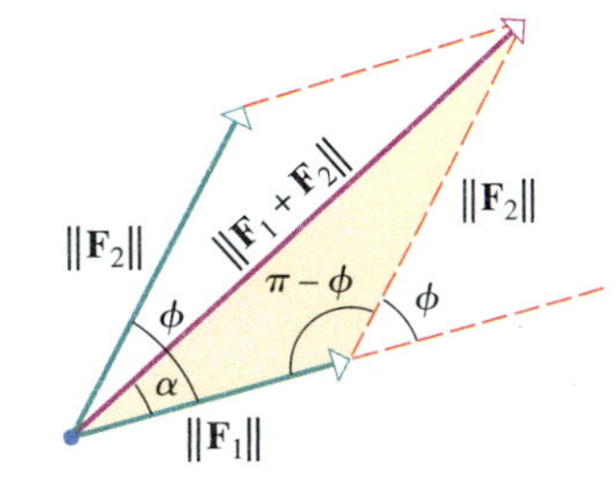

Figure 12.2.19

RESULTANT OF TWO CONCURRENT FORCES

The effect that a force has on an object depends on the magnitude and direction of the force and the point at which it is applied. Thus, forces are regarded to be vector quantities and, indeed, the algebraic operations on vectors that we have defined in this section have their origin in the study of forces. For example, it is a fact of physics that if two forces $\mathbf{F}_1$ and $\mathbf{F}_2$ are applied at the same point on an object, then the two forces have the same effect on the object as the single force $\mathbf{F}_1 + \mathbf{F}_2$ applied at the point (Figure 12.2.16). Physicists and engineers call $\mathbf{F}_1 + \mathbf{F}_2$ the ***resultant*** of $\mathbf{F}_1$ and $\mathbf{F}_2$, and they say that the forces $\mathbf{F}_1$ and $\mathbf{F}_2$ are ***concurrent*** to indicate that they are applied at the same point.

In many applications, the magnitudes of two concurrent forces and the angle between them are known, and the problem is to find the magnitude and direction of the resultant. For example, referring to Figure 12.2.17, suppose that we know the magnitudes of the forces $\mathbf{F}_1$ and $\mathbf{F}_2$ and the angle ϕ between them, and we are interested in finding the magnitude of the resultant $\mathbf{F}_1 + \mathbf{F}_2$ and the angle α that the resultant makes with the force $\mathbf{F}_1$. This can be done by trigonometric methods based on the laws of sines and cosines. For this purpose, recall that the law of sines applied to the triangle in Figure 12.2.18 states that

$$\frac{a}{\sin\alpha} = \frac{b}{\sin\beta} = \frac{c}{\sin\gamma}$$

and the law of cosines implies that

$$c^2 = a^2 + b^2 - 2ab\cos\gamma$$

Referring to Figure 12.2.19, and using the fact that $\cos(\pi - \phi) = -\cos\phi$, it follows from the law of cosines that

$$\|\mathbf{F}_1 + \mathbf{F}_2\|^2 = \|\mathbf{F}_1\|^2 + \|\mathbf{F}_2\|^2 + 2\|\mathbf{F}_1\|\|\mathbf{F}_2\|\cos\phi \tag{14}$$

Moreover, it follows from the law of sines that

$$\frac{\|\mathbf{F}_2\|}{\sin\alpha} = \frac{\|\mathbf{F}_1 + \mathbf{F}_2\|}{\sin(\pi - \phi)}$$

which, with the help of the identity $\sin(\pi - \phi) = \sin\phi$, can be expressed as

$$\sin\alpha = \frac{\|\mathbf{F}_2\|}{\|\mathbf{F}_1 + \mathbf{F}_2\|}\sin\phi \tag{15}$$

▶ **Example 8** Suppose that two forces are applied to an eye bracket, as shown in Figure 12.2.20. Find the magnitude of the resultant and the angle θ that it makes with the positive x-axis.

Solution. We are given that $\|\mathbf{F}_1\| = 200$ N and $\|\mathbf{F}_2\| = 300$ N and that the angle between the vectors $\mathbf{F}_1$ and $\mathbf{F}_2$ is $\phi = 40°$. Thus, it follows from (14) that the magnitude of the resultant is

$$\begin{aligned}\|\mathbf{F}_1 + \mathbf{F}_2\| &= \sqrt{\|\mathbf{F}_1\|^2 + \|\mathbf{F}_2\|^2 + 2\|\mathbf{F}_1\|\|\mathbf{F}_2\|\cos\phi} \\ &= \sqrt{(200)^2 + (300)^2 + 2(200)(300)\cos 40°} \\ &\approx 471 \text{ N}\end{aligned}$$

Moreover, it follows from (15) that the angle α between $\mathbf{F}_1$ and the resultant is

$$\alpha = \sin^{-1}\left(\frac{\|\mathbf{F}_2\|}{\|\mathbf{F}_1 + \mathbf{F}_2\|}\sin\phi\right) \approx \sin^{-1}\left(\frac{300}{471}\sin 40°\right) \approx 24.2°$$

Thus, the angle θ that the resultant makes with the positive x-axis is

$$\theta = \alpha + 30^\circ \approx 24.2^\circ + 30^\circ = 54.2^\circ$$

(Figure 12.2.21). ◀

The resultant of three or more concurrent forces can be found by working in pairs. For example, the resultant of three forces can be found by finding the resultant of any two of the forces and then finding the resultant of that resultant with the third force.

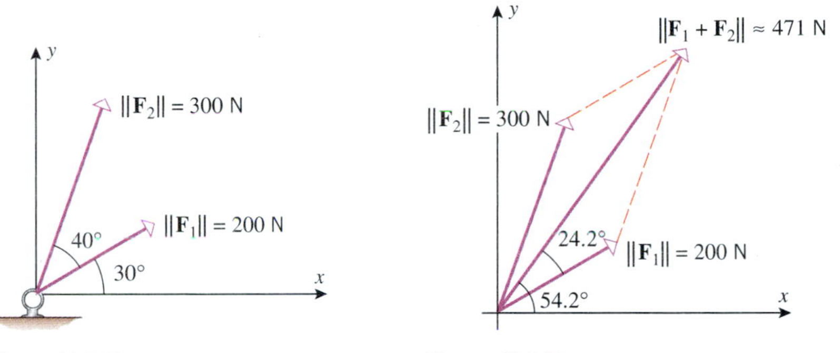

Figure 12.2.20

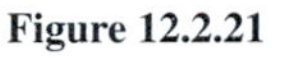

Figure 12.2.21

✔QUICK CHECK EXERCISES 12.2 *(See page 804 for answers.)*

1. If $\mathbf{v} = \langle 3, -1, 7\rangle$ and $\mathbf{w} = \langle 4, 10, -5\rangle$, then
   (a) $\|\mathbf{v}\| =$ ________ (b) $\mathbf{v} + \mathbf{w} =$ ________
   (c) $\mathbf{v} - \mathbf{w} =$ ________ (d) $2\mathbf{v} =$ ________.
2. The unit vector in the direction of $\mathbf{v} = \langle 3, -1, 7\rangle$ is ________.
3. The unit vector in 2-space that makes an angle of $\pi/3$ with the positive x-axis is ________.
4. Consider points $A(3, 4, 0)$ and $B(0, 0, 5)$.
   (a) $\overrightarrow{AB} =$ ________
   (b) If $\mathbf{v}$ is a vector in the same direction as $\overrightarrow{AB}$ and the length of $\mathbf{v}$ is $\sqrt{2}$, then $\mathbf{v} =$ ________.

EXERCISE SET 12.2

1–4 Sketch the vectors with their initial points at the origin.

1. (a) $\langle 2, 5\rangle$ (b) $\langle -5, -4\rangle$ (c) $\langle 2, 0\rangle$
   (d) $-5\mathbf{i} + 3\mathbf{j}$ (e) $3\mathbf{i} - 2\mathbf{j}$ (f) $-6\mathbf{j}$
2. (a) $\langle -3, 7\rangle$ (b) $\langle 6, -2\rangle$ (c) $\langle 0, -8\rangle$
   (d) $4\mathbf{i} + 2\mathbf{j}$ (e) $-2\mathbf{i} - \mathbf{j}$ (f) $4\mathbf{i}$
3. (a) $\langle 1, -2, 2\rangle$ (b) $\langle 2, 2, -1\rangle$
   (c) $-\mathbf{i} + 2\mathbf{j} + 3\mathbf{k}$ (d) $2\mathbf{i} + 3\mathbf{j} - \mathbf{k}$
4. (a) $\langle -1, 3, 2\rangle$ (b) $\langle 3, 4, 2\rangle$
   (c) $2\mathbf{j} - \mathbf{k}$ (d) $\mathbf{i} - \mathbf{j} + 2\mathbf{k}$

5–6 Find the components of the vector, and sketch an equivalent vector with its initial point at the origin.

5. (a) (b)

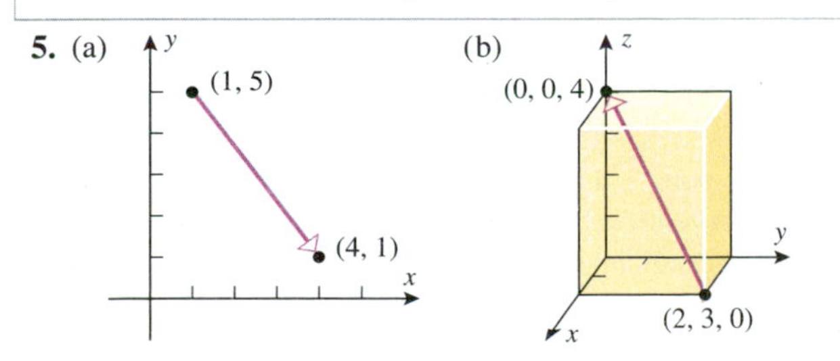

6. (a) (b)

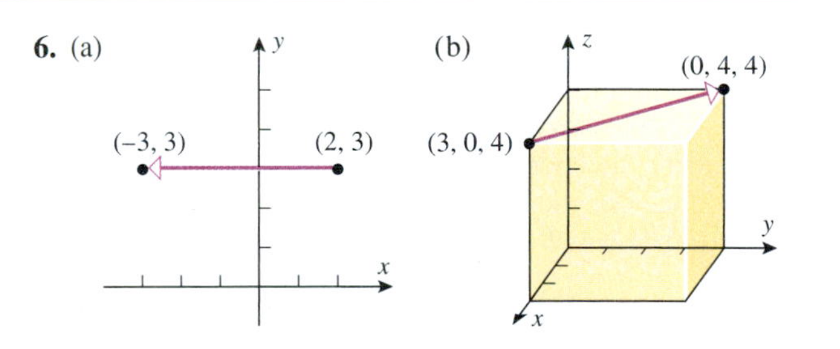

7–8 Find the components of the vector $\overrightarrow{P_1P_2}$.

7. (a) $P_1(3, 5)$, $P_2(2, 8)$ (b) $P_1(7, -2)$, $P_2(0, 0)$
   (c) $P_1(5, -2, 1)$, $P_2(2, 4, 2)$
8. (a) $P_1(-6, -2)$, $P_2(-4, -1)$
   (b) $P_1(0, 0, 0)$, $P_2(-1, 6, 1)$
   (c) $P_1(4, 1, -3)$, $P_2(9, 1, -3)$
9. (a) Find the terminal point of $\mathbf{v} = 3\mathbf{i} - 2\mathbf{j}$ if the initial point is $(1, -2)$.
   (b) Find the initial point of $\mathbf{v} = \langle -3, 1, 2\rangle$ if the terminal point is $(5, 0, -1)$.
10. (a) Find the terminal point of $\mathbf{v} = \langle 7, 6\rangle$ if the initial point is $(2, -1)$.

(b) Find the terminal point of $\mathbf{v} = \mathbf{i} + 2\mathbf{j} - 3\mathbf{k}$ if the initial point is $(-2, 1, 4)$.

11–12 Perform the stated operations on the vectors **u**, **v**, and **w**.

11. $\mathbf{u} = 3\mathbf{i} - \mathbf{k}$, $\mathbf{v} = \mathbf{i} - \mathbf{j} + 2\mathbf{k}$, $\mathbf{w} = 3\mathbf{j}$
(a) $\mathbf{w} - \mathbf{v}$ (b) $6\mathbf{u} + 4\mathbf{w}$
(c) $-\mathbf{v} - 2\mathbf{w}$ (d) $4(3\mathbf{u} + \mathbf{v})$
(e) $-8(\mathbf{v} + \mathbf{w}) + 2\mathbf{u}$ (f) $3\mathbf{w} - (\mathbf{v} - \mathbf{w})$

12. $\mathbf{u} = \langle 2, -1, 3\rangle$, $\mathbf{v} = \langle 4, 0, -2\rangle$, $\mathbf{w} = \langle 1, 1, 3\rangle$
(a) $\mathbf{u} - \mathbf{w}$ (b) $7\mathbf{v} + 3\mathbf{w}$ (c) $-\mathbf{w} + \mathbf{v}$
(d) $3(\mathbf{u} - 7\mathbf{v})$ (e) $-3\mathbf{v} - 8\mathbf{w}$ (f) $2\mathbf{v} - (\mathbf{u} + \mathbf{w})$

13–14 Find the norm of **v**.

13. (a) $\mathbf{v} = \langle 1, -1\rangle$ (b) $\mathbf{v} = -\mathbf{i} + 7\mathbf{j}$
(c) $\mathbf{v} = \langle -1, 2, 4\rangle$ (d) $\mathbf{v} = -3\mathbf{i} + 2\mathbf{j} + \mathbf{k}$

14. (a) $\mathbf{v} = \langle 3, 4\rangle$ (b) $\mathbf{v} = \sqrt{2}\mathbf{i} - \sqrt{7}\mathbf{j}$
(c) $\mathbf{v} = \langle 0, -3, 0\rangle$ (d) $\mathbf{v} = \mathbf{i} + \mathbf{j} + \mathbf{k}$

15. Let $\mathbf{u} = \mathbf{i} - 3\mathbf{j} + 2\mathbf{k}$, $\mathbf{v} = \mathbf{i} + \mathbf{j}$, and $\mathbf{w} = 2\mathbf{i} + 2\mathbf{j} - 4\mathbf{k}$. Find
(a) $\|\mathbf{u} + \mathbf{v}\|$ (b) $\|\mathbf{u}\| + \|\mathbf{v}\|$
(c) $\|-2\mathbf{u}\| + 2\|\mathbf{v}\|$ (d) $\|3\mathbf{u} - 5\mathbf{v} + \mathbf{w}\|$
(e) $\dfrac{1}{\|\mathbf{w}\|}\mathbf{w}$ (f) $\left\|\dfrac{1}{\|\mathbf{w}\|}\mathbf{w}\right\|$.

16. Is it possible to have $\|\mathbf{u}\| + \|\mathbf{v}\| = \|\mathbf{u} + \mathbf{v}\|$ if **u** and **v** are nonzero vectors? Justify your conclusion geometrically.

17–18 Find the unit vectors that satisfy the stated conditions.

17. (a) Same direction as $-\mathbf{i} + 4\mathbf{j}$.
(b) Oppositely directed to $6\mathbf{i} - 4\mathbf{j} + 2\mathbf{k}$.
(c) Same direction as the vector from the point $A(-1, 0, 2)$ to the point $B(3, 1, 1)$.

18. (a) Oppositely directed to $3\mathbf{i} - 4\mathbf{j}$.
(b) Same direction as $2\mathbf{i} - \mathbf{j} - 2\mathbf{k}$.
(c) Same direction as the vector from the point $A(-3, 2)$ to the point $B(1, -1)$.

19–20 Find the vectors that satisfy the stated conditions.

19. (a) Oppositely directed to $\mathbf{v} = \langle 3, -4\rangle$ and half the length of **v**.
(b) Length $\sqrt{17}$ and same direction as $\mathbf{v} = \langle 7, 0, -6\rangle$.

20. (a) Same direction as $\mathbf{v} = -2\mathbf{i} + 3\mathbf{j}$ and three times the length of **v**.
(b) Length 2 and oppositely directed to $\mathbf{v} = -3\mathbf{i} + 4\mathbf{j} + \mathbf{k}$.

21. In each part, find the component form of the vector **v** in 2-space that has the stated length and makes the stated angle θ with the positive x-axis.
(a) $\|\mathbf{v}\| = 3$; $\theta = \pi/4$ (b) $\|\mathbf{v}\| = 2$; $\theta = 90°$
(c) $\|\mathbf{v}\| = 5$; $\theta = 120°$ (d) $\|\mathbf{v}\| = 1$; $\theta = \pi$

22. Find the component forms of $\mathbf{v} + \mathbf{w}$ and $\mathbf{v} - \mathbf{w}$ in 2-space, given that $\|\mathbf{v}\| = 1$, $\|\mathbf{w}\| = 1$, **v** makes an angle of $\pi/6$ with the positive x-axis, and **w** makes an angle of $3\pi/4$ with the positive x-axis.

23–24 Find the component form of $\mathbf{v} + \mathbf{w}$, given that **v** and **w** are unit vectors.

23. **24.**

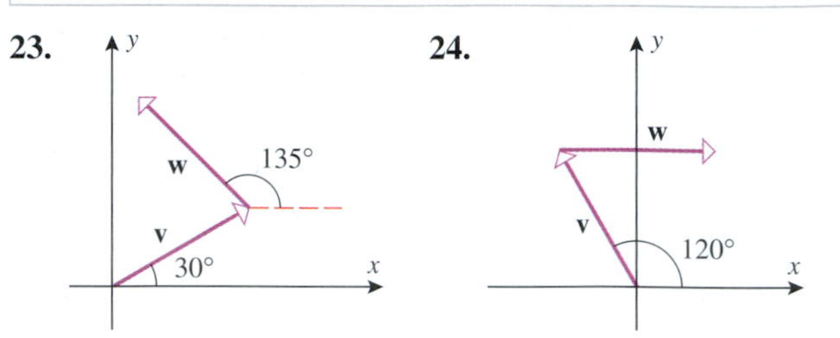

25. In each part, sketch the vector $\mathbf{u} + \mathbf{v} + \mathbf{w}$ and express it in component form.

(a) (b)

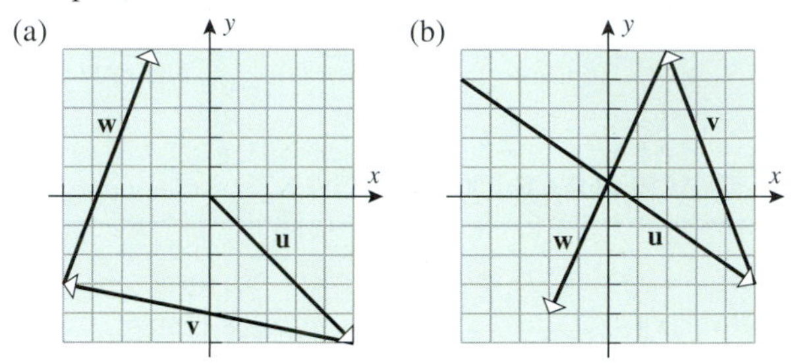

26. In each part of Exercise 25, sketch the vector $\mathbf{u} - \mathbf{v} + \mathbf{w}$ and express it in component form.

27. Let $\mathbf{u} = \langle 1, 3\rangle$, $\mathbf{v} = \langle 2, 1\rangle$, $\mathbf{w} = \langle 4, -1\rangle$. Find the vector **x** that satisfies $2\mathbf{u} - \mathbf{v} + \mathbf{x} = 7\mathbf{x} + \mathbf{w}$.

28. Let $\mathbf{u} = \langle -1, 1\rangle$, $\mathbf{v} = \langle 0, 1\rangle$, and $\mathbf{w} = \langle 3, 4\rangle$. Find the vector **x** that satisfies $\mathbf{u} - 2\mathbf{x} = \mathbf{x} - \mathbf{w} + 3\mathbf{v}$.

29. Find **u** and **v** if $\mathbf{u} + 2\mathbf{v} = 3\mathbf{i} - \mathbf{k}$ and $3\mathbf{u} - \mathbf{v} = \mathbf{i} + \mathbf{j} + \mathbf{k}$.

30. Find **u** and **v** if $\mathbf{u} + \mathbf{v} = \langle 2, -3\rangle$ and $3\mathbf{u} + 2\mathbf{v} = \langle -1, 2\rangle$.

31. Use vectors to find the lengths of the diagonals of the parallelogram that has $\mathbf{i} + \mathbf{j}$ and $\mathbf{i} - 2\mathbf{j}$ as adjacent sides.

32. Use vectors to find the fourth vertex of a parallelogram, three of whose vertices are $(0, 0)$, $(1, 3)$, and $(2, 4)$. [*Note:* There is more than one answer.]

33. (a) Given that $\|\mathbf{v}\| = 3$, find all values of k such that $\|k\mathbf{v}\| = 5$.
(b) Given that $k = -2$ and $\|k\mathbf{v}\| = 6$, find $\|\mathbf{v}\|$.

34. What do you know about k and **v** if $\|k\mathbf{v}\| = 0$?

35. In each part, find two unit vectors in 2-space that satisfy the stated condition.
(a) Parallel to the line $y = 3x + 2$
(b) Parallel to the line $x + y = 4$
(c) Perpendicular to the line $y = -5x + 1$

36. In each part, find two unit vectors in 3-space that satisfy the stated condition.
(a) Perpendicular to the xy-plane
(b) Perpendicular to the xz-plane
(c) Perpendicular to the yz-plane

FOCUS ON CONCEPTS

37. Let $\mathbf{r} = \langle x, y \rangle$ be an arbitrary vector. In each part, describe the set of all points (x, y) in 2-space that satisfy the stated condition.
(a) $\|\mathbf{r}\| = 1$ (b) $\|\mathbf{r}\| \leq 1$ (c) $\|\mathbf{r}\| > 1$

38. Let $\mathbf{r} = \langle x, y \rangle$ and $\mathbf{r}_0 = \langle x_0, y_0 \rangle$. In each part, describe the set of all points (x, y) in 2-space that satisfy the stated condition.
(a) $\|\mathbf{r} - \mathbf{r}_0\| = 1$ (b) $\|\mathbf{r} - \mathbf{r}_0\| \leq 1$ (c) $\|\mathbf{r} - \mathbf{r}_0\| > 1$

39. Let $\mathbf{r} = \langle x, y, z \rangle$ be an arbitrary vector. In each part, describe the set of all points (x, y, z) in 3-space that satisfy the stated condition.
(a) $\|\mathbf{r}\| = 1$ (b) $\|\mathbf{r}\| \leq 1$ (c) $\|\mathbf{r}\| > 1$

40. Let $\mathbf{r}_1 = \langle x_1, y_1 \rangle$, $\mathbf{r}_2 = \langle x_2, y_2 \rangle$, and $\mathbf{r} = \langle x, y \rangle$. Assuming that $k > \|\mathbf{r}_2 - \mathbf{r}_1\|$, describe the set of all points (x, y) for which $\|\mathbf{r} - \mathbf{r}_1\| + \|\mathbf{r} - \mathbf{r}_2\| = k$.

41–46 Find the magnitude of the resultant force and the angle that it makes with the positive x-axis.

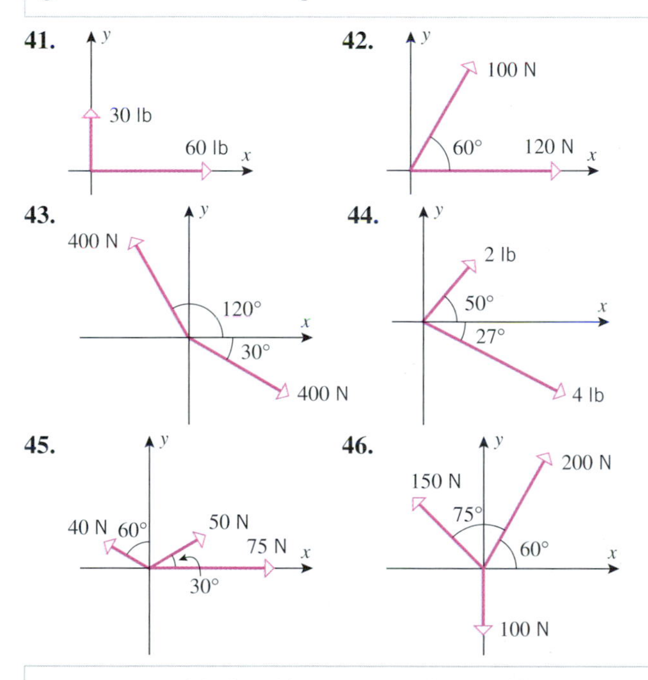

47–48 A particle is said to be in ***static equilibrium*** if the resultant of all forces applied to it is zero. In these exercises, find the force $\mathbf{F}$ that must be applied to the point to produce static equilibrium. Describe $\mathbf{F}$ by specifying its magnitude and the angle that it makes with the positive x-axis.

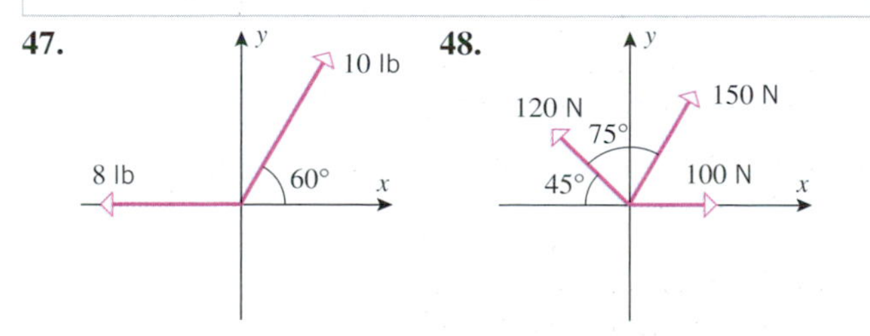

49. The accompanying figure shows a 250-lb traffic light supported by two flexible cables. The magnitudes of the forces that the cables apply to the eye ring are called the cable ***tensions***. Find the tensions in the cables if the traffic light is in static equilibrium (defined above Exercise 47).

50. Find the tensions in the cables shown in the accompanying figure if the block is in static equilibrium (see Exercise 49).

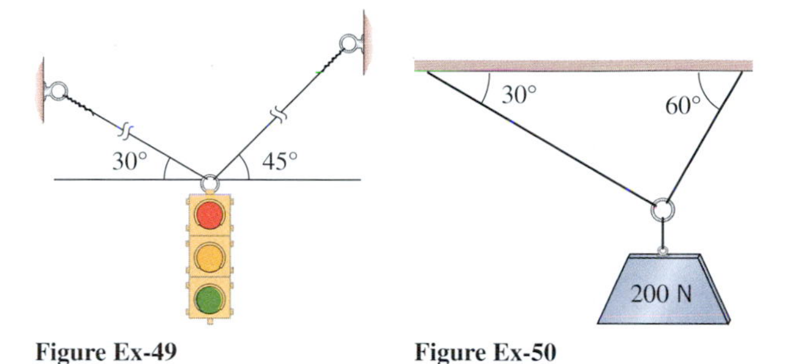

Figure Ex-49 **Figure Ex-50**

51. A vector $\mathbf{w}$ is said to be a ***linear combination*** of the vectors $\mathbf{v}_1$ and $\mathbf{v}_2$ if $\mathbf{w}$ can be expressed as $\mathbf{w} = c_1\mathbf{v}_1 + c_2\mathbf{v}_2$, where c_1 and c_2 are scalars.
(a) Find scalars c_1 and c_2 to express the vector $4\mathbf{j}$ as a linear combination of the vectors $\mathbf{v}_1 = 2\mathbf{i} - \mathbf{j}$ and $\mathbf{v}_2 = 4\mathbf{i} + 2\mathbf{j}$.
(b) Show that the vector $\langle 3, 5 \rangle$ cannot be expressed as a linear combination of the vectors $\mathbf{v}_1 = \langle 1, -3 \rangle$ and $\mathbf{v}_2 = \langle -2, 6 \rangle$.

52. A vector $\mathbf{w}$ is a ***linear combination*** of the vectors $\mathbf{v}_1$, $\mathbf{v}_2$, and $\mathbf{v}_3$ if $\mathbf{w}$ can be expressed as $\mathbf{w} = c_1\mathbf{v}_1 + c_2\mathbf{v}_2 + c_3\mathbf{v}_3$, where c_1, c_2, and c_3 are scalars.
(a) Find scalars c_1, c_2, and c_3 to express $\langle -1, 1, 5 \rangle$ as a linear combination of $\mathbf{v}_1 = \langle 1, 0, 1 \rangle$, $\mathbf{v}_2 = \langle 3, 2, 0 \rangle$, and $\mathbf{v}_3 = \langle 0, 1, 1 \rangle$.
(b) Show that the vector $2\mathbf{i} + \mathbf{j} - \mathbf{k}$ cannot be expressed as a linear combination of $\mathbf{v}_1 = \mathbf{i} - \mathbf{j}$, $\mathbf{v}_2 = 3\mathbf{i} + \mathbf{k}$, and $\mathbf{v}_3 = 4\mathbf{i} - \mathbf{j} + \mathbf{k}$.

53. Use a theorem from plane geometry to show that if $\mathbf{u}$ and $\mathbf{v}$ are vectors in 2-space or 3-space, then

$$\|\mathbf{u} + \mathbf{v}\| \leq \|\mathbf{u}\| + \|\mathbf{v}\|$$

which is called the ***triangle inequality for vectors***. Give some examples to illustrate this inequality.

54. Prove parts (a), (c), and (e) of Theorem 12.2.6 algebraically in 2-space.

55. Prove parts (d), (g), and (h) of Theorem 12.2.6 algebraically in 2-space.

56. Prove part (f) of Theorem 12.2.6 geometrically.

FOCUS ON CONCEPTS

57. Use vectors to prove that the line segment joining the midpoints of two sides of a triangle is parallel to the third side and half as long.

58. Use vectors to prove that the midpoints of the sides of a quadrilateral are the vertices of a parallelogram.

✔ QUICK CHECK ANSWERS 12.2

1. (a) $\sqrt{59}$ (b) $\langle 7, 9, 2\rangle$ (c) $\langle -1, -11, 12\rangle$ (d) $\langle 6, -2, 14\rangle$ **2.** $\frac{1}{\sqrt{59}}\mathbf{v} = \left\langle \frac{3}{\sqrt{59}}, -\frac{1}{\sqrt{59}}, \frac{7}{\sqrt{59}}\right\rangle$ **3.** $\left\langle \frac{1}{2}, \frac{\sqrt{3}}{2}\right\rangle = \frac{1}{2}\mathbf{i} + \frac{\sqrt{3}}{2}\mathbf{j}$
4. (a) $\langle -3, -4, 5\rangle$ (b) $\frac{1}{5}\overrightarrow{AB} = \langle -\frac{3}{5}, -\frac{4}{5}, 1\rangle$

12.3 DOT PRODUCT; PROJECTIONS

In the last section we defined three operations on vectors—addition, subtraction, and scalar multiplication. In scalar multiplication a vector is multiplied by a scalar and the result is a vector. In this section we will define a new kind of multiplication in which two vectors are multiplied to produce a scalar. This multiplication operation has many uses, some of which we will also discuss in this section.

DEFINITION OF THE DOT PRODUCT

12.3.1 DEFINITION. If $\mathbf{u} = \langle u_1, u_2\rangle$ and $\mathbf{v} = \langle v_1, v_2\rangle$ are vectors in 2-space, then the ***dot product*** of $\mathbf{u}$ and $\mathbf{v}$ is written as $\mathbf{u}\cdot\mathbf{v}$ and is defined as

$$\mathbf{u}\cdot\mathbf{v} = u_1v_1 + u_2v_2$$

Similarly, if $\mathbf{u} = \langle u_1, u_2, u_3\rangle$ and $\mathbf{v} = \langle v_1, v_2, v_3\rangle$ are vectors in 3-space, then their dot product is defined as

$$\mathbf{u}\cdot\mathbf{v} = u_1v_1 + u_2v_2 + u_3v_3$$

In words, the dot product of two vectors is formed by multiplying their corresponding components and adding the resulting products. Note that the dot product of two vectors is a scalar.

► Example 1

$$\langle 3, 5\rangle\cdot\langle -1, 2\rangle = 3(-1) + 5(2) = 7$$
$$\langle 2, 3\rangle\cdot\langle -3, 2\rangle = 2(-3) + 3(2) = 0$$
$$\langle 1, -3, 4\rangle\cdot\langle 1, 5, 2\rangle = 1(1) + (-3)(5) + 4(2) = -6$$

Here are the same computations expressed another way:

$$(3\mathbf{i} + 5\mathbf{j})\cdot(-\mathbf{i} + 2\mathbf{j}) = 3(-1) + 5(2) = 7$$
$$(2\mathbf{i} + 3\mathbf{j})\cdot(-3\mathbf{i} + 2\mathbf{j}) = 2(-3) + 3(2) = 0$$
$$(\mathbf{i} - 3\mathbf{j} + 4\mathbf{k})\cdot(\mathbf{i} + 5\mathbf{j} + 2\mathbf{k}) = 1(1) + (-3)(5) + 4(2) = -6 \;◄$$

TECHNOLOGY MASTERY

Many calculating utilities have a built-in dot product operation. If your calculating utility has this capability, use it to check the computations in Example 1.

ALGEBRAIC PROPERTIES OF THE DOT PRODUCT

The following theorem provides some of the basic algebraic properties of the dot product.

12.3.2 THEOREM. *If* $\mathbf{u}$, $\mathbf{v}$, *and* $\mathbf{w}$ *are vectors in 2- or 3-space and k is a scalar, then*

(a) $\mathbf{u}\cdot\mathbf{v} = \mathbf{v}\cdot\mathbf{u}$

(b) $\mathbf{u}\cdot(\mathbf{v} + \mathbf{w}) = \mathbf{u}\cdot\mathbf{v} + \mathbf{u}\cdot\mathbf{w}$

(c) $k(\mathbf{u}\cdot\mathbf{v}) = (k\mathbf{u})\cdot\mathbf{v} = \mathbf{u}\cdot(k\mathbf{v})$

(d) $\mathbf{v}\cdot\mathbf{v} = \|\mathbf{v}\|^2$

(e) $\mathbf{0}\cdot\mathbf{v} = 0$

Note the difference between the two zeros that appear in part (e) of Theorem 12.3.2–the zero on the left side is the *zero vector* (boldface), whereas the zero on the right side is the *zero scalar* (lightface).

We will prove parts (*c*) and (*d*) for vectors in 3-space and leave some of the others as exercises.

PROOF (*c*). Let $\mathbf{u} = \langle u_1, u_2, u_3 \rangle$ and $\mathbf{v} = \langle v_1, v_2, v_3 \rangle$. Then

$$k(\mathbf{u} \cdot \mathbf{v}) = k(u_1v_1 + u_2v_2 + u_3v_3) = (ku_1)v_1 + (ku_2)v_2 + (ku_3)v_3 = (k\mathbf{u}) \cdot \mathbf{v}$$

Similarly, $k(\mathbf{u} \cdot \mathbf{v}) = \mathbf{u} \cdot (k\mathbf{v})$.

PROOF (*d*). $\mathbf{v} \cdot \mathbf{v} = v_1v_1 + v_2v_2 + v_3v_3 = v_1^2 + v_2^2 + v_3^2 = \|\mathbf{v}\|^2$. ■

The following alternative form of the formula in part (*d*) of Theorem 12.3.2 provides a useful way of expressing the norm of a vector in terms of a dot product:

$$\|\mathbf{v}\| = \sqrt{\mathbf{v} \cdot \mathbf{v}} \tag{1}$$

ANGLE BETWEEN VECTORS

Suppose that **u** and **v** are nonzero vectors in 2-space or 3-space that are positioned so their initial points coincide. We define the ***angle between* u *and* v** to be the angle θ determined by the vectors that satisfies the condition $0 \le \theta \le \pi$ (Figure 12.3.1). In 2-space, θ is the smallest counterclockwise angle through which one of the vectors can be rotated until it aligns with the other.

The next theorem provides a way of calculating the angle between two vectors from their components.

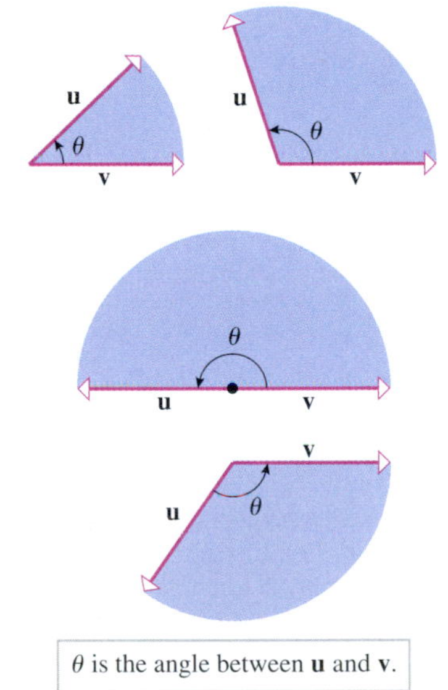

Figure 12.3.1

12.3.3 THEOREM. *If* **u** *and* **v** *are nonzero vectors in 2-space or 3-space, and if* θ *is the angle between them, then*

$$\cos\theta = \frac{\mathbf{u} \cdot \mathbf{v}}{\|\mathbf{u}\|\,\|\mathbf{v}\|} \tag{2}$$

PROOF. Suppose that the vectors **u**, **v**, and $\mathbf{v} - \mathbf{u}$ are positioned to form three sides of a triangle, as shown in Figure 12.3.2. It follows from the law of cosines that

$$\|\mathbf{v} - \mathbf{u}\|^2 = \|\mathbf{u}\|^2 + \|\mathbf{v}\|^2 - 2\|\mathbf{u}\|\,\|\mathbf{v}\| \cos\theta \tag{3}$$

Using the properties of the dot product in Theorem 12.3.2, we can rewrite the left side of this equation as

$$\begin{aligned} \|\mathbf{v} - \mathbf{u}\|^2 &= (\mathbf{v} - \mathbf{u}) \cdot (\mathbf{v} - \mathbf{u}) \\ &= (\mathbf{v} - \mathbf{u}) \cdot \mathbf{v} - (\mathbf{v} - \mathbf{u}) \cdot \mathbf{u} \\ &= \mathbf{v} \cdot \mathbf{v} - \mathbf{u} \cdot \mathbf{v} - \mathbf{v} \cdot \mathbf{u} + \mathbf{u} \cdot \mathbf{u} \\ &= \|\mathbf{v}\|^2 - 2\mathbf{u} \cdot \mathbf{v} + \|\mathbf{u}\|^2 \end{aligned}$$

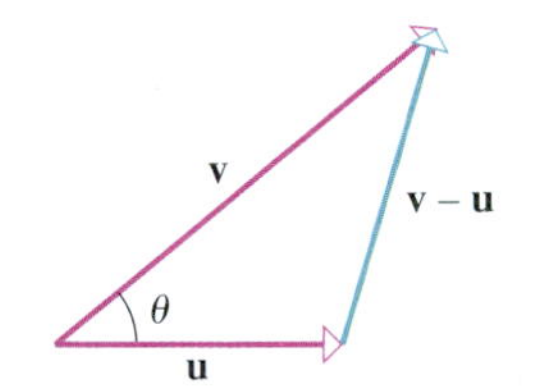

Figure 12.3.2

Substituting this back into (3) yields

$$\|\mathbf{v}\|^2 - 2\mathbf{u} \cdot \mathbf{v} + \|\mathbf{u}\|^2 = \|\mathbf{u}\|^2 + \|\mathbf{v}\|^2 - 2\|\mathbf{u}\|\,\|\mathbf{v}\| \cos\theta$$

which we can simplify and rewrite as

$$\mathbf{u} \cdot \mathbf{v} = \|\mathbf{u}\|\,\|\mathbf{v}\| \cos\theta$$

Finally, dividing both sides of this equation by $\|\mathbf{u}\|\,\|\mathbf{v}\|$ yields (2). ■

▶ **Example 2** Find the angle between the vector $\mathbf{u} = \mathbf{i} - 2\mathbf{j} + 2\mathbf{k}$ and

(a) $\mathbf{v} = -3\mathbf{i} + 6\mathbf{j} + 2\mathbf{k}$ (b) $\mathbf{w} = 2\mathbf{i} + 7\mathbf{j} + 6\mathbf{k}$ (c) $\mathbf{z} = -3\mathbf{i} + 6\mathbf{j} - 6\mathbf{k}$

Solution (a).

$$\cos\theta = \frac{\mathbf{u} \cdot \mathbf{v}}{\|\mathbf{u}\|\,\|\mathbf{v}\|} = \frac{-11}{(3)(7)} = -\frac{11}{21}$$

Thus,

$$\theta = \cos^{-1}\left(-\tfrac{11}{21}\right) \approx 2.12 \text{ radians} \approx 121.6^\circ$$

Solution (b).

$$\cos\theta = \frac{\mathbf{u}\cdot\mathbf{w}}{\|\mathbf{u}\|\,\|\mathbf{w}\|} = \frac{0}{\|\mathbf{u}\|\,\|\mathbf{w}\|} = 0$$

Thus, $\theta = \pi/2$, which means that the vectors are perpendicular.

Solution (c).

$$\cos\theta = \frac{\mathbf{u}\cdot\mathbf{z}}{\|\mathbf{u}\|\,\|\mathbf{z}\|} = \frac{-27}{(3)(9)} = -1$$

Thus, $\theta = \pi$, which means that the vectors are oppositely directed. In retrospect, we could have seen this without computing θ, since $\mathbf{z} = -3\mathbf{u}$. ◀

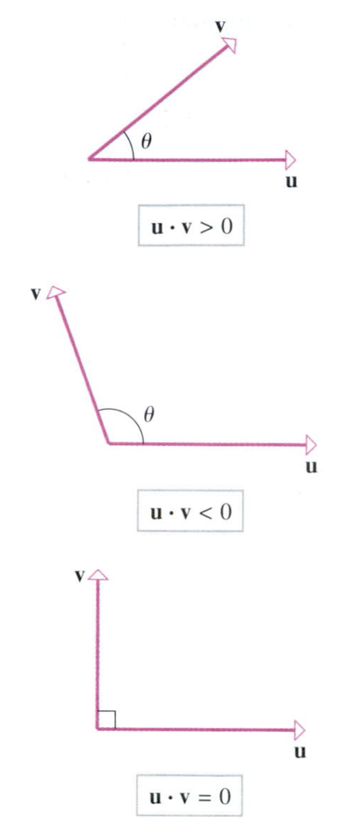

Figure 12.3.3

INTERPRETING THE SIGN OF THE DOT PRODUCT

It will often be convenient to express Formula (2) as

$$\mathbf{u}\cdot\mathbf{v} = \|\mathbf{u}\|\,\|\mathbf{v}\|\cos\theta \tag{4}$$

which expresses the dot product of $\mathbf{u}$ and $\mathbf{v}$ in terms of the lengths of these vectors and the angle between them. Since $\mathbf{u}$ and $\mathbf{v}$ are assumed to be nonzero vectors, this version of the formula makes it clear that the sign of $\mathbf{u}\cdot\mathbf{v}$ is the same as the sign of $\cos\theta$. Thus, we can tell from the dot product whether the angle between two vectors is acute or obtuse or whether the vectors are perpendicular (Figure 12.3.3).

> The terms "perpendicular," "orthogonal," and "normal" are all commonly used to describe geometric objects that meet at right angles. For consistency, we will say that two vectors are *orthogonal*, a vector is *normal* to a plane, and two planes are *perpendicular*. Moreover, although the zero vector does not make a well-defined angle with other vectors, we will consider $\mathbf{0}$ to be orthogonal to *all* vectors. This convention allows us to say that $\mathbf{u}$ and $\mathbf{v}$ are orthogonal vectors if and only if $\mathbf{u}\cdot\mathbf{v} = 0$, and makes Formula (4) valid if $\mathbf{u}$ or $\mathbf{v}$ (or both) is zero.

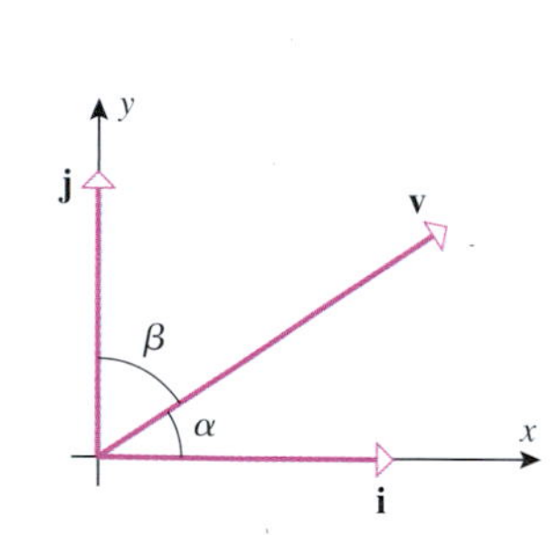

Figure 12.3.4

DIRECTION ANGLES

In an xy-coordinate system, the direction of a nonzero vector $\mathbf{v}$ is completely determined by the angles α and β between $\mathbf{v}$ and the unit vectors $\mathbf{i}$ and $\mathbf{j}$ (Figure 12.3.4), and in an xyz-coordinate system the direction is completely determined by the angles α, β, and γ between $\mathbf{v}$ and the unit vectors $\mathbf{i}$, $\mathbf{j}$, and $\mathbf{k}$ (Figure 12.3.5). In both 2-space and 3-space the angles between a nonzero vector $\mathbf{v}$ and the vectors $\mathbf{i}$, $\mathbf{j}$, and $\mathbf{k}$ are called the ***direction angles*** of $\mathbf{v}$, and the cosines of those angles are called the ***direction cosines*** of $\mathbf{v}$. Formulas for the direction cosines of a vector can be obtained from Formula (2). For example, if $\mathbf{v} = v_1\mathbf{i} + v_2\mathbf{j} + v_3\mathbf{k}$, then

$$\cos\alpha = \frac{\mathbf{v}\cdot\mathbf{i}}{\|\mathbf{v}\|\,\|\mathbf{i}\|} = \frac{v_1}{\|\mathbf{v}\|}, \quad \cos\beta = \frac{\mathbf{v}\cdot\mathbf{j}}{\|\mathbf{v}\|\,\|\mathbf{j}\|} = \frac{v_2}{\|\mathbf{v}\|}, \quad \cos\gamma = \frac{\mathbf{v}\cdot\mathbf{k}}{\|\mathbf{v}\|\,\|\mathbf{k}\|} = \frac{v_3}{\|\mathbf{v}\|}$$

Thus, we have the following theorem.

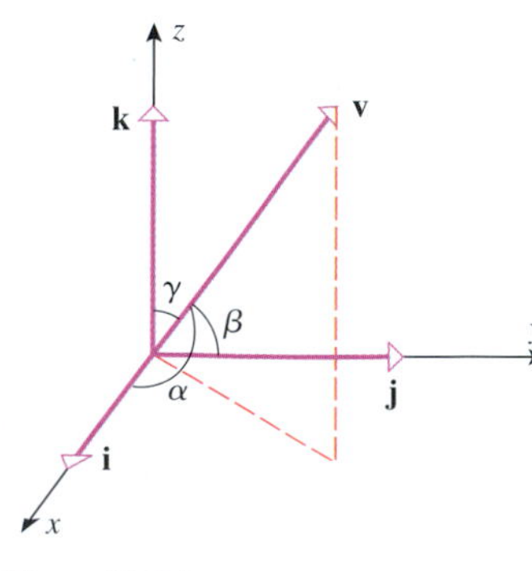

Figure 12.3.5

12.3.4 THEOREM. *The direction cosines of a nonzero vector* $\mathbf{v} = v_1\mathbf{i} + v_2\mathbf{j} + v_3\mathbf{k}$ *are*

$$\cos\alpha = \frac{v_1}{\|\mathbf{v}\|}, \quad \cos\beta = \frac{v_2}{\|\mathbf{v}\|}, \quad \cos\gamma = \frac{v_3}{\|\mathbf{v}\|}$$

The direction cosines of a vector $\mathbf{v} = v_1\mathbf{i} + v_2\mathbf{j} + v_3\mathbf{k}$ can be computed by normalizing $\mathbf{v}$ and reading off the components of $\mathbf{v}/\|\mathbf{v}\|$, since

$$\frac{\mathbf{v}}{\|\mathbf{v}\|} = \frac{v_1}{\|\mathbf{v}\|}\mathbf{i} + \frac{v_2}{\|\mathbf{v}\|}\mathbf{j} + \frac{v_3}{\|\mathbf{v}\|}\mathbf{k} = (\cos\alpha)\mathbf{i} + (\cos\beta)\mathbf{j} + (\cos\gamma)\mathbf{k}$$

We leave it as an exercise for you to show that the direction cosines of a vector satisfy the equation

$$\cos^2\alpha + \cos^2\beta + \cos^2\gamma = 1 \tag{5}$$

► **Example 3** Find the direction cosines of the vector $\mathbf{v} = 2\mathbf{i} - 4\mathbf{j} + 4\mathbf{k}$, and approximate the direction angles to the nearest degree.

Solution. First we will normalize the vector $\mathbf{v}$ and then read off the components. We have $\|\mathbf{v}\| = \sqrt{4+16+16} = 6$, so that $\mathbf{v}/\|\mathbf{v}\| = \frac{1}{3}\mathbf{i} - \frac{2}{3}\mathbf{j} + \frac{2}{3}\mathbf{k}$. Thus,

$$\cos\alpha = \tfrac{1}{3}, \quad \cos\beta = -\tfrac{2}{3}, \quad \cos\gamma = \tfrac{2}{3}$$

With the help of a calculating utility we obtain

$$\alpha = \cos^{-1}\left(\tfrac{1}{3}\right) \approx 71^\circ, \quad \beta = \cos^{-1}\left(-\tfrac{2}{3}\right) \approx 132^\circ, \quad \gamma = \cos^{-1}\left(\tfrac{2}{3}\right) \approx 48^\circ$$ ◄

► **Example 4** Find the angle between a diagonal of a cube and one of its edges.

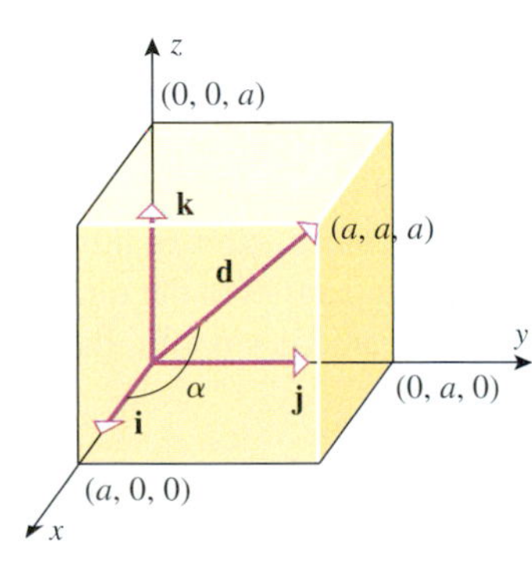

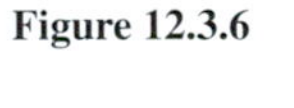
Figure 12.3.6

Solution. Assume that the cube has side a, and introduce a coordinate system as shown in Figure 12.3.6. In this coordinate system the vector

$$\mathbf{d} = a\mathbf{i} + a\mathbf{j} + a\mathbf{k}$$

is a diagonal of the cube and the unit vectors $\mathbf{i}$, $\mathbf{j}$, and $\mathbf{k}$ run along the edges. By symmetry, the diagonal makes the same angle with each edge, so it is sufficient to find the angle between $\mathbf{d}$ and $\mathbf{i}$ (the direction angle α). Thus,

$$\cos\alpha = \frac{\mathbf{d}\cdot\mathbf{i}}{\|\mathbf{d}\|\,\|\mathbf{i}\|} = \frac{a}{\|\mathbf{d}\|} = \frac{a}{\sqrt{3a^2}} = \frac{1}{\sqrt{3}}$$

and hence

$$\alpha = \cos^{-1}\left(\frac{1}{\sqrt{3}}\right) \approx 0.955 \text{ radian} \approx 54.7^\circ$$ ◄

DECOMPOSING VECTORS INTO ORTHOGONAL COMPONENTS

In many applications it is desirable to "decompose" a vector into a sum of two orthogonal vectors with convenient specified directions. For example, Figure 12.3.7 shows a block on an inclined plane. The downward force $\mathbf{F}$ that gravity exerts on the block can be decomposed into the sum

$$\mathbf{F} = \mathbf{F}_1 + \mathbf{F}_2$$

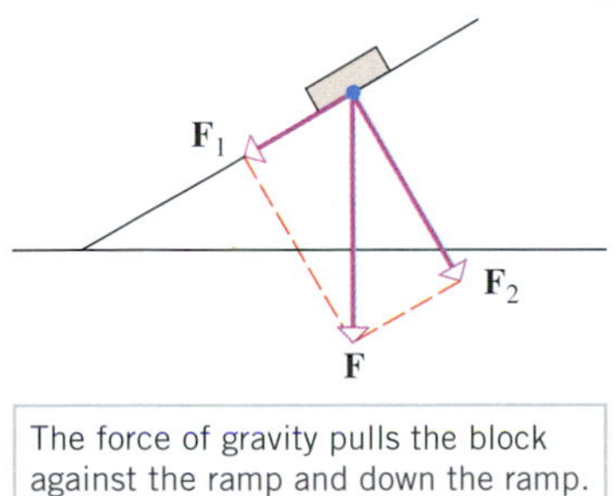

The force of gravity pulls the block against the ramp and down the ramp.

Figure 12.3.7

where the force $\mathbf{F}_1$ is parallel to the ramp and the force $\mathbf{F}_2$ is perpendicular to the ramp. The forces $\mathbf{F}_1$ and $\mathbf{F}_2$ are useful because $\mathbf{F}_1$ is the force that pulls the block *along* the ramp, and $\mathbf{F}_2$ is the force that the block exerts *against* the ramp.

Thus, our next objective is to develop a computational procedure for decomposing a vector into a sum of orthogonal vectors. For this purpose, suppose that $\mathbf{e}_1$ and $\mathbf{e}_2$ are two orthogonal *unit* vectors in 2-space, and suppose that we want to express a given vector $\mathbf{v}$ as a sum

$$\mathbf{v} = \mathbf{w}_1 + \mathbf{w}_2$$

so that $\mathbf{w}_1$ is a scalar multiple of $\mathbf{e}_1$ and $\mathbf{w}_2$ is a scalar multiple of $\mathbf{e}_2$ (Figure 12.3.8a). That is, we want to find scalars k_1 and k_2 such that

$$\mathbf{v} = k_1\mathbf{e}_1 + k_2\mathbf{e}_2 \tag{6}$$

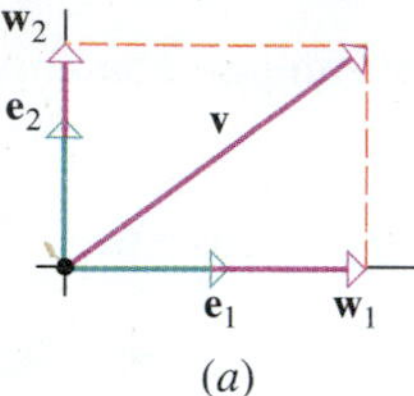

(a)

(‖v‖ sin θ)e2
‖v‖ cos θ
e2
‖v‖
‖v‖ sin θ
θ
e1 (‖v‖ cos θ)e1

(b)

Figure 12.3.8

We can find k_1 by taking the dot product of $\mathbf{v}$ with $\mathbf{e}_1$. This yields

$$\begin{aligned}\mathbf{v}\cdot\mathbf{e}_1 &= (k_1\mathbf{e}_1 + k_2\mathbf{e}_2)\cdot\mathbf{e}_1\\ &= k_1(\mathbf{e}_1\cdot\mathbf{e}_1) + k_2(\mathbf{e}_2\cdot\mathbf{e}_1)\\ &= k_1\|\mathbf{e}_1\|^2 + 0 = k_1\end{aligned}$$

Similarly,

$$\mathbf{v}\cdot\mathbf{e}_2 = (k_1\mathbf{e}_1 + k_2\mathbf{e}_2)\cdot\mathbf{e}_2 = k_1(\mathbf{e}_1\cdot\mathbf{e}_2) + k_2(\mathbf{e}_2\cdot\mathbf{e}_2) = 0 + k_2\|\mathbf{e}_2\|^2 = k_2$$

Substituting these expressions for k_1 and k_2 in (6) yields

$$\mathbf{v} = (\mathbf{v}\cdot\mathbf{e}_1)\mathbf{e}_1 + (\mathbf{v}\cdot\mathbf{e}_2)\mathbf{e}_2 \tag{7}$$

In this formula we call $(\mathbf{v}\cdot\mathbf{e}_1)\mathbf{e}_1$ and $(\mathbf{v}\cdot\mathbf{e}_2)\mathbf{e}_2$ the ***vector components*** of $\mathbf{v}$ along $\mathbf{e}_1$ and $\mathbf{e}_2$, respectively; and we call $\mathbf{v}\cdot\mathbf{e}_1$ and $\mathbf{v}\cdot\mathbf{e}_2$ the ***scalar components*** of $\mathbf{v}$ along $\mathbf{e}_1$ and $\mathbf{e}_2$, respectively. If θ denotes the angle between $\mathbf{v}$ and $\mathbf{e}_1$, then the scalar components of $\mathbf{v}$ can be written in trigonometric form as

$$\mathbf{v}\cdot\mathbf{e}_1 = \|\mathbf{v}\|\cos\theta \quad\text{and}\quad \mathbf{v}\cdot\mathbf{e}_2 = \|\mathbf{v}\|\sin\theta \tag{8}$$

(Figure 12.3.8*b*). Moreover, the vector components of $\mathbf{v}$ can be expressed as

$$(\mathbf{v}\cdot\mathbf{e}_1)\mathbf{e}_1 = (\|\mathbf{v}\|\cos\theta)\mathbf{e}_1 \quad\text{and}\quad (\mathbf{v}\cdot\mathbf{e}_2)\mathbf{e}_2 = (\|\mathbf{v}\|\sin\theta)\mathbf{e}_2 \tag{9}$$

and the decomposition (6) can be expressed as

$$\mathbf{v} = (\|\mathbf{v}\|\cos\theta)\mathbf{e}_1 + (\|\mathbf{v}\|\sin\theta)\mathbf{e}_2 \tag{10}$$

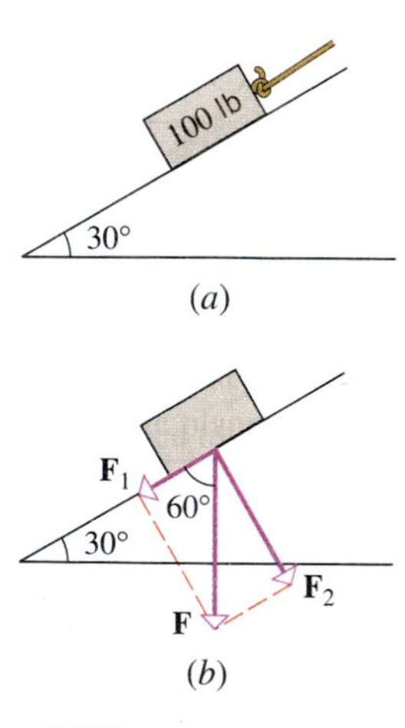

(a)

(b)

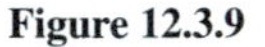
Figure 12.3.9

▸ **Example 5** A rope is attached to a 100-lb block on a ramp that is inclined at an angle of 30° with the ground (Figure 12.3.9*a*). How much force does the block exert against the ramp, and how much force must be applied to the rope in a direction parallel to the ramp to prevent the block from sliding down the ramp? (Assume that the ramp is smooth, that is, exerts no frictional forces.)

Solution. Let $\mathbf{F}$ denote the downward force of gravity on the block (so $\|\mathbf{F}\| = 100$ lb), and let $\mathbf{F}_1$ and $\mathbf{F}_2$ be the vector components of $\mathbf{F}$ parallel and perpendicular to the ramp (as shown in Figure 12.3.9*b*). The lengths of $\mathbf{F}_1$ and $\mathbf{F}_2$ are

$$\|\mathbf{F}_1\| = \|\mathbf{F}\|\cos 60^\circ = 100\left(\frac{1}{2}\right) = 50 \text{ lb}$$

$$\|\mathbf{F}_2\| = \|\mathbf{F}\|\sin 60^\circ = 100\left(\frac{\sqrt{3}}{2}\right) \approx 86.6 \text{ lb}$$

Thus, the block exerts a force of approximately 86.6 lb against the ramp, and it requires a force of 50 lb to prevent the block from sliding down the ramp. ◂

ORTHOGONAL PROJECTIONS

The vector components of $\mathbf{v}$ along $\mathbf{e}_1$ and $\mathbf{e}_2$ in (7) are also called the *orthogonal projections* of $\mathbf{v}$ on $\mathbf{e}_1$ and $\mathbf{e}_2$ and are commonly denoted by

$$\text{proj}_{\mathbf{e}_1}\mathbf{v} = (\mathbf{v}\cdot\mathbf{e}_1)\mathbf{e}_1 \quad\text{and}\quad \text{proj}_{\mathbf{e}_2}\mathbf{v} = (\mathbf{v}\cdot\mathbf{e}_2)\mathbf{e}_2$$

In general, if $\mathbf{e}$ is a unit vector, then we define the ***orthogonal projection of*** $\mathbf{v}$ ***on*** $\mathbf{e}$ to be

$$\text{proj}_{\mathbf{e}}\mathbf{v} = (\mathbf{v}\cdot\mathbf{e})\mathbf{e} \tag{11}$$

The orthogonal projection of $\mathbf{v}$ on an arbitrary nonzero vector $\mathbf{b}$ can be obtained by normalizing $\mathbf{b}$ and then applying Formula (11); that is,

$$\text{proj}_{\mathbf{b}}\mathbf{v} = \left(\mathbf{v} \cdot \frac{\mathbf{b}}{\|\mathbf{b}\|}\right)\left(\frac{\mathbf{b}}{\|\mathbf{b}\|}\right)$$

which can be rewritten as

$$\text{proj}_{\mathbf{b}}\mathbf{v} = \frac{\mathbf{v} \cdot \mathbf{b}}{\|\mathbf{b}\|^2}\mathbf{b} \tag{12}$$

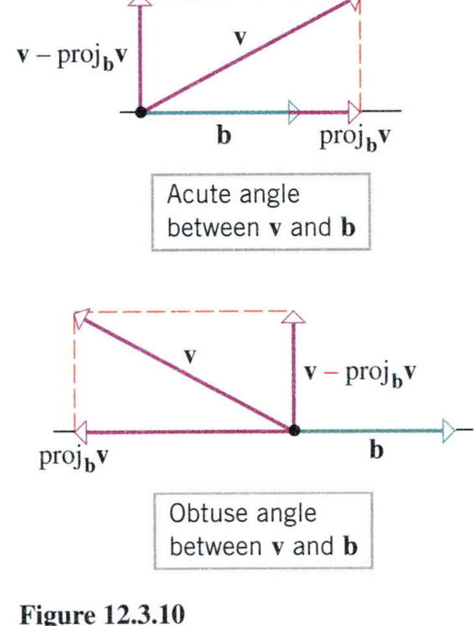

Figure 12.3.10

Geometrically, if $\mathbf{b}$ and $\mathbf{v}$ have a common initial point, then $\text{proj}_{\mathbf{b}}\mathbf{v}$ is the vector that is determined when a perpendicular is dropped from the terminal point of $\mathbf{v}$ to the line through $\mathbf{b}$ (illustrated in Figure 12.3.10 in two cases). Moreover, it is evident from Figure 12.3.10 that if we subtract $\text{proj}_{\mathbf{b}}\mathbf{v}$ from $\mathbf{v}$, then the resulting vector

$$\mathbf{v} - \text{proj}_{\mathbf{b}}\mathbf{v}$$

will be orthogonal to $\mathbf{b}$; we call this the ***vector component of* $\mathbf{v}$ *orthogonal to* $\mathbf{b}$**.

► Example 6 Find the orthogonal projection of $\mathbf{v} = \mathbf{i} + \mathbf{j} + \mathbf{k}$ on $\mathbf{b} = 2\mathbf{i} + 2\mathbf{j}$, and then find the vector component of $\mathbf{v}$ orthogonal to $\mathbf{b}$.

Solution. We have

$$\mathbf{v} \cdot \mathbf{b} = (\mathbf{i} + \mathbf{j} + \mathbf{k}) \cdot (2\mathbf{i} + 2\mathbf{j}) = 2 + 2 + 0 = 4$$
$$\|\mathbf{b}\|^2 = 2^2 + 2^2 = 8$$

Thus, the orthogonal projection of $\mathbf{v}$ on $\mathbf{b}$ is

$$\text{proj}_{\mathbf{b}}\mathbf{v} = \frac{\mathbf{v} \cdot \mathbf{b}}{\|\mathbf{b}\|^2}\mathbf{b} = \frac{4}{8}(2\mathbf{i} + 2\mathbf{j}) = \mathbf{i} + \mathbf{j}$$

and the vector component of $\mathbf{v}$ orthogonal to $\mathbf{b}$ is

$$\mathbf{v} - \text{proj}_{\mathbf{b}}\mathbf{v} = (\mathbf{i} + \mathbf{j} + \mathbf{k}) - (\mathbf{i} + \mathbf{j}) = \mathbf{k}$$

These results are consistent with Figure 12.3.11. ◄

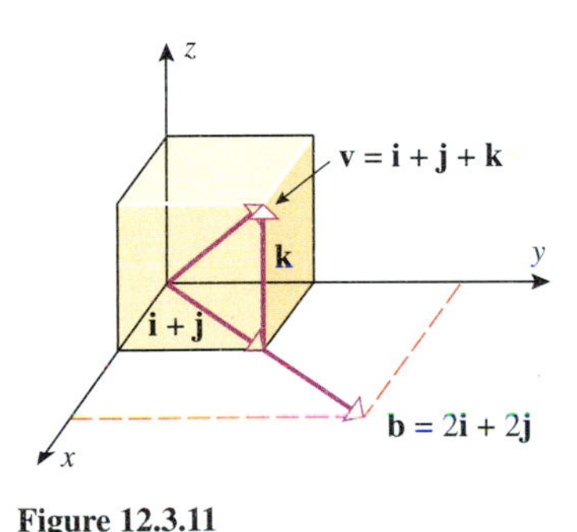

Figure 12.3.11

WORK

In Section 7.7 we discussed the work done by a constant force acting on an object that moves along a line. We defined the work W done on the object by a constant force of magnitude F acting in the direction of motion over a distance d to be

$$W = Fd = \text{force} \times \text{distance} \tag{13}$$

If we let $\mathbf{F}$ denote a force vector of magnitude $\|\mathbf{F}\| = F$ *acting in the direction of motion*, then we can write (13) as

$$W = \|\mathbf{F}\|d$$

Furthermore, if we assume that the object moves along a line from point P to point Q, then $d = \|\overrightarrow{PQ}\|$, so that the work can be expressed entirely in vector form as

$$W = \|\mathbf{F}\|\|\overrightarrow{PQ}\|$$

(Figure 12.3.12*a*). The vector $\overrightarrow{PQ}$ is called the ***displacement vector*** for the object. In the case where a constant force $\mathbf{F}$ is not in the direction of motion, but rather makes an angle θ with the displacement vector, then we *define* the work W done by $\mathbf{F}$ to be

$$W = (\|\mathbf{F}\|\cos\theta)\|\overrightarrow{PQ}\| = \mathbf{F} \cdot \overrightarrow{PQ} \tag{14}$$

(Figure 12.3.12*b*).

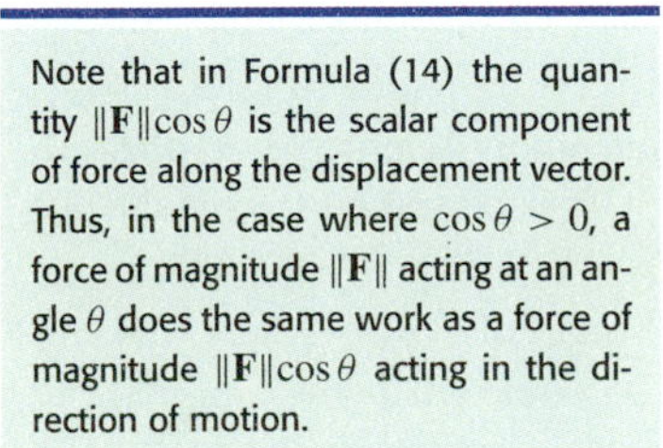
Note that in Formula (14) the quantity $\|\mathbf{F}\|\cos\theta$ is the scalar component of force along the displacement vector. Thus, in the case where $\cos\theta > 0$, a force of magnitude $\|\mathbf{F}\|$ acting at an angle θ does the same work as a force of magnitude $\|\mathbf{F}\|\cos\theta$ acting in the direction of motion.

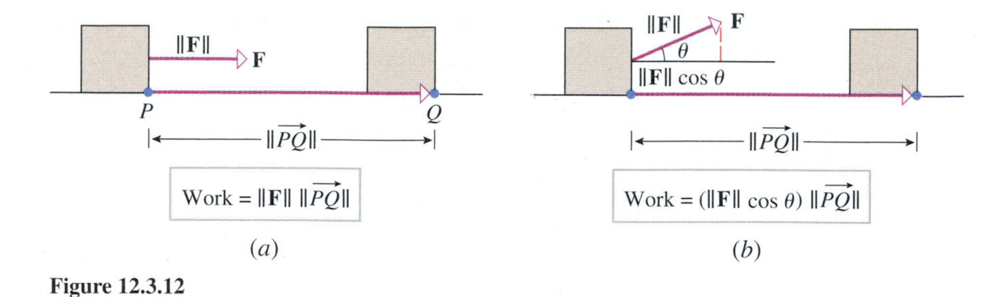

Figure 12.3.12

Example 7 A wagon is pulled horizontally by exerting a constant force of 10 lb on the handle at an angle of 60° with the horizontal. How much work is done in moving the wagon 50 ft?

Solution. Introduce an xy-coordinate system so that the wagon moves from $P(0, 0)$ to $Q(50, 0)$ along the x-axis (Figure 12.3.13). In this coordinate system

$$\overrightarrow{PQ} = 50\mathbf{i}$$

and

$$\mathbf{F} = (10\cos 60°)\mathbf{i} + (10\sin 60°)\mathbf{j} = 5\mathbf{i} + 5\sqrt{3}\mathbf{j}$$

so the work done is

$$W = \mathbf{F} \cdot \overrightarrow{PQ} = (5\mathbf{i} + 5\sqrt{3}\mathbf{j}) \cdot (50\mathbf{i}) = 250 \text{ (foot-pounds)}$$

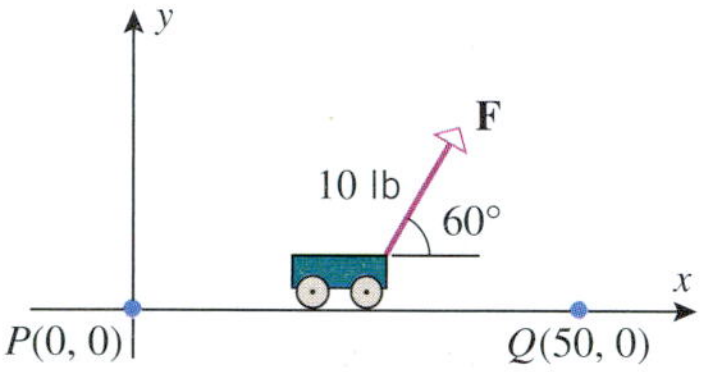

Figure 12.3.13

✔ QUICK CHECK EXERCISES 12.3 *(See page 813 for answers.)*

1. $\langle 3, 1, -2\rangle \cdot \langle 6, 0, 5\rangle =$ ______

2. Suppose that $\mathbf{u}$, $\mathbf{v}$, and $\mathbf{w}$ are vectors in 3-space such that $\|\mathbf{u}\| = 5$, $\mathbf{u}\cdot\mathbf{v} = 7$, and $\mathbf{u}\cdot\mathbf{w} = -3$.
   (a) $\mathbf{u}\cdot\mathbf{u} =$ ______ (b) $\mathbf{v}\cdot\mathbf{u} =$ ______
   (c) $\mathbf{u}\cdot(\mathbf{v}-\mathbf{w}) =$ ______ (d) $\mathbf{u}\cdot(2\mathbf{w}) =$ ______

3. For the vectors $\mathbf{u}$ and $\mathbf{v}$ in the preceding exercise, if the angle between $\mathbf{u}$ and $\mathbf{v}$ is $\pi/3$, then $\|\mathbf{v}\| =$ ______.

4. The direction cosines of $\langle 2, -1, 3\rangle$ are $\cos\alpha =$ ______, $\cos\beta =$ ______, and $\cos\gamma =$ ______.

5. The orthogonal projection of $\mathbf{v} = 10\mathbf{i}$ on $\mathbf{b} = -3\mathbf{i} + \mathbf{j}$ is ______.

EXERCISE SET 12.3 Graphing Utility C CAS

1. In each part, find the dot product of the vectors and the cosine of the angle between them.
   (a) $\mathbf{u} = \mathbf{i} + 2\mathbf{j}$, $\mathbf{v} = 6\mathbf{i} - 8\mathbf{j}$
   (b) $\mathbf{u} = \langle -7, -3\rangle$, $\mathbf{v} = \langle 0, 1\rangle$
   (c) $\mathbf{u} = \mathbf{i} - 3\mathbf{j} + 7\mathbf{k}$, $\mathbf{v} = 8\mathbf{i} - 2\mathbf{j} - 2\mathbf{k}$
   (d) $\mathbf{u} = \langle -3, 1, 2\rangle$, $\mathbf{v} = \langle 4, 2, -5\rangle$
   (c) $\mathbf{u} = \langle 1, 1, 1\rangle$, $\mathbf{v} = \langle -1, 0, 0\rangle$
   (d) $\mathbf{u} = \langle 4, 1, 6\rangle$, $\mathbf{v} = \langle -3, 0, 2\rangle$

2. In each part use the given information to find $\mathbf{u}\cdot\mathbf{v}$.
   (a) $\|\mathbf{u}\| = 1$, $\|\mathbf{v}\| = 2$, the angle between $\mathbf{u}$ and $\mathbf{v}$ is $\pi/6$.
   (b) $\|\mathbf{u}\| = 2$, $\|\mathbf{v}\| = 3$, the angle between $\mathbf{u}$ and $\mathbf{v}$ is 135°.

3. In each part, determine whether $\mathbf{u}$ and $\mathbf{v}$ make an acute angle, an obtuse angle, or are orthogonal.
   (a) $\mathbf{u} = 7\mathbf{i} + 3\mathbf{j} + 5\mathbf{k}$, $\mathbf{v} = -8\mathbf{i} + 4\mathbf{j} + 2\mathbf{k}$
   (b) $\mathbf{u} = 6\mathbf{i} + \mathbf{j} + 3\mathbf{k}$, $\mathbf{v} = 4\mathbf{i} - 6\mathbf{k}$

FOCUS ON CONCEPTS

4. Does the triangle in 3-space with vertices $(-1, 2, 3)$, $(2, -2, 0)$, and $(3, 1, -4)$ have an obtuse angle? Justify your answer.

5. The accompanying figure shows eight vectors that are equally spaced around a circle of radius 1. Find the dot product of $\mathbf{v}_0$ with each of the other seven vectors.

6. The accompanying figure shows six vectors that are equally spaced around a circle of radius 5. Find the dot product of $\mathbf{v}_0$ with each of the other five vectors.

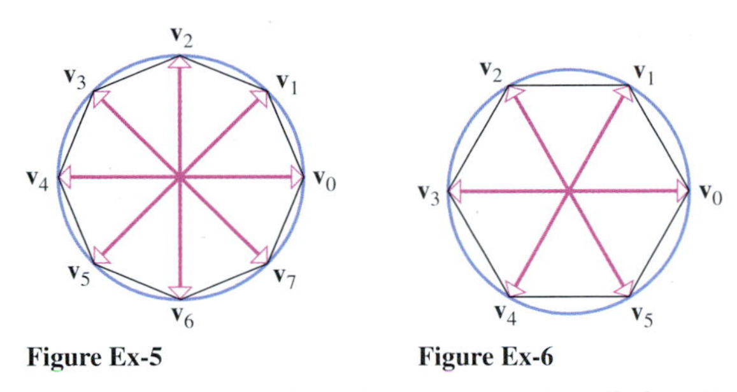

Figure Ex-5 **Figure Ex-6**

7. (a) Use vectors to show that $A(2, -1, 1)$, $B(3, 2, -1)$, and $C(7, 0, -2)$ are vertices of a right triangle. At which vertex is the right angle?
(b) Use vectors to find the interior angles of the triangle with vertices $(-1, 0)$, $(2, -1)$, and $(1, 4)$. Express your answers to the nearest degree.

8. (a) Show that if $\mathbf{v} = a\mathbf{i} + b\mathbf{j}$ is a vector in 2-space, then the vectors

$$\mathbf{v}_1 = -b\mathbf{i} + a\mathbf{j} \quad \text{and} \quad \mathbf{v}_2 = b\mathbf{i} - a\mathbf{j}$$

are both orthogonal to $\mathbf{v}$.
(b) Use the result in part (a) to find two unit vectors that are orthogonal to the vector $\mathbf{v} = 3\mathbf{i} - 2\mathbf{j}$. Sketch the vectors $\mathbf{v}$, $\mathbf{v}_1$, and $\mathbf{v}_2$.

9. Explain why each of the following expressions makes no sense.
(a) $\mathbf{u} \cdot (\mathbf{v} \cdot \mathbf{w})$ (b) $(\mathbf{u} \cdot \mathbf{v}) + \mathbf{w}$
(c) $\|\mathbf{u} \cdot \mathbf{v}\|$ (d) $k \cdot (\mathbf{u} + \mathbf{v})$

10. True or false? If $\mathbf{a} \cdot \mathbf{b} = \mathbf{a} \cdot \mathbf{c}$ and if $\mathbf{a} \neq \mathbf{0}$, then $\mathbf{b} = \mathbf{c}$. Justify your conclusion.

11. Verify parts (*b*) and (*c*) of Theorem 12.3.2 for the vectors $\mathbf{u} = 6\mathbf{i} - \mathbf{j} + 2\mathbf{k}$, $\mathbf{v} = 2\mathbf{i} + 7\mathbf{j} + 4\mathbf{k}$, $\mathbf{w} = \mathbf{i} + \mathbf{j} - 3\mathbf{k}$ and $k = -5$.

12. Let $\mathbf{u} = \langle 1, 2\rangle$, $\mathbf{v} = \langle 4, -2\rangle$, and $\mathbf{w} = \langle 6, 0\rangle$. Find
(a) $\mathbf{u} \cdot (7\mathbf{v} + \mathbf{w})$ (b) $\|(\mathbf{u} \cdot \mathbf{w})\mathbf{w}\|$
(c) $\|\mathbf{u}\|(\mathbf{v} \cdot \mathbf{w})$ (d) $(\|\mathbf{u}\|\mathbf{v}) \cdot \mathbf{w}$.

13. Find r so that the vector from the point $A(1, -1, 3)$ to the point $B(3, 0, 5)$ is orthogonal to the vector from A to the point $P(r, r, r)$.

14. Find two unit vectors in 2-space that make an angle of 45° with $4\mathbf{i} + 3\mathbf{j}$.

15–16 Find the direction cosines of $\mathbf{v}$ and confirm that they satisfy Equation (5). Then use the direction cosines to approximate the direction angles to the nearest degree.

15. (a) $\mathbf{v} = \mathbf{i} + \mathbf{j} - \mathbf{k}$ (b) $\mathbf{v} = 2\mathbf{i} - 2\mathbf{j} + \mathbf{k}$

16. (a) $\mathbf{v} = 3\mathbf{i} - 2\mathbf{j} - 6\mathbf{k}$ (b) $\mathbf{v} = 3\mathbf{i} - 4\mathbf{k}$

FOCUS ON CONCEPTS

17. Show that the direction cosines of a vector satisfy

$$\cos^2\alpha + \cos^2\beta + \cos^2\gamma = 1$$

18. Let θ and λ be the angles shown in the accompanying figure. Show that the direction cosines of $\mathbf{v}$ can be expressed as

$$\cos\alpha = \cos\lambda\cos\theta$$
$$\cos\beta = \cos\lambda\sin\theta$$
$$\cos\gamma = \sin\lambda$$

[*Hint:* Express $\mathbf{v}$ in component form and normalize.]

19. The accompanying figure shows a cube.
(a) Find the angle between the vectors $\mathbf{d}$ and $\mathbf{u}$ to the nearest degree.
(b) Make a conjecture about the angle between the vectors $\mathbf{d}$ and $\mathbf{v}$, and confirm your conjecture by computing the angle.

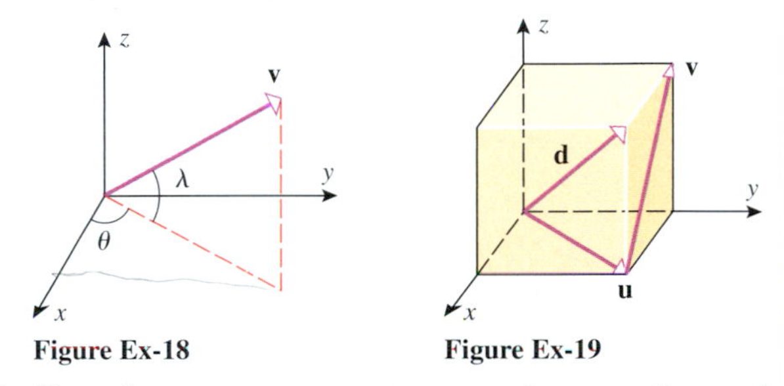

Figure Ex-18 **Figure Ex-19**

20. Show that two nonzero vectors $\mathbf{v}_1$ and $\mathbf{v}_2$ are orthogonal if and only if their direction cosines satisfy

$$\cos\alpha_1\cos\alpha_2 + \cos\beta_1\cos\beta_2 + \cos\gamma_1\cos\gamma_2 = 0$$

21. Use the result in Exercise 18 to find the direction angles of the vector shown in the accompanying figure to the nearest degree.

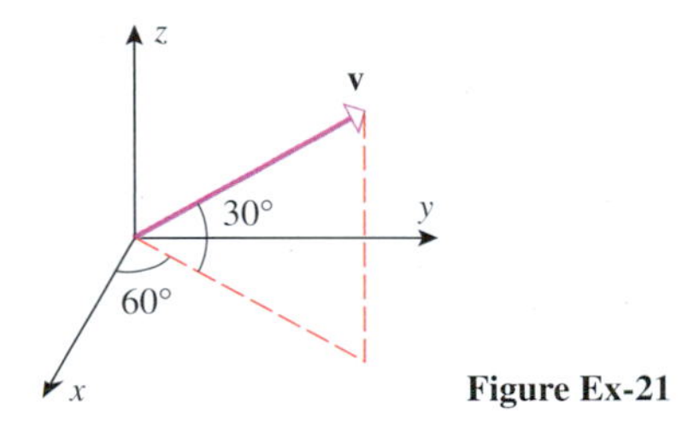

Figure Ex-21

22. Find, to the nearest degree, the acute angle formed by two diagonals of a cube.

23. Find, to the nearest degree, the angles that a diagonal of a box with dimensions 10 cm by 15 cm by 25 cm makes with the edges of the box.

24. In each part, find the vector component of $\mathbf{v}$ along $\mathbf{b}$ and the vector component of $\mathbf{v}$ orthogonal to $\mathbf{b}$. Then sketch the vectors $\mathbf{v}$, $\text{proj}_{\mathbf{b}}\mathbf{v}$, and $\mathbf{v} - \text{proj}_{\mathbf{b}}\mathbf{v}$.
(a) $\mathbf{v} = 2\mathbf{i} - \mathbf{j}$, $\mathbf{b} = 3\mathbf{i} + 4\mathbf{j}$
(b) $\mathbf{v} = \langle 4, 5\rangle$, $\mathbf{b} = \langle 1, -2\rangle$
(c) $\mathbf{v} = -3\mathbf{i} - 2\mathbf{j}$, $\mathbf{b} = 2\mathbf{i} + \mathbf{j}$

25. In each part, find the vector component of $\mathbf{v}$ along $\mathbf{b}$ and the vector component of $\mathbf{v}$ orthogonal to $\mathbf{b}$.
(a) $\mathbf{v} = 2\mathbf{i} - \mathbf{j} + 3\mathbf{k}$, $\mathbf{b} = \mathbf{i} + 2\mathbf{j} + 2\mathbf{k}$
(b) $\mathbf{v} = \langle 4, -1, 7\rangle$, $\mathbf{b} = \langle 2, 3, -6\rangle$

26–27 Express the vector $\mathbf{v}$ as the sum of a vector parallel to $\mathbf{b}$ and a vector orthogonal to $\mathbf{b}$.

26. (a) $\mathbf{v} = 2\mathbf{i} - 4\mathbf{j},\ \mathbf{b} = \mathbf{i} + \mathbf{j}$
(b) $\mathbf{v} = 3\mathbf{i} + \mathbf{j} - 2\mathbf{k},\ \mathbf{b} = 2\mathbf{i} - \mathbf{k}$
(c) $\mathbf{v} = 4\mathbf{i} - 2\mathbf{j} + 6\mathbf{k},\ \mathbf{b} = -2\mathbf{i} + \mathbf{j} - 3\mathbf{k}$

27. (a) $\mathbf{v} = \langle -3, 5\rangle,\ \mathbf{b} = \langle 1, 1\rangle$
(b) $\mathbf{v} = \langle -2, 1, 6\rangle,\ \mathbf{b} = \langle 0, -2, 1\rangle$
(c) $\mathbf{v} = \langle 1, 4, 1\rangle,\ \mathbf{b} = \langle 3, -2, 5\rangle$

28. If L is a line in 2-space or 3-space that passes through the points A and B, then the distance from a point P to the line L is equal to the length of the component of the vector $\overrightarrow{AP}$ that is orthogonal to the vector $\overrightarrow{AB}$ (see the accompanying figure). Use this result to find the distance from the point $P(1, 0)$ to the line through $A(2, -3)$ and $B(5, 1)$.

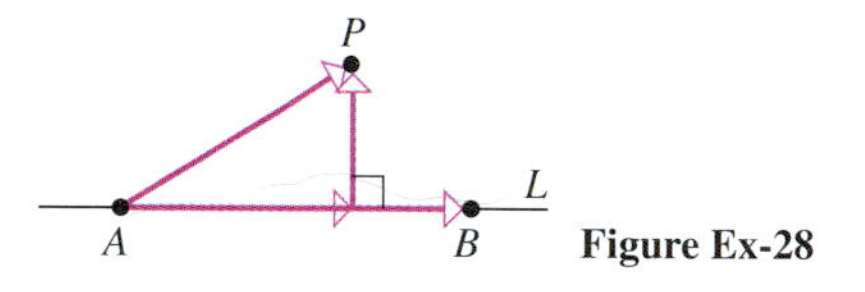

Figure Ex-28

29. Use the method of Exercise 28 to find the distance from the point $P(-3, 1, 2)$ to the line through $A(1, 1, 0)$ and $B(-2, 3, -4)$.

30. As shown in the accompanying figure, a child with mass 34 kg is seated on a smooth (frictionless) playground slide that is inclined at an angle of 27° with the horizontal. How much force does the child exert on the slide, and how much force must be applied in the direction of $\mathbf{P}$ to prevent the child from sliding down the slide? Take the acceleration due to gravity to be 9.8 m/s^2.

31. For the child in Exercise 30, how much force must be applied in the direction of $\mathbf{Q}$ (shown in the accompanying figure) to prevent the child from sliding down the slide?

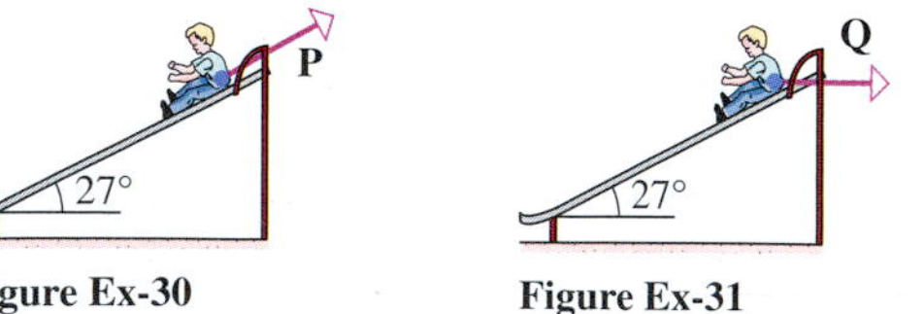

Figure Ex-30 Figure Ex-31

32. A block weighing 300 lb is suspended by cables A and B, as shown in the accompanying figure. Determine the forces that the block exerts along the cables.

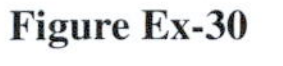

33. A block weighing 100 N is suspended by cables A and B, as shown in the accompanying figure.
(a) Use a graphing utility to graph the forces that the block exerts along cables A and B as functions of the "sag" d.
(b) Does increasing the sag increase or decrease the forces on the cables?
(c) How much sag is required if the cables cannot tolerate forces in excess of 150 N?

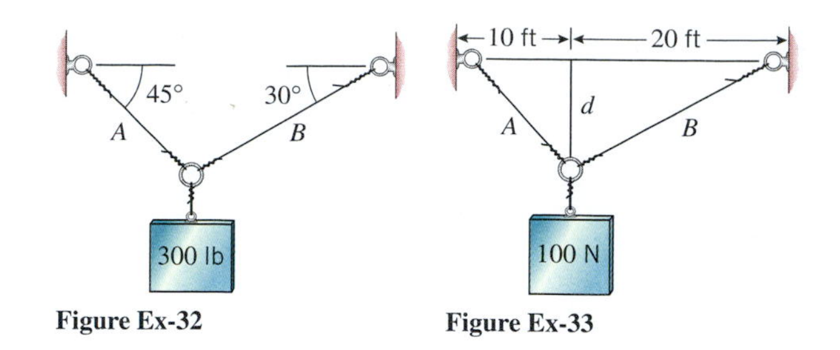

Figure Ex-32 Figure Ex-33

34. Find the work done by a force $\mathbf{F} = -3\mathbf{j}$ (pounds) applied to a point that moves on a line from $(1, 3)$ to $(4, 7)$. Assume that distance is measured in feet.

35. A force of $\mathbf{F} = 4\mathbf{i} - 6\mathbf{j} + \mathbf{k}$ newtons is applied to a point that moves a distance of 15 meters in the direction of the vector $\mathbf{i} + \mathbf{j} + \mathbf{k}$. How much work is done?

36. A boat travels 100 meters due north while the wind exerts a force of 500 newtons toward the northeast. How much work does the wind do?

37. A box is dragged along the floor by a rope that applies a force of 50 lb at an angle of 60° with the floor. How much work is done in moving the box 15 ft?

38. As shown in the accompanying figure, a force of 250 N is applied to a boat at an angle of 38° with the positive x-axis. What force $\mathbf{F}$ should be applied to the boat to produce a resultant force of 1000 N acting in the positive x-direction? State your answer by giving the magnitude of the force and its angle with the positive x-axis to the nearest degree.

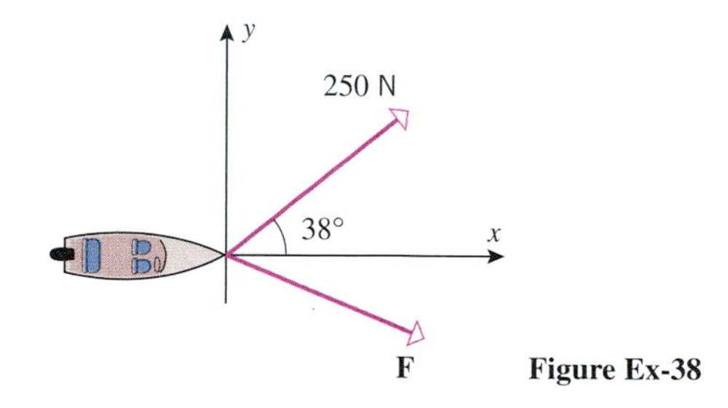

Figure Ex-38

FOCUS ON CONCEPTS

39. Let $\mathbf{u}$ and $\mathbf{v}$ be adjacent sides of a parallelogram. Use vectors to prove that the diagonals of the parallelogram are perpendicular if the sides are equal in length.

40. Let $\mathbf{u}$ and $\mathbf{v}$ be adjacent sides of a parallelogram. Use vectors to prove that the parallelogram is a rectangle if the diagonals are equal in length.

41. Prove that

$$\|\mathbf{u} + \mathbf{v}\|^2 + \|\mathbf{u} - \mathbf{v}\|^2 = 2\|\mathbf{u}\|^2 + 2\|\mathbf{v}\|^2$$

and interpret the result geometrically by translating it into a theorem about parallelograms.

42. Prove: $\mathbf{u} \cdot \mathbf{v} = \frac{1}{4}\|\mathbf{u} + \mathbf{v}\|^2 - \frac{1}{4}\|\mathbf{u} - \mathbf{v}\|^2$.

43. Show that if $\mathbf{v}_1$, $\mathbf{v}_2$, and $\mathbf{v}_3$ are mutually orthogonal nonzero vectors in 3-space, and if a vector $\mathbf{v}$ in 3-space is expressed as

$$\mathbf{v} = c_1\mathbf{v}_1 + c_2\mathbf{v}_2 + c_3\mathbf{v}_3$$

then the scalars c_1, c_2, and c_3 are given by the formulas

$$c_i = (\mathbf{v} \cdot \mathbf{v}_i)/\|\mathbf{v}_i\|^2, \quad i = 1, 2, 3$$

44. Show that the three vectors

$$\mathbf{v}_1 = 3\mathbf{i} - \mathbf{j} + 2\mathbf{k}, \quad \mathbf{v}_2 = \mathbf{i} + \mathbf{j} - \mathbf{k}, \quad \mathbf{v}_3 = \mathbf{i} - 5\mathbf{j} - 4\mathbf{k}$$

are mutually orthogonal, and then use the result of Exercise 43 to find scalars c_1, c_2, and c_3 so that

$$c_1\mathbf{v}_1 + c_2\mathbf{v}_2 + c_3\mathbf{v}_3 = \mathbf{i} - \mathbf{j} + \mathbf{k}$$

C **45.** For each x in $(-\infty, +\infty)$, let $\mathbf{u}(x)$ be the vector from the origin to the point $P(x, y)$ on the curve $y = x^2 + 1$, and $\mathbf{v}(x)$ the vector from the origin to the point $Q(x, y)$ on the line $y = -x - 1$.

(a) Use a CAS to find, to the nearest degree, the minimum angle between $\mathbf{u}(x)$ and $\mathbf{v}(x)$ for x in $(-\infty, +\infty)$.

(b) Determine whether there are any real values of x for which $\mathbf{u}(x)$ and $\mathbf{v}(x)$ are orthogonal.

C **46.** Let $\mathbf{u}$ be a unit vector in the xy-plane of an xyz-coordinate system, and let $\mathbf{v}$ be a unit vector in the yz-plane. Let θ_1 be the angle between $\mathbf{u}$ and $\mathbf{i}$, let θ_2 be the angle between $\mathbf{v}$ and $\mathbf{k}$, and let θ be the angle between $\mathbf{u}$ and $\mathbf{v}$.

(a) Show that $\cos\theta = \pm\sin\theta_1 \sin\theta_2$.

(b) Find θ if θ is acute and $\theta_1 = \theta_2 = 45°$.

(c) Use a CAS to find, to the nearest degree, the maximum and minimum values of θ if θ is acute and $\theta_2 = 2\theta_1$.

47. Prove parts (*b*) and (*e*) of Theorem 12.3.2 for vectors in 3-space.

✔ QUICK CHECK ANSWERS 12.3

1. 8 **2.** (a) 25 (b) 7 (c) 10 (d) −6 **3.** $\frac{14}{5}$ **4.** $\frac{2}{\sqrt{14}}; -\frac{1}{\sqrt{14}}; \frac{3}{\sqrt{14}}$ **5.** $9\mathbf{i} - 3\mathbf{j}$

12.4 CROSS PRODUCT

In many applications of vectors in mathematics, physics, and engineering, there is a need to find a vector that is orthogonal to two given vectors. In this section we will discuss a new type of vector multiplication that can be used for this purpose.

DETERMINANTS

Some of the concepts that we will develop in this section require basic ideas about ***determinants***, which are functions that assign numerical values to square arrays of numbers. For example, if a_1, a_2, b_1, and b_2 are real numbers, then we define a **2 × 2 *determinant*** by

$$\begin{vmatrix} a_1 & a_2 \\ b_1 & b_2 \end{vmatrix} = a_1b_2 - a_2b_1 \tag{1}$$

The purpose of the arrows is to help you remember the formula—the determinant is the product of the entries on the rightward arrow minus the product of the entries on the leftward arrow. For example,

$$\begin{vmatrix} 3 & -2 \\ 4 & 5 \end{vmatrix} = (3)(5) - (-2)(4) = 15 + 8 = 23$$

A **3 × 3 *determinant*** is defined in terms of 2 × 2 determinants by

$$\begin{vmatrix} a_1 & a_2 & a_3 \\ b_1 & b_2 & b_3 \\ c_1 & c_2 & c_3 \end{vmatrix} = a_1\begin{vmatrix} b_2 & b_3 \\ c_2 & c_3 \end{vmatrix} - a_2\begin{vmatrix} b_1 & b_3 \\ c_1 & c_3 \end{vmatrix} + a_3\begin{vmatrix} b_1 & b_2 \\ c_1 & c_2 \end{vmatrix} \tag{2}$$

The right side of this formula is easily remembered by noting that a_1, a_2, and a_3 are the entries in the first "row" of the left side, and the 2 × 2 determinants on the right side arise by deleting the first row and an appropriate column from the left side. The pattern is as follows:

$$\begin{vmatrix} a_1 & a_2 & a_3 \\ b_1 & b_2 & b_3 \\ c_1 & c_2 & c_3 \end{vmatrix} = a_1 \begin{vmatrix} a_1 & a_2 & a_3 \\ b_1 & b_2 & b_3 \\ c_1 & c_2 & c_3 \end{vmatrix} - a_2 \begin{vmatrix} a_1 & a_2 & a_3 \\ b_1 & b_2 & b_3 \\ c_1 & c_2 & c_3 \end{vmatrix} + a_3 \begin{vmatrix} a_1 & a_2 & a_3 \\ b_1 & b_2 & b_3 \\ c_1 & c_2 & c_3 \end{vmatrix}$$

For example,

$$\begin{vmatrix} 3 & -2 & -5 \\ 1 & 4 & -4 \\ 0 & 3 & 2 \end{vmatrix} = 3 \begin{vmatrix} 4 & -4 \\ 3 & 2 \end{vmatrix} - (-2) \begin{vmatrix} 1 & -4 \\ 0 & 2 \end{vmatrix} + (-5) \begin{vmatrix} 1 & 4 \\ 0 & 3 \end{vmatrix}$$

$$= 3(20) + 2(2) - 5(3) = 49$$

There are also definitions of 4×4 determinants, 5×5 determinants, and higher, but we will not need them in this text. Properties of determinants are studied in a branch of mathematics called ***linear algebra***, but we will only need the two properties stated in the following theorem.

12.4.1 THEOREM.

(*a*) *If two rows in the array of a determinant are the same, then the value of the determinant is* 0.

(*b*) *Interchanging two rows in the array of a determinant multiplies its value by* -1.

We will give the proofs of parts (*a*) and (*b*) for 2×2 determinants and leave the proofs for 3×3 determinants as exercises.

PROOF (*a*).

$$\begin{vmatrix} a_1 & a_2 \\ a_1 & a_2 \end{vmatrix} = a_1a_2 - a_2a_1 = 0$$

PROOF (*b*).

$$\begin{vmatrix} b_1 & b_2 \\ a_1 & a_2 \end{vmatrix} = b_1a_2 - b_2a_1 = -(a_1b_2 - a_2b_1) = -\begin{vmatrix} a_1 & a_2 \\ b_1 & b_2 \end{vmatrix}$$ ■

CROSS PRODUCT

We now turn to the main concept in this section.

12.4.2 DEFINITION. If $\mathbf{u} = \langle u_1, u_2, u_3 \rangle$ and $\mathbf{v} = \langle v_1, v_2, v_3 \rangle$ are vectors in 3-space, then the ***cross product*** $\mathbf{u} \times \mathbf{v}$ is the vector defined by

$$\mathbf{u} \times \mathbf{v} = \begin{vmatrix} u_2 & u_3 \\ v_2 & v_3 \end{vmatrix} \mathbf{i} - \begin{vmatrix} u_1 & u_3 \\ v_1 & v_3 \end{vmatrix} \mathbf{j} + \begin{vmatrix} u_1 & u_2 \\ v_1 & v_2 \end{vmatrix} \mathbf{k} \quad (3)$$

or, equivalently,

$$\mathbf{u} \times \mathbf{v} = (u_2v_3 - u_3v_2)\mathbf{i} - (u_1v_3 - u_3v_1)\mathbf{j} + (u_1v_2 - u_2v_1)\mathbf{k} \quad (4)$$

Observe that the right side of Formula (3) has the same form as the right side of Formula (2), the difference being notation and the order of the factors in the three terms. Thus, we can rewrite (3) as

$$\mathbf{u} \times \mathbf{v} = \begin{vmatrix} \mathbf{i} & \mathbf{j} & \mathbf{k} \\ u_1 & u_2 & u_3 \\ v_1 & v_2 & v_3 \end{vmatrix} \tag{5}$$

However, this is just a mnemonic device and not a true determinant since the entries in a determinant are numbers, not vectors.

Example 1 Let $\mathbf{u} = \langle 1, 2, -2 \rangle$ and $\mathbf{v} = \langle 3, 0, 1 \rangle$. Find

(a) $\mathbf{u} \times \mathbf{v}$ (b) $\mathbf{v} \times \mathbf{u}$

Solution (a).

$$\mathbf{u} \times \mathbf{v} = \begin{vmatrix} \mathbf{i} & \mathbf{j} & \mathbf{k} \\ 1 & 2 & -2 \\ 3 & 0 & 1 \end{vmatrix}$$

$$= \begin{vmatrix} 2 & -2 \\ 0 & 1 \end{vmatrix}\mathbf{i} - \begin{vmatrix} 1 & -2 \\ 3 & 1 \end{vmatrix}\mathbf{j} + \begin{vmatrix} 1 & 2 \\ 3 & 0 \end{vmatrix}\mathbf{k} = 2\mathbf{i} - 7\mathbf{j} - 6\mathbf{k}$$

Solution (b). We could use the method of part (a), but it is really not necessary to perform any computations. We need only observe that reversing $\mathbf{u}$ and $\mathbf{v}$ interchanges the second and third rows in (5), which in turn interchanges the rows in the arrays for the 2×2 determinants in (3). But interchanging the rows in the array of a 2×2 determinant reverses its sign, so the net effect of reversing the factors in a cross product is to reverse the signs of the components. Thus, by inspection

$$\mathbf{v} \times \mathbf{u} = -(\mathbf{u} \times \mathbf{v}) = -2\mathbf{i} + 7\mathbf{j} + 6\mathbf{k}$$ ◀

Example 2 Show that $\mathbf{u} \times \mathbf{u} = \mathbf{0}$ for any vector $\mathbf{u}$ in 3-space.

Solution. We could let $\mathbf{u} = u_1\mathbf{i} + u_2\mathbf{j} + u_3\mathbf{k}$ and apply the method in part (a) of Example 1 to show that

$$\mathbf{u} \times \mathbf{u} = \begin{vmatrix} \mathbf{i} & \mathbf{j} & \mathbf{k} \\ u_1 & u_2 & u_3 \\ u_1 & u_2 & u_3 \end{vmatrix} = 0$$

However, the actual computations are unnecessary. We need only observe that if the two factors in a cross product are the same, then each 2×2 determinant in (3) is zero because its array has identical rows. Thus, $\mathbf{u} \times \mathbf{u} = \mathbf{0}$ by inspection. ◀

ALGEBRAIC PROPERTIES OF THE CROSS PRODUCT

Our next goal is to establish some of the basic algebraic properties of the cross product. As you read the discussion, keep in mind the essential differences between the cross product and the dot product:

- The cross product is defined only for vectors in 3-space, whereas the dot product is defined for vectors in 2-space and 3-space.
- The cross product of two vectors is a vector, whereas the dot product of two vectors is a scalar.

The main algebraic properties of the cross product are listed in the next theorem.

12.4.3 THEOREM. *If* $\mathbf{u}$, $\mathbf{v}$, *and* $\mathbf{w}$ *are any vectors in 3-space and* k *is any scalar, then*

(*a*) $\mathbf{u} \times \mathbf{v} = -(\mathbf{v} \times \mathbf{u})$

(*b*) $\mathbf{u} \times (\mathbf{v} + \mathbf{w}) = (\mathbf{u} \times \mathbf{v}) + (\mathbf{u} \times \mathbf{w})$

(*c*) $(\mathbf{u} + \mathbf{v}) \times \mathbf{w} = (\mathbf{u} \times \mathbf{w}) + (\mathbf{v} \times \mathbf{w})$

(*d*) $k(\mathbf{u} \times \mathbf{v}) = (k\mathbf{u}) \times \mathbf{v} = \mathbf{u} \times (k\mathbf{v})$

(*e*) $\mathbf{u} \times \mathbf{0} = \mathbf{0} \times \mathbf{u} = \mathbf{0}$

(*f*) $\mathbf{u} \times \mathbf{u} = \mathbf{0}$

Unlike for ordinary multiplication for dot products, the order of the factors matters for cross products. Specifically, part (*a*) of Theorem 12.4.3 shows that reversing the order of the factors in a cross product reverses the direction of the resulting vector.

Parts (*a*) and (*f*) were addressed in Examples 1 and 2. The other proofs are left as exercises.

The following cross products occur so frequently that it is helpful to be familiar with them:

$$\begin{array}{lll} \mathbf{i} \times \mathbf{j} = \mathbf{k} & \mathbf{j} \times \mathbf{k} = \mathbf{i} & \mathbf{k} \times \mathbf{i} = \mathbf{j} \\ \mathbf{j} \times \mathbf{i} = -\mathbf{k} & \mathbf{k} \times \mathbf{j} = -\mathbf{i} & \mathbf{i} \times \mathbf{k} = -\mathbf{j} \end{array} \tag{6}$$

These results are easy to obtain; for example,

$$\mathbf{i} \times \mathbf{j} = \begin{vmatrix} \mathbf{i} & \mathbf{j} & \mathbf{k} \\ 1 & 0 & 0 \\ 0 & 1 & 0 \end{vmatrix} = \begin{vmatrix} 0 & 0 \\ 1 & 0 \end{vmatrix} \mathbf{i} - \begin{vmatrix} 1 & 0 \\ 0 & 0 \end{vmatrix} \mathbf{j} + \begin{vmatrix} 1 & 0 \\ 0 & 1 \end{vmatrix} \mathbf{k} = \mathbf{k}$$

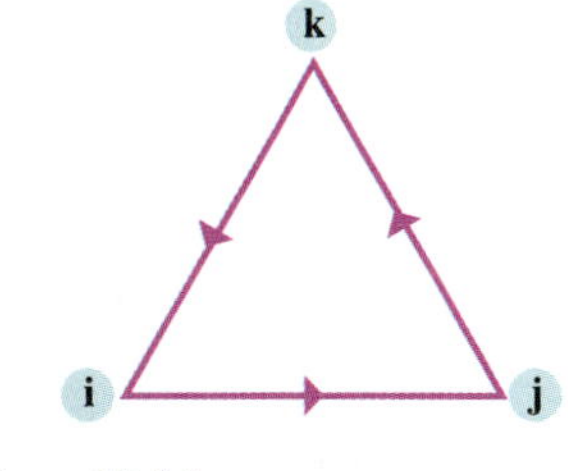

Figure 12.4.1

However, rather than computing these cross products each time you need them, you can use the diagram in Figure 12.4.1. In this diagram, the cross product of two consecutive vectors in the counterclockwise direction is the next vector around, and the cross product of two consecutive vectors in the clockwise direction is the negative of the next vector around.

WARNING We can write a product of three real numbers as uvw since the associative law $u(vw) = (uv)w$ ensures that the same value for the product results no matter how the factors are grouped. However, the associative law *does not* hold for cross products. For example,

$$\mathbf{i} \times (\mathbf{j} \times \mathbf{j}) = \mathbf{i} \times \mathbf{0} = \mathbf{0} \quad \text{and} \quad (\mathbf{i} \times \mathbf{j}) \times \mathbf{j} = \mathbf{k} \times \mathbf{j} = -\mathbf{i}$$

so that $\mathbf{i} \times (\mathbf{j} \times \mathbf{j}) \neq (\mathbf{i} \times \mathbf{j}) \times \mathbf{j}$. Thus, we cannot write a cross product with three vectors as $\mathbf{u} \times \mathbf{v} \times \mathbf{w}$, since this expression is ambiguous without parentheses.

GEOMETRIC PROPERTIES OF THE CROSS PRODUCT

The following theorem shows that the cross product of two vectors is orthogonal to both factors.

12.4.4 THEOREM. *If* $\mathbf{u}$ *and* $\mathbf{v}$ *are vectors in 3-space, then:*

(*a*) $\mathbf{u} \cdot (\mathbf{u} \times \mathbf{v}) = 0$ $\quad$ ($\mathbf{u} \times \mathbf{v}$ *is orthogonal to* $\mathbf{u}$)

(*b*) $\mathbf{v} \cdot (\mathbf{u} \times \mathbf{v}) = 0$ $\quad$ ($\mathbf{u} \times \mathbf{v}$ *is orthogonal to* $\mathbf{v}$)

We will prove part (a). The proof of part (b) is similar.

PROOF (a). Let $\mathbf{u} = \langle u_1, u_2, u_3 \rangle$ and $\mathbf{v} = \langle v_1, v_2, v_3 \rangle$. Then from (4)

$$\mathbf{u} \times \mathbf{v} = \langle u_2v_3 - u_3v_2, u_3v_1 - u_1v_3, u_1v_2 - u_2v_1 \rangle \tag{7}$$

so that

$$\mathbf{u} \cdot (\mathbf{u} \times \mathbf{v}) = u_1(u_2v_3 - u_3v_2) + u_2(u_3v_1 - u_1v_3) + u_3(u_1v_2 - u_2v_1) = 0 \quad \blacksquare$$

▶ **Example 3** In Example 1 we showed that the cross product $\mathbf{u} \times \mathbf{v}$ of $\mathbf{u} = \langle 1, 2, -2 \rangle$ and $\mathbf{v} = \langle 3, 0, 1 \rangle$ is

$$\mathbf{u} \times \mathbf{v} = 2\mathbf{i} - 7\mathbf{j} - 6\mathbf{k} = \langle 2, -7, -6 \rangle$$

Theorem 12.4.4 guarantees that this vector is orthogonal to both $\mathbf{u}$ and $\mathbf{v}$; this is confirmed by the computations

$$\mathbf{u} \cdot (\mathbf{u} \times \mathbf{v}) = \langle 1, 2, -2 \rangle \cdot \langle 2, -7, -6 \rangle = (1)(2) + (2)(-7) + (-2)(-6) = 0$$
$$\mathbf{v} \cdot (\mathbf{u} \times \mathbf{v}) = \langle 3, 0, 1 \rangle \cdot \langle 2, -7, -6 \rangle = (3)(2) + (0)(-7) + (1)(-6) = 0 \quad ◀$$

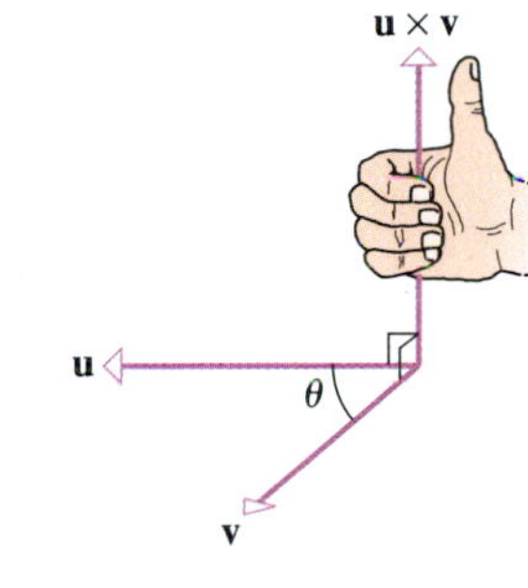

Figure 12.4.2

It can be proved that if $\mathbf{u}$ and $\mathbf{v}$ are nonzero and nonparallel vectors, then the direction of $\mathbf{u} \times \mathbf{v}$ relative to $\mathbf{u}$ and $\mathbf{v}$ is determined by a right-hand rule;* that is, if the fingers of the right hand are cupped so they curl from $\mathbf{u}$ toward $\mathbf{v}$ in the direction of rotation that takes $\mathbf{u}$ into $\mathbf{v}$ in less than 180°, then the thumb will point (roughly) in the direction of $\mathbf{u} \times \mathbf{v}$ (Figure 12.4.2). For example, we stated in (6) that

$$\mathbf{i} \times \mathbf{j} = \mathbf{k}, \quad \mathbf{j} \times \mathbf{k} = \mathbf{i}, \quad \mathbf{k} \times \mathbf{i} = \mathbf{j}$$

all of which are consistent with the right-hand rule (verify).

The next theorem lists some more important geometric properties of the cross product.

12.4.5 THEOREM. *Let* $\mathbf{u}$ *and* $\mathbf{v}$ *be nonzero vectors in 3-space, and let* θ *be the angle between these vectors when they are positioned so their initial points coincide.*

(a) $\|\mathbf{u} \times \mathbf{v}\| = \|\mathbf{u}\|\|\mathbf{v}\| \sin\theta$

(b) *The area A of the parallelogram that has* $\mathbf{u}$ *and* $\mathbf{v}$ *as adjacent sides is*

$$A = \|\mathbf{u} \times \mathbf{v}\| \tag{8}$$

(c) $\mathbf{u} \times \mathbf{v} = \mathbf{0}$ *if and only if* $\mathbf{u}$ *and* $\mathbf{v}$ *are parallel vectors, that is, if and only if they are scalar multiples of one another.*

PROOF (a).

$$\begin{aligned}
\|\mathbf{u}\|\|\mathbf{v}\|\sin\theta &= \|\mathbf{u}\|\|\mathbf{v}\|\sqrt{1 - \cos^2\theta} \\
&= \|\mathbf{u}\|\|\mathbf{v}\|\sqrt{1 - \frac{(\mathbf{u} \cdot \mathbf{v})^2}{\|\mathbf{u}\|^2\|\mathbf{v}\|^2}} \qquad \text{Theorem 12.3.3} \\
&= \sqrt{\|\mathbf{u}\|^2\|\mathbf{v}\|^2 - (\mathbf{u} \cdot \mathbf{v})^2} \\
&= \sqrt{(u_1^2 + u_2^2 + u_3^2)(v_1^2 + v_2^2 + v_3^2) - (u_1v_1 + u_2v_2 + u_3v_3)^2} \\
&= \sqrt{(u_2v_3 - u_3v_2)^2 + (u_1v_3 - u_3v_1)^2 + (u_1v_2 - u_2v_1)^2} \\
&= \|\mathbf{u} \times \mathbf{v}\| \qquad \text{See Formula (4).}
\end{aligned}$$

*Recall that we agreed to consider only right-handed coordinate systems in this text. Had we used left-handed systems instead, a "left-hand rule" would apply here.

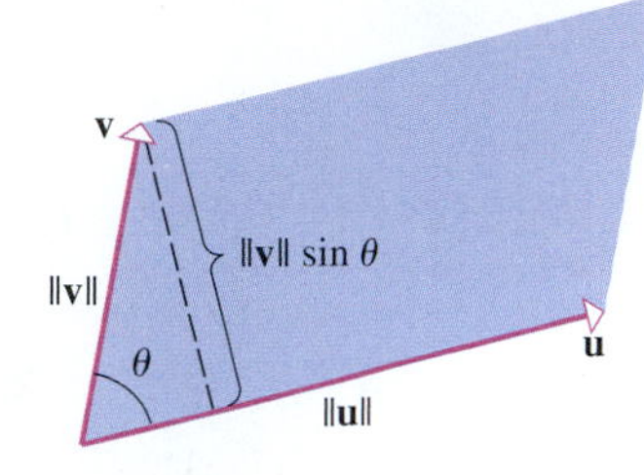

Figure 12.4.3

PROOF (b). Referring to Figure 12.4.3, the parallelogram that has $\mathbf{u}$ and $\mathbf{v}$ as adjacent sides can be viewed as having base $\|\mathbf{u}\|$ and altitude $\|\mathbf{v}\|\sin\theta$. Thus, its area A is

$$A = (\text{base})(\text{altitude}) = \|\mathbf{u}\|\|\mathbf{v}\|\sin\theta = \|\mathbf{u}\times\mathbf{v}\|$$

PROOF (c). Since $\mathbf{u}$ and $\mathbf{v}$ are assumed to be nonzero vectors, it follows from part (a) that $\mathbf{u}\times\mathbf{v} = \mathbf{0}$ if and only if $\sin\theta = 0$; this is true if and only if $\theta = 0$ or $\theta = \pi$ (since $0 \le \theta \le \pi$). Geometrically, this means that $\mathbf{u}\times\mathbf{v} = \mathbf{0}$ if and only if $\mathbf{u}$ and $\mathbf{v}$ are parallel vectors. ■

▶ **Example 4** Find the area of the triangle that is determined by the points $P_1(2, 2, 0)$, $P_2(-1, 0, 2)$, and $P_3(0, 4, 3)$.

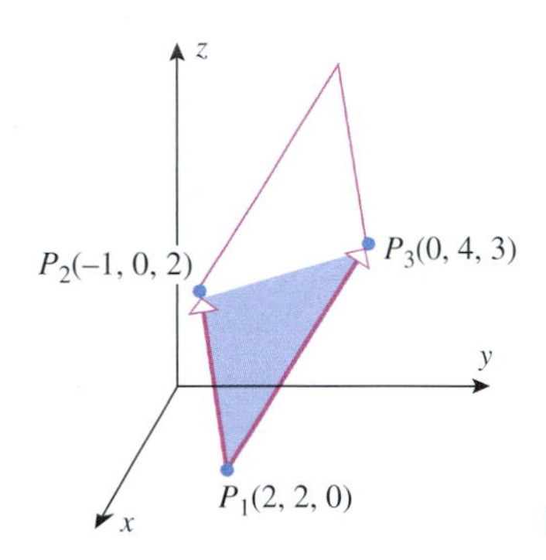

Figure 12.4.4

Solution. The area A of the triangle is half the area of the parallelogram determined by the vectors $\overrightarrow{P_1P_2}$ and $\overrightarrow{P_1P_3}$ (Figure 12.4.4). But $\overrightarrow{P_1P_2} = \langle -3, -2, 2\rangle$ and $\overrightarrow{P_1P_3} = \langle -2, 2, 3\rangle$, so

$$\overrightarrow{P_1P_2}\times\overrightarrow{P_1P_3} = \langle -10, 5, -10\rangle$$

(verify), and consequently

$$A = \tfrac{1}{2}\|\overrightarrow{P_1P_2}\times\overrightarrow{P_1P_3}\| = \tfrac{15}{2} \blacktriangleleft$$

SCALAR TRIPLE PRODUCTS

If $\mathbf{u} = \langle u_1, u_2, u_3\rangle$, $\mathbf{v} = \langle v_1, v_2, v_3\rangle$, and $\mathbf{w} = \langle w_1, w_2, w_3\rangle$ are vectors in 3-space, then the number

$$\mathbf{u}\cdot(\mathbf{v}\times\mathbf{w})$$

is called the ***scalar triple product*** of $\mathbf{u}$, $\mathbf{v}$, and $\mathbf{w}$. It is not necessary to compute the dot product and cross product to evaluate a scalar triple product—the value can be obtained directly from the formula

$$\mathbf{u}\cdot(\mathbf{v}\times\mathbf{w}) = \begin{vmatrix} u_1 & u_2 & u_3 \\ v_1 & v_2 & v_3 \\ w_1 & w_2 & w_3 \end{vmatrix} \tag{9}$$

the validity of which can be seen by writing

$$\begin{aligned}
\mathbf{u}\cdot(\mathbf{v}\times\mathbf{w}) &= \mathbf{u}\cdot\left(\begin{vmatrix} v_2 & v_3 \\ w_2 & w_3 \end{vmatrix}\mathbf{i} - \begin{vmatrix} v_1 & v_3 \\ w_1 & w_3 \end{vmatrix}\mathbf{j} + \begin{vmatrix} v_1 & v_2 \\ w_1 & w_2 \end{vmatrix}\mathbf{k}\right) \\
&= u_1\begin{vmatrix} v_2 & v_3 \\ w_2 & w_3 \end{vmatrix} - u_2\begin{vmatrix} v_1 & v_3 \\ w_1 & w_3 \end{vmatrix} + u_3\begin{vmatrix} v_1 & v_2 \\ w_1 & w_2 \end{vmatrix} \\
&= \begin{vmatrix} u_1 & u_2 & u_3 \\ v_1 & v_2 & v_3 \\ w_1 & w_2 & w_3 \end{vmatrix}
\end{aligned}$$

▶ **Example 5** Calculate the scalar triple product $\mathbf{u}\cdot(\mathbf{v}\times\mathbf{w})$ of the vectors

$$\mathbf{u} = 3\mathbf{i} - 2\mathbf{j} - 5\mathbf{k}, \quad \mathbf{v} = \mathbf{i} + 4\mathbf{j} - 4\mathbf{k}, \quad \mathbf{w} = 3\mathbf{j} + 2\mathbf{k}$$

Solution.

$$\mathbf{u}\cdot(\mathbf{v}\times\mathbf{w}) = \begin{vmatrix} 3 & -2 & -5 \\ 1 & 4 & -4 \\ 0 & 3 & 2 \end{vmatrix} = 49 \blacktriangleleft$$

TECHNOLOGY MASTERY

Many calculating utilities have built-in cross product and determinant operations. If your calculating utility has these capabilities, use it to check the computations in Examples 1 and 5.

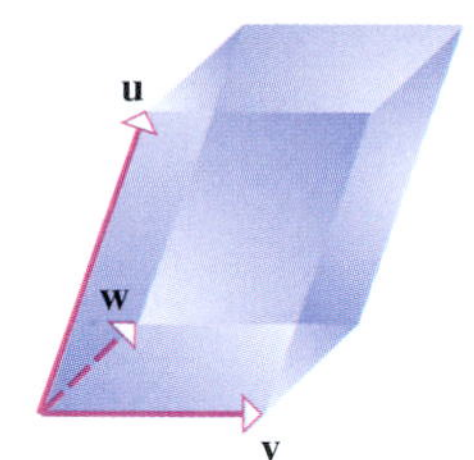

Figure 12.4.5

GEOMETRIC PROPERTIES OF THE SCALAR TRIPLE PRODUCT

If **u**, **v**, and **w** are nonzero vectors in 3-space that are positioned so their initial points coincide, then these vectors form the adjacent sides of a parallelepiped (Figure 12.4.5). The following theorem establishes a relationship between the volume of this parallelepiped and the scalar triple product of the sides.

12.4.6 THEOREM. *Let* **u**, **v**, *and* **w** *be nonzero vectors in 3-space.*

(*a*) *The volume V of the parallelepiped that has* **u**, **v**, *and* **w** *as adjacent edges is*

$$V = |\mathbf{u} \cdot (\mathbf{v} \times \mathbf{w})| \tag{10}$$

(*b*) $\mathbf{u} \cdot (\mathbf{v} \times \mathbf{w}) = 0$ *if and only if* **u**, **v**, *and* **w** *lie in the same plane.*

It follows from Formula (10) that

$$\mathbf{u} \cdot (\mathbf{v} \times \mathbf{w}) = \pm V$$

The + occurs when **u** makes an acute angle with $\mathbf{v} \times \mathbf{w}$ and the − occurs when it makes an obtuse angle.

PROOF (*a*). Referring to Figure 12.4.6, let us regard the base of the parallelepiped with **u**, **v**, and **w** as adjacent sides to be the parallelogram determined by **v** and **w**. Thus, the area of the base is $\|\mathbf{v} \times \mathbf{w}\|$, and the altitude h of the parallelepiped (shown in the figure) is the length of the orthogonal projection of **u** on the vector $\mathbf{v} \times \mathbf{w}$. Therefore, from Formula (12) of Section 12.3 we have

$$h = \|\text{proj}_{\mathbf{v}\times\mathbf{w}}\mathbf{u}\| = \frac{|\mathbf{u} \cdot (\mathbf{v} \times \mathbf{w})|}{\|\mathbf{v} \times \mathbf{w}\|^2}\|\mathbf{v} \times \mathbf{w}\| = \frac{|\mathbf{u} \cdot (\mathbf{v} \times \mathbf{w})|}{\|\mathbf{v} \times \mathbf{w}\|}$$

It now follows that the volume of the parallelepiped is

$$V = (\text{area of base})(\text{height}) = \|\mathbf{v} \times \mathbf{w}\|h = |\mathbf{u} \cdot (\mathbf{v} \times \mathbf{w})|$$

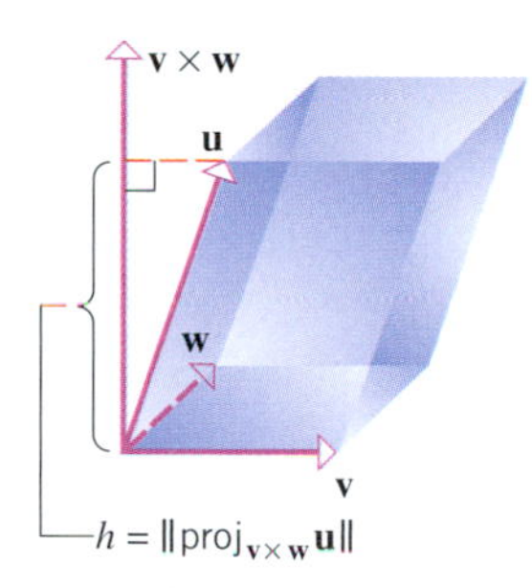

Figure 12.4.6

PROOF (*b*). The vectors **u**, **v**, and **w** lie in the same plane if and only if the parallelepiped with these vectors as adjacent sides has volume zero (why?). Thus, from part (*a*) the vectors lie in the same plane if and only if $\mathbf{u} \cdot (\mathbf{v} \times \mathbf{w}) = 0$. ■

ALGEBRAIC PROPERTIES OF THE SCALAR TRIPLE PRODUCT

We observed earlier in this section that the expression $\mathbf{u} \times \mathbf{v} \times \mathbf{w}$ must be avoided because it is ambiguous without parentheses. However, the expression $\mathbf{u} \cdot \mathbf{v} \times \mathbf{w}$ is not ambiguous—it has to mean $\mathbf{u} \cdot (\mathbf{v} \times \mathbf{w})$ and not $(\mathbf{u} \cdot \mathbf{v}) \times \mathbf{w}$ because we cannot form the cross product of a scalar and a vector. Similarly, the expression $\mathbf{u} \times \mathbf{v} \cdot \mathbf{w}$ must mean $(\mathbf{u} \times \mathbf{v}) \cdot \mathbf{w}$ and not $\mathbf{u} \times (\mathbf{v} \cdot \mathbf{w})$. Thus, when you see an expression of the form $\mathbf{u} \cdot \mathbf{v} \times \mathbf{w}$ or $\mathbf{u} \times \mathbf{v} \cdot \mathbf{w}$, the cross product is formed first and the dot product second.

Since interchanging two rows of a determinant multiplies its value by −1, making two row interchanges in a determinant has no effect on its value. This being the case, it follows that

$$\mathbf{u} \cdot (\mathbf{v} \times \mathbf{w}) = \mathbf{w} \cdot (\mathbf{u} \times \mathbf{v}) = \mathbf{v} \cdot (\mathbf{w} \times \mathbf{u}) \tag{11}$$

since the 3×3 determinants that are used to compute these scalar triple products can be obtained from one another by two row interchanges (verify).

A good way to remember Formula (11) is to observe that the second expression in the formula can be obtained from the first by leaving the dot, cross, and parentheses fixed, moving the first two vectors to the right, and bringing the third vector to the first position. The same procedure produces the third expression from the second and the first expression from the third (verify).

Another useful formula can be obtained by rewriting the first equality in (11) as

$$\mathbf{u} \cdot (\mathbf{v} \times \mathbf{w}) = (\mathbf{u} \times \mathbf{v}) \cdot \mathbf{w}$$

and then omitting the superfluous parentheses to obtain

$$\mathbf{u} \cdot \mathbf{v} \times \mathbf{w} = \mathbf{u} \times \mathbf{v} \cdot \mathbf{w} \tag{12}$$

In words, this formula states that the dot and cross in a scalar triple product can be interchanged (provided the factors are grouped appropriately).

DOT AND CROSS PRODUCTS ARE COORDINATE INDEPENDENT

In Definitions 12.3.1 and 12.4.2 we defined the dot product and the cross product of two vectors in terms of the components of those vectors in a coordinate system. Thus, it is theoretically possible that changing the coordinate system might change $\mathbf{u} \cdot \mathbf{v}$ or $\mathbf{u} \times \mathbf{v}$, since the components of a vector depend on the coordinate system that is chosen. However, the relationships

$$\mathbf{u} \cdot \mathbf{v} = \|\mathbf{u}\|\|\mathbf{v}\|\cos\theta \tag{13}$$

$$\|\mathbf{u} \times \mathbf{v}\| = \|\mathbf{u}\|\|\mathbf{v}\|\sin\theta \tag{14}$$

that were obtained in Theorems 12.3.3 and 12.4.5 show that this is not the case. Formula (13) shows that the value of $\mathbf{u} \cdot \mathbf{v}$ depends only on the lengths of the vectors and the angle between them—not on the coordinate system. Similarly, Formula (14), in combination with the right-hand rule and Theorem 12.4.4, shows that $\mathbf{u} \times \mathbf{v}$ does not depend on the coordinate system (as long as it is right-handed). These facts are important in applications because they allow us to choose any convenient coordinate system for solving a problem with full confidence that the choice will not affect computations that involve dot products or cross products.

MOMENTS AND ROTATIONAL MOTION IN 3-SPACE

Astronauts use tools that are designed to limit forces that would impart unintended rotational motion to a satellite.

Cross products play an important role in describing rotational motion in 3-space. For example, suppose that an astronaut on a satellite repair mission in space applies a force $\mathbf{F}$ at a point Q on the surface of a spherical satellite. If the force is directed along a line that passes through the center P of the satellite, then Newton's Second Law of Motion implies that the force will accelerate the satellite in the direction of $\mathbf{F}$. However, if the astronaut applies the same force at an angle θ with the vector $\overrightarrow{PQ}$, then $\mathbf{F}$ will tend to cause a rotation, as well as an acceleration in the direction of $\mathbf{F}$. To see why this is so, let us resolve $\mathbf{F}$ into a sum of orthogonal components $\mathbf{F} = \mathbf{F}_1 + \mathbf{F}_2$, where $\mathbf{F}_1$ is the orthogonal projection of $\mathbf{F}$ on the vector $\overrightarrow{PQ}$ and $\mathbf{F}_2$ is the component of $\mathbf{F}$ orthogonal to $\overrightarrow{PQ}$ (Figure 12.4.7). Since the force $\mathbf{F}_1$ acts along the line through the center of the satellite, it contributes to the linear acceleration of the satellite but does not cause any rotation. However, the force $\mathbf{F}_2$ is tangent to the circle around the satellite in the plane of $\mathbf{F}$ and $\overrightarrow{PQ}$, so it causes the satellite to rotate about an axis that is perpendicular to that plane.

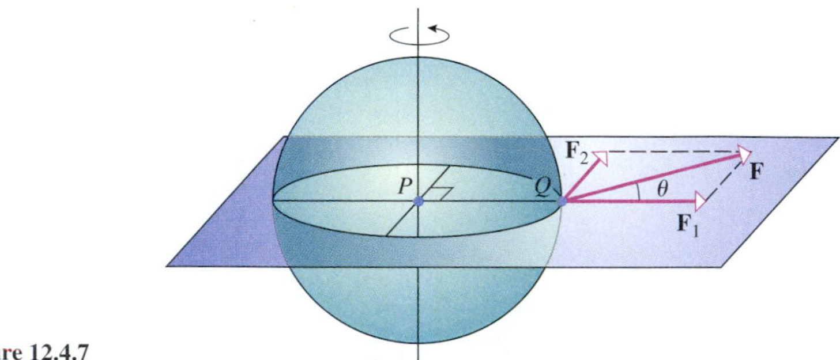

Figure 12.4.7

You know from your own experience that the "tendency" for rotation about an axis depends both on the amount of force and how far from the axis it is applied. For example, it is easier to close a door by pushing on its outer edge than applying the same force close to the hinges. In fact, the tendency of rotation of the satellite can be measured by

$$\|\overrightarrow{PQ}\|\|\mathbf{F}_2\| \quad \boxed{\text{distance from the center} \times \text{magnitude of the force}} \tag{15}$$

However, $\|\mathbf{F}_2\| = \|\mathbf{F}\|\sin\theta$, so we can rewrite (15) as

$$\|\overrightarrow{PQ}\|\|\mathbf{F}\|\sin\theta = \|\overrightarrow{PQ} \times \mathbf{F}\|$$

This is called the ***scalar moment*** or ***torque*** of $\mathbf{F}$ about the point P. Scalar moments have units of force times distance—pound-feet or newton-meters, for example. The vector $\overrightarrow{PQ} \times \mathbf{F}$ is called the ***vector moment*** or ***torque vector*** of $\mathbf{F}$ about P.

Recalling that the direction of $\overrightarrow{PQ} \times \mathbf{F}$ is determined by the right-hand rule, it follows that the direction of rotation about P that results by applying the force $\mathbf{F}$ at the point Q is counterclockwise looking down the axis of $\overrightarrow{PQ} \times \mathbf{F}$ (Figure 12.4.7). Thus, the vector moment $\overrightarrow{PQ} \times \mathbf{F}$ captures the essential information about the rotational effect of the force—the magnitude of the cross product provides the scalar moment of the force, and the cross product vector itself provides the axis and direction of rotation.

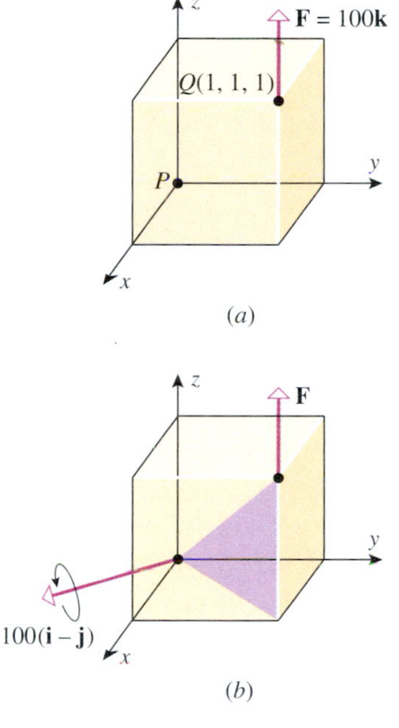

Figure 12.4.8

▶ **Example 6** Figure 12.4.8a shows a force $\mathbf{F}$ of 100 N applied in the positive z-direction at the point $Q(1, 1, 1)$ of a cube whose sides have a length of 1 m. Assuming that the cube is free to rotate about the point $P(0, 0, 0)$ (the origin), find the scalar moment of the force about P, and describe the direction of rotation.

Solution. The force vector is $\mathbf{F} = 100\mathbf{k}$, and the vector from P to Q is $\overrightarrow{PQ} = \mathbf{i} + \mathbf{j} + \mathbf{k}$, so the vector moment of $\mathbf{F}$ about P is

$$\overrightarrow{PQ} \times \mathbf{F} = \begin{vmatrix} \mathbf{i} & \mathbf{j} & \mathbf{k} \\ 1 & 1 & 1 \\ 0 & 0 & 100 \end{vmatrix} = 100\mathbf{i} - 100\mathbf{j}$$

Thus, the scalar moment of $\mathbf{F}$ about P is $\|100\mathbf{i} - 100\mathbf{j}\| = 100\sqrt{2} \approx 141$ N·m, and the direction of rotation is counterclockwise looking along the vector $100\mathbf{i} - 100\mathbf{j} = 100(\mathbf{i} - \mathbf{j})$ toward its initial point (Figure 12.4.8b). ◀

✔QUICK CHECK EXERCISES 12.4 *(See page 823 for answers.)*

1. (a) $\begin{vmatrix} 3 & 2 \\ 4 & 5 \end{vmatrix} =$ ________ (b) $\begin{vmatrix} 3 & 2 & 1 \\ 3 & 2 & 1 \\ 5 & 5 & 5 \end{vmatrix} =$ ________

2. $\langle 1, 2, 0\rangle \times \langle 3, 0, 4\rangle =$ ________

3. Suppose that $\mathbf{u}$, $\mathbf{v}$, and $\mathbf{w}$ are vectors in 3-space such that $\mathbf{u} \times \mathbf{v} = \langle 2, 7, 3\rangle$ and $\mathbf{u} \times \mathbf{w} = \langle -5, 4, 0\rangle$.
   (a) $\mathbf{u} \times \mathbf{u} =$ ________ (b) $\mathbf{v} \times \mathbf{u} =$ ________
   (c) $\mathbf{u} \times (\mathbf{v} + \mathbf{w}) =$ ________
   (d) $\mathbf{u} \times (2\mathbf{w}) =$ ________

4. Let $\mathbf{u} = \mathbf{i} - 5\mathbf{k}$, $\mathbf{v} = 2\mathbf{i} - 4\mathbf{j} + \mathbf{k}$, and $\mathbf{w} = 3\mathbf{i} - 2\mathbf{j} + 5\mathbf{k}$.
   (a) $\mathbf{u} \cdot (\mathbf{v} \times \mathbf{w}) =$ ________
   (b) The volume of the parallelepiped that has $\mathbf{u}$, $\mathbf{v}$, and $\mathbf{w}$ as adjacent edges is $V =$ ________.

EXERCISE SET 12.4 [C] CAS

1. (a) Use a determinant to find the cross product
$$\mathbf{i} \times (\mathbf{i} + \mathbf{j} + \mathbf{k})$$
   (b) Check your answer in part (a) by rewriting the cross product as
$$\mathbf{i} \times (\mathbf{i} + \mathbf{j} + \mathbf{k}) = (\mathbf{i} \times \mathbf{i}) + (\mathbf{i} \times \mathbf{j}) + (\mathbf{i} \times \mathbf{k})$$
   and evaluating each term.

2. In each part, use the two methods in Exercise 1 to find
   (a) $\mathbf{j} \times (\mathbf{i} + \mathbf{j} + \mathbf{k})$ (b) $\mathbf{k} \times (\mathbf{i} + \mathbf{j} + \mathbf{k})$.

3–6 Find $\mathbf{u} \times \mathbf{v}$ and check that it is orthogonal to both $\mathbf{u}$ and $\mathbf{v}$.

3. $\mathbf{u} = \langle 1, 2, -3\rangle$, $\mathbf{v} = \langle -4, 1, 2\rangle$

4. $\mathbf{u} = 3\mathbf{i} + 2\mathbf{j} - \mathbf{k}$, $\mathbf{v} = -\mathbf{i} - 3\mathbf{j} + \mathbf{k}$

5. $\mathbf{u} = \langle 0, 1, -2 \rangle$, $\mathbf{v} = \langle 3, 0, -4 \rangle$

6. $\mathbf{u} = 4\mathbf{i} + \mathbf{k}$, $\mathbf{v} = 2\mathbf{i} - \mathbf{j}$

7. Let $\mathbf{u} = \langle 2, -1, 3 \rangle$, $\mathbf{v} = \langle 0, 1, 7 \rangle$, and $\mathbf{w} = \langle 1, 4, 5 \rangle$. Find
(a) $\mathbf{u} \times (\mathbf{v} \times \mathbf{w})$ (b) $(\mathbf{u} \times \mathbf{v}) \times \mathbf{w}$
(c) $(\mathbf{u} \times \mathbf{v}) \times (\mathbf{v} \times \mathbf{w})$ (d) $(\mathbf{v} \times \mathbf{w}) \times (\mathbf{u} \times \mathbf{v})$.

C **8.** Use a CAS or a calculating utility that can compute determinants or cross products to solve Exercise 7.

9. Find the direction cosines of $\mathbf{u} \times \mathbf{v}$ for the vectors $\mathbf{u}$ and $\mathbf{v}$ in the accompanying figure.

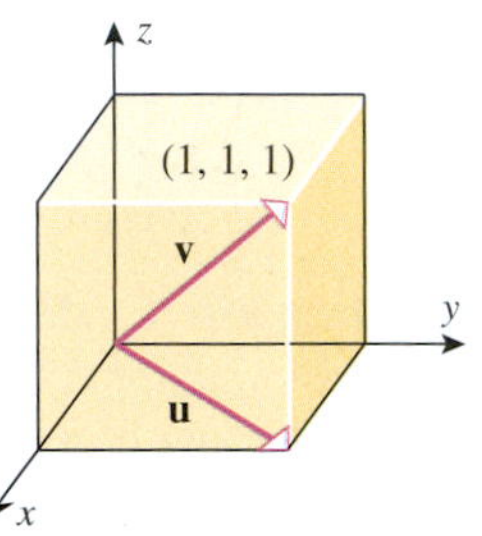

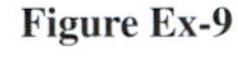
Figure Ex-9

10. Find two unit vectors that are orthogonal to both
$$\mathbf{u} = -7\mathbf{i} + 3\mathbf{j} + \mathbf{k}, \quad \mathbf{v} = 2\mathbf{i} + 4\mathbf{k}$$

11. Find two unit vectors that are normal to the plane determined by the points $A(0, -2, 1)$, $B(1, -1, -2)$, and $C(-1, 1, 0)$.

12. Find two unit vectors that are parallel to the yz-plane and are orthogonal to the vector $3\mathbf{i} - \mathbf{j} + 2\mathbf{k}$.

13–14 Find the area of the parallelogram that has $\mathbf{u}$ and $\mathbf{v}$ as adjacent sides.

13. $\mathbf{u} = \mathbf{i} - \mathbf{j} + 2\mathbf{k}$, $\mathbf{v} = 3\mathbf{j} + \mathbf{k}$

14. $\mathbf{u} = 2\mathbf{i} + 3\mathbf{j}$, $\mathbf{v} = -\mathbf{i} + 2\mathbf{j} - 2\mathbf{k}$

15–16 Find the area of the triangle with vertices P, Q, and R.

15. $P(1, 5, -2)$, $Q(0, 0, 0)$, $R(3, 5, 1)$

16. $P(2, 0, -3)$, $Q(1, 4, 5)$, $R(7, 2, 9)$

17–20 Find $\mathbf{u} \cdot (\mathbf{v} \times \mathbf{w})$.

17. $\mathbf{u} = 2\mathbf{i} - 3\mathbf{j} + \mathbf{k}$, $\mathbf{v} = 4\mathbf{i} + \mathbf{j} - 3\mathbf{k}$, $\mathbf{w} = \mathbf{j} + 5\mathbf{k}$

18. $\mathbf{u} = \langle 1, -2, 2 \rangle$, $\mathbf{v} = \langle 0, 3, 2 \rangle$, $\mathbf{w} = \langle -4, 1, -3 \rangle$

19. $\mathbf{u} = \langle 2, 1, 0 \rangle$, $\mathbf{v} = \langle 1, -3, 1 \rangle$, $\mathbf{w} = \langle 4, 0, 1 \rangle$

20. $\mathbf{u} = \mathbf{i}$, $\mathbf{v} = \mathbf{i} + \mathbf{j}$, $\mathbf{w} = \mathbf{i} + \mathbf{j} + \mathbf{k}$

21–22 Use a scalar triple product to find the volume of the parallelepiped that has $\mathbf{u}$, $\mathbf{v}$, and $\mathbf{w}$ as adjacent edges.

21. $\mathbf{u} = \langle 2, -6, 2 \rangle$, $\mathbf{v} = \langle 0, 4, -2 \rangle$, $\mathbf{w} = \langle 2, 2, -4 \rangle$

22. $\mathbf{u} = 3\mathbf{i} + \mathbf{j} + 2\mathbf{k}$, $\mathbf{v} = 4\mathbf{i} + 5\mathbf{j} + \mathbf{k}$, $\mathbf{w} = \mathbf{i} + 2\mathbf{j} + 4\mathbf{k}$

23. In each part, use a scalar triple product to determine whether the vectors lie in the same plane.
(a) $\mathbf{u} = \langle 1, -2, 1 \rangle$, $\mathbf{v} = \langle 3, 0, -2 \rangle$, $\mathbf{w} = \langle 5, -4, 0 \rangle$
(b) $\mathbf{u} = 5\mathbf{i} - 2\mathbf{j} + \mathbf{k}$, $\mathbf{v} = 4\mathbf{i} - \mathbf{j} + \mathbf{k}$, $\mathbf{w} = \mathbf{i} - \mathbf{j}$
(c) $\mathbf{u} = \langle 4, -8, 1 \rangle$, $\mathbf{v} = \langle 2, 1, -2 \rangle$, $\mathbf{w} = \langle 3, -4, 12 \rangle$

24. Suppose that $\mathbf{u} \cdot (\mathbf{v} \times \mathbf{w}) = 3$. Find
(a) $\mathbf{u} \cdot (\mathbf{w} \times \mathbf{v})$ (b) $(\mathbf{v} \times \mathbf{w}) \cdot \mathbf{u}$
(c) $\mathbf{w} \cdot (\mathbf{u} \times \mathbf{v})$ (d) $\mathbf{v} \cdot (\mathbf{u} \times \mathbf{w})$
(e) $(\mathbf{u} \times \mathbf{w}) \cdot \mathbf{v}$ (f) $\mathbf{v} \cdot (\mathbf{w} \times \mathbf{w})$.

25. Consider the parallelepiped with adjacent edges
$$\mathbf{u} = 3\mathbf{i} + 2\mathbf{j} + \mathbf{k}$$
$$\mathbf{v} = \mathbf{i} + \mathbf{j} + 2\mathbf{k}$$
$$\mathbf{w} = \mathbf{i} + 3\mathbf{j} + 3\mathbf{k}$$
(a) Find the volume.
(b) Find the area of the face determined by $\mathbf{u}$ and $\mathbf{w}$.
(c) Find the angle between $\mathbf{u}$ and the plane containing the face determined by $\mathbf{v}$ and $\mathbf{w}$.

26. Show that in 3-space the distance d from a point P to the line L through points A and B can be expressed as
$$d = \frac{\|\overrightarrow{AP} \times \overrightarrow{AB}\|}{\|\overrightarrow{AB}\|}$$

27. Use the result in Exercise 26 to find the distance between the point P and the line through the points A and B.
(a) $P(-3, 1, 2)$, $A(1, 1, 0)$, $B(-2, 3, -4)$
(b) $P(4, 3)$, $A(2, 1)$, $B(0, 2)$

28. It is a theorem of solid geometry that the volume of a tetrahedron is $\frac{1}{3}$(area of base) · (height). Use this result to prove that the volume of a tetrahedron with adjacent edges given by the vectors $\mathbf{u}$, $\mathbf{v}$, and $\mathbf{w}$ is $\frac{1}{6}|\mathbf{u} \cdot (\mathbf{v} \times \mathbf{w})|$.

29. Use the result of Exercise 28 to find the volume of the tetrahedron with vertices
$$P(-1, 2, 0), \quad Q(2, 1, -3), \quad R(1, 0, 1), \quad S(3, -2, 3)$$

30. Let θ be the angle between the vectors $\mathbf{u} = 2\mathbf{i} + 3\mathbf{j} - 6\mathbf{k}$ and $\mathbf{v} = 2\mathbf{i} + 3\mathbf{j} + 6\mathbf{k}$.
(a) Use the dot product to find $\cos\theta$.
(b) Use the cross product to find $\sin\theta$.
(c) Confirm that $\sin^2\theta + \cos^2\theta = 1$.

FOCUS ON CONCEPTS

31. Let A, B, C and D be four distinct points in 3-space. If $\overrightarrow{AB} \times \overrightarrow{CD} \neq \mathbf{0}$ and $\overrightarrow{AC} \cdot (\overrightarrow{AB} \times \overrightarrow{CD}) = 0$, explain why the line through A and B must intersect the line through C and D.

32. Let A, B, and C be three distinct noncollinear points in 3-space. Describe the set of all points P that satisfy the vector equation $\overrightarrow{AP} \cdot (\overrightarrow{AB} \times \overrightarrow{AC}) = 0$.

33. What can you say about the angle between nonzero vectors $\mathbf{u}$ and $\mathbf{v}$ if $\mathbf{u} \cdot \mathbf{v} = \|\mathbf{u} \times \mathbf{v}\|$?

34. Show that if $\mathbf{u}$ and $\mathbf{v}$ are vectors in 3-space, then
$$\|\mathbf{u} \times \mathbf{v}\|^2 = \|\mathbf{u}\|^2\|\mathbf{v}\|^2 - (\mathbf{u} \cdot \mathbf{v})^2$$
[*Note:* This result is sometimes called ***Lagrange's identity***.]

35. The accompanying figure shows a force $\mathbf{F}$ of 10 lb applied in the positive y-direction to the point $Q(1, 1, 1)$ of a cube whose sides have a length of 1 ft. In each part, find the scalar moment of $\mathbf{F}$ about the point P, and describe the direction of rotation, if any, if the cube is free to rotate about P.
(a) P is the point $(0, 0, 0)$. (b) P is the point $(1, 0, 0)$.
(c) P is the point $(1, 0, 1)$.

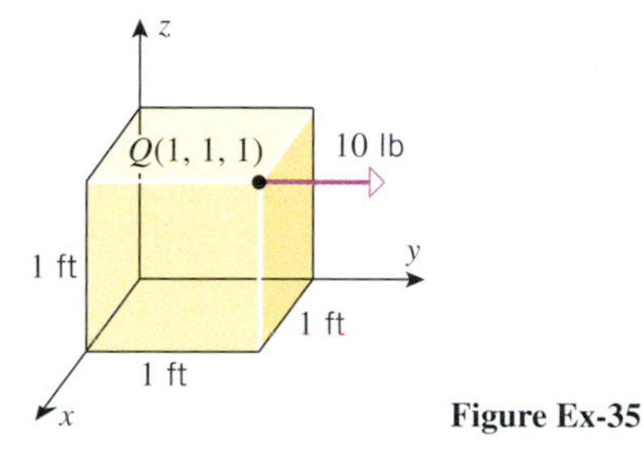

Figure Ex-35

36. The accompanying figure shows a force $\mathbf{F}$ of 1000 N applied to the corner of a box.
(a) Find the scalar moment of $\mathbf{F}$ about the point P.
(b) Find the direction angles of the vector moment of $\mathbf{F}$ about the point P to the nearest degree.

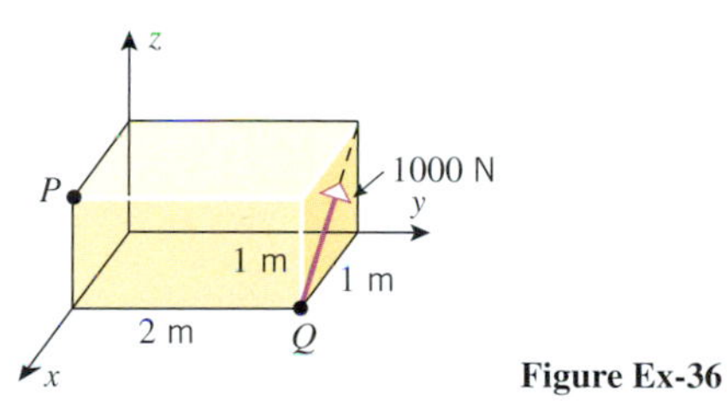

Figure Ex-36

37. As shown in the accompanying figure, a force of 200 N is applied at an angle of 18° to a point near the end of a monkey wrench. Find the scalar moment of the force about the center of the bolt. [Treat this as a problem in two dimensions.]

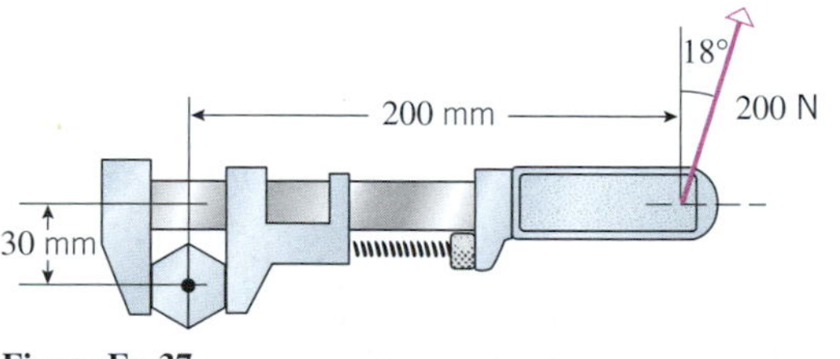

Figure Ex-37

38. Prove parts (b) and (c) of Theorem 12.4.3.

39. Prove parts (d) and (e) of Theorem 12.4.3.

40. Prove part (b) of Theorem 12.4.1 for 3×3 determinants. [Just give the proof for the first two rows.] Then use (b) to prove (a).

FOCUS ON CONCEPTS

41. Expressions of the form

$$\mathbf{u} \times (\mathbf{v} \times \mathbf{w}) \quad \text{and} \quad (\mathbf{u} \times \mathbf{v}) \times \mathbf{w}$$

are called ***vector triple products***. It can be proved with some effort that

$$\mathbf{u} \times (\mathbf{v} \times \mathbf{w}) = (\mathbf{u} \cdot \mathbf{w})\mathbf{v} - (\mathbf{u} \cdot \mathbf{v})\mathbf{w}$$
$$(\mathbf{u} \times \mathbf{v}) \times \mathbf{w} = (\mathbf{w} \cdot \mathbf{u})\mathbf{v} - (\mathbf{w} \cdot \mathbf{v})\mathbf{u}$$

These expressions can be summarized with the following mnemonic rule:

vector triple product = (outer · remote)adjacent
− (outer · adjacent)remote

See if you can figure out what the expressions "outer," "remote," and "adjacent" mean in this rule, and then use the rule to find the two vector triple products of the vectors

$$\mathbf{u} = \mathbf{i} + 3\mathbf{j} - \mathbf{k}, \quad \mathbf{v} = \mathbf{i} + \mathbf{j} + 2\mathbf{k}, \quad \mathbf{w} = 3\mathbf{i} - \mathbf{j} + 2\mathbf{k}$$

42. (a) Use the result in Exercise 41 to show that $\mathbf{u} \times (\mathbf{v} \times \mathbf{w})$ lies in the same plane as $\mathbf{v}$ and $\mathbf{w}$, and $(\mathbf{u} \times \mathbf{v}) \times \mathbf{w}$ lies in the same plane as $\mathbf{u}$ and $\mathbf{v}$.
(b) Use a geometrical argument to justify the results in part (a).

43. In each part, use the result in Exercise 41 to prove the vector identity.
(a) $(\mathbf{a} \times \mathbf{b}) \times (\mathbf{c} \times \mathbf{d}) = (\mathbf{a} \times \mathbf{b} \cdot \mathbf{d})\mathbf{c} - (\mathbf{a} \times \mathbf{b} \cdot \mathbf{c})\mathbf{d}$
(b) $(\mathbf{a} \times \mathbf{b}) \times \mathbf{c} + (\mathbf{b} \times \mathbf{c}) \times \mathbf{a} + (\mathbf{c} \times \mathbf{a}) \times \mathbf{b} = 0$

44. Prove: If $\mathbf{a}$, $\mathbf{b}$, $\mathbf{c}$, and $\mathbf{d}$ lie in the same plane when positioned with a common initial point, then

$$(\mathbf{a} \times \mathbf{b}) \times (\mathbf{c} \times \mathbf{d}) = \mathbf{0}$$

C **45.** Use a CAS to approximate the minimum area of a triangle if two of its vertices are $(2, -1, 0)$ and $(3, 2, 2)$ and its third vertex is on the curve $y = \ln x$ in the xy-plane.

46. If a force $\mathbf{F}$ is applied to an object at a point Q, then the line through Q parallel to $\mathbf{F}$ is called the ***line of action*** of the force. We defined the vector moment of $\mathbf{F}$ about a point P to be $\overrightarrow{PQ} \times \mathbf{F}$. Show that if Q' is any point on the line of action of $\mathbf{F}$, then $\overrightarrow{PQ} \times \mathbf{F} = \overrightarrow{PQ'} \times \mathbf{F}$; that is, it is not essential to use the point of application to compute the vector moment—any point on the line of action will do. [*Hint:* Write $\overrightarrow{PQ'} = \overrightarrow{PQ} + \overrightarrow{QQ'}$ and use properties of the cross product.]

✔QUICK CHECK ANSWERS 12.4

1. (a) 7 (b) 0 **2.** $8\mathbf{i} - 4\mathbf{j} - 6\mathbf{k}$ **3.** (a) $\langle 0, 0, 0\rangle$ (b) $\langle -2, -7, -3\rangle$ (c) $\langle -3, 11, 3\rangle$ (d) $\langle -10, 8, 0\rangle$ **4.** (a) −58 (b) 58

12.5 PARAMETRIC EQUATIONS OF LINES

In this section we will discuss parametric equations of lines in 2-space and 3-space. In 3-space, parametric equations of lines are especially important because they generally provide the most convenient form for representing lines algebraically.

LINES DETERMINED BY A POINT AND A VECTOR

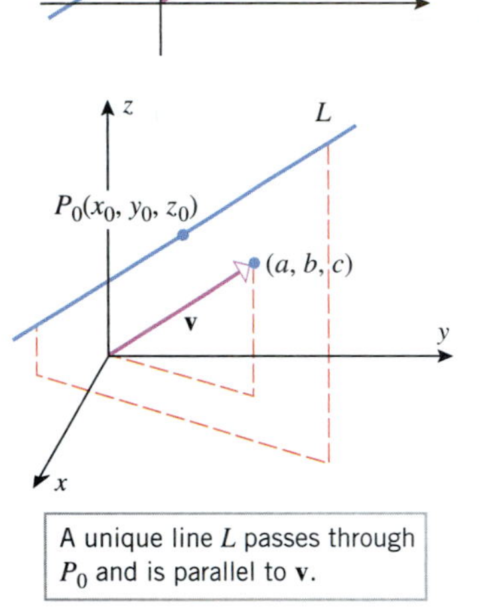

A unique line L passes through P_0 and is parallel to $\mathbf{v}$.

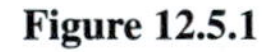

Figure 12.5.1

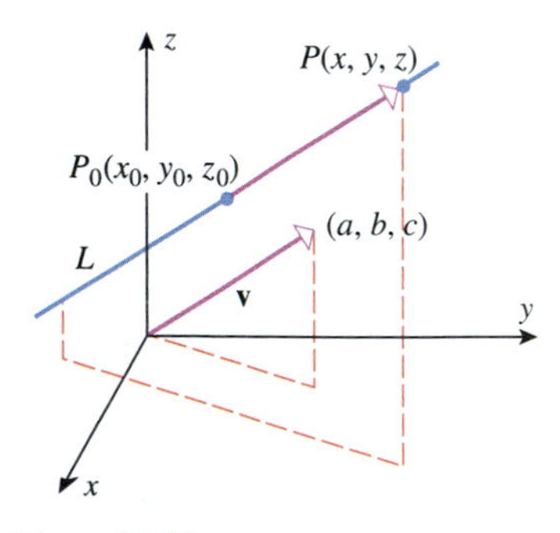

Figure 12.5.2

A line in 2-space or 3-space can be determined uniquely by specifying a point on the line and a nonzero vector parallel to the line (Figure 12.5.1). For example, consider a line L in 3-space that passes through the point $P_0(x_0, y_0, z_0)$ and is parallel to the nonzero vector $\mathbf{v} = \langle a, b, c \rangle$. Then L consists precisely of those points $P(x, y, z)$ for which the vector $\overrightarrow{P_0P}$ is parallel to $\mathbf{v}$ (Figure 12.5.2). In other words, the point $P(x, y, z)$ is on L if and only if $\overrightarrow{P_0P}$ is a scalar multiple of $\mathbf{v}$, say

$$\overrightarrow{P_0P} = t\mathbf{v}$$

This equation can be written as

$$\langle x - x_0, y - y_0, z - z_0 \rangle = \langle ta, tb, tc \rangle$$

which implies that

$$x - x_0 = ta, \quad y - y_0 = tb, \quad z - z_0 = tc$$

Thus, L can be described by the parametric equations

$$x = x_0 + at, \quad y = y_0 + bt, \quad z = z_0 + ct$$

A similar description applies to lines in 2-space. We summarize these descriptions in the following theorem.

12.5.1 THEOREM.

(*a*) *The line in 2-space that passes through the point $P_0(x_0, y_0)$ and is parallel to the nonzero vector $\mathbf{v} = \langle a, b \rangle = a\mathbf{i} + b\mathbf{j}$ has parametric equations*

$$x = x_0 + at, \quad y = y_0 + bt \tag{1}$$

(*b*) *The line in 3-space that passes through the point $P_0(x_0, y_0, z_0)$ and is parallel to the nonzero vector $\mathbf{v} = \langle a, b, c \rangle = a\mathbf{i} + b\mathbf{j} + c\mathbf{k}$ has parametric equations*

$$x = x_0 + at, \quad y = y_0 + bt, \quad z = z_0 + ct \tag{2}$$

Although it is not stated explicitly, it is understood in Equations (1) and (2) that $-\infty < t < +\infty$, which reflects the fact that lines extend indefinitely.

▶ Example 1 Find parametric equations of the line

(a) passing through (4, 2) and parallel to $\mathbf{v} = \langle -1, 5 \rangle$;

(b) passing through (1, 2, −3) and parallel to $\mathbf{v} = 4\mathbf{i} + 5\mathbf{j} - 7\mathbf{k}$;

(c) passing through the origin in 3-space and parallel to $\mathbf{v} = \langle 1, 1, 1 \rangle$.

Solution (a). From (1) with $x_0 = 4$, $y_0 = 2$, $a = -1$, and $b = 5$ we obtain

$$x = 4 - t, \quad y = 2 + 5t$$

Solution (b). From (2) we obtain

$$x = 1 + 4t, \quad y = 2 + 5t, \quad z = -3 - 7t$$

Solution (c). From (2) with $x_0 = 0$, $y_0 = 0$, $z_0 = 0$, $a = 1$, $b = 1$, and $c = 1$ we obtain

$$x = t, \quad y = t, \quad z = t \blacktriangleleft$$

▶ Example 2

(a) Find parametric equations of the line L passing through the points $P_1(2, 4, -1)$ and $P_2(5, 0, 7)$.

(b) Where does the line intersect the xy-plane?

Solution (a). The vector $\overrightarrow{P_1P_2} = \langle 3, -4, 8 \rangle$ is parallel to L and the point $P_1(2, 4, -1)$ lies on L, so it follows from (2) that L has parametric equations

$$x = 2 + 3t, \quad y = 4 - 4t, \quad z = -1 + 8t \tag{3}$$

Had we used P_2 as the point on L rather than P_1, we would have obtained the equations

$$x = 5 + 3t, \quad y = -4t, \quad z = 7 + 8t$$

Although these equations look different from those obtained using P_1, the two sets of equations are actually equivalent in that both generate L as t varies from $-\infty$ to $+\infty$. To see this, note that if t_1 gives a point

$$(x, y, z) = (2 + 3t_1, 4 - 4t_1, -1 + 8t_1)$$

on L using the first set of equations, then $t_2 = t_1 - 1$ gives the *same* point

$$\begin{aligned}(x, y, z) &= (5 + 3t_2, -4t_2, 7 + 8t_2)\\ &= (5 + 3(t_1 - 1), -4(t_1 - 1), 7 + 8(t_1 - 1))\\ &= (2 + 3t_1, 4 - 4t_1, -1 + 8t_1)\end{aligned}$$

on L using the second set of equations. Conversely, if t_2 gives a point on L using the second set of equations, then $t_1 = t_2 + 1$ gives the same point using the first set.

Solution (b). It follows from (3) in part (a) that the line intersects the xy-plane at the point where $z = -1 + 8t = 0$, that is, when $t = \frac{1}{8}$. Substituting this value of t in (3) yields the point of intersection $(x, y, z) = \left(\frac{19}{8}, \frac{7}{2}, 0\right)$. ◀

▶ Example 3

Let L_1 and L_2 be the lines

$$\begin{aligned} L_1\colon x &= 1 + 4t, \quad y = 5 - 4t, \quad z = -1 + 5t\\ L_2\colon x &= 2 + 8t, \quad y = 4 - 3t, \quad z = 5 + t\end{aligned}$$

(a) Are the lines parallel?

(b) Do the lines intersect?

Solution (a). The line L_1 is parallel to the vector $4\mathbf{i} - 4\mathbf{j} + 5\mathbf{k}$, and the line L_2 is parallel to the vector $8\mathbf{i} - 3\mathbf{j} + \mathbf{k}$. These vectors are not parallel since neither is a scalar multiple of the other. Thus, the lines are not parallel.

Solution (b). For L_1 and L_2 to intersect at some point (x_0, y_0, z_0) these coordinates would have to satisfy the equations of both lines. In other words, there would have to exist values t_1 and t_2 for the parameters such that

$$x_0 = 1 + 4t_1, \quad y_0 = 5 - 4t_1, \quad z_0 = -1 + 5t_1$$

and

$$x_0 = 2 + 8t_2, \quad y_0 = 4 - 3t_2, \quad z_0 = 5 + t_2$$

This leads to three conditions on t_1 and t_2,

$$\begin{aligned} 1 + 4t_1 &= 2 + 8t_2 \\ 5 - 4t_1 &= 4 - 3t_2 \\ -1 + 5t_1 &= 5 + t_2 \end{aligned} \tag{4}$$

Thus, the lines intersect if there are values of t_1 and t_2 that satisfy all three equations, and the lines do not intersect if there are no such values. You should be familiar with methods for solving systems of two linear equations in two unknowns; however, this is a system of three linear equations in two unknowns. To determine whether this system has a solution we will solve the first two equations for t_1 and t_2 and then check whether these values satisfy the third equation.

We will solve the first two equations by the method of elimination. We can eliminate the unknown t_1 by adding the equations. This yields the equation

$$6 = 6 + 5t_2$$

from which we obtain $t_2 = 0$. We can now find t_1 by substituting this value of t_2 in either the first or second equation. This yields $t_1 = \frac{1}{4}$. However, the values $t_1 = \frac{1}{4}$ and $t_2 = 0$ do not satisfy the third equation in (4), so the lines do not intersect. ◀

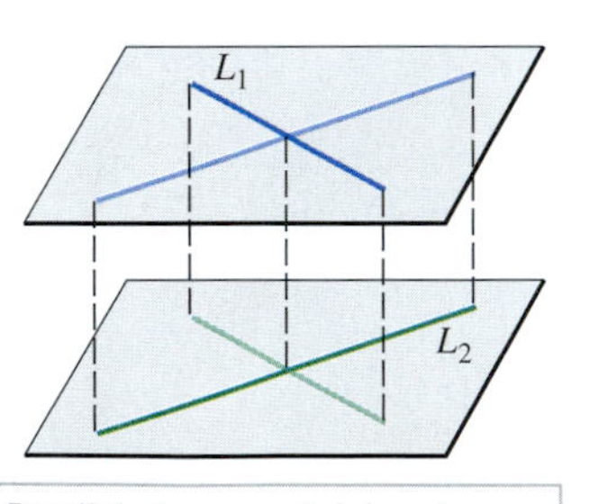

Parallel planes containing skew lines L_1 and L_2 can be determined by translating each line until it intersects the other.

Figure 12.5.3

Two lines in 3-space that are not parallel and do not intersect (such as those in Example 3) are called ***skew*** lines. As illustrated in Figure 12.5.3, any two skew lines lie in parallel planes.

LINE SEGMENTS

Sometimes one is not interested in an entire line, but rather some *segment* of a line. Parametric equations of a line segment can be obtained by finding parametric equations for the entire line, then restricting the parameter appropriately so that only the desired segment is generated.

▶ Example 4 Find parametric equations describing the line segment joining the points $P_1(2, 4, -1)$ and $P_2(5, 0, 7)$.

Solution. From Example 2, the line through the points P_1 and P_2 has parametric equations $x = 2 + 3t$, $y = 4 - 4t$, $z = -1 + 8t$. With these equations, the point P_1 corresponds to $t = 0$ and P_2 to $t = 1$. Thus, the line segment that joins P_1 and P_2 is given by

$$x = 2 + 3t, \quad y = 4 - 4t, \quad z = -1 + 8t \qquad (0 \le t \le 1)$$ ◀

VECTOR EQUATIONS OF LINES

We will now show how vector notation can be used to express the parametric equations of a line more compactly. Because two vectors are equal if and only if their components are equal, (1) and (2) can be written in vector form as

$$\langle x, y\rangle = \langle x_0 + at, y_0 + bt\rangle$$
$$\langle x, y, z\rangle = \langle x_0 + at, y_0 + bt, z_0 + ct\rangle$$

or, equivalently, as

$$\langle x, y\rangle = \langle x_0, y_0\rangle + t\langle a, b\rangle \tag{5}$$
$$\langle x, y, z\rangle = \langle x_0, y_0, z_0\rangle + t\langle a, b, c\rangle \tag{6}$$

For the equation in 2-space we define the vectors $\mathbf{r}$, $\mathbf{r}_0$, and $\mathbf{v}$ as

$$\mathbf{r} = \langle x, y\rangle, \quad \mathbf{r}_0 = \langle x_0, y_0\rangle, \quad \mathbf{v} = \langle a, b\rangle \tag{7}$$

and for the equation in 3-space we define them as

$$\mathbf{r} = \langle x, y, z\rangle, \quad \mathbf{r}_0 = \langle x_0, y_0, z_0\rangle, \quad \mathbf{v} = \langle a, b, c\rangle \tag{8}$$

Substituting (7) and (8) in (5) and (6), respectively, yields the equation

$$\mathbf{r} = \mathbf{r}_0 + t\mathbf{v} \tag{9}$$

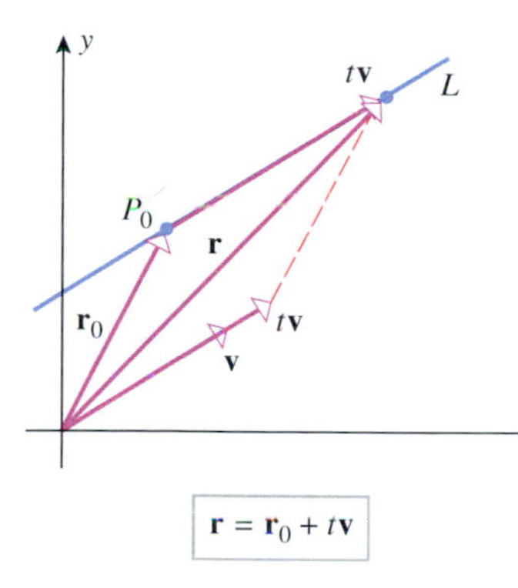

Figure 12.5.4

in both cases. We call this the ***vector equation of a line*** in 2-space or 3-space. In this equation, $\mathbf{v}$ is a nonzero vector parallel to the line, and $\mathbf{r}_0$ is a vector whose components are the coordinates of a point on the line.

We can interpret Equation (9) geometrically by positioning the vectors $\mathbf{r}_0$ and $\mathbf{v}$ with their initial points at the origin and the vector $t\mathbf{v}$ with its initial point at P_0 (Figure 12.5.4). The vector $t\mathbf{v}$ is a scalar multiple of $\mathbf{v}$ and hence is parallel to $\mathbf{v}$ and L. Moreover, since the initial point of $t\mathbf{v}$ is at the point P_0 on L, this vector actually runs along L; hence, the vector $\mathbf{r} = \mathbf{r}_0 + t\mathbf{v}$ can be interpreted as the vector from the origin to a point on L. As the parameter t varies from 0 to $+\infty$, the terminal point of $\mathbf{r}$ traces out the portion of L that extends from P_0 in the direction of $\mathbf{v}$, and as t varies from 0 to $-\infty$, the terminal point of $\mathbf{r}$ traces out the portion of L that extends from P_0 in the direction that is opposite to $\mathbf{v}$. Thus, the entire line is traced as t varies over the interval $(-\infty, +\infty)$, and it is traced in the direction of $\mathbf{v}$ as t increases.

▶ **Example 5** The equation

$$\langle x, y, z\rangle = \langle -1, 0, 2\rangle + t\langle 1, 5, -4\rangle$$

is of form (9) with

$$\mathbf{r}_0 = \langle -1, 0, 2\rangle \quad \text{and} \quad \mathbf{v} = \langle 1, 5, -4\rangle$$

Thus, the equation represents the line in 3-space that passes through the point $(-1, 0, 2)$ and is parallel to the vector $\langle 1, 5, -4\rangle$. ◀

▶ **Example 6** Find an equation of the line in 3-space that passes through the points $P_1(2, 4, -1)$ and $P_2(5, 0, 7)$.

Solution. The vector

$$\overrightarrow{P_1P_2} = \langle 3, -4, 8\rangle$$

is parallel to the line, so it can be used as $\mathbf{v}$ in (9). For $\mathbf{r}_0$ we can use either the vector from the origin to P_1 or the vector from the origin to P_2. Using the former yields

$$\mathbf{r}_0 = \langle 2, 4, -1\rangle$$

Thus, a vector equation of the line through P_1 and P_2 is

$$\langle x, y, z\rangle = \langle 2, 4, -1\rangle + t\langle 3, -4, 8\rangle$$

If needed, we can express the line parametrically by equating corresponding components on the two sides of this vector equation, in which case we obtain the parametric equations in Example 2 (verify). ◀

✔ QUICK CHECK EXERCISES 12.5 (See page 830 for answers.)

1. Let L be the line through (2, 5) and parallel to $\mathbf{v} = \langle 3, -1\rangle$.
(a) Parametric equations of L are

$x =$ ________ $y =$ ________

(b) A vector equation of L is $\langle x, y\rangle =$ ________.

2. Parametric equations for the line through (5, 3, 7) and parallel to the line $x = 3 - t$, $y = 2$, $z = 8 + 4t$ are

$x =$ ________, $y =$ ________, $z =$ ________

3. Parametric equations for the line segment joining the points (3, 0, 11) and (2, 6, 7) are

$x =$ ________, $y =$ ________, $z =$ ________ (________)

4. The line through the points $(-3, 8, -4)$ and $(1, 0, 8)$ intersects the yz-plane at ________.

EXERCISE SET 12.5 Graphing Utility C CAS

1. (a) Find parametric equations for the lines through the corner of the unit square shown in part (a) of the accompanying figure.
(b) Find parametric equations for the lines through the corner of the unit cube shown in part (b) of the accompanying figure.

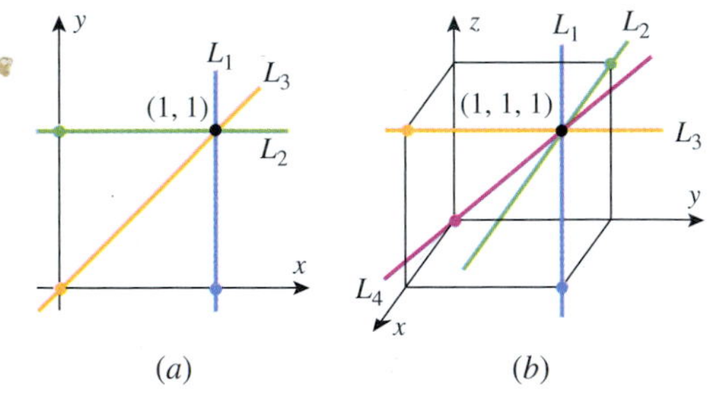

Figure Ex-1

2. (a) Find parametric equations for the line segments in the unit square in part (a) of the accompanying figure.
(b) Find parametric equations for the line segments in the unit cube shown in part (b) of the accompanying figure.

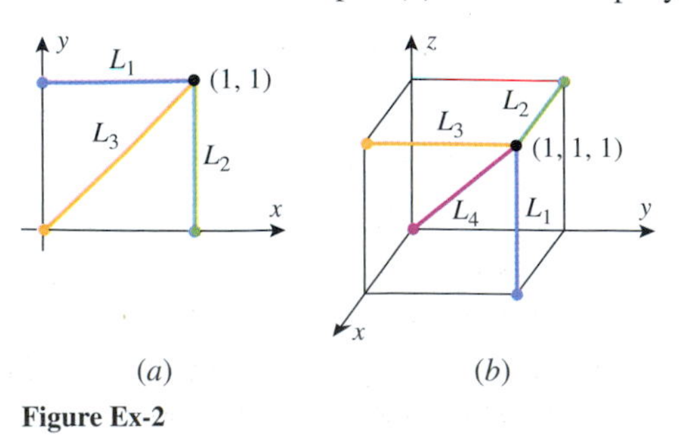

Figure Ex-2

3–4 Find parametric equations for the line through P_1 and P_2 and also for the line segment joining those points.

3. (a) $P_1(3, -2)$, $P_2(5, 1)$ (b) $P_1(5, -2, 1)$, $P_2(2, 4, 2)$

4. (a) $P_1(0, 1)$, $P_2(-3, -4)$
(b) $P_1(-1, 3, 5)$, $P_2(-1, 3, 2)$

5–6 Find parametric equations for the line whose vector equation is given.

5. (a) $\langle x, y\rangle = \langle 2, -3\rangle + t\langle 1, -4\rangle$
(b) $x\mathbf{i} + y\mathbf{j} + z\mathbf{k} = \mathbf{k} + t(\mathbf{i} - \mathbf{j} + \mathbf{k})$

6. (a) $x\mathbf{i} + y\mathbf{j} = (3\mathbf{i} - 4\mathbf{j}) + t(2\mathbf{i} + \mathbf{j})$
(b) $\langle x, y, z\rangle = \langle -1, 0, 2\rangle + t\langle -1, 3, 0\rangle$

7–8 Find a point P on the line and a vector $\mathbf{v}$ parallel to the line by inspection.

7. (a) $x\mathbf{i} + y\mathbf{j} = (2\mathbf{i} - \mathbf{j}) + t(4\mathbf{i} - \mathbf{j})$
(b) $\langle x, y, z\rangle = \langle -1, 2, 4\rangle + t\langle 5, 7, -8\rangle$

8. (a) $\langle x, y\rangle = \langle -1, 5\rangle + t\langle 2, 3\rangle$
(b) $x\mathbf{i} + y\mathbf{j} + z\mathbf{k} = (\mathbf{i} + \mathbf{j} - 2\mathbf{k}) + t\mathbf{j}$

9–10 Express the given parametric equations of a line using bracket notation and also using $\mathbf{i}, \mathbf{j}, \mathbf{k}$ notation.

9. (a) $x = -3 + t$, $y = 4 + 5t$
(b) $x = 2 - t$, $y = -3 + 5t$, $z = t$

10. (a) $x = t$, $y = -2 + t$
(b) $x = 1 + t$, $y = -7 + 3t$, $z = 4 - 5t$

11–18 Find parametric equations of the line that satisfies the stated conditions.

11. The line through $(-5, 2)$ that is parallel to $2\mathbf{i} - 3\mathbf{j}$.

12. The line through $(0, 3)$ that is parallel to the line $x = -5 + t$, $y = 1 - 2t$.

13. The line that is tangent to the circle $x^2 + y^2 = 25$ at the point $(3, -4)$.

14. The line that is tangent to the parabola $y = x^2$ at the point $(-2, 4)$.

15. The line through $(-1, 2, 4)$ that is parallel to $3\mathbf{i} - 4\mathbf{j} + \mathbf{k}$.

16. The line through $(2, -1, 5)$ that is parallel to $\langle -1, 2, 7\rangle$.

17. The line through $(-2, 0, 5)$ that is parallel to the line given by $x = 1 + 2t$, $y = 4 - t$, $z = 6 + 2t$.

18. The line through the origin that is parallel to the line given by $x = t$, $y = -1 + t$, $z = 2$.

19. Where does the line $x = 1 + 3t$, $y = 2 - t$ intersect
(a) the x-axis (b) the y-axis
(c) the parabola $y = x^2$?

20. Where does the line $\langle x, y\rangle = \langle 4t, 3t\rangle$ intersect the circle $x^2 + y^2 = 25$?

21–22 Find the intersections of the lines with the xy-plane, the xz-plane, and the yz-plane.

21. $x = -2,\ y = 4 + 2t,\ z = -3 + t$

22. $x = -1 + 2t,\ y = 3 + t,\ z = 4 - t$

23. Where does the line $x = 1 + t$, $y = 3 - t$, $z = 2t$ intersect the cylinder $x^2 + y^2 = 16$?

24. Where does the line $x = 2 - t$, $y = 3t$, $z = -1 + 2t$ intersect the plane $2y + 3z = 6$?

25–26 Show that the lines L_1 and L_2 intersect, and find their point of intersection.

25. L_1: $x = 2 + t,\ y = 2 + 3t,\ z = 3 + t$
L_2: $x = 2 + t,\ y = 3 + 4t,\ z = 4 + 2t$

26. L_1: $x + 1 = 4t,\ y - 3 = t,\ z - 1 = 0$
L_2: $x + 13 = 12t,\ y - 1 = 6t,\ z - 2 = 3t$

27–28 Show that the lines L_1 and L_2 are skew.

27. L_1: $x = 1 + 7t,\ y = 3 + t,\ z = 5 - 3t$
L_2: $x = 4 - t,\ y = 6,\ z = 7 + 2t$

28. L_1: $x = 2 + 8t,\ y = 6 - 8t,\ z = 10t$
L_2: $x = 3 + 8t,\ y = 5 - 3t,\ z = 6 + t$

29–30 Determine whether the lines L_1 and L_2 are parallel.

29. L_1: $x = 3 - 2t,\ y = 4 + t,\ z = 6 - t$
L_2: $x = 5 - 4t,\ y = -2 + 2t,\ z = 7 - 2t$

30. L_1: $x = 5 + 3t,\ y = 4 - 2t,\ z = -2 + 3t$
L_2: $x = -1 + 9t,\ y = 5 - 6t,\ z = 3 + 8t$

31–32 Determine whether the points P_1, P_2, and P_3 lie on the same line.

31. $P_1(6, 9, 7),\ P_2(9, 2, 0),\ P_3(0, -5, -3)$

32. $P_1(1, 0, 1),\ P_2(3, -4, -3),\ P_3(4, -6, -5)$

33–34 Show that the lines L_1 and L_2 are the same.

33. L_1: $x = 3 - t,\ y = 1 + 2t$
L_2: $x = -1 + 3t,\ y = 9 - 6t$

34. L_1: $x = 1 + 3t,\ y = -2 + t,\ z = 2t$
L_2: $x = 4 - 6t,\ y = -1 - 2t,\ z = 2 - 4t$

FOCUS ON CONCEPTS

35. Sketch the vectors $\mathbf{r}_0 = \langle -1, 2\rangle$ and $\mathbf{v} = \langle 1, 1\rangle$, and then sketch the six vectors $\mathbf{r}_0 \pm \mathbf{v}$, $\mathbf{r}_0 \pm 2\mathbf{v}$, $\mathbf{r}_0 \pm 3\mathbf{v}$. Draw the line $L: x = -1 + t,\ y = 2 + t$, and describe the relationship between L and the vectors you sketched. What is the vector equation of L?

36. Sketch the vectors $\mathbf{r}_0 = \langle 0, 2, 1\rangle$ and $\mathbf{v} = \langle 1, 0, 1\rangle$, and then sketch the vectors $\mathbf{r}_0 + \mathbf{v}$, $\mathbf{r}_0 + 2\mathbf{v}$, and $\mathbf{r}_0 + 3\mathbf{v}$. Draw the line $L: x = t,\ y = 2,\ z = 1 + t$, and describe the relationship between L and the vectors you sketched. What is the vector equation of L?

37. Sketch the vectors $\mathbf{r}_0 = \langle -2, 0\rangle$ and $\mathbf{r}_1 = \langle 1, 3\rangle$, and then sketch the vectors

$$\tfrac{1}{3}\mathbf{r}_0 + \tfrac{2}{3}\mathbf{r}_1,\quad \tfrac{1}{2}\mathbf{r}_0 + \tfrac{1}{2}\mathbf{r}_1,\quad \tfrac{2}{3}\mathbf{r}_0 + \tfrac{1}{3}\mathbf{r}_1$$

Draw the line segment $(1 - t)\mathbf{r}_0 + t\mathbf{r}_1$ $(0 \le t \le 1)$. If n is a positive integer, what is the position of the point on this line segment corresponding to $t = 1/n$, relative to the points $(-2, 0)$ and $(1, 3)$?

38. Sketch the vectors $\mathbf{r}_0 = \langle 2, 0, 4\rangle$ and $\mathbf{r}_1 = \langle 0, 4, 0\rangle$, and then sketch the vectors

$$\tfrac{1}{4}\mathbf{r}_0 + \tfrac{3}{4}\mathbf{r}_1,\quad \tfrac{1}{2}\mathbf{r}_0 + \tfrac{1}{2}\mathbf{r}_1,\quad \tfrac{3}{4}\mathbf{r}_0 + \tfrac{1}{4}\mathbf{r}_1$$

Draw the line segment $(1 - t)\mathbf{r}_0 + t\mathbf{r}_1$ $(0 \le t \le 1)$. If n is a positive integer, what is the position of the point on this line segment corresponding to $t = 1/n$, relative to the points $(2, 0, 4)$ and $(0, 4, 0)$?

39–40 Describe the line segment represented by the vector equation.

39. $\langle x, y\rangle = \langle 1, 0\rangle + t\langle -2, 3\rangle \quad (0 \le t \le 2)$

40. $\langle x, y, z\rangle = \langle -2, 1, 4\rangle + t\langle 3, 0, -1\rangle \quad (0 \le t \le 3)$

41. Find the point on the line segment joining $P_1(3, 6)$ and $P_2(8, -4)$ that is $\frac{2}{5}$ of the way from P_1 to P_2.

42. Find the point on the line segment joining $P_1(1, 4, -3)$ and $P_2(1, 5, -1)$ that is $\frac{2}{3}$ of the way from P_1 to P_2.

43–44 Use the method in Exercise 28 of Section 12.3 to find the distance from the point P to the line L, and then check your answer using the method in Exercise 26 of Section 12.4.

43. $P(-2, 1, 1)$
$L: x = 3 - t,\ y = t,\ z = 1 + 2t$

44. $P(1, 4, -3)$
$L: x = 2 + t,\ y = -1 - t,\ z = 3t$

45–46 Show that the lines L_1 and L_2 are parallel, and find the distance between them.

45. $L_1: x = 2 - t,\ y = 2t,\ z = 1 + t$
$L_2: x = 1 + 2t,\ y = 3 - 4t,\ z = 5 - 2t$

46. $L_1: x = 2t,\ y = 3 + 4t,\ z = 2 - 6t$
$L_2: x = 1 + 3t,\ y = 6t,\ z = -9t$

47. (a) Find parametric equations for the line through the points (x_0, y_0, z_0) and (x_1, y_1, z_1).
(b) Find parametric equations for the line through the point (x_1, y_1, z_1) and parallel to the line

$$x = x_0 + at, \quad y = y_0 + bt, \quad z = z_0 + ct$$

48. Let L be the line that passes through the point (x_0, y_0, z_0) and is parallel to the vector $\mathbf{v} = \langle a, b, c\rangle$, where a, b, and c are nonzero. Show that a point (x, y, z) lies on the line L if and only if

$$\frac{x - x_0}{a} = \frac{y - y_0}{b} = \frac{z - z_0}{c}$$

These equations, which are called the ***symmetric equations*** of L, provide a nonparametric representation of L.

49. (a) Describe the line whose symmetric equations are

$$\frac{x - 1}{2} = \frac{y + 3}{4} = z - 5$$

[See Exercise 48.]
(b) Find parametric equations for the line in part (a).

50. Consider the lines L_1 and L_2 whose symmetric equations are

$$L_1: \frac{x - 1}{2} = \frac{y + \frac{3}{2}}{1} = \frac{z + 1}{2}$$

$$L_2: \frac{x - 4}{-1} = \frac{y - 3}{-2} = \frac{z + 4}{2}$$

[See Exercise 48.]
(a) Are L_1 and L_2 parallel? Perpendicular?
(b) Find parametric equations for L_1 and L_2.
(c) Do L_1 and L_2 intersect? If so, where?

51. Let L_1 and L_2 be the lines whose parametric equations are

$$L_1: x = 1 + 2t, \quad y = 2 - t, \quad z = 4 - 2t$$
$$L_2: x = 9 + t, \quad y = 5 + 3t, \quad z = -4 - t$$

(a) Show that L_1 and L_2 intersect at the point $(7, -1, -2)$.
(b) Find, to the nearest degree, the acute angle between L_1 and L_2 at their intersection.
(c) Find parametric equations for the line that is perpendicular to L_1 and L_2 and passes through their point of intersection.

52. Let L_1 and L_2 be the lines whose parametric equations are

$$L_1: x = 4t, \quad y = 1 - 2t, \quad z = 2 + 2t$$
$$L_2: x = 1 + t, \quad y = 1 - t, \quad z = -1 + 4t$$

(a) Show that L_1 and L_2 intersect at the point $(2, 0, 3)$.
(b) Find, to the nearest degree, the acute angle between L_1 and L_2 at their intersection.
(c) Find parametric equations for the line that is perpendicular to L_1 and L_2 and passes through their point of intersection.

53–54 Find parametric equations of the line that contains the point P and intersects the line L at a right angle, and find the distance between P and L.

53. $P(0, 2, 1)$
$L: x = 2t,\ y = 1 - t,\ z = 2 + t$

54. $P(3, 1, -2)$
$L: x = -2 + 2t,\ y = 4 + 2t,\ z = 2 + t$

55. Two bugs are walking along lines in 3-space. At time t bug 1 is at the point (x, y, z) on the line

$$x = 4 - t, \quad y = 1 + 2t, \quad z = 2 + t$$

and at the same time t bug 2 is at the point (x, y, z) on the line

$$x = t, \quad y = 1 + t, \quad z = 1 + 2t$$

Assume that distance is in centimeters and that time is in minutes.
(a) Find the distance between the bugs at time $t = 0$.
(b) Use a graphing utility to graph the distance between the bugs as a function of time from $t = 0$ to $t = 5$.
(c) What does the graph tell you about the distance between the bugs?
(d) How close do the bugs get?

C **56.** Suppose that the temperature T at a point (x, y, z) on the line $x = t$, $y = 1 + t$, $z = 3 - 2t$ is $T = 25x^2yz$. Use a CAS or a calculating utility with a root-finding capability to approximate the maximum temperature on that portion of the line that extends from the xz-plane to the xy-plane.

1. (a) $2 + 3t$; $5 - t$ (b) $\langle 2, 5\rangle + t\langle 3, -1\rangle$ **2.** $5 - t$; 3; $7 + 4t$ **3.** $3 - t$; $6t$; $11 - 4t$; $0 \le t \le 1$ **4.** $(0, 2, 5)$

12.6 PLANES IN 3-SPACE

In this section we will use vectors to derive equations of planes in 3-space, and then we will use these equations to solve various geometric problems.

PLANES PARALLEL TO THE COORDINATE PLANES

The graph of the equation $x = a$ in an xyz-coordinate system consists of all points of the form (a, y, z), where y and z are arbitrary. One such point is $(a, 0, 0)$, and all others are in the plane that passes through this point and is parallel to the yz-plane (Figure 12.6.1). Similarly, the graph of $y = b$ is the plane through $(0, b, 0)$ that is parallel to the xz-plane, and the graph of $z = c$ is the plane through $(0, 0, c)$ that is parallel to the xy-plane.

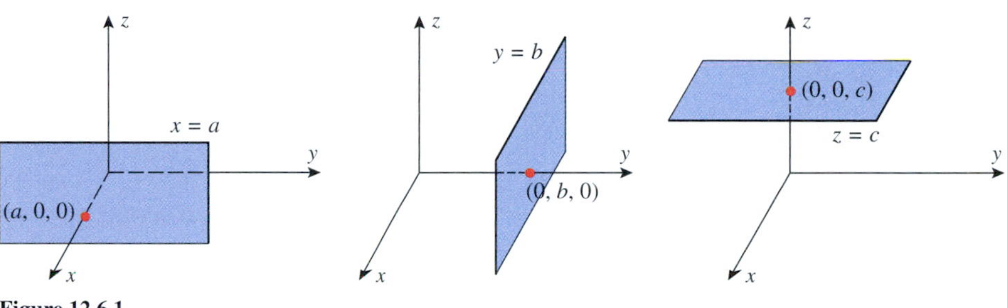

Figure 12.6.1

PLANES DETERMINED BY A POINT AND A NORMAL VECTOR

A plane in 3-space can be determined uniquely by specifying a point in the plane and a vector perpendicular to the plane (Figure 12.6.2). A vector perpendicular to a plane is called a ***normal*** to the plane.

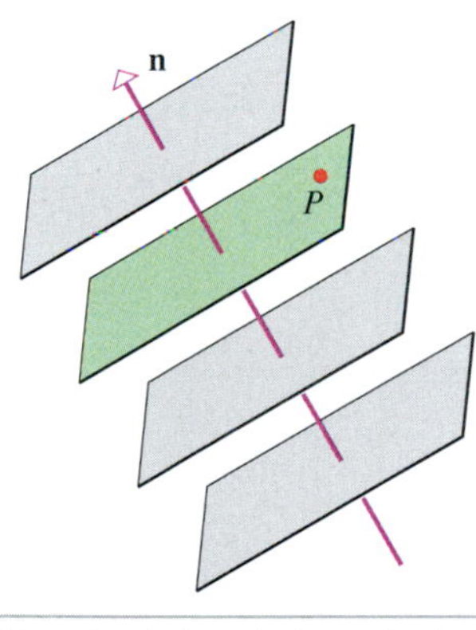

The colored plane is determined uniquely by the point P and the vector $\mathbf{n}$ perpendicular to the plane.

Figure 12.6.2

Suppose that we want to find an equation of the plane passing through $P_0(x_0, y_0, z_0)$ and perpendicular to the vector $\mathbf{n} = \langle a, b, c\rangle$. Define the vectors $\mathbf{r}_0$ and $\mathbf{r}$ as

$$\mathbf{r}_0 = \langle x_0, y_0, z_0\rangle \quad \text{and} \quad \mathbf{r} = \langle x, y, z\rangle$$

It should be evident from Figure 12.6.3 that the plane consists precisely of those points $P(x, y, z)$ for which the vector $\mathbf{r} - \mathbf{r}_0$ is orthogonal to $\mathbf{n}$; or, expressed as an equation,

$$\mathbf{n} \cdot (\mathbf{r} - \mathbf{r}_0) = 0 \tag{1}$$

If preferred, we can express this vector equation in terms of components as

$$\langle a, b, c\rangle \cdot \langle x - x_0, y - y_0, z - z_0\rangle = 0 \tag{2}$$

from which we obtain

$$a(x - x_0) + b(y - y_0) + c(z - z_0) = 0 \tag{3}$$

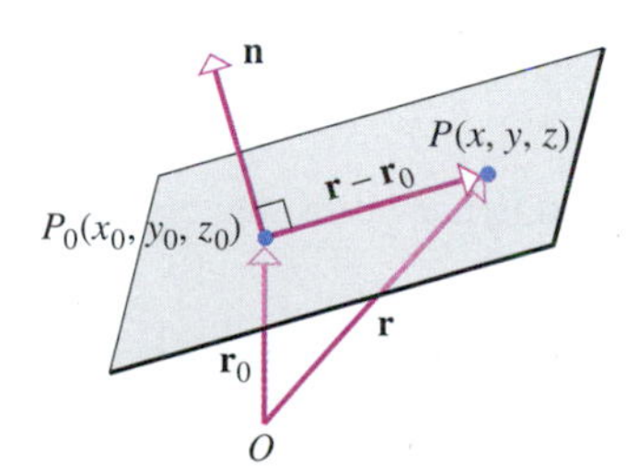

Figure 12.6.3

This is called the ***point-normal form*** of the equation of a plane. Formulas (1) and (2) are vector versions of this formula.

► **Example 1** Find an equation of the plane passing through the point $(3, -1, 7)$ and perpendicular to the vector $\mathbf{n} = \langle 4, 2, -5\rangle$.

Solution. From (3), a point-normal form of the equation is

$$4(x - 3) + 2(y + 1) - 5(z - 7) = 0 \tag{4}$$

What does Equation (1) represent if

$\mathbf{n} = \langle a, b \rangle$, $\mathbf{r}_0 = \langle x_0, y_0 \rangle$, $\mathbf{r} = \langle x, y \rangle$

are vectors in an xy-plane in 2-space? Draw a picture.

If preferred, this equation can be written in vector form as

$$\langle 4, 2, -5 \rangle \cdot \langle x - 3, y + 1, z - 7 \rangle = 0 \blacktriangleleft$$

Observe that if we multiply out the terms in (3) and simplify, we obtain an equation of the form

$$ax + by + cz + d = 0 \qquad (5)$$

For example, Equation (4) in Example 1 can be rewritten as

$$4x + 2y - 5z + 25 = 0$$

The following theorem shows that every equation of form (5) represents a plane in 3-space.

12.6.1 THEOREM. *If a, b, c, and d are constants, and a, b, and c are not all zero, then the graph of the equation*

$$ax + by + cz + d = 0 \qquad (6)$$

is a plane that has the vector $\mathbf{n} = \langle a, b, c \rangle$ as a normal.

PROOF. Since a, b, and c are not all zero, there is at least one point (x_0, y_0, z_0) whose coordinates satisfy Equation (6). For example, if $a \neq 0$, then such a point is $(-d/a, 0, 0)$, and similarly if $b \neq 0$ or $c \neq 0$ (verify). Thus, let (x_0, y_0, z_0) be any point whose coordinates satisfy (6); that is,

$$ax_0 + by_0 + cz_0 + d = 0$$

Subtracting this equation from (6) yields

$$a(x - x_0) + b(y - y_0) + c(z - z_0) = 0$$

which is the point-normal form of a plane with normal $\mathbf{n} = \langle a, b, c \rangle$. ■

Equation (6) is called the ***general form*** of the equation of a plane.

▶ **Example 2** Determine whether the planes

$$3x - 4y + 5z = 0 \quad \text{and} \quad -6x + 8y - 10z - 4 = 0$$

are parallel.

Solution. It is clear geometrically that two planes are parallel if and only if their normals are parallel vectors. A normal to the first plane is

$$\mathbf{n}_1 = \langle 3, -4, 5 \rangle$$

and a normal to the second plane is

$$\mathbf{n}_2 = \langle -6, 8, -10 \rangle$$

Since $\mathbf{n}_2$ is a scalar multiple of $\mathbf{n}_1$, the normals are parallel, and hence so are the planes. ◀

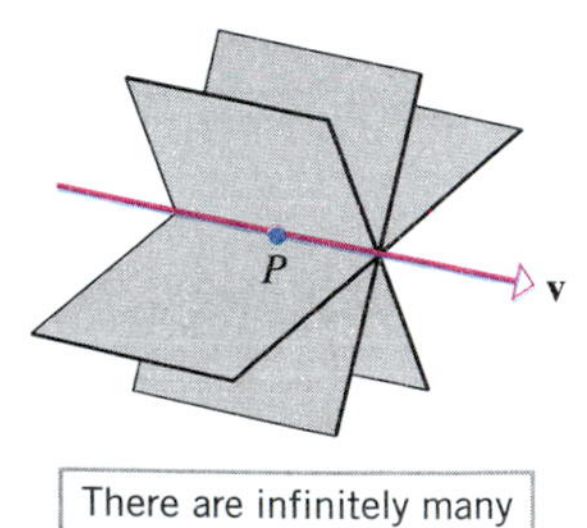

There are infinitely many planes containing P and parallel to $\mathbf{v}$.

Figure 12.6.4

We have seen that a unique plane is determined by a point in the plane and a nonzero vector normal to the plane. In contrast, a unique plane is not determined by a point in the plane and a nonzero vector *parallel* to the plane (Figure 12.6.4). However, a unique plane is determined by a point in the plane and two nonparallel vectors that are parallel to the

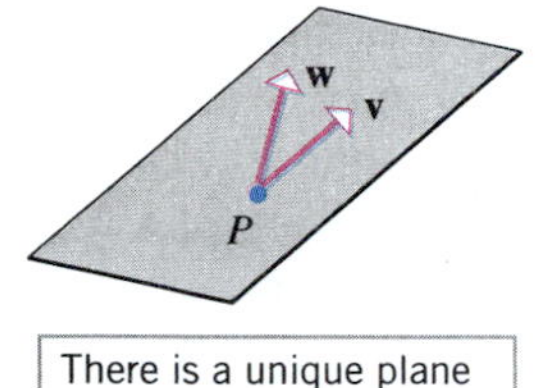

There is a unique plane through P that is parallel to both $\mathbf{v}$ and $\mathbf{w}$.

Figure 12.6.5

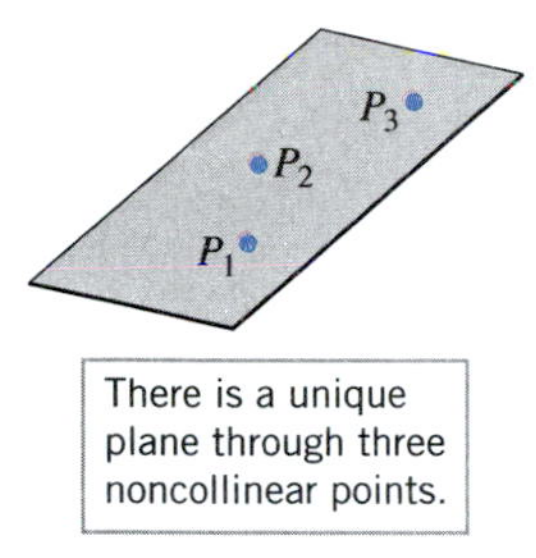

There is a unique plane through three noncollinear points.

Figure 12.6.6

plane (Figure 12.6.5). A unique plane is also determined by three noncollinear points that lie in the plane (Figure 12.6.6).

▶ Example 3 Find an equation of the plane through the points $P_1(1, 2, -1)$, $P_2(2, 3, 1)$, and $P_3(3, -1, 2)$.

Solution. Since the points P_1, P_2, and P_3 lie in the plane, the vectors $\overrightarrow{P_1P_2} = \langle 1, 1, 2\rangle$ and $\overrightarrow{P_1P_3} = \langle 2, -3, 3\rangle$ are parallel to the plane. Therefore,

$$\overrightarrow{P_1P_2} \times \overrightarrow{P_1P_3} = \begin{vmatrix} \mathbf{i} & \mathbf{j} & \mathbf{k} \\ 1 & 1 & 2 \\ 2 & -3 & 3 \end{vmatrix} = 9\mathbf{i} + \mathbf{j} - 5\mathbf{k}$$

is normal to the plane, since it is orthogonal to both $\overrightarrow{P_1P_2}$ and $\overrightarrow{P_1P_3}$. By using this normal and the point $P_1(1, 2, -1)$ in the plane, we obtain the point-normal form

$$9(x-1) + (y-2) - 5(z+1) = 0$$

which can be rewritten as

$$9x + y - 5z - 16 = 0 \blacktriangleleft$$

▶ Example 4 Determine whether the line

$$x = 3 + 8t, \quad y = 4 + 5t, \quad z = -3 - t$$

is parallel to the plane $x - 3y + 5z = 12$.

Solution. The vector $\mathbf{v} = \langle 8, 5, -1\rangle$ is parallel to the line and the vector $\mathbf{n} = \langle 1, -3, 5\rangle$ is normal to the plane. For the line and plane to be parallel, the vectors $\mathbf{v}$ and $\mathbf{n}$ must be orthogonal. But this is not so, since the dot product

$$\mathbf{v} \cdot \mathbf{n} = (8)(1) + (5)(-3) + (-1)(5) = -12$$

is nonzero. Thus, the line and plane are not parallel. ◀

▶ Example 5 Find the intersection of the line and plane in Example 4.

Solution. If we let (x_0, y_0, z_0) be the point of intersection, then the coordinates of this point satisfy both the equation of the plane and the parametric equations of the line. Thus,

$$x_0 - 3y_0 + 5z_0 = 12 \tag{7}$$

and for some value of t, say $t = t_0$,

$$x_0 = 3 + 8t_0, \quad y_0 = 4 + 5t_0, \quad z_0 = -3 - t_0 \tag{8}$$

Substituting (8) in (7) yields

$$(3 + 8t_0) - 3(4 + 5t_0) + 5(-3 - t_0) = 12$$

Solving for t_0 yields $t_0 = -3$ and on substituting this value in (8), we obtain

$$(x_0, y_0, z_0) = (-21, -11, 0) \blacktriangleleft$$

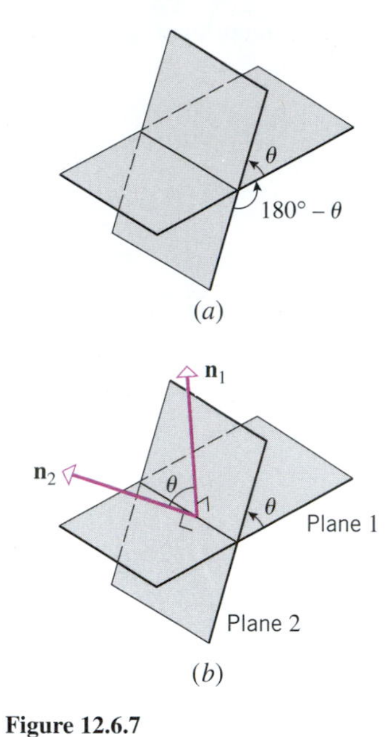

Figure 12.6.7

INTERSECTING PLANES

Two distinct intersecting planes determine two positive angles of intersection—an (acute) angle θ that satisfies the condition $0 \leq \theta \leq \pi/2$ and the supplement of that angle (Figure 12.6.7a). If $\mathbf{n}_1$ and $\mathbf{n}_2$ are normals to the planes, then depending on the directions of $\mathbf{n}_1$ and $\mathbf{n}_2$, the angle θ is either the angle between $\mathbf{n}_1$ and $\mathbf{n}_2$ or the angle between $\mathbf{n}_1$ and $-\mathbf{n}_2$ (Figure 12.6.7b). In both cases, Theorem 12.3.3 yields the following formula for the acute angle θ between the planes:

$$\cos\theta = \frac{|\mathbf{n}_1 \cdot \mathbf{n}_2|}{\|\mathbf{n}_1\|\|\mathbf{n}_2\|} \tag{9}$$

Example 6 Find the acute angle of intersection between the two planes

$$2x - 4y + 4z = 6 \quad \text{and} \quad 6x + 2y - 3z = 4$$

Solution. The given equations yield the normals $\mathbf{n}_1 = \langle 2, -4, 4\rangle$ and $\mathbf{n}_2 = \langle 6, 2, -3\rangle$. Thus, Formula (9) yields

$$\cos\theta = \frac{|\mathbf{n}_1 \cdot \mathbf{n}_2|}{\|\mathbf{n}_1\|\,\|\mathbf{n}_2\|} = \frac{|-8|}{\sqrt{36}\sqrt{49}} = \frac{4}{21}$$

from which we obtain

$$\theta = \cos^{-1}\left(\frac{4}{21}\right) \approx 79^\circ \blacktriangleleft$$

Example 7 Find an equation for the line L of intersection of the planes in Example 6.

Solution. First compute $\mathbf{v} = \mathbf{n}_1 \times \mathbf{n}_2 = \langle 2, -4, 4\rangle \times \langle 6, 2, -3\rangle = \langle 4, 30, 28\rangle$. Since $\mathbf{v}$ is orthogonal to $\mathbf{n}_1$, it is parallel to the first plane, and since $\mathbf{v}$ is orthogonal to $\mathbf{n}_2$, it is parallel to the second plane. That is, $\mathbf{v}$ is parallel to L, the intersection of the two planes. To find a point on L we observe that L must intersect the xy-plane, $z = 0$, since $\mathbf{v} \cdot \langle 0, 0, 1\rangle = 28 \neq 0$. Substituting $z = 0$ in the equations of both planes yields

$$\begin{aligned} 2x - 4y &= 6 \\ 6x + 2y &= 4 \end{aligned}$$

with solution $x = 1$, $y = -1$. Thus, $P(1, -1, 0)$ is a point on L. A vector equation for L is

$$\langle x, y, z\rangle = \langle 1, -1, 0\rangle + t\langle 4, 30, 28\rangle \blacktriangleleft$$

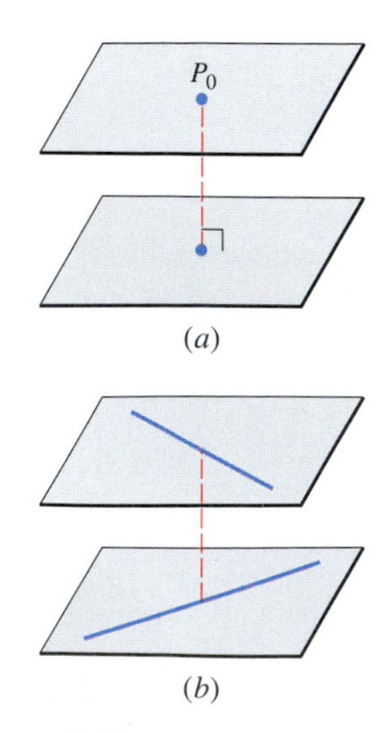

Figure 12.6.8

DISTANCE PROBLEMS INVOLVING PLANES

Next we will consider three basic "distance problems" in 3-space:

- Find the distance between a point and a plane.
- Find the distance between two parallel planes.
- Find the distance between two skew lines.

The three problems are related. If we can find the distance between a point and a plane, then we can find the distance between parallel planes by computing the distance between one of the planes and an arbitrary point P_0 in the other plane (Figure 12.6.8a). Moreover, we can find the distance between two skew lines by computing the distance between parallel planes containing them (Figure 12.6.8b).

12.6.2 THEOREM. *The distance D between a point $P_0(x_0, y_0, z_0)$ and the plane $ax + by + cz + d = 0$ is*

$$D = \frac{|ax_0 + by_0 + cz_0 + d|}{\sqrt{a^2 + b^2 + c^2}} \tag{10}$$

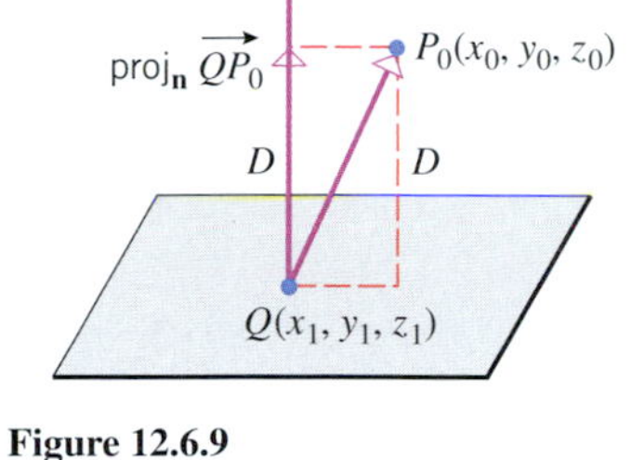

Figure 12.6.9

PROOF. Let $Q(x_1, y_1, z_1)$ be any point in the plane, and position the normal $\mathbf{n} = \langle a, b, c\rangle$ so that its initial point is at Q. As illustrated in Figure 12.6.9, the distance D is equal to the length of the orthogonal projection of $\overrightarrow{QP_0}$ on $\mathbf{n}$. Thus, from (12) of Section 12.3,

$$D = \|\text{proj}_{\mathbf{n}} \overrightarrow{QP_0}\| = \left\| \frac{\overrightarrow{QP_0} \cdot \mathbf{n}}{\|\mathbf{n}\|^2}\mathbf{n} \right\| = \frac{|\overrightarrow{QP_0} \cdot \mathbf{n}|}{\|\mathbf{n}\|^2}\|\mathbf{n}\| = \frac{|\overrightarrow{QP_0} \cdot \mathbf{n}|}{\|\mathbf{n}\|}$$

But

$$\overrightarrow{QP_0} = \langle x_0 - x_1, y_0 - y_1, z_0 - z_1\rangle$$
$$\overrightarrow{QP_0} \cdot \mathbf{n} = a(x_0 - x_1) + b(y_0 - y_1) + c(z_0 - z_1)$$
$$\|\mathbf{n}\| = \sqrt{a^2 + b^2 + c^2}$$

Thus,

$$D = \frac{|a(x_0 - x_1) + b(y_0 - y_1) + c(z_0 - z_1)|}{\sqrt{a^2 + b^2 + c^2}} \tag{11}$$

Since the point $Q(x_1, y_1, z_1)$ lies in the plane, its coordinates satisfy the equation of the plane; that is,

$$ax_1 + by_1 + cz_1 + d = 0$$

or

$$d = -ax_1 - by_1 - cz_1$$

Combining this expression with (11) yields (10). ■

▶ Example 8 Find the distance D between the point $(1, -4, -3)$ and the plane

$$2x - 3y + 6z = -1$$

There is an analog of Formula (10) in 2-space that can be used to compute the distance between a point and a line. (See Exercise 50).

Solution. Formula (10) requires the plane be rewritten in the form $ax + by + cz + d = 0$. Thus, we rewrite the equation of the given plane as

$$2x - 3y + 6z + 1 = 0$$

from which we obtain $a = 2$, $b = -3$, $c = 6$, and $d = 1$. Substituting these values and the coordinates of the given point in (10), we obtain

$$D = \frac{|(2)(1) + (-3)(-4) + 6(-3) + 1|}{\sqrt{2^2 + (-3)^2 + 6^2}} = \frac{|-3|}{7} = \frac{3}{7}$$ ◀

▶ Example 9 The planes

$$x + 2y - 2z = 3 \quad \text{and} \quad 2x + 4y - 4z = 7$$

are parallel since their normals, $\langle 1, 2, -2\rangle$ and $\langle 2, 4, -4\rangle$, are parallel vectors. Find the distance between these planes.

Solution. To find the distance D between the planes, we can select an arbitrary point in one of the planes and compute its distance to the other plane. By setting $y = z = 0$ in the equation $x + 2y - 2z = 3$, we obtain the point $P_0(3, 0, 0)$ in this plane. From (10), the distance from P_0 to the plane $2x + 4y - 4z = 7$ is

$$D = \frac{|(2)(3) + 4(0) + (-4)(0) - 7|}{\sqrt{2^2 + 4^2 + (-4)^2}} = \frac{1}{6} \blacktriangleleft$$

▶ Example 10 It was shown in Example 3 of Section 12.5 that the lines

$$L_1\colon x = 1 + 4t, \quad y = 5 - 4t, \quad z = -1 + 5t$$

$$L_2\colon x = 2 + 8t, \quad y = 4 - 3t, \quad z = 5 + t$$

are skew. Find the distance between them.

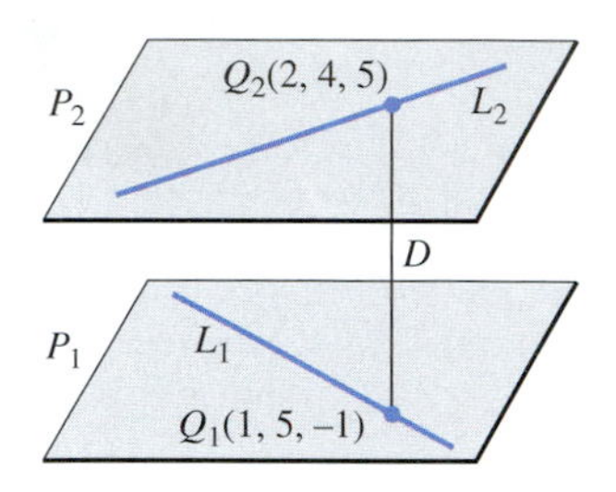

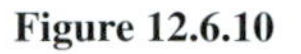
Figure 12.6.10

Solution. Let P_1 and P_2 denote parallel planes containing L_1 and L_2, respectively (Figure 12.6.10). To find the distance D between L_1 and L_2, we will calculate the distance from a point in P_1 to the plane P_2. Since L_1 lies in plane P_1, we can find a point in P_1 by finding a point on the line L_1; we can do this by substituting any convenient value of t in the parametric equations of L_1. The simplest choice is $t = 0$, which yields the point $Q_1(1, 5, -1)$.

The next step is to find an equation for the plane P_2. For this purpose, observe that the vector $\mathbf{u}_1 = \langle 4, -4, 5\rangle$ is parallel to line L_1, and therefore also parallel to planes P_1 and P_2. Similarly, $\mathbf{u}_2 = \langle 8, -3, 1\rangle$ is parallel to L_2 and hence parallel to P_1 and P_2. Therefore, the cross product

$$\mathbf{n} = \mathbf{u}_1 \times \mathbf{u}_2 = \begin{vmatrix} \mathbf{i} & \mathbf{j} & \mathbf{k} \\ 4 & -4 & 5 \\ 8 & -3 & 1 \end{vmatrix} = 11\mathbf{i} + 36\mathbf{j} + 20\mathbf{k}$$

is normal to both P_1 and P_2. Using this normal and the point $Q_2(2, 4, 5)$ found by setting $t = 0$ in the equations of L_2, we obtain an equation for P_2:

$$11(x - 2) + 36(y - 4) + 20(z - 5) = 0$$

or

$$11x + 36y + 20z - 266 = 0$$

The distance between $Q_1(1, 5, -1)$ and this plane is

$$D = \frac{|(11)(1) + (36)(5) + (20)(-1) - 266|}{\sqrt{11^2 + 36^2 + 20^2}} = \frac{95}{\sqrt{1817}}$$

which is also the distance between L_1 and L_2. ◀

✔QUICK CHECK EXERCISES 12.6 *(See page 839 for answers.)*

1. The point-normal form of the equation of the plane through $(0, 3, 5)$ and perpendicular to $\langle -4, 1, 7\rangle$ is ________.
2. A normal vector for the plane $4x - 2y + 7z - 11 = 0$ is ________.
3. A normal vector for the plane through the points $(2, 5, 1)$, $(3, 7, 0)$, and $(2, 5, 2)$ is ________.
4. The acute angle of intersection of the planes $x + y - 2z = 5$ and $3y - 4z = 6$ is ________.
5. The distance between the point $(9, 8, 3)$ and the plane $x + y - 2z = 5$ is ________.

EXERCISE SET 12.6

1. Find equations of the planes P_1, P_2, and P_3 that are parallel to the coordinate planes and pass through the corner $(3, 4, 5)$ of the box shown in the accompanying figure.

2. Find equations of the planes P_1, P_2, and P_3 that are parallel to the coordinate planes and pass through the corner (x_0, y_0, z_0) of the box shown in the accompanying figure.

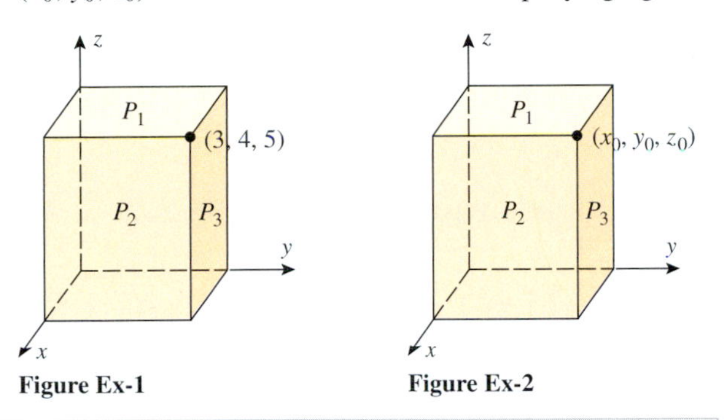

Figure Ex-1 **Figure Ex-2**

3–6 Find an equation of the plane that passes through the point P and has the vector $\mathbf{n}$ as a normal.

3. $P(2, 6, 1)$; $\mathbf{n} = \langle 1, 4, 2\rangle$

4. $P(-1, -1, 2)$; $\mathbf{n} = \langle -1, 7, 6\rangle$

5. $P(1, 0, 0)$; $\mathbf{n} = \langle 0, 0, 1\rangle$

6. $P(0, 0, 0)$; $\mathbf{n} = \langle 2, -3, -4\rangle$

7–10 Find an equation of the plane indicated in the figure.

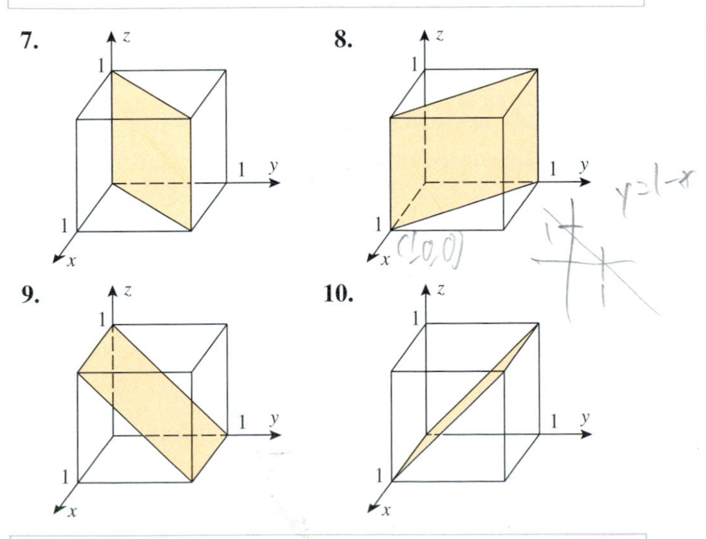

11–12 Find an equation of the plane that passes through the given points.

11. $(-2, 1, 1)$, $(0, 2, 3)$, and $(1, 0, -1)$

12. $(3, 2, 1)$, $(2, 1, -1)$, and $(-1, 3, 2)$

13–14 Determine whether the planes are parallel, perpendicular, or neither.

13. (a) $2x - 8y - 6z - 2 = 0$, $-x + 4y + 3z - 5 = 0$
(b) $3x - 2y + z = 1$, $4x + 5y - 2z = 4$
(c) $x - y + 3z - 2 = 0$, $2x + z = 1$

14. (a) $3x - 2y + z = 4$, $6x - 4y + 3z = 7$
(b) $y = 4x - 2z + 3$, $x = \frac{1}{4}y + \frac{1}{2}z$
(c) $x + 4y + 7z = 3$, $5x - 3y + z = 0$

15–16 Determine whether the line and plane are parallel, perpendicular, or neither.

15. (a) $x = 4 + 2t,\ y = -t,\ z = -1 - 4t$; $3x + 2y + z - 7 = 0$
(b) $x = t,\ y = 2t,\ z = 3t$; $x - y + 2z = 5$
(c) $x = -1 + 2t,\ y = 4 + t,\ z = 1 - t$; $4x + 2y - 2z = 7$

16. (a) $x = 3 - t,\ y = 2 + t,\ z = 1 - 3t$; $2x + 2y - 5 = 0$
(b) $x = 1 - 2t,\ y = t,\ z = -t$; $6x - 3y + 3z = 1$
(c) $x = t,\ y = 1 - t,\ z = 2 + t$; $x + y + z = 1$

17–18 Determine whether the line and plane intersect; if so, find the coordinates of the intersection.

17. (a) $x = t,\ y = t,\ z = t$; $3x - 2y + z - 5 = 0$
(b) $x = 2 - t,\ y = 3 + t,\ z = t$; $2x + y + z = 1$

18. (a) $x = 3t,\ y = 5t,\ z = -t$; $2x - y + z + 1 = 0$
(b) $x = 1 + t,\ y = -1 + 3t,\ z = 2 + 4t$; $x - y + 4z = 7$

19–20 Find the acute angle of intersection of the planes to the nearest degree.

19. $x = 0$ and $2x - y + z - 4 = 0$

20. $x + 2y - 2z = 5$ and $6x - 3y + 2z = 8$

21–30 Find an equation of the plane that satisfies the stated conditions.

21. The plane through the origin that is parallel to the plane $4x - 2y + 7z + 12 = 0$.

22. The plane that contains the line $x = -2 + 3t$, $y = 4 + 2t$, $z = 3 - t$ and is perpendicular to the plane $x - 2y + z = 5$.

23. The plane through the point $(-1, 4, 2)$ that contains the line of intersection of the planes $4x - y + z - 2 = 0$ and $2x + y - 2z - 3 = 0$.

24. The plane through $(-1, 4, -3)$ that is perpendicular to the line $x - 2 = t$, $y + 3 = 2t$, $z = -t$.

25. The plane through $(1, 2, -1)$ that is perpendicular to the line of intersection of the planes $2x + y + z = 2$ and $x + 2y + z = 3$.

26. The plane through the points $P_1(-2, 1, 4)$, $P_2(1, 0, 3)$ that is perpendicular to the plane $4x - y + 3z = 2$.

27. The plane through $(-1, 2, -5)$ that is perpendicular to the planes $2x - y + z = 1$ and $x + y - 2z = 3$.

28. The plane that contains the point $(2, 0, 3)$ and the line $x = -1 + t$, $y = t$, $z = -4 + 2t$.

29. The plane whose points are equidistant from $(2, -1, 1)$ and $(3, 1, 5)$.

30. The plane that contains the line $x = 3t$, $y = 1 + t$, $z = 2t$ and is parallel to the intersection of the planes $y + z = -1$ and $2x - y + z = 0$.

31. Find parametric equations of the line through the point $(5, 0, -2)$ that is parallel to the planes $x - 4y + 2z = 0$ and $2x + 3y - z + 1 = 0$.

32. Let L be the line $x = 3t + 1$, $y = -5t$, $z = t$.
(a) Show that L lies in the plane $2x + y - z = 2$.
(b) Show that L is parallel to the plane $x + y + 2z = 0$. Is the line above, below, or on this plane?

33. Show that the lines

$$x = -2 + t, \quad y = 3 + 2t, \quad z = 4 - t$$
$$x = 3 - t, \quad y = 4 - 2t, \quad z = t$$

are parallel and find an equation of the plane they determine.

34. Show that the lines

$$L_1: x + 1 = 4t, \quad y - 3 = t, \quad z - 1 = 0$$
$$L_2: x + 13 = 12t, \quad y - 1 = 6t, \quad z - 2 = 3t$$

intersect and find an equation of the plane they determine.

FOCUS ON CONCEPTS

35. Do the points $(1, 0, -1)$, $(0, 2, 3)$, $(-2, 1, 1)$, and $(4, 2, 3)$ lie in the same plane? Justify your answer two different ways.

36. Show that if a, b, and c are nonzero, then the plane whose intercepts with the coordinate axes are $x = a$, $y = b$, and $z = c$ is given by the equation

$$\frac{x}{a} + \frac{y}{b} + \frac{z}{c} = 1$$

37. If L is a line in 3-space, must L lie in some vertical plane? Explain.

38. If L is a line in 3-space, must L lie in some horizontal plane? Explain.

39–40 Find parametric equations of the line of intersection of the planes.

39. $-2x + 3y + 7z + 2 = 0$
$x + 2y - 3z + 5 = 0$

40. $3x - 5y + 2z = 0$
$z = 0$

41–42 Find the distance between the point and the plane.

41. $(1, -2, 3)$; $2x - 2y + z = 4$

42. $(0, 1, 5)$; $3x + 6y - 2z - 5 = 0$

43–44 Find the distance between the given parallel planes.

43. $-2x + y + z = 0$
$6x - 3y - 3z - 5 = 0$

44. $x + y + z = 1$
$x + y + z = -1$

45–46 Find the distance between the given skew lines.

45. $x = 1 + 7t$, $y = 3 + t$, $z = 5 - 3t$
$x = 4 - t$, $y = 6$, $z = 7 + 2t$

46. $x = 3 - t$, $y = 4 + 4t$, $z = 1 + 2t$
$x = t$, $y = 3$, $z = 2t$

47. Find an equation of the sphere with center $(2, 1, -3)$ that is tangent to the plane $x - 3y + 2z = 4$.

48. Locate the point of intersection of the plane $2x + y - z = 0$ and the line through $(3, 1, 0)$ that is perpendicular to the plane.

49. Show that the line $x = -1 + t$, $y = 3 + 2t$, $z = -t$ and the plane $2x - 2y - 2z + 3 = 0$ are parallel, and find the distance between them.

FOCUS ON CONCEPTS

50. Formulas (1), (2), (3), (5), and (10), which apply to planes in 3-space, have analogs for lines in 2-space.
(a) Draw an analog of Figure 12.6.3 in 2-space to illustrate that the equation of the line that passes through the point $P(x_0, y_0)$ and is perpendicular to the vector $\mathbf{n} = \langle a, b \rangle$ can be expressed as

$$\mathbf{n} \cdot (\mathbf{r} - \mathbf{r}_0) = 0$$

where $\mathbf{r} = \langle x, y \rangle$ and $\mathbf{r}_0 = \langle x_0, y_0 \rangle$.
(b) Show that the vector equation in part (a) can be expressed as

$$a(x - x_0) + b(y - y_0) = 0$$

This is called the ***point-normal form of a line***.
(c) Using the proof of Theorem 12.6.1 as a guide, show that if a and b are not both zero, then the graph of the equation

$$ax + by + c = 0$$

is a line that has $\mathbf{n} = \langle a, b \rangle$ as a normal.

(d) Using the proof of Theorem 12.6.2 as a guide, show that the distance D between a point $P(x_0, y_0)$ and the line $ax + by + c = 0$ is

$$D = \frac{|ax_0 + by_0 + c|}{\sqrt{a^2 + b^2}}$$

(e) Use the formula in part (d) to find the distance between the point $P(-3, 5)$ and the line $y = -2x + 1$.

51. (a) Show that the distance D between parallel planes

$$ax + by + cz + d_1 = 0$$
$$ax + by + cz + d_2 = 0$$

is

$$D = \frac{|d_1 - d_2|}{\sqrt{a^2 + b^2 + c^2}}$$

(b) Use the formula in part (a) to solve Exercise 43.

✔QUICK CHECK ANSWERS 12.6

1. $-4x + (y - 3) + 7(z - 5) = 0$ **2.** $\langle 4, -2, 7\rangle$ **3.** $\langle 2, -1, 0\rangle$ **4.** $\cos^{-1}\dfrac{11}{5\sqrt{6}} \approx 26°$ **5.** $\sqrt{6}$

12.7 QUADRIC SURFACES

In this section we will study an important class of surfaces that are the three-dimensional analogs of the conic sections.

TRACES OF SURFACES

Although the general shape of a curve in 2-space can be obtained by plotting points, this method is not usually helpful for surfaces in 3-space because too many points are required. It is more common to build up the shape of a surface with a network of ***mesh lines***, which are curves obtained by cutting the surface with well-chosen planes. For example, Figure 12.7.1, which was generated by a CAS, shows the graph of $z = x^3 - 3xy^2$ rendered with a combination of mesh lines and colorization to produce the surface detail. This surface is called a "monkey saddle" because a monkey sitting astride the surface has a place for its two legs and tail.

A monkey saddle

Figure 12.7.1

The mesh line that results when a surface is cut by a plane is called the ***trace*** of the surface in the plane (Figure 12.7.2). Usually, surfaces are built up from traces in planes that are parallel to the coordinate planes, so we will begin by showing how the equations of such traces can be obtained. For this purpose, we will consider the surface

$$z = x^2 + y^2 \qquad (1)$$

shown in Figure 12.7.3*a*.

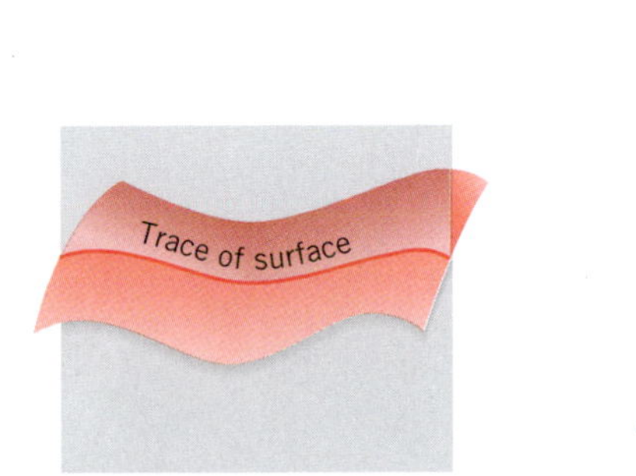

Figure 12.7.2

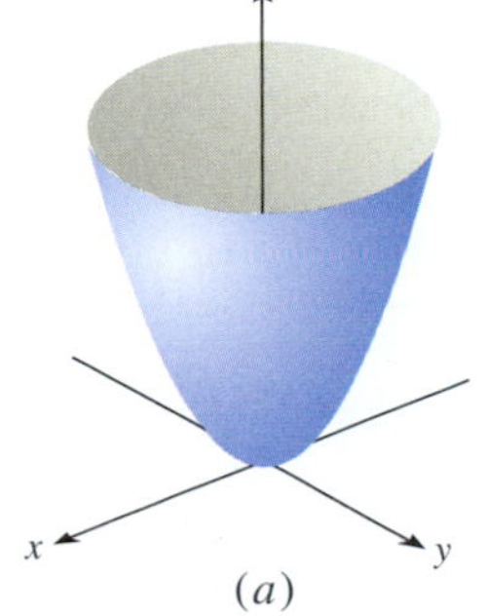

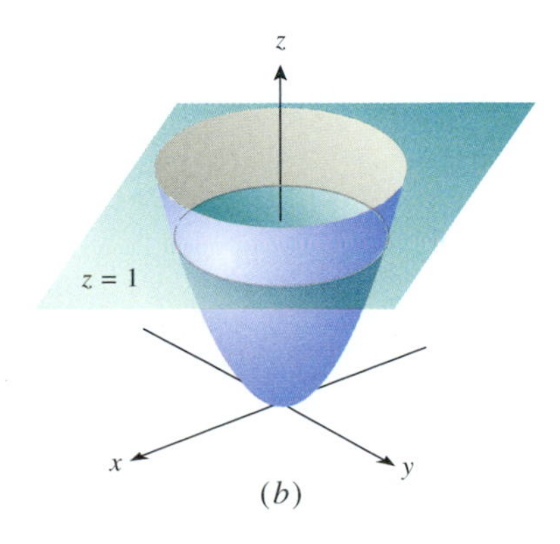

Figure 12.7.3

The parenthetical part of Equation (2) is a reminder that the z-coordinate of each point in the trace is $z = 1$. This needs to be stated explicitly because the variable z does not appear in the equation $x^2 + y^2 = 1$.

The basic procedure for finding the equation of a trace is to substitute the equation of the plane into the equation of the surface. For example, to find the trace of the surface $z = x^2 + y^2$ in the plane $z = 1$, we substitute $z = 1$ in (1), which yields

$$x^2 + y^2 = 1 \qquad (z = 1) \tag{2}$$

This is a circle of radius 1 centered at the point (0, 0, 1) (Figure 12.7.3*b*).

Figure 12.7.4*a* suggests that the traces of (1) in planes that are parallel to and above the xy-plane form a family of circles that are centered on the z-axis and whose radii increase with z. To confirm this, let us consider the trace in a general plane $z = k$ that is parallel to the xy-plane. The equation of the trace is

$$x^2 + y^2 = k \qquad (z = k)$$

If $k \geq 0$, then the trace is a circle of radius $\sqrt{k}$ centered at the point $(0, 0, k)$. In particular, if $k = 0$, then the radius is zero, so the trace in the xy-plane is the single point (0, 0, 0). Thus, for nonnegative values of k the traces parallel to the xy-plane form a family of circles, centered on the z-axis, whose radii start at zero and increase with k. This confirms our conjecture. If $k < 0$, then the equation $x^2 + y^2 = k$ has no graph, which means that there is no trace.

Now let us examine the traces of (1) in planes parallel to the yz-plane. Such planes have equations of the form $x = k$, so we substitute this in (1) to obtain

$$z = k^2 + y^2 \qquad (x = k)$$

which we can rewrite as

$$z - k^2 = y^2 \qquad (x = k) \tag{3}$$

For simplicity, let us start with the case where $k = 0$ (the trace in the yz-plane), in which case the trace has the equation

$$z = y^2 \qquad (x = 0)$$

You should be able to recognize that this is a parabola that has its vertex at the origin, opens in the positive z-direction, and is symmetric about the z-axis (Figure 12.7.4*b* shows a two-dimensional view). You should also be able to recognize that the $-k^2$ term in (3) has the effect of translating the parabola $z = y^2$ in the positive z-direction, so the new vertex falls at $(k, 0, k^2)$. Thus, the traces parallel to the yz-plane form a family of parabolas whose vertices move upward as k^2 increases. This is consistent with Figure 12.7.4*c*. Similarly, the traces in planes parallel to the xz-plane have equations of the form

$$z - k^2 = x^2 \qquad (y = k)$$

which again is a family of parabolas whose vertices move upward as k^2 increases (Figure 12.7.4*d*).

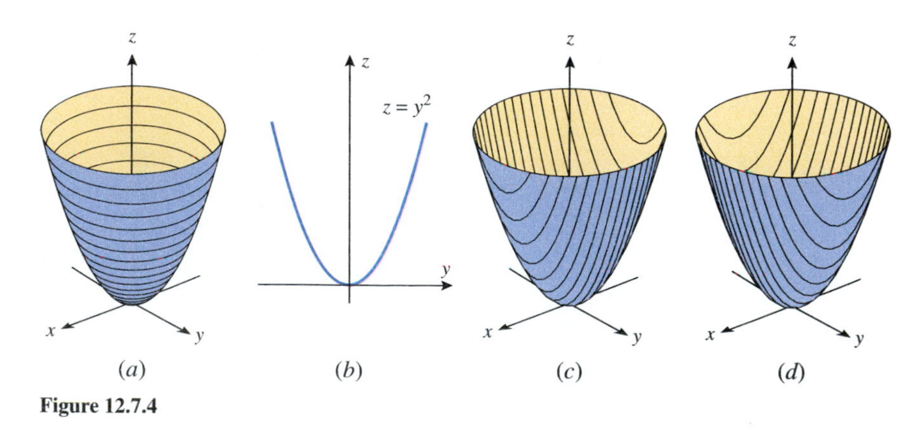

Figure 12.7.4

THE QUADRIC SURFACES

In the discussion of Formula (2) in Section 11.5 we noted that a second-degree equation

$$Ax^2 + Bxy + Cy^2 + Dx + Ey + F = 0$$

represents a conic section (possibly degenerate). The analog of this equation in an xyz-coordinate system is

$$Ax^2 + By^2 + Cz^2 + Dxy + Exz + Fyz + Gx + Hy + Iz + J = 0 \tag{4}$$

which is called a ***second-degree equation in x, y, and z***. The graphs of such equations are called ***quadric surfaces*** or sometimes ***quadrics***.

Six common types of quadric surfaces are shown in Table 12.7.1—*ellipsoids, hyperboloids of one sheet, hyperboloids of two sheets, elliptic cones, elliptic paraboloids,* and *hyperbolic paraboloids.* (The constants a, b, and c that appear in the equations in the table are assumed to be positive.) Observe that none of the quadric surfaces in the table have cross-product terms in their equations. This is because of their orientations relative to the coordinate axes. Later in this section we will discuss other possible orientations that produce equations of the quadric surfaces with no cross-product terms. In the special case where the elliptic cross sections of an elliptic cone or an elliptic paraboloid are circles, the terms *circular cone* and *circular paraboloid* are used.

TECHNIQUES FOR GRAPHING QUADRIC SURFACES

Accurate graphs of quadric surfaces are best left for graphing utilities. However, the techniques that we will now discuss can be used to generate rough sketches of these surfaces that are useful for various purposes.

A rough sketch of an ellipsoid

$$\frac{x^2}{a^2} + \frac{y^2}{b^2} + \frac{z^2}{c^2} = 1 \qquad (a > 0, b > 0, c > 0) \tag{5}$$

can be obtained by first plotting the intersections with the coordinate axes, then sketching the elliptical traces in the coordinate planes, and then sketching the surface itself using the traces as a guide. Example 1 illustrates this technique.

► **Example 1** Sketch the ellipsoid

$$\frac{x^2}{4} + \frac{y^2}{16} + \frac{z^2}{9} = 1 \tag{6}$$

Solution. The x-intercepts can be obtained by setting $y = 0$ and $z = 0$ in (6). This yields $x = \pm 2$. Similarly, the y-intercepts are $y = \pm 4$, and the z-intercepts are $z = \pm 3$. From these intercepts we obtain the elliptical traces and the ellipsoid sketched in Figure 12.7.5. ◄

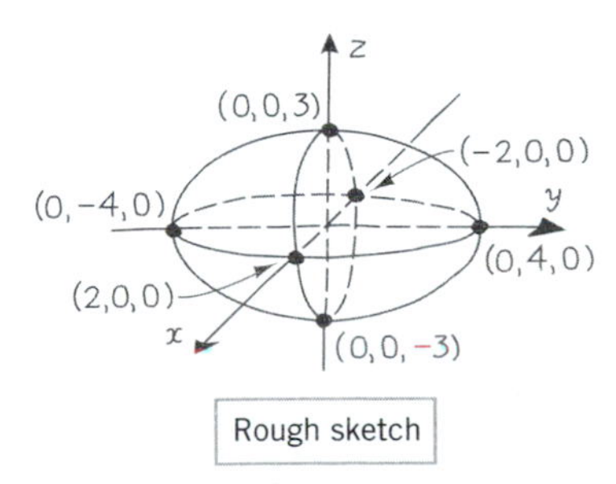

Figure 12.7.5

A rough sketch of a hyperboloid of one sheet

$$\frac{x^2}{a^2} + \frac{y^2}{b^2} - \frac{z^2}{c^2} = 1 \qquad (a > 0, b > 0, c > 0) \tag{7}$$

can be obtained by first sketching the elliptical trace in the xy-plane, then the elliptical traces in the planes $z = \pm c$, and then the hyperbolic curves that join the endpoints of the axes of these ellipses. The next example illustrates this technique.

Table 12.7.1

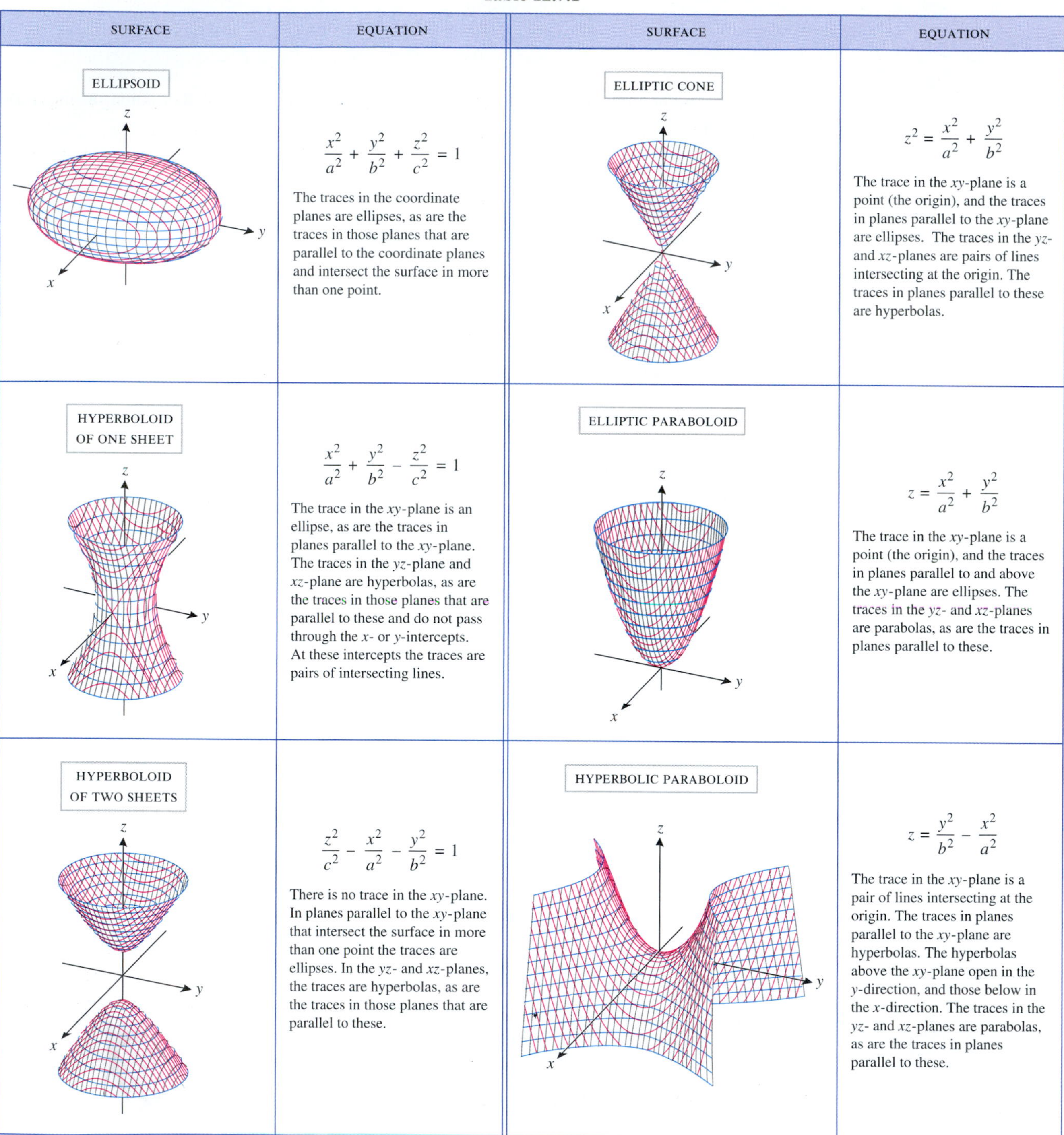

SURFACE	EQUATION	SURFACE	EQUATION
ELLIPSOID	$\dfrac{x^2}{a^2} + \dfrac{y^2}{b^2} + \dfrac{z^2}{c^2} = 1$ The traces in the coordinate planes are ellipses, as are the traces in those planes that are parallel to the coordinate planes and intersect the surface in more than one point.	ELLIPTIC CONE	$z^2 = \dfrac{x^2}{a^2} + \dfrac{y^2}{b^2}$ The trace in the xy-plane is a point (the origin), and the traces in planes parallel to the xy-plane are ellipses. The traces in the yz- and xz-planes are pairs of lines intersecting at the origin. The traces in planes parallel to these are hyperbolas.
HYPERBOLOID OF ONE SHEET	$\dfrac{x^2}{a^2} + \dfrac{y^2}{b^2} - \dfrac{z^2}{c^2} = 1$ The trace in the xy-plane is an ellipse, as are the traces in planes parallel to the xy-plane. The traces in the yz-plane and xz-plane are hyperbolas, as are the traces in those planes that are parallel to these and do not pass through the x- or y-intercepts. At these intercepts the traces are pairs of intersecting lines.	ELLIPTIC PARABOLOID	$z = \dfrac{x^2}{a^2} + \dfrac{y^2}{b^2}$ The trace in the xy-plane is a point (the origin), and the traces in planes parallel to and above the xy-plane are ellipses. The traces in the yz- and xz-planes are parabolas, as are the traces in planes parallel to these.
HYPERBOLOID OF TWO SHEETS	$\dfrac{z^2}{c^2} - \dfrac{x^2}{a^2} - \dfrac{y^2}{b^2} = 1$ There is no trace in the xy-plane. In planes parallel to the xy-plane that intersect the surface in more than one point the traces are ellipses. In the yz- and xz-planes, the traces are hyperbolas, as are the traces in those planes that are parallel to these.	HYPERBOLIC PARABOLOID	$z = \dfrac{y^2}{b^2} - \dfrac{x^2}{a^2}$ The trace in the xy-plane is a pair of lines intersecting at the origin. The traces in planes parallel to the xy-plane are hyperbolas. The hyperbolas above the xy-plane open in the y-direction, and those below in the x-direction. The traces in the yz- and xz-planes are parabolas, as are the traces in planes parallel to these.

▶ **Example 2** Sketch the graph of the hyperboloid of one sheet

$$x^2 + y^2 - \frac{z^2}{4} = 1 \qquad (8)$$

Solution. The trace in the xy-plane, obtained by setting $z = 0$ in (8), is

$$x^2 + y^2 = 1 \qquad (z = 0)$$

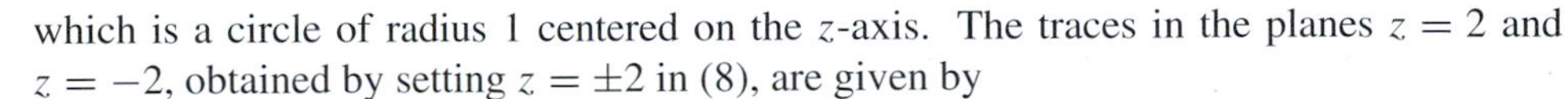

which is a circle of radius 1 centered on the z-axis. The traces in the planes $z = 2$ and $z = -2$, obtained by setting $z = \pm 2$ in (8), are given by

$$x^2 + y^2 = 2 \qquad (z = \pm 2)$$

which are circles of radius $\sqrt{2}$ centered on the z-axis. Joining these circles by the hyperbolic traces in the vertical coordinate planes yields the graph in Figure 12.7.6. ◀

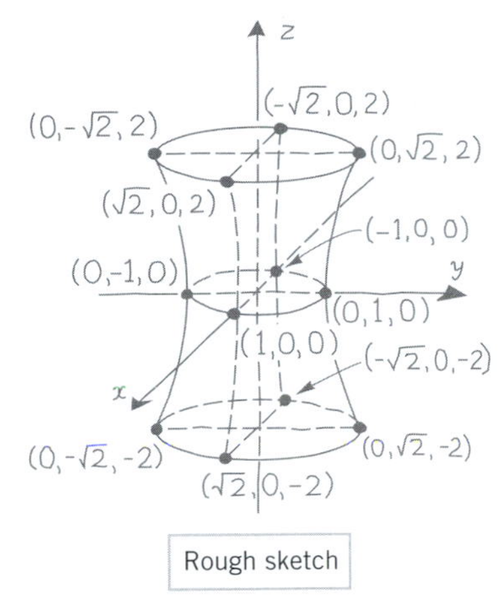

Figure 12.7.6

A rough sketch of the hyperboloid of two sheets

$$\frac{z^2}{c^2} - \frac{x^2}{a^2} - \frac{y^2}{b^2} = 1 \qquad (a > 0, b > 0, c > 0) \tag{9}$$

can be obtained by first plotting the intersections with the z-axis, then sketching the elliptical traces in the planes $z = \pm 2c$, and then sketching the hyperbolic traces that connect the z-axis intersections and the endpoints of the axes of the ellipses. (It is not essential to use the planes $z = \pm 2c$, but these are good choices since they simplify the calculations slightly and have the right spacing for a good sketch.) The next example illustrates this technique.

▶ **Example 3** Sketch the graph of the hyperboloid of two sheets

$$z^2 - x^2 - \frac{y^2}{4} = 1 \tag{10}$$

Solution. The z-intercepts, obtained by setting $x = 0$ and $y = 0$ in (10), are $z = \pm 1$. The traces in the planes $z = 2$ and $z = -2$, obtained by setting $z = \pm 2$ in (10), are given by

$$\frac{x^2}{3} + \frac{y^2}{12} = 1 \qquad (z = \pm 2)$$

Sketching these ellipses and the hyperbolic traces in the vertical coordinate planes yields Figure 12.7.7. ◀

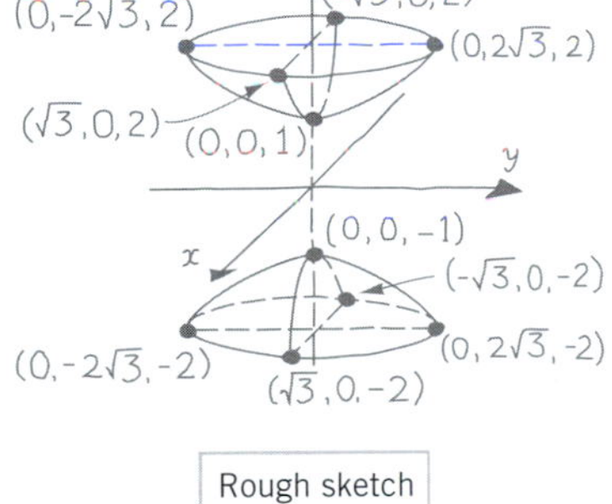

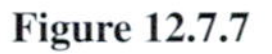
Figure 12.7.7

A rough sketch of the elliptic cone

$$z^2 = \frac{x^2}{a^2} + \frac{y^2}{b^2} \qquad (a > 0, b > 0) \tag{11}$$

can be obtained by first sketching the elliptical traces in the planes $z = \pm 1$ and then sketching the linear traces that connect the endpoints of the axes of the ellipses. The next example illustrates this technique.

▶ **Example 4** Sketch the graph of the elliptic cone

$$z^2 = x^2 + \frac{y^2}{4} \tag{12}$$

Solution. The traces of (12) in the planes $z = \pm 1$ are given by

$$x^2 + \frac{y^2}{4} = 1 \qquad (z = \pm 1)$$

Sketching these ellipses and the linear traces in the vertical coordinate planes yields the graph in Figure 12.7.8. ◀

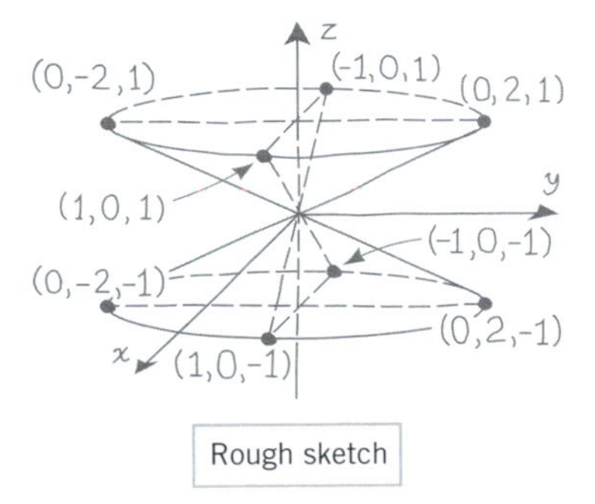

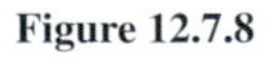
Figure 12.7.8

In the special cases of (11) and (13) where $a = b$, the traces parallel to the xy-plane are circles. In these cases, we call (11) a ***circular cone*** and (13) a ***circular paraboloid.***

A rough sketch of the elliptic paraboloid

$$z = \frac{x^2}{a^2} + \frac{y^2}{b^2} \qquad (a > 0, b > 0) \tag{13}$$

can be obtained by first sketching the elliptical trace in the plane $z = 1$ and then sketching the parabolic traces in the vertical coordinate planes to connect the origin to the ends of the axes of the ellipse. The next example illustrates this technique.

▶ Example 5 Sketch the graph of the elliptic paraboloid

$$z = \frac{x^2}{4} + \frac{y^2}{9} \tag{14}$$

Solution. The trace of (14) in the plane $z = 1$ is

$$\frac{x^2}{4} + \frac{y^2}{9} = 1 \qquad (z = 1)$$

Sketching this ellipse and the parabolic traces in the vertical coordinate planes yields the graph in Figure 12.7.9. ◀

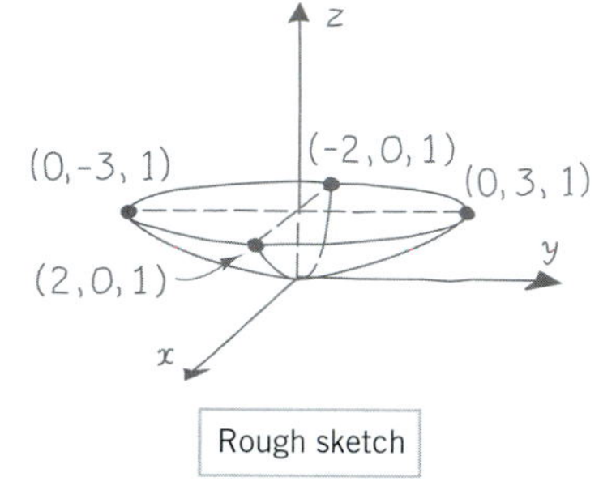

Figure 12.7.9

A rough sketch of the hyperbolic paraboloid

$$z = \frac{y^2}{b^2} - \frac{x^2}{a^2} \qquad (a > 0, b > 0) \tag{15}$$

can be obtained by first sketching the two parabolic traces that pass through the origin (one in the plane $x = 0$ and the other in the plane $y = 0$). After the parabolic traces are drawn, sketch the hyperbolic traces in the planes $z = \pm 1$ and then fill in any missing edges. The next example illustrates this technique.

▶ Example 6 Sketch the graph of the hyperbolic paraboloid

$$z = \frac{y^2}{4} - \frac{x^2}{9} \tag{16}$$

Solution. Setting $x = 0$ in (16) yields

$$z = \frac{y^2}{4} \qquad (x = 0)$$

which is a parabola in the yz-plane with vertex at the origin and opening in the positive z-direction (since $z \geq 0$), and setting $y = 0$ yields

$$z = -\frac{x^2}{9} \qquad (y = 0)$$

which is a parabola in the xz-plane with vertex at the origin and opening in the negative z-direction.

The trace in the plane $z = 1$ is

$$\frac{y^2}{4} - \frac{x^2}{9} = 1 \qquad (z = 1)$$

which is a hyperbola that opens along a line parallel to the y-axis (verify), and the trace in the plane $z = -1$ is

$$\frac{x^2}{9} - \frac{y^2}{4} = 1 \qquad (z = -1)$$

which is a hyperbola that opens along a line parallel to the x-axis. Combining all of the above information leads to the sketch in Figure 12.7.10. ◀

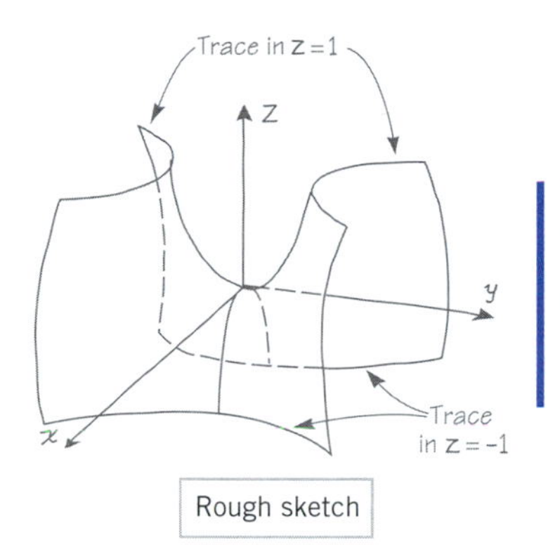

Figure 12.7.10

The hyperbolic paraboloid in Figure 12.7.10 has an interesting behavior at the origin—the trace in the xz-plane has a relative maximum at $(0, 0, 0)$, and the trace in the yz-plane has a relative minimum at $(0, 0, 0)$. Thus, a bug walking on the surface may view the origin as a highest point if traveling along one path, or may view the origin as a lowest point if traveling along a different path. A point with this property is commonly called a ***saddle point*** or a ***minimax point***.

Figure 12.7.11 shows two computer-generated views of the hyperbolic paraboloid in Example 6. The first view, which is much like our rough sketch in Figure 12.7.10, has cuts at the top and bottom that are hyperbolic traces parallel to the xy-plane. In the second view the top horizontal cut has been omitted; this helps to emphasize the parabolic traces parallel to the xz-plane.

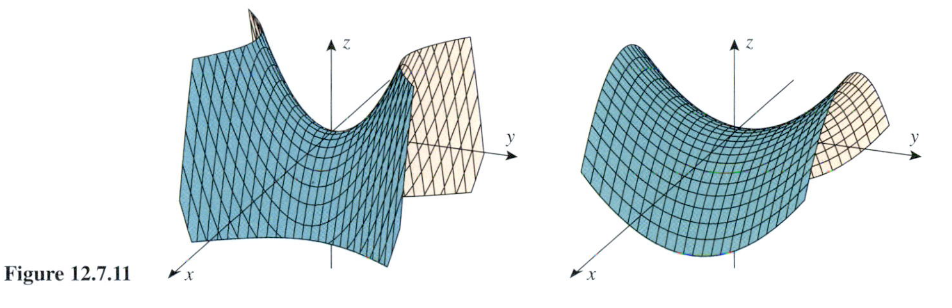

Figure 12.7.11

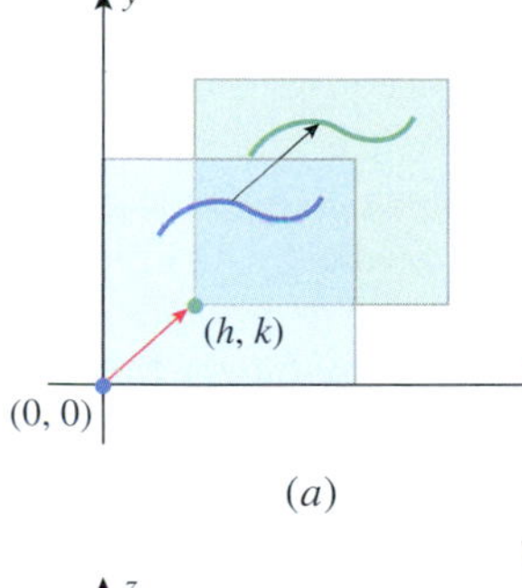

(*a*)

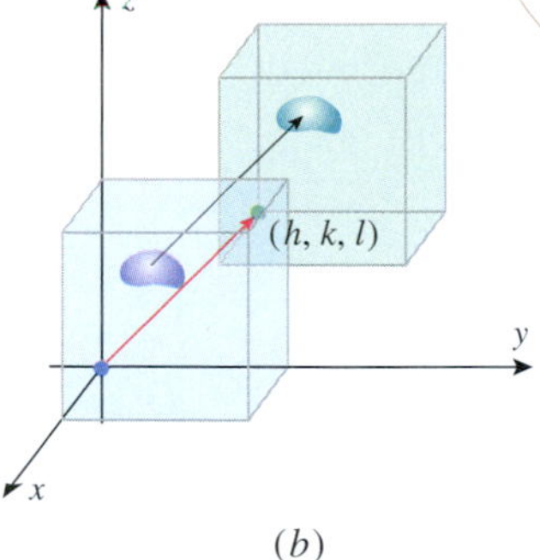

(*b*)

Figure 12.7.12

TRANSLATIONS OF QUADRIC SURFACES

In Section 11.4 we saw that a conic in an xy-coordinate system can be translated by substituting $x - h$ for x and $y - k$ for y in its equation. To understand why this works, think of the xy-axes as fixed and think of the plane as a transparent sheet of plastic on which all graphs are drawn. When the coordinates of points are modified by substituting $(x - h, y - k)$ for (x, y), the geometric effect is to translate the sheet of plastic (and hence all curves) so that the point on the plastic that was initially at $(0, 0)$ is moved to the point (h, k) (see Figure 12.7.12*a*).

For the analog in three dimensions, think of the xyz-axes as fixed and think of 3-space as a transparent block of plastic in which all surfaces are embedded. When the coordinates of points are modified by substituting $(x - h, y - k, z - l)$ for (x, y, z), the geometric effect is to translate the block of plastic (and hence all surfaces) so that the point in the plastic block that was initially at $(0, 0, 0)$ is moved to the point (h, k, l) (see Figure 12.7.12*b*).

▶ Example 7 Describe the surface $z = (x - 1)^2 + (y + 2)^2 + 3$.

Solution. The equation can be rewritten as

$$z - 3 = (x - 1)^2 + (y + 2)^2$$

This surface is the paraboloid that results by translating the paraboloid

$$z = x^2 + y^2$$

in Figure 12.7.3 so that the new "vertex" is at the point $(1, -2, 3)$. A rough sketch of this paraboloid is shown in Figure 12.7.13. ◀

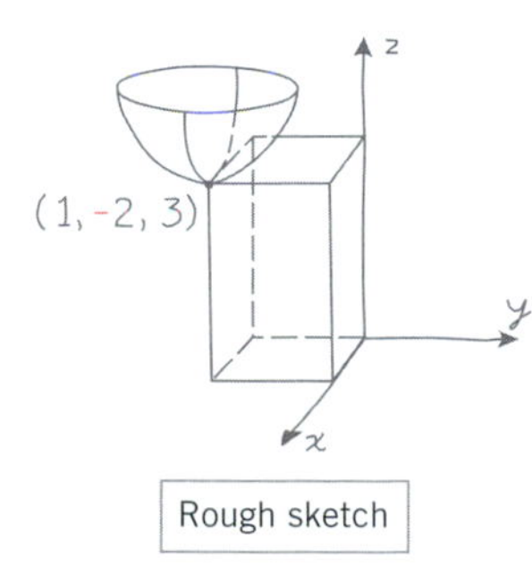

Figure 12.7.13

▶ Example 8 Describe the surface

$$4x^2 + 4y^2 + z^2 + 8y - 4z = -4$$

Solution. Completing the squares yields

$$4x^2 + 4(y + 1)^2 + (z - 2)^2 = -4 + 4 + 4$$

or

$$x^2 + (y + 1)^2 + \frac{(z - 2)^2}{4} = 1$$

Thus, the surface is the ellipsoid that results when the ellipsoid

$$x^2 + y^2 + \frac{z^2}{4} = 1$$

is translated so that the new "center" is at the point $(0, -1, 2)$. A rough sketch of this ellipsoid is shown in Figure 12.7.14. ◀

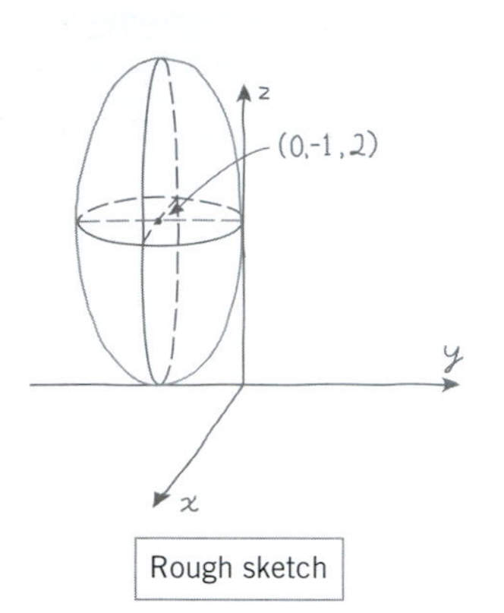

Figure 12.7.14

In Figure 12.7.14, the cross section in the yz-plane is shown tangent to both the y- and z-axes. Confirm that this is correct.

REFLECTIONS OF SURFACES IN 3-SPACE

Recall that in an xy-coordinate system a point (x, y) is reflected about the x-axis if y is replaced by $-y$, and it is reflected about the y-axis if x is replaced by $-x$. In an xyz-coordinate system, a point (x, y, z) is reflected about the xy-plane if z is replaced by $-z$, it is reflected about the yz-plane if x is replaced by $-x$, and it is reflected about the xz-plane if y is replaced by $-y$ (Figure 12.7.15). It follows that *replacing a variable by its negative in the equation of a surface causes that surface to be reflected about a coordinate plane.*

Recall also that in an xy-coordinate system a point (x, y) is reflected about the line $y = x$ if x and y are interchanged. However, in an xyz-coordinate system, interchanging x and y reflects the point (x, y, z) about the plane $y = x$ (Figure 12.7.16). Similarly, interchanging x and z reflects the point about the plane $x = z$, and interchanging y and z reflects it about the plane $y = z$. Thus, it follows that *interchanging two variables in the equation of a surface reflects that surface about a plane that makes a* 45° *angle with two of the coordinate planes.*

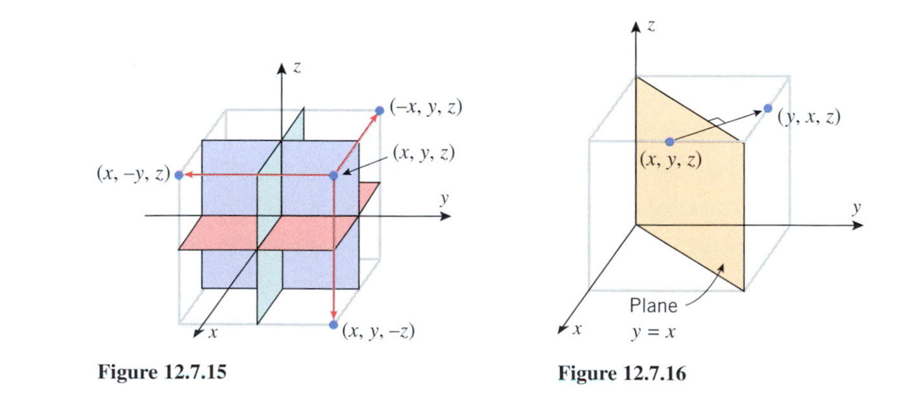

Figure 12.7.15 **Figure 12.7.16**

▶ Example 9 Describe the surfaces

(a) $y^2 = x^2 + z^2$ (b) $z = -(x^2 + y^2)$

Solution (a). The graph of the equation $y^2 = x^2 + z^2$ results from interchanging y and z in the equation $z^2 = x^2 + y^2$. Thus, the graph of the equation $y^2 = x^2 + z^2$ can be obtained by reflecting the graph of $z^2 = x^2 + y^2$ about the plane $y = z$. Since the graph of

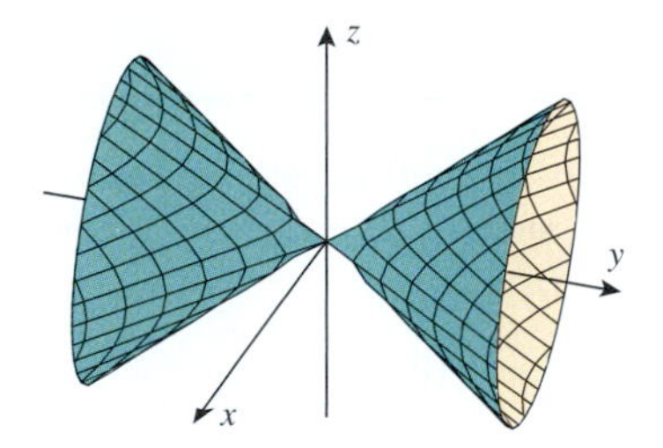

Figure 12.7.17

$z^2 = x^2 + y^2$ is a circular cone opening along the z-axis (see Table 12.7.1), it follows that the graph of $y^2 = x^2 + z^2$ is a circular cone opening along the y-axis (Figure 12.7.17).

Solution (b). The graph of the equation $z = -(x^2 + y^2)$ can be written as $-z = x^2 + y^2$, which can be obtained by replacing z with $-z$ in the equation $z = x^2 + y^2$. Since the graph of $z = x^2 + y^2$ is a circular paraboloid opening in the positive z-direction (see Table 12.7.1), it follows that the graph of $z = -(x^2 + y^2)$ is a circular paraboloid opening in the negative z-direction (Figure 12.7.18). ◄

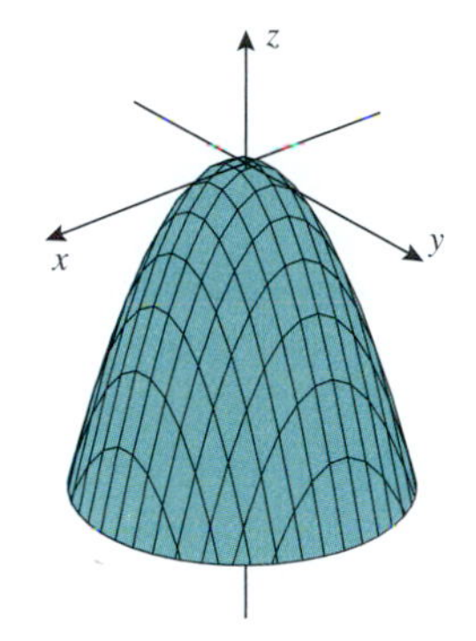

Figure 12.7.18

A TECHNIQUE FOR IDENTIFYING QUADRIC SURFACES

The equations of the quadric surfaces in Table 12.7.1 have certain characteristics that make it possible to identify quadric surfaces that are derived from these equations by reflections. These identifying characteristics, which are shown in Table 12.7.2, are based on writing the equation of the quadric surface so that all of the variable terms are on the left side of the equation and there is a 1 or a 0 on the right side. When there is a 1 on the right side the surface is an ellipsoid, hyperboloid of one sheet, or a hyperboloid of two sheets, and when there is a 0 on the right side it is an elliptic cone, an elliptic paraboloid, or a hyperbolic paraboloid. Within the group with a 1 on the right side, ellipsoids have no minus signs, hyperboloids of one sheet have one minus sign, and hyperboloids of two sheets have two minus signs. Within the group with a 0 on the right side, elliptic cones have no linear terms, elliptic paraboloids have one linear term and two quadratic terms with the same sign, and hyperbolic paraboloids have one linear term and two quadratic terms with opposite signs. These characteristics do not change when the surface is reflected about a coordinate plane or planes of the form $x = y$, $x = z$, or $y = z$, thereby making it possible to identify the reflected quadric surface from the form of its equation.

Table 12.7.2

EQUATION	$\frac{x^2}{a^2} + \frac{y^2}{b^2} + \frac{z^2}{c^2} = 1$	$\frac{x^2}{a^2} + \frac{y^2}{b^2} - \frac{z^2}{c^2} = 1$	$\frac{z^2}{c^2} - \frac{x^2}{a^2} - \frac{y^2}{b^2} = 1$	$z^2 - \frac{x^2}{a^2} - \frac{y^2}{b^2} = 0$	$z - \frac{x^2}{a^2} - \frac{y^2}{b^2} = 0$	$z - \frac{y^2}{b^2} + \frac{x^2}{a^2} = 0$
CHARACTERISTIC	No minus signs	One minus sign	Two minus signs	No linear terms	One linear term; two quadratic terms with the same sign	One linear term; two quadratic terms with opposite signs
CLASSIFICATION	Ellipsoid	Hyperboloid of one sheet	Hyperboloid of two sheets	Elliptic cone	Elliptic paraboloid	Hyperbolic paraboloid

► **Example 10** Identify the surfaces

$$\text{(a) } 3x^2 - 4y^2 + 12z^2 + 12 = 0 \qquad \text{(b) } 4x^2 - 4y + z^2 = 0$$

Solution (a). The equation can be rewritten as

$$\frac{y^2}{3} - \frac{x^2}{4} - z^2 = 1$$

This equation has a 1 on the right side and two negative terms on the left side, so its graph is a hyperboloid of two sheets.

Solution (b). The equation has one linear term and two quadratic terms with the same sign, so its graph is an elliptic paraboloid. ◄

✔ QUICK CHECK EXERCISES 12.7 (See page 850 for answers.)

1. For the surface $4x^2 + y^2 + z^2 = 9$, classify the indicated trace as an ellipse, hyperbola, or parabola.
(a) $x = 0$ (b) $y = 0$ (c) $z = 1$

2. For the surface $4x^2 + z^2 - y^2 = 9$, classify the indicated trace as an ellipse, hyperbola, or parabola.
(a) $x = 0$ (b) $y = 0$ (c) $z = 1$

3. For the surface $4x^2 + y^2 - z = 0$, classify the indicated trace as an ellipse, hyperbola, or parabola.
(a) $x = 0$ (b) $y = 0$ (c) $z = 1$

4. Classify each surface as an ellipsoid, hyperboloid of one sheet, hyperboloid of two sheets, elliptic cone, elliptic paraboloid, or hyperbolic paraboloid.
(a) $\dfrac{x^2}{36} + \dfrac{y^2}{25} - z = 0$ (b) $\dfrac{x^2}{36} + \dfrac{y^2}{25} + z^2 = 1$
(c) $\dfrac{x^2}{36} - \dfrac{y^2}{25} + z = 0$ (d) $\dfrac{x^2}{36} + \dfrac{y^2}{25} - z^2 = 1$
(e) $\dfrac{x^2}{36} + \dfrac{y^2}{25} - z^2 = 0$ (f) $z^2 - \dfrac{x^2}{36} - \dfrac{y^2}{25} = 1$

EXERCISE SET 12.7

1–2 Identify the quadric surface as an ellipsoid, hyperboloid of one sheet, hyperboloid of two sheets, elliptic cone, elliptic paraboloid, or hyperbolic paraboloid by matching the equation with one of the forms given in Table 12.7.1. State the values of a, b, and c in each case.

1. (a) $z = \dfrac{x^2}{4} + \dfrac{y^2}{9}$ (b) $z = \dfrac{y^2}{25} - x^2$
(c) $x^2 + y^2 - z^2 = 16$ (d) $x^2 + y^2 - z^2 = 0$
(e) $4z = x^2 + 4y^2$ (f) $z^2 - x^2 - y^2 = 1$

2. (a) $6x^2 + 3y^2 + 4z^2 = 12$ (b) $y^2 - x^2 - z = 0$
(c) $9x^2 + y^2 - 9z^2 = 9$ (d) $4x^2 + y^2 - 4z^2 = -4$
(e) $2z - x^2 - 4y^2 = 0$ (f) $12z^2 - 3x^2 = 4y^2$

3. Find an equation for and sketch the surface that results when the circular paraboloid $z = x^2 + y^2$ is reflected about the plane
(a) $z = 0$ (b) $x = 0$ (c) $y = 0$
(d) $y = x$ (e) $x = z$ (f) $y = z$.

4. Find an equation for and sketch the surface that results when the hyperboloid of one sheet $x^2 + y^2 - z^2 = 1$ is reflected about the plane
(a) $z = 0$ (b) $x = 0$ (c) $y = 0$
(d) $y = x$ (e) $x = z$ (f) $y = z$.

FOCUS ON CONCEPTS

5. The given equations represent quadric surfaces whose orientations are different from those in Table 12.7.1. In each part, identify the quadric surface, and give a verbal description of its orientation (e.g., an elliptic cone opening along the z-axis or a hyperbolic paraboloid straddling the y-axis).
(a) $\dfrac{z^2}{c^2} - \dfrac{y^2}{b^2} + \dfrac{x^2}{a^2} = 1$ (b) $\dfrac{x^2}{a^2} - \dfrac{y^2}{b^2} - \dfrac{z^2}{c^2} = 1$
(c) $x = \dfrac{y^2}{b^2} + \dfrac{z^2}{c^2}$ (d) $x^2 = \dfrac{y^2}{b^2} + \dfrac{z^2}{c^2}$
(e) $y = \dfrac{z^2}{c^2} - \dfrac{x^2}{a^2}$ (f) $y = -\left(\dfrac{x^2}{a^2} + \dfrac{z^2}{c^2}\right)$

6. For each of the surfaces in Exercise 5, find the equation of the surface that results if the given surface is reflected about the xz-plane and that surface is then reflected about the plane $z = 0$.

7–8 Find equations of the traces in the coordinate planes and sketch the traces in an xyz-coordinate system. [*Suggestion:* If you have trouble sketching a trace directly in three dimensions, start with a sketch in two dimensions by placing the coordinate plane in the plane of the paper, then transfer the sketch to three dimensions.]

7. (a) $\dfrac{x^2}{9} + \dfrac{y^2}{25} + \dfrac{z^2}{4} = 1$ (b) $z = x^2 + 4y^2$
(c) $\dfrac{x^2}{9} + \dfrac{y^2}{16} - \dfrac{z^2}{4} = 1$

8. (a) $y^2 + 9z^2 = x$ (b) $4x^2 - y^2 + 4z^2 = 4$
(c) $z^2 = x^2 + \dfrac{y^2}{4}$

9–10 In these exercises, traces of the surfaces in the planes are conic sections. In each part, find an equation of the trace, and state whether it is an ellipse, a parabola, or a hyperbola.

9. (a) $4x^2 + y^2 + z^2 = 4$; $y = 1$
(b) $4x^2 + y^2 + z^2 = 4$; $x = \frac{1}{2}$
(c) $9x^2 - y^2 - z^2 = 16$; $x = 2$
(d) $9x^2 - y^2 - z^2 = 16$; $z = 2$
(e) $z = 9x^2 + 4y^2$; $y = 2$
(f) $z = 9x^2 + 4y^2$; $z = 4$

10. (a) $9x^2 - y^2 + 4z^2 = 9$; $x = 2$
(b) $9x^2 - y^2 + 4z^2 = 9$; $y = 4$
(c) $x^2 + 4y^2 - 9z^2 = 0$; $y = 1$
(d) $x^2 + 4y^2 - 9z^2 = 0$; $z = 1$
(e) $z = x^2 - 4y^2$; $x = 1$
(f) $z = x^2 - 4y^2$; $z = 4$

11–22 Identify and sketch the quadric surface.

11. $x^2 + \frac{y^2}{4} + \frac{z^2}{9} = 1$

12. $x^2 + 4y^2 + 9z^2 = 36$

13. $\frac{x^2}{4} + \frac{y^2}{9} - \frac{z^2}{16} = 1$

14. $x^2 + y^2 - z^2 = 9$

15. $4z^2 = x^2 + 4y^2$

16. $9x^2 + 4y^2 - 36z^2 = 0$

17. $9z^2 - 4y^2 - 9x^2 = 36$

18. $y^2 - \frac{x^2}{4} - \frac{z^2}{9} = 1$

19. $z = y^2 - x^2$

20. $16z = y^2 - x^2$

21. $4z = x^2 + 2y^2$

22. $z - 3x^2 - 3y^2 = 0$

23–28 The given equation represents a quadric surface whose orientation is different from that in Table 12.7.1. Identify and sketch the surface.

23. $x^2 - 3y^2 - 3z^2 = 0$

24. $x - y^2 - 4z^2 = 0$

25. $2y^2 - x^2 + 2z^2 = 8$

26. $x^2 - 3y^2 - 3z^2 = 9$

27. $z = \frac{x^2}{4} - \frac{y^2}{9}$

28. $4x^2 - y^2 + 4z^2 = 16$

29–32 Sketch the surface.

29. $z = \sqrt{x^2 + y^2}$

30. $z = \sqrt{1 - x^2 - y^2}$

31. $z = \sqrt{x^2 + y^2 - 1}$

32. $z = \sqrt{1 + x^2 + y^2}$

33–36 Identify the surface and make a rough sketch that shows its position and orientation.

33. $z = (x + 2)^2 + (y - 3)^2 - 9$

34. $4x^2 - y^2 + 16(z - 2)^2 = 100$

35. $9x^2 + y^2 + 4z^2 - 18x + 2y + 16z = 10$

36. $z^2 = 4x^2 + y^2 + 8x - 2y + 4z$

37–38 Use the ellipsoid $4x^2 + 9y^2 + 18z^2 = 72$ in these exercises.

37. (a) Find an equation of the elliptical trace in the plane $z = \sqrt{2}$.
(b) Find the lengths of the major and minor axes of the ellipse in part (a).
(c) Find the coordinates of the foci of the ellipse in part (a).
(d) Describe the orientation of the focal axis of the ellipse in part (a) relative to the coordinate axes.

38. (a) Find an equation of the elliptical trace in the plane $x = 3$.
(b) Find the lengths of the major and minor axes of the ellipse in part (a).
(c) Find the coordinates of the foci of the ellipse in part (a).
(d) Describe the orientation of the focal axis of the ellipse in part (a) relative to the coordinate axes.

39–42 These exercises refer to the hyperbolic paraboloid $z = y^2 - x^2$.

39. (a) Find an equation of the hyperbolic trace in the plane $z = 4$.
(b) Find the vertices of the hyperbola in part (a).
(c) Find the foci of the hyperbola in part (a).
(d) Describe the orientation of the focal axis of the hyperbola in part (a) relative to the coordinate axes.

40. (a) Find an equation of the hyperbolic trace in the plane $z = -4$.
(b) Find the vertices of the hyperbola in part (a).
(c) Find the foci of the hyperbola in part (a).
(d) Describe the orientation of the focal axis of the hyperbola in part (a) relative to the coordinate axes.

41. (a) Find an equation of the parabolic trace in the plane $x = 2$.
(b) Find the vertices of the parabola in part (a).
(c) Find the focus of the parabola in part (a).
(d) Describe the orientation of the focal axis of the parabola in part (a) relative to the coordinate axes.

42. (a) Find an equation of the parabolic trace in the plane $y = 2$.
(b) Find the vertex of the parabola in part (a).
(c) Find the focus of the parabola in part (a).
(d) Describe the orientation of the focal axis of the parabola in part (a) relative to the coordinate axes.

43–44 Sketch the region enclosed between the surfaces and describe their curve of intersection.

43. The paraboloids $z = x^2 + y^2$ and $z = 4 - x^2 - y^2$

44. The hyperbolic paraboloid $x^2 = y^2 + z$ and the ellipsoid $x^2 = 4 - 2y^2 - 2z$

45–46 Find an equation for the surface generated by revolving the curve about the y-axis.

45. $y = 4x^2$ $(z = 0)$

46. $y = 2x$ $(z = 0)$

47. Find an equation of the surface consisting of all points $P(x, y, z)$ that are equidistant from the point $(0, 0, 1)$ and the plane $z = -1$. Identify the surface.

48. Find an equation of the surface consisting of all points $P(x, y, z)$ that are twice as far from the plane $z = -1$ as from the point $(0, 0, 1)$. Identify the surface.

49. If a sphere

$$\frac{x^2}{a^2} + \frac{y^2}{a^2} + \frac{z^2}{a^2} = 1$$

of radius a is compressed in the z-direction, then the resulting surface, called an ***oblate spheroid***, has an equation of the form

$$\frac{x^2}{a^2} + \frac{y^2}{a^2} + \frac{z^2}{c^2} = 1$$

where $c < a$. Show that the oblate spheroid has a circular trace of radius a in the xy-plane and an elliptical trace in the

xz-plane with major axis of length $2a$ along the x-axis and minor axis of length $2c$ along the z-axis.

50. The Earth's rotation causes a flattening at the poles, so its shape is often modeled as an oblate spheroid rather than a sphere (see Exercise 49 for terminology). One of the models used by global positioning satellites is the ***World Geodetic System of 1984*** (WGS-84), which treats the Earth as an oblate spheroid whose equatorial radius is 6378.1370 km and whose polar radius (the distance from the Earth's center to the poles) is 6356.5231 km. Use the WGS-84 model to find an equation for the surface of the Earth relative to the coordinate system shown in the accompanying figure.

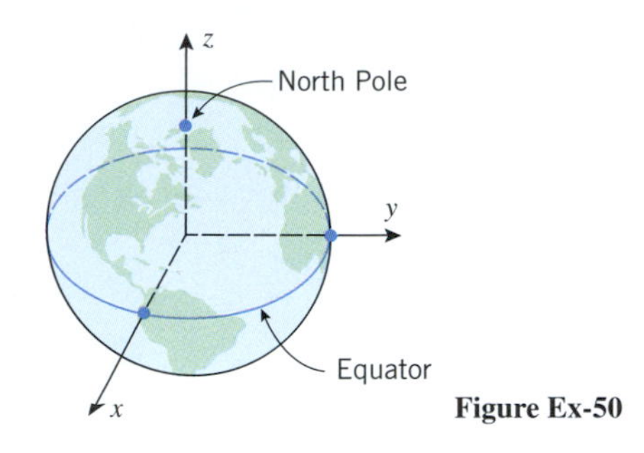

Figure Ex-50

51. Use the method of slicing to show that the volume of the ellipsoid

$$\frac{x^2}{a^2}+\frac{y^2}{b^2}+\frac{z^2}{c^2}=1$$

is $\frac{4}{3}\pi abc$.

✔ QUICK CHECK ANSWERS 12.7

1. (a) ellipse (b) ellipse (c) ellipse **2.** (a) hyperbola (b) ellipse (c) hyperbola **3.** (a) parabola (b) parabola (c) ellipse **4.** (a) elliptic paraboloid (b) ellipsoid (c) hyperbolic paraboloid (d) hyperboloid of one sheet (e) elliptic cone (f) hyperboloid of two sheets

12.8 CYLINDRICAL AND SPHERICAL COORDINATES

In this section we will discuss two new types of coordinate systems in 3-space that are often more useful than rectangular coordinate systems for studying surfaces with symmetries. These new coordinate systems also have important applications in navigation, astronomy, and the study of rotational motion about an axis.

CYLINDRICAL AND SPHERICAL COORDINATE SYSTEMS

Three coordinates are required to establish the location of a point in 3-space. We have already done this using rectangular coordinates. However, Figure 12.8.1 shows two other possibilities: part (a) of the figure shows the ***rectangular coordinates*** (x, y, z) of a point P, part (b) shows the ***cylindrical coordinates*** (r, θ, z) of P, and part (c) shows the ***spherical coordinates*** (ρ, θ, ϕ) of P. In a rectangular coordinate system the coordinates can be any real numbers, but in cylindrical and spherical coordinate systems there are restrictions on the allowable values of the coordinates (as indicated in Figure 12.8.1).

CONSTANT SURFACES

In rectangular coordinates the surfaces represented by equations of the form

$$x = x_0, \quad y = y_0, \quad \text{and} \quad z = z_0$$

where x_0, y_0, and z_0 are constants, are planes parallel to the yz-plane, xz-plane, and xy-plane, respectively (Figure 12.8.2). In cylindrical coordinates the surfaces represented by equations of the form

$$r = r_0, \quad \theta = \theta_0, \quad \text{and} \quad z = z_0$$

where r_0, θ_0, and z_0 are constants, are shown in Figure 12.8.3:

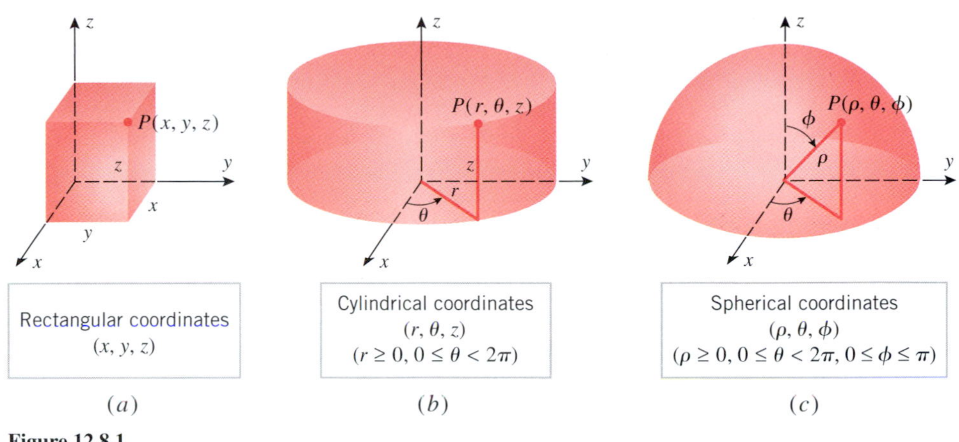

Figure 12.8.1

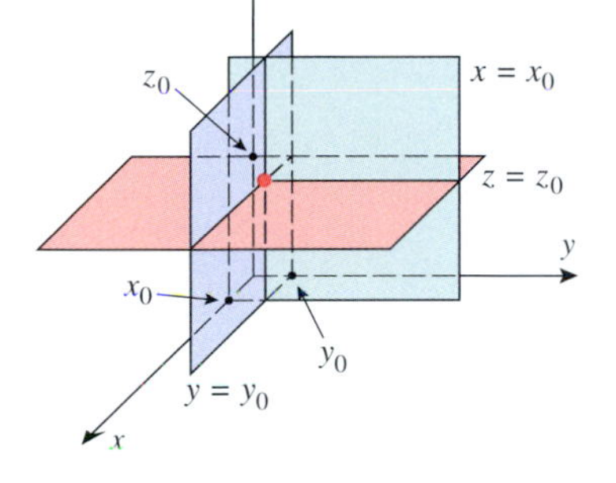

Figure 12.8.2

- The surface $r = r_0$ is a right circular cylinder of radius r_0 centered on the z-axis. At each point (r, θ, z) on this cylinder, r has the value r_0, but θ and z are unrestricted except for our general restriction that $0 \leq \theta < 2\pi$.
- The surface $\theta = \theta_0$ is a half-plane attached along the z-axis and making an angle θ_0 with the positive x-axis. At each point (r, θ, z) on this surface, θ has the value θ_0, but r and z are unrestricted except for our general restriction that $r \geq 0$.
- The surface $z = z_0$ is a horizontal plane. At each point (r, θ, z) on this plane, z has the value z_0, but r and θ are unrestricted except for the general restrictions.

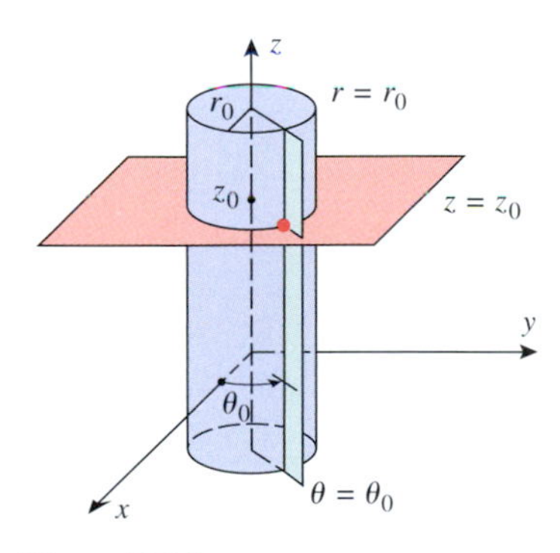

Figure 12.8.3

In spherical coordinates the surfaces represented by equations of the form

$$\rho = \rho_0, \quad \theta = \theta_0, \quad \text{and} \quad \phi = \phi_0$$

where ρ_0, θ_0, and ϕ_0 are constants, are shown in Figure 12.8.4:

- The surface $\rho = \rho_0$ consists of all points whose distance ρ from the origin is ρ_0. Assuming ρ_0 to be nonnegative, this is a sphere of radius ρ_0 centered at the origin.
- As in cylindrical coordinates, the surface $\theta = \theta_0$ is a half-plane attached along the z-axis, making an angle of θ_0 with the positive x-axis.
- The surface $\phi = \phi_0$ consists of all points from which a line segment to the origin makes an angle of ϕ_0 with the positive z-axis. Depending on whether $0 < \phi_0 < \pi/2$ or $\pi/2 < \phi_0 < \pi$, this will be the nappe of a cone opening up or opening down. (If $\phi_0 = \pi/2$, then the cone is flat, and the surface is the xy-plane.)

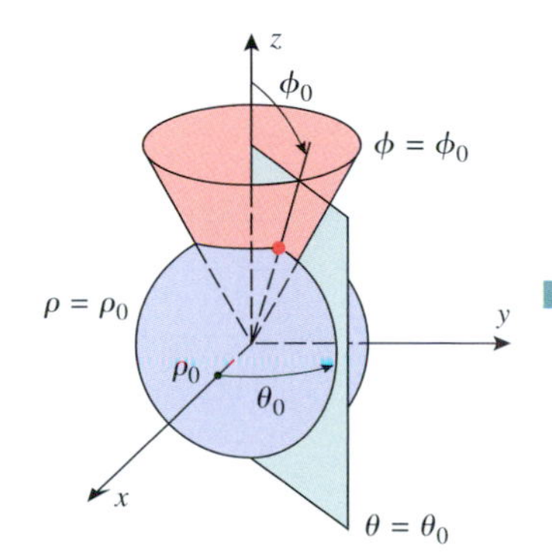

Figure 12.8.4

CONVERTING COORDINATES

Just as we needed to convert between rectangular and polar coordinates in 2-space, so we will need to be able to convert between rectangular, cylindrical, and spherical coordinates in 3-space. Table 12.8.1 provides formulas for making these conversions.

The diagrams in Figure 12.8.5 will help you to understand how the formulas in Table 12.8.1 are derived. For example, part (a) of the figure shows that in converting between rectangular coordinates (x, y, z) and cylindrical coordinates (r, θ, z), we can interpret (r, θ) as polar coordinates of (x, y). Thus, the polar-to-rectangular and rectangular-to-polar conversion formulas (1) and (2) of Section 11.1 provide the conversion formulas between rectangular and cylindrical coordinates in the table.

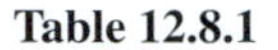

Table 12.8.1

CONVERSION		FORMULAS	RESTRICTIONS
Cylindrical to rectangular	$(r, \theta, z) \to (x, y, z)$	$x = r\cos\theta,\quad y = r\sin\theta,\quad z = z$	
Rectangular to cylindrical	$(x, y, z) \to (r, \theta, z)$	$r = \sqrt{x^2 + y^2},\quad \tan\theta = y/x,\quad z = z$	
Spherical to cylindrical	$(\rho, \theta, \phi) \to (r, \theta, z)$	$r = \rho\sin\phi,\quad \theta = \theta,\quad z = \rho\cos\phi$	$r \ge 0, \rho \ge 0$
Cylindrical to spherical	$(r, \theta, z) \to (\rho, \theta, \phi)$	$\rho = \sqrt{r^2 + z^2},\quad \theta = \theta,\quad \tan\phi = r/z$	$0 \le \theta < 2\pi$
Spherical to rectangular	$(\rho, \theta, \phi) \to (x, y, z)$	$x = \rho\sin\phi\cos\theta,\quad y = \rho\sin\phi\sin\theta,\quad z = \rho\cos\phi$	$0 \le \phi \le \pi$
Rectangular to spherical	$(x, y, z) \to (\rho, \theta, \phi)$	$\rho = \sqrt{x^2 + y^2 + z^2},\quad \tan\theta = y/x,\quad \cos\phi = z/\sqrt{x^2 + y^2 + z^2}$	

(*a*)

(*b*)

Figure 12.8.5

Part (*b*) of Figure 12.8.5 suggests that the spherical coordinates (ρ, θ, ϕ) of a point P can be converted to cylindrical coordinates (r, θ, z) by the conversion formulas

$$r = \rho\sin\phi, \qquad \theta = \theta, \qquad z = \rho\cos\phi \tag{1}$$

Moreover, since the cylindrical coordinates (r, θ, z) of P can be converted to rectangular coordinates (x, y, z) by the conversion formulas

$$x = r\cos\theta, \qquad y = r\sin\theta, \qquad z = z \tag{2}$$

we can obtain direct conversion formulas from spherical coordinates to rectangular coordinates by substituting (1) in (2). This yields

$$x = \rho\sin\phi\cos\theta, \qquad y = \rho\sin\phi\sin\theta, \qquad z = \rho\cos\phi \tag{3}$$

The other conversion formulas in Table 12.8.1 are left as exercises.

▶ Example 1

(a) Find the rectangular coordinates of the point with cylindrical coordinates

$$(r, \theta, z) = (4, \pi/3, -3)$$

(b) Find the rectangular coordinates of the point with spherical coordinates

$$(\rho, \theta, \phi) = (4, \pi/3, \pi/4)$$

Solution (a). Applying the cylindrical-to-rectangular conversion formulas in Table 12.8.1 yields

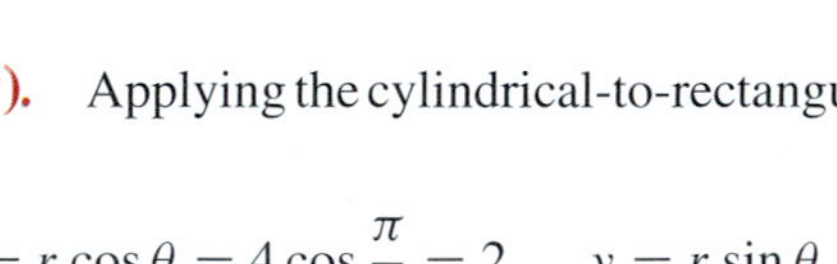

$$x = r\cos\theta = 4\cos\frac{\pi}{3} = 2, \qquad y = r\sin\theta = 4\sin\frac{\pi}{3} = 2\sqrt{3}, \qquad z = -3$$

Thus, the rectangular coordinates of the point are $(x, y, z) = (2, 2\sqrt{3}, -3)$ (Figure 12.8.6).

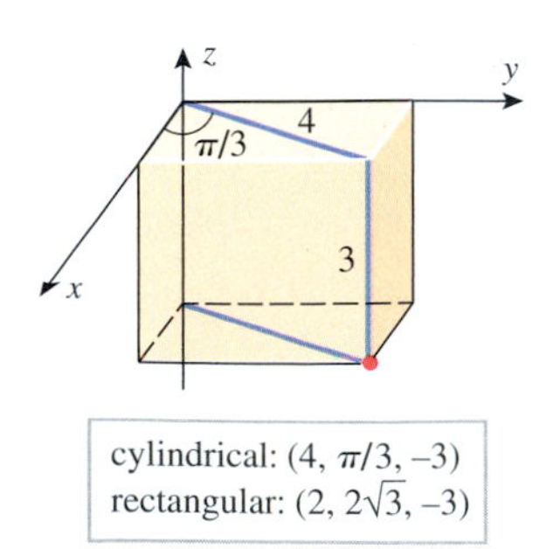

Figure 12.8.6

Solution (b). Applying the spherical-to-rectangular conversion formulas in Table 12.8.1 yields

$$x = \rho\sin\phi\cos\theta = 4\sin\frac{\pi}{4}\cos\frac{\pi}{3} = \sqrt{2}$$

$$y = \rho\sin\phi\sin\theta = 4\sin\frac{\pi}{4}\sin\frac{\pi}{3} = \sqrt{6}$$

$$z = \rho\cos\phi = 4\cos\frac{\pi}{4} = 2\sqrt{2}$$

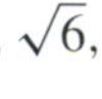

The rectangular coordinates of the point are $(x, y, z) = (\sqrt{2}, \sqrt{6}, 2\sqrt{2})$ (Figure 12.8.7). ◀

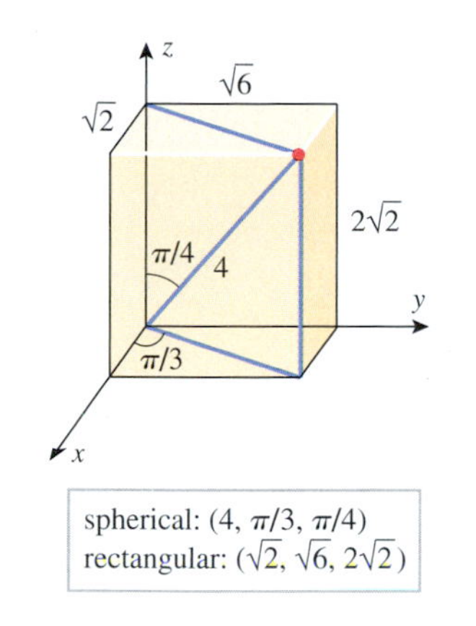

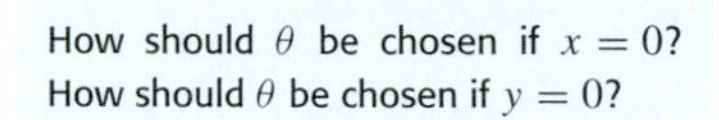

Figure 12.8.7

How should θ be chosen if $x = 0$? How should θ be chosen if $y = 0$?

Since the interval $0 \leq \theta < 2\pi$ covers two periods of the tangent function, the conversion formula $\tan\theta = y/x$ does not completely determine θ. When converting from rectangular to cylindrical or spherical coordinates, it is evident from parts (*b*) and (*c*) of Figure 12.8.1 that we should select θ so that

$$0 < \theta < \pi \quad \text{if } y > 0 \quad \text{and} \quad \pi < \theta < 2\pi \quad \text{if } y < 0$$

This is illustrated in the following example.

▶ **Example 2** Find the spherical coordinates of the point that has rectangular coordinates

$$(x, y, z) = (4, -4, 4\sqrt{6})$$

Solution. From the rectangular-to-spherical conversion formulas in Table 12.8.1 we obtain

$$\rho = \sqrt{x^2 + y^2 + z^2} = \sqrt{16 + 16 + 96} = \sqrt{128} = 8\sqrt{2}$$

$$\tan\theta = \frac{y}{x} = -1$$

$$\cos\phi = \frac{z}{\sqrt{x^2 + y^2 + z^2}} = \frac{4\sqrt{6}}{8\sqrt{2}} = \frac{\sqrt{3}}{2}$$

From the restriction $0 \leq \theta < 2\pi$ and the computed value of $\tan\theta$, the possibilities for θ are $\theta = 3\pi/4$ and $\theta = 7\pi/4$. However, the given point has a negative y-coordinate, so we must have $\theta = 7\pi/4$. Moreover, from the restriction $0 \leq \phi \leq \pi$ and the computed value of $\cos\phi$, the only possibility for ϕ is $\phi = \pi/6$. Thus, the spherical coordinates of the point are $(\rho, \theta, \phi) = (8\sqrt{2}, 7\pi/4, \pi/6)$ (Figure 12.8.8). ◀

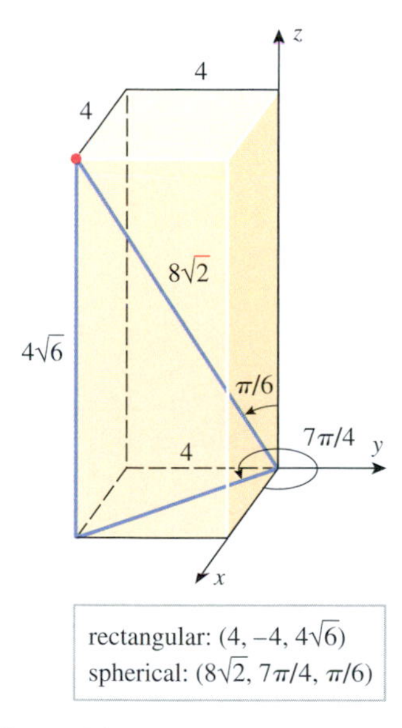

Figure 12.8.8

EQUATIONS OF SURFACES IN CYLINDRICAL AND SPHERICAL COORDINATES

Surfaces of revolution about the z-axis of a rectangular coordinate system usually have simpler equations in cylindrical coordinates than in rectangular coordinates, and the equations of surfaces with symmetry about the origin are usually simpler in spherical coordinates than in rectangular coordinates. For example, consider the upper nappe of the circular cone whose equation in rectangular coordinates is

$$z = \sqrt{x^2 + y^2}$$

(Table 12.8.2). The corresponding equation in cylindrical coordinates can be obtained from the cylindrical-to-rectangular conversion formulas in Table 12.8.1. This yields

$$z = \sqrt{(r\cos\theta)^2 + (r\sin\theta)^2} = \sqrt{r^2} = |r| = r$$

so the equation of the cone in cylindrical coordinates is $z = r$. Going a step further, the equation of the cone in spherical coordinates can be obtained from the spherical-to-cylindrical conversion formulas from Table 12.8.1. This yields

$$\rho\cos\phi = \rho\sin\phi$$

which, if $\rho \neq 0$, can be rewritten as

$$\tan\phi = 1 \quad \text{or} \quad \phi = \frac{\pi}{4}$$

Geometrically, this tells us that the radial line from the origin to any point on the cone makes an angle of $\pi/4$ with the z-axis.

Table 12.8.2

	CONE	CYLINDER	SPHERE	PARABOLOID	HYPERBOLOID
RECTANGULAR	$z = \sqrt{x^2 + y^2}$	$x^2 + y^2 = 1$	$x^2 + y^2 + z^2 = 1$	$z = x^2 + y^2$	$x^2 + y^2 - z^2 = 1$
CYLINDRICAL	$z = r$	$r = 1$	$z^2 = 1 - r^2$	$z = r^2$	$z^2 = r^2 - 1$
SPHERICAL	$\phi = \pi/4$	$\rho = \csc\phi$	$\rho = 1$	$\rho = \cos\phi\csc^2\phi$	$\rho^2 = -\sec 2\phi$

▶ **Example 3** Find equations of the paraboloid $z = x^2 + y^2$ in cylindrical and spherical coordinates.

Verify the equations given in Table 12.8.2 for the cylinder and hyperboloid in cylindrical and spherical coordinates.

Solution. The rectangular-to-cylindrical conversion formulas in Table 12.8.1 yield

$$z = r^2 \tag{4}$$

which is the equation in cylindrical coordinates. Now applying the spherical-to-cylindrical conversion formulas to (4) yields

$$\rho\cos\phi = \rho^2\sin^2\phi$$

which we can rewrite as

$$\rho = \cos\phi\csc^2\phi$$

Alternatively, we could have obtained this equation directly from the equation in rectangular coordinates by applying the spherical-to-rectangular conversion formulas (verify). ◀

SPHERICAL COORDINATES IN NAVIGATION

Spherical coordinates are related to longitude and latitude coordinates used in navigation. To see why this is so, let us construct a right-hand rectangular coordinate system with its origin at the center of the Earth, its positive z-axis passing through the North Pole, and its positive x-axis passing through the prime meridian (Figure 12.8.9). If we assume the Earth to be a sphere of radius $\rho = 4000$ miles, then each point on the Earth has spherical coordinates of the form $(4000, \theta, \phi)$, where ϕ and θ determine the latitude and longitude of the point. It is common to specify longitudes in degrees east or west of the prime meridian and latitudes in degrees north or south of the equator. However, the next example shows that it is a simple matter to determine ϕ and θ from such data.

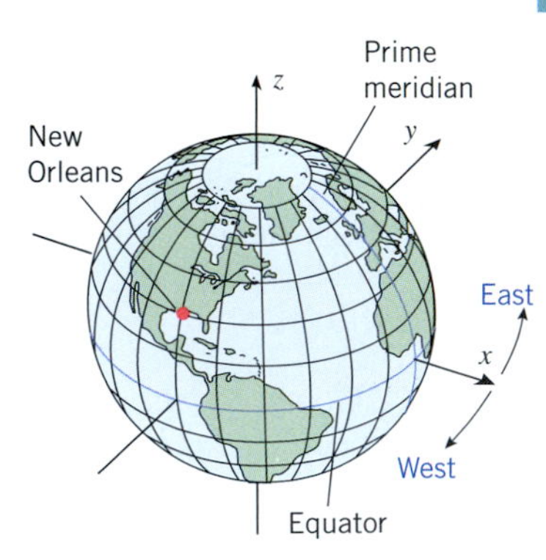

Figure 12.8.9

▶ **Example 4** The city of New Orleans is located at 90° west longitude and 30° north latitude. Find its spherical and rectangular coordinates relative to the coordinate axes of Figure 12.8.9. (Assume that distance is in miles.)

Solution. A longitude of 90° west corresponds to $\theta = 360° - 90° = 270°$ or $\theta = 3\pi/2$ radians; and a latitude of 30° north corresponds to $\phi = 90° - 30° = 60°$ or $\phi = \pi/3$ radians. Thus, the spherical coordinates (ρ, θ, ϕ) of New Orleans are $(4000, 3\pi/2, \pi/3)$.

To find the rectangular coordinates we apply the spherical-to-rectangular conversion formulas in Table 12.8.1. This yields

$$x = 4000 \sin\frac{\pi}{3}\cos\frac{3\pi}{2} = 4000\frac{\sqrt{3}}{2}(0) = 0 \text{ mi}$$

$$y = 4000 \sin\frac{\pi}{3}\sin\frac{3\pi}{2} = 4000\frac{\sqrt{3}}{2}(-1) = -2000\sqrt{3} \text{ mi}$$

$$z = 4000 \cos\frac{\pi}{3} = 4000\left(\frac{1}{2}\right) = 2000 \text{ mi} \blacktriangleleft$$

✔ QUICK CHECK EXERCISES 12.8 (See page 856 for answers.)

1. The conversion formulas from cylindrical coordinates (r, θ, z) to rectangular coordinates (x, y, z) are

$x =$ ________, $y =$ ________, $z =$ ________

2. The conversion formulas from spherical coordinates (ρ, θ, ϕ) to rectangular coordinates (x, y, z) are

$x =$ ________, $y =$ ________, $z =$ ________

3. The conversion formulas from spherical coordinates (ρ, θ, ϕ) to cylindrical coordinates (r, θ, z) are

$r =$ ________, $\theta =$ ________, $z =$ ________

4. Let P be the point in 3-space with rectangular coordinates $(\sqrt{2}, -\sqrt{2}, 2\sqrt{3})$.
(a) Cylindrical coordinates for P are $(r, \theta, z) =$ ________.
(b) Spherical coordinates for P are $(\rho, \theta, \phi) =$ ________.

5. Give an equation of a sphere of radius 5, centered at the origin, in
(a) rectangular coordinates
(b) cylindrical coordinates
(c) spherical coordinates.

EXERCISE SET 12.8 Graphing Utility C CAS

1–2 Convert from rectangular to cylindrical coordinates.

1. (a) $(4\sqrt{3}, 4, -4)$ (b) $(-5, 5, 6)$
(c) $(0, 2, 0)$ (d) $(4, -4\sqrt{3}, 6)$

2. (a) $(\sqrt{2}, -\sqrt{2}, 1)$ (b) $(0, 1, 1)$
(c) $(-4, 4, -7)$ (d) $(2, -2, -2)$

3–4 Convert from cylindrical to rectangular coordinates.

3. (a) $(4, \pi/6, 3)$ (b) $(8, 3\pi/4, -2)$
(c) $(5, 0, 4)$ (d) $(7, \pi, -9)$

4. (a) $(6, 5\pi/3, 7)$ (b) $(1, \pi/2, 0)$
(c) $(3, \pi/2, 5)$ (d) $(4, \pi/2, -1)$

5–6 Convert from rectangular to spherical coordinates.

5. (a) $(1, \sqrt{3}, -2)$ (b) $(1, -1, \sqrt{2})$
(c) $(0, 3\sqrt{3}, 3)$ (d) $(-5\sqrt{3}, 5, 0)$

6. (a) $(4, 4, 4\sqrt{6})$ (b) $(1, -\sqrt{3}, -2)$
(c) $(2, 0, 0)$ (d) $(\sqrt{3}, 1, 2\sqrt{3})$

7–8 Convert from spherical to rectangular coordinates.

7. (a) $(5, \pi/6, \pi/4)$ (b) $(7, 0, \pi/2)$
(c) $(1, \pi, 0)$ (d) $(2, 3\pi/2, \pi/2)$

8. (a) $(1, 2\pi/3, 3\pi/4)$ (b) $(3, 7\pi/4, 5\pi/6)$
(c) $(8, \pi/6, \pi/4)$ (d) $(4, \pi/2, \pi/3)$

9–10 Convert from cylindrical to spherical coordinates.

9. (a) $(\sqrt{3}, \pi/6, 3)$ (b) $(1, \pi/4, -1)$
(c) $(2, 3\pi/4, 0)$ (d) $(6, 1, -2\sqrt{3})$

10. (a) $(4, 5\pi/6, 4)$ (b) $(2, 0, -2)$
(c) $(4, \pi/2, 3)$ (d) $(6, \pi, 2)$

11–12 Convert from spherical to cylindrical coordinates.

11. (a) $(5, \pi/4, 2\pi/3)$ (b) $(1, 7\pi/6, \pi)$
(c) $(3, 0, 0)$ (d) $(4, \pi/6, \pi/2)$

12. (a) $(5, \pi/2, 0)$ (b) $(6, 0, 3\pi/4)$
(c) $(\sqrt{2}, 3\pi/4, \pi)$ (d) $(5, 2\pi/3, 5\pi/6)$

C **13.** Use a CAS or a programmable calculating utility to set up the conversion formulas in Table 12.8.1, and then use the CAS or calculating utility to solve the problems in Exercises 1, 3, 5, 7, 9, and 11.

C **14.** Use a CAS or a programmable calculating utility to set up the conversion formulas in Table 12.8.1, and then use the CAS or calculating utility to solve the problems in Exercises 2, 4, 6, 8, 10, and 12.

15–22 An equation is given in cylindrical coordinates. Express the equation in rectangular coordinates and sketch the graph.

15. $r = 3$ **16.** $\theta = \pi/4$ **17.** $z = r^2$
18. $z = r\cos\theta$ **19.** $r = 4\sin\theta$ **20.** $r = 2\sec\theta$
21. $r^2 + z^2 = 1$ **22.** $r^2\cos 2\theta = z$

23–30 An equation is given in spherical coordinates. Express the equation in rectangular coordinates and sketch the graph.

23. $\rho = 3$ **24.** $\theta = \pi/3$ **25.** $\phi = \pi/4$
26. $\rho = 2\sec\phi$ **27.** $\rho = 4\cos\phi$ **28.** $\rho\sin\phi = 1$
29. $\rho\sin\phi = 2\cos\theta$ **30.** $\rho - 2\sin\phi\cos\theta = 0$

31–42 An equation of a surface is given in rectangular coordinates. Find an equation of the surface in (a) cylindrical coordinates and (b) spherical coordinates.

31. $z = 3$ **32.** $y = 2$
33. $z = 3x^2 + 3y^2$ **34.** $z = \sqrt{3x^2 + 3y^2}$
35. $x^2 + y^2 = 4$ **36.** $x^2 + y^2 - 6y = 0$
37. $x^2 + y^2 + z^2 = 9$ **38.** $z^2 = x^2 - y^2$
39. $2x + 3y + 4z = 1$ **40.** $x^2 + y^2 - z^2 = 1$
41. $x^2 = 16 - z^2$ **42.** $x^2 + y^2 + z^2 = 2z$

FOCUS ON CONCEPTS

43–46 Describe the region in 3-space that satisfies the given inequalities.

43. $r^2 \le z \le 4$ **44.** $0 \le r \le 2\sin\theta,\quad 0 \le z \le 3$
45. $1 \le \rho \le 3$ **46.** $0 \le \phi \le \pi/6,\quad 0 \le \rho \le 2$

47. St. Petersburg (Leningrad), Russia, is located at 30° east longitude and 60° north latitude. Find its spherical and rectangular coordinates relative to the coordinate axes of Figure 12.8.9. Take miles as the unit of distance and assume the Earth to be a sphere of radius 4000 miles.

48. (a) Show that the curve of intersection of the surfaces $z = \sin\theta$ and $r = a$ (cylindrical coordinates) is an ellipse.
(b) Sketch the surface $z = \sin\theta$ for $0 \le \theta \le \pi/2$.

49. The accompanying figure shows a right circular cylinder of radius 10 cm spinning at 3 revolutions per minute about the z-axis. At time $t = 0$ s, a bug at the point (0, 10, 0) begins walking straight up the face of the cylinder at the rate of 0.5 cm/min.
(a) Find the cylindrical coordinates of the bug after 2 min.
(b) Find the rectangular coordinates of the bug after 2 min.
(c) Find the spherical coordinates of the bug after 2 min.

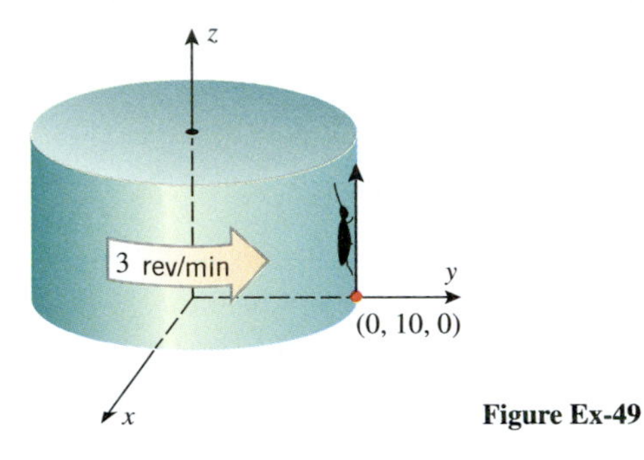

Figure Ex-49

50. Referring to Exercise 49, use a graphing utility to graph the bug's distance from the origin as a function of time.

51. A ship at sea is at point A that is 60° west longitude and 40° north latitude. The ship travels to point B that is 40° west longitude and 20° north latitude. Assuming that the Earth is a sphere with radius 6370 kilometers, find the shortest distance the ship can travel in going from A to B, given that the shortest distance between two points on a sphere is along the arc of the great circle joining the points. [*Suggestion:* Introduce an xyz-coordinate system as in Figure 12.8.9, and consider the angle between the vectors from the center of the Earth to the points A and B. If you are not familiar with the term "great circle," consult a dictionary.]

✔ QUICK CHECK ANSWERS 12.8

1. $r\cos\theta$; $r\sin\theta$; z **2.** $\rho\sin\phi\cos\theta$; $\rho\sin\phi\sin\theta$; $\rho\cos\phi$ **3.** $\rho\sin\phi$; θ; $\rho\cos\theta$
4. (a) $(2, 7\pi/4, 2\sqrt{3})$ (b) $(4, 7\pi/4, \pi/6)$ **5.** (a) $x^2 + y^2 + z^2 = 25$ (b) $r^2 + z^2 = 25$ (c) $\rho = 5$

CHAPTER REVIEW EXERCISES

1. (a) What is the difference between a vector and a scalar? Give a physical example of each.
(b) How can you determine whether or not two vectors are orthogonal?
(c) How can you determine whether or not two vectors are parallel?
(d) How can you determine whether or not three vectors with a common initial point lie in the same plane in 3-space?

2. (a) Sketch vectors $\mathbf{u}$ and $\mathbf{v}$ for which $\mathbf{u}+\mathbf{v}$ and $\mathbf{u}-\mathbf{v}$ are orthogonal.
(b) How can you use vectors to determine whether four points in 3-space lie in the same plane?
(c) If forces $\mathbf{F}_1 = \mathbf{i}$ and $\mathbf{F}_2 = \mathbf{j}$ are applied at a point in 2-space, what force would you apply at that point to cancel the combined effect of $\mathbf{F}_1$ and $\mathbf{F}_2$?
(d) Write an equation of the sphere with center $(1, -2, 2)$ that passes through the origin.

3. (a) Draw a picture that shows the direction angles α, β, and γ of a vector.
(b) What are the components of a unit vector in 2-space that makes an angle of $120°$ with the vector $\mathbf{i}$ (two answers)?
(c) How can you use vectors to determine whether a triangle with known vertices P_1, P_2, and P_3 has an obtuse angle?
(d) True or false: The cross product of orthogonal unit vectors is a unit vector. Explain your reasoning.

4. (a) Make a table that shows all possible cross products of the vectors $\mathbf{i}$, $\mathbf{j}$, and $\mathbf{k}$.
(b) Give a geometric interpretation of $\|\mathbf{u} \times \mathbf{v}\|$.
(c) Give a geometric interpretation of $|\mathbf{u} \cdot (\mathbf{v} \times \mathbf{w})|$.
(d) Write an equation of the plane that passes through the origin and is perpendicular to the line $x = t$, $y = 2t$, $z = -t$.

5. In each part, find an equation of the sphere with center $(-3, 5, -4)$ and satisfying the given condition.
(a) Tangent to the xy-plane
(b) Tangent to the xz-plane
(c) Tangent to the yz-plane

6. Find the largest and smallest distances between the point $P(1, 1, 1)$ and the sphere

$$x^2 + y^2 + z^2 - 2y + 6z - 6 = 0$$

7. Given the points $P(3, 4)$, $Q(1, 1)$, and $R(5, 2)$, use vector methods to find the coordinates of the fourth vertex of the parallelogram whose adjacent sides are $\overrightarrow{PQ}$ and $\overrightarrow{QR}$.

8. Let $\mathbf{u} = \langle 3, 5, -1 \rangle$ and $\mathbf{v} = \langle 2, -2, 3 \rangle$. Find
(a) $2\mathbf{u} + 5\mathbf{v}$ (b) $\dfrac{1}{\|\mathbf{v}\|}\mathbf{v}$
(c) $\|\mathbf{u}\|$ (d) $\|\mathbf{u} - \mathbf{v}\|$.

9. Let $\mathbf{a} = c\mathbf{i} + \mathbf{j}$ and $\mathbf{b} = 4\mathbf{i} + 3\mathbf{j}$. Find c so that
(a) $\mathbf{a}$ and $\mathbf{b}$ are orthogonal
(b) the angle between $\mathbf{a}$ and $\mathbf{b}$ is $\pi/4$
(c) the angle between $\mathbf{a}$ and $\mathbf{b}$ is $\pi/6$
(d) $\mathbf{a}$ and $\mathbf{b}$ are parallel.

10. Let $\mathbf{r}_0 = \langle x_0, y_0, z_0 \rangle$ and $\mathbf{r} = \langle x, y, z \rangle$. Describe the set of all points (x, y, z) for which
(a) $\mathbf{r} \cdot \mathbf{r}_0 = 0$ (b) $(\mathbf{r} - \mathbf{r}_0) \cdot \mathbf{r}_0 = 0$.

11. Show that if $\mathbf{u}$ and $\mathbf{v}$ are unit vectors and θ is the angle between them, then $\|\mathbf{u} - \mathbf{v}\| = 2 \sin \frac{1}{2}\theta$.

12. Find the vector with length 5 and direction angles $\alpha = 60°$, $\beta = 120°$, $\gamma = 135°$.

13. Assuming that force is in pounds and distance is in feet, find the work done by a constant force $\mathbf{F} = 3\mathbf{i} - 4\mathbf{j} + \mathbf{k}$ acting on a particle that moves on a straight line from $P(5, 7, 0)$ to $Q(6, 6, 6)$.

14. Assuming that force is in newtons and distance is in meters, find the work done by the resultant of the constant forces $\mathbf{F}_1 = \mathbf{i} - 3\mathbf{j} + \mathbf{k}$ and $\mathbf{F}_2 = \mathbf{i} + 2\mathbf{j} + 2\mathbf{k}$ acting on a particle that moves on a straight line from $P(-1, -2, 3)$ to $Q(0, 2, 0)$.

15. (a) Find the area of the triangle with vertices $A(1, 0, 1)$, $B(0, 2, 3)$, and $C(2, 1, 0)$.
(b) Use the result in part (a) to find the length of the altitude from vertex C to side AB.

16. True or false? Explain your reasoning.
(a) If $\mathbf{u} \cdot \mathbf{v} = 0$, then $\mathbf{u} = \mathbf{0}$ or $\mathbf{v} = \mathbf{0}$.
(b) If $\mathbf{u} \times \mathbf{v} = \mathbf{0}$, then $\mathbf{u} = \mathbf{0}$ or $\mathbf{v} = \mathbf{0}$.
(c) If $\mathbf{u} \cdot \mathbf{v} = 0$ and $\mathbf{u} \times \mathbf{v} = \mathbf{0}$, then $\mathbf{u} = \mathbf{0}$ or $\mathbf{v} = \mathbf{0}$.

17. Consider the points

$$A(1, -1, 2), \quad B(2, -3, 0), \quad C(-1, -2, 0), \quad D(2, 1, -1)$$

(a) Find the volume of the parallelepiped that has the vectors $\overrightarrow{AB}$, $\overrightarrow{AC}$, $\overrightarrow{AD}$ as adjacent edges.
(b) Find the distance from D to the plane containing A, B, and C.

18. Suppose that a force $\mathbf{F}$ with a magnitude of 9 lb is applied to the lever–shaft assembly shown in the accompanying figure.
(a) Express the force $\mathbf{F}$ in component form.
(b) Find the vector moment of $\mathbf{F}$ about the origin.

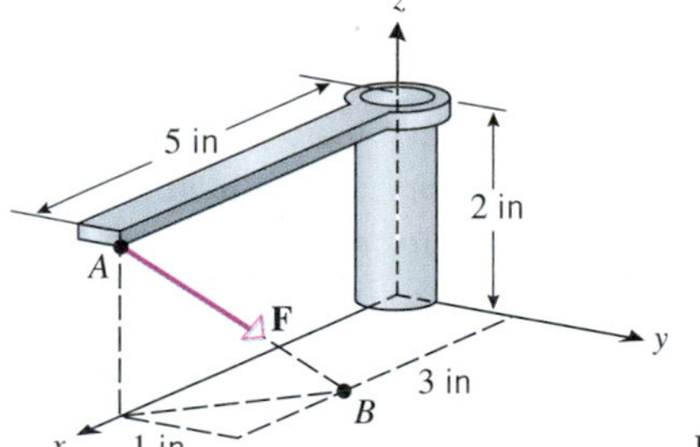

Figure Ex-18

19. Let P be the point $(4, 1, 2)$. Find parametric equations for the line through P and parallel to the vector $\langle 1, -1, 0 \rangle$.

20. (a) Find parametric equations for the intersection of the planes $2x + y - z = 3$ and $x + 2y + z = 3$.
(b) Find the acute angle between the two planes.

21. Find an equation of the plane that is parallel to the plane $x + 5y - z + 8 = 0$ and contains the point $(1, 1, 4)$.

22. Find an equation of the plane through the point $(4, 3, 0)$ and parallel to the vectors $\mathbf{i} + \mathbf{k}$ and $2\mathbf{j} - \mathbf{k}$.

23. What condition must the constants satisfy for the planes

$$a_1x + b_1y + c_1z = d_1 \quad \text{and} \quad a_2x + b_2y + c_2z = d_2$$

to be perpendicular?

24. (a) List six common types of quadric surfaces, and describe their traces in planes parallel to the coordinate planes.
(b) Give the coordinates of the points that result when the point (x, y, z) is reflected about the plane $y = x$, the plane $y = z$, and the plane $x = z$.
(c) Describe the intersection of the surfaces $r = 5$ and $z = 1$ in cylindrical coordinates.
(d) Describe the intersection of the surfaces $\phi = \pi/4$ and $\theta = 0$ in spherical coordinates.

25. In each part, identify the surface by completing the squares.
(a) $x^2 + 4y^2 - z^2 - 6x + 8y + 4z = 0$
(b) $x^2 + y^2 + z^2 + 6x - 4y + 12z = 0$
(c) $x^2 + y^2 - z^2 - 2x + 4y + 5 = 0$

26. In each part, express the equation in cylindrical and spherical coordinates.
(a) $x^2 + y^2 = z$ (b) $x^2 - y^2 - z^2 = 0$

27. In each part, express the equation in rectangular coordinates.
(a) $z = r^2 \cos 2\theta$ (b) $\rho^2 \sin\phi \cos\phi \cos\theta = 1$

28–29 Sketch the solid in 3-space that is described in cylindrical coordinates by the stated inequalities.

28. (a) $1 \le r \le 2$ (b) $2 \le z \le 3$ (c) $\pi/6 \le \theta \le \pi/3$
(d) $1 \le r \le 2$, $2 \le z \le 3$, and $\pi/6 \le \theta \le \pi/3$

29. (a) $r^2 + z^2 \le 4$ (b) $r \le 1$
(c) $r^2 + z^2 \le 4$ and $r > 1$

30–31 Sketch the solid in 3-space that is described in spherical coordinates by the stated inequalities.

30. (a) $0 \le \rho \le 2$ (b) $0 \le \phi \le \pi/6$
(c) $0 \le \rho \le 2$ and $0 \le \phi \le \pi/6$

31. (a) $0 \le \rho \le 5$, $0 \le \phi \le \pi/2$, and $0 \le \theta \le \pi/2$
(b) $0 \le \phi \le \pi/3$ and $0 \le \rho \le 2\sec\phi$
(c) $0 \le \rho \le 2$ and $\pi/6 \le \phi \le \pi/3$

32. Sketch the surface whose equation in spherical coordinates is $\rho = a(1 - \cos\phi)$. [*Hint:* The surface is shaped like a familiar fruit.]

chapter thirteen

VECTOR-VALUED FUNCTIONS

Everyone knows what a curve is, until he has studied enough mathematics to become confused through the countless number of possible exceptions.

—Felix Klein
Mathematician

In this chapter we will consider functions whose values are vectors. Such functions provide a unified way of studying parametric curves in 2-space and 3-space and are a basic tool for analyzing the motion of particles along curved paths. We will begin by developing the calculus of vector-valued functions—we will show how to differentiate and integrate such functions, and we will develop some of the basic properties of these operations. We will then apply these calculus tools to define three fundamental vectors that can be used to describe such basic characteristics of curves as curvature and twisting tendencies. Once this is done, we will develop the concepts of velocity and acceleration for such motion, and we will apply these concepts to explain various physical phenomena. Finally, we will use the calculus of vector-valued functions to develop basic principles of gravitational attraction and to derive Kepler's laws of planetary motion.

Photo: *A roller coaster moves with variable velocity and direction. We will study this kind of motion in this chapter.*

13.1 INTRODUCTION TO VECTOR-VALUED FUNCTIONS

In Section 12.5 we discussed parametric equations of lines in 3-space. In this section we will discuss more general parametric curves in 3-space, and we will show how vector notation can be used to express parametric equations in 2-space and 3-space in a more compact form. This will lead us to consider a new kind of function—namely, functions that associate vectors with real numbers. Such functions have many important applications in physics and engineering.

PARAMETRIC CURVES IN 3-SPACE

Recall from Section 1.8 that if f and g are well-behaved functions, then the pair of parametric equations

$$x = f(t), \quad y = g(t) \tag{1}$$

generates a curve in 2-space that is traced in a specific direction as the parameter t increases. We defined this direction to be the *orientation* of the curve or the *direction of increasing parameter*, and we called the curve together with its orientation the *graph* of the parametric equations or the *parametric curve* represented by the equations. Analogously, if f, g, and h are three well-behaved functions, then the parametric equations

$$x = f(t), \quad y = g(t), \quad z = h(t) \tag{2}$$

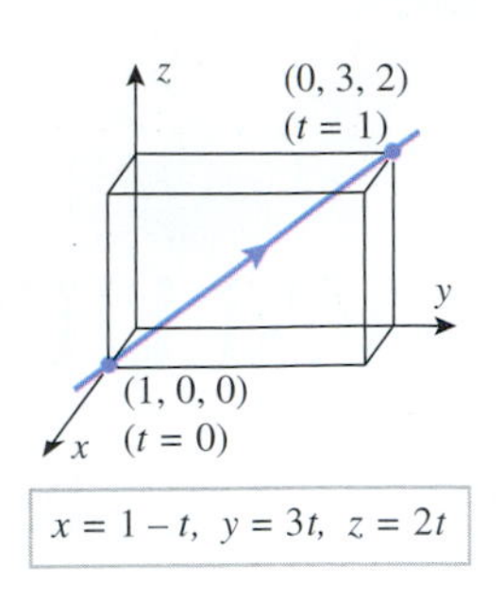

Figure 13.1.1

generate a curve in 3-space that is traced in a specific direction as t increases. As in 2-space, this direction is called the ***orientation*** or ***direction of increasing parameter***, and the curve together with its orientation is called the ***graph*** of the parametric equations or the ***parametric curve*** represented by the equations. If no restrictions are stated explicitly or are implied by the equations, then it will be understood that t varies over the interval $(-\infty, +\infty)$.

Example 1 The parametric equations

$$x = 1 - t, \quad y = 3t, \quad z = 2t$$

represent a line in 3-space that passes through the point $(1, 0, 0)$ and is parallel to the vector $\langle -1, 3, 2\rangle$. Since x decreases as t increases, the line has the orientation shown in Figure 13.1.1. ◄

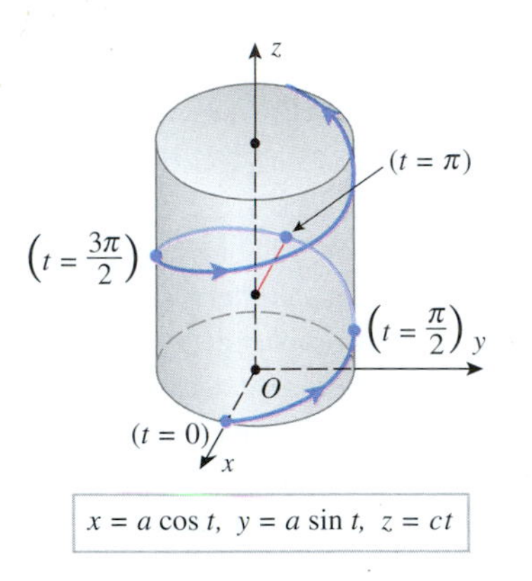

Figure 13.1.2

Example 2 Describe the parametric curve represented by the equations

$$x = a\cos t, \quad y = a\sin t, \quad z = ct$$

where a and c are positive constants.

Solution. As the parameter t increases, the value of $z = ct$ also increases, so the point (x, y, z) moves upward. However, as t increases, the point (x, y, z) also moves in a path directly over the circle

$$x = a\cos t, \quad y = a\sin t$$

in the xy-plane. The combination of these upward and circular motions produces a corkscrew-shaped curve that wraps around a right circular cylinder of radius a centered on the z-axis (Figure 13.1.2). This curve is called a ***circular helix***. ◄

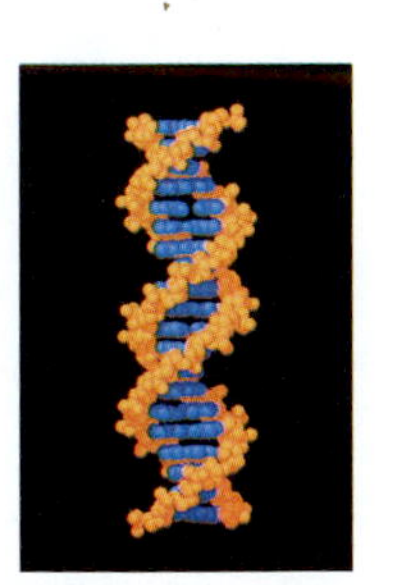

The circular helix described in Example 2 occurs in nature. Above is a computer representation of the twin helix DNA molecule (deoxyribonucleic acid). This structure contains all the inherited instructions necessary for the development of a living organism.

PARAMETRIC CURVES GENERATED WITH TECHNOLOGY

Except in the simplest cases, parametric curves in 3-space can be difficult to visualize and draw without the help of a graphing utility. For example, Figure 13.1.3a shows the graph of the parametric curve called a *torus knot* that was produced by a CAS. However, even this computer rendering is difficult to visualize because it is unclear whether the points of overlap are intersections or whether one portion of the curve is in front of the other. To resolve this visualization problem, some graphing utilities provide the capability of enclosing the curve within a thin tube, as in Figure 13.1.3b. Such graphs are called ***tube plots***.

TECHNOLOGY MASTERY

If you have a CAS, use it to generate the line in Example 1 and the helix

$$x = 4\cos t, \quad y = 4\sin t, \quad z = t \qquad (0 \le t \le 3\pi)$$

in Figure 13.1.4.

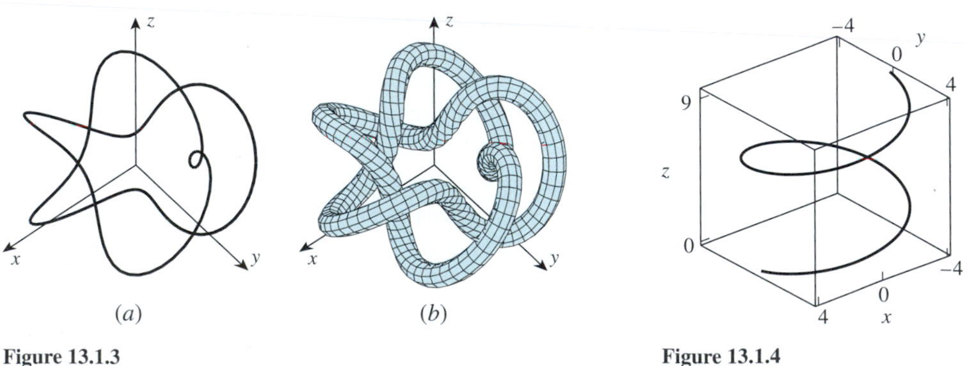

Figure 13.1.3

Figure 13.1.4

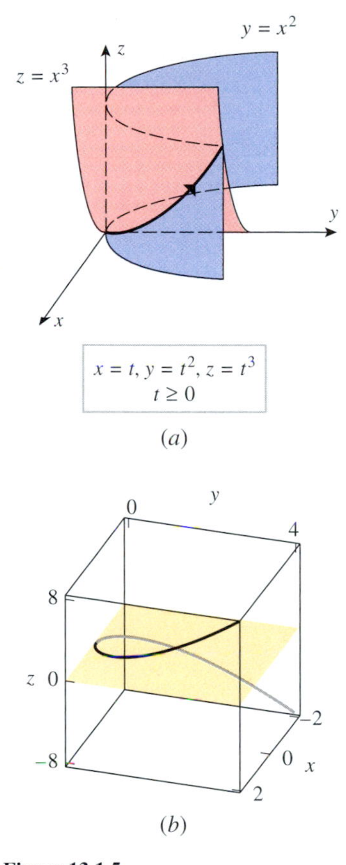

Figure 13.1.5

Find the vector-valued function in 2-space whose component functions are $x(t) = t$ and $y(t) = t^2$.

PARAMETRIC EQUATIONS FOR INTERSECTIONS OF SURFACES

Curves in 3-space often arise as intersections of surfaces. For example, Figure 13.1.5a shows a portion of the intersection of the cylinders $z = x^3$ and $y = x^2$. One method for finding parametric equations for the curve of intersection is to choose one of the variables as the parameter and use the two equations to express the remaining two variables in terms of that parameter. In particular, if we choose $x = t$ as the parameter and substitute this into the equations $z = x^3$ and $y = x^2$, we obtain the parametric equations

$$x = t, \quad y = t^2, \quad z = t^3 \tag{3}$$

This curve is called a ***twisted cubic***. The portion of the twisted cubic shown in Figure 13.1.5a corresponds to $t \geq 0$; a computer-generated graph of the twisted cubic for positive and negative values of t is shown in Figure 13.1.5b. Some other examples and techniques for finding intersections of surfaces are discussed in the exercises.

VECTOR-VALUED FUNCTIONS

The twisted cubic defined by the equations in (3) is the set of points of the form (t, t^2, t^3) for real values of t. If we view each of these points as a terminal point for a vector $\mathbf{r}$ whose initial point is at the origin,

$$\mathbf{r} = \langle x, y, z\rangle = \langle t, t^2, t^3\rangle = t\mathbf{i} + t^2\mathbf{j} + t^3\mathbf{k}$$

then we obtain $\mathbf{r}$ as a function of the parameter t, that is, $\mathbf{r} = \mathbf{r}(t)$. Since this function produces a *vector*, we say that $\mathbf{r} = \mathbf{r}(t)$ defines $\mathbf{r}$ as a ***vector-valued function of a real variable***, or more simply, a ***vector-valued function***. The vectors that we will consider in this text are either in 2-space or 3-space, so we will say that a vector-valued function is in 2-space or in 3-space according to the kind of vectors that it produces.

If $\mathbf{r}(t)$ is a vector-valued function in 2-space, then for each allowable value of t the vector $\mathbf{r} = \mathbf{r}(t)$ can be represented in terms of components as

$$\mathbf{r} = \mathbf{r}(t) = \langle x(t), y(t)\rangle = x(t)\mathbf{i} + y(t)\mathbf{j}$$

The functions $x(t)$ and $y(t)$ are called the ***component functions*** or the ***components*** of $\mathbf{r}(t)$. Similarly, the component functions of a vector-valued function

$$\mathbf{r}(t) = \langle x(t), y(t), z(t)\rangle = x(t)\mathbf{i} + y(t)\mathbf{j} + z(t)\mathbf{k}$$

in 3-space are $x(t)$, $y(t)$, and $z(t)$.

▶ Example 3 The component functions of

$$\mathbf{r}(t) = \langle t, t^2, t^3\rangle = t\mathbf{i} + t^2\mathbf{j} + t^3\mathbf{k}$$

are

$$x(t) = t, \quad y(t) = t^2, \quad z(t) = t^3 \quad ◀$$

The ***domain*** of a vector-valued function $\mathbf{r}(t)$ is the set of allowable values for t. If $\mathbf{r}(t)$ is defined in terms of component functions and the domain is not specified explicitly, then it will be understood that the domain is the intersection of the natural domains of the component functions; this is called the ***natural domain*** of $\mathbf{r}(t)$.

▶ Example 4 Find the natural domain of

$$\mathbf{r}(t) = \langle \ln|t-1|, e^t, \sqrt{t}\rangle = (\ln|t-1|)\mathbf{i} + e^t\mathbf{j} + \sqrt{t}\mathbf{k}$$

Solution. The natural domains of the component functions

$$x(t) = \ln|t-1|, \quad y(t) = e^t, \quad z(t) = \sqrt{t}$$

are

$$(-\infty, 1) \cup (1, +\infty), \quad (-\infty, +\infty), \quad [0, +\infty)$$

respectively. The intersection of these sets is

$$[0, 1) \cup (1, +\infty)$$

(verify), so the natural domain of $\mathbf{r}(t)$ consists of all values of t such that

$$0 \leq t < 1 \quad \text{or} \quad t > 1 \ ◀$$

GRAPHS OF VECTOR-VALUED FUNCTIONS

If $\mathbf{r}(t)$ is a vector-valued function in 2-space or 3-space, then we define the ***graph*** of $\mathbf{r}(t)$ to be the parametric curve described by the component functions for $\mathbf{r}(t)$. For example, if

$$\mathbf{r}(t) = \langle 1-t, 3t, 2t \rangle = (1-t)\mathbf{i} + 3t\mathbf{j} + 2t\mathbf{k} \tag{4}$$

then the graph of $\mathbf{r} = \mathbf{r}(t)$ is the graph of the parametric equations

$$x = 1-t, \quad y = 3t, \quad z = 2t$$

Thus, the graph of (4) is the line in Figure 13.1.1.

Strictly speaking, we should write $(\cos t)\mathbf{i}$ and $(\sin t)\mathbf{j}$ rather than $\cos t\mathbf{i}$ and $\sin t\mathbf{j}$ for clarity. However, it is a common practice to omit the parentheses in such cases, since no misinterpretation is possible. Why?

▶ Example 5 Describe the graph of the vector-valued function

$$\mathbf{r}(t) = \langle \cos t, \sin t, t \rangle = \cos t\mathbf{i} + \sin t\mathbf{j} + t\mathbf{k}$$

Solution. The corresponding parametric equations are

$$x = \cos t, \quad y = \sin t, \quad z = t$$

Thus, as we saw in Example 2, the graph is a circular helix wrapped around a cylinder of radius 1. ◀

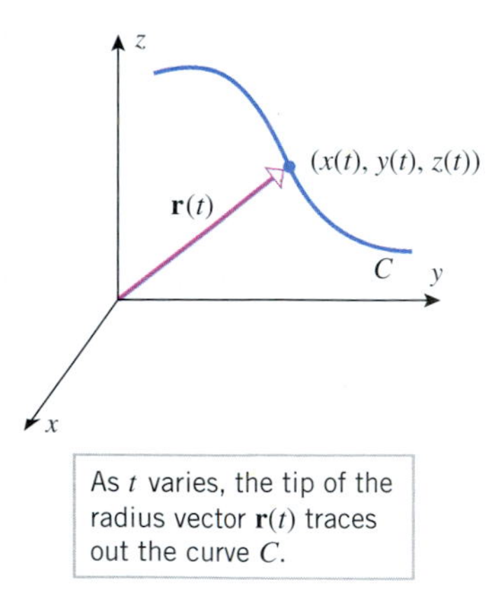

As t varies, the tip of the radius vector $\mathbf{r}(t)$ traces out the curve C.

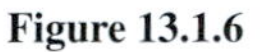

Figure 13.1.6

Up to now we have considered parametric curves to be paths traced by moving points. However, if a parametric curve is viewed as the graph of a vector-valued function, then we can also imagine the graph to be traced by the tip of a moving vector. For example, if the curve C in 3-space is the graph of

$$\mathbf{r}(t) = x(t)\mathbf{i} + y(t)\mathbf{j} + z(t)\mathbf{k}$$

and if we position $\mathbf{r}(t)$ so its initial point is at the origin, then its terminal point will fall on the curve C (as shown in Figure 13.1.6). Thus, when $\mathbf{r}(t)$ is positioned with its initial point at the origin, its terminal point will trace out the curve C as the parameter t varies, in which case we call $\mathbf{r}(t)$ the ***radius vector*** or the ***position vector*** for C. For simplicity, we will sometimes let the dependence on t be understood and write $\mathbf{r}$ rather than $\mathbf{r}(t)$ for a radius vector.

▶ Example 6 Sketch the graph and a radius vector of

(a) $\mathbf{r}(t) = \cos t\mathbf{i} + \sin t\mathbf{j}, \quad 0 \leq t \leq 2\pi$

(b) $\mathbf{r}(t) = \cos t\mathbf{i} + \sin t\mathbf{j} + 2\mathbf{k}, \quad 0 \leq t \leq 2\pi$

Solution (a). The corresponding parametric equations are

$$x = \cos t, \quad y = \sin t \qquad (0 \leq t \leq 2\pi)$$

so the graph is a circle of radius 1, centered at the origin, and oriented counterclockwise. The graph and a radius vector are shown in Figure 13.1.7.

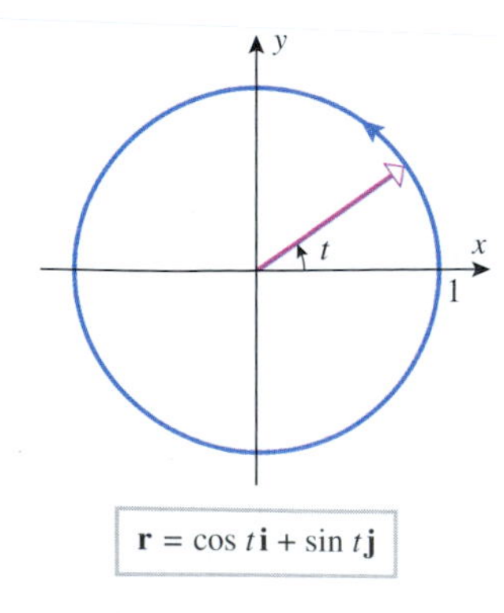

$\mathbf{r} = \cos t\mathbf{i} + \sin t\mathbf{j}$

Figure 13.1.7

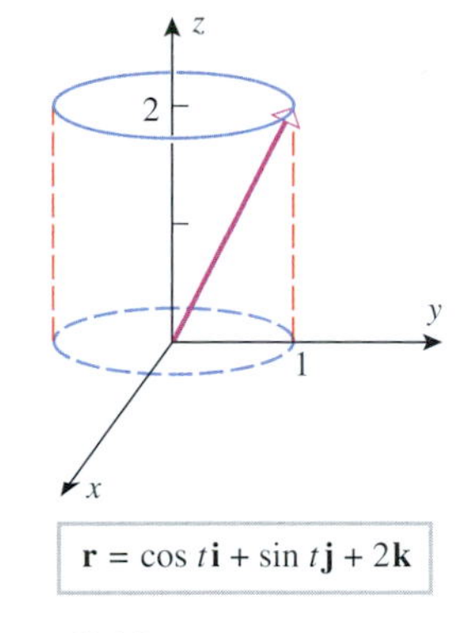

Figure 13.1.8

Solution (b). The corresponding parametric equations are

$$x = \cos t, \quad y = \sin t, \quad z = 2 \qquad (0 \le t \le 2\pi)$$

From the third equation, the tip of the radius vector traces a curve in the plane $z = 2$, and from the first two equations, the curve is a circle of radius 1 centered at the point (0, 0, 2) and traced counterclockwise looking down the z-axis. The graph and a radius vector are shown in Figure 13.1.8. ◀

VECTOR FORM OF A LINE SEGMENT

Recall from Formula (9) of Section 12.5 that if $\mathbf{r}_0$ is a vector in 2-space or 3-space with its initial point at the origin, then the line that passes through the terminal point of $\mathbf{r}_0$ and is parallel to the vector $\mathbf{v}$ can be expressed in vector form as

$$\mathbf{r} = \mathbf{r}_0 + t\mathbf{v}$$

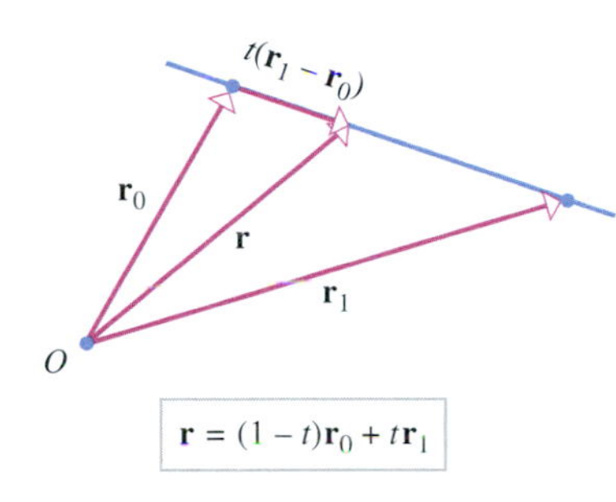

Figure 13.1.9

In particular, if $\mathbf{r}_0$ and $\mathbf{r}_1$ are vectors in 2-space or 3-space with their initial points at the origin, then the line that passes through the terminal points of these vectors can be expressed in vector form as

$$\mathbf{r} = \mathbf{r}_0 + t(\mathbf{r}_1 - \mathbf{r}_0) \qquad \text{or} \qquad \mathbf{r} = (1-t)\mathbf{r}_0 + t\mathbf{r}_1 \tag{5–6}$$

as indicated in Figure 13.1.9.

It is common to call either (5) or (6) the ***two-point vector form of a line*** and to say, for simplicity, that the line passes through the *points* $\mathbf{r}_0$ and $\mathbf{r}_1$ (as opposed to saying that it passes through the *terminal points* of $\mathbf{r}_0$ and $\mathbf{r}_1$).

It is understood in (5) and (6) that t varies from $-\infty$ to $+\infty$. However, if we restrict t to vary over the interval $0 \le t \le 1$, then $\mathbf{r}$ will vary from $\mathbf{r}_0$ to $\mathbf{r}_1$. Thus, the equation

$$\mathbf{r} = (1-t)\mathbf{r}_0 + t\mathbf{r}_1 \qquad (0 \le t \le 1) \tag{7}$$

represents the line segment in 2-space or 3-space that is traced from $\mathbf{r}_0$ to $\mathbf{r}_1$.

✔ QUICK CHECK EXERCISES 13.1 (See page 865 for answers.)

1. (a) Express the parametric equations

$$x = \frac{1}{t}, \quad y = \sqrt{t}, \quad z = \sin^{-1} t$$

as a single vector equation of the form

$$\mathbf{r} = x(t)\mathbf{i} + y(t)\mathbf{j} + z(t)\mathbf{k}$$

(b) The vector equation in part (a) defines $\mathbf{r} = \mathbf{r}(t)$ as a vector-valued function. The domain of $\mathbf{r}(t)$ is ________ and $\mathbf{r}\left(\frac{1}{2}\right) =$ ________.

2. Describe the graph of $\mathbf{r}(t) = \langle 1+2t, -1+3t \rangle$.
3. Describe the graph of $\mathbf{r}(t) = \sin^2 t\mathbf{i} + \cos^2 t\mathbf{j}$.
4. Find a vector equation for the curve of intersection of the surfaces $y = x^2$ and $z = y$ in terms of the parameter $x = t$.

EXERCISE SET 13.1 Graphing Utility

1–4 Find the domain of $\mathbf{r}(t)$ and the value of $\mathbf{r}(t_0)$.

1. $\mathbf{r}(t) = \cos t\mathbf{i} - 3t\mathbf{j}; \ t_0 = \pi$
2. $\mathbf{r}(t) = \langle \sqrt{3t+1}, t^2 \rangle; \ t_0 = 1$
3. $\mathbf{r}(t) = \cos \pi t\mathbf{i} - \ln t\mathbf{j} + \sqrt{t-2}\mathbf{k}; \ t_0 = 3$
4. $\mathbf{r}(t) = \langle 2e^{-t}, \sin^{-1} t, \ln(1-t) \rangle; \ t_0 = 0$

5–8 Express the parametric equations as a single vector equation of the form

$$\mathbf{r} = x(t)\mathbf{i} + y(t)\mathbf{j} \quad \text{or} \quad \mathbf{r} = x(t)\mathbf{i} + y(t)\mathbf{j} + z(t)\mathbf{k}$$

5. $x = 3\cos t, \ y = t + \sin t$
6. $x = t^2 + 1, \ y = e^{-2t}$
7. $x = 2t, \ y = 2\sin 3t, \ z = 5\cos 3t$

8. $x = t\sin t,\ \ y = \ln t,\ \ z = \cos^2 t$

9–12 Find the parametric equations that correspond to the given vector equation.

9. $\mathbf{r} = 3t^2\mathbf{i} - 2\mathbf{j}$ **10.** $\mathbf{r} = \sin^2 t\mathbf{i} + (1 - \cos 2t)\mathbf{j}$

11. $\mathbf{r} = (2t - 1)\mathbf{i} - 3\sqrt{t}\,\mathbf{j} + \sin 3t\mathbf{k}$

12. $\mathbf{r} = te^{-t}\mathbf{i} - 5t^2\mathbf{k}$

13–18 Describe the graph of the equation.

13. $\mathbf{r} = (3 - 2t)\mathbf{i} + 5t\mathbf{j}$ **14.** $\mathbf{r} = 2\sin 3t\mathbf{i} - 2\cos 3t\,\mathbf{j}$

15. $\mathbf{r} = 2t\mathbf{i} - 3\mathbf{j} + (1 + 3t)\mathbf{k}$

16. $\mathbf{r} = 3\mathbf{i} + 2\cos t\,\mathbf{j} + 2\sin t\mathbf{k}$

17. $\mathbf{r} = 2\cos t\mathbf{i} - 3\sin t\,\mathbf{j} + \mathbf{k}$

18. $\mathbf{r} = -3\mathbf{i} + (1 - t^2)\mathbf{j} + t\mathbf{k}$

19. (a) Find the slope of the line in 2-space that is represented by the vector equation $\mathbf{r} = (1 - 2t)\mathbf{i} - (2 - 3t)\mathbf{j}$.
(b) Find the coordinates of the point where the line

$$\mathbf{r} = (2 + t)\mathbf{i} + (1 - 2t)\mathbf{j} + 3t\mathbf{k}$$

intersects the xz-plane.

20. (a) Find the y-intercept of the line in 2-space that is represented by the vector equation $\mathbf{r} = (3 + 2t)\mathbf{i} + 5t\mathbf{j}$.
(b) Find the coordinates of the point where the line

$$\mathbf{r} = t\mathbf{i} + (1 + 2t)\mathbf{j} - 3t\mathbf{k}$$

intersects the plane $3x - y - z = 2$.

21–22 Sketch the line segment represented by the vector equation.

21. (a) $\mathbf{r} = (1 - t)\mathbf{i} + t\mathbf{j};\ \ 0 \le t \le 1$
(b) $\mathbf{r} = (1 - t)(\mathbf{i} + \mathbf{j}) + t(\mathbf{i} - \mathbf{j});\ \ 0 \le t \le 1$

22. (a) $\mathbf{r} = (1 - t)(\mathbf{i} + \mathbf{j}) + t\mathbf{k};\ \ 0 \le t \le 1$
(b) $\mathbf{r} = (1 - t)(\mathbf{i} + \mathbf{j} + \mathbf{k}) + t(\mathbf{i} + \mathbf{j});\ \ 0 \le t \le 1$

23–24 Write a vector equation for the line segment from P to Q.

23. **24.**

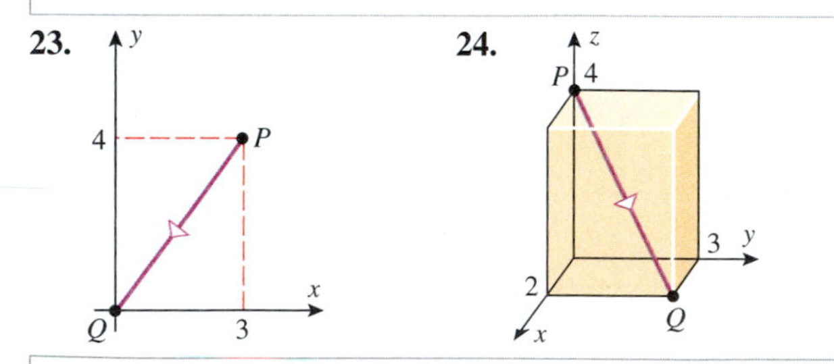

25–34 Sketch the graph of $\mathbf{r}(t)$ and show the direction of increasing t.

25. $\mathbf{r}(t) = 2\mathbf{i} + t\mathbf{j}$ **26.** $\mathbf{r}(t) = \langle 3t - 4, 6t + 2\rangle$

27. $\mathbf{r}(t) = (1 + \cos t)\mathbf{i} + (3 - \sin t)\mathbf{j};\ \ 0 \le t \le 2\pi$

28. $\mathbf{r}(t) = \langle 2\cos t, 5\sin t\rangle;\ \ 0 \le t \le 2\pi$

29. $\mathbf{r}(t) = \cosh t\mathbf{i} + \sinh t\,\mathbf{j}$ **30.** $\mathbf{r}(t) = \sqrt{t}\,\mathbf{i} + (2t + 4)\mathbf{j}$

31. $\mathbf{r}(t) = 2\cos t\mathbf{i} + 2\sin t\,\mathbf{j} + t\mathbf{k}$

32. $\mathbf{r}(t) = 9\cos t\mathbf{i} + 4\sin t\,\mathbf{j} + t\mathbf{k}$

33. $\mathbf{r}(t) = t\mathbf{i} + t^2\mathbf{j} + 2\mathbf{k}$

34. $\mathbf{r}(t) = t\mathbf{i} + t\mathbf{j} + \sin t\mathbf{k};\ \ 0 \le t \le 2\pi$

35–36 Sketch the curve of intersection of the surfaces, and find parametric equations for the intersection in terms of parameter $x = t$. Check your work with a graphing utility by generating the parametric curve over the interval $-1 \le t \le 1$.

35. $z = x^2 + y^2,\ \ x - y = 0$

36. $y + x = 0,\ \ z = \sqrt{2 - x^2 - y^2}$

37–38 Sketch the curve of intersection of the surfaces, and find a vector equation for the curve in terms of the parameter $x = t$.

37. $9x^2 + y^2 + 9z^2 = 81,\ \ y = x^2\ \ \ (z > 0)$

38. $y = x,\ \ x + y + z = 1$

39. Show that the graph of

$$\mathbf{r} = t\sin t\mathbf{i} + t\cos t\,\mathbf{j} + t^2\mathbf{k}$$

lies on the paraboloid $z = x^2 + y^2$.

40. Show that the graph of

$$\mathbf{r} = t\mathbf{i} + \frac{1 + t}{t}\mathbf{j} + \frac{1 - t^2}{t}\mathbf{k},\quad t > 0$$

lies in the plane $x - y + z + 1 = 0$.

FOCUS ON CONCEPTS

41. Show that the graph of

$$\mathbf{r} = \sin t\mathbf{i} + 2\cos t\,\mathbf{j} + \sqrt{3}\sin t\mathbf{k}$$

is a circle, and find its center and radius. [*Hint:* Show that the curve lies on both a sphere and a plane.]

42. Show that the graph of

$$\mathbf{r} = 3\cos t\mathbf{i} + 3\sin t\,\mathbf{j} + 3\sin t\mathbf{k}$$

is an ellipse, and find the lengths of the major and minor axes. [*Hint:* Show that the graph lies on both a circular cylinder and a plane and use the result in Exercise 60 of Section 11.4.]

43. For the helix $\mathbf{r} = a\cos t\mathbf{i} + a\sin t\,\mathbf{j} + ct\mathbf{k}$, find the value of c ($c > 0$) so that the helix will make one complete turn in a distance of 3 units measured along the z-axis.

44. How many revolutions will the circular helix

$$\mathbf{r} = a\cos t\mathbf{i} + a\sin t\,\mathbf{j} + 0.2t\mathbf{k}$$

make in a distance of 10 units measured along the z-axis?

45. Show that the curve $\mathbf{r} = t\cos t\mathbf{i} + t\sin t\,\mathbf{j} + t\mathbf{k}$, $t \ge 0$, lies on the cone $z = \sqrt{x^2 + y^2}$. Describe the curve.

46. Describe the curve $\mathbf{r} = a\cos t\mathbf{i} + b\sin t\,\mathbf{j} + ct\mathbf{k}$, where a, b, and c are positive constants such that $a \ne b$.

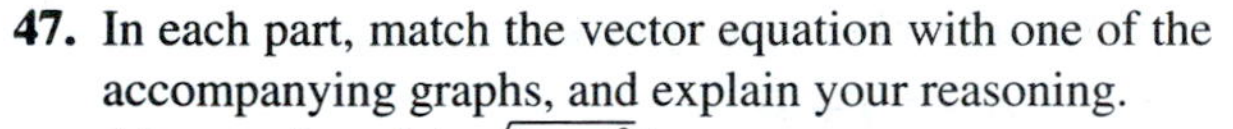

47. In each part, match the vector equation with one of the accompanying graphs, and explain your reasoning.

(a) $\mathbf{r} = t\mathbf{i} - t\mathbf{j} + \sqrt{2 - t^2}\,\mathbf{k}$

(b) $\mathbf{r} = \sin \pi t\mathbf{i} - t\mathbf{j} + t\mathbf{k}$

(c) $\mathbf{r} = \sin t\mathbf{i} + \cos t\,\mathbf{j} + \sin 2t\mathbf{k}$

(d) $\mathbf{r} = \frac{1}{2}t\mathbf{i} + \cos 3t\,\mathbf{j} + \sin 3t\mathbf{k}$

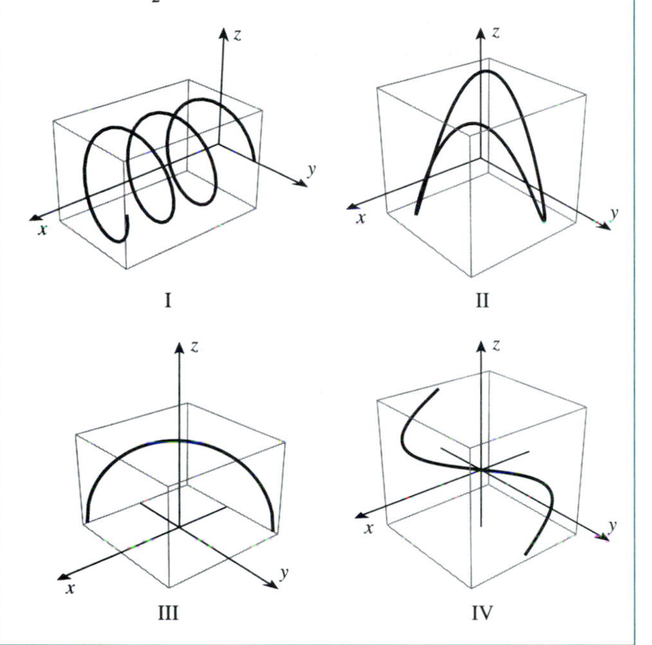

48. Check your conclusions in Exercise 47 by generating the curves with a graphing utility. [*Note:* Your graphing utility may look at the curve from a different viewpoint. Read the documentation for your graphing utility to determine how to control the viewpoint, and see if you can generate a reasonable facsimile of the graphs shown in the figure by adjusting the viewpoint and choosing the interval of t-values appropriately.]

49. (a) Find parametric equations for the curve of intersection of the circular cylinder $x^2 + y^2 = 9$ and the parabolic cylinder $z = x^2$ in terms of a parameter t for which $x = 3\cos t$.

(b) Use a graphing utility to generate the curve of intersection in part (a).

50. Use a graphing utility to generate the intersection of the cone $z = \sqrt{x^2 + y^2}$ and the plane $z = y + 2$. Identify the curve and explain your reasoning.

51. (a) Sketch the graph of

$$\mathbf{r}(t) = \left\langle 2t, \frac{2}{1+t^2} \right\rangle$$

(b) Prove that the curve in part (a) is also the graph of the function

$$y = \frac{8}{4 + x^2}$$

[The graphs of $y = a^3/(a^2 + x^2)$, where a denotes a constant, were first studied by the French mathematician Pierre de Fermat, and later by the Italian mathematicians Guido Grandi and Maria Agnesi. Any such curve is now known as a "witch of Agnesi." There are a number of theories for the origin of this name. Some suggest there was a mistranslation by either Grandi or Agnesi of some less colorful Latin name into Italian. Others lay the blame on a translation into English of Agnesi's 1748 treatise, *Analytical Institutions*.]

✔ QUICK CHECK ANSWERS 13.1

1. (a) $\mathbf{r} = \frac{1}{t}\mathbf{i} + \sqrt{t}\mathbf{j} + \sin^{-1} t\mathbf{k}$ (b) $0 < t \leq 1$; $2\mathbf{i} + \frac{\sqrt{2}}{2}\mathbf{j} + \frac{\pi}{6}\mathbf{k}$ **2.** The graph is a line through $(1, -1)$ with direction vector $2\mathbf{i} + 3\mathbf{j}$. **3.** The graph is the line segment in the xy-plane from $(0, 1)$ to $(1, 0)$. **4.** $\mathbf{r} = \langle t, t^2, t^2 \rangle$

13.2 CALCULUS OF VECTOR-VALUED FUNCTIONS

In this section we will define limits, derivatives, and integrals of vector-valued functions and discuss their properties.

LIMITS AND CONTINUITY

Our first goal in this section is to develop a notion of what it means for a vector-valued function $\mathbf{r}(t)$ in 2-space or 3-space to approach a limiting vector $\mathbf{L}$ as t approaches a number a. That is, we want to define

$$\lim_{t \to a} \mathbf{r}(t) = \mathbf{L} \tag{1}$$

One way to motivate a reasonable definition of (1) is to position $\mathbf{r}(t)$ and $\mathbf{L}$ with their initial points at the origin and interpret this limit to mean that the terminal point of $\mathbf{r}(t)$ approaches

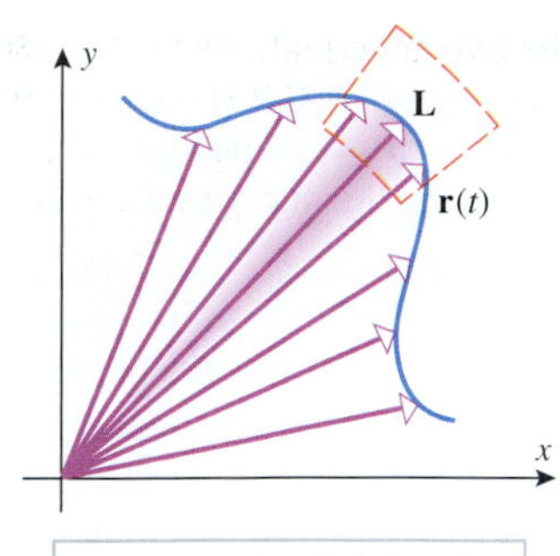

$\mathbf{r}(t)$ approaches $\mathbf{L}$ in length and direction if $\lim_{t \to a} \mathbf{r}(t) = \mathbf{L}$.

Figure 13.2.1

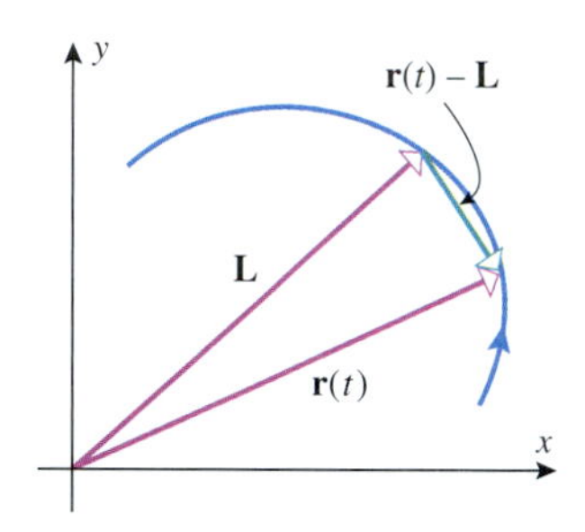

$\|\mathbf{r}(t) - \mathbf{L}\|$ is the distance between terminal points for vectors $\mathbf{r}(t)$ and $\mathbf{L}$ when positioned with the same initial points.

Figure 13.2.2

Note that $\|\mathbf{r}(t) - \mathbf{L}\|$ is a real number for each value of t, so even though this expression involves a vector-valued function, the limit

$$\lim_{t \to a} \|\mathbf{r}(t) - \mathbf{L}\|$$

is an ordinary limit of a real-valued function.

How would you define the one-sided limits

$$\lim_{t \to a^+} \mathbf{r}(t) \quad \text{and} \quad \lim_{t \to a^-} \mathbf{r}(t)?$$

Limits of vector-valued functions have many of the same properties as limits of real-valued functions. For example, assuming that the limits exist, the limit of a sum is the sum of the limits, the limit of a difference is the difference of the limits, and a constant scalar factor can be moved through a limit symbol.

the terminal point of $\mathbf{L}$ as t approaches a or, equivalently, that the vector $\mathbf{r}(t)$ approaches the vector $\mathbf{L}$ in both length and direction at t approaches a (Figure 13.2.1). Algebraically, this is equivalent to stating that

$$\lim_{t \to a} \|\mathbf{r}(t) - \mathbf{L}\| = 0 \tag{2}$$

(Figure 13.2.2). Thus, we make the following definition.

13.2.1 DEFINITION. Let $\mathbf{r}(t)$ be a vector-valued function that is defined for all t in some open interval containing the number a, except that $\mathbf{r}(t)$ need not be defined at a. We will write

$$\lim_{t \to a} \mathbf{r}(t) = \mathbf{L}$$

if and only if

$$\lim_{t \to a} \|\mathbf{r}(t) - \mathbf{L}\| = 0$$

It is clear intuitively that $\mathbf{r}(t)$ will approach a limiting vector $\mathbf{L}$ as t approaches a if and only if the component functions of $\mathbf{r}(t)$ approach the corresponding components of $\mathbf{L}$. This suggests the following theorem, whose formal proof is omitted.

13.2.2 THEOREM.

(*a*) *If* $\mathbf{r}(t) = \langle x(t), y(t) \rangle = x(t)\mathbf{i} + y(t)\mathbf{j}$, *then*

$$\lim_{t \to a} \mathbf{r}(t) = \left\langle \lim_{t \to a} x(t), \lim_{t \to a} y(t) \right\rangle = \lim_{t \to a} x(t)\mathbf{i} + \lim_{t \to a} y(t)\mathbf{j}$$

provided the limits of the component functions exist. Conversely, the limits of the component functions exist provided $\mathbf{r}(t)$ *approaches a limiting vector as* t *approaches* a.

(*b*) *If* $\mathbf{r}(t) = \langle x(t), y(t), z(t) \rangle = x(t)\mathbf{i} + y(t)\mathbf{j} + z(t)\mathbf{k}$, *then*

$$\begin{aligned} \lim_{t \to a} \mathbf{r}(t) &= \left\langle \lim_{t \to a} x(t), \lim_{t \to a} y(t), \lim_{t \to a} z(t) \right\rangle \\ &= \lim_{t \to a} x(t)\mathbf{i} + \lim_{t \to a} y(t)\mathbf{j} + \lim_{t \to a} z(t)\mathbf{k} \end{aligned}$$

provided the limits of the component functions exist. Conversely, the limits of the component functions exist provided $\mathbf{r}(t)$ *approaches a limiting vector as* t *approaches* a.

▶ **Example 1** Let $\mathbf{r}(t) = t^2\mathbf{i} + e^t\mathbf{j} - (2\cos \pi t)\mathbf{k}$. Then

$$\lim_{t \to 0} \mathbf{r}(t) = \left(\lim_{t \to 0} t^2\right)\mathbf{i} + \left(\lim_{t \to 0} e^t\right)\mathbf{j} - \left(\lim_{t \to 0} 2\cos \pi t\right)\mathbf{k} = \mathbf{j} - 2\mathbf{k}$$

Alternatively, using the angle bracket notation for vectors,

$$\lim_{t \to 0} \mathbf{r}(t) = \lim_{t \to 0} \langle t^2, e^t, -2\cos \pi t \rangle = \left\langle \lim_{t \to 0} t^2, \lim_{t \to 0} e^t, \lim_{t \to 0} (-2\cos \pi t) \right\rangle = \langle 0, 1, -2 \rangle$$ ◀

Motivated by the definition of continuity for real-valued functions, we define a vector-valued function $\mathbf{r}(t)$ to be ***continuous*** at $t = a$ if

$$\lim_{t \to a} \mathbf{r}(t) = \mathbf{r}(a) \tag{3}$$

That is, $\mathbf{r}(a)$ is defined, the limit of $\mathbf{r}(t)$ as $t \to a$ exists, and the two are equal. As in the case for real-valued functions, we say that $\mathbf{r}(t)$ is ***continuous on an interval*** I if it is continuous at each point of I [with the understanding that at an endpoint in I the two-sided limit in (3) is replaced by the appropriate one-sided limit]. It follows from Theorem 13.2.2 that a vector-valued function is continuous at $t = a$ if and only if its component functions are continuous at $t = a$.

DERIVATIVES

The derivative of a vector-valued function is defined by a limit similar to that for the derivative of a real-valued function.

13.2.3 DEFINITION. If $\mathbf{r}(t)$ is a vector-valued function, we define the ***derivative of*** $\mathbf{r}$ ***with respect to t*** to be the vector-valued function $\mathbf{r}'$ given by

$$\mathbf{r}'(t) = \lim_{h \to 0} \frac{\mathbf{r}(t+h) - \mathbf{r}(t)}{h} \tag{4}$$

The domain of $\mathbf{r}'$ consists of all values of t in the domain of $\mathbf{r}(t)$ for which the limit exists.

The function $\mathbf{r}(t)$ is ***differentiable*** at t if the limit in (4) exists. All of the standard notations for derivatives continue to apply. For example, the derivative of $\mathbf{r}(t)$ can be expressed as

$$\frac{d}{dt}[\mathbf{r}(t)], \quad \frac{d\mathbf{r}}{dt}, \quad \mathbf{r}'(t), \quad \text{or} \quad \mathbf{r}'$$

It is important to keep in mind that $\mathbf{r}'(t)$ is a vector, not a number, and hence has a magnitude and a direction for each value of t [except if $\mathbf{r}'(t) = \mathbf{0}$, in which case $\mathbf{r}'(t)$ has magnitude zero but no specific direction]. In the next section we will consider the significance of the magnitude of $\mathbf{r}'(t)$, but for now our goal is to obtain a geometric interpretation of the direction of $\mathbf{r}'(t)$. For this purpose, consider parts (a) and (b) of Figure 13.2.3. These illustrations show the graph C of $\mathbf{r}(t)$ (with its orientation) and the vectors $\mathbf{r}(t)$, $\mathbf{r}(t+h)$, and $\mathbf{r}(t+h) - \mathbf{r}(t)$ for positive h and for negative h. In both cases, the vector $\mathbf{r}(t+h) - \mathbf{r}(t)$ runs along the secant line joining the terminal points of $\mathbf{r}(t+h)$ and $\mathbf{r}(t)$, but with opposite directions in the two cases. In the case where h is positive the vector $\mathbf{r}(t+h) - \mathbf{r}(t)$ points in the direction of increasing parameter, and in the case where h is negative it points in the opposite direction. However, in the case where h is negative the direction gets reversed when we multiply by $1/h$, so in both cases the vector

$$\frac{1}{h}[\mathbf{r}(t+h) - \mathbf{r}(t)] = \frac{\mathbf{r}(t+h) - \mathbf{r}(t)}{h}$$

points in the direction of increasing parameter and runs along the secant line. As $h \to 0$, the secant line approaches the tangent line at the terminal point of $\mathbf{r}(t)$, so we can conclude that the limit

$$\mathbf{r}'(t) = \lim_{h \to 0} \frac{\mathbf{r}(t+h) - \mathbf{r}(t)}{h}$$

(if it exists and is nonzero) is a vector that is tangent to the curve C at the tip of $\mathbf{r}(t)$ and points in the direction of increasing parameter (Figure 13:2.3).

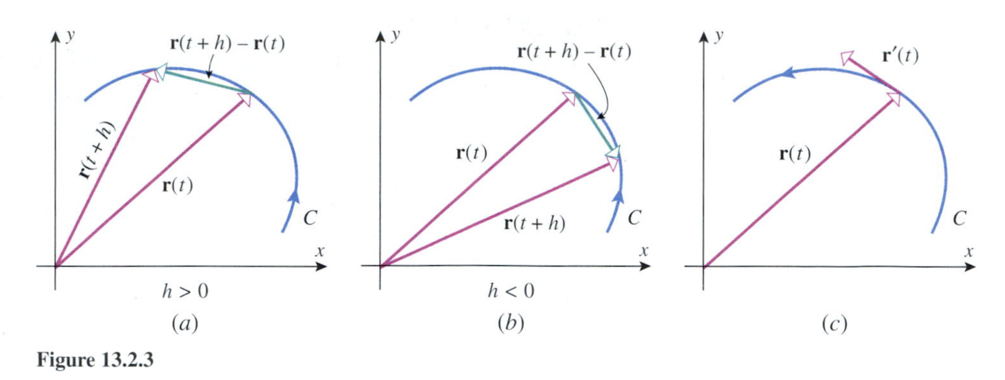

Figure 13.2.3

We can summarize all of this as follows.

13.2.4 GEOMETRIC INTERPRETATION OF THE DERIVATIVE. Suppose that C is the graph of a vector-valued function $\mathbf{r}(t)$ in 2-space or 3-space and that $\mathbf{r}'(t)$ exists and is nonzero for a given value of t. If the vector $\mathbf{r}'(t)$ is positioned with its initial point at the terminal point of the radius vector $\mathbf{r}(t)$, then $\mathbf{r}'(t)$ is tangent to C and points in the direction of increasing parameter.

Since limits of vector-valued functions can be computed componentwise, it seems reasonable that we should be able to compute derivatives in terms of component functions as well. This is the result of the next theorem.

13.2.5 THEOREM. *If $\mathbf{r}(t)$ is a vector-valued function, then $\mathbf{r}$ is differentiable at t if and only if each of its component functions is differentiable at t, in which case the component functions of $\mathbf{r}'(t)$ are the derivatives of the corresponding component functions of $\mathbf{r}(t)$.*

PROOF. For simplicity, we give the proof in 2-space; the proof in 3-space is identical, except for the additional component. Assume that $\mathbf{r}(t) = x(t)\mathbf{i} + y(t)\mathbf{j}$. Then

$$\begin{aligned}
\mathbf{r}'(t) &= \lim_{h\to 0} \frac{\mathbf{r}(t+h) - \mathbf{r}(t)}{h} \\
&= \lim_{h\to 0} \frac{[x(t+h)\mathbf{i} + y(t+h)\mathbf{j}] - [x(t)\mathbf{i} + y(t)\mathbf{j}]}{h} \\
&= \left(\lim_{h\to 0} \frac{x(t+h) - x(t)}{h}\right)\mathbf{i} + \left(\lim_{h\to 0} \frac{y(t+h) - y(t)}{h}\right)\mathbf{j} \\
&= x'(t)\mathbf{i} + y'(t)\mathbf{j}
\end{aligned}$$

■

Example 2 Let $\mathbf{r}(t) = t^2\mathbf{i} + e^t\mathbf{j} - (2\cos \pi t)\mathbf{k}$. Then

$$\begin{aligned}
\mathbf{r}'(t) &= \frac{d}{dt}(t^2)\mathbf{i} + \frac{d}{dt}(e^t)\mathbf{j} - \frac{d}{dt}(2\cos \pi t)\mathbf{k} \\
&= 2t\mathbf{i} + e^t\mathbf{j} + (2\pi \sin \pi t)\mathbf{k}
\end{aligned}$$

◂

DERIVATIVE RULES

Many of the rules for differentiating real-valued functions have analogs in the context of differentiating vector-valued functions. We state some of these in the following theorem.

13.2.6 THEOREM (*Rules of Differentiation*). *Let $\mathbf{r}(t)$, $\mathbf{r}_1(t)$, and $\mathbf{r}_2(t)$ be vector-valued functions that are all in 2-space or all in 3-space, and let $f(t)$ be a real-valued function, k a scalar, and $\mathbf{c}$ a constant vector (that is, a vector whose value does not depend on t). Then the following rules of differentiation hold:*

(a) $\dfrac{d}{dt}[\mathbf{c}] = \mathbf{0}$

(b) $\dfrac{d}{dt}[k\mathbf{r}(t)] = k\dfrac{d}{dt}[\mathbf{r}(t)]$

(c) $\dfrac{d}{dt}[\mathbf{r}_1(t) + \mathbf{r}_2(t)] = \dfrac{d}{dt}[\mathbf{r}_1(t)] + \dfrac{d}{dt}[\mathbf{r}_2(t)]$

(d) $\dfrac{d}{dt}[\mathbf{r}_1(t) - \mathbf{r}_2(t)] = \dfrac{d}{dt}[\mathbf{r}_1(t)] - \dfrac{d}{dt}[\mathbf{r}_2(t)]$

(e) $\dfrac{d}{dt}[f(t)\mathbf{r}(t)] = f(t)\dfrac{d}{dt}[\mathbf{r}(t)] + \dfrac{d}{dt}[f(t)]\mathbf{r}(t)$

The proofs of most of these rules are immediate consequences of Definition 13.2.3, although the last rule can be seen more easily by application of the product rule for real-valued functions to the component functions. The proof of Theorem 13.2.6 is left as an exercise.

TANGENT LINES TO GRAPHS OF VECTOR-VALUED FUNCTIONS

Motivated by the discussion of the geometric interpretation of the derivative of a vector-valued function, we make the following definition.

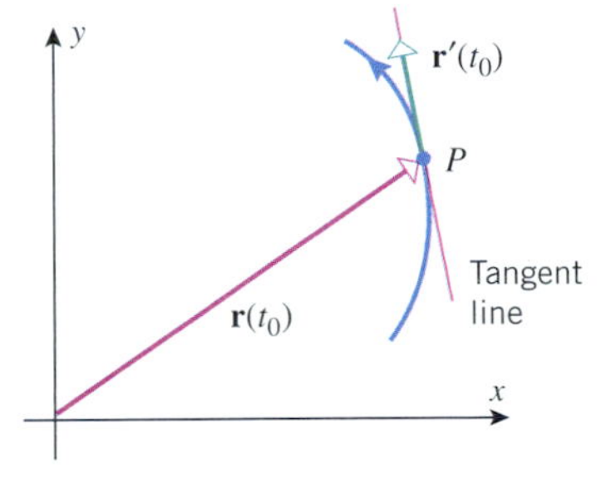

Figure 13.2.4

13.2.7 DEFINITION. Let P be a point on the graph of a vector-valued function $\mathbf{r}(t)$, and let $\mathbf{r}(t_0)$ be the radius vector from the origin to P (Figure 13.2.4). If $\mathbf{r}'(t_0)$ exists and $\mathbf{r}'(t_0) \neq \mathbf{0}$, then we call $\mathbf{r}'(t_0)$ a ***tangent vector*** to the graph of $\mathbf{r}(t)$ at $\mathbf{r}(t_0)$, and we call the line through P that is parallel to the tangent vector the ***tangent line*** to the graph of $\mathbf{r}(t)$ at $\mathbf{r}(t_0)$.

Let $\mathbf{r}_0 = \mathbf{r}(t_0)$ and $\mathbf{v}_0 = \mathbf{r}'(t_0)$. It follows from Formula (9) of Section 12.5 that the tangent line to the graph of $\mathbf{r}(t)$ at $\mathbf{r}_0$ is given by the vector equation

$$\mathbf{r} = \mathbf{r}_0 + t\mathbf{v}_0 \tag{5}$$

Example 3 Find parametric equations of the tangent line to the circular helix

$$x = \cos t, \quad y = \sin t, \quad z = t$$

where $t = t_0$, and use that result to find parametric equations for the tangent line at the point where $t = \pi$.

Solution. The vector equation of the helix is

$$\mathbf{r}(t) = \cos t\mathbf{i} + \sin t\mathbf{j} + t\mathbf{k}$$

so we have

$$\mathbf{r}_0 = \mathbf{r}(t_0) = \cos t_0\mathbf{i} + \sin t_0\mathbf{j} + t_0\mathbf{k}$$

$$\mathbf{v}_0 = \mathbf{r}'(t_0) = (-\sin t_0)\mathbf{i} + \cos t_0\mathbf{j} + \mathbf{k}$$

It follows from (5) that the vector equation of the tangent line at $t = t_0$ is

$$\begin{aligned}\mathbf{r} &= \cos t_0\mathbf{i} + \sin t_0\mathbf{j} + t_0\mathbf{k} + t[(-\sin t_0)\mathbf{i} + \cos t_0\mathbf{j} + \mathbf{k}]\\ &= (\cos t_0 - t\sin t_0)\mathbf{i} + (\sin t_0 + t\cos t_0)\mathbf{j} + (t_0 + t)\mathbf{k}\end{aligned}$$

Thus, the parametric equations of the tangent line at $t = t_0$ are

$$x = \cos t_0 - t\sin t_0, \quad y = \sin t_0 + t\cos t_0, \quad z = t_0 + t$$

In particular, the tangent line at the point where $t = \pi$ has parametric equations

$$x = -1, \quad y = -t, \quad z = \pi + t$$

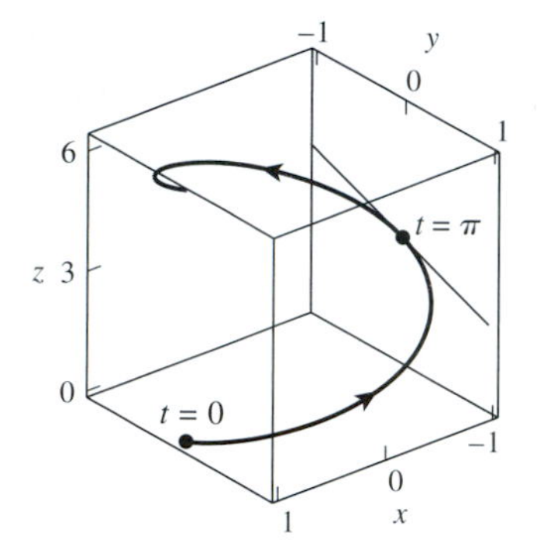

Figure 13.2.5

The graph of the helix and this tangent line are shown in Figure 13.2.5. ◀

▶ Example 4 Let

$$\mathbf{r}_1(t) = (\tan^{-1} t)\mathbf{i} + (\sin t)\mathbf{j} + t^2\mathbf{k}$$

and

$$\mathbf{r}_2(t) = (t^2 - t)\mathbf{i} + (2t - 2)\mathbf{j} + (\ln t)\mathbf{k}$$

The graphs of $\mathbf{r}_1(t)$ and $\mathbf{r}_2(t)$ intersect at the origin. Find the degree measure of the acute angle between the tangent lines to the graphs of $\mathbf{r}_1(t)$ and $\mathbf{r}_2(t)$ at the origin.

Solution. The graph of $\mathbf{r}_1(t)$ passes through the origin at $t = 0$, where its tangent vector is

$$\mathbf{r}_1'(0) = \left\langle \frac{1}{1+t^2}, \cos t, 2t \right\rangle\bigg|_{t=0} = \langle 1, 1, 0\rangle$$

The graph of $\mathbf{r}_2(t)$ passes through the origin at $t = 1$ (verify), where its tangent vector is

$$\mathbf{r}_2'(1) = \left\langle 2t - 1, 2, \frac{1}{t} \right\rangle\bigg|_{t=1} = \langle 1, 2, 1\rangle$$

By Theorem 12.3.3, the angle θ between these two tangent vectors satisfies

$$\cos\theta = \frac{\langle 1, 1, 0\rangle \cdot \langle 1, 2, 1\rangle}{\|\langle 1, 1, 0\rangle\| \, \|\langle 1, 2, 1\rangle\|} = \frac{3}{\sqrt{12}} = \frac{\sqrt{3}}{2}$$

It follows that $\theta = \pi/6$ radians, or $30°$. ◀

DERIVATIVES OF DOT AND CROSS PRODUCTS

The following rules, which are derived in the exercises, provide a method for differentiating dot products in 2-space and 3-space and cross products in 3-space.

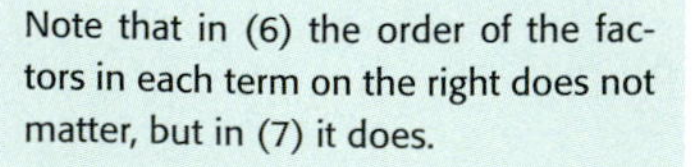
Note that in (6) the order of the factors in each term on the right does not matter, but in (7) it does.

$$\frac{d}{dt}[\mathbf{r}_1(t) \cdot \mathbf{r}_2(t)] = \mathbf{r}_1(t) \cdot \frac{d\mathbf{r}_2}{dt} + \frac{d\mathbf{r}_1}{dt} \cdot \mathbf{r}_2(t) \tag{6}$$

$$\frac{d}{dt}[\mathbf{r}_1(t) \times \mathbf{r}_2(t)] = \mathbf{r}_1(t) \times \frac{d\mathbf{r}_2}{dt} + \frac{d\mathbf{r}_1}{dt} \times \mathbf{r}_2(t) \tag{7}$$

In plane geometry one learns that a tangent line to a circle is perpendicular to the radius at the point of tangency. Consequently, if a point moves along a circle in 2-space that is centered at the origin, then one would expect the radius vector and the tangent vector at any point on the circle to be orthogonal. This is the motivation for the following useful theorem, which is applicable in both 2-space and 3-space.

13.2.8 THEOREM. *If* $\mathbf{r}(t)$ *is a vector-valued function in 2-space or 3-space and* $\|\mathbf{r}(t)\|$ *is constant for all* t*, then*

$$\mathbf{r}(t) \cdot \mathbf{r}'(t) = 0 \tag{8}$$

that is, $\mathbf{r}(t)$ *and* $\mathbf{r}'(t)$ *are orthogonal vectors for all* t*.*

PROOF. It follows from (6) with $\mathbf{r}_1(t) = \mathbf{r}_2(t) = \mathbf{r}(t)$ that

$$\frac{d}{dt}[\mathbf{r}(t) \cdot \mathbf{r}(t)] = \mathbf{r}(t) \cdot \frac{d\mathbf{r}}{dt} + \frac{d\mathbf{r}}{dt} \cdot \mathbf{r}(t)$$

or, equivalently,

$$\frac{d}{dt}[\|\mathbf{r}(t)\|^2] = 2\mathbf{r}(t) \cdot \frac{d\mathbf{r}}{dt} \tag{9}$$

But $\|\mathbf{r}(t)\|^2$ is constant, so its derivative is zero. Thus

$$2\mathbf{r}(t) \cdot \frac{d\mathbf{r}}{dt} = 0$$

from which (8) follows. ■

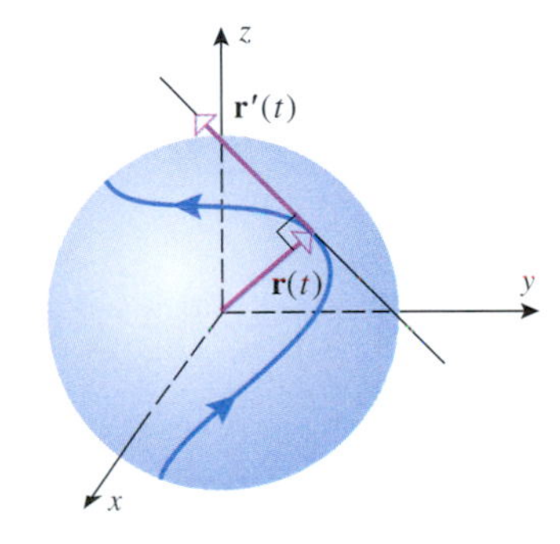

Figure 13.2.6

▶ **Example 5** Just as a tangent line to a circle in 2-space is perpendicular to the radius at the point of tangency, so a tangent vector to a curve on the surface of a sphere in 3-space that is centered at the origin is orthogonal to the radius vector at the point of tangency (Figure 13.2.6). To see that this is so, suppose that the graph of $\mathbf{r}(t)$ lies on the surface of a sphere of positive radius k centered at the origin. For each value of t we have $\|\mathbf{r}(t)\| = k$, so by Theorem 13.2.8

$$\mathbf{r}(t) \cdot \mathbf{r}'(t) = 0$$

and hence the radius vector $\mathbf{r}(t)$ and the tangent vector $\mathbf{r}'(t)$ are orthogonal. ◀

DEFINITE INTEGRALS OF VECTOR-VALUED FUNCTIONS

If $\mathbf{r}(t)$ is a vector-valued function that is continuous on the interval $a \le t \le b$, then we define the ***definite integral*** of $\mathbf{r}(t)$ over this interval as a limit of Riemann sums, just as in Definition 6.5.1, except here the integrand is a vector-valued function. Specifically, we define

$$\int_a^b \mathbf{r}(t)\,dt = \lim_{\max \Delta t_k \to 0} \sum_{k=1}^{n} \mathbf{r}(t_k^*)\Delta t_k \tag{10}$$

It follows from (10) that the definite integral of $\mathbf{r}(t)$ over the interval $a \le t \le b$ can be expressed as a vector whose components are the definite integrals of the component functions of $\mathbf{r}(t)$. For example, if $\mathbf{r}(t) = x(t)\mathbf{i} + y(t)\mathbf{j}$, then

$$\begin{aligned}
\int_a^b \mathbf{r}(t)\,dt &= \lim_{\max \Delta t_k \to 0} \sum_{k=1}^{n} \mathbf{r}(t_k^*)\Delta t_k \\
&= \lim_{\max \Delta t_k \to 0} \left[\left(\sum_{k=1}^{n} x(t_k^*)\Delta t_k \right)\mathbf{i} + \left(\sum_{k=1}^{n} y(t_k^*)\Delta t_k \right)\mathbf{j} \right] \\
&= \left(\lim_{\max \Delta t_k \to 0} \sum_{k=1}^{n} x(t_k^*)\Delta t_k \right)\mathbf{i} + \left(\lim_{\max \Delta t_k \to 0} \sum_{k=1}^{n} y(t_k^*)\Delta t_k \right)\mathbf{j} \\
&= \left(\int_a^b x(t)\,dt \right)\mathbf{i} + \left(\int_a^b y(t)\,dt \right)\mathbf{j}
\end{aligned}$$

In general, we have

$$\int_a^b \mathbf{r}(t)\,dt = \left(\int_a^b x(t)\,dt \right)\mathbf{i} + \left(\int_a^b y(t)\,dt \right)\mathbf{j} \qquad \text{2-space} \qquad (11)$$

$$\int_a^b \mathbf{r}(t)\,dt = \left(\int_a^b x(t)\,dt \right)\mathbf{i} + \left(\int_a^b y(t)\,dt \right)\mathbf{j} + \left(\int_a^b z(t)\,dt \right)\mathbf{k} \qquad \text{3-space} \qquad (12)$$

Rewrite Formulas (11) and (12) in bracket notation with

$$\mathbf{r}(t) = \langle x(t), y(t) \rangle$$

and

$$\mathbf{r}(t) = \langle x(t), y(t), z(t) \rangle$$

respectively.

► **Example 6** Let $\mathbf{r}(t) = t^2\mathbf{i} + e^t\mathbf{j} - (2\cos \pi t)\mathbf{k}$. Then

$$\begin{aligned}
\int_0^1 \mathbf{r}(t)\,dt &= \left(\int_0^1 t^2\,dt \right)\mathbf{i} + \left(\int_0^1 e^t\,dt \right)\mathbf{j} - \left(\int_0^1 2\cos \pi t\,dt \right)\mathbf{k} \\
&= \frac{t^3}{3}\bigg]_0^1 \mathbf{i} + e^t\bigg]_0^1 \mathbf{j} - \frac{2}{\pi}\sin \pi t\bigg]_0^1 \mathbf{k} = \frac{1}{3}\mathbf{i} + (e-1)\mathbf{j}
\end{aligned}$$ ◄

RULES OF INTEGRATION

As with differentiation, many of the rules for integrating real-valued functions have analogs for vector-valued functions.

13.2.9 THEOREM (*Rules of Integration*). *Let $\mathbf{r}(t)$, $\mathbf{r}_1(t)$, and $\mathbf{r}_2(t)$ be vector-valued functions in 2-space or 3-space that are continuous on the interval $a \le t \le b$, and let k be a scalar. Then the following rules of integration hold:*

(*a*) $$\int_a^b k\mathbf{r}(t)\,dt = k\int_a^b \mathbf{r}(t)\,dt$$

(*b*) $$\int_a^b [\mathbf{r}_1(t) + \mathbf{r}_2(t)]\,dt = \int_a^b \mathbf{r}_1(t)\,dt + \int_a^b \mathbf{r}_2(t)\,dt$$

(*c*) $$\int_a^b [\mathbf{r}_1(t) - \mathbf{r}_2(t)]\,dt = \int_a^b \mathbf{r}_1(t)\,dt - \int_a^b \mathbf{r}_2(t)\,dt$$

We omit the proof.

ANTIDERIVATIVES OF VECTOR-VALUED FUNCTIONS

An ***antiderivative*** for a vector-valued function $\mathbf{r}(t)$ is a vector-valued function $\mathbf{R}(t)$ such that

$$\mathbf{R}'(t) = \mathbf{r}(t) \tag{13}$$

As in Chapter 6, we express Equation (13) using integral notation as

$$\int \mathbf{r}(t)\,dt = \mathbf{R}(t) + \mathbf{C} \tag{14}$$

where $\mathbf{C}$ represents an arbitrary constant *vector*.

Since differentiation of vector-valued functions can be performed componentwise, it follows that antidifferentiation can be done this way as well. This is illustrated in the next example.

Example 7

$$\begin{aligned}\int (2t\mathbf{i} + 3t^2\mathbf{j})\,dt &= \left(\int 2t\,dt\right)\mathbf{i} + \left(\int 3t^2\,dt\right)\mathbf{j}\\ &= (t^2 + C_1)\mathbf{i} + (t^3 + C_2)\mathbf{j}\\ &= (t^2\mathbf{i} + t^3\mathbf{j}) + (C_1\mathbf{i} + C_2\mathbf{j}) = (t^2\mathbf{i} + t^3\mathbf{j}) + \mathbf{C}\end{aligned}$$

where $\mathbf{C} = C_1\mathbf{i} + C_2\mathbf{j}$ is an arbitrary vector constant of integration. ◄

Most of the familiar integration properties have vector counterparts. For example, vector differentiation and integration are inverse operations in the sense that

$$\frac{d}{dt}\left[\int \mathbf{r}(t)\,dt\right] = \mathbf{r}(t) \quad \text{and} \quad \int \mathbf{r}'(t)\,dt = \mathbf{r}(t) + \mathbf{C} \tag{15–16}$$

Moreover, if $\mathbf{R}(t)$ is an antiderivative of $\mathbf{r}(t)$ on an interval containing $t = a$ and $t = b$, then we have the following vector form of the Fundamental Theorem of Calculus:

$$\int_a^b \mathbf{r}(t)\,dt = \mathbf{R}(t)\Big]_a^b = \mathbf{R}(b) - \mathbf{R}(a) \tag{17}$$

Example 8 Evaluate the definite integral $\int_0^2 (2t\mathbf{i} + 3t^2\mathbf{j})\,dt$.

Solution. Integrating the components yields

$$\int_0^2 (2t\mathbf{i} + 3t^2\mathbf{j})\,dt = t^2\Big]_0^2\mathbf{i} + t^3\Big]_0^2\mathbf{j} = 4\mathbf{i} + 8\mathbf{j}$$

Alternative Solution. The function $\mathbf{R}(t) = t^2\mathbf{i} + t^3\mathbf{j}$ is an antiderivative of the integrand since $\mathbf{R}'(t) = 2t\mathbf{i} + 3t^2\mathbf{j}$. Thus, it follows from (17) that

$$\int_0^2 (2t\mathbf{i} + 3t^2\mathbf{j})\,dt = \mathbf{R}(t)\Big]_0^2 = t^2\mathbf{i} + t^3\mathbf{j}\Big]_0^2 = (4\mathbf{i} + 8\mathbf{j}) - (0\mathbf{i} + 0\mathbf{j}) = 4\mathbf{i} + 8\mathbf{j}$$ ◄

Example 9 Find $\mathbf{r}(t)$ given that $\mathbf{r}'(t) = \langle 3, 2t\rangle$ and $\mathbf{r}(1) = \langle 2, 5\rangle$.

Solution. Integrating $\mathbf{r}'(t)$ to obtain $\mathbf{r}(t)$ yields

$$\mathbf{r}(t) = \int \mathbf{r}'(t)\,dt = \int \langle 3, 2t\rangle\,dt = \langle 3t, t^2\rangle + \mathbf{C}$$

where $\mathbf{C}$ is a vector constant of integration. To find the value of $\mathbf{C}$ we substitute $t = 1$ and use the given value of $\mathbf{r}(1)$ to obtain

$$\mathbf{r}(1) = \langle 3, 1\rangle + \mathbf{C} = \langle 2, 5\rangle$$

so that $\mathbf{C} = \langle -1, 4\rangle$. Thus,

$$\mathbf{r}(t) = \langle 3t, t^2\rangle + \langle -1, 4\rangle = \langle 3t-1, t^2+4\rangle$$ ◄

✔ QUICK CHECK EXERCISES 13.2 (See page 876 for answers.)

1. (a) $\lim_{t\to 3}(t^2\mathbf{i} + 2t\mathbf{j}) =$ ______
   (b) $\lim_{t\to \pi/4}\langle \cos t, \sin t\rangle =$ ______
2. Find $\mathbf{r}'(t)$.
   (a) $\mathbf{r}(t) = (4+5t)\mathbf{i} + (t - t^2)\mathbf{j}$
   (b) $\mathbf{r}(t) = \left\langle \frac{1}{t}, \tan t, e^{2t}\right\rangle$
3. Suppose that $\mathbf{r}_1(0) = \langle 3, 2, 1\rangle$, $\mathbf{r}_2(0) = \langle 1, 2, 3\rangle$, $\mathbf{r}_1'(0) = \langle 0, 0, 0\rangle$, and $\mathbf{r}_2'(0) = \langle -6, -4, -2\rangle$. Use this information to evaluate the derivative of each function at $t = 0$.
   (a) $\mathbf{r}(t) = 2\mathbf{r}_1(t) - \mathbf{r}_2(t)$
   (b) $\mathbf{r}(t) = \cos t\,\mathbf{r}_1(t) + e^{2t}\mathbf{r}_2(t)$
   (c) $\mathbf{r}(t) = \mathbf{r}_1(t) \times \mathbf{r}_2(t)$
   (d) $f(t) = \mathbf{r}_1(t)\cdot \mathbf{r}_2(t)$
4. (a) $\int_0^1 \langle 2t, t^2, \sin \pi t\rangle\,dt =$ ______
   (b) $\int (t\mathbf{i} - 3t^2\mathbf{j} + e^t\mathbf{k})\,dt =$ ______

EXERCISE SET 13.2 Graphing Utility

1–4 Find the limit.

1. $\lim_{t\to+\infty}\left\langle \frac{t^2+1}{3t^2+2}, \frac{1}{t}\right\rangle$
2. $\lim_{t\to 0^+}\left(\sqrt{t}\,\mathbf{i} + \frac{\sin t}{t}\mathbf{j}\right)$
3. $\lim_{t\to 2}(t\mathbf{i} - 3\mathbf{j} + t^2\mathbf{k})$
4. $\lim_{t\to 1}\left\langle \frac{3}{t^2}, \frac{\ln t}{t^2-1}, \sin 2t\right\rangle$

5–6 Determine whether $\mathbf{r}(t)$ is continuous at $t = 0$. Explain your reasoning.

5. (a) $\mathbf{r}(t) = 3\sin t\,\mathbf{i} - 2t\mathbf{j}$ (b) $\mathbf{r}(t) = t^2\mathbf{i} + \frac{1}{t}\mathbf{j} + t\mathbf{k}$
6. (a) $\mathbf{r}(t) = e^t\mathbf{i} + \mathbf{j} + \csc t\,\mathbf{k}$
   (b) $\mathbf{r}(t) = 5\mathbf{i} - \sqrt{3t+1}\,\mathbf{j} + e^{2t}\mathbf{k}$
7. Sketch the circle $\mathbf{r}(t) = \cos t\,\mathbf{i} + \sin t\,\mathbf{j}$, and in each part draw the vector with its correct length.
   (a) $\mathbf{r}'(\pi/4)$ (b) $\mathbf{r}''(\pi)$ (c) $\mathbf{r}(2\pi) - \mathbf{r}(3\pi/2)$
8. Sketch the circle $\mathbf{r}(t) = \cos t\,\mathbf{i} - \sin t\,\mathbf{j}$, and in each part draw the vector with its correct length.
   (a) $\mathbf{r}'(\pi/4)$ (b) $\mathbf{r}''(\pi)$ (c) $\mathbf{r}(2\pi) - \mathbf{r}(3\pi/2)$

9–10 Find $\mathbf{r}'(t)$.

9. $\mathbf{r}(t) = 4\mathbf{i} - \cos t\,\mathbf{j}$
10. $\mathbf{r}(t) = (\tan^{-1} t)\mathbf{i} + t\cos t\,\mathbf{j} - \sqrt{t}\,\mathbf{k}$

11–14 Find the vector $\mathbf{r}'(t_0)$; then sketch the graph of $\mathbf{r}(t)$ in 2-space and draw the tangent vector $\mathbf{r}'(t_0)$.

11. $\mathbf{r}(t) = \langle t, t^2\rangle$; $t_0 = 2$
12. $\mathbf{r}(t) = t^3\mathbf{i} + t^2\mathbf{j}$; $t_0 = 1$
13. $\mathbf{r}(t) = \sec t\,\mathbf{i} + \tan t\,\mathbf{j}$; $t_0 = 0$
14. $\mathbf{r}(t) = 2\sin t\,\mathbf{i} + 3\cos t\,\mathbf{j}$; $t_0 = \pi/6$

15–16 Find the vector $\mathbf{r}'(t_0)$; then sketch the graph of $\mathbf{r}(t)$ in 3-space and draw the tangent vector $\mathbf{r}'(t_0)$.

15. $\mathbf{r}(t) = 2\sin t\,\mathbf{i} + \mathbf{j} + 2\cos t\,\mathbf{k}$; $t_0 = \pi/2$
16. $\mathbf{r}(t) = \cos t\,\mathbf{i} + \sin t\,\mathbf{j} + t\mathbf{k}$; $t_0 = \pi/4$

17–18 Use a graphing utility to generate the graph of $\mathbf{r}(t)$ and the graph of the tangent line at t_0 on the same screen.

17. $\mathbf{r}(t) = \sin \pi t\,\mathbf{i} + t^2\mathbf{j}$; $t_0 = \frac{1}{2}$

18. $\mathbf{r}(t) = 3\sin t\mathbf{i} + 4\cos t\mathbf{j};\ t_0 = \pi/4$

19–22 Find parametric equations of the line tangent to the graph of $\mathbf{r}(t)$ at the point where $t = t_0$.

19. $\mathbf{r}(t) = t^2\mathbf{i} + (2 - \ln t)\mathbf{j};\ t_0 = 1$
20. $\mathbf{r}(t) = e^{2t}\mathbf{i} - 2\cos 3t\,\mathbf{j};\ t_0 = 0$
21. $\mathbf{r}(t) = 2\cos \pi t\mathbf{i} + 2\sin \pi t\mathbf{j} + 3t\mathbf{k};\ t_0 = \frac{1}{3}$
22. $\mathbf{r}(t) = \ln t\mathbf{i} + e^{-t}\mathbf{j} + t^3\mathbf{k};\ t_0 = 2$

23–26 Find a vector equation of the line tangent to the graph of $\mathbf{r}(t)$ at the point P_0 on the curve.

23. $\mathbf{r}(t) = (2t - 1)\mathbf{i} + \sqrt{3t + 4}\mathbf{j};\ P_0(-1, 2)$
24. $\mathbf{r}(t) = 4\cos t\mathbf{i} - 3t\mathbf{j};\ P_0(2, -\pi)$
25. $\mathbf{r}(t) = t^2\mathbf{i} - \dfrac{1}{t+1}\mathbf{j} + (4 - t^2)\mathbf{k};\ P_0(4, 1, 0)$
26. $\mathbf{r}(t) = \sin t\mathbf{i} + \cosh t\mathbf{j} + (\tan^{-1} t)\mathbf{k};\ P_0(0, 1, 0)$
27. Let $\mathbf{r}(t) = \cos t\mathbf{i} + \sin t\mathbf{j} + \mathbf{k}$. Find
    (a) $\lim_{t\to 0} (\mathbf{r}(t) - \mathbf{r}'(t))$
    (b) $\lim_{t\to 0} (\mathbf{r}(t) \times \mathbf{r}'(t))$
    (c) $\lim_{t\to 0} (\mathbf{r}(t) \cdot \mathbf{r}'(t))$.
28. Let $\mathbf{r}(t) = t\mathbf{i} + t^2\mathbf{j} + t^3\mathbf{k}$. Find
$$\lim_{t\to 1} \mathbf{r}(t) \cdot (\mathbf{r}'(t) \times \mathbf{r}''(t))$$

29–30 Calculate
$$\frac{d}{dt}[\mathbf{r}_1(t) \cdot \mathbf{r}_2(t)] \quad \text{and} \quad \frac{d}{dt}[\mathbf{r}_1(t) \times \mathbf{r}_2(t)]$$
first by differentiating the product directly and then by applying Formulas (6) and (7).

29. $\mathbf{r}_1(t) = 2t\mathbf{i} + 3t^2\mathbf{j} + t^3\mathbf{k},\ \mathbf{r}_2(t) = t^4\mathbf{k}$
30. $\mathbf{r}_1(t) = \cos t\mathbf{i} + \sin t\mathbf{j} + t\mathbf{k},\ \mathbf{r}_2(t) = \mathbf{i} + t\mathbf{k}$

31–36 Evaluate the indefinite integral.

31. $\int (3\mathbf{i} + 4t\mathbf{j})\,dt$
32. $\int (\sin t\mathbf{i} - \cos t\mathbf{j})\,dt$
33. $\int (t\sin t\mathbf{i} + \mathbf{j})\,dt$
34. $\int \langle te^t, \ln t\rangle\,dt$
35. $\int \left(t^2\mathbf{i} - 2t\mathbf{j} + \frac{1}{t}\mathbf{k}\right)dt$
36. $\int \langle e^{-t}, e^t, 3t^2\rangle\,dt$

37–42 Evaluate the definite integral.

37. $\int_0^{\pi/2} \langle \cos 2t, \sin 2t\rangle\,dt$
38. $\int_0^1 (t^2\mathbf{i} + t^3\mathbf{j})\,dt$
39. $\int_0^2 \|t\mathbf{i} + t^2\mathbf{j}\|\,dt$
40. $\int_{-3}^3 \langle (3-t)^{3/2}, (3+t)^{3/2}, 1\rangle\,dt$
41. $\int_1^9 (t^{1/2}\mathbf{i} + t^{-1/2}\mathbf{j})\,dt$
42. $\int_0^1 (e^{2t}\mathbf{i} + e^{-t}\mathbf{j} + t\mathbf{k})\,dt$

43–46 Solve the vector initial-value problem for $\mathbf{y}(t)$ by integrating and using the initial conditions to find the constants of integration.

43. $\mathbf{y}'(t) = 2t\mathbf{i} + 3t^2\mathbf{j},\ \mathbf{y}(0) = \mathbf{i} - \mathbf{j}$
44. $\mathbf{y}'(t) = \cos t\mathbf{i} + \sin t\mathbf{j},\ \mathbf{y}(0) = \mathbf{i} - \mathbf{j}$
45. $\mathbf{y}''(t) = \mathbf{i} + e^t\mathbf{j},\ \mathbf{y}(0) = 2\mathbf{i},\ \mathbf{y}'(0) = \mathbf{j}$
46. $\mathbf{y}''(t) = 12t^2\mathbf{i} - 2t\mathbf{j},\ \mathbf{y}(0) = 2\mathbf{i} - 4\mathbf{j},\ \mathbf{y}'(0) = \mathbf{0}$

47–48 Let $\theta(t)$ be the angle between $\mathbf{r}(t)$ and $\mathbf{r}'(t)$. Use a graphing calculator to generate the graph of θ versus t, and make rough estimates of the t-values at which t-intercepts or relative extrema occur. What do these values tell you about the vectors $\mathbf{r}(t)$ and $\mathbf{r}'(t)$?

47. $\mathbf{r}(t) = 4\cos t\mathbf{i} + 3\sin t\mathbf{j};\ 0 \le t \le 2\pi$
48. $\mathbf{r}(t) = t^2\mathbf{i} + t^3\mathbf{j};\ 0 \le t \le 1$
49. (a) Find the points where the curve
$$\mathbf{r} = t\mathbf{i} + t^2\mathbf{j} - 3t\mathbf{k}$$
intersects the plane $2x - y + z = -2$.
    (b) For the curve and plane in part (a), find, to the nearest degree, the acute angle that the tangent line to the curve makes with a line normal to the plane at each point of intersection.
50. Find where the tangent line to the curve
$$\mathbf{r} = e^{-2t}\mathbf{i} + \cos t\mathbf{j} + 3\sin t\mathbf{k}$$
at the point $(1, 1, 0)$ intersects the yz-plane.

51–52 Show that the graphs of $\mathbf{r}_1(t)$ and $\mathbf{r}_2(t)$ intersect at the point P. Find, to the nearest degree, the acute angle between the tangent lines to the graphs of $\mathbf{r}_1(t)$ and $\mathbf{r}_2(t)$ at the point P.

51. $\mathbf{r}_1(t) = t^2\mathbf{i} + t\mathbf{j} + 3t^3\mathbf{k}$
    $\mathbf{r}_2(t) = (t - 1)\mathbf{i} + \frac{1}{4}t^2\mathbf{j} + (5 - t)\mathbf{k};\ P(1, 1, 3)$
52. $\mathbf{r}_1(t) = 2e^{-t}\mathbf{i} + \cos t\mathbf{j} + (t^2 + 3)\mathbf{k}$
    $\mathbf{r}_2(t) = (1 - t)\mathbf{i} + t^2\mathbf{j} + (t^3 + 4)\mathbf{k};\ P(2, 1, 3)$

FOCUS ON CONCEPTS

53. Use Formula (7) to derive the differentiation formula
$$\frac{d}{dt}[\mathbf{r}(t) \times \mathbf{r}'(t)] = \mathbf{r}(t) \times \mathbf{r}''(t)$$
54. Let $\mathbf{u} = \mathbf{u}(t)$, $\mathbf{v} = \mathbf{v}(t)$, and $\mathbf{w} = \mathbf{w}(t)$ be differentiable vector-valued functions. Use Formulas (6) and (7) to show that
$$\frac{d}{dt}[\mathbf{u} \cdot (\mathbf{v} \times \mathbf{w})] = \frac{d\mathbf{u}}{dt} \cdot [\mathbf{v} \times \mathbf{w}] + \mathbf{u} \cdot \left[\frac{d\mathbf{v}}{dt} \times \mathbf{w}\right] + \mathbf{u} \cdot \left[\mathbf{v} \times \frac{d\mathbf{w}}{dt}\right]$$

55. Let u_1, u_2, u_3, v_1, v_2, v_3, w_1, w_2, and w_3 be differentiable functions of t. Use Exercise 54 to show that

$$\frac{d}{dt}\begin{vmatrix} u_1 & u_2 & u_3 \\ v_1 & v_2 & v_3 \\ w_1 & w_2 & w_3 \end{vmatrix} = \begin{vmatrix} u_1' & u_2' & u_3' \\ v_1 & v_2 & v_3 \\ w_1 & w_2 & w_3 \end{vmatrix} + \begin{vmatrix} u_1 & u_2 & u_3 \\ v_1' & v_2' & v_3' \\ w_1 & w_2 & w_3 \end{vmatrix} + \begin{vmatrix} u_1 & u_2 & u_3 \\ v_1 & v_2 & v_3 \\ w_1' & w_2' & w_3' \end{vmatrix}$$

56. Prove Theorem 13.2.6 for 2-space.

57. Derive Formulas (6) and (7) for 3-space.

58. Prove Theorem 13.2.9 for 2-space.

✔QUICK CHECK ANSWERS 13.2

1. (a) $9\mathbf{i}+6\mathbf{j}$ (b) $\left\langle \frac{\sqrt{2}}{2}, \frac{\sqrt{2}}{2} \right\rangle$ **2.** (a) $\mathbf{r}'(t) = 5\mathbf{i} + (1-2t)\mathbf{j}$ (b) $\mathbf{r}'(t) = \left\langle -\frac{1}{t^2}, \sec^2 t, 2e^{2t} \right\rangle$ **3.** (a) $\langle 6, 4, 2\rangle$ (b) $\langle -4, 0, 4\rangle$ (c) $\mathbf{0}$ (d) -28 **4.** (a) $\left\langle 1, \frac{1}{3}, \frac{2}{\pi} \right\rangle$ (b) $\frac{t^2}{2}\mathbf{i} - t^3\mathbf{j} + e^t\mathbf{k} + \mathbf{C}$

13.3 CHANGE OF PARAMETER; ARC LENGTH

We observed in earlier sections that a curve in 2-space or 3-space can be represented parametrically in more than one way. For example, in Section 1.8 we gave two parametric representations of a circle—one in which the circle was traced clockwise and the other in which it was traced counterclockwise. Sometimes it will be desirable to change the parameter for a parametric curve to a different parameter that is better suited for the problem at hand. In this section we will investigate issues associated with changes of parameter, and we will show that arc length plays a special role in parametric representations of curves.

SMOOTH PARAMETRIZATIONS

Graphs of vector-valued functions range from continuous and smooth to discontinuous and wildly erratic. In this text we will not be concerned with graphs of the latter type, so we will need to impose restrictions to eliminate the unwanted behavior. We will say that $\mathbf{r}(t)$ is ***smoothly parametrized*** or that $\mathbf{r}(t)$ is a ***smooth function*** of t if $\mathbf{r}'(t)$ is continuous and $\mathbf{r}'(t) \neq \mathbf{0}$ for any allowable value of t. Algebraically, smoothness implies that the components of $\mathbf{r}(t)$ have continuous derivatives that are not all zero for the same value of t, and geometrically, it implies that the tangent vector $\mathbf{r}'(t)$ varies continuously along the curve. For this reason a smoothly parametrized function is said to have a ***continuously turning tangent vector***.

▶ **Example 1** Determine whether the following vector-valued functions have continuously turning tangent vectors.

(a) $\mathbf{r}(t) = a\cos t\mathbf{i} + a\sin t\mathbf{j} + ct\mathbf{k} \quad (a > 0, c > 0)$

(b) $\mathbf{r}(t) = t^2\mathbf{i} + t^3\mathbf{j}$

Solution (a). We have

$$\mathbf{r}'(t) = -a\sin t\mathbf{i} + a\cos t\mathbf{j} + c\mathbf{k}$$

The components are continuous functions, and there is no value of t for which all three of them are zero (verify), so $\mathbf{r}(t)$ has a continuously turning tangent vector. The graph of $\mathbf{r}(t)$ is the circular helix in Figure 13.1.2.

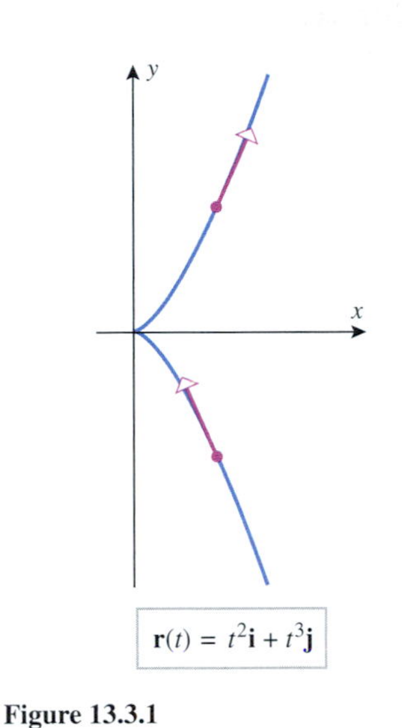

Figure 13.3.1

Solution (b). We have

$$\mathbf{r}'(t) = 2t\mathbf{i} + 3t^2\mathbf{j}$$

Although the components are continuous functions, they are both equal to zero if $t = 0$, so $\mathbf{r}(t)$ does not have a continuously turning tangent vector. The graph of $\mathbf{r}(t)$, which is shown in Figure 13.3.1, is a semicubical parabola traced in the upward direction (see Example 3 of Section 11.2). Observe that for values of t slightly less than zero the angle between $\mathbf{r}'(t)$ and $\mathbf{i}$ is near π, and for values of t slightly larger than zero the angle is near 0; hence there is a sudden reversal in the direction of the tangent vector as t increases through $t = 0$. ◀

ARC LENGTH FROM THE VECTOR VIEWPOINT

Recall from Theorem 7.4.3 that the arc length L of a parametric curve

$$x = x(t), \quad y = y(t) \qquad (a \le t \le b) \tag{1}$$

is given by the formula

$$L = \int_a^b \sqrt{\left(\frac{dx}{dt}\right)^2 + \left(\frac{dy}{dt}\right)^2}\, dt \tag{2}$$

Analogously, the arc length L of a parametric curve

$$x = x(t), \quad y = y(t), \quad z = z(t) \qquad (a \le t \le b) \tag{3}$$

in 3-space is given by the formula

$$L = \int_a^b \sqrt{\left(\frac{dx}{dt}\right)^2 + \left(\frac{dy}{dt}\right)^2 + \left(\frac{dz}{dt}\right)^2}\, dt \tag{4}$$

Formulas (2) and (4) have vector forms that we can obtain by letting

$$\underbrace{\mathbf{r}(t) = x(t)\mathbf{i} + y(t)\mathbf{j}}_{\text{2-space}} \quad \text{or} \quad \underbrace{\mathbf{r}(t) = x(t)\mathbf{i} + y(t)\mathbf{j} + z(t)\mathbf{k}}_{\text{3-space}}$$

It follows that

$$\underbrace{\frac{d\mathbf{r}}{dt} = \frac{dx}{dt}\mathbf{i} + \frac{dy}{dt}\mathbf{j}}_{\text{2-space}} \quad \text{or} \quad \underbrace{\frac{d\mathbf{r}}{dt} = \frac{dx}{dt}\mathbf{i} + \frac{dy}{dt}\mathbf{j} + \frac{dz}{dt}\mathbf{k}}_{\text{3-space}}$$

and hence

$$\underbrace{\left\|\frac{d\mathbf{r}}{dt}\right\| = \sqrt{\left(\frac{dx}{dt}\right)^2 + \left(\frac{dy}{dt}\right)^2}}_{\text{2-space}} \quad \text{or} \quad \underbrace{\left\|\frac{d\mathbf{r}}{dt}\right\| = \sqrt{\left(\frac{dx}{dt}\right)^2 + \left(\frac{dy}{dt}\right)^2 + \left(\frac{dz}{dt}\right)^2}}_{\text{3-space}}$$

Substituting these expressions in (2) and (4) leads us to the following theorem.

13.3.1 THEOREM. *If C is the graph in 2-space or 3-space of a smooth vector-valued function $\mathbf{r}(t)$, then its arc length L from $t = a$ to $t = b$ is*

$$L = \int_a^b \left\| \frac{d\mathbf{r}}{dt} \right\| dt \tag{5}$$

▶ Example 2 Find the arc length of that portion of the circular helix

$$x = \cos t, \quad y = \sin t, \quad z = t$$

from $t = 0$ to $t = \pi$.

Solution. Set $\mathbf{r}(t) = (\cos t)\mathbf{i} + (\sin t)\mathbf{j} + t\mathbf{k} = \langle \cos t, \sin t, t \rangle$. Then

$$\mathbf{r}'(t) = \langle -\sin t, \cos t, 1 \rangle \quad \text{and} \quad \|\mathbf{r}'(t)\| = \sqrt{(-\sin t)^2 + (\cos t)^2 + 1} = \sqrt{2}$$

From Theorem 13.3.1 the arc length of the helix is

$$L = \int_0^\pi \left\| \frac{d\mathbf{r}}{dt} \right\| dt = \int_0^\pi \sqrt{2}\, dt = \sqrt{2}\pi \quad ◀$$

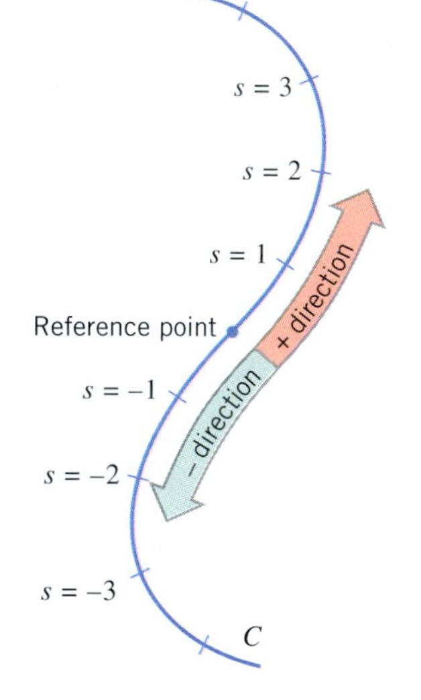

Figure 13.3.2

ARC LENGTH AS A PARAMETER

For many purposes the best parameter to use for representing a curve in 2-space or 3-space parametrically is the length of arc measured along the curve from some fixed reference point. This can be done as follows:

Using Arc Length as a Parameter

Step 1. Select an arbitrary point on the curve C to serve as a ***reference point***.

Step 2. Starting from the reference point, choose one direction along the curve to be the ***positive direction*** and the other to be the ***negative direction***.

Step 3. If P is a point on the curve, let s be the "signed" arc length along C from the reference point to P, where s is positive if P is in the positive direction from the reference point and s is negative if P is in the negative direction. Figure 13.3.2 illustrates this idea.

By this procedure, a unique point P on the curve is determined when a value for s is given. For example, $s = 2$ determines the point that is 2 units along the curve in the positive direction from the reference point, and $s = -\frac{3}{2}$ determines the point that is $\frac{3}{2}$ units along the curve in the negative direction from the reference point.

Let us now treat s as a variable. As the value of s changes, the corresponding point P moves along C and the coordinates of P become functions of s. Thus, in 2-space the coordinates of P are $(x(s), y(s))$, and in 3-space they are $(x(s), y(s), z(s))$. Therefore, in 2-space or 3-space the curve C is given by the parametric equations

$$x = x(s), \quad y = y(s) \quad \text{or} \quad x = x(s), \quad y = y(s), \quad z = z(s)$$

A parametric representation of a curve with arc length as the parameter is called an ***arc length parametrization*** of the curve. Note that a given curve will generally have infinitely many different arc length parametrizations, since the reference point and orientation can be chosen arbitrarily.

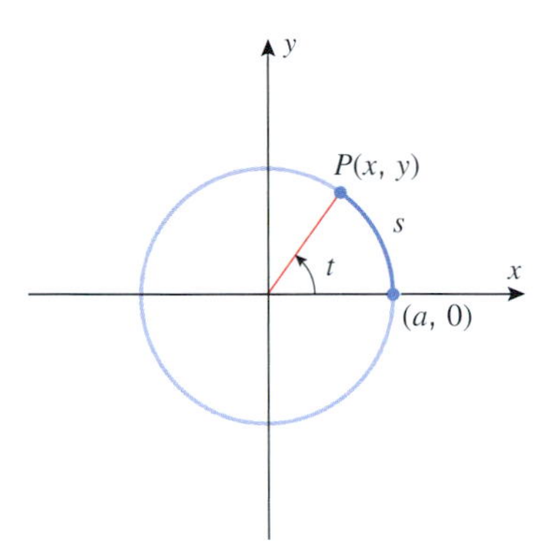

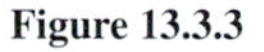
Figure 13.3.3

▶ **Example 3** Find the arc length parametrization of the circle $x^2 + y^2 = a^2$ with counterclockwise orientation and $(a, 0)$ as the reference point.

Solution. The circle with counterclockwise orientation can be represented by the parametric equations

$$x = a\cos t, \quad y = a\sin t \qquad (0 \le t \le 2\pi) \tag{6}$$

in which t can be interpreted as the angle in radian measure from the positive x-axis to the radius from the origin to the point $P(x, y)$ (Figure 13.3.3). If we take the positive direction for measuring the arc length to be counterclockwise, and we take $(a, 0)$ to be the reference point, then s and t are related by

$$s = at \quad \text{or} \quad t = s/a$$

Making this change of variable in (6) and noting that s increases from 0 to $2\pi a$ as t increases from 0 to 2π yields the following arc length parametrization of the circle:

$$x = a\cos(s/a), \quad y = a\sin(s/a) \qquad (0 \le s \le 2\pi a)$$ ◀

CHANGE OF PARAMETER

In many situations the solution of a problem can be simplified by choosing the parameter in a vector-valued function or a parametric curve in the right way. The two most common parameters for curves in 2-space or 3-space are time and arc length. However, there are other useful possibilities as well. For example, in analyzing the motion of a particle in 2-space, it is often desirable to parametrize its trajectory in terms of the angle ϕ between the tangent vector and the positive x-axis (Figure 13.3.4). Thus, our next objective is to develop methods for changing the parameter in a vector-valued function or parametric curve. This will allow us to move freely between different possible parametrizations.

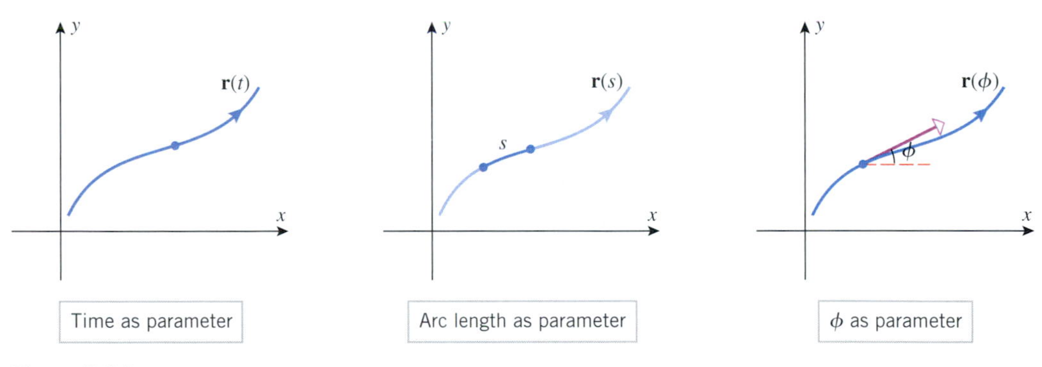

Figure 13.3.4

A ***change of parameter*** in a vector-valued function $\mathbf{r}(t)$ is a substitution $t = g(\tau)$ that produces a new vector-valued function $\mathbf{r}(g(\tau))$ having the same graph as $\mathbf{r}(t)$, but possibly traced differently as the parameter τ increases.

▶ **Example 4** Find a change of parameter $t = g(\tau)$ for the circle

$$\mathbf{r}(t) = \cos t\,\mathbf{i} + \sin t\,\mathbf{j} \qquad (0 \le t \le 2\pi)$$

such that

(a) the circle is traced counterclockwise as τ increases over the interval $[0, 1]$;

(b) the circle is traced clockwise as τ increases over the interval $[0, 1]$.

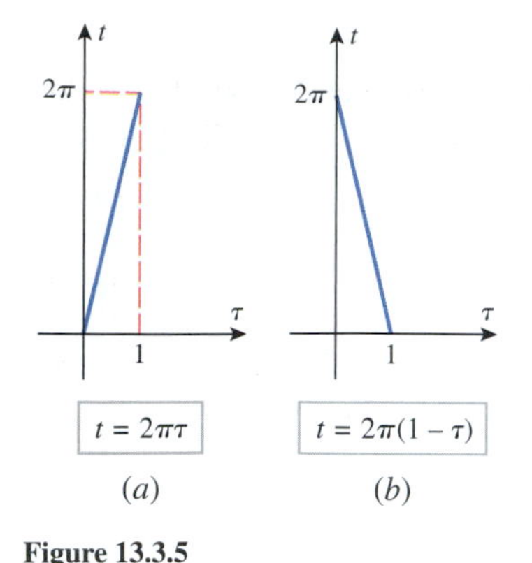

Figure 13.3.5

Solution (a). The given circle is traced counterclockwise as t increases. Thus, if we choose g to be an increasing function, then it will follow from the relationship $t = g(\tau)$ that t increases when τ increases, thereby ensuring that the circle will be traced counterclockwise as τ increases. We also want to choose g so that t increases from 0 to 2π as τ increases from 0 to 1. A simple choice of g that satisfies all of the required criteria is the linear function graphed in Figure 13.3.5*a*. The equation of this line is

$$t = g(\tau) = 2\pi\tau \tag{7}$$

which is the desired change of parameter. The resulting representation of the circle in terms of the parameter τ is

$$\mathbf{r}(g(\tau)) = \cos 2\pi\tau\,\mathbf{i} + \sin 2\pi\tau\,\mathbf{j} \qquad (0 \le \tau \le 1)$$

Solution (b). To ensure that the circle is traced clockwise, we will choose g to be a decreasing function such that t decreases from 2π to 0 as τ increases from 0 to 1. A simple choice of g that achieves this is the linear function

$$t = g(\tau) = 2\pi(1 - \tau) \tag{8}$$

graphed in Figure 13.3.5*b*. The resulting representation of the circle in terms of the parameter τ is

$$\mathbf{r}(g(\tau)) = \cos(2\pi(1 - \tau))\mathbf{i} + \sin(2\pi(1 - \tau))\mathbf{j} \qquad (0 \le \tau \le 1)$$

which simplifies to (verify)

$$\mathbf{r}(g(\tau)) = \cos 2\pi\tau\,\mathbf{i} - \sin 2\pi\tau\,\mathbf{j} \qquad (0 \le \tau \le 1)$$ ◂

When making a change of parameter $t = g(\tau)$ in a vector-valued function $\mathbf{r}(t)$, it will be important to ensure that the new vector-valued function $\mathbf{r}(g(\tau))$ is smooth if $\mathbf{r}(t)$ is smooth. To establish conditions under which this happens, we will need the following version of the chain rule for vector-valued functions. The proof is left as an exercise.

Strictly speaking, since $d\mathbf{r}/dt$ is a vector and $dt/d\tau$ is a scalar, Formula (9) should be written in the form

$$\frac{d\mathbf{r}}{d\tau} = \frac{dt}{d\tau}\frac{d\mathbf{r}}{dt}$$

However, reversing the order of the factors makes the formula easier to remember, and we will continue to do so.

13.3.2 THEOREM (*Chain Rule*). *Let $\mathbf{r}(t)$ be a vector-valued function in 2-space or 3-space that is differentiable with respect to t. If $t = g(\tau)$ is a change of parameter in which g is differentiable with respect to τ, then $\mathbf{r}(g(\tau))$ is differentiable with respect to τ and*

$$\frac{d\mathbf{r}}{d\tau} = \frac{d\mathbf{r}}{dt}\frac{dt}{d\tau} \tag{9}$$

A change of parameter $t = g(\tau)$ in which $\mathbf{r}(g(\tau))$ is smooth if $\mathbf{r}(t)$ is smooth is called a ***smooth change of parameter***. It follows from (9) that $t = g(\tau)$ will be a smooth change of parameter if $dt/d\tau$ is continuous and $dt/d\tau \ne 0$ for all values of τ, since these conditions imply that $d\mathbf{r}/d\tau$ is continuous and nonzero if $d\mathbf{r}/dt$ is continuous and nonzero. Smooth changes of parameter fall into two categories—those for which $dt/d\tau > 0$ for all τ (called ***positive changes of parameter***) and those for which $dt/d\tau < 0$ for all τ (called ***negative changes of parameter***). A positive change of parameter preserves the orientation of a parametric curve, and a negative change of parameter reverses it.

▸ **Example 5** In Example 4 the change of parameter in Formula (7) is positive since $dt/d\tau = 2\pi > 0$, and the change of parameter given by Formula (8) is negative since $dt/d\tau = -2\pi < 0$. The positive change of parameter preserved the orientation of the circle, and the negative change of parameter reversed it. ◂

FINDING ARC LENGTH PARAMETRIZATIONS

Next we will consider the problem of finding an arc length parametrization of a vector-valued function that is expressed initially in terms of some other parameter t. The following theorem will provide a general method for doing this.

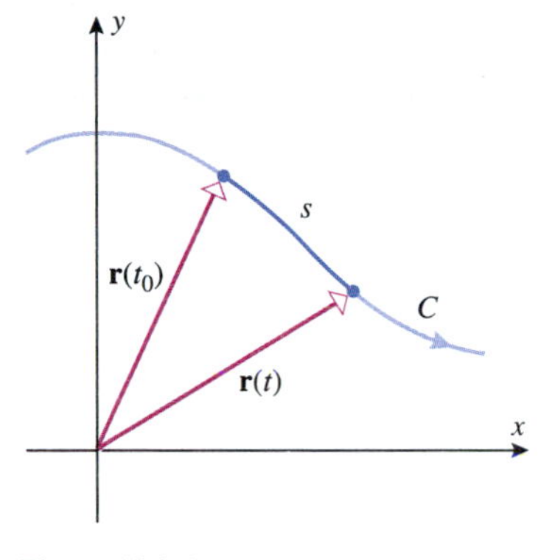

Figure 13.3.6

13.3.3 THEOREM. *Let C be the graph of a smooth vector-valued function $\mathbf{r}(t)$ in 2-space or 3-space, and let $\mathbf{r}(t_0)$ be any point on C. Then the following formula defines a positive change of parameter from t to s, where s is an arc length parameter having $\mathbf{r}(t_0)$ as its reference point* (Figure 13.3.6):

$$s = \int_{t_0}^{t} \left\| \frac{d\mathbf{r}}{du} \right\| du \tag{10}$$

PROOF. From (5) with u as the variable of integration instead of t, the integral represents the arc length of that portion of C between $\mathbf{r}(t_0)$ and $\mathbf{r}(t)$ if $t > t_0$ and the negative of that arc length if $t < t_0$. Thus, s is the arc length parameter with $\mathbf{r}(t_0)$ as its reference point and its positive direction in the direction of increasing t. ■

When needed, Formula (10) can be expressed in component form as

$$s = \int_{t_0}^{t} \sqrt{\left(\frac{dx}{du}\right)^2 + \left(\frac{dy}{du}\right)^2}\, du \qquad \text{2-space} \tag{11}$$

$$s = \int_{t_0}^{t} \sqrt{\left(\frac{dx}{du}\right)^2 + \left(\frac{dy}{du}\right)^2 + \left(\frac{dz}{du}\right)^2}\, du \qquad \text{3-space} \tag{12}$$

▶ **Example 6** Find the arc length parametrization of the circular helix

$$\mathbf{r} = \cos t\mathbf{i} + \sin t\mathbf{j} + t\mathbf{k} \tag{13}$$

that has reference point $\mathbf{r}(0) = (1, 0, 0)$ and the same orientation as the given helix.

Solution. Replacing t by u in $\mathbf{r}$ for integration purposes and taking $t_0 = 0$ in Formula (10), we obtain

$$\mathbf{r} = \cos u\mathbf{i} + \sin u\mathbf{j} + u\mathbf{k}$$

$$\frac{d\mathbf{r}}{du} = (-\sin u)\mathbf{i} + \cos u\mathbf{j} + \mathbf{k}$$

$$\left\| \frac{d\mathbf{r}}{du} \right\| = \sqrt{(-\sin u)^2 + \cos^2 u + 1} = \sqrt{2}$$

$$s = \int_0^t \left\| \frac{d\mathbf{r}}{du} \right\| du = \int_0^t \sqrt{2}\, du = \sqrt{2}u\Big]_0^t = \sqrt{2}t$$

Thus, $t = s/\sqrt{2}$, so (13) can be reparametrized in terms of s as

$$\mathbf{r} = \cos\left(\frac{s}{\sqrt{2}}\right)\mathbf{i} + \sin\left(\frac{s}{\sqrt{2}}\right)\mathbf{j} + \frac{s}{\sqrt{2}}\mathbf{k}$$

We are guaranteed that this reparametrization preserves the orientation of the helix since Formula (10) produces a positive change of parameter. ◀

▶ **Example 7** A bug, starting at the reference point (1, 0, 0) of the helix in Example 6, walks up the helix for a distance of 10 units. What are the bug's final coordinates?

Solution. From Example 6, the arc length parametrization of the helix relative to the reference point (1, 0, 0) is

$$\mathbf{r} = \cos\left(\frac{s}{\sqrt{2}}\right)\mathbf{i} + \sin\left(\frac{s}{\sqrt{2}}\right)\mathbf{j} + \frac{s}{\sqrt{2}}\mathbf{k}$$

or, expressed parametrically,

$$x = \cos\left(\frac{s}{\sqrt{2}}\right), \quad y = \sin\left(\frac{s}{\sqrt{2}}\right), \quad z = \frac{s}{\sqrt{2}}$$

Thus, at $s = 10$ the coordinates are

$$\left(\cos\left(\frac{10}{\sqrt{2}}\right), \sin\left(\frac{10}{\sqrt{2}}\right), \frac{10}{\sqrt{2}}\right) \approx (0.705, 0.709, 7.07)$$ ◀

▶ **Example 8** Recall from Formula (9) of Section 12.5 that the equation

$$\mathbf{r} = \mathbf{r}_0 + t\mathbf{v} \tag{14}$$

is the vector form of the line that passes through the terminal point of $\mathbf{r}_0$ and is parallel to the vector $\mathbf{v}$. Find the arc length parametrization of the line that has reference point $\mathbf{r}_0$ and the same orientation as the given line.

Solution. Replacing t by u in (14) for integration purposes and taking $t_0 = 0$ in Formula (10), we obtain

$$\mathbf{r} = \mathbf{r}_0 + u\mathbf{v} \quad \text{and} \quad \frac{d\mathbf{r}}{du} = \mathbf{v}$$

Since $\mathbf{r}_0$ is constant

It follows from this that

$$s = \int_0^t \left\|\frac{d\mathbf{r}}{du}\right\| du = \int_0^t \|\mathbf{v}\|\, du = \|\mathbf{v}\|u\Big]_0^t = t\|\mathbf{v}\|$$

This implies that $t = s/\|\mathbf{v}\|$, so (14) can be reparametrized in terms of s as

$$\mathbf{r} = \mathbf{r}_0 + s\left(\frac{\mathbf{v}}{\|\mathbf{v}\|}\right) \tag{15}$$ ◀

In words, Formula (15) tells us that the line represented by Equation (14) can be reparametrized in terms of arc length with $\mathbf{r}_0$ as the reference point by normalizing $\mathbf{v}$ and then replacing t by s.

▶ **Example 9** Find the arc length parametrization of the line

$$x = 2t + 1, \quad y = 3t - 2$$

that has the same orientation as the given line and uses (1, −2) as the reference point.

Solution. The line passes through the point (1, −2) and is parallel to $\mathbf{v} = 2\mathbf{i} + 3\mathbf{j}$. To find the arc length parametrization of the line, we need only rewrite the given equations using $\mathbf{v}/\|\mathbf{v}\|$ rather than $\mathbf{v}$ to determine the direction and replace t by s. Since

$$\frac{\mathbf{v}}{\|\mathbf{v}\|} = \frac{2\mathbf{i} + 3\mathbf{j}}{\sqrt{13}} = \frac{2}{\sqrt{13}}\mathbf{i} + \frac{3}{\sqrt{13}}\mathbf{j}$$

it follows that the parametric equations for the line in terms of s are

$$x = \frac{2}{\sqrt{13}}s + 1, \quad y = \frac{3}{\sqrt{13}}s - 2$$

PROPERTIES OF ARC LENGTH PARAMETRIZATIONS

Because arc length parameters for a curve C are intimately related to the geometric characteristics of C, arc length parametrizations have properties that are not enjoyed by other parametrizations. For example, the following theorem shows that if a smooth curve is represented parametrically using an arc length parameter, then the tangent vectors all have length 1.

13.3.4 THEOREM.

(*a*) *If C is the graph of a smooth vector-valued function $\mathbf{r}(t)$ in 2-space or 3-space, where t is a general parameter, and if s is the arc length parameter for C defined by Formula (10), then for every value of t the tangent vector has length*

$$\left\| \frac{d\mathbf{r}}{dt} \right\| = \frac{ds}{dt} \tag{16}$$

(*b*) *If C is the graph of a smooth vector-valued function $\mathbf{r}(s)$ in 2-space or 3-space, where s is an arc length parameter, then for every value of s the tangent vector to C has length*

$$\left\| \frac{d\mathbf{r}}{ds} \right\| = 1 \tag{17}$$

(*c*) *If C is the graph of a smooth vector-valued function $\mathbf{r}(t)$ in 2-space or 3-space, and if $\|d\mathbf{r}/dt\| = 1$ for every value of t, then for any value of t_0 in the domain of $\mathbf{r}$, the parameter $s = t - t_0$ is an arc length parameter that has its reference point at the point on C where $t = t_0$.*

PROOF (*a*). This result follows by applying the Fundamental Theorem of Calculus (Theorem 6.6.3) to Formula (10).

PROOF (*b*). Let $t = s$ in part (*a*).

PROOF (*c*). It follows from Theorem 13.3.3 that the formula

$$s = \int_{t_0}^{t} \left\| \frac{d\mathbf{r}}{du} \right\| du$$

defines an arc length parameter for C with reference point $\mathbf{r}(0)$. However, $\|d\mathbf{r}/du\| = 1$ by hypothesis, so we can rewrite the formula for s as

$$s = \int_{t_0}^{t} du = u \Big]_{t_0}^{t} = t - t_0$$

The component forms of Formulas (16) and (17) will be of sufficient interest in later sections that we provide them here for reference:

$$\frac{ds}{dt} = \left\| \frac{d\mathbf{r}}{dt} \right\| = \sqrt{\left(\frac{dx}{dt}\right)^2 + \left(\frac{dy}{dt}\right)^2} \qquad \text{2-space} \tag{18}$$

Note that Formulas (18) and (19) do not involve t_0, and hence do not depend on where the reference point for s is chosen. This is to be expected since changing the reference point shifts s by a constant (the arc length between the two reference points), and this constant drops out on differentiating.

$$\frac{ds}{dt} = \left\| \frac{d\mathbf{r}}{dt} \right\| = \sqrt{\left(\frac{dx}{dt}\right)^2 + \left(\frac{dy}{dt}\right)^2 + \left(\frac{dz}{dt}\right)^2} \quad \text{3-space} \tag{19}$$

$$\left\| \frac{d\mathbf{r}}{ds} \right\| = \sqrt{\left(\frac{dx}{ds}\right)^2 + \left(\frac{dy}{ds}\right)^2} = 1 \quad \text{2-space} \tag{20}$$

$$\left\| \frac{d\mathbf{r}}{ds} \right\| = \sqrt{\left(\frac{dx}{ds}\right)^2 + \left(\frac{dy}{ds}\right)^2 + \left(\frac{dz}{ds}\right)^2} = 1 \quad \text{3-space} \tag{21}$$

✔ QUICK CHECK EXERCISES 13.3 *(See page 886 for answers.)*

1. If $\mathbf{r}(t)$ is a smooth vector-valued function, then the integral
$$\int_a^b \left\| \frac{d\mathbf{r}}{dt} \right\| dt$$
may be interpreted geometrically as the ________.

2. If $\mathbf{r}(s)$ is a smooth vector-valued function parametrized by arc length s, then
$$\left\| \frac{d\mathbf{r}}{ds} \right\| = \text{\_\_\_\_\_\_\_\_}$$
and the arc length of the graph of $\mathbf{r}$ over the interval $a \le s \le b$ is ________.

3. If $\mathbf{r}(t)$ is a smooth vector-valued function, then the arc length parameter s having $\mathbf{r}(t_0)$ as the reference point may be defined by the integral
$$s = \int_{t_0}^{t} \text{\_\_\_\_\_\_\_\_}\, du$$

4. Suppose that $\mathbf{r}(t)$ is a smooth vector-valued function of t with $\mathbf{r}'(1) = \langle \sqrt{3}, -\sqrt{3}, -1 \rangle$, and let $\mathbf{r}_1(t)$ be defined by the equation $\mathbf{r}_1(t) = \mathbf{r}(2\cos t)$. Then $\mathbf{r}_1'(\pi/3) =$ ________.

EXERCISE SET 13.3

1–4 Determine whether $\mathbf{r}(t)$ is a smooth function of the parameter t.

1. $\mathbf{r}(t) = t^3\mathbf{i} + (3t^2 - 2t)\mathbf{j} + t^2\mathbf{k}$
2. $\mathbf{r}(t) = \cos t^2\mathbf{i} + \sin t^2\mathbf{j} + e^{-t}\mathbf{k}$
3. $\mathbf{r}(t) = te^{-t}\mathbf{i} + (t^2 - 2t)\mathbf{j} + \cos \pi t\mathbf{k}$
4. $\mathbf{r}(t) = \sin \pi t\mathbf{i} + (2t - \ln t)\mathbf{j} + (t^2 - t)\mathbf{k}$

5–8 Find the arc length of the parametric curve.

5. $x = \cos^3 t,\ y = \sin^3 t,\ z = 2;\ 0 \le t \le \pi/2$
6. $x = 3\cos t,\ y = 3\sin t,\ z = 4t;\ 0 \le t \le \pi$
7. $x = e^t,\ y = e^{-t},\ z = \sqrt{2}t;\ 0 \le t \le 1$
8. $x = \frac{1}{2}t,\ y = \frac{1}{3}(1-t)^{3/2},\ z = \frac{1}{3}(1+t)^{3/2};\ -1 \le t \le 1$

9–12 Find the arc length of the graph of $\mathbf{r}(t)$.

9. $\mathbf{r}(t) = t^3\mathbf{i} + t\mathbf{j} + \frac{1}{2}\sqrt{6}t^2\mathbf{k};\ 1 \le t \le 3$
10. $\mathbf{r}(t) = (4 + 3t)\mathbf{i} + (2 - 2t)\mathbf{j} + (5 + t)\mathbf{k};\ 3 \le t \le 4$
11. $\mathbf{r}(t) = 3\cos t\mathbf{i} + 3\sin t\mathbf{j} + t\mathbf{k};\ 0 \le t \le 2\pi$
12. $\mathbf{r}(t) = t^2\mathbf{i} + (\cos t + t\sin t)\mathbf{j} + (\sin t - t\cos t)\mathbf{k};$ $0 \le t \le \pi$

13–16 Calculate $d\mathbf{r}/d\tau$ by the chain rule, and then check your result by expressing $\mathbf{r}$ in terms of τ and differentiating.

13. $\mathbf{r} = t\mathbf{i} + t^2\mathbf{j};\ t = 4\tau + 1$
14. $\mathbf{r} = \langle 3\cos t, 3\sin t \rangle;\ t = \pi\tau$
15. $\mathbf{r} = e^t\mathbf{i} + 4e^{-t}\mathbf{j};\ t = \tau^2$
16. $\mathbf{r} = \mathbf{i} + 3t^{3/2}\mathbf{j} + t\mathbf{k};\ t = 1/\tau$

FOCUS ON CONCEPTS

17. The accompanying figure shows the graph of the *four-cusped hypocycloid*
$$\mathbf{r}(t) = \cos^3 t\mathbf{i} + \sin^3 t\mathbf{j} \qquad (0 \le t \le 2\pi)$$
(a) Give an informal explanation of why $\mathbf{r}(t)$ is not smooth.
(b) Confirm that $\mathbf{r}(t)$ is not smooth by examining $\mathbf{r}'(t)$.

18. The accompanying figure shows the graph of the vector-valued function
$$\mathbf{r}(t) = \sin t\mathbf{i} + \sin^2 t\mathbf{j} \qquad (0 \le t \le 2\pi)$$
Show that this parametric curve is not smooth, even though it has no corners. Give an informal explanation of what causes the lack of smoothness.

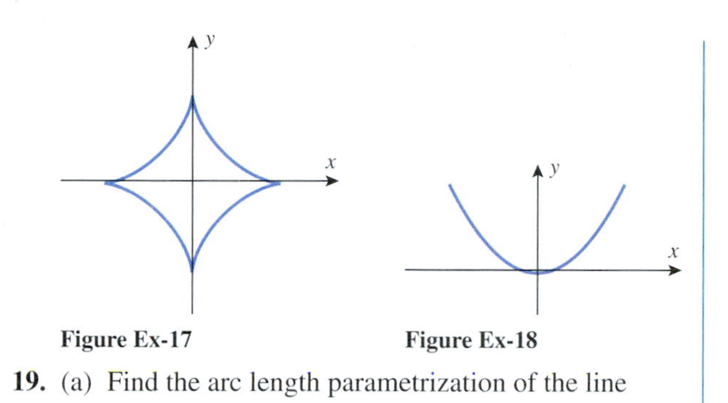

Figure Ex-17 Figure Ex-18

19. (a) Find the arc length parametrization of the line

$$x = t, \quad y = t$$

that has the same orientation as the given line and has reference point $(0, 0)$.

(b) Find the arc length parametrization of the line

$$x = t, \quad y = t, \quad z = t$$

that has the same orientation as the given line and has reference point $(0, 0, 0)$.

20. Find arc length parametrizations of the lines in Exercise 19 that have the stated reference points but are oriented opposite to the given lines.

21. (a) Find the arc length parametrization of the line

$$x = 1 + t, \quad y = 3 - 2t, \quad z = 4 + 2t$$

that has the same direction as the given line and has reference point $(1, 3, 4)$.

(b) Use the parametric equations obtained in part (a) to find the point on the line that is 25 units from the reference point in the direction of increasing parameter.

22. (a) Find the arc length parametrization of the line

$$x = -5 + 3t, \quad y = 2t, \quad z = 5 + t$$

that has the same direction as the given line and has reference point $(-5, 0, 5)$.

(b) Use the parametric equations obtained in part (a) to find the point on the line that is 10 units from the reference point in the direction of increasing parameter.

23–28 Find an arc length parametrization of the curve that has the same orientation as the given curve and has $t = 0$ as the reference point.

23. $\mathbf{r}(t) = (3 + \cos t)\mathbf{i} + (2 + \sin t)\mathbf{j};\ 0 \le t \le 2\pi$

24. $\mathbf{r}(t) = \cos^3 t\mathbf{i} + \sin^3 t\mathbf{j};\ 0 \le t \le \pi/2$

25. $\mathbf{r}(t) = \frac{1}{3}t^3\mathbf{i} + \frac{1}{2}t^2\mathbf{j};\ t \ge 0$

26. $\mathbf{r}(t) = (1 + t)^2\mathbf{i} + (1 + t)^3\mathbf{j};\ 0 \le t \le 1$

27. $\mathbf{r}(t) = e^t \cos t\mathbf{i} + e^t \sin t\mathbf{j};\ 0 \le t \le \pi/2$

28. $\mathbf{r}(t) = \sin e^t\mathbf{i} + \cos e^t\mathbf{j} + \sqrt{3}e^t\mathbf{k};\ t \ge 0$

29. Show that the arc length of the circular helix $x = a\cos t$, $y = a\sin t$, $z = ct$ for $0 \le t \le t_0$ is $t_0\sqrt{a^2 + c^2}$.

30. Use the result in Exercise 29 to show the circular helix

$$\mathbf{r} = a\cos t\mathbf{i} + a\sin t\mathbf{j} + ct\mathbf{k}$$

can be expressed as

$$\mathbf{r} = \left(a\cos\frac{s}{w}\right)\mathbf{i} + \left(a\sin\frac{s}{w}\right)\mathbf{j} + \frac{cs}{w}\mathbf{k}$$

where $w = \sqrt{a^2 + c^2}$ and s is an arc length parameter with reference point at $(a, 0, 0)$.

31. Find an arc length parametrization of the cycloid

$$x = at - a\sin t, \quad y = a - a\cos t \qquad (0 \le t \le 2\pi)$$

with $(0, 0)$ as the reference point.

32. Show that in cylindrical coordinates a curve given by the parametric equations $r = r(t)$, $\theta = \theta(t)$, $z = z(t)$ for $a \le t \le b$ has arc length

$$L = \int_a^b \sqrt{\left(\frac{dr}{dt}\right)^2 + r^2\left(\frac{d\theta}{dt}\right)^2 + \left(\frac{dz}{dt}\right)^2}\, dt$$

[*Hint:* Use the relationships $x = r\cos\theta$, $y = r\sin\theta$.]

33. In each part, use the formula in Exercise 32 to find the arc length of the curve.

(a) $r = e^{2t}, \theta = t, z = e^{2t};\ 0 \le t \le \ln 2$

(b) $r = t^2, \theta = \ln t, z = \frac{1}{3}t^3;\ 1 \le t \le 2$

34. Show that in spherical coordinates a curve given by the parametric equations $\rho = \rho(t)$, $\theta = \theta(t)$, $\phi = \phi(t)$ for $a \le t \le b$ has arc length

$$L = \int_a^b \sqrt{\left(\frac{d\rho}{dt}\right)^2 + \rho^2\sin^2\phi\left(\frac{d\theta}{dt}\right)^2 + \rho^2\left(\frac{d\phi}{dt}\right)^2}\, dt$$

[*Hint:* $x = \rho\sin\phi\cos\theta$, $y = \rho\sin\phi\sin\theta$, $z = \rho\cos\phi$.]

35. In each part, use the formula in Exercise 34 to find the arc length of the curve.

(a) $\rho = e^{-t}, \theta = 2t, \phi = \pi/4;\ 0 \le t \le 2$

(b) $\rho = 2t, \theta = \ln t, \phi = \pi/6;\ 1 \le t \le 5$

FOCUS ON CONCEPTS

36. (a) Show that $\mathbf{r}(t) = t\mathbf{i} + t^2\mathbf{j}\ (-1 \le t \le 1)$ is a smooth vector-valued function, but the change of parameter $t = \tau^3$ produces a vector-valued function that is not smooth, yet has the same graph as $\mathbf{r}(t)$.

(b) Examine how the two vector-valued functions are traced and see if you can explain what causes the problem.

37. Find a change of parameter $t = g(\tau)$ for the semicircle

$$\mathbf{r}(t) = \cos t\mathbf{i} + \sin t\mathbf{j} \quad (0 \le t \le \pi)$$

such that

(a) the semicircle is traced counterclockwise as τ varies over the interval $[0, 1]$

(b) the semicircle is traced clockwise as τ varies over the interval $[0, 1]$.

38. What change of parameter $t = g(\tau)$ would you make if you wanted to trace the graph of $\mathbf{r}(t)$ $(0 \le t \le 1)$ in the opposite direction with τ varying from 0 to 1?

39. As illustrated in the accompanying figure, copper cable with a diameter of $\frac{1}{2}$ inch is to be wrapped in a circular helix around a cylinder that has a 12-inch diameter. What length of cable (measured along its centerline) will make one complete turn around the cylinder in a distance of 20 inches (between centerlines) measured parallel to the axis of the cylinder?

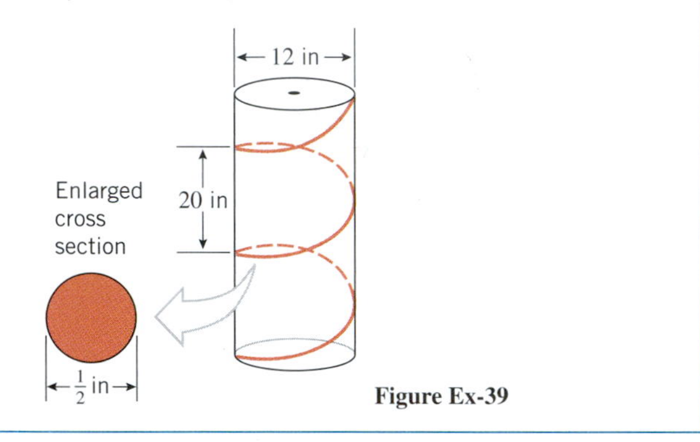

Figure Ex-39

40. Let $x = \cos t$, $y = \sin t$, $z = t^{3/2}$. Find
(a) $\|\mathbf{r}'(t)\|$ (b) $\dfrac{ds}{dt}$ (c) $\displaystyle\int_0^2 \|\mathbf{r}'(t)\|\,dt$.

41. Let $\mathbf{r}(t) = \ln t\mathbf{i} + 2t\mathbf{j} + t^2\mathbf{k}$. Find
(a) $\|\mathbf{r}'(t)\|$ (b) $\dfrac{ds}{dt}$ (c) $\displaystyle\int_1^3 \|\mathbf{r}'(t)\|\,dt$.

42. Prove: If $\mathbf{r}(t)$ is a smoothly parametrized function, then the angles between $\mathbf{r}'(t)$ and the vectors $\mathbf{i}$, $\mathbf{j}$, and $\mathbf{k}$ are continuous functions of t.

43. Prove the vector form of the chain rule for 2-space (Theorem 13.3.2) by expressing $\mathbf{r}(t)$ in terms of components.

✔ QUICK CHECK ANSWERS 13.3

1. arc length of the graph of $\mathbf{r}(t)$ from $t = a$ to $t = b$ **2.** 1; $b - a$ **3.** $\left\|\dfrac{d\mathbf{r}}{du}\right\|$ **4.** $\langle -3, 3, \sqrt{3}\rangle$

13.4 UNIT TANGENT, NORMAL, AND BINORMAL VECTORS

In this section we will discuss some of the fundamental geometric properties of vector-valued functions. Our work here will have important applications to the study of motion along a curved path in 2-space or 3-space and to the study of the geometric properties of curves and surfaces.

UNIT TANGENT VECTORS

Recall that if C is the graph of a *smooth* vector-valued function $\mathbf{r}(t)$ in 2-space or 3-space, then the vector $\mathbf{r}'(t)$ is nonzero, tangent to C, and points in the direction of increasing parameter. Thus, by normalizing $\mathbf{r}'(t)$ we obtain a unit vector

$$\mathbf{T}(t) = \frac{\mathbf{r}'(t)}{\|\mathbf{r}'(t)\|} \tag{1}$$

that is tangent to C and points in the direction of increasing parameter. We call $\mathbf{T}(t)$ the ***unit tangent vector*** to C at t.

Unless stated otherwise, we will assume that $\mathbf{T}(t)$ is positioned with its initial point at the terminal point of $\mathbf{r}(t)$, as in Figure 13.4.1. This will ensure that $\mathbf{T}(t)$ is actually tangent to the graph of $\mathbf{r}(t)$ and not simply parallel to the tangent line.

▶ **Example 1** Find the unit tangent vector to the graph of $\mathbf{r}(t) = t^2\mathbf{i} + t^3\mathbf{j}$ at the point where $t = 2$.

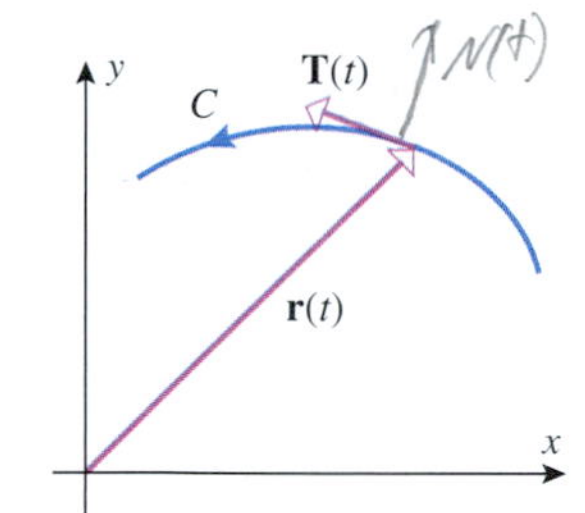

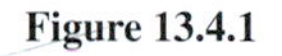
Figure 13.4.1

Solution. Since

$$\mathbf{r}'(t) = 2t\mathbf{i} + 3t^2\mathbf{j}$$

we obtain

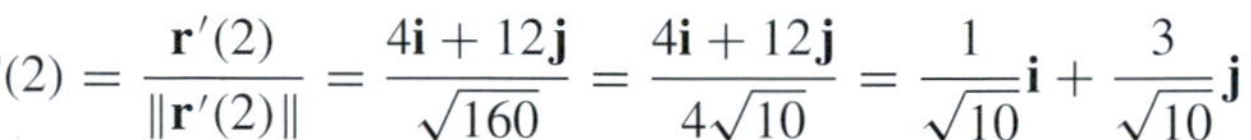

$$\mathbf{T}(2) = \frac{\mathbf{r}'(2)}{\|\mathbf{r}'(2)\|} = \frac{4\mathbf{i} + 12\mathbf{j}}{\sqrt{160}} = \frac{4\mathbf{i} + 12\mathbf{j}}{4\sqrt{10}} = \frac{1}{\sqrt{10}}\mathbf{i} + \frac{3}{\sqrt{10}}\mathbf{j}$$

The graph of $\mathbf{r}(t)$ and the vector $\mathbf{T}(2)$ are shown in Figure 13.4.2. ◀

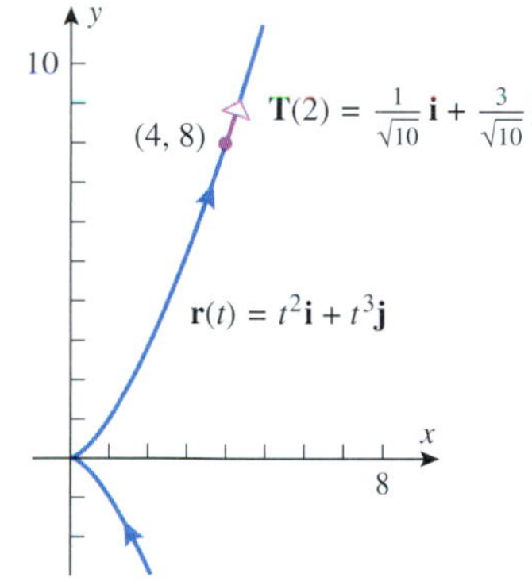
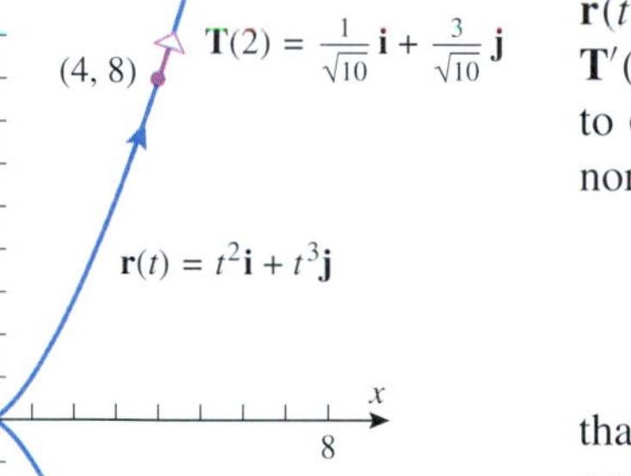

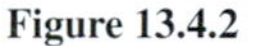
Figure 13.4.2

UNIT NORMAL VECTORS

Recall from Theorem 13.2.8 that if a vector-valued function $\mathbf{r}(t)$ has constant norm, then $\mathbf{r}(t)$ and $\mathbf{r}'(t)$ are orthogonal vectors. In particular, $\mathbf{T}(t)$ has constant norm 1, so $\mathbf{T}(t)$ and $\mathbf{T}'(t)$ are orthogonal vectors. This implies that $\mathbf{T}'(t)$ is perpendicular to the tangent line to C at t, so we say that $\mathbf{T}'(t)$ is ***normal*** to C at t. It follows that if $\mathbf{T}'(t) \neq \mathbf{0}$, and if we normalize $\mathbf{T}'(t)$, then we obtain a unit vector

$$\mathbf{N}(t) = \frac{\mathbf{T}'(t)}{\|\mathbf{T}'(t)\|} \tag{2}$$

that is normal to C and points in the same direction as $\mathbf{T}'(t)$. We call $\mathbf{N}(t)$ the ***principal unit normal vector*** to C at t, or more simply, the ***unit normal vector***. Observe that the unit normal vector is only defined at points where $\mathbf{T}'(t) \neq \mathbf{0}$. Unless stated otherwise, we will assume that this condition is satisfied. In particular, this *excludes* straight lines.

In 2-space there are two unit vectors that are orthogonal to $\mathbf{T}(t)$, and in 3-space there are infinitely many such vectors (Figure 13.4.3). In both cases the principal unit normal is that particular normal that points in the direction of $\mathbf{T}'(t)$. After the next example we will show that for a nonlinear parametric curve in 2-space the principal unit normal is the one that points "inward" toward the concave side of the curve.

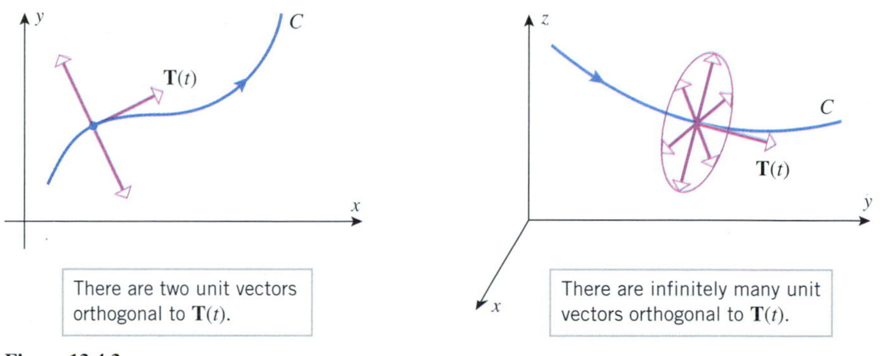

Figure 13.4.3

▶ Example 2 Find $\mathbf{T}(t)$ and $\mathbf{N}(t)$ for the circular helix

$$x = a\cos t, \quad y = a\sin t, \quad z = ct$$

where $a > 0$.

Solution. The radius vector for the helix is

$$\mathbf{r}(t) = a\cos t\,\mathbf{i} + a\sin t\,\mathbf{j} + ct\mathbf{k}$$

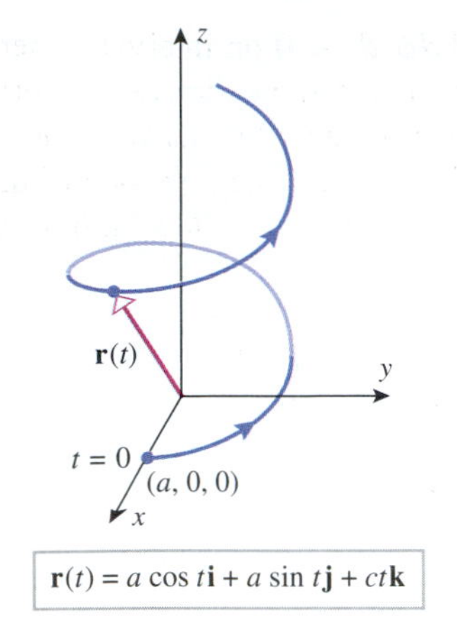

Figure 13.4.4

(Figure 13.4.4). Thus,

$$\mathbf{r}'(t) = (-a \sin t)\mathbf{i} + a \cos t\,\mathbf{j} + c\mathbf{k}$$

$$\|\mathbf{r}'(t)\| = \sqrt{(-a \sin t)^2 + (a \cos t)^2 + c^2} = \sqrt{a^2 + c^2}$$

$$\mathbf{T}(t) = \frac{\mathbf{r}'(t)}{\|\mathbf{r}'(t)\|} = -\frac{a \sin t}{\sqrt{a^2 + c^2}}\mathbf{i} + \frac{a \cos t}{\sqrt{a^2 + c^2}}\mathbf{j} + \frac{c}{\sqrt{a^2 + c^2}}\mathbf{k}$$

$$\mathbf{T}'(t) = -\frac{a \cos t}{\sqrt{a^2 + c^2}}\mathbf{i} - \frac{a \sin t}{\sqrt{a^2 + c^2}}\mathbf{j}$$

$$\|\mathbf{T}'(t)\| = \sqrt{\left(-\frac{a \cos t}{\sqrt{a^2 + c^2}}\right)^2 + \left(-\frac{a \sin t}{\sqrt{a^2 + c^2}}\right)^2} = \sqrt{\frac{a^2}{a^2 + c^2}} = \frac{a}{\sqrt{a^2 + c^2}}$$

$$\mathbf{N}(t) = \frac{\mathbf{T}'(t)}{\|\mathbf{T}'(t)\|} = (-\cos t)\mathbf{i} - (\sin t)\mathbf{j} = -(\cos t\,\mathbf{i} + \sin t\,\mathbf{j})$$

Note that the $\mathbf{k}$ component of the principal unit normal $\mathbf{N}(t)$ is zero for every value of t, so this vector always lies in a horizontal plane, as illustrated in Figure 13.4.5. We leave it as an exercise to show that this vector actually always points toward the z-axis. ◀

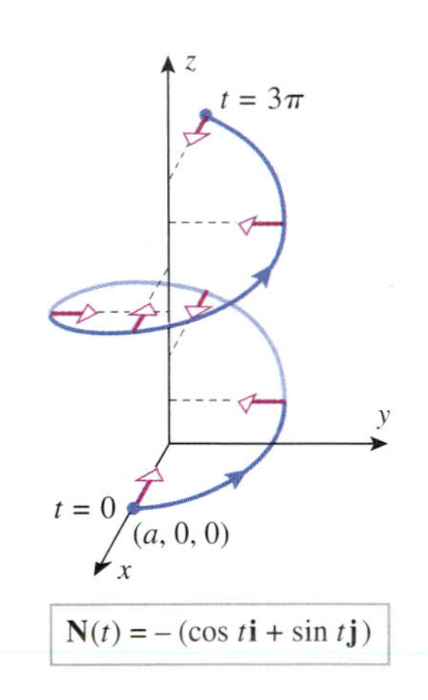

Figure 13.4.5

INWARD UNIT NORMAL VECTORS IN 2-SPACE

Our next objective is to show that for a nonlinear parametric curve C in 2-space the unit normal vector always points toward the concave side of C. For this purpose, let $\phi(t)$ be the angle from the positive x-axis to $\mathbf{T}(t)$, and let $\mathbf{n}(t)$ be the unit vector that results when $\mathbf{T}(t)$ is rotated counterclockwise through an angle of $\pi/2$ (Figure 13.4.6). Since $\mathbf{T}(t)$ and $\mathbf{n}(t)$ are unit vectors, it follows from Formula (12) of Section 12.2 that these vectors can be expressed as

$$\mathbf{T}(t) = \cos\phi(t)\mathbf{i} + \sin\phi(t)\mathbf{j} \tag{3}$$

and

$$\mathbf{n}(t) = \cos[\phi(t) + \pi/2]\mathbf{i} + \sin[\phi(t) + \pi/2]\mathbf{j} = -\sin\phi(t)\mathbf{i} + \cos\phi(t)\mathbf{j} \tag{4}$$

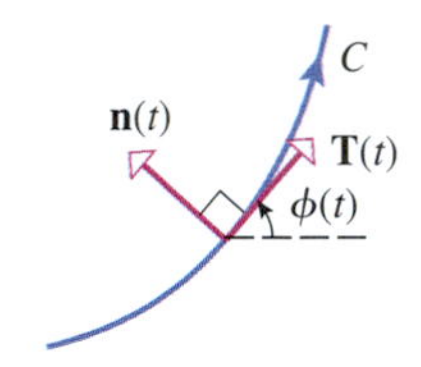

Figure 13.4.6

Observe that on intervals where $\phi(t)$ is increasing the vector $\mathbf{n}(t)$ points *toward* the concave side of C, and on intervals where $\phi(t)$ is decreasing it points *away* from the concave side (Figure 13.4.7).

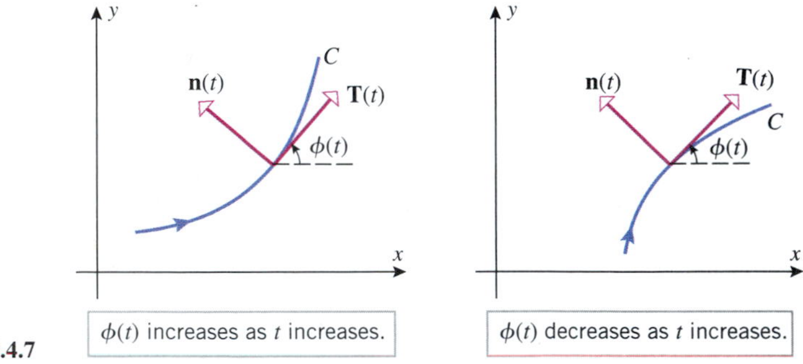

Figure 13.4.7

Now let us differentiate $\mathbf{T}(t)$ by using Formula (3) and applying the chain rule. This yields

$$\frac{d\mathbf{T}}{dt} = \frac{d\mathbf{T}}{d\phi}\frac{d\phi}{dt} = [(-\sin\phi)\mathbf{i} + (\cos\phi)\mathbf{j}]\frac{d\phi}{dt}$$

and thus from (4)

$$\frac{d\mathbf{T}}{dt} = \mathbf{n}(t)\frac{d\phi}{dt} \tag{5}$$

But $d\phi/dt > 0$ on intervals where $\phi(t)$ is increasing and $d\phi/dt < 0$ on intervals where $\phi(t)$ is decreasing. Thus, it follows from (5) that $d\mathbf{T}/dt$ has the same direction as $\mathbf{n}(t)$ on intervals where $\phi(t)$ is increasing and the opposite direction on intervals where $\phi(t)$ is decreasing. Therefore, $\mathbf{T}'(t) = d\mathbf{T}/dt$ points "inward" toward the concave side of the curve in all cases, and hence so does $\mathbf{N}(t)$. For this reason, $\mathbf{N}(t)$ is also called the ***inward unit normal*** when applied to curves in 2-space.

COMPUTING T AND N FOR CURVES PARAMETRIZED BY ARC LENGTH

In the case where $\mathbf{r}(s)$ is parametrized by arc length, the procedures for computing the unit tangent vector $\mathbf{T}(s)$ and the unit normal vector $\mathbf{N}(s)$ are simpler than in the general case. For example, we showed in Theorem 13.3.4 that if s is an arc length parameter, then $\|\mathbf{r}'(s)\| = 1$. Thus, Formula (1) for the unit tangent vector simplifies to

WARNING

Formulas (6) and (7) are only applicable when the curve is parametrized by an arc length parameter s. For other parametrizations Formulas (1) and (2) can be used.

$$\mathbf{T}(s) = \mathbf{r}'(s) \tag{6}$$

and consequently Formula (2) for the unit normal vector simplifies to

$$\mathbf{N}(s) = \frac{\mathbf{r}''(s)}{\|\mathbf{r}''(s)\|} \tag{7}$$

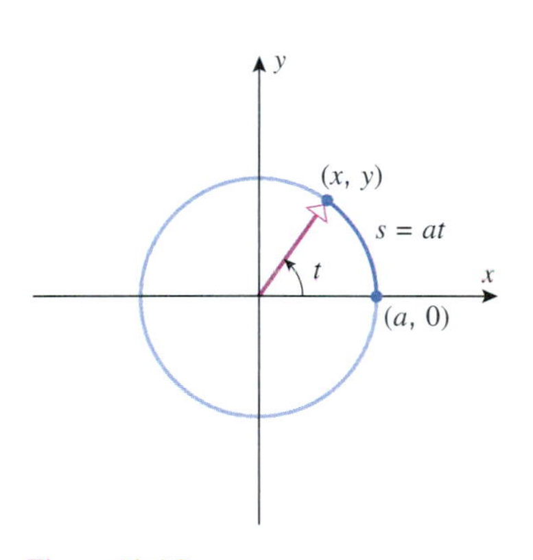

Figure 13.4.8

Example 3 The circle of radius a with counterclockwise orientation and centered at the origin can be represented by the vector-valued function

$$\mathbf{r} = a\cos t\mathbf{i} + a\sin t\mathbf{j} \qquad (0 \le t \le 2\pi) \tag{8}$$

In this representation we can interpret t as the angle in radian measure from the positive x-axis to the radius vector (Figure 13.4.8). This angle subtends an arc of length $s = at$ on the circle, so we can reparametrize the circle in terms of s by substituting s/a for t in (8). This yields

$$\mathbf{r}(s) = a\cos(s/a)\mathbf{i} + a\sin(s/a)\mathbf{j} \qquad (0 \le s \le 2\pi a)$$

To find $\mathbf{T}(s)$ and $\mathbf{N}(s)$ from Formulas (6) and (7), we must compute $\mathbf{r}'(s)$, $\mathbf{r}''(s)$, and $\|\mathbf{r}''(s)\|$. Doing so, we obtain

$$\mathbf{r}'(s) = -\sin(s/a)\mathbf{i} + \cos(s/a)\mathbf{j}$$

$$\mathbf{r}''(s) = -(1/a)\cos(s/a)\mathbf{i} - (1/a)\sin(s/a)\mathbf{j}$$

$$\|\mathbf{r}''(s)\| = \sqrt{(-1/a)^2\cos^2(s/a) + (-1/a)^2\sin^2(s/a)} = 1/a$$

Thus,

$$\mathbf{T}(s) = \mathbf{r}'(s) = -\sin(s/a)\mathbf{i} + \cos(s/a)\mathbf{j}$$

$$\mathbf{N}(s) = \mathbf{r}''(s)/\|\mathbf{r}''(s)\| = -\cos(s/a)\mathbf{i} - \sin(s/a)\mathbf{j}$$

so $\mathbf{N}(s)$ points toward the center of the circle for all s (Figure 13.4.9). This makes sense geometrically and is also consistent with our earlier observation that in 2-space the unit normal vector is the inward normal. ◂

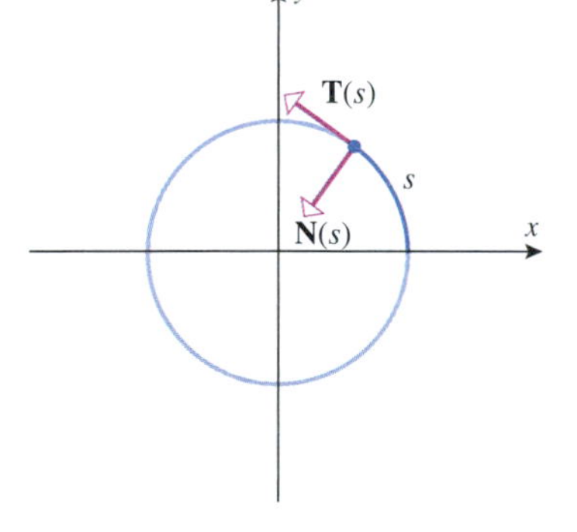

Figure 13.4.9

BINORMAL VECTORS IN 3-SPACE

If C is the graph of a vector-valued function $\mathbf{r}(t)$ in 3-space, then we define the ***binormal vector*** to C at t to be

$$\mathbf{B}(t) = \mathbf{T}(t) \times \mathbf{N}(t) \tag{9}$$

It follows from properties of the cross product that $\mathbf{B}(t)$ is orthogonal to both $\mathbf{T}(t)$ and $\mathbf{N}(t)$ and is oriented relative to $\mathbf{T}(t)$ and $\mathbf{N}(t)$ by the right-hand rule. Moreover, $\mathbf{T}(t) \times \mathbf{N}(t)$ is a unit vector since

$$\|\mathbf{T}(t) \times \mathbf{N}(t)\| = \|\mathbf{T}(t)\|\|\mathbf{N}(t)\|\sin(\pi/2) = 1$$

Thus, $\{\mathbf{T}(t), \mathbf{N}(t), \mathbf{B}(t)\}$ is a set of three mutually orthogonal unit vectors.

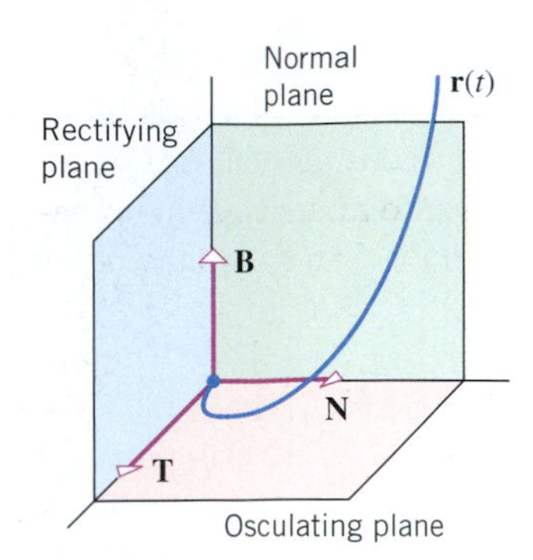

Figure 13.4.10

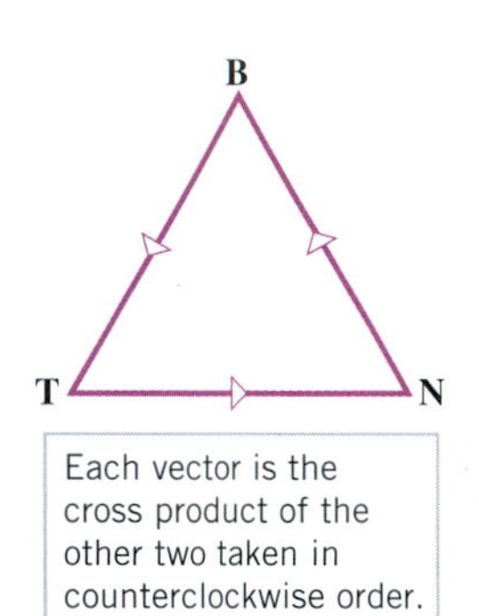

Figure 13.4.11

Just as the vectors **i**, **j**, and **k** determine a right-handed coordinate system in 3-space, so do the vectors $\mathbf{T}(t)$, $\mathbf{N}(t)$, and $\mathbf{B}(t)$. At each point on a smooth parametric curve C in 3-space, these vectors determine three mutually perpendicular planes that pass through the point—the **TB**-plane (called the ***rectifying plane***), the **TN**-plane (called the ***osculating plane***), and the **NB**-plane (called the ***normal plane***) (Figure 13.4.10). Moreover, one can show that a coordinate system determined by $\mathbf{T}(t)$, $\mathbf{N}(t)$, and $\mathbf{B}(t)$ is right-handed in the sense that each of these vectors is related to the other two by the right-hand rule (Figure 13.4.11):

$$\mathbf{B}(t) = \mathbf{T}(t) \times \mathbf{N}(t), \quad \mathbf{N}(t) = \mathbf{B}(t) \times \mathbf{T}(t), \quad \mathbf{T}(t) = \mathbf{N}(t) \times \mathbf{B}(t) \tag{10}$$

The coordinate system determined by $\mathbf{T}(t)$, $\mathbf{N}(t)$, and $\mathbf{B}(t)$ is called the **TNB*-frame*** or sometimes the ***Frenet frame*** in honor of the French mathematician Jean Frédéric Frenet (1816–1900) who pioneered its application to the study of space curves. Typically, the xyz-coordinate system determined by the unit vectors **i**, **j**, and **k** remains fixed, whereas the **TNB**-frame changes as its origin moves along the curve C (Figure 13.4.12).

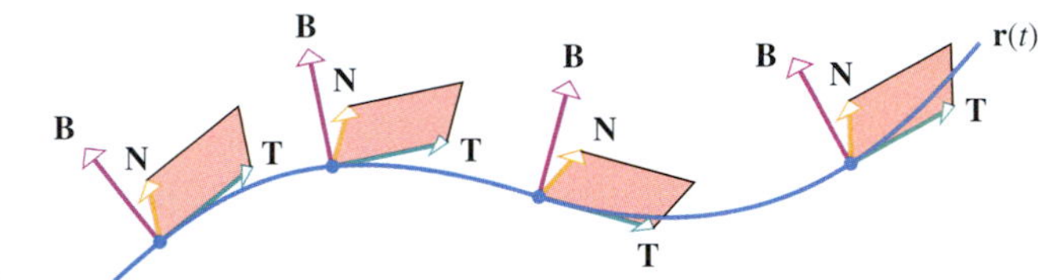

Figure 13.4.12

Formula (9) expresses $\mathbf{B}(t)$ in terms of $\mathbf{T}(t)$ and $\mathbf{N}(t)$. Alternatively, the binormal $\mathbf{B}(t)$ can be expressed directly in terms of $\mathbf{r}(t)$ as

$$\mathbf{B}(t) = \frac{\mathbf{r}'(t) \times \mathbf{r}''(t)}{\|\mathbf{r}'(t) \times \mathbf{r}''(t)\|} \tag{11}$$

and in the case where the parameter is arc length it can be expressed in terms of $\mathbf{r}(s)$ as

$$\mathbf{B}(s) = \frac{\mathbf{r}'(s) \times \mathbf{r}''(s)}{\|\mathbf{r}''(s)\|} \tag{12}$$

We omit the proof.

✔QUICK CHECK EXERCISES 13.4 (See page 892 for answers.)

1. If C is the graph of a smooth vector-valued function $\mathbf{r}(t)$, then the unit tangent, unit normal, and binormal to C at t are defined, respectively, by

   $\mathbf{T}(t) =$ ________, $\mathbf{N}(t) =$ ________, $\mathbf{B}(t) =$ ________

2. If C is the graph of a smooth vector-valued function $\mathbf{r}(s)$ parametrized by arc length, then the definitions of the unit tangent and unit normal to C at s simplify, respectively, to

   $\mathbf{T}(s) =$ ________ and $\mathbf{N}(s) =$ ________

3. If C is the graph of a smooth vector-valued function $\mathbf{r}(t)$, then the unit binormal vector to C at t may be computed directly in terms of $\mathbf{r}'(t)$ and $\mathbf{r}''(t)$ by the formula $\mathbf{B}(t) =$ ________. When $t = s$ is the arc length parameter, this formula simplifies to $\mathbf{B}(s) =$ ________.

4. Suppose that C is the graph of a smooth vector-valued function $\mathbf{r}(s)$ parametrized by arc length with $\mathbf{r}'(0) = \langle \frac{2}{3}, \frac{1}{3}, \frac{2}{3} \rangle$ and $\mathbf{r}''(0) = \langle -3, 12, -3 \rangle$. Then

   $\mathbf{T}(0) =$ ________, $\mathbf{N}(0) =$ ________, $\mathbf{B}(0) =$ ________

EXERCISE SET 13.4

FOCUS ON CONCEPTS

1. In each part, sketch the unit tangent and normal vectors at the points P, Q, and R, taking into account the orientation of the curve C.

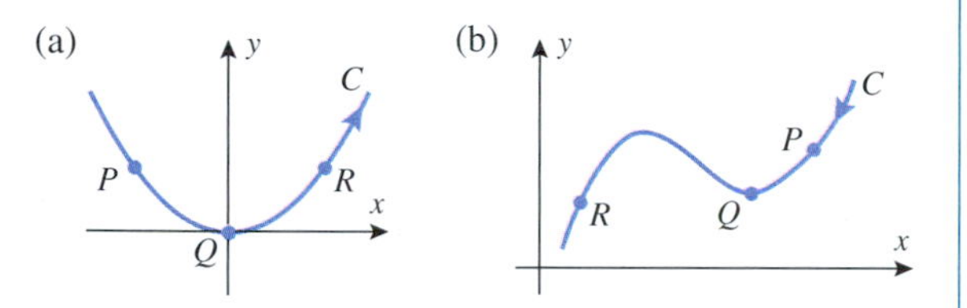

2. Make a rough sketch that shows the ellipse
$$\mathbf{r}(t) = 3\cos t\,\mathbf{i} + 2\sin t\,\mathbf{j}$$
for $0 \le t \le 2\pi$ and the unit tangent and normal vectors at the points $t = 0$, $t = \pi/4$, $t = \pi/2$, and $t = \pi$.

3. In the marginal note associated with Example 8 of Section 13.3, we observed that a line $\mathbf{r} = \mathbf{r}_0 + t\mathbf{v}$ can be parametrized in terms of an arc length parameter s with reference point $\mathbf{r}_0$ by normalizing $\mathbf{v}$. Use this result to show that the tangent line to the graph of $\mathbf{r}(t)$ at the point t_0 can be expressed as
$$\mathbf{r} = \mathbf{r}(t_0) + s\mathbf{T}(t_0)$$
where s is an arc length parameter with reference point $\mathbf{r}(t_0)$.

4. Use the result in Exercise 3 to show that the tangent line to the parabola
$$x = t, \quad y = t^2$$
at the point $(1, 1)$ can be expressed parametrically as
$$x = 1 + \frac{s}{\sqrt{5}}, \quad y = 1 + \frac{2s}{\sqrt{5}}$$

5–12 Find $\mathbf{T}(t)$ and $\mathbf{N}(t)$ at the given point.

5. $\mathbf{r}(t) = (t^2 - 1)\mathbf{i} + t\mathbf{j};\ t = 1$

6. $\mathbf{r}(t) = \frac{1}{2}t^2\mathbf{i} + \frac{1}{3}t^3\mathbf{j};\ t = 1$

7. $\mathbf{r}(t) = 5\cos t\,\mathbf{i} + 5\sin t\,\mathbf{j};\ t = \pi/3$

8. $\mathbf{r}(t) = \ln t\,\mathbf{i} + t\mathbf{j};\ t = e$

9. $\mathbf{r}(t) = 4\cos t\,\mathbf{i} + 4\sin t\,\mathbf{j} + t\mathbf{k};\ t = \pi/2$

10. $\mathbf{r}(t) = t\mathbf{i} + \frac{1}{2}t^2\mathbf{j} + \frac{1}{3}t^3\mathbf{k};\ t = 0$

11. $x = e^t\cos t,\ y = e^t\sin t,\ z = e^t;\ t = 0$

12. $x = \cosh t,\ y = \sinh t,\ z = t;\ t = \ln 2$

13–14 Use the result in Exercise 3 to find parametric equations for the tangent line to the graph of $\mathbf{r}(t)$ at t_0 in terms of an arc length parameter s.

13. $\mathbf{r}(t) = \sin t\,\mathbf{i} + \cos t\,\mathbf{j} + \frac{1}{2}t^2\mathbf{k};\ t_0 = 0$

14. $\mathbf{r}(t) = t\mathbf{i} + t\mathbf{j} + \sqrt{9 - t^2}\,\mathbf{k};\ t_0 = 1$

15–18 Use the formula $\mathbf{B}(t) = \mathbf{T}(t) \times \mathbf{N}(t)$ to find $\mathbf{B}(t)$, and then check your answer by using Formula (11) to find $\mathbf{B}(t)$ directly from $\mathbf{r}(t)$.

15. $\mathbf{r}(t) = 3\sin t\,\mathbf{i} + 3\cos t\,\mathbf{j} + 4t\mathbf{k}$

16. $\mathbf{r}(t) = e^t\sin t\,\mathbf{i} + e^t\cos t\,\mathbf{j} + 3\mathbf{k}$

17. $\mathbf{r}(t) = (\sin t - t\cos t)\mathbf{i} + (\cos t + t\sin t)\mathbf{j} + \mathbf{k}$

18. $\mathbf{r}(t) = a\cos t\,\mathbf{i} + a\sin t\,\mathbf{j} + ct\mathbf{k} \quad (a \ne 0, c \ne 0)$

19–20 Find $\mathbf{T}(t)$, $\mathbf{N}(t)$, and $\mathbf{B}(t)$ for the given value of t. Then find equations for the osculating, normal, and rectifying planes at the point that corresponds to that value of t.

19. $\mathbf{r}(t) = \cos t\,\mathbf{i} + \sin t\,\mathbf{j} + \mathbf{k};\ t = \pi/4$

20. $\mathbf{r}(t) = e^t\mathbf{i} + e^t\cos t\,\mathbf{j} + e^t\sin t\,\mathbf{k};\ t = 0$

21. (a) Use the formula $\mathbf{N}(t) = \mathbf{B}(t) \times \mathbf{T}(t)$ and Formulas (1) and (11) to show that $\mathbf{N}(t)$ can be expressed in terms of $\mathbf{r}(t)$ as
$$\mathbf{N}(t) = \frac{\mathbf{r}'(t) \times \mathbf{r}''(t)}{\|\mathbf{r}'(t) \times \mathbf{r}''(t)\|} \times \frac{\mathbf{r}'(t)}{\|\mathbf{r}'(t)\|}$$
(b) Use properties of cross products to show that the formula in part (a) can be expressed as
$$\mathbf{N}(t) = \frac{(\mathbf{r}'(t) \times \mathbf{r}''(t)) \times \mathbf{r}'(t)}{\|(\mathbf{r}'(t) \times \mathbf{r}''(t)) \times \mathbf{r}'(t)\|}$$
(c) Use the result in part (b) and Exercise 41 of Section 12.4 to show that $\mathbf{N}(t)$ can be expressed directly in terms of $\mathbf{r}(t)$ as
$$\mathbf{N}(t) = \frac{\mathbf{u}(t)}{\|\mathbf{u}(t)\|}$$
where
$$\mathbf{u}(t) = \|\mathbf{r}'(t)\|^2\mathbf{r}''(t) - (\mathbf{r}'(t) \cdot \mathbf{r}''(t))\mathbf{r}'(t)$$

22. Use the result in part (b) of Exercise 21 to find the unit normal vector requested in
(a) Exercise 5 (b) Exercise 9.

23–24 Use the result in part (c) of Exercise 21 to find $\mathbf{N}(t)$.

23. $\mathbf{r}(t) = \sin t\,\mathbf{i} + \cos t\,\mathbf{j} + t\mathbf{k}$

24. $\mathbf{r}(t) = t\mathbf{i} + t^2\mathbf{j} + t^3\mathbf{k}$

✔ QUICK CHECK ANSWERS 13.4

1. $\dfrac{\mathbf{r}'(t)}{\|\mathbf{r}'(t)\|}$; $\dfrac{\mathbf{T}'(t)}{\|\mathbf{T}'(t)\|}$; $\mathbf{T}(t)\times\mathbf{N}(t)$ **2.** $\mathbf{r}'(s)$; $\dfrac{\mathbf{r}''(s)}{\|\mathbf{r}''(s)\|}$ **3.** $\dfrac{\mathbf{r}'(t)\times\mathbf{r}''(t)}{\|\mathbf{r}'(t)\times\mathbf{r}''(t)\|}$; $\dfrac{\mathbf{r}'(s)\times\mathbf{r}''(s)}{\|\mathbf{r}''(s)\|}$
4. $\left\langle \dfrac{2}{3}, \dfrac{1}{3}, \dfrac{2}{3}\right\rangle$; $\left\langle -\dfrac{1}{3\sqrt{2}}, \dfrac{4}{3\sqrt{2}}, -\dfrac{1}{3\sqrt{2}}\right\rangle$; $\left\langle -\dfrac{1}{\sqrt{2}}, 0, \dfrac{1}{\sqrt{2}}\right\rangle$

13.5 CURVATURE

In this section we will consider the problem of obtaining a numerical measure of how sharply a curve in 2-space or 3-space bends. Our results will have applications in geometry and in the study of motion along a curved path.

DEFINITION OF CURVATURE

Suppose that C is the graph of a smooth vector-valued function in 2-space or 3-space that is parametrized in terms of arc length. Figure 13.5.1 suggests that for a curve in 2-space the "sharpness" of the bend in C is closely related to $d\mathbf{T}/ds$, which is the rate of change of the unit tangent vector $\mathbf{T}$ with respect to s. (Keep in mind that $\mathbf{T}$ has constant length, so only its direction changes.) If C is a straight line (no bend), then the direction of $\mathbf{T}$ remains constant (Figure 13.5.1*a*); if C bends slightly, then $\mathbf{T}$ undergoes a gradual change of direction (Figure 13.5.1*b*); and if C bends sharply, then $\mathbf{T}$ undergoes a rapid change of direction (Figure 13.5.1*c*).

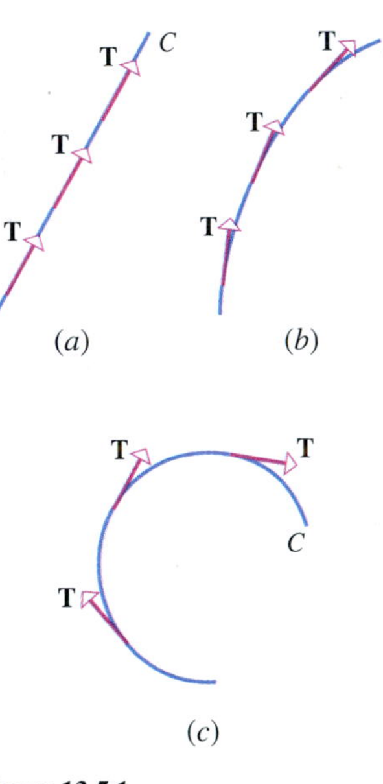

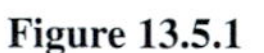
Figure 13.5.1

The situation in 3-space is more complicated because bends in a curve are not limited to a single plane—they can occur in all directions, as illustrated by the complicated tube plot in Figure 13.1.3. To describe the bending characteristics of a curve in 3-space completely, one must take into account $d\mathbf{T}/ds$, $d\mathbf{N}/ds$, and $d\mathbf{B}/ds$. A complete study of this topic would take us too far afield, so we will limit our discussion to $d\mathbf{T}/ds$, which is the most important of these derivatives in applications.

13.5.1 DEFINITION. If C is a smooth curve in 2-space or 3-space that is parametrized by arc length, then the ***curvature*** of C, denoted by $\kappa = \kappa(s)$ (κ = Greek "kappa"), is defined by

$$\kappa(s) = \left\|\frac{d\mathbf{T}}{ds}\right\| = \|\mathbf{r}''(s)\| \tag{1}$$

Observe that $\kappa(s)$ is a real-valued function of s, since it is the *length* of $d\mathbf{T}/ds$ that measures the curvature. In general, the curvature will vary from point to point along a curve; however, the following example shows that the curvature is constant for circles in 2-space, as you might expect.

▶ **Example 1** In Example 3 of Section 13.4 we showed that the circle of radius a, centered at the origin, can be parametrized in terms of arc length as

$$\mathbf{r}(s) = a\cos\left(\frac{s}{a}\right)\mathbf{i} + a\sin\left(\frac{s}{a}\right)\mathbf{j} \qquad (0 \le s \le 2\pi a)$$

Thus,

$$\mathbf{r}''(s) = -\frac{1}{a}\cos\left(\frac{s}{a}\right)\mathbf{i} - \frac{1}{a}\sin\left(\frac{s}{a}\right)\mathbf{j}$$

and hence from (1)

$$\kappa(s) = \|\mathbf{r}''(s)\| = \sqrt{\left[-\frac{1}{a}\cos\left(\frac{s}{a}\right)\right]^2 + \left[-\frac{1}{a}\sin\left(\frac{s}{a}\right)\right]^2} = \frac{1}{a}$$

so the circle has constant curvature $1/a$. ◀

The next example shows that lines have zero curvature, which is consistent with the fact that they do not bend.

▶ **Example 2** Recall from Formula (15) of Section 13.3 that a line in 2-space or 3-space can be parametrized in terms of arc length as

$$\mathbf{r} = \mathbf{r}_0 + s\mathbf{u}$$

where the terminal point of $\mathbf{r}_0$ is a point on the line and $\mathbf{u}$ is a unit vector parallel to the line. Since $\mathbf{u}$ and $\mathbf{r}_0$ are constant, their derivatives with respect to s are zero, and hence

$$\mathbf{r}'(s) = \frac{d\mathbf{r}}{ds} = \frac{d}{ds}[\mathbf{r}_0 + s\mathbf{u}] = \mathbf{0} + \mathbf{u} = \mathbf{u}$$

$$\mathbf{r}''(s) = \frac{d\mathbf{r}'}{ds} = \frac{d}{ds}[\mathbf{u}] = \mathbf{0}$$

Thus,

$$\kappa(s) = \|\mathbf{r}''(s)\| = 0 \text{ ◀}$$

FORMULAS FOR CURVATURE

Formula (1) is only applicable if the curve is parametrized in terms of arc length. The following theorem provides two formulas for curvature in terms of a general parameter t.

13.5.2 THEOREM. *If $\mathbf{r}(t)$ is a smooth vector-valued function in 2-space or 3-space, then for each value of t at which $\mathbf{T}'(t)$ and $\mathbf{r}''(t)$ exist, the curvature κ can be expressed as*

(a) $$\kappa(t) = \frac{\|\mathbf{T}'(t)\|}{\|\mathbf{r}'(t)\|} \tag{2}$$

(b) $$\kappa(t) = \frac{\|\mathbf{r}'(t) \times \mathbf{r}''(t)\|}{\|\mathbf{r}'(t)\|^3} \tag{3}$$

PROOF (a). It follows from Formula (1) and Formulas (16) and (17) of Section 13.3 that

$$\kappa(t) = \left\|\frac{d\mathbf{T}}{ds}\right\| = \left\|\frac{d\mathbf{T}/dt}{ds/dt}\right\| = \left\|\frac{d\mathbf{T}/dt}{\|d\mathbf{r}/dt\|}\right\| = \frac{\|\mathbf{T}'(t)\|}{\|\mathbf{r}'(t)\|}$$

PROOF (b). It follows from Formula (1) of Section 13.4 that

$$\mathbf{r}'(t) = \|\mathbf{r}'(t)\|\mathbf{T}(t) \tag{4}$$

$$\mathbf{r}''(t) = \|\mathbf{r}'(t)\|'\mathbf{T}(t) + \|\mathbf{r}'(t)\|\mathbf{T}'(t) \tag{5}$$

But from Formula (2) of Section 13.4 and part (a) of this theorem we have

$$\mathbf{T}'(t) = \|\mathbf{T}'(t)\|\mathbf{N}(t) \quad \text{and} \quad \|\mathbf{T}'(t)\| = \kappa(t)\|\mathbf{r}'(t)\|$$

so

$$\mathbf{T}'(t) = \kappa(t)\|\mathbf{r}'(t)\|\mathbf{N}(t)$$

Substituting this into (5) yields

$$\mathbf{r}''(t) = \|\mathbf{r}'(t)\|'\mathbf{T}(t) + \kappa(t)\|\mathbf{r}'(t)\|^2\mathbf{N}(t) \tag{6}$$

Thus, from (4) and (6)

$$\mathbf{r}'(t) \times \mathbf{r}''(t) = \|\mathbf{r}'(t)\|\,\|\mathbf{r}'(t)\|'(\mathbf{T}(t) \times \mathbf{T}(t)) + \kappa(t)\|\mathbf{r}'(t)\|^3(\mathbf{T}(t) \times \mathbf{N}(t))$$

But the cross product of a vector with itself is zero, so this equation simplifies to

$$\mathbf{r}'(t) \times \mathbf{r}''(t) = \kappa(t)\|\mathbf{r}'(t)\|^3(\mathbf{T}(t) \times \mathbf{N}(t)) = \kappa(t)\|\mathbf{r}'(t)\|^3\mathbf{B}(t)$$

It follows from this equation and the fact that $\mathbf{B}(t)$ is a unit vector that

$$\|\mathbf{r}'(t) \times \mathbf{r}''(t)\| = \kappa(t)\|\mathbf{r}'(t)\|^3$$

Formula (3) now follows. ■

> Formula (2) is useful if $\mathbf{T}(t)$ is known or is easy to obtain; however, Formula (3) will usually be easier to apply, since it involves only $\mathbf{r}(t)$ and its derivatives. We also note that cross products were defined only for vectors in 3-space, so to use Formula (3) in 2-space we must first write the 2-space function $\mathbf{r}(t) = x(t)\mathbf{i} + y(t)\mathbf{j}$ as the 3-space function $\mathbf{r}(t) = x(t)\mathbf{i} + y(t)\mathbf{j} + 0\mathbf{k}$ with a zero $\mathbf{k}$ component.

▶ Example 3 Find $\kappa(t)$ for the circular helix

$$x = a\cos t, \quad y = a\sin t, \quad z = ct$$

where $a > 0$.

Solution. The radius vector for the helix is

$$\mathbf{r}(t) = a\cos t\mathbf{i} + a\sin t\mathbf{j} + ct\mathbf{k}$$

Thus,

$$\mathbf{r}'(t) = (-a\sin t)\mathbf{i} + a\cos t\mathbf{j} + c\mathbf{k}$$

$$\mathbf{r}''(t) = (-a\cos t)\mathbf{i} + (-a\sin t)\mathbf{j}$$

$$\mathbf{r}'(t) \times \mathbf{r}''(t) = \begin{vmatrix} \mathbf{i} & \mathbf{j} & \mathbf{k} \\ -a\sin t & a\cos t & c \\ -a\cos t & -a\sin t & 0 \end{vmatrix} = (ac\sin t)\mathbf{i} - (ac\cos t)\mathbf{j} + a^2\mathbf{k}$$

Therefore,

$$\|\mathbf{r}'(t)\| = \sqrt{(-a\sin t)^2 + (a\cos t)^2 + c^2} = \sqrt{a^2 + c^2}$$

and

$$\|\mathbf{r}'(t) \times \mathbf{r}''(t)\| = \sqrt{(ac\sin t)^2 + (-ac\cos t)^2 + a^4}$$
$$= \sqrt{a^2c^2 + a^4} = a\sqrt{a^2 + c^2}$$

so

$$\kappa(t) = \frac{\|\mathbf{r}'(t) \times \mathbf{r}''(t)\|}{\|\mathbf{r}'(t)\|^3} = \frac{a\sqrt{a^2 + c^2}}{(\sqrt{a^2 + c^2})^3} = \frac{a}{a^2 + c^2}$$

Note that κ does not depend on t, which tells us that the helix has constant curvature. ◀

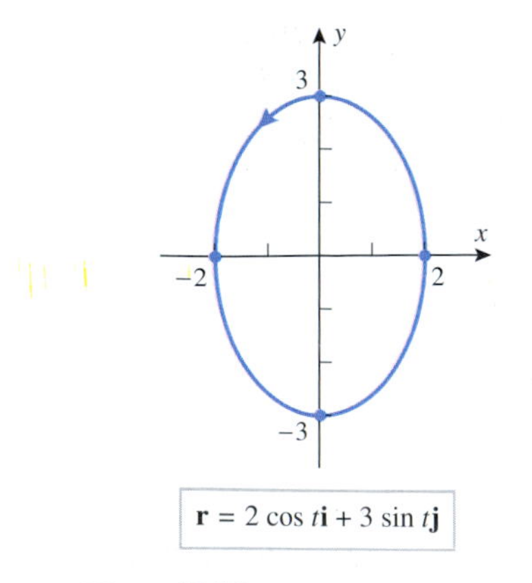

Figure 13.5.2

▶ Example 4 The graph of the vector equation

$$\mathbf{r} = 2\cos t\mathbf{i} + 3\sin t\mathbf{j} \quad (0 \le t \le 2\pi)$$

is the ellipse in Figure 13.5.2. Find the curvature of the ellipse at the endpoints of the major and minor axes, and use a graphing utility to generate the graph of $\kappa(t)$.

Solution. To apply Formula (3), we must treat the ellipse as a curve in the xy-plane of an xyz-coordinate system by adding a zero $\mathbf{k}$ component and writing its equation as

$$\mathbf{r} = 2\cos t\mathbf{i} + 3\sin t\mathbf{j} + 0\mathbf{k}$$

It is not essential to write the zero $\mathbf{k}$ component explicitly as long as you assume it to be there when you calculate a cross product. Thus,

$$\mathbf{r}'(t) = (-2\sin t)\mathbf{i} + 3\cos t\mathbf{j}$$

$$\mathbf{r}''(t) = (-2\cos t)\mathbf{i} + (-3\sin t)\mathbf{j}$$

$$\mathbf{r}'(t) \times \mathbf{r}''(t) = \begin{vmatrix} \mathbf{i} & \mathbf{j} & \mathbf{k} \\ -2\sin t & 3\cos t & 0 \\ -2\cos t & -3\sin t & 0 \end{vmatrix} = [(6\sin^2 t) + (6\cos^2 t)]\mathbf{k} = 6\mathbf{k}$$

Therefore,

$$\|\mathbf{r}'(t)\| = \sqrt{(-2\sin t)^2 + (3\cos t)^2} = \sqrt{4\sin^2 t + 9\cos^2 t}$$

$$\|\mathbf{r}'(t) \times \mathbf{r}''(t)\| = 6$$

so

$$\kappa(t) = \frac{\|\mathbf{r}'(t) \times \mathbf{r}''(t)\|}{\|\mathbf{r}'(t)\|^3} = \frac{6}{[4\sin^2 t + 9\cos^2 t]^{3/2}} \tag{7}$$

The endpoints of the minor axis are $(2, 0)$ and $(-2, 0)$, which correspond to $t = 0$ and $t = \pi$, respectively. Substituting these values in (7) yields the same curvature at both points, namely,

$$\kappa = \kappa(0) = \kappa(\pi) = \frac{6}{9^{3/2}} = \frac{6}{27} = \frac{2}{9}$$

The endpoints of the major axis are $(0, 3)$ and $(0, -3)$, which correspond to $t = \pi/2$ and $t = 3\pi/2$, respectively; from (7) the curvature at these points is

$$\kappa = \kappa\left(\frac{\pi}{2}\right) = \kappa\left(\frac{3\pi}{2}\right) = \frac{6}{4^{3/2}} = \frac{3}{4}$$

Observe that the curvature is greater at the ends of the major axis than at the ends of the minor axis, as you might expect. Figure 13.5.3 shows the graph of κ versus t. This graph illustrates clearly that the curvature is minimum at $t = 0$ (the right end of the minor axis), increases to a maximum at $t = \pi/2$ (the top of the major axis), decreases to a minimum again at $t = \pi$ (the left end of the minor axis), and continues cyclically in this manner. Figure 13.5.4 provides another way of picturing the curvature. ◀

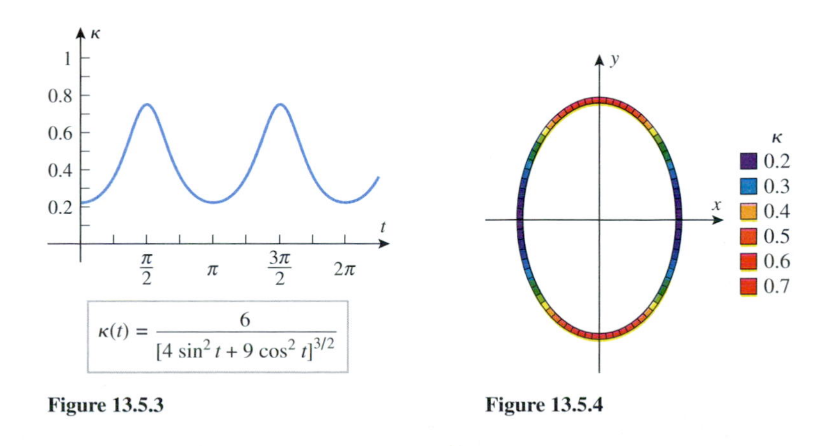

Figure 13.5.3

Figure 13.5.4

RADIUS OF CURVATURE

In the last example we found the curvature at the ends of the minor axis to be $\frac{2}{9}$ and the curvature at the ends of the major axis to be $\frac{3}{4}$. To obtain a better understanding of the meaning of these numbers, recall from Example 1 that a circle of radius a has a constant

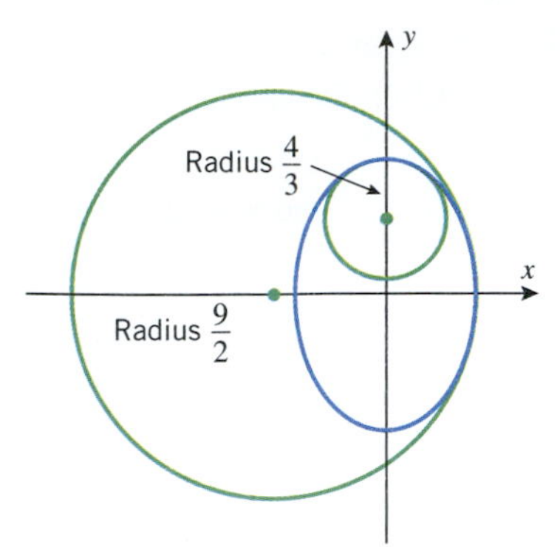

Figure 13.5.5

curvature of $1/a$; thus, the curvature of the ellipse at the ends of the minor axis is the same as that of a circle of radius $\frac{9}{2}$, and the curvature at the ends of the major axis is the same as that of a circle of radius $\frac{4}{3}$ (Figure 13.5.5).

In general, if a curve C in 2-space has nonzero curvature κ at a point P, then the circle of radius $\rho = 1/\kappa$ sharing a common tangent with C at P, and centered on the concave side of the curve at P, is called the ***circle of curvature*** or ***osculating circle*** at P (Figure 13.5.6). The osculating circle and the curve C not only touch at P but they have equal curvatures at that point. In this sense, the osculating circle is the circle that best approximates the curve C near P. The radius ρ of the osculating circle at P is called the ***radius of curvature*** at P, and the center of the circle is called the ***center of curvature*** at P (Figure 13.5.6).

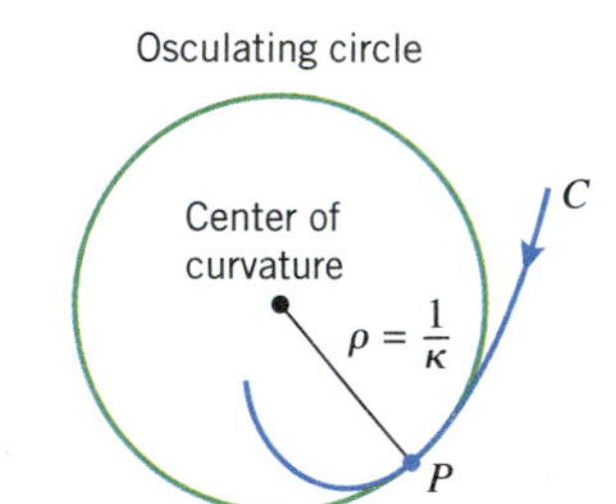

Figure 13.5.6

AN INTERPRETATION OF CURVATURE IN 2-SPACE

A useful geometric interpretation of curvature in 2-space can be obtained by considering the angle ϕ measured counterclockwise from the direction of the positive x-axis to the unit tangent vector $\mathbf{T}$ (Figure 13.5.7). By Formula (12) of Section 12.2, we can express $\mathbf{T}$ in terms of ϕ as

$$\mathbf{T}(\phi) = \cos\phi\mathbf{i} + \sin\phi\mathbf{j}$$

Thus,

$$\frac{d\mathbf{T}}{d\phi} = (-\sin\phi)\mathbf{i} + \cos\phi\mathbf{j}$$

$$\frac{d\mathbf{T}}{ds} = \frac{d\mathbf{T}}{d\phi}\frac{d\phi}{ds}$$

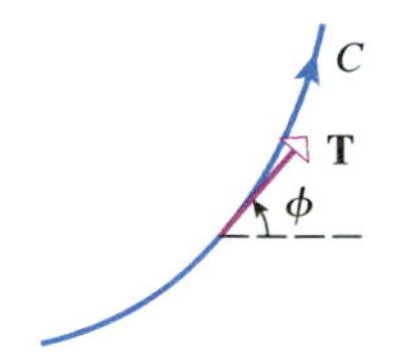

Figure 13.5.7

from which we obtain

$$\kappa(s) = \left\|\frac{d\mathbf{T}}{ds}\right\| = \left|\frac{d\phi}{ds}\right|\left\|\frac{d\mathbf{T}}{d\phi}\right\| = \left|\frac{d\phi}{ds}\right|\sqrt{(-\sin\phi)^2 + \cos^2\phi} = \left|\frac{d\phi}{ds}\right|$$

In summary, we have shown that

$$\kappa(s) = \left|\frac{d\phi}{ds}\right| \tag{8}$$

which tells us that curvature in 2-space can be interpreted as the magnitude of the rate of change of ϕ with respect to s—the greater the curvature, the more rapidly ϕ changes with s (Figure 13.5.8). In the case of a straight line, the angle ϕ is constant (Figure 13.5.9) and consequently $\kappa(s) = |d\phi/ds| = 0$, which is consistent with the fact that a straight line has zero curvature at every point.

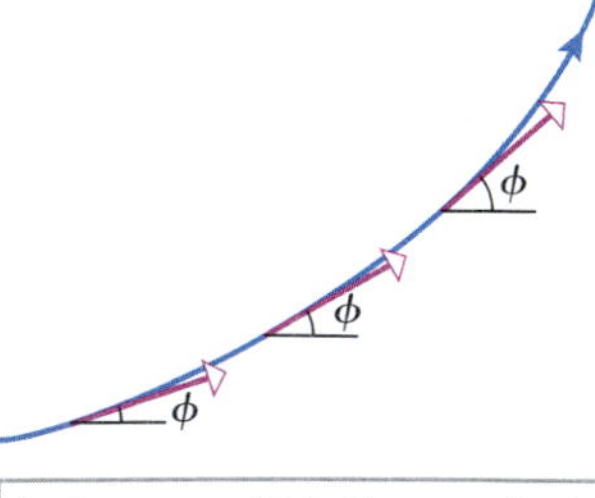

In 2-space, $\kappa(s)$ is the magnitude of the rate of change of ϕ with respect to s.

Figure 13.5.8

FORMULA SUMMARY

We conclude this section with a summary of formulas for $\mathbf{T}$, $\mathbf{N}$, and $\mathbf{B}$. These formulas have either been derived in the text or are easily derivable from formulas we have already established.

$$\mathbf{T}(s) = \mathbf{r}'(s) \tag{9}$$

$$\mathbf{N}(s) = \frac{1}{\kappa(s)}\frac{d\mathbf{T}}{ds} = \frac{\mathbf{r}''(s)}{\|\mathbf{r}''(s)\|} = \frac{\mathbf{r}''(s)}{\kappa(s)} \tag{10}$$

$$\mathbf{B}(s) = \frac{\mathbf{r}'(s)\times\mathbf{r}''(s)}{\|\mathbf{r}''(s)\|} = \frac{\mathbf{r}'(s)\times\mathbf{r}''(s)}{\kappa(s)} \tag{11}$$

ϕ is constant, so the line has zero curvature.

Figure 13.5.9

$$\mathbf{T}(t) = \frac{\mathbf{r}'(t)}{\|\mathbf{r}'(t)\|} \tag{12}$$

$$\mathbf{B}(t) = \frac{\mathbf{r}'(t) \times \mathbf{r}''(t)}{\|\mathbf{r}'(t) \times \mathbf{r}''(t)\|} \tag{13}$$

$$\mathbf{N}(t) = \mathbf{B}(t) \times \mathbf{T}(t) \tag{14}$$

✔ QUICK CHECK EXERCISES 13.5 (See page 900 for answers.)

1. If C is a smooth curve parametrized by arc length, then the curvature is defined by $\kappa(s) =$ ________.

2. Let $\mathbf{r}(t)$ be a smooth vector-valued function with curvature $\kappa(t)$.
(a) The curvature may be expressed in terms of $\mathbf{T}'(t)$ and $\mathbf{r}'(t)$ as $\kappa(t) =$ ________.
(b) The curvature may be expressed directly in terms of $\mathbf{r}'(t)$ and $\mathbf{r}''(t)$ as $\kappa(t) =$ ________.

3. Suppose that C is the graph of a smooth vector-valued function $\mathbf{r}(s) = \langle x(s), y(s)\rangle$ parametrized by arc length and that the unit tangent $\mathbf{T}(s) = \langle \cos\phi(s), \sin\phi(s)\rangle$. Then the curvature may be expressed in terms of $\phi(s)$ as $\kappa(s) =$ ________.

4. Suppose that C is a smooth curve and that $x^2 + y^2 = 4$ is the osculating circle to C at $P(1, \sqrt{3})$. Then the curvature of C at P is ________.

EXERCISE SET 13.5 Graphing Utility C CAS

FOCUS ON CONCEPTS

1–2 Use the osculating circle shown in the figure to estimate the curvature at the indicated point.

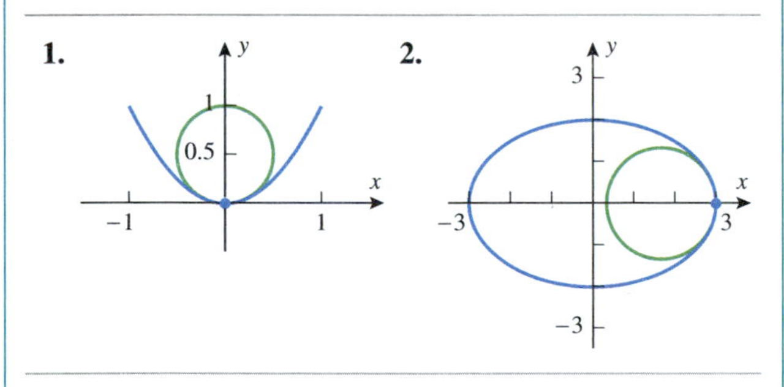

3–4 For a plane curve $y = f(x)$ the curvature at $(x, f(x))$ is a function $\kappa(x)$. In these exercises the graphs of $f(x)$ and $\kappa(x)$ are shown. Determine which is which and explain your reasoning.

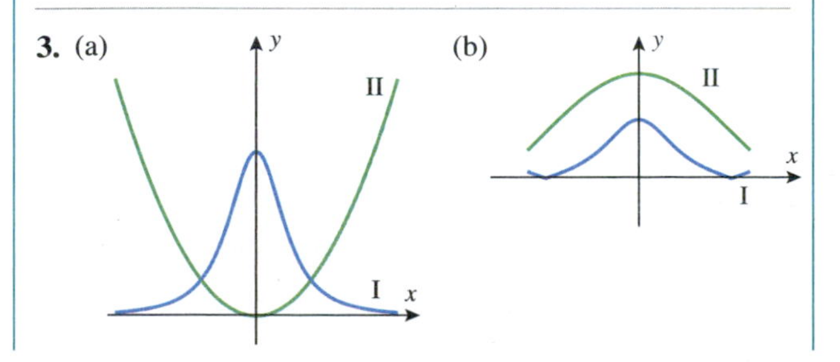

4. (a) (b)

5–12 Use Formula (3) to find $\kappa(t)$.

5. $\mathbf{r}(t) = t^2\mathbf{i} + t^3\mathbf{j}$

6. $\mathbf{r}(t) = 4\cos t\mathbf{i} + \sin t\mathbf{j}$

7. $\mathbf{r}(t) = e^{3t}\mathbf{i} + e^{-t}\mathbf{j}$

8. $x = 1 - t^3, \quad y = t - t^2$

9. $\mathbf{r}(t) = 4\cos t\mathbf{i} + 4\sin t\mathbf{j} + t\mathbf{k}$

10. $\mathbf{r}(t) = t\mathbf{i} + \frac{1}{2}t^2\mathbf{j} + \frac{1}{3}t^3\mathbf{k}$

11. $x = \cosh t, \quad y = \sinh t, \quad z = t$

12. $\mathbf{r}(t) = \mathbf{i} + t\mathbf{j} + t^2\mathbf{k}$

13–16 Find the curvature and the radius of curvature at the stated point.

13. $\mathbf{r}(t) = 3\cos t\mathbf{i} + 4\sin t\mathbf{j} + t\mathbf{k}; \quad t = \pi/2$

14. $\mathbf{r}(t) = e^t\mathbf{i} + e^{-t}\mathbf{j} + t\mathbf{k}; \quad t = 0$

15. $x = e^t\cos t, y = e^t\sin t, z = e^t; \quad t = 0$

16. $x = \sin t, y = \cos t, z = \frac{1}{2}t^2; \quad t = 0$

17–18 Confirm that s is an arc length parameter by showing that $\|d\mathbf{r}/ds\| = 1$, and then apply Formula (1) to find $\kappa(s)$.

17. $\mathbf{r} = \sin\left(1 + \frac{s}{2}\right)\mathbf{i} + \cos\left(1 + \frac{s}{2}\right)\mathbf{j} + \sqrt{3}\left(1 + \frac{s}{2}\right)\mathbf{k}$

18. $\mathbf{r} = \left(1 - \frac{2}{3}s\right)^{3/2}\mathbf{i} + \left(\frac{2}{3}s\right)^{3/2}\mathbf{j} \quad \left(0 \le s \le \frac{3}{2}\right)$

19. (a) Use Formula (3) to show that in 2-space the curvature of a smooth parametric curve

$$x = x(t), \quad y = y(t)$$

is

$$\kappa(t) = \frac{|x'y'' - y'x''|}{(x'^2 + y'^2)^{3/2}}$$

where primes denote differentiation with respect to t.

(b) Use the result in part (a) to show that in 2-space the curvature of the plane curve given by $y = f(x)$ is

$$\kappa(x) = \frac{|d^2y/dx^2|}{[1 + (dy/dx)^2]^{3/2}}$$

[*Hint:* Express $y = f(x)$ parametrically with $x = t$ as the parameter.]

20. Use part (b) of Exercise 19 to show that the curvature of $y = f(x)$ can be expressed in terms of the angle of inclination of the tangent line as

$$\kappa(\phi) = \left|\frac{d^2y}{dx^2}\cos^3\phi\right|$$

[*Hint:* $\tan\phi = dy/dx$.]

21–26 Use the result in Exercise 19(b) to find the curvature at the stated point.

21. $y = \sin x;\ x = \pi/2$

22. $y = x^3/3;\ x = 0$

23. $y = 1/x;\ x = 1$

24. $y = e^{-x};\ x = 1$

25. $y = \tan x;\ x = \pi/4$

26. $y^2 - 4x^2 = 9;\ (2, 5)$

27–32 Use the result in Exercise 19(a) to find the curvature at the stated point.

27. $x = t^2,\ y = t^3;\ t = \frac{1}{2}$

28. $x = 4\cos t,\ y = \sin t;\ t = \pi/2$

29. $x = e^{3t},\ y = e^{-t};\ t = 0$

30. $x = 1 - t^3,\ y = t - t^2;\ t = 1$

31. $x = t,\ y = 1/t;\ t = 1$

32. $x = 2\sin 2t,\ y = 3\sin t;\ t = \pi/2$

33. In each part, use the formulas in Exercise 19 to help find the radius of curvature at the stated points. Then sketch the graph together with the osculating circles at those points.
(a) $y = \cos x$ at $x = 0$ and $x = \pi$
(b) $x = 2\cos t,\ y = \sin t\ \ (0 \le t \le 2\pi)$ at $t = 0$ and $t = \pi/2$

34. Use the formula in Exercise 19(a) to find $\kappa(t)$ for the curve $x = e^{-t}\cos t,\ y = e^{-t}\sin t$. Then sketch the graph of $\kappa(t)$.

35–36 Generate the graph of $y = f(x)$ using a graphing utility, and then make a conjecture about the shape of the graph of $y = \kappa(x)$. Check your conjecture by generating the graph of $y = \kappa(x)$.

35. $f(x) = xe^{-x}$ for $0 \le x \le 5$

36. $f(x) = x^3 - x$ for $-1 \le x \le 1$

37. (a) If you have a CAS, read the documentation on calculating higher-order derivatives. Then use the CAS and part (b) of Exercise 19 to find $\kappa(x)$ for $f(x) = x^4 - 2x^2$.
(b) Use the CAS to generate the graphs of $f(x) = x^4 - 2x^2$ and $\kappa(x)$ on the same screen for $-2 \le x \le 2$.
(c) Find the radius of curvature at each relative extremum.
(d) Make a reasonably accurate hand-drawn sketch that shows the graph of $f(x) = x^4 - 2x^2$ and the osculating circles in their correct proportions at the relative extrema.

38. (a) Use a CAS to graph the parametric curve $x = t\cos t$, $y = t\sin t$ for $t \ge 0$.
(b) Make a conjecture about the behavior of the curvature $\kappa(t)$ as $t \to +\infty$.
(c) Use the CAS and part (a) of Exercise 19 to find $\kappa(t)$.
(d) Check your conjecture by finding the limit of $\kappa(t)$ as $t \to +\infty$.

39. Use the formula in Exercise 19(a) to show that for a curve in polar coordinates described by $r = f(\theta)$ the curvature is

$$\kappa(\theta) = \frac{\left|r^2 + 2\left(\frac{dr}{d\theta}\right)^2 - r\frac{d^2r}{d\theta^2}\right|}{\left[r^2 + \left(\frac{dr}{d\theta}\right)^2\right]^{3/2}}$$

[*Hint:* Let θ be the parameter and use the relationships $x = r\cos\theta$, $y = r\sin\theta$.]

40. Use the result in Exercise 39 to show that a circle has constant curvature.

41–44 Use the formula in Exercise 39 to find the curvature at the indicated point.

41. $r = 1 + \cos\theta;\ \theta = \pi/2$

42. $r = e^{2\theta};\ \theta = 1$

43. $r = \sin 3\theta;\ \theta = 0$

44. $r = \theta;\ \theta = 1$

45. The accompanying figure is the graph of the radius of curvature versus θ in rectangular coordinates for the cardioid $r = 1 + \cos\theta$. In words, explain what the graph tells you about the cardioid.

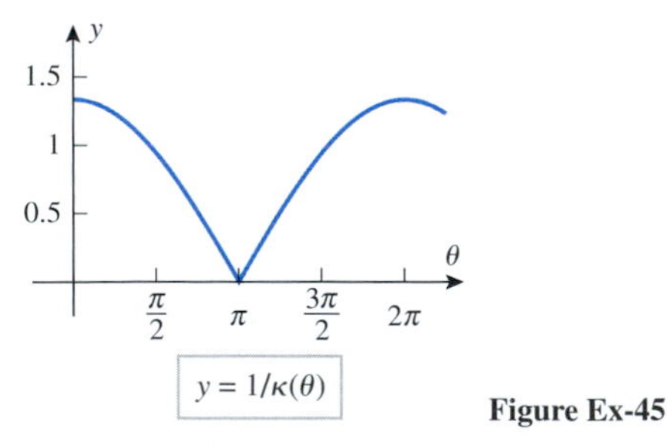

Figure Ex-45

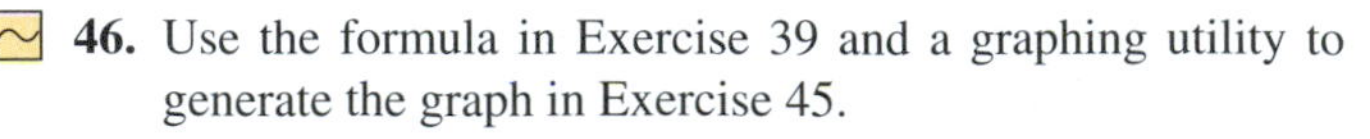

46. Use the formula in Exercise 39 and a graphing utility to generate the graph in Exercise 45.

47. Find the radius of curvature of the parabola $y^2 = 4px$ at $(0, 0)$.

48. At what point(s) does $y = e^x$ have maximum curvature?

49. At what point(s) does $4x^2 + 9y^2 = 36$ have minimum radius of curvature?

50. Find the value of $x, x > 0$, where $y = x^3$ has maximum curvature.

51. Find the maximum and minimum values of the radius of curvature for the curve $x = \cos t$, $y = \sin t$, $z = \cos t$.

52. Find the minimum value of the radius of curvature for the curve $x = e^t$, $y = e^{-t}$, $z = \sqrt{2}t$.

53. Use the formula in Exercise 39 to show that the curvature of the polar curve $r = e^{a\theta}$ is inversely proportional to r.

54. Use the formula in Exercise 39 and a CAS to show that the curvature of the lemniscate $r = \sqrt{a \cos 2\theta}$ is directly proportional to r.

55. (a) Use the result in Exercise 20 to show that for the parabola $y = x^2$ the curvature $\kappa(\phi)$ at points where the tangent line has an angle of inclination of ϕ is

$$\kappa(\phi) = |2 \cos^3 \phi|$$

(b) Use the result in part (a) to find the radius of curvature of the parabola at the point on the parabola where the tangent line has slope 1.

(c) Make a sketch with reasonably accurate proportions that shows the osculating circle at the point on the parabola where the tangent line has slope 1.

56. The ***evolute*** of a smooth parametric curve C in 2-space is the curve formed from the centers of curvature of C. The accompanying figure shows the ellipse $x = 3\cos t$, $y = 2\sin t$ $(0 \le t \le 2\pi)$ and its evolute graphed together.

(a) Which points on the evolute correspond to $t = 0$ and $t = \pi/2$?

(b) In what direction is the evolute traced as t increases from 0 to 2π?

(c) What does the evolute of a circle look like? Explain your reasoning.

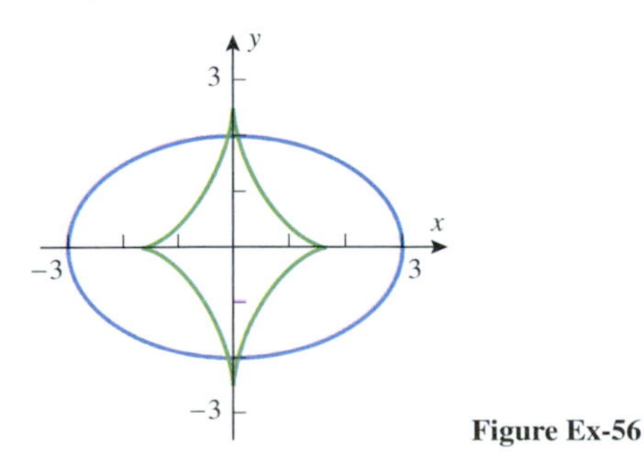

Figure Ex-56

FOCUS ON CONCEPTS

57–62 These exercises are concerned with the problem of creating a single smooth curve by piecing together two separate smooth curves. If two smooth curves C_1 and C_2 are joined at a point P to form a curve C, then we will say that C_1 and C_2 make a ***smooth transition*** at P if the curvature of C is continuous at P.

57. Show that the transition at $x = 0$ from the horizontal line $y = 0$ for $x \le 0$ to the parabola $y = x^2$ for $x > 0$ is not smooth, whereas the transition to $y = x^3$ for $x > 0$ is smooth.

58. (a) Sketch the graph of the curve defined piecewise by $y = x^2$ for $x < 0$, $y = x^4$ for $x \ge 0$.

(b) Show that for the curve in part (a) the transition at $x = 0$ is not smooth.

59. The accompanying figure shows the arc of a circle of radius r with center at $(0, r)$. Find the value of a so that there is a smooth transition from the circle to the parabola $y = ax^2$ at the point where $x = 0$.

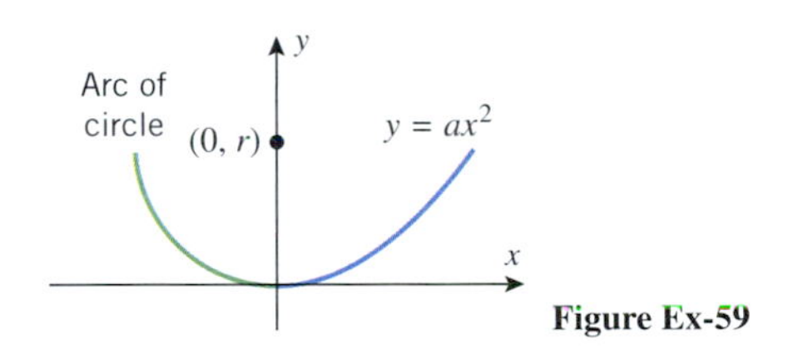

Figure Ex-59

60. Find a, b, and c so that there is a smooth transition at $x = 0$ from the curve $y = e^x$ for $x \le 0$ to the parabola $y = ax^2 + bx + c$ for $x > 0$. [*Hint:* The curvature is continuous at those points where y'' is continuous.]

61. Assume that f is a function for which $f'''(x)$ is defined for all $x \le 0$. Explain why it is always possible to find numbers a, b, and c such that there is a smooth transition at $x = 0$ from the curve $y = f(x)$, $x \le 0$, to the parabola $y = ax^2 + bx + c$.

62. In Exercise 60 of Section 11.2 we defined the Cornu spiral parametrically as

$$x = \int_0^t \cos\left(\frac{\pi u^2}{2}\right) du, \quad y = \int_0^t \sin\left(\frac{\pi u^2}{2}\right) du$$

This curve, which is graphed in the accompanying figure (next page), is used in highway design to create a gradual transition from a straight road (zero curvature) to an exit ramp with positive curvature.

(a) Express the Cornu spiral as a vector-valued function $\mathbf{r}(t)$, and then use Theorem 13.3.4 to show that $s = t$ is the arc length parameter with reference point $(0, 0)$.

(b) Replace t by s and use Formula (1) to show that $\kappa(s) = \pi|s|$. [*Note:* If $s \ge 0$, then the curvature $\kappa(s) = \pi s$ increases from 0 at a constant rate with respect to s. This makes the spiral ideal for joining a curved road to a straight road.]

(c) What happens to the curvature of the Cornu spiral as $s \to +\infty$? In words, explain why this is consistent with the graph.

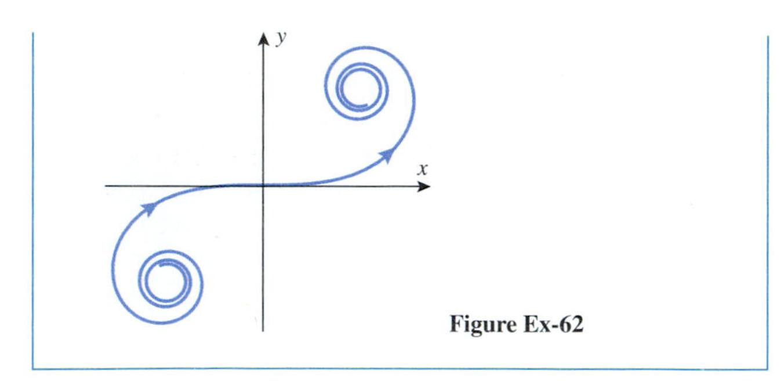

Figure Ex-62

63–66 Assume that s is an arc length parameter for a smooth vector-valued function $\mathbf{r}(s)$ in 3-space and that $d\mathbf{T}/ds$ and $d\mathbf{N}/ds$ exist at each point on the curve. (This implies that $d\mathbf{B}/ds$ exists as well, since $\mathbf{B} = \mathbf{T} \times \mathbf{N}$.)

63. Show that
$$\frac{d\mathbf{T}}{ds} = \kappa(s)\mathbf{N}(s)$$
and use this result to obtain the formulas in (10).

64. (a) Show that $d\mathbf{B}/ds$ is perpendicular to $\mathbf{B}(s)$.
(b) Show that $d\mathbf{B}/ds$ is perpendicular to $\mathbf{T}(s)$. [*Hint:* Use the fact that $\mathbf{B}(s)$ is perpendicular to both $\mathbf{T}(s)$ and $\mathbf{N}(s)$, and differentiate $\mathbf{B} \cdot \mathbf{T}$ with respect to s.]
(c) Use the results in parts (a) and (b) to show that $d\mathbf{B}/ds$ is a scalar multiple of $\mathbf{N}(s)$. The *negative* of this scalar is called the ***torsion*** of $\mathbf{r}(s)$ and is denoted by $\tau(s)$. Thus,
$$\frac{d\mathbf{B}}{ds} = -\tau(s)\mathbf{N}(s)$$
(d) Show that $\tau(s) = 0$ for all s if the graph of $\mathbf{r}(s)$ lies in a plane. [*Note:* For reasons that we cannot discuss here, the torsion is related to the "twisting" properties of the curve, and $\tau(s)$ is regarded as a numerical measure of the tendency for the curve to twist out of the osculating plane.]

65. Let κ be the curvature of C and τ the torsion (defined in Exercise 64). By differentiating $\mathbf{N} = \mathbf{B} \times \mathbf{T}$ with respect to s, show that $d\mathbf{N}/ds = -\kappa\mathbf{T} + \tau\mathbf{B}$.

66. The following derivatives, known as the ***Frenet–Serret formulas***, are fundamental in the theory of curves in 3-space:

$d\mathbf{T}/ds = \kappa\mathbf{N}$ [Exercise 63]
$d\mathbf{N}/ds = -\kappa\mathbf{T} + \tau\mathbf{B}$ [Exercise 65]
$d\mathbf{B}/ds = -\tau\mathbf{N}$ [Exercise 64(c)]

Use the first two Frenet–Serret formulas and the fact that $\mathbf{r}'(s) = \mathbf{T}$ if $\mathbf{r} = \mathbf{r}(s)$ to show that
$$\tau = \frac{[\mathbf{r}'(s) \times \mathbf{r}''(s)] \cdot \mathbf{r}'''(s)}{\|\mathbf{r}''(s)\|^2} \quad \text{and} \quad \mathbf{B} = \frac{\mathbf{r}'(s) \times \mathbf{r}''(s)}{\|\mathbf{r}''(s)\|}$$

67. Use the results in Exercise 66 and the results in Exercise 30 of Section 13.3 to show that for the circular helix
$$\mathbf{r} = a\cos t\,\mathbf{i} + a\sin t\,\mathbf{j} + ct\mathbf{k}$$
with $a > 0$ the torsion and the binormal vector are
$$\tau = \frac{c}{w^2}$$
and
$$\mathbf{B} = \left(\frac{c}{w}\sin\frac{s}{w}\right)\mathbf{i} - \left(\frac{c}{w}\cos\frac{s}{w}\right)\mathbf{j} + \left(\frac{a}{w}\right)\mathbf{k}$$
where $w = \sqrt{a^2 + c^2}$ and s has reference point $(a, 0, 0)$.

68. (a) Use the chain rule and the first two Frenet–Serret formulas in Exercise 66 to show that
$$\mathbf{T}' = \kappa s'\mathbf{N} \quad \text{and} \quad \mathbf{N}' = -\kappa s'\mathbf{T} + \tau s'\mathbf{B}$$
where primes denote differentiation with respect to t.
(b) Show that Formulas (4) and (6) can be written in the form
$$\mathbf{r}'(t) = s'\mathbf{T} \quad \text{and} \quad \mathbf{r}''(t) = s''\mathbf{T} + \kappa(s')^2\mathbf{N}$$
(c) Use the results in parts (a) and (b) to show that
$$\mathbf{r}'''(t) = [s''' - \kappa^2(s')^3]\mathbf{T} + [3\kappa s's'' + \kappa'(s')^2]\mathbf{N} + \kappa\tau(s')^3\mathbf{B}$$
(d) Use the results in parts (b) and (c) to show that
$$\tau(t) = \frac{[\mathbf{r}'(t) \times \mathbf{r}''(t)] \cdot \mathbf{r}'''(t)}{\|\mathbf{r}'(t) \times \mathbf{r}''(t)\|^2}$$

69–72 Use the formula in Exercise 68(d) to find the torsion $\tau = \tau(t)$.

69. The twisted cubic $\mathbf{r}(t) = 2t\mathbf{i} + t^2\mathbf{j} + \frac{1}{3}t^3\mathbf{k}$

70. The circular helix $\mathbf{r}(t) = a\cos t\,\mathbf{i} + a\sin t\,\mathbf{j} + ct\mathbf{k}$

71. $\mathbf{r}(t) = e^t\mathbf{i} + e^{-t}\mathbf{j} + \sqrt{2}t\mathbf{k}$

72. $\mathbf{r}(t) = (t - \sin t)\mathbf{i} + (1 - \cos t)\mathbf{j} + t\mathbf{k}$

✔ QUICK CHECK ANSWERS 13.5

1. $\left\|\frac{d\mathbf{T}}{ds}\right\| = \|\mathbf{r}''(s)\|$ **2.** (a) $\frac{\|\mathbf{T}'(t)\|}{\|\mathbf{r}'(t)\|}$ (b) $\frac{\|\mathbf{r}'(t) \times \mathbf{r}''(t)\|}{\|\mathbf{r}'(t)\|^3}$ **3.** $\left|\frac{d\phi}{ds}\right|$ **4.** $\frac{1}{2}$

13.6 MOTION ALONG A CURVE

In earlier sections we considered the motion of a particle along a line. In that situation there are only two directions in which the particle can move—the positive direction or the negative direction. Motion in 2-space or 3-space is more complicated because there are infinitely many directions in which a particle can move. In this section we will show how vectors can be used to analyze motion along curves in 2-space or 3-space.

VELOCITY, ACCELERATION, AND SPEED

Let us assume that the motion of a particle in 2-space or 3-space is described by a smooth vector-valued function $\mathbf{r}(t)$ in which the parameter t denotes time; we will call this the ***position function*** or ***trajectory*** of the particle. As the particle moves along its trajectory, its direction of motion and its speed can vary from instant to instant. Thus, before we can undertake any analysis of such motion, we must have clear answers to the following questions:

- What is the direction of motion of the particle at an instant of time?
- What is the speed of the particle at an instant of time?

We will define the direction of motion at time t to be the direction of the unit tangent vector $\mathbf{T}(t)$, and we will define the speed to be ds/dt—the instantaneous rate of change of the arc length traveled by the particle from an arbitrary reference point. Taking this a step further, we will combine the speed and the direction of motion to form the vector

$$\mathbf{v}(t) = \frac{ds}{dt}\mathbf{T}(t) \tag{1}$$

which we call the ***velocity*** of the particle at time t. Thus, at each instant of time the velocity vector $\mathbf{v}(t)$ points in the direction of motion and has a magnitude that is equal to the speed of the particle (Figure 13.6.1).

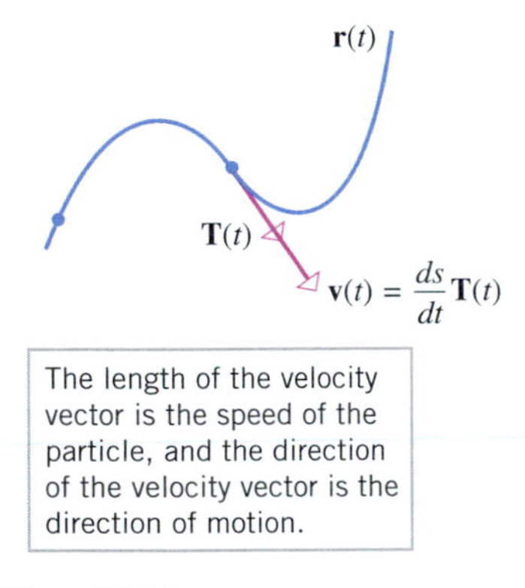

The length of the velocity vector is the speed of the particle, and the direction of the velocity vector is the direction of motion.

Figure 13.6.1

Recall that for motion along a coordinate line the velocity function is the derivative of the position function. The same is true for motion along a curve, since

$$\frac{d\mathbf{r}}{dt} = \frac{d\mathbf{r}}{ds}\frac{ds}{dt} = \frac{ds}{dt}\mathbf{T}(t) = \mathbf{v}(t)$$

For motion along a coordinate line, the acceleration function was defined to be the derivative of the velocity function. The definition is the same for motion along a curve.

13.6.1 DEFINITION. If $\mathbf{r}(t)$ is the position function of a particle moving along a curve in 2-space or 3-space, then the ***instantaneous velocity***, ***instantaneous acceleration***, and ***instantaneous speed*** of the particle at time t are defined by

$$\text{velocity} = \mathbf{v}(t) = \frac{d\mathbf{r}}{dt} \tag{2}$$

$$\text{acceleration} = \mathbf{a}(t) = \frac{d\mathbf{v}}{dt} = \frac{d^2\mathbf{r}}{dt^2} \tag{3}$$

$$\text{speed} = \|\mathbf{v}(t)\| = \frac{ds}{dt} \tag{4}$$

As shown in Table 13.6.1, the position, velocity, acceleration, and speed can also be expressed in component form.

Table 13.6.1

	2-SPACE	3-SPACE
POSITION	$\mathbf{r}(t) = x(t)\mathbf{i} + y(t)\mathbf{j}$	$\mathbf{r}(t) = x(t)\mathbf{i} + y(t)\mathbf{j} + z(t)\mathbf{k}$
VELOCITY	$\mathbf{v}(t) = \dfrac{dx}{dt}\mathbf{i} + \dfrac{dy}{dt}\mathbf{j}$	$\mathbf{v}(t) = \dfrac{dx}{dt}\mathbf{i} + \dfrac{dy}{dt}\mathbf{j} + \dfrac{dz}{dt}\mathbf{k}$
ACCELERATION	$\mathbf{a}(t) = \dfrac{d^2x}{dt^2}\mathbf{i} + \dfrac{d^2y}{dt^2}\mathbf{j}$	$\mathbf{a}(t) = \dfrac{d^2x}{dt^2}\mathbf{i} + \dfrac{d^2y}{dt^2}\mathbf{j} + \dfrac{d^2z}{dt^2}\mathbf{k}$
SPEED	$\lVert\mathbf{v}(t)\rVert = \sqrt{\left(\dfrac{dx}{dt}\right)^2 + \left(\dfrac{dy}{dt}\right)^2}$	$\lVert\mathbf{v}(t)\rVert = \sqrt{\left(\dfrac{dx}{dt}\right)^2 + \left(\dfrac{dy}{dt}\right)^2 + \left(\dfrac{dz}{dt}\right)^2}$

▶ **Example 1** A particle moves along a circular path in such a way that its x- and y-coordinates at time t are

$$x = 2\cos t, \quad y = 2\sin t$$

(a) Find the instantaneous velocity and speed of the particle at time t.

(b) Sketch the path of the particle, and show the position and velocity vectors at time $t = \pi/4$ with the velocity vector drawn so that its initial point is at the tip of the position vector.

(c) Show that at each instant the acceleration vector is perpendicular to the velocity vector.

Solution (a). At time t, the position vector is

$$\mathbf{r}(t) = 2\cos t\mathbf{i} + 2\sin t\mathbf{j}$$

so the instantaneous velocity and speed are

$$\mathbf{v}(t) = \frac{d\mathbf{r}}{dt} = -2\sin t\mathbf{i} + 2\cos t\mathbf{j}$$

$$\lVert\mathbf{v}(t)\rVert = \sqrt{(-2\sin t)^2 + (2\cos t)^2} = 2$$

Solution (b). The graph of the parametric equations is a circle of radius 2 centered at the origin. At time $t = \pi/4$ the position and velocity vectors of the particle are

$$\mathbf{r}(\pi/4) = 2\cos(\pi/4)\mathbf{i} + 2\sin(\pi/4)\mathbf{j} = \sqrt{2}\mathbf{i} + \sqrt{2}\mathbf{j}$$

$$\mathbf{v}(\pi/4) = -2\sin(\pi/4)\mathbf{i} + 2\cos(\pi/4)\mathbf{j} = -\sqrt{2}\mathbf{i} + \sqrt{2}\mathbf{j}$$

These vectors and the circle are shown in Figure 13.6.2.

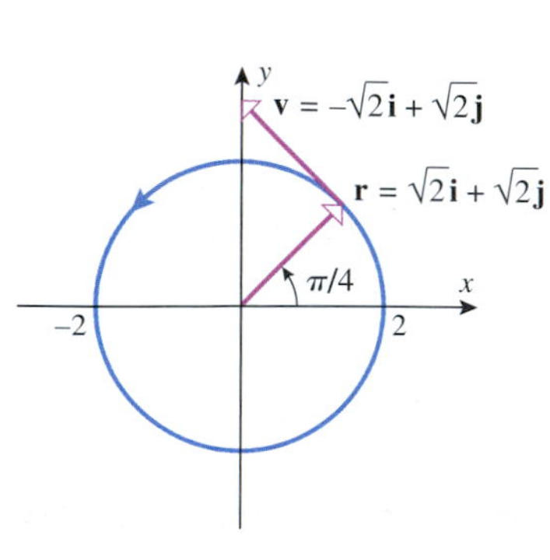

Figure 13.6.2

Solution (c). At time t, the acceleration vector is

$$\mathbf{a}(t) = \frac{d\mathbf{v}}{dt} = -2\cos t\mathbf{i} - 2\sin t\mathbf{j}$$

One way of showing that $\mathbf{v}(t)$ and $\mathbf{a}(t)$ are perpendicular is to show that their dot product is zero (try it). However, it is easier to observe that $\mathbf{a}(t)$ is the negative of $\mathbf{r}(t)$, which implies that $\mathbf{v}(t)$ and $\mathbf{a}(t)$ are perpendicular, since at each point on a circle the radius and tangent line are perpendicular. ◀

Since $\mathbf{v}(t)$ can be obtained by differentiating $\mathbf{r}(t)$, and since $\mathbf{a}(t)$ can be obtained by differentiating $\mathbf{v}(t)$, it follows that $\mathbf{r}(t)$ can be obtained by integrating $\mathbf{v}(t)$, and $\mathbf{v}(t)$ can be obtained by integrating $\mathbf{a}(t)$. However, such integrations do not produce unique functions because constants of integration occur. Typically, initial conditions are required to determine these constants.

Example 2 A particle moves through 3-space in such a way that its velocity is

$$\mathbf{v}(t) = \mathbf{i} + t\mathbf{j} + t^2\mathbf{k}$$

Find the coordinates of the particle at time $t = 1$ given that the particle is at the point $(-1, 2, 4)$ at time $t = 0$.

Solution. Integrating the velocity function to obtain the position function yields

$$\mathbf{r}(t) = \int \mathbf{v}(t)\,dt = \int (\mathbf{i} + t\mathbf{j} + t^2\mathbf{k})\,dt = t\mathbf{i} + \frac{t^2}{2}\mathbf{j} + \frac{t^3}{3}\mathbf{k} + \mathbf{C} \tag{5}$$

where $\mathbf{C}$ is a vector constant of integration. Since the coordinates of the particle at time $t = 0$ are $(-1, 2, 4)$, the position vector at time $t = 0$ is

$$\mathbf{r}(0) = -\mathbf{i} + 2\mathbf{j} + 4\mathbf{k} \tag{6}$$

It follows on substituting $t = 0$ in (5) and equating the result with (6) that

$$\mathbf{C} = -\mathbf{i} + 2\mathbf{j} + 4\mathbf{k}$$

Substituting this value of $\mathbf{C}$ in (5) and simplifying yields

$$\mathbf{r}(t) = (t-1)\mathbf{i} + \left(\frac{t^2}{2} + 2\right)\mathbf{j} + \left(\frac{t^3}{3} + 4\right)\mathbf{k}$$

Thus, at time $t = 1$ the position vector of the particle is

$$\mathbf{r}(1) = 0\mathbf{i} + \frac{5}{2}\mathbf{j} + \frac{13}{3}\mathbf{k}$$

so its coordinates at that instant are $\left(0, \frac{5}{2}, \frac{13}{3}\right)$. ◀

DISPLACEMENT AND DISTANCE TRAVELED

If a particle travels along a curve C in 2-space or 3-space, the ***displacement*** of the particle over the time interval $t_1 \le t \le t_2$ is commonly denoted by $\Delta\mathbf{r}$ and is defined as

$$\Delta\mathbf{r} = \mathbf{r}(t_2) - \mathbf{r}(t_1) \tag{7}$$

(Figure 13.6.3). The displacement vector, which describes the change in position of the particle during the time interval, can be obtained by integrating the velocity function from t_1 to t_2:

$$\Delta\mathbf{r} = \int_{t_1}^{t_2} \mathbf{v}(t)\,dt = \int_{t_1}^{t_2} \frac{d\mathbf{r}}{dt}\,dt = \mathbf{r}(t)\Big]_{t_1}^{t_2} = \mathbf{r}(t_2) - \mathbf{r}(t_1) \tag{8}$$

Displacement

It follows from Theorem 13.3.1 that we can find the distance s traveled by a particle over a time interval $t_1 \le t \le t_2$ by integrating the speed over that interval, since

$$s = \int_{t_1}^{t_2} \left\|\frac{d\mathbf{r}}{dt}\right\| dt = \int_{t_1}^{t_2} \|\mathbf{v}(t)\|\,dt \tag{9}$$

Distance traveled

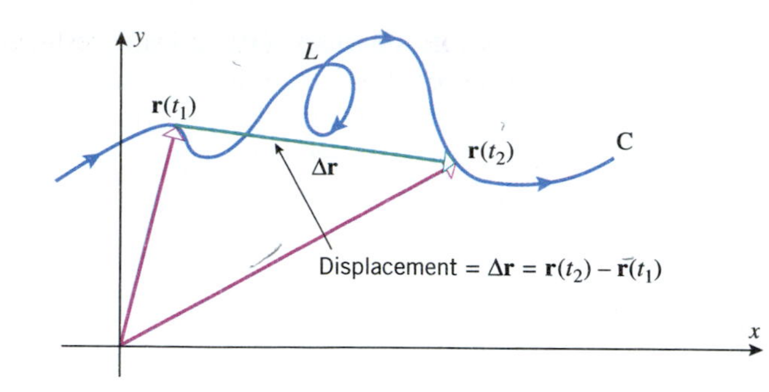

Figure 13.6.3

▶ **Example 3** Suppose that a particle moves along a circular helix in 3-space so that its position vector at time t is

$$\mathbf{r}(t) = (4\cos \pi t)\mathbf{i} + (4\sin \pi t)\mathbf{j} + t\mathbf{k}$$

Find the distance traveled and the displacement of the particle during the time interval $1 \le t \le 5$.

Solution. We have

$$\mathbf{v}(t) = \frac{d\mathbf{r}}{dt} = (-4\pi \sin \pi t)\mathbf{i} + (4\pi \cos \pi t)\mathbf{j} + \mathbf{k}$$

$$\|\mathbf{v}(t)\| = \sqrt{(-4\pi \sin \pi t)^2 + (4\pi \cos \pi t)^2 + 1} = \sqrt{16\pi^2 + 1}$$

Thus, it follows from (9) that the distance traveled by the particle from time $t = 1$ to $t = 5$ is

$$s = \int_1^5 \sqrt{16\pi^2 + 1}\, dt = 4\sqrt{16\pi^2 + 1}$$

Moreover, it follows from (8) that the displacement over the time interval is

$$\begin{aligned} \Delta\mathbf{r} &= \mathbf{r}(5) - \mathbf{r}(1) \\ &= (4\cos 5\pi\mathbf{i} + 4\sin 5\pi\mathbf{j} + 5\mathbf{k}) - (4\cos \pi\mathbf{i} + 4\sin \pi\mathbf{j} + \mathbf{k}) \\ &= (-4\mathbf{i} + 5\mathbf{k}) - (-4\mathbf{i} + \mathbf{k}) = 4\mathbf{k} \end{aligned}$$

which tells us that the change in the position of the particle over the time interval was 4 units straight up. ◀

NORMAL AND TANGENTIAL COMPONENTS OF ACCELERATION

You know from your experience as an automobile passenger that if a car speeds up rapidly, then your body is thrown back against the backrest of the seat. You also know that if the car rounds a turn in the road, then your body is thrown toward the outside of the curve—the greater the curvature in the road, the greater this effect. The explanation of these effects can be understood by resolving the velocity and acceleration components of the motion into vector components that are parallel to the unit tangent and unit normal vectors. The following theorem explains how to do this.

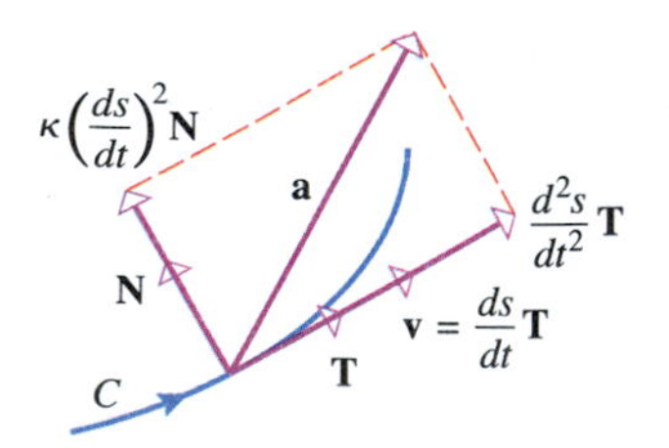

Figure 13.6.4

13.6.2 THEOREM. *If a particle moves along a smooth curve C in 2-space or 3-space, then at each point on the curve velocity and acceleration vectors can be written as*

$$\mathbf{v} = \frac{ds}{dt}\mathbf{T} \qquad \mathbf{a} = \frac{d^2s}{dt^2}\mathbf{T} + \kappa\left(\frac{ds}{dt}\right)^2\mathbf{N} \tag{10–11}$$

where s is an arc length parameter for the curve, and **T**, **N**, *and* κ *denote the unit tangent vector, unit normal vector, and curvature at the point* (Figure 13.6.4).

PROOF. Formula (10) is just a restatement of (1). To obtain (11), we differentiate both sides of (10) with respect to t; this yields

$$\begin{aligned}
\mathbf{a} = \frac{d}{dt}\left(\frac{ds}{dt}\mathbf{T}\right) &= \frac{d^2s}{dt^2}\mathbf{T} + \frac{ds}{dt}\frac{d\mathbf{T}}{dt} \\
&= \frac{d^2s}{dt^2}\mathbf{T} + \frac{ds}{dt}\frac{d\mathbf{T}}{ds}\frac{ds}{dt} \\
&= \frac{d^2s}{dt^2}\mathbf{T} + \left(\frac{ds}{dt}\right)^2\frac{d\mathbf{T}}{ds} \\
&= \frac{d^2s}{dt^2}\mathbf{T} + \left(\frac{ds}{dt}\right)^2\kappa\mathbf{N}
\end{aligned}$$

Formula (10) of Section 13.5

from which (11) follows. ■

The coefficients of **T** and **N** in (11) are commonly denoted by

$$a_T = \frac{d^2s}{dt^2} \qquad a_N = \kappa\left(\frac{ds}{dt}\right)^2 \qquad (12\text{–}13)$$

Formula (14) applies to motion in both 2-space and 3-space. What is interesting is that the 3-space formula does not involve the binormal vector **B**, so the acceleration vector always lies in the plane of **T** and **N** (the osculating plane), even for highly twisting paths of motion (Figure 13.6.5).

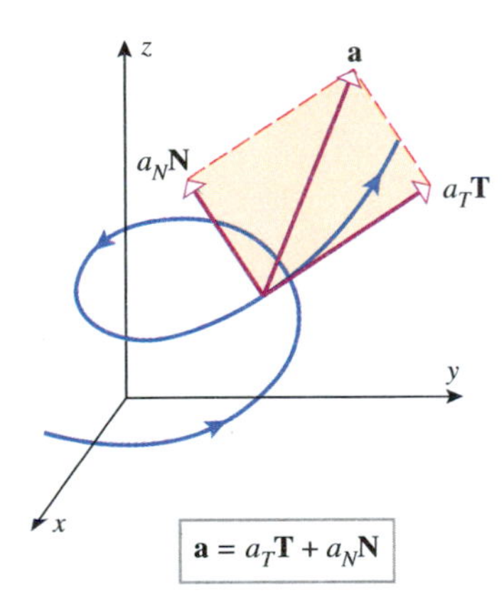

Figure 13.6.5

in which case Formula (11) is expressed as

$$\mathbf{a} = a_T\mathbf{T} + a_N\mathbf{N} \qquad (14)$$

In this formula the scalars a_T and a_N are called the ***tangential scalar component of acceleration*** and the ***normal scalar component of acceleration***, and the vectors $a_T\mathbf{T}$ and $a_N\mathbf{N}$ are called the ***tangential vector component of acceleration*** and the ***normal vector component of acceleration***.

The scalar components of acceleration explain the effect that you experience when a car speeds up rapidly or rounds a turn. The rapid increase in speed produces a large value for d^2s/dt^2, which results in a large tangential scalar component of acceleration; and by Newton's second law this corresponds to a large tangential force on the car in the direction of motion. To understand the effect of rounding a turn, observe that the normal scalar component of acceleration has the curvature κ and the square of the speed ds/dt as factors. Thus, sharp turns or turns taken at high speed both correspond to large normal forces on the car.

Although Formulas (12) and (13) provide useful insight into the behavior of particles moving along curved paths, they are not always the best formulas for computations. The following theorem provides some more useful formulas that relate a_T, a_N, and κ to the velocity **v** and acceleration **a**.

Theorem 13.6.3 applies to motion in 2-space and 3-space, but for motion in 2-space you will have to add a zero **k** component to **v** to calculate the cross product.

13.6.3 THEOREM. *If a particle moves along a smooth curve C in 2-space or 3-space, then at each point on the curve the velocity* **v** *and the acceleration* **a** *are related to a_T, a_N, and κ by the formulas*

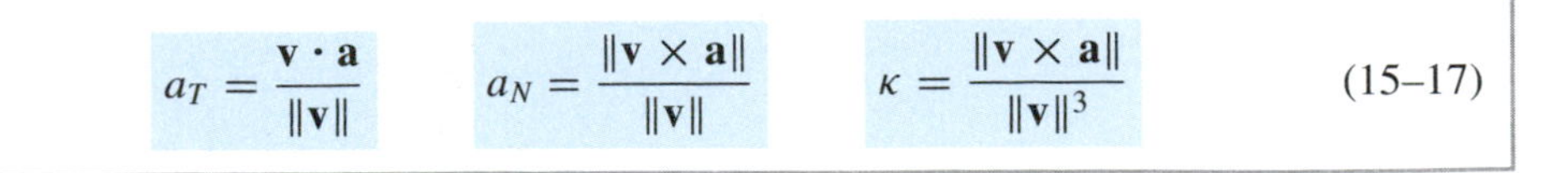

$$a_T = \frac{\mathbf{v}\cdot\mathbf{a}}{\|\mathbf{v}\|} \qquad a_N = \frac{\|\mathbf{v}\times\mathbf{a}\|}{\|\mathbf{v}\|} \qquad \kappa = \frac{\|\mathbf{v}\times\mathbf{a}\|}{\|\mathbf{v}\|^3} \qquad (15\text{–}17)$$

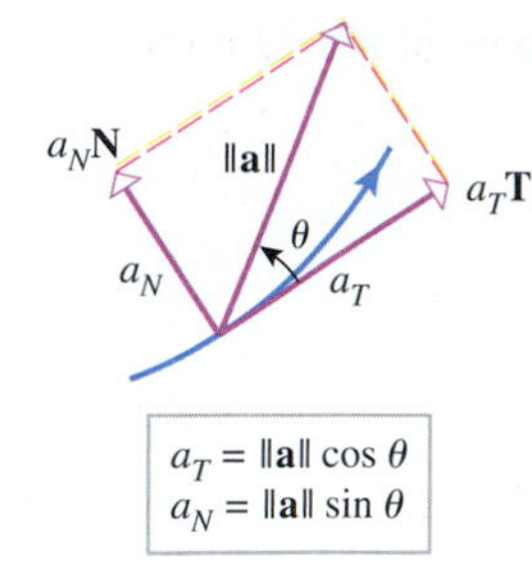

Figure 13.6.6

PROOF. As illustrated in Figure 13.6.6, let θ be the angle between the vector $\mathbf{a}$ and the vector $a_T\mathbf{T}$. Thus,

$$a_T = \|\mathbf{a}\| \cos\theta \quad \text{and} \quad a_N = \|\mathbf{a}\| \sin\theta$$

from which we obtain

$$a_T = \|\mathbf{a}\| \cos\theta = \frac{\|\mathbf{v}\|\,\|\mathbf{a}\| \cos\theta}{\|\mathbf{v}\|} = \frac{\mathbf{v} \cdot \mathbf{a}}{\|\mathbf{v}\|}$$

$$a_N = \|\mathbf{a}\| \sin\theta = \frac{\|\mathbf{v}\|\,\|\mathbf{a}\| \sin\theta}{\|\mathbf{v}\|} = \frac{\|\mathbf{v} \times \mathbf{a}\|}{\|\mathbf{v}\|}$$

$$\kappa = \frac{a_N}{(ds/dt)^2} = \frac{a_N}{\|\mathbf{v}\|^2} = \frac{1}{\|\mathbf{v}\|^2}\frac{\|\mathbf{v} \times \mathbf{a}\|}{\|\mathbf{v}\|} = \frac{\|\mathbf{v} \times \mathbf{a}\|}{\|\mathbf{v}\|^3}$$ ■

Recall that for nonlinear smooth curves in 2-space the unit normal vector $\mathbf{N}$ is the inward normal (points toward the concave side of the curve). Explain why the same is true for $a_N\mathbf{N}$.

► **Example 4** Suppose that a particle moves through 3-space so that its position vector at time t is

$$\mathbf{r}(t) = t\mathbf{i} + t^2\mathbf{j} + t^3\mathbf{k}$$

(The path is the twisted cubic shown in Figure 13.1.5.)

(a) Find the scalar tangential and normal components of acceleration at time t.

(b) Find the scalar tangential and normal components of acceleration at time $t = 1$.

(c) Find the vector tangential and normal components of acceleration at time $t = 1$.

(d) Find the curvature of the path at the point where the particle is located at time $t = 1$.

Solution (a). We have

$$\mathbf{v}(t) = \mathbf{r}'(t) = \mathbf{i} + 2t\mathbf{j} + 3t^2\mathbf{k}$$
$$\mathbf{a}(t) = \mathbf{v}'(t) = 2\mathbf{j} + 6t\mathbf{k}$$
$$\|\mathbf{v}(t)\| = \sqrt{1 + 4t^2 + 9t^4}$$
$$\mathbf{v}(t) \cdot \mathbf{a}(t) = 4t + 18t^3$$
$$\mathbf{v}(t) \times \mathbf{a}(t) = \begin{vmatrix} \mathbf{i} & \mathbf{j} & \mathbf{k} \\ 1 & 2t & 3t^2 \\ 0 & 2 & 6t \end{vmatrix} = 6t^2\mathbf{i} - 6t\mathbf{j} + 2\mathbf{k}$$

Thus, from (15) and (16)

$$a_T = \frac{\mathbf{v} \cdot \mathbf{a}}{\|\mathbf{v}\|} = \frac{4t + 18t^3}{\sqrt{1 + 4t^2 + 9t^4}}$$

$$a_N = \frac{\|\mathbf{v} \times \mathbf{a}\|}{\|\mathbf{v}\|} = \frac{\sqrt{36t^4 + 36t^2 + 4}}{\sqrt{1 + 4t^2 + 9t^4}} = 2\sqrt{\frac{9t^4 + 9t^2 + 1}{9t^4 + 4t^2 + 1}}$$

Solution (b). At time $t = 1$, the components a_T and a_N in part (a) are

$$a_T = \frac{22}{\sqrt{14}} \approx 5.88 \quad \text{and} \quad a_N = 2\sqrt{\frac{19}{14}} \approx 2.33$$

Solution (c). Since $\mathbf{T}$ and $\mathbf{v}$ have the same direction, $\mathbf{T}$ can be obtained by normalizing $\mathbf{v}$, that is,

$$\mathbf{T}(t) = \frac{\mathbf{v}(t)}{\|\mathbf{v}(t)\|}$$

At time $t = 1$ we have

$$\mathbf{T}(1) = \frac{\mathbf{v}(1)}{\|\mathbf{v}(1)\|} = \frac{\mathbf{i} + 2\mathbf{j} + 3\mathbf{k}}{\|\mathbf{i} + 2\mathbf{j} + 3\mathbf{k}\|} = \frac{1}{\sqrt{14}}(\mathbf{i} + 2\mathbf{j} + 3\mathbf{k})$$

From this and part (b) we obtain the vector tangential component of acceleration:

$$a_T(1)\mathbf{T}(1) = \frac{22}{\sqrt{14}}\mathbf{T}(1) = \frac{11}{7}(\mathbf{i} + 2\mathbf{j} + 3\mathbf{k}) = \frac{11}{7}\mathbf{i} + \frac{22}{7}\mathbf{j} + \frac{33}{7}\mathbf{k}$$

To find the normal vector component of acceleration, we rewrite $\mathbf{a} = a_T\mathbf{T} + a_N\mathbf{N}$ as

$$a_N\mathbf{N} = \mathbf{a} - a_T\mathbf{T}$$

Thus, at time $t = 1$ the normal vector component of acceleration is

$$\begin{aligned} a_N(1)\mathbf{N}(1) &= \mathbf{a}(1) - a_T(1)\mathbf{T}(1) \\ &= (2\mathbf{j} + 6\mathbf{k}) - \left(\frac{11}{7}\mathbf{i} + \frac{22}{7}\mathbf{j} + \frac{33}{7}\mathbf{k}\right) \\ &= -\frac{11}{7}\mathbf{i} - \frac{8}{7}\mathbf{j} + \frac{9}{7}\mathbf{k} \end{aligned}$$

Solution (d). We will apply Formula (17) with $t = 1$. From part (a)

$$\|\mathbf{v}(1)\| = \sqrt{14} \quad \text{and} \quad \mathbf{v}(1) \times \mathbf{a}(1) = 6\mathbf{i} - 6\mathbf{j} + 2\mathbf{k}$$

Thus, at time $t = 1$

$$\kappa = \frac{\|\mathbf{v} \times \mathbf{a}\|}{\|\mathbf{v}\|^3} = \frac{\sqrt{76}}{(\sqrt{14})^3} = \frac{1}{14}\sqrt{\frac{38}{7}} \approx 0.17 \blacktriangleleft$$

In the case where $\|\mathbf{a}\|$ and a_T are known, there is a useful alternative to Formula (16) for a_N that does not require the calculation of a cross product. It follows algebraically from Formula (14) or geometrically from Figure 13.6.6 and the Theorem of Pythagoras that

Use Formula (18) to confirm the value of a_N found in Example 4.

$$a_N = \sqrt{\|\mathbf{a}\|^2 - a_T^2} \tag{18}$$

A MODEL OF PROJECTILE MOTION

Earlier in this text we examined various problems concerned with objects moving *vertically* in the Earth's gravitational field (see the subsection of Section 6.7 entitled Free-Fall Motion and the subsection of Section 9.1 entitled A Model of Free-Fall Motion Retarded by Air Resistance). Now we will consider the motion of a projectile launched along a *curved* path in the Earth's gravitational field. For this purpose we will need the following *vector version* of Newton's Second Law of Motion (9.1.1)

$$\mathbf{F} = m\mathbf{a} \tag{19}$$

and we will need to make three modeling assumptions:

- The mass m of the object is constant.
- The only force acting on the object after it is launched is the force of the Earth's gravity. (Thus, air resistance and the gravitational effect of other planets and celestial objects are ignored.)
- The object remains sufficiently close to the Earth that we can assume the force of gravity to be constant.

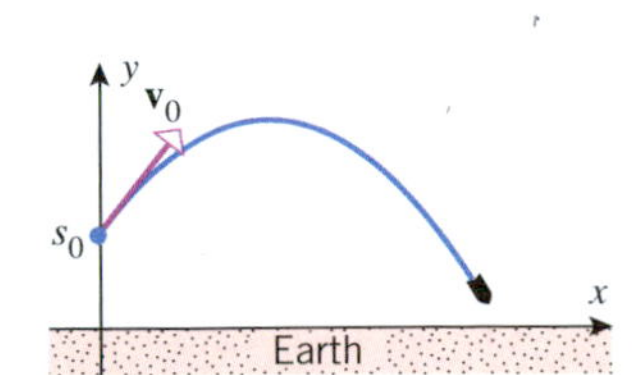

Figure 13.6.7

Let us assume that at time $t = 0$ an object of mass m is launched from a height of s_0 above the Earth with an initial velocity vector of $\mathbf{v}_0$. Furthermore, let us introduce an xy-coordinate system as shown in Figure 13.6.7. In this coordinate system the positive y-direction is up, the origin is at the surface of the Earth, and the initial location of the object is $(0, s_0)$. Our objective is to use basic principles of physics to derive the velocity function $\mathbf{v}(t)$ and the position function $\mathbf{r}(t)$ from the acceleration function $\mathbf{a}(t)$ of the object. Our starting point is the physical observation that the downward force $\mathbf{F}$ of the Earth's gravity on an object of mass m is

$$\mathbf{F} = -mg\mathbf{j}$$

where g is the acceleration due to gravity (see 9.4.3). It follows from this fact and Newton's second law (19) that

$$m\mathbf{a} = -mg\mathbf{j}$$

or on canceling m from both sides

$$\mathbf{a} = -g\mathbf{j} \tag{20}$$

Observe that this acceleration function does not involve t and hence is constant. We can now obtain the velocity function $\mathbf{v}(t)$ by integrating this acceleration function and using the initial condition $\mathbf{v}(0) = \mathbf{v}_0$ to find the constant of integration. Integrating (20) with respect to t and keeping in mind that $-g\mathbf{j}$ is constant yields

$$\mathbf{v}(t) = \int -g\mathbf{j}\,dt = -gt\mathbf{j} + \mathbf{c}_1$$

where $\mathbf{c}_1$ is a vector constant of integration. Substituting $t = 0$ in this equation and using the initial condition $\mathbf{v}(0) = \mathbf{v}_0$ yields $\mathbf{v}_0 = \mathbf{c}_1$. Thus, the velocity function of the object is

$$\mathbf{v}(t) = -gt\mathbf{j} + \mathbf{v}_0 \tag{21}$$

To obtain the position function $\mathbf{r}(t)$ of the object, we will integrate the velocity function and use the known initial position of the object to find the constant of integration. For this purpose observe that the object has coordinates $(0, s_0)$ at time $t = 0$, so the position vector at that time is

$$\mathbf{r}(0) = 0\mathbf{i} + s_0\mathbf{j} = s_0\mathbf{j} \tag{22}$$

This is the initial condition that we will need to find the constant of integration. Integrating (21) with respect to t yields

$$\mathbf{r}(t) = \int (-gt\mathbf{j} + \mathbf{v}_0)\,dt = -\tfrac{1}{2}gt^2\mathbf{j} + t\mathbf{v}_0 + \mathbf{c}_2 \tag{23}$$

where $\mathbf{c}_2$ is another vector constant of integration. Substituting $t = 0$ in (23) and using initial condition (22) yields

$$s_0\mathbf{j} = \mathbf{c}_2$$

so that (23) can be written as

$$\mathbf{r}(t) = \left(-\tfrac{1}{2}gt^2 + s_0\right)\mathbf{j} + t\mathbf{v}_0 \tag{24}$$

This formula expresses the position function of the object in terms of its known initial position and velocity.

Observe that the mass m does not appear in Formulas (21) and (24) and hence has no influence on the velocity or the trajectory of the object. This explains the famous observation of Galileo that two objects of different mass that are released from the same height reach the ground at the same time if air resistance is neglected.

PARAMETRIC EQUATIONS OF PROJECTILE MOTION

Formulas (21) and (24) can be used to obtain parametric equations for the position and velocity in terms of the initial speed of the object and the angle that the initial velocity vector makes with the positive x-axis. For this purpose, let $v_0 = \|\mathbf{v}_0\|$ be the initial speed, let α be the angle that the initial velocity vector $\mathbf{v}_0$ makes with the positive x-axis, let v_x and v_y be the horizontal and vertical scalar components of $\mathbf{v}(t)$ at time t, and let x and y be the horizontal and vertical components of $\mathbf{r}(t)$ at time t. As illustrated in Figure 13.6.8, the initial velocity vector can be expressed as

$$\mathbf{v}_0 = (v_0 \cos\alpha)\mathbf{i} + (v_0 \sin\alpha)\mathbf{j} \tag{25}$$

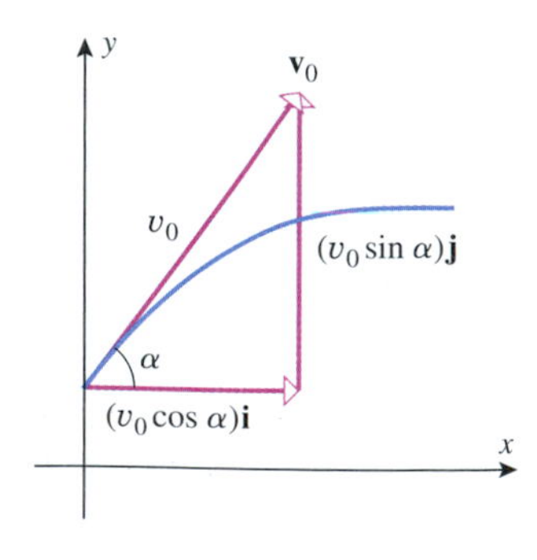

Figure 13.6.8

Substituting this expression in (24) and combining like components yields (verify)

$$\mathbf{r}(t) = (v_0\cos\alpha)t\mathbf{i} + \left(s_0 + (v_0\sin\alpha)t - \tfrac{1}{2}gt^2\right)\mathbf{j} \tag{26}$$

which is equivalent to the parametric equations

$$x = (v_0 \cos \alpha)t, \quad y = s_0 + (v_0 \sin \alpha)t - \tfrac{1}{2}gt^2 \tag{27}$$

Similarly, substituting (25) in (21) and combining like components yields

$$\mathbf{v}(t) = (v_0 \cos \alpha)\mathbf{i} + (v_0 \sin \alpha - gt)\mathbf{j}$$

which is equivalent to the parametric equations

$$v_x = v_0 \cos \alpha, \quad v_y = v_0 \sin \alpha - gt \tag{28}$$

The parameter t can be eliminated in (27) by solving the first equation for t and substituting in the second equation. We leave it for you to show that this yields

$$y = s_0 + (\tan \alpha)x - \left(\frac{g}{2v_0^2 \cos^2 \alpha}\right)x^2 \tag{29}$$

which is the equation of a parabola, since the right side is a quadratic polynomial in x. Thus, we have shown that the trajectory of the projectile is a parabolic arc.

Example 5 A shell, fired from a cannon, has a muzzle speed (the speed as it leaves the barrel) of 800 ft/s. The barrel makes an angle of 45° with the horizontal and, for simplicity, the barrel opening is assumed to be at ground level.

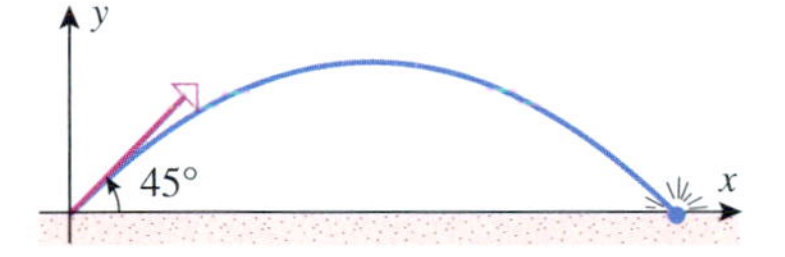

Figure 13.6.9

(a) Find parametric equations for the shell's trajectory relative to the coordinate system in Figure 13.6.9.

(b) How high does the shell rise?

(c) How far does the shell travel horizontally?

(d) What is the speed of the shell at its point of impact with the ground?

Solution (a). From (27) with $v_0 = 800$ ft/s, $\alpha = 45°$, $s_0 = 0$ ft (since the shell starts at ground level), and $g = 32$ ft/s^2, we obtain the parametric equations

$$x = (800 \cos 45°)t, \quad y = (800 \sin 45°)t - 16t^2 \quad (t \geq 0)$$

which simplify to

$$x = 400\sqrt{2}t, \quad y = 400\sqrt{2}t - 16t^2 \quad (t \geq 0) \tag{30}$$

Solution (b). The maximum height of the shell is the maximum value of y in (30), which occurs when $dy/dt = 0$, that is, when

$$400\sqrt{2} - 32t = 0 \quad \text{or} \quad t = \frac{25\sqrt{2}}{2}$$

Substituting this value of t in (30) yields

$$y = 5000 \text{ ft}$$

as the maximum height of the shell.

Solution (c). The shell will hit the ground when $y = 0$. From (30), this occurs when

$$400\sqrt{2}t - 16t^2 = 0 \quad \text{or} \quad t(400\sqrt{2} - 16t) = 0$$

The solution $t = 0$ corresponds to the initial position of the shell and the solution $t = 25\sqrt{2}$ to the time of impact. Substituting the latter value in the equation for x in (30) yields

$$x = 20{,}000 \text{ ft}$$

as the horizontal distance traveled by the shell.

Solution (d). From (30), the position function of the shell is

$$\mathbf{r}(t) = 400\sqrt{2}t\mathbf{i} + (400\sqrt{2}t - 16t^2)\mathbf{j}$$

so that the velocity function is

$$\mathbf{v}(t) = \mathbf{r}'(t) = 400\sqrt{2}\mathbf{i} + (400\sqrt{2} - 32t)\mathbf{j}$$

From part (c), impact occurs when $t = 25\sqrt{2}$, so the velocity vector at this point is

$$\mathbf{v}(25\sqrt{2}) = 400\sqrt{2}\mathbf{i} + [400\sqrt{2} - 32(25\sqrt{2})]\mathbf{j} = 400\sqrt{2}\mathbf{i} - 400\sqrt{2}\mathbf{j}$$

Thus, the speed at impact is

$$\|\mathbf{v}(25\sqrt{2})\| = \sqrt{(400\sqrt{2})^2 + (-400\sqrt{2})^2} = 800 \text{ ft/s} \blacktriangleleft$$

✔ QUICK CHECK EXERCISES 13.6 (See page 914 for answers.)

1. If $\mathbf{r}(t)$ is the position function of a particle, then the velocity, acceleration, and speed of the particle at time t are given, respectively, by

$$\mathbf{v}(t) = \_\_\_\_\_\_, \quad \mathbf{a}(t) = \_\_\_\_\_\_, \quad \frac{ds}{dt} = \_\_\_\_\_\_$$

2. If $\mathbf{r}(t)$ is the position function of a particle, then the displacement of the particle over the time interval $t_1 \le t \le t_2$ is ______, and the distance s traveled by the particle during this time interval is given by the integral ______.

3. The tangential scalar component of acceleration is given by the formula ______, and the normal scalar component of acceleration is given by the formula ______.

4. The projectile motion model

$$\mathbf{r}(t) = \left(-\tfrac{1}{2}gt^2 + s_0\right)\mathbf{j} + t\mathbf{v}_0$$

describes the motion of an object with constant acceleration $\mathbf{a} =$ ______ and velocity function $\mathbf{v}(t) =$ ______. The initial position of the object is ______ and its initial velocity is ______.

EXERCISE SET 13.6 Graphing Utility C CAS

1–4 In these exercises $\mathbf{r}(t)$ is the position vector of a particle moving in the plane. Find the velocity, acceleration, and speed at an arbitrary time t. Then sketch the path of the particle together with the velocity and acceleration vectors at the indicated time t.

1. $\mathbf{r}(t) = 3\cos t\mathbf{i} + 3\sin t\mathbf{j};\ t = \pi/3$

2. $\mathbf{r}(t) = t\mathbf{i} + t^2\mathbf{j};\ t = 2$

3. $\mathbf{r}(t) = e^t\mathbf{i} + e^{-t}\mathbf{j};\ t = 0$

4. $\mathbf{r}(t) = (2 + 4t)\mathbf{i} + (1 - t)\mathbf{j};\ t = 1$

5–8 Find the velocity, speed, and acceleration at the given time t of a particle moving along the given curve.

5. $\mathbf{r}(t) = t\mathbf{i} + \frac{1}{2}t^2\mathbf{j} + \frac{1}{3}t^3\mathbf{k};\ t = 1$

6. $x = 1 + 3t,\ y = 2 - 4t,\ z = 7 + t;\ t = 2$

7. $x = 2\cos t,\ y = 2\sin t,\ z = t;\ t = \pi/4$

8. $\mathbf{r}(t) = e^t \sin t\mathbf{i} + e^t \cos t\mathbf{j} + t\mathbf{k};\ t = \pi/2$

FOCUS ON CONCEPTS

9. As illustrated in the accompanying figure, suppose that the equations of motion of a particle moving along an elliptic path are $x = a\cos\omega t$, $y = b\sin\omega t$.
   (a) Show that the acceleration is directed toward the origin.
   (b) Show that the magnitude of the acceleration is proportional to the distance from the particle to the origin.

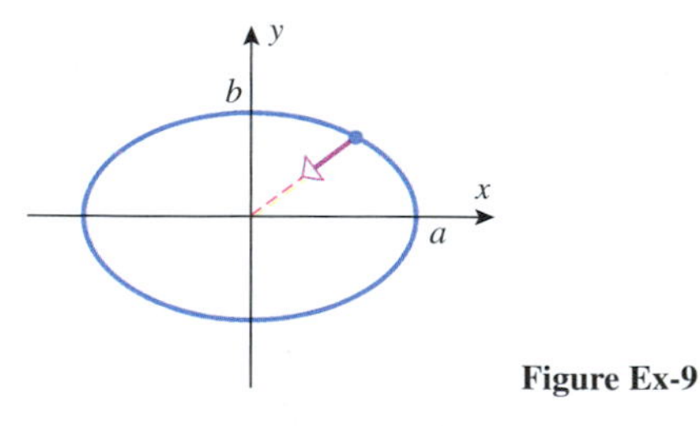

Figure Ex-9

10. Suppose that a particle vibrates in such a way that its position function is $\mathbf{r}(t) = 16\sin\pi t\mathbf{i} + 4\cos 2\pi t\mathbf{j}$, where distance is in millimeters and t is in seconds.

(a) Find the velocity and acceleration at time $t = 1$ s.
(b) Show that the particle moves along a parabolic curve.
(c) Show that the particle moves back and forth along the curve.

11. What can you say about the trajectory of a particle that moves in 2-space or 3-space with zero acceleration? Justify your answer.

12. Recall from Theorem 13.2.8 that if $\mathbf{r}(t)$ is a vector-valued function in 2-space or 3-space, and if $\|\mathbf{r}(t)\|$ is constant for all t, then $\mathbf{r}(t) \cdot \mathbf{r}'(t) = 0$.
(a) Translate this theorem into a statement about the motion of a particle in 2-space or 3-space.
(b) Replace $\mathbf{r}(t)$ by $\mathbf{r}'(t)$ in the theorem, and translate the result into a statement about the motion of a particle in 2-space or 3-space.

13. Suppose that the position vector of a particle moving in the plane is $\mathbf{r} = 12\sqrt{t}\,\mathbf{i} + t^{3/2}\mathbf{j}$, $t > 0$. Find the minimum speed of the particle and its location when it has this speed.

14. Suppose that the motion of a particle is described by the position vector $\mathbf{r} = (t - t^2)\mathbf{i} - t^2\mathbf{j}$. Find the minimum speed of the particle and its location when it has this speed.

15. Suppose that the position function of a particle moving in 2-space is $\mathbf{r} = \sin 3t\mathbf{i} - 2\cos 3t\mathbf{j}$.
(a) Use a graphing utility to graph the speed of the particle versus time from $t = 0$ to $t = 2\pi/3$.
(b) What are the maximum and minimum speeds of the particle?
(c) Use the graph to estimate the time at which the maximum speed first occurs.
(d) Find the exact time at which the maximum speed first occurs.

16. Suppose that the position function of a particle moving in 3-space is $\mathbf{r} = 3\cos 2t\mathbf{i} + \sin 2t\mathbf{j} + 4t\mathbf{k}$.
(a) Use a graphing utility to graph the speed of the particle versus time from $t = 0$ to $t = \pi$.
(b) Use the graph to estimate the maximum and minimum speeds of the particle.
(c) Use the graph to estimate the time at which the maximum speed first occurs.
(d) Find the exact values of the maximum and minimum speeds and the exact time at which the maximum speed first occurs.

17–20 Use the given information to find the position and velocity vectors of the particle.

17. $\mathbf{a}(t) = -\cos t\mathbf{i} - \sin t\mathbf{j}$; $\mathbf{v}(0) = \mathbf{i}$; $\mathbf{r}(0) = \mathbf{j}$

18. $\mathbf{a}(t) = \mathbf{i} + e^{-t}\mathbf{j}$; $\mathbf{v}(0) = 2\mathbf{i} + \mathbf{j}$; $\mathbf{r}(0) = \mathbf{i} - \mathbf{j}$

19. $\mathbf{a}(t) = \sin t\mathbf{i} + \cos t\mathbf{j} + e^t\mathbf{k}$; $\mathbf{v}(0) = \mathbf{k}$; $\mathbf{r}(0) = -\mathbf{i} + \mathbf{k}$

20. $\mathbf{a}(t) = (t+1)^{-2}\mathbf{j} - e^{-2t}\mathbf{k}$; $\mathbf{v}(0) = 3\mathbf{i} - \mathbf{j}$; $\mathbf{r}(0) = 2\mathbf{k}$

21. Find, to the nearest degree, the angle between $\mathbf{v}$ and $\mathbf{a}$ for $\mathbf{r} = t^3\mathbf{i} + t^2\mathbf{j}$ when $t = 1$.

22. Show that the angle between $\mathbf{v}$ and $\mathbf{a}$ is constant for the position vector $\mathbf{r} = e^t\cos t\mathbf{i} + e^t\sin t\mathbf{j}$. Find the angle.

23. (a) Suppose that at time $t = t_0$ an electron has a position vector of $\mathbf{r} = 3.5\mathbf{i} - 1.7\mathbf{j} + \mathbf{k}$, and at a later time $t = t_1$ it has a position vector of $\mathbf{r} = 4.2\mathbf{i} + \mathbf{j} - 2.4\mathbf{k}$. What is the displacement of the electron during the time interval from t_0 to t_1?
(b) Suppose that during a certain time interval a proton has a displacement of $\Delta\mathbf{r} = 0.7\mathbf{i} + 2.9\mathbf{j} - 1.2\mathbf{k}$ and its final position vector is known to be $\mathbf{r} = 3.6\mathbf{k}$. What was the initial position vector of the proton?

24. Suppose that the position function of a particle moving along a circle in the xy-plane is $\mathbf{r} = 5\cos 2\pi t\mathbf{i} + 5\sin 2\pi t\mathbf{j}$.
(a) Sketch some typical displacement vectors over the time interval from $t = 0$ to $t = 1$.
(b) What is the distance traveled by the particle during the time interval?

25–28 Find the displacement and the distance traveled over the indicated time interval.

25. $\mathbf{r} = t^2\mathbf{i} + \frac{1}{3}t^3\mathbf{j}$; $1 \le t \le 3$

26. $\mathbf{r} = (1 - 3\sin t)\mathbf{i} + 3\cos t\mathbf{j}$; $0 \le t \le 3\pi/2$

27. $\mathbf{r} = e^t\mathbf{i} + e^{-t}\mathbf{j} + \sqrt{2}t\mathbf{k}$; $0 \le t \le \ln 3$

28. $\mathbf{r} = \cos 2t\mathbf{i} + (1 - \cos 2t)\mathbf{j} + \left(3 + \frac{1}{2}\cos 2t\right)\mathbf{k}$; $0 \le t \le \pi$

29–30 The position vectors of two particles are given. Show that the particles move along the same path but the speed of the first is constant and the speed of the second is not.

29. $\mathbf{r}_1 = 2\cos 3t\mathbf{i} + 2\sin 3t\mathbf{j}$
$\mathbf{r}_2 = 2\cos(t^2)\mathbf{i} + 2\sin(t^2)\mathbf{j} \quad (t \ge 0)$

30. $\mathbf{r}_1 = (3 + 2t)\mathbf{i} + t\mathbf{j} + (1 - t)\mathbf{k}$
$\mathbf{r}_2 = (5 - 2t^3)\mathbf{i} + (1 - t^3)\mathbf{j} + t^3\mathbf{k}$

31–38 The position function of a particle is given. Use Theorem 13.6.3 to find
(a) the scalar tangential and normal components of acceleration at the stated time t;
(b) the vector tangential and normal components of acceleration at the stated time t;
(c) the curvature of the path at the point where the particle is located at the stated time t.

31. $\mathbf{r} = e^{-t}\mathbf{i} + e^t\mathbf{j}$; $t = 0$

32. $\mathbf{r} = \cos(t^2)\mathbf{i} + \sin(t^2)\mathbf{j}$; $t = \sqrt{\pi}/2$

33. $\mathbf{r} = (t^3 - 2t)\mathbf{i} + (t^2 - 4)\mathbf{j}$; $t = 1$

34. $\mathbf{r} = e^t\cos t\mathbf{i} + e^t\sin t\mathbf{j}$; $t = \pi/4$

35. $\mathbf{r} = (1/t)\mathbf{i} + t^2\mathbf{j} + t^3\mathbf{k}$; $t = 1$

36. $\mathbf{r} = e^t\mathbf{i} + e^{-2t}\mathbf{j} + t\mathbf{k}$; $t = 0$

37. $\mathbf{r} = 3\sin t\mathbf{i} + 2\cos t\mathbf{j} - \sin 2t\mathbf{k}$; $t = \pi/2$

38. $\mathbf{r} = 2\mathbf{i} + t^3\mathbf{j} - 16\ln t\mathbf{k}$; $t = 1$

39–42 In these exercises **v** and **a** are given at a certain instant of time. Find a_T, a_N, **T**, and **N** at this instant.

39. $\mathbf{v} = -4\mathbf{j},\ \mathbf{a} = 2\mathbf{i} + 3\mathbf{j}$ **40.** $\mathbf{v} = \mathbf{i} + 2\mathbf{j},\ \mathbf{a} = 3\mathbf{i}$

41. $\mathbf{v} = 2\mathbf{i} + 2\mathbf{j} + \mathbf{k},\ \mathbf{a} = \mathbf{i} + 2\mathbf{k}$

42. $\mathbf{v} = 3\mathbf{i} - 4\mathbf{k},\ \mathbf{a} = \mathbf{i} - \mathbf{j} + 2\mathbf{k}$

43–46 The speed $\|\mathbf{v}\|$ of a particle at an arbitrary time t is given. Find the scalar tangential component of acceleration at the indicated time.

43. $\|\mathbf{v}\| = \sqrt{3t^2 + 4};\ t = 2$ **44.** $\|\mathbf{v}\| = \sqrt{t^2 + e^{-3t}};\ t = 0$

45. $\|\mathbf{v}\| = \sqrt{(4t - 1)^2 + \cos^2 \pi t};\ t = \frac{1}{4}$

46. $\|\mathbf{v}\| = \sqrt{t^4 + 5t^2 + 3};\ t = 1$

47. The nuclear accelerator at the Enrico Fermi Laboratory is circular with a radius of 1 km. Find the scalar normal component of acceleration of a proton moving around the accelerator with a constant speed of 2.9×10^5 km/s.

48. Suppose that a particle moves with nonzero acceleration along the curve $y = f(x)$. Use part (b) of Exercise 19 in Section 13.5 to show that the acceleration vector is tangent to the curve at each point where $f''(x) = 0$.

49–50 Use the given information and Exercise 19 of Section 13.5 to find the normal scalar component of acceleration as a function of x.

49. A particle moves along the parabola $y = x^2$ with a constant speed of 3 units per second.

50. A particle moves along the curve $x = \ln y$ with a constant speed of 2 units per second.

51–52 Use the given information to find the normal scalar component of acceleration at time $t = 1$.

51. $\mathbf{a}(1) = \mathbf{i} + 2\mathbf{j} - 2\mathbf{k};\ a_T(1) = 3$

52. $\|\mathbf{a}(1)\| = 9;\ a_T(1)\mathbf{T}(1) = 2\mathbf{i} - 2\mathbf{j} + \mathbf{k}$

53. An automobile travels at a constant speed around a curve whose radius of curvature is 1000 m. What is the maximum allowable speed if the maximum acceptable value for the normal scalar component of acceleration is 1.5 m/s^2?

54. If an automobile of mass m rounds a curve, then its inward vector component of acceleration $a_N\mathbf{N}$ is caused by the frictional force **F** of the road. Thus, it follows from the vector form of Newton's second law [Equation (19)] that the frictional force and the normal scalar component of acceleration are related by the equation $\mathbf{F} = ma_N\mathbf{N}$. Thus,

$$\|\mathbf{F}\| = m\kappa\left(\frac{ds}{dt}\right)^2$$

Use this result to find the magnitude of the frictional force in newtons exerted by the road on a 500-kg go-cart driven at a speed of 10 km/h around a circular track of radius 15 m. [*Note:* 1 N = 1 kg·m/s^2]

55. A shell is fired from ground level with a muzzle speed of 320 ft/s and elevation angle of 60°. Find
(a) parametric equations for the shell's trajectory
(b) the maximum height reached by the shell
(c) the horizontal distance traveled by the shell
(d) the speed of the shell at impact.

56. Solve Exercise 55 assuming that the muzzle speed is 980 m/s and the elevation angle is 45°.

57. A rock is thrown downward from the top of a building, 168 ft high, at an angle of 60° with the horizontal. How far from the base of the building will the rock land if its initial speed is 80 ft/s?

58. Solve Exercise 57 assuming that the rock is thrown horizontally at a speed of 80 ft/s.

59. A shell is to be fired from ground level at an elevation angle of 30°. What should the muzzle speed be in order for the maximum height of the shell to be 2500 ft?

60. A shell, fired from ground level at an elevation angle of 45°, hits the ground 24,500 m away. Calculate the muzzle speed of the shell.

61. Find two elevation angles that will enable a shell, fired from ground level with a muzzle speed of 800 ft/s, to hit a ground-level target 10,000 ft away.

62. A ball rolls off a table 4 ft high while moving at a constant speed of 5 ft/s.
(a) How long does it take for the ball to hit the floor after it leaves the table?
(b) At what speed does the ball hit the floor?
(c) If a ball were dropped from rest at table height just as the rolling ball leaves the table, which ball would hit the ground first? Justify your answer.

63. As illustrated in the accompanying figure, a fire hose sprays water with an initial velocity of 40 ft/s at an angle of 60° with the horizontal.
(a) Confirm that the water will clear corner point A.
(b) Confirm that the water will hit the roof.
(c) How far from corner point A will the water hit the roof?

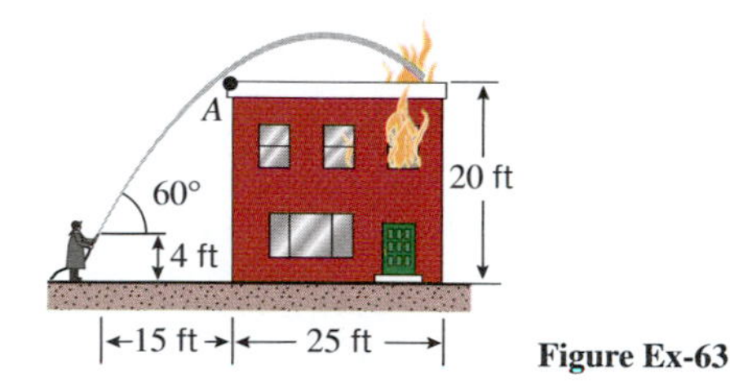

Figure Ex-63

64. What is the minimum initial velocity that will allow the water in Exercise 63 to hit the roof?

65. As shown in the accompanying figure, water is sprayed from a hose with an initial velocity of 35 m/s at an angle of 45° with the horizontal.
(a) What is the radius of curvature of the stream at the point where it leaves the hose?

(b) What is the maximum height of the stream above the nozzle of the hose?

66. As illustrated in the accompanying figure, a train is traveling on a curved track. At a point where the train is traveling at a speed of 132 ft/s and the radius of curvature of the track is 3000 ft, the engineer hits the brakes to make the train slow down at a constant rate of 7.5 ft/s^2.

(a) Find the magnitude of the acceleration vector at the instant the engineer hits the brakes.

(b) Approximate the angle between the acceleration vector and the unit tangent vector **T** at the instant the engineer hits the brakes.

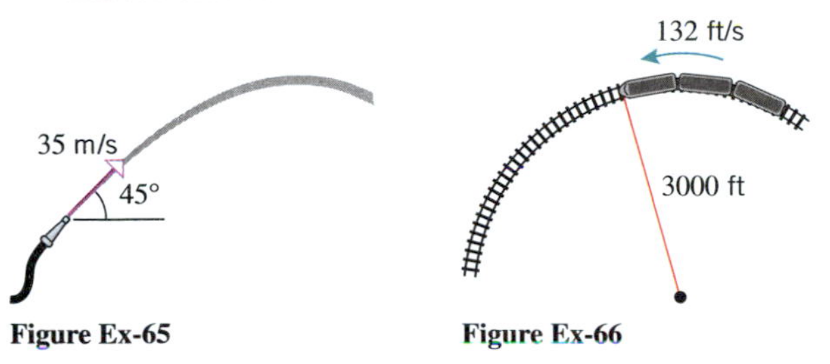

Figure Ex-65 **Figure Ex-66**

67. A shell is fired from ground level at an elevation angle of α and a muzzle speed of v_0.

(a) Show that the maximum height reached by the shell is

$$\text{maximum height} = \frac{(v_0 \sin\alpha)^2}{2g}$$

(b) The ***horizontal range*** R of the shell is the horizontal distance traveled when the shell returns to ground level. Show that $R = (v_0^2 \sin 2\alpha)/g$. For what elevation angle will the range be maximum? What is the maximum range?

68. A shell is fired from ground level with an elevation angle α and a muzzle speed of v_0. Find two angles that can be used to hit a target at ground level that is a distance of three-fourths the maximum range of the shell. Express your answer to the nearest tenth of a degree. [*Hint:* See Exercise 67(b).]

69. At time $t = 0$ a baseball that is 5 ft above the ground is hit with a bat. The ball leaves the bat with a speed of 80 ft/s at an angle of 30° above the horizontal.

(a) How long will it take for the baseball to hit the ground? Express your answer to the nearest hundredth of a second.

(b) Use the result in part (a) to find the horizontal distance traveled by the ball. Express your answer to the nearest tenth of a foot.

70. Repeat Exercise 69, assuming that the ball leaves the bat with a speed of 70 ft/s at an angle of 60° above the horizontal.

C **71.** At time $t = 0$ a skier leaves the end of a ski jump with a speed of v_0 ft/s at an angle α with the horizontal (see the accompanying figure). The skier lands 259 ft down the incline 2.9 s later.

(a) Approximate v_0 to the nearest ft/s and α to the nearest degree.

(b) Use a CAS or a calculating utility with a numerical integration capability to approximate the distance traveled by the skier.

(Use $g = 32$ ft/s^2 as the acceleration due to gravity.)

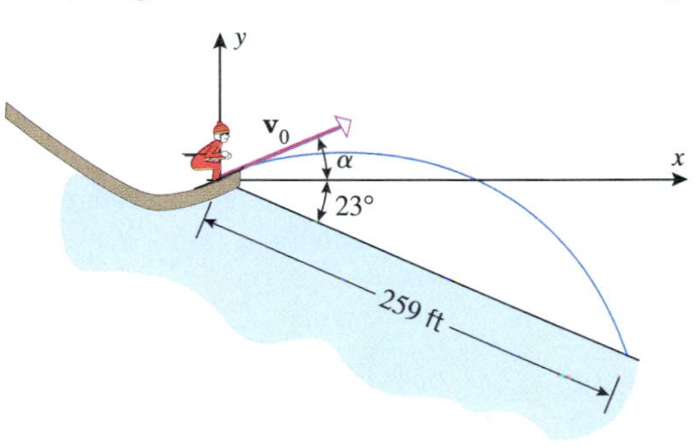

Figure Ex-71

FOCUS ON CONCEPTS

72. At time $t = 0$ a projectile is fired from a height h above level ground at an elevation angle of α with a speed v. Let R be the horizontal distance to the point where the projectile hits the ground.

(a) Show that α and R must satisfy the equation

$$g(\sec^2\alpha)R^2 - 2v^2(\tan\alpha)R - 2v^2h = 0$$

(b) If g, h, and v are constant, then the equation in part (a) defines R implicitly as a function of α. Let R_0 be the maximum value of R and α_0 the value of α when $R = R_0$. Use implicit differentiation to find $dR/d\alpha$ and show that

$$\tan\alpha_0 = \frac{v^2}{gR_0}$$

[*Hint:* Assume that $dR/d\alpha = 0$ when R attains a maximum.]

(c) Use the results in parts (a) and (b) to show that

$$R_0 = \frac{v}{g}\sqrt{v^2 + 2gh}$$

and

$$\alpha_0 = \tan^{-1}\frac{v}{\sqrt{v^2 + 2gh}}$$

73. Suppose that the position function of a point moving in the xy-plane is

$$\mathbf{r} = x(t)\mathbf{i} + y(t)\mathbf{j}$$

This equation can be expressed in polar coordinates by making the substitution

$$x(t) = r(t)\cos\theta(t), \quad y(t) = r(t)\sin\theta(t)$$

This yields

$$\mathbf{r} = r(t)\cos\theta(t)\mathbf{i} + r(t)\sin\theta(t)\mathbf{j}$$

which can be expressed as

$$\mathbf{r} = r(t)\mathbf{e}_r(t)$$

where $\mathbf{e}_r(t) = \cos\theta(t)\mathbf{i} + \sin\theta(t)\mathbf{j}$.

(a) Show that $\mathbf{e}_r(t)$ is a unit vector that has the same direction as the radius vector $\mathbf{r}$ if $r(t) > 0$ and that $\mathbf{e}_\theta(t) = -\sin\theta(t)\mathbf{i} + \cos\theta(t)\mathbf{j}$ is the unit vector that results when $\mathbf{e}_r(t)$ is rotated counterclockwise through an angle of $\pi/2$. The vector $\mathbf{e}_r(t)$ is called the ***radial unit vector*** and the vector $\mathbf{e}_\theta(t)$ is called the ***transverse unit vector*** (see the accompanying figure).

(b) Show that the velocity function $\mathbf{v} = \mathbf{v}(t)$ can be expressed in terms of radial and transverse components as

$$\mathbf{v} = \frac{dr}{dt}\mathbf{e}_r + r\frac{d\theta}{dt}\mathbf{e}_\theta$$

(c) Show that the acceleration function $\mathbf{a} = \mathbf{a}(t)$ can be expressed in terms of radial and transverse components as

$$\mathbf{a} = \left[\frac{d^2r}{dt^2} - r\left(\frac{d\theta}{dt}\right)^2\right]\mathbf{e}_r + \left[r\frac{d^2\theta}{dt^2} + 2\frac{dr}{dt}\frac{d\theta}{dt}\right]\mathbf{e}_\theta$$

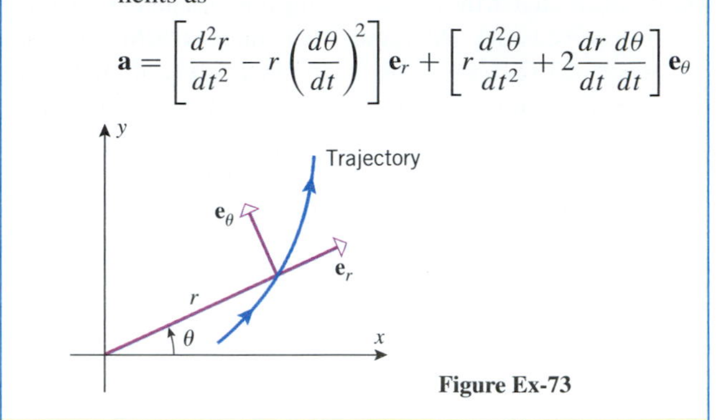

Figure Ex-73

✔QUICK CHECK ANSWERS 13.6

1. $\dfrac{d\mathbf{r}}{dt}$; $\dfrac{d\mathbf{v}}{dt} = \dfrac{d^2\mathbf{r}}{dt^2}$; $\|\mathbf{v}(t)\|$ **2.** $\mathbf{r}(t_2) - \mathbf{r}(t_1)$; $\displaystyle\int_{t_1}^{t_2} \|\mathbf{v}(t)\|\, dt$ **3.** $\dfrac{d^2s}{dt^2}$; $\kappa(ds/dt)^2$ **4.** $-g\mathbf{j}$; $-gt\mathbf{j} + \mathbf{v}_0$; $s_0\mathbf{j}$; $\mathbf{v}_0$

13.7 KEPLER'S LAWS OF PLANETARY MOTION

One of the great advances in the history of astronomy occurred in the early 1600s when Johannes Kepler deduced from empirical data that all planets in our solar system move in elliptical orbits with the Sun at a focus. Subsequently, Isaac Newton showed mathematically that such planetary motion is the consequence of an inverse-square law of gravitational attraction. In this section we will use the concepts developed in the preceding sections of this chapter to derive three basic laws of planetary motion, known as **Kepler's laws**.*

KEPLER'S LAWS

In Section 11.6 we stated the following laws of planetary motion that were published by Johannes Kepler in 1609 in his book known as *Astronomia Nova*.

13.7.1 KEPLER'S LAWS.

- First law (***Law of Orbits***). Each planet moves in an elliptical orbit with the Sun at a focus.
- Second law (***Law of Areas***). Equal areas are swept out in equal times by the line from the Sun to a planet.
- Third law (***Law of Periods***). The square of a planet's period (the time it takes the planet to complete one orbit about the Sun) is proportional to the cube of the semimajor axis of its orbit.

*See biography on p. 776.

CENTRAL FORCES

If a particle moves under the influence of a *single* force that is always directed toward a fixed point O, then the particle is said to be moving in a ***central force field***. The force is called a ***central force***, and the point O is called the ***center of force***. For example, in the simplest model of planetary motion, it is assumed that the only force acting on a planet is the force of the Sun's gravity, directed toward the center of the Sun. This model, which produces Kepler's laws, ignores the forces that other celestial objects exert on the planet as well as the minor effect that the planet's gravity has on the Sun. Central force models are also used to study the motion of comets, asteroids, planetary moons, and artificial satellites. They also have important applications in electromagnetics. Our objective in this section is to develop some basic principles about central force fields and then use those results to derive Kepler's laws.

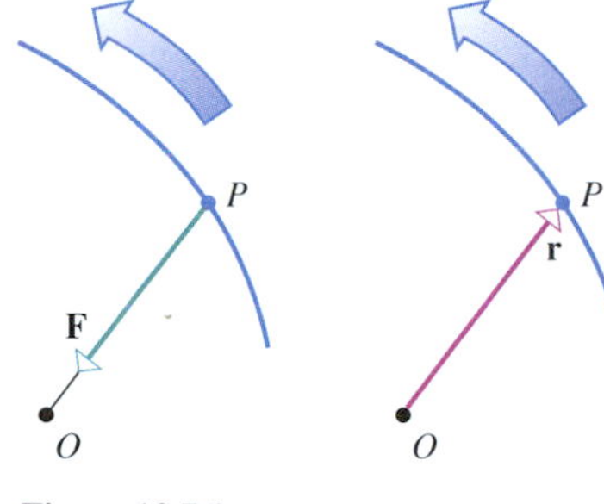

Figure 13.7.1

Suppose that a particle P of mass m moves in a central force field due to a force $\mathbf{F}$ that is directed toward a fixed point O, and let $\mathbf{r} = \mathbf{r}(t)$ be the position vector from O to P (Figure 13.7.1). Let $\mathbf{v} = \mathbf{v}(t)$ and $\mathbf{a} = \mathbf{a}(t)$ be the velocity and acceleration functions of the particle, and assume that $\mathbf{F}$ and $\mathbf{a}$ are related by Newton's second law ($\mathbf{F} = m\mathbf{a}$).

Our first objective is to show that the particle P moves in a plane containing the point O. For this purpose observe that $\mathbf{a}$ has the same direction as $\mathbf{F}$ by Newton's second law, and this implies that $\mathbf{a}$ and $\mathbf{r}$ are oppositely directed vectors. Thus, it follows from part (*c*) of Theorem 12.4.5 that

$$\mathbf{r} \times \mathbf{a} = \mathbf{0}$$

Since the velocity and acceleration of the particle are given by $\mathbf{v} = d\mathbf{r}/dt$ and $\mathbf{a} = d\mathbf{v}/dt$, respectively, we have

$$\frac{d}{dt}(\mathbf{r} \times \mathbf{v}) = \mathbf{r} \times \frac{d\mathbf{v}}{dt} + \frac{d\mathbf{r}}{dt} \times \mathbf{v} = (\mathbf{r} \times \mathbf{a}) + (\mathbf{v} \times \mathbf{v}) = \mathbf{0} + \mathbf{0} = \mathbf{0} \tag{1}$$

Integrating the left and right sides of this equation with respect to t yields

$$\mathbf{r} \times \mathbf{v} = \mathbf{b} \tag{2}$$

Astronomers call the plane containing the orbit of a planet the ***ecliptic*** of the planet.

where $\mathbf{b}$ is a constant (independent of t). However, $\mathbf{b}$ is orthogonal to both $\mathbf{r}$ and $\mathbf{v}$, so we can conclude that $\mathbf{r} = \mathbf{r}(t)$ and $\mathbf{v} = \mathbf{v}(t)$ lie in a fixed plane containing the point O.

NEWTON'S LAW OF UNIVERSAL GRAVITATION

Our next objective is to derive the position function of a particle moving under a central force in a polar coordinate system. For this purpose we will need the following result, known as ***Newton's Law of Universal Gravitation***.

13.7.2 NEWTON'S LAW OF UNIVERSAL GRAVITATION. Every particle of matter in the Universe attracts every other particle of matter in the Universe with a force that is proportional to the product of their masses and inversely proportional to the square of the distance between them. Specifically, if a particle of mass M and a particle of mass m are at a distance r from one another, then they attract each other with equal and opposite forces, $\mathbf{F}$ and $-\mathbf{F}$, of magnitude

$$\|\mathbf{F}\| = \frac{GMm}{r^2} \tag{3}$$

where G is a constant called the ***universal gravitational constant***.

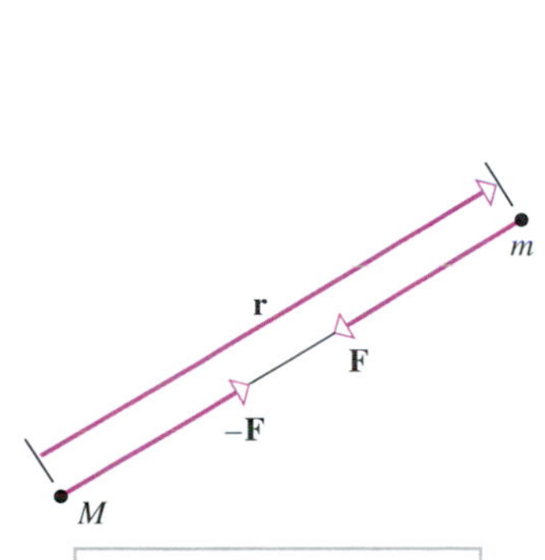

Figure 13.7.2

To obtain a formula for the vector force $\mathbf{F}$ that mass M exerts on mass m, we will let $\mathbf{r}$ be the radius vector from mass M to mass m (Figure 13.7.2). Thus, the distance r between the masses is $\|\mathbf{r}\|$, and the force $\mathbf{F}$ can be expressed in terms of $\mathbf{r}$ as

$$\mathbf{F} = \|\mathbf{F}\|\left(-\frac{\mathbf{r}}{\|\mathbf{r}\|}\right) = \|\mathbf{F}\|\left(-\frac{\mathbf{r}}{r}\right)$$

which from (3) can be expressed as

$$\mathbf{F} = -\frac{GMm}{r^3}\mathbf{r} \tag{4}$$

We start by finding a formula for the acceleration function. To do this we use Formula (4) and Newton's second law to obtain

Observe in Formula (5) that the acceleration $\mathbf{a}$ does not involve m. Thus, the mass of a planet has no effect on its acceleration.

$$m\mathbf{a} = -\frac{GMm}{r^3}\mathbf{r}$$

from which we obtain

$$\mathbf{a} = -\frac{GM}{r^3}\mathbf{r} \tag{5}$$

To obtain a formula for the position function of the mass m, we will need to introduce a coordinate system and make some assumptions about the initial conditions. Let us assume:

- The distance r from m to M is minimum at time $t = 0$.
- The mass m has nonzero position and velocity vectors $\mathbf{r}_0$ and $\mathbf{v}_0$ at time $t = 0$.
- A polar coordinate system is introduced with its pole at mass M and oriented so $\theta = 0$ at time $t = 0$.
- The vector $\mathbf{v}_0$ is perpendicular to the polar axis at time $t = 0$.

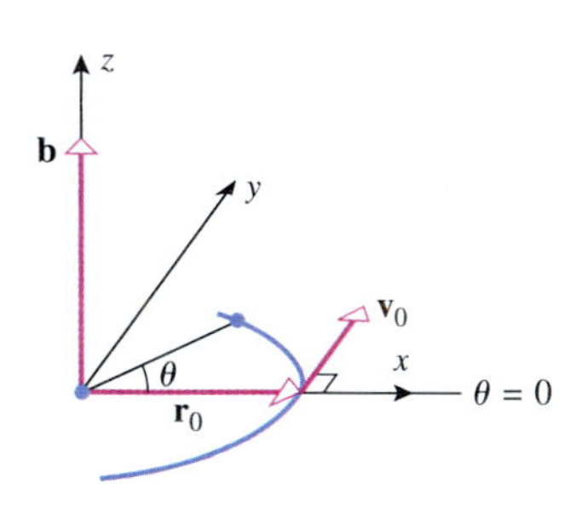

Figure 13.7.3

Moreover, to ensure that the polar angle θ increases with t, let us agree to observe this polar coordinate system looking toward the pole from the terminal point of the vector $\mathbf{b} = \mathbf{r}_0 \times \mathbf{v}_0$. We will also find it useful to superimpose an xyz-coordinate system on the polar coordinate system with the positive z-axis in the direction of $\mathbf{b}$ (Figure 13.7.3).

For computational purposes, it will be helpful to denote $\|\mathbf{r}_0\|$ by r_0 and $\|\mathbf{v}_0\|$ by v_0, in which case we can express the vectors $\mathbf{r}_0$ and $\mathbf{v}_0$ in xyz-coordinates as

$$\mathbf{r}_0 = r_0\mathbf{i} \quad \text{and} \quad \mathbf{v}_0 = v_0\mathbf{j}$$

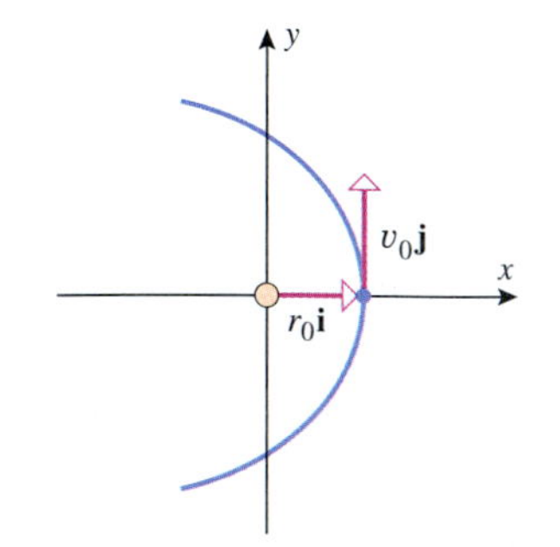

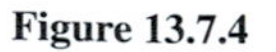
Figure 13.7.4

and the vector $\mathbf{b}$ as

$$\mathbf{b} = \mathbf{r}_0 \times \mathbf{v}_0 = r_0\mathbf{i} \times v_0\mathbf{j} = r_0v_0\mathbf{k} \tag{6}$$

(Figure 13.7.4). It will also be useful to introduce the unit vector

$$\mathbf{u} = \cos\theta\mathbf{i} + \sin\theta\mathbf{j} \tag{7}$$

which will allow us to express the polar form of the position vector $\mathbf{r}$ as

$$\mathbf{r} = r\cos\theta\mathbf{i} + r\sin\theta\mathbf{j} = r(\cos\theta\mathbf{i} + \sin\theta\mathbf{j}) = r\mathbf{u} \tag{8}$$

and to express the acceleration vector $\mathbf{a}$ in terms of $\mathbf{u}$ by rewriting (5) as

$$\mathbf{a} = -\frac{GM}{r^2}\mathbf{u} \tag{9}$$

We are now ready to derive the position function of the mass m in polar coordinates. For this purpose, recall from (2) that the vector $\mathbf{b} = \mathbf{r} \times \mathbf{v}$ is constant, so it follows from (6) that the relationship

$$\mathbf{b} = \mathbf{r} \times \mathbf{v} = r_0v_0\mathbf{k} \tag{10}$$

holds for *all* values of t. Now let us examine $\mathbf{b}$ from another point of view. It follows from (8) that

$$\mathbf{v} = \frac{d\mathbf{r}}{dt} = \frac{d}{dt}(r\mathbf{u}) = r\frac{d\mathbf{u}}{dt} + \frac{dr}{dt}\mathbf{u}$$

and hence

$$\mathbf{b} = \mathbf{r} \times \mathbf{v} = (r\mathbf{u}) \times \left(r\frac{d\mathbf{u}}{dt} + \frac{dr}{dt}\mathbf{u}\right) = r^2\mathbf{u} \times \frac{d\mathbf{u}}{dt} + r\frac{dr}{dt}\mathbf{u} \times \mathbf{u} = r^2\mathbf{u} \times \frac{d\mathbf{u}}{dt} \tag{11}$$

But (7) implies that

$$\frac{d\mathbf{u}}{dt} = \frac{d\mathbf{u}}{d\theta}\frac{d\theta}{dt} = (-\sin\theta\mathbf{i} + \cos\theta\mathbf{j})\frac{d\theta}{dt}$$

so

$$\mathbf{u} \times \frac{d\mathbf{u}}{dt} = \frac{d\theta}{dt}\mathbf{k} \tag{12}$$

Substituting (12) in (11) yields

$$\mathbf{b} = r^2\frac{d\theta}{dt}\mathbf{k} \tag{13}$$

Thus, it follows from (7), (9), and (13) that

$$\begin{aligned}\mathbf{a} \times \mathbf{b} &= -\frac{GM}{r^2}(\cos\theta\mathbf{i} + \sin\theta\mathbf{j}) \times \left(r^2\frac{d\theta}{dt}\mathbf{k}\right) \\ &= GM(-\sin\theta\mathbf{i} + \cos\theta\mathbf{j})\frac{d\theta}{dt} = GM\frac{d\mathbf{u}}{dt}\end{aligned} \tag{14}$$

From this formula and the fact that $d\mathbf{b}/dt = \mathbf{0}$ (since $\mathbf{b}$ is constant), we obtain

$$\frac{d}{dt}(\mathbf{v} \times \mathbf{b}) = \mathbf{v} \times \frac{d\mathbf{b}}{dt} + \frac{d\mathbf{v}}{dt} \times \mathbf{b} = \mathbf{a} \times \mathbf{b} = GM\frac{d\mathbf{u}}{dt}$$

Integrating both sides of this equation with respect to t yields

$$\mathbf{v} \times \mathbf{b} = GM\mathbf{u} + \mathbf{C} \tag{15}$$

where $\mathbf{C}$ is a vector constant of integration. This constant can be obtained by evaluating both sides of the equation at $t = 0$. We leave it as an exercise to show that

$$\mathbf{C} = (r_0v_0^2 - GM)\mathbf{i} \tag{16}$$

from which it follows that

$$\mathbf{v} \times \mathbf{b} = GM\mathbf{u} + (r_0v_0^2 - GM)\mathbf{i} \tag{17}$$

We can now obtain the position function by computing the scalar triple product $\mathbf{r} \cdot (\mathbf{v} \times \mathbf{b})$ in two ways. First we use (10) and property (11) of Section 12.4 to obtain

$$\mathbf{r} \cdot (\mathbf{v} \times \mathbf{b}) = (\mathbf{r} \times \mathbf{v}) \cdot \mathbf{b} = \mathbf{b} \cdot \mathbf{b} = r_0^2v_0^2 \tag{18}$$

and next we use (17) to obtain

$$\begin{aligned}\mathbf{r} \cdot (\mathbf{v} \times \mathbf{b}) &= \mathbf{r} \cdot (GM\mathbf{u}) + \mathbf{r} \cdot (r_0v_0^2 - GM)\mathbf{i} \\ &= \mathbf{r} \cdot \left(GM\frac{\mathbf{r}}{r}\right) + r\mathbf{u} \cdot (r_0v_0^2 - GM)\mathbf{i} \\ &= GMr + r(r_0v_0^2 - GM)\cos\theta\end{aligned}$$

If we now equate this to (18), we obtain

$$r_0^2v_0^2 = GMr + r(r_0v_0^2 - GM)\cos\theta$$

which when solved for r gives

$$r = \frac{r_0^2v_0^2}{GM + (r_0v_0^2 - GM)\cos\theta} = \frac{\dfrac{r_0^2v_0^2}{GM}}{1 + \left(\dfrac{r_0v_0^2}{GM} - 1\right)\cos\theta} \tag{19}$$

or more simply

$$r = \frac{k}{1 + e\cos\theta} \tag{20}$$

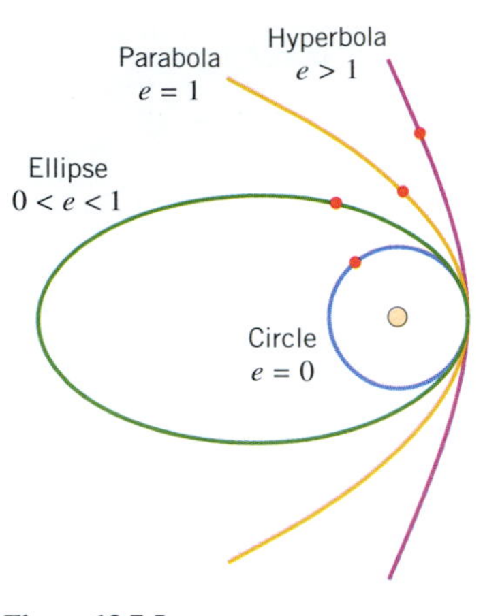

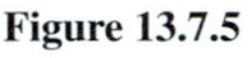
Figure 13.7.5

where

$$k = \frac{r_0^2 v_0^2}{GM} \quad \text{and} \quad e = \frac{r_0 v_0^2}{GM} - 1 \tag{21–22}$$

We will leave it as an exercise to show that $e \geq 0$. Accepting this to be so, it follows by comparing (20) to Formula (3) of Section 11.6 that the trajectory is a conic section with eccentricity e, the focus at the pole, and $d = k/e$. Thus, depending on whether $e < 1$, $e = 1$, or $e > 1$, the trajectory will be, respectively, an ellipse, a parabola, or a hyperbola (Figure 13.7.5).

Note from Formula (22) that e depends on r_0 and v_0, so the exact form of the trajectory is determined by the mass M and the initial conditions. If the initial conditions are such that $e < 1$, then the mass m becomes trapped in an elliptical orbit; otherwise the mass m "escapes" and never returns to its initial position. Accordingly, the initial velocity that produces an eccentricity of $e = 1$ is called the ***escape speed*** and is denoted by v_{esc}. Thus, it follows from (22) that

$$v_{\text{esc}} = \sqrt{\frac{2GM}{r_0}} \tag{23}$$

(verify).

KEPLER'S FIRST AND SECOND LAWS

It follows from our general discussion of central force fields that the planets have elliptical orbits with the Sun at the focus, which is Kepler's first law. To derive Kepler's second law, we begin by equating (10) and (13) to obtain

$$r^2 \frac{d\theta}{dt} = r_0 v_0 \tag{24}$$

To prove that the radial line from the center of the Sun to the center of a planet sweeps out equal areas in equal times, let $r = f(\theta)$ denote the polar equation of the planet, and let A denote the area swept out by the radial line as it varies from any fixed angle θ_0 to an angle θ. It follows from the area formula in 11.3.2 that A can be expressed as

$$A = \int_{\theta_0}^{\theta} \frac{1}{2}[f(\phi)]^2 \, d\phi$$

where the dummy variable ϕ is introduced for the integration to reserve θ for the upper limit. It now follows from Part 2 of the Fundamental Theorem of Calculus and the chain rule that

$$\frac{dA}{dt} = \frac{dA}{d\theta}\frac{d\theta}{dt} = \frac{1}{2}[f(\theta)]^2 \frac{d\theta}{dt} = \frac{1}{2} r^2 \frac{d\theta}{dt}$$

Thus, it follows from (24) that

$$\frac{dA}{dt} = \frac{1}{2} r_0 v_0 \tag{25}$$

which shows that A changes at a constant rate. This implies that equal areas are swept out in equal times.

KEPLER'S THIRD LAW

To derive Kepler's third law, we let a and b be the semimajor and semiminor axes of the elliptical orbit, and we recall that the area of this ellipse is πab. It follows by integrating (25) that in t units of time the radial line will sweep out an area of $A = \frac{1}{2} r_0 v_0 t$. Thus, if T denotes the time required for the planet to make one revolution around the Sun (the period), then the radial line will sweep out the area of the entire ellipse during that time and hence

$$\pi ab = \frac{1}{2} r_0 v_0 T$$

from which we obtain

$$T^2 = \frac{4\pi^2 a^2 b^2}{r_0^2 v_0^2} \tag{26}$$

However, it follows from Formula (1) of Section 11.6 and the relationship $c^2 = a^2 - b^2$ for an ellipse that

$$e = \frac{c}{a} = \frac{\sqrt{a^2 - b^2}}{a}$$

Thus, $b^2 = a^2(1 - e^2)$ and hence (26) can be written as

$$T^2 = \frac{4\pi^2 a^4 (1 - e^2)}{r_0^2 v_0^2} \tag{27}$$

But comparing Equation (20) to Equation (17) of Section 11.6 shows that

$$k = a(1 - e^2)$$

Finally, substituting this expression and (21) in (27) yields

$$T^2 = \frac{4\pi^2 a^3}{r_0^2 v_0^2} k = \frac{4\pi^2 a^3}{r_0^2 v_0^2} \frac{r_0^2 v_0^2}{GM} = \frac{4\pi^2}{GM} a^3 \tag{28}$$

Thus, we have proved that T^2 is proportional to a^3, which is Kepler's third law. When convenient, Formula (28) can also be expressed as

$$T = \frac{2\pi}{\sqrt{GM}} a^{3/2} \tag{29}$$

ARTIFICIAL SATELLITES

Kepler's second and third laws and Formula (23) also apply to satellites that orbit a celestial body; we need only interpret M to be the mass of the body exerting the force and m to be the mass of the satellite. Values of GM that are required in many of the formulas in this section have been determined experimentally for various attracting bodies (Table 13.7.1).

Table 13.7.1

ATTRACTING BODY	INTERNATIONAL SYSTEM	BRITISH ENGINEERING SYSTEM
Earth	$GM = 3.99 \times 10^{14}\ \text{m}^3/\text{s}^2$ $GM = 3.99 \times 10^{5}\ \text{km}^3/\text{s}^2$	$GM = 1.41 \times 10^{16}\ \text{ft}^3/\text{s}^2$ $GM = 1.24 \times 10^{12}\ \text{mi}^3/\text{h}^2$
Sun	$GM = 1.33 \times 10^{20}\ \text{m}^3/\text{s}^2$ $GM = 1.33 \times 10^{11}\ \text{km}^3/\text{s}^2$	$GM = 4.69 \times 10^{21}\ \text{ft}^3/\text{s}^2$ $GM = 4.13 \times 10^{17}\ \text{mi}^3/\text{h}^2$
Moon	$GM = 4.90 \times 10^{12}\ \text{m}^3/\text{s}^2$ $GM = 4.90 \times 10^{3}\ \text{km}^3/\text{s}^2$	$GM = 1.73 \times 10^{14}\ \text{ft}^3/\text{s}^2$ $GM = 1.53 \times 10^{10}\ \text{mi}^3/\text{h}^2$

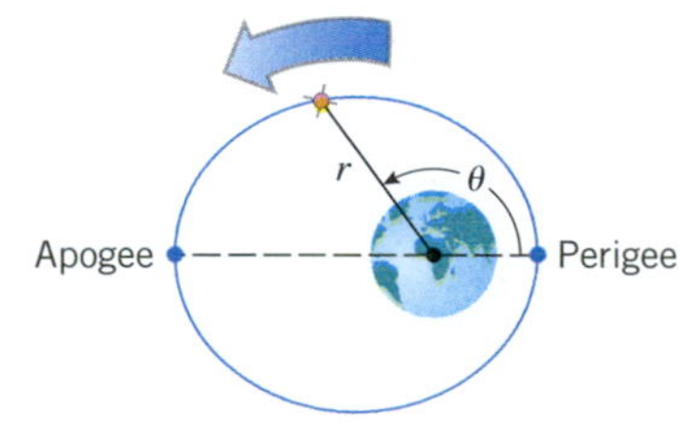

Figure 13.7.6

Recall that for orbits of planets around the Sun, the point at which the distance between the center of the planet and the center of the Sun is maximum is called the *aphelion* and the point at which it is minimum the *perihelion*. For satellites around the Earth the point at which the maximum distance occurs is called the ***apogee*** and the point at which the minimum distance occurs is called the ***perigee*** (Figure 13.7.6). The actual distances between the centers at apogee and perigee are called the ***apogee distance*** and the ***perigee distance***.

▶ **Example 1** A geosynchronous orbit for a satellite is a circular orbit about the equator of the Earth in which the satellite stays fixed over a point on the equator. Use the fact that the Earth makes one revolution about its axis every 24 hours to find the altitude in miles of a communications satellite in geosynchronous orbit. Assume the Earth to be a sphere of radius 4000 mi.

Solution. To remain fixed over a point on the equator, the satellite must have a period of $T = 24$ h. It follows from (28) or (29) and the Earth value of $GM = 1.24 \times 10^{12}$ mi^3/h^2 from Table 13.7.1 that

$$a = \sqrt[3]{\frac{GMT^2}{4\pi^2}} = \sqrt[3]{\frac{(1.24 \times 10^{12})(24)^2}{4\pi^2}} \approx 26{,}250 \text{ mi}$$

and hence the altitude h of the satellite is

$$h \approx 26{,}250 - 4000 = 22{,}250 \text{ mi} \blacktriangleleft$$

✔QUICK CHECK EXERCISES 13.7 (See page 921 for answers.)

1. Let G denote the universal gravitational constant and let M and m denote masses a distance r apart.
   (a) According to Newton's Law of Universal Gravitation, M and m attract each other with a force of magnitude ________.
   (b) If $\mathbf{r}$ is the radius vector from M to m, then the force of attraction that mass M exerts on mass m is ________.
2. Suppose that a mass m is in an orbit about a mass M and that r_0 is the minimum distance from m to M. If G is the universal gravitational constant, then the "escape" speed of m is ________.
3. For a planet in an elliptical orbit about the Sun, the square of the planet's period is proportional to what power of the semimajor axis of its orbit?
4. Suppose that a mass m is in an orbit about a mass M and that r_0 is the minimum distance from m to M. If v_0 is the speed of mass m when it is a distance r_0 from M, and if G denotes the universal gravitational constant, then the eccentricity of the orbit is ________.

EXERCISE SET 13.7

In exercises that require numerical values, use Table 13.7.1 and the following values, where needed:

radius of Earth = 4000 mi = 6440 km
radius of Moon = 1080 mi = 1740 km
1 year (Earth year) = 365 days

FOCUS ON CONCEPTS

1. (a) Obtain the value of $\mathbf{C}$ given in Formula (16) by setting $t = 0$ in (15).
   (b) Use Formulas (7), (17), and (22) to show that
   $$\mathbf{v} \times \mathbf{b} = GM[(e + \cos\theta)\mathbf{i} + \sin\theta\,\mathbf{j}]$$
   (c) Show that $\|\mathbf{v} \times \mathbf{b}\| = \|\mathbf{v}\|\|\mathbf{b}\|$.
   (d) Use the results in parts (b) and (c) to show that the speed of a particle in an elliptical orbit is
   $$v = \frac{v_0}{1+e}\sqrt{e^2 + 2e\cos\theta + 1}$$
   (e) Suppose that a particle is in an elliptical orbit. Use part (d) to conclude that the distance from the particle to the center of force is a minimum if and only if the speed of the particle is a maximum. Similarly, argue that the distance from the particle to the center of force is a maximum if and only if the speed of the particle is a minimum.
2. Use the result in Exercise 1(d) to show that when a particle in an elliptical orbit with eccentricity e reaches an end of the minor axis, its speed is
   $$v = v_0\sqrt{\frac{1-e}{1+e}}$$
3. Use the result in Exercise 1(d) to show that for a particle in an elliptical orbit with eccentricity e, the maximum and minimum speeds are related by
   $$v_{\max} = v_{\min}\frac{1+e}{1-e}$$
4. Use Formula (22) and the result in Exercise 1(d) to show that the speed v of a particle in a circular orbit of radius r_0 is constant and is given by
   $$v = \sqrt{\frac{GM}{r_0}}$$

5. Suppose that a particle is in an elliptical orbit in a central force field in which the center of force is at a focus, and let $\mathbf{r} = \mathbf{r}(t)$ and $\mathbf{v} = \mathbf{v}(t)$ be the position and velocity functions of the particle respectively. Let $r_{\min}$ and $r_{\max}$ denote the minimum and maximum distances from the particle to the center of force, and let $v_{\min}$ and $v_{\max}$ denote the minimum and maximum speeds of the particle.
   (a) Review the discussion of ellipses in polar coordinates in Section 11.6, and show that if the ellipse has eccentricity e and semimajor axis a, then $r_{\min} = a(1-e)$ and $r_{\max} = a(1+e)$.
   (b) Explain why $r_{\min}$ and $r_{\max}$ occur at points at which $\mathbf{r}$ and $\mathbf{v}$ are orthogonal. [*Hint:* First argue that the extreme values of $\|\mathbf{r}\|$ occur at critical points of the function $\|\mathbf{r}\|^2 = \mathbf{r}\cdot\mathbf{r}$.]
   (c) Explain why $v_{\min}$ and $v_{\max}$ occur at points at which $\mathbf{r}$ and $\mathbf{v}$ are orthogonal. [*Hint:* First argue that the extreme values of $\|\mathbf{v}\|$ occur at critical points of the function $\|\mathbf{v}\|^2 = \mathbf{v}\cdot\mathbf{v}$. Then use Equation (5).]
   (d) Use Equation (2) and parts (b) and (c) to conclude that $r_{\max}v_{\min} = r_{\min}v_{\max}$.
6. Use the results in parts (a) and (d) of Exercise 5 to give a derivation of the equation in Exercise 3.
7. Use the result in Exercise 4 to find the speed in km/s of a satellite in a circular orbit that is 200 km above the surface of the Earth.
8. Use the result in Exercise 4 to find the speed in mi/h of a communications satellite that is in geosynchronous orbit around the Earth. [See Example 1.]
9. Find the escape speed in km/s for a space probe in a circular orbit that is 300 km above the surface of the Earth.
10. The universal gravitational constant is approximately

$$G = 6.67 \times 10^{-11}\ \text{m}^3/\text{kg}\cdot\text{s}^2$$

and the semimajor axis of the Earth's orbit is approximately

$$a = 149.6 \times 10^6\ \text{km}$$

Estimate the mass of the Sun in kg.
11. (a) The eccentricity of the Moon's orbit around the Earth is 0.055, and its semimajor axis is $a = 238{,}900$ mi. Find the maximum and minimum distances between the surface of the Earth and the surface of the Moon.
   (b) Find the period of the Moon's orbit in days.
12. (a) *Vanguard 1* was launched in March 1958 with perigee and apogee altitudes above the Earth of 649 km and 4340 km, respectively. Find the length of the semimajor axis of its orbit.
   (b) Use the result in part (a) of Exercise 19 in Section 11.6 to find the eccentricity of its orbit.
   (c) Find the period of *Vanguard 1* in minutes.
13. (a) Suppose that a space probe is in a circular orbit at an altitude of 180 mi above the surface of the Earth. Use the result in Exercise 4 to find its speed.
   (b) During a very short period of time, a thruster rocket on the space probe is fired to increase the speed of the probe by 600 mi/h in its direction of motion. Find the eccentricity of the resulting elliptical orbit, and use the result in part (a) of Exercise 5 to find the apogee altitude.
14. Show that the quantity e defined by Formula (22) is nonnegative. [*Hint:* The polar axis was chosen so that r is minimum when $\theta = 0$.]

✔QUICK CHECK ANSWERS 13.7

1. (a) $\dfrac{GMm}{r^2}$ (b) $-\dfrac{GMm}{r^3}\mathbf{r}$ 2. $\sqrt{\dfrac{2GM}{r_0}}$ 3. 3 4. $e = \dfrac{r_0 v_0^2}{GM} - 1$

CHAPTER REVIEW EXERCISES

1. In words, what is meant by the graph of a vector-valued function?

2–5 Describe the graph of the equation.

2. $\mathbf{r} = (2-3t)\mathbf{i} - 4t\mathbf{j}$
3. $\mathbf{r} = 3\sin 2t\mathbf{i} + 3\cos 2t\mathbf{j}$
4. $\mathbf{r} = 3\cos t\mathbf{i} + 2\sin t\mathbf{j} - \mathbf{k}$
5. $\mathbf{r} = -2\mathbf{i} + t\mathbf{j} + (t^2-1)\mathbf{k}$
6. Describe the graph of the vector-valued function.
   (a) $\mathbf{r} = \mathbf{r}_0 + t(\mathbf{r}_1 - \mathbf{r}_0)$
   (b) $\mathbf{r} = \mathbf{r}_0 + t(\mathbf{r}_1 - \mathbf{r}_0)\quad (0 \le t \le 1)$
   (c) $\mathbf{r} = \mathbf{r}_0 + t\mathbf{r}'(t_0)$
7. Show that the graph of $\mathbf{r}(t) = t\sin\pi t\mathbf{i} + t\mathbf{j} + t\cos\pi t\mathbf{k}$ lies on the surface of a cone, and sketch the cone.
8. Find parametric equations for the intersection of the surfaces

$$y = x^2 \quad \text{and} \quad 2x^2 + y^2 + 6z^2 = 24$$

and sketch the intersection.
9. In words, give a geometric description of the statement $\lim_{t\to a}\mathbf{r}(t) = \mathbf{L}$.

10. Evaluate $\lim_{t\to 0}\left(e^{-t}\mathbf{i}+\frac{1-\cos t}{t}\mathbf{j}+t^2\mathbf{k}\right)$.

11. Find parametric equations of the line tangent to the graph of

$$\mathbf{r}(t)=(t+\cos 2t)\mathbf{i}-(t^2+t)\mathbf{j}+\sin t\mathbf{k}$$

at the point where $t=0$.

12. Suppose that $\mathbf{r}_1(t)$ and $\mathbf{r}_2(t)$ are smooth vector-valued functions such that $\mathbf{r}_1(0)=\langle -1,1,2\rangle$, $\mathbf{r}_2(0)=\langle 1,2,1\rangle$, $\mathbf{r}_1'(0)=\langle 1,0,1\rangle$, and $\mathbf{r}_2'(0)=\langle 4,0,2\rangle$. Use this information to evaluate the derivative at $t=0$ of each function.
(a) $\mathbf{r}(t)=3\mathbf{r}_1(t)+2\mathbf{r}_2(t)$ (b) $\mathbf{r}(t)=[\ln(t+1)]\mathbf{r}_1(t)$
(c) $\mathbf{r}(t)=\mathbf{r}_1(t)\times\mathbf{r}_2(t)$ (d) $f(t)=\mathbf{r}_1(t)\cdot\mathbf{r}_2(t)$

13. Evaluate $\int(\cos t\mathbf{i}+\sin t\mathbf{j})\,dt$.

14. Evaluate $\int_0^{\pi/3}\langle\cos 3t,-\sin 3t\rangle\,dt$.

15. Solve the vector initial-value problem

$$\mathbf{y}'(t)=t^2\mathbf{i}+2t\mathbf{j},\quad \mathbf{y}(0)=\mathbf{i}+\mathbf{j}$$

16. Solve the vector initial-value problem

$$\frac{d\mathbf{r}}{dt}=\mathbf{r},\quad \mathbf{r}(0)=\mathbf{r}_0$$

for the unknown vector-valued function $\mathbf{r}(t)$.

17. Find the arc length of the graph of

$$\mathbf{r}(t)=e^{\sqrt{2}t}\mathbf{i}+e^{-\sqrt{2}t}\mathbf{j}+2t\mathbf{k}\quad (0\le t\le\sqrt{2}\ln 2)$$

18. Suppose that $\mathbf{r}(t)$ is a smooth vector-valued function of t with $\mathbf{r}'(0)=3\mathbf{i}-\mathbf{j}+\mathbf{k}$ and that $\mathbf{r}_1(t)=\mathbf{r}(2-e^{t\ln 2})$. Find $\mathbf{r}_1'(1)$.

19. Find the arc length parametrization of the line through $P(-1,4,3)$ and $Q(0,2,5)$ that has reference point P and orients the line in the direction from P to Q.

20. Find an arc length parametrization of the curve

$$\mathbf{r}(t)=\langle e^t\cos t,-e^t\sin t\rangle\quad (0\le t\le\pi/2)$$

which has the same orientation and has $\mathbf{r}(0)$ as the reference point.

21. Suppose that $\mathbf{r}(t)$ is a smooth vector-valued function. State the definitions of $\mathbf{T}(t)$, $\mathbf{N}(t)$, and $\mathbf{B}(t)$.

22. Find $\mathbf{T}(0)$, $\mathbf{N}(0)$, and $\mathbf{B}(0)$ for the curve

$$\mathbf{r}(t)=\left\langle 2\cos t,2\cos t+\frac{3}{\sqrt{5}}\sin t,\cos t-\frac{6}{\sqrt{5}}\sin t\right\rangle$$

23. State the definition of "curvature" and explain what it means geometrically.

24. Suppose that $\mathbf{r}(t)$ is a smooth curve with $\mathbf{r}'(0)=\mathbf{i}$ and $\mathbf{r}''(0)=\mathbf{i}+2\mathbf{j}$. Find the curvature at $t=0$.

25–28 Find the curvature of the curve at the stated point.

25. $\mathbf{r}(t)=2\cos t\mathbf{i}+3\sin t\mathbf{j}-t\mathbf{k};\ t=\pi/2$

26. $\mathbf{r}(t)=\langle 2t,e^{2t},e^{-2t}\rangle;\ t=0$

27. $y=\cos x;\ x=\pi/2$ **28.** $y=\ln x;\ x=1$

29. Suppose that $\mathbf{r}(t)$ is the position function of a particle moving in 2-space or 3-space. In each part, explain what the given quantity represents physically.
(a) $\left\|\frac{d\mathbf{r}}{dt}\right\|$ (b) $\int_{t_0}^{t_1}\left\|\frac{d\mathbf{r}}{dt}\right\|dt$ (c) $\|\mathbf{r}(t)\|$

30. (a) What does Theorem 13.2.8 tell you about the velocity vector of a particle that moves over a sphere?
(b) What does Theorem 13.2.8 tell you about the acceleration vector of a particle that moves with constant speed?
(c) Show that the particle with position function

$$\mathbf{r}(t)=\sqrt{1-\tfrac{1}{4}\cos^2 t}\cos t\mathbf{i}+\sqrt{1-\tfrac{1}{4}\cos^2 t}\sin t\mathbf{j}+\tfrac{1}{2}\cos t\mathbf{k}$$

moves over a sphere.

31. As illustrated in the accompanying figure, suppose that a particle moves counterclockwise around a circle of radius R centered at the origin at a constant rate of ω radians per second. This is called ***uniform circular motion***. If we assume that the particle is at the point $(R,0)$ at time $t=0$, then its position function will be

$$\mathbf{r}(t)=R\cos\omega t\mathbf{i}+R\sin\omega t\mathbf{j}$$

(a) Show that the velocity vector $\mathbf{v}(t)$ is always tangent to the circle and that the particle has constant speed v given by

$$v=R\omega$$

(b) Show that the acceleration vector $\mathbf{a}(t)$ is always directed toward the center of the circle and has constant magnitude a given by

$$a=R\omega^2$$

(c) Show that the time T required for the particle to make one complete revolution is

$$T=\frac{2\pi}{\omega}=\frac{2\pi R}{v}$$

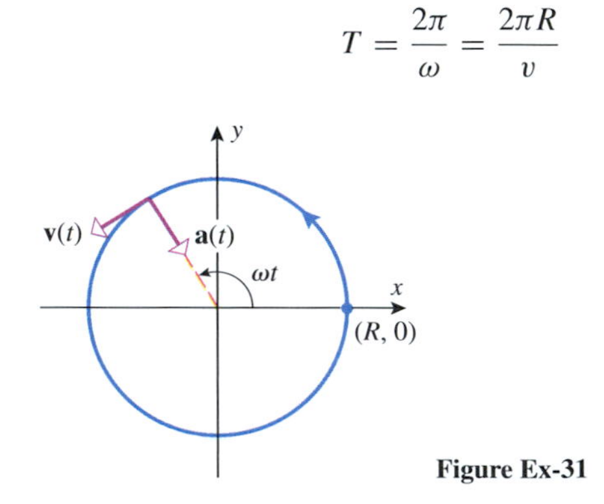

Figure Ex-31

32. If a particle of mass m has uniform circular motion (see Exercise 31), then the acceleration vector $\mathbf{a}(t)$ is called the ***centripetal acceleration***. According to Newton's second law, this acceleration must be produced by some force $\mathbf{F}(t)$, called the ***centripetal force***, that is related to $\mathbf{a}(t)$ by the equation $\mathbf{F}(t)=m\mathbf{a}(t)$. If this force is not present, then the particle cannot undergo uniform circular motion.

(a) Show that the direction of the centripetal force varies with time but that it has constant magnitude F given by

$$F = \frac{mv^2}{R}$$

(b) An astronaut with a mass of $m = 60$ kg orbits the Earth at an altitude of $h = 3200$ km with a constant speed of $v = 6.43$ km/s. Find her centripetal acceleration assuming that the radius of the Earth is 6440 km.

(c) What centripetal gravitational force in newtons does the Earth exert on the astronaut?

33. At time $t = 0$ a particle at the origin of an xyz-coordinate system has a velocity vector of $\mathbf{v}_0 = \mathbf{i} + 2\mathbf{j} - \mathbf{k}$. The acceleration function of the particle is $\mathbf{a}(t) = 2t^2\mathbf{i} + \mathbf{j} + \cos 2t\mathbf{k}$.
(a) Find the position function of the particle.
(b) Find the speed of the particle at time $t = 1$.

34. As illustrated in the accompanying figure, the polar coordinates of a rocket are tracked by radar from a point that is b units from the launching pad. Show that the speed v of the rocket can be expressed in terms b, θ, and $d\theta/dt$ as

$$v = b \sec^2 \theta \frac{d\theta}{dt}$$

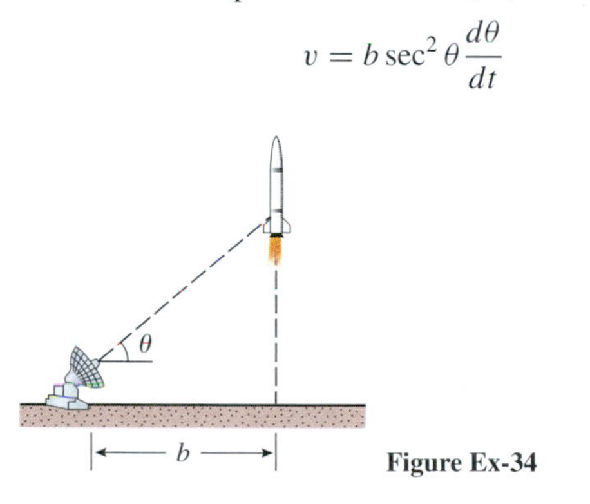

Figure Ex-34

35. A player throws a ball with an initial speed of 60 ft/s at an unknown angle α with the horizontal from a point that is 4 ft above the floor of a gymnasium. Given that the ceiling of the gymnasium is 25 ft high, determine the maximum height h at which the ball can hit a wall that is 60 ft away (see the accompanying figure).

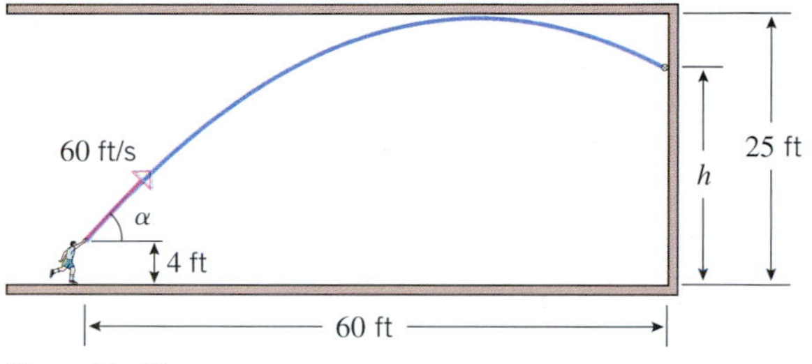

Figure Ex-35

36. Let $\mathbf{v} = \mathbf{v}(t)$ and $\mathbf{a} = \mathbf{a}(t)$ be the velocity and acceleration vectors for a particle moving in 2-space or 3-space. Show that the rate of change of its speed can be expressed as

$$\frac{d}{dt}(\|\mathbf{v}\|) = \frac{1}{\|\mathbf{v}\|}(\mathbf{v} \cdot \mathbf{a})$$

37. Use Formula (23) in Section 13.7 to find the escape speed (in km/s) for a space probe in a circular orbit 600 km above the surface of the Earth.

38. The universal gravitational constant is approximately 6.67×10^{-11} m^3/kg·s^2 and the semimajor axis of the Moon's orbit about the Earth is $a = 384{,}629$ km. Estimate the mass of the Earth in kilograms.

PARTIAL DERIVATIVES

Let no one ignorant of mathematics enter here.

—Plato
Ancient Greek Philosopher and Scholar

In this chapter we will extend many of the basic concepts of calculus to functions of two or more variables, commonly called functions of several variables. We will begin by discussing limits and continuity for functions of two and three variables, then we will define derivatives of such functions, and then we will use these derivatives to study tangent planes, rates of change, slopes of surfaces, and maximization and minimization problems. Although many of the basic ideas that we developed for functions of one variable will carry over in a natural way, functions of several variables are intrinsically more complicated than functions of one variable, so we will need to develop new tools and new ideas to deal with such functions.

Photo: *Three-dimensional surfaces are analogous to mountain ranges. In this chapter we will use derivatives to analyze steepness and other features of such surfaces.*

14.1 FUNCTIONS OF TWO OR MORE VARIABLES

In previous sections we studied real-valued functions of a real variable and vector-valued functions of a real variable. In this section we will consider real-valued functions of two or more real variables.

NOTATION AND TERMINOLOGY

There are many familiar formulas in which a given variable depends on two or more other variables. For example, the area A of a triangle depends on the base length b and height h by the formula $A = \frac{1}{2}bh$; the volume V of a rectangular box depends on the length l, the width w, and the height h by the formula $V = lwh$; and the arithmetic average $\bar{x}$ of n real numbers, $x_1, x_2, \ldots, x_n$, depends on those numbers by the formula

$$\bar{x} = \frac{1}{n}(x_1 + x_2 + \cdots + x_n)$$

Thus, we say that

A is a function of the two variables b and h;

V is a function of the three variables l, w, and h;

$\bar{x}$ is a function of the n variables $x_1, x_2, \ldots, x_n$.

The terminology and notation for functions of two or more variables is similar to that for functions of one variable. For example, the expression

$$z = f(x, y)$$

means that z is a function of x and y in the sense that a unique value of the dependent variable z is determined by specifying values for the independent variables x and y. Similarly,

$$w = f(x, y, z)$$

expresses w as a function of x, y, and z, and

$$u = f(x_1, x_2, \dots, x_n)$$

expresses u as a function of $x_1, x_2, \dots, x_n$.

We will find it useful to think of functions of two or three independent variables in geometric terms. For example, if $z = f(x, y)$, then we can view (x, y) as a point in the xy-plane and think of f as a rule that associates a unique numerical value z with the point (x, y); similarly, we can think of $w = f(x, y, z)$ as a rule that associates a unique numerical value w with a point (x, y, z) in an xyz-coordinate system (Figure 14.1.1).

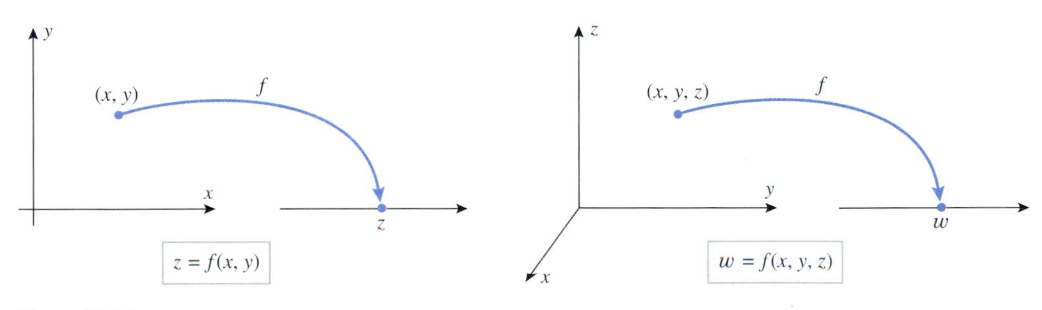

Figure 14.1.1

As with functions of one variable, the independent variables of a function of two or more variables may be restricted to lie in some set D, which we call the ***domain*** of f. Sometimes the domain will be determined by physical restrictions on the variables. If the function is defined by a formula and if there are no physical restrictions or other restrictions stated explicitly, then it is understood that the domain consists of all points for which the formula yields a real value for the dependent variable. We call this the ***natural domain*** of the function. The following definitions summarize this discussion.

14.1.1 DEFINITION. A ***function f of two variables***, x and y, is a rule that assigns a unique real number $f(x, y)$ to each point (x, y) in some set D in the xy-plane.

By extension, one can define the notion of "n-dimensional space" in which a "point" is a sequence of n real numbers $(x_1, x_2, \dots, x_n)$. Then a function of n real variables is a rule that assigns a unique real number $f(x_1, x_2, \dots, x_n)$ to each point in some set in this space.

14.1.2 DEFINITION. A ***function f of three variables***, x, y, and z, is a rule that assigns a unique real number $f(x, y, z)$ to each point (x, y, z) in some set D in three-dimensional space.

▶ Example 1 Let

$$f(x, y) = 3x^2\sqrt{y} - 1$$

Find $f(1, 4)$, $f(0, 9)$, $f(t^2, t)$, $f(ab, 9b)$, and the natural domain of f.

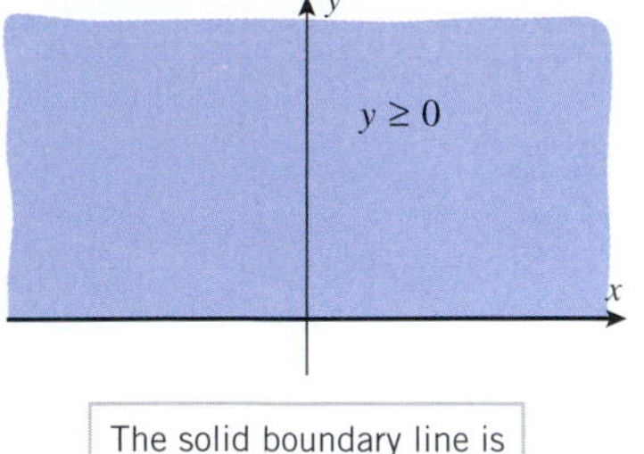

The solid boundary line is included in the domain.

Figure 14.1.2

Solution. By substitution

$$f(1, 4) = 3(1)^2\sqrt{4} - 1 = 5$$
$$f(0, 9) = 3(0)^2\sqrt{9} - 1 = -1$$
$$f(t^2, t) = 3(t^2)^2\sqrt{t} - 1 = 3t^4\sqrt{t} - 1$$
$$f(ab, 9b) = 3(ab)^2\sqrt{9b} - 1 = 9a^2b^2\sqrt{b} - 1$$

Because of the radical $\sqrt{y}$ in the formula for f, we must have $y \geq 0$ to avoid non-real values for $f(x, y)$. Thus, the natural domain of f consists of all points in the xy-plane that are on or above the x-axis. (See Figure 14.1.2.) ◀

▶ Example 2 Sketch the natural domain of the function $f(x, y) = \ln(x^2 - y)$.

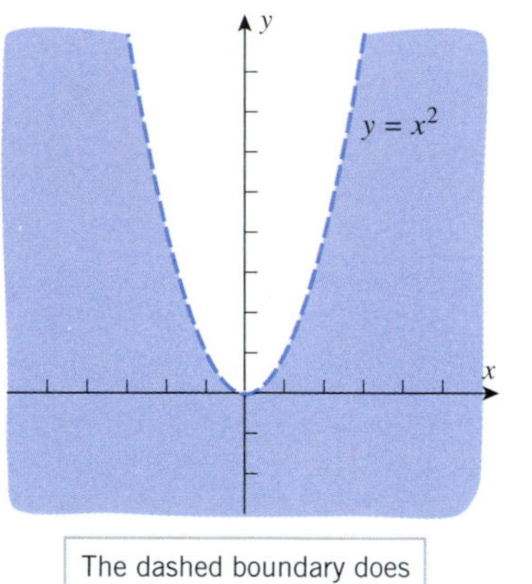

The dashed boundary does not belong to the domain.

Figure 14.1.3

Solution. $\ln(x^2 - y)$ is defined only when $0 < x^2 - y$ or $y < x^2$. We first sketch the parabola $y = x^2$ as a "dashed" curve. The region $y < x^2$ then consists of all points below this curve (Figure 14.1.3). ◀

▶ Example 3 Let

$$f(x, y, z) = \sqrt{1 - x^2 - y^2 - z^2}$$

Find $f\left(0, \frac{1}{2}, -\frac{1}{2}\right)$ and the natural domain of f.

Solution. By substitution,

$$f\left(0, \tfrac{1}{2}, -\tfrac{1}{2}\right) = \sqrt{1 - (0)^2 - \left(\tfrac{1}{2}\right)^2 - \left(-\tfrac{1}{2}\right)^2} = \sqrt{\tfrac{1}{2}}$$

Because of the square root sign, we must have $0 \leq 1 - x^2 - y^2 - z^2$ in order to have a real value for $f(x, y, z)$. Rewriting this inequality in the form

$$x^2 + y^2 + z^2 \leq 1$$

we see that the natural domain of f consists of all points on or within the sphere

$$x^2 + y^2 + z^2 = 1 \quad ◀$$

FUNCTIONS DESCRIBED BY TABLES

Sometimes it is either desirable or necessary to represent a function of two variables in table form, rather than as an explicit formula. For example, the U.S. National Weather Service uses the formula

$$W = 35.74 + 0.6215T + (0.4275T - 35.75)v^{0.16} \tag{1}$$

to model the wind chill index W (in °F) as a function of the temperature T (in °F) and the wind speed v (in mi/h) for wind speeds greater than 3 mi/h. This formula is sufficiently complex that it is difficult to get an intuitive feel for the relationship between the variables. One can get a clearer sense of the relationship by selecting sample values of T and v and constructing a table, such as Table 14.1.1, in which we have rounded the values of W to the

The wind chill index is that temperature (in °F) which would produce the same sensation on exposed skin at a wind speed of 3 mi/h as the temperature and wind speed combination in current weather conditions.

Table 14.1.1

TEMPERATURE T (°F)

WIND SPEED v (mi/h)	20	25	30	35
5	13	19	25	31
15	6	13	19	25
25	3	9	16	23
35	0	7	14	21
45	−2	5	12	19

nearest integer. For example, if the temperature is 30°F and the wind speed is 5 mi/h, it feels as if the temperature is 25°F. If the wind speed increases to 15 mi/h, the temperature then feels as if it has dropped to 19°F. Note that in this case, an increase in wind speed of 10 mi/h causes a 6°F decrease in the wind chill index. To estimate wind chill values not displayed in the table, we can use ***linear interpolation***. For example, suppose that the temperature is 30°F and the wind speed is 7 mi/h. A reasonable estimate for the drop in the wind chill index from its value when the wind speed is 5 mi/h would be $\frac{2}{10} \cdot 6°\text{F} = 1.2°\text{F}$. (Why?) The resulting estimate in wind chill would then be $25° - 1.2° = 23.8°\text{F}$.

In some cases, tables for functions of two variables arise directly from experimental data, in which case one must either work directly with the table or else use some technique to construct a formula that models the data in the table. We will not discuss such modeling techniques in this text.

GRAPHS OF FUNCTIONS OF TWO VARIABLES

Recall that for a function f of one variable, the graph of $f(x)$ in the xy-plane was defined to be the graph of the equation $y = f(x)$. Similarly, if f is a function of two variables, we define the ***graph*** of $f(x, y)$ in xyz-space to be the graph of the equation $z = f(x, y)$. In general, such a graph will be a surface in 3-space.

Example 4 In each part, describe the graph of the function in an xyz-coordinate system.

(a) $f(x, y) = 1 - x - \frac{1}{2}y$ (b) $f(x, y) = \sqrt{1 - x^2 - y^2}$

(c) $f(x, y) = -\sqrt{x^2 + y^2}$

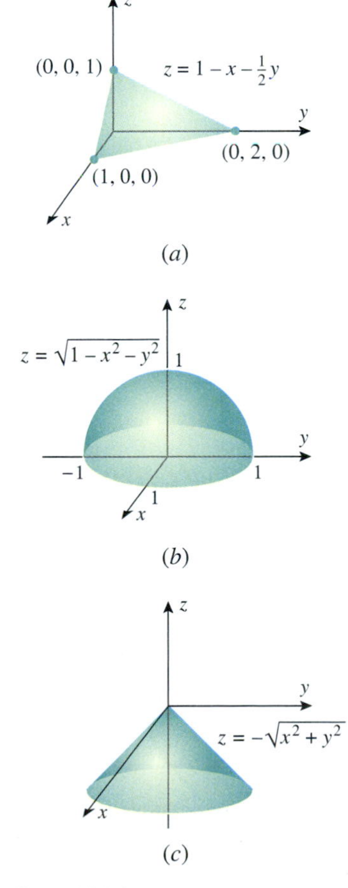

Figure 14.1.4

Solution (a). By definition, the graph of the given function is the graph of the equation

$$z = 1 - x - \tfrac{1}{2}y$$

which is a plane. A triangular portion of the plane can be sketched by plotting the intersections with the coordinate axes and joining them with line segments (Figure 14.1.4*a*).

Solution (b). By definition, the graph of the given function is the graph of the equation

$$z = \sqrt{1 - x^2 - y^2} \tag{2}$$

After squaring both sides, this can be rewritten as

$$x^2 + y^2 + z^2 = 1$$

which represents a sphere of radius 1, centered at the origin. Since (2) imposes the added condition that $z \geq 0$, the graph is just the upper hemisphere (Figure 14.1.4*b*).

Solution (c). The graph of the given function is the graph of the equation

$$z = -\sqrt{x^2 + y^2} \tag{3}$$

After squaring, we obtain

$$z^2 = x^2 + y^2$$

which is the equation of a circular cone (see Table 12.7.1). Since (3) imposes the condition that $z \leq 0$, the graph is just the lower nappe of the cone (Figure 14.1.4*c*). ◄

LEVEL CURVES

We are all familiar with the topographic (or contour) maps in which a three-dimensional landscape, such as a mountain range, is represented by two-dimensional contour lines or curves of constant elevation. Consider, for example, the model hill and its contour map shown in Figure 14.1.5. The contour map is constructed by passing planes of constant elevation through the hill, projecting the resulting contours onto a flat surface, and labeling the contours with their elevations. In Figure 14.1.5, note how the two gullies appear as indentations in the contour lines and how the curves are close together on the contour map where the hill has a steep slope and become more widely spaced where the slope is gradual.

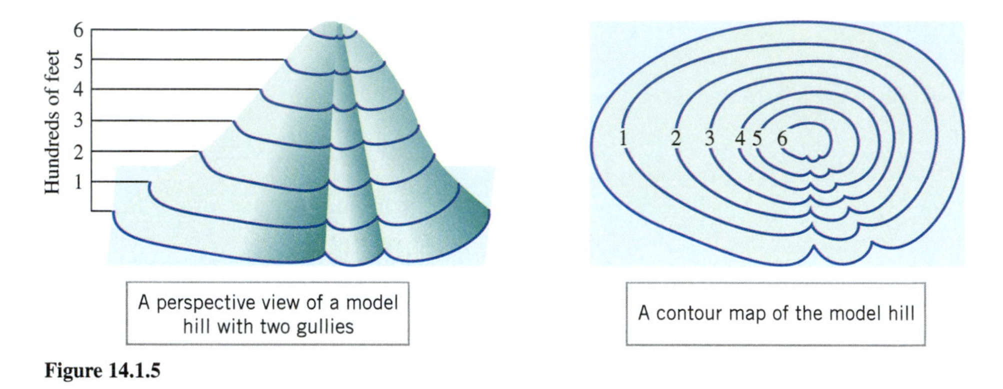

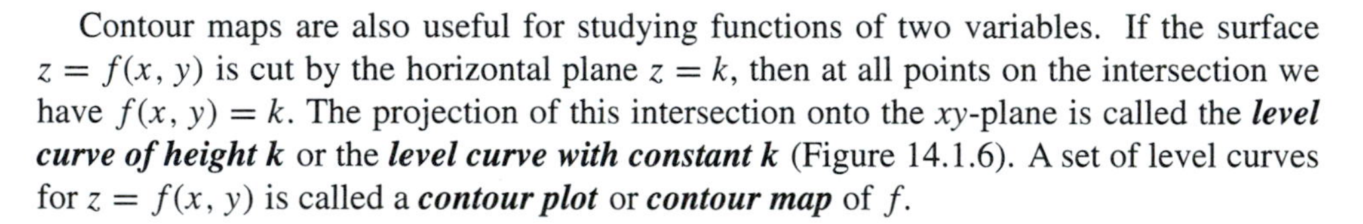

Figure 14.1.5

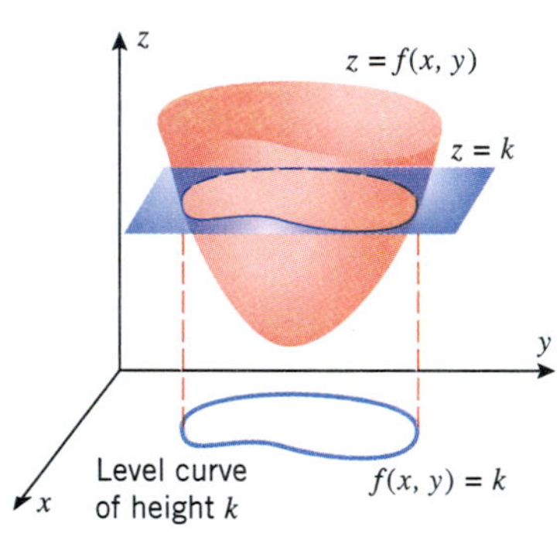

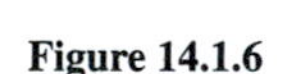

Figure 14.1.6

Contour maps are also useful for studying functions of two variables. If the surface $z = f(x, y)$ is cut by the horizontal plane $z = k$, then at all points on the intersection we have $f(x, y) = k$. The projection of this intersection onto the xy-plane is called the ***level curve of height k*** or the ***level curve with constant k*** (Figure 14.1.6). A set of level curves for $z = f(x, y)$ is called a ***contour plot*** or ***contour map*** of f.

Example 5 The graph of the function $f(x, y) = y^2 - x^2$ in xyz-space is the hyperbolic paraboloid (saddle surface) shown in Figure 14.1.7a. The level curves have equations of the form $y^2 - x^2 = k$. For $k > 0$ these curves are hyperbolas opening along lines parallel to the y-axis; for $k < 0$ they are hyperbolas opening along lines parallel to the x-axis; and for $k = 0$ the level curve consists of the intersecting lines $y + x = 0$ and $y - x = 0$ (Figure 14.1.7b). ◀

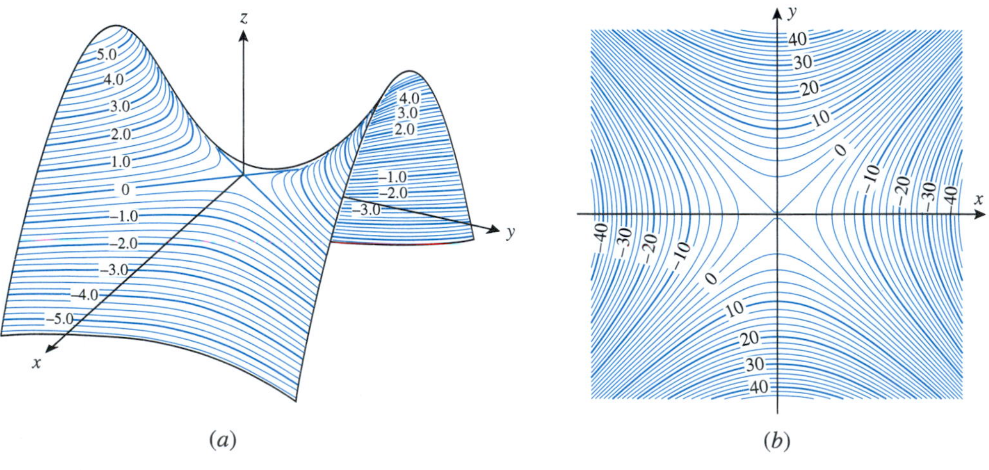

Figure 14.1.7

► **Example 6** Sketch the contour plot of $f(x, y) = 4x^2 + y^2$ using level curves of height $k = 0, 1, 2, 3, 4, 5$.

Solution. The graph of the surface $z = 4x^2 + y^2$ is the paraboloid shown in the left part of Figure 14.1.8, so we can reasonably expect the contour plot to be a family of ellipses centered at the origin. The level curve of height k has the equation $4x^2 + y^2 = k$. If $k = 0$, then the graph is the single point $(0, 0)$. For $k > 0$ we can rewrite the equation as

$$\frac{x^2}{k/4} + \frac{y^2}{k} = 1$$

which represents a family of ellipses with x-intercepts $\pm\sqrt{k}/2$ and y-intercepts $\pm\sqrt{k}$. The contour plot for the specified values of k is shown in the right part of Figure 14.1.8. ◄

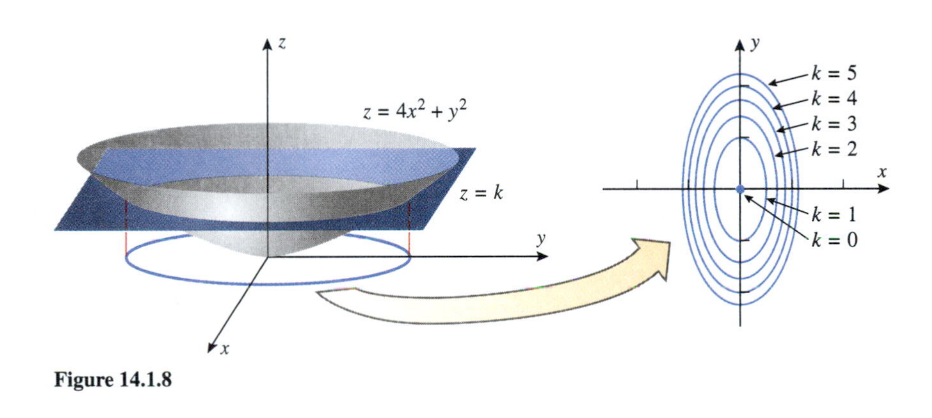

Figure 14.1.8

► **Example 7** Sketch the contour plot of $f(x, y) = 2 - x - y$ using level curves of height $k = -6, -4, -2, 0, 2, 4, 6$.

Solution. The graph of the surface $z = 2 - x - y$ is the plane shown in the left part of Figure 14.1.9, so we can reasonably expect the contour plot to be a family of parallel lines. The level curve of height k has the equation $2 - x - y = k$, which we can rewrite as

$$y = -x + (2 - k)$$

This represents a family of parallel lines of slope -1. The contour plot for the specified values of k is shown in the right part of Figure 14.1.9. ◄

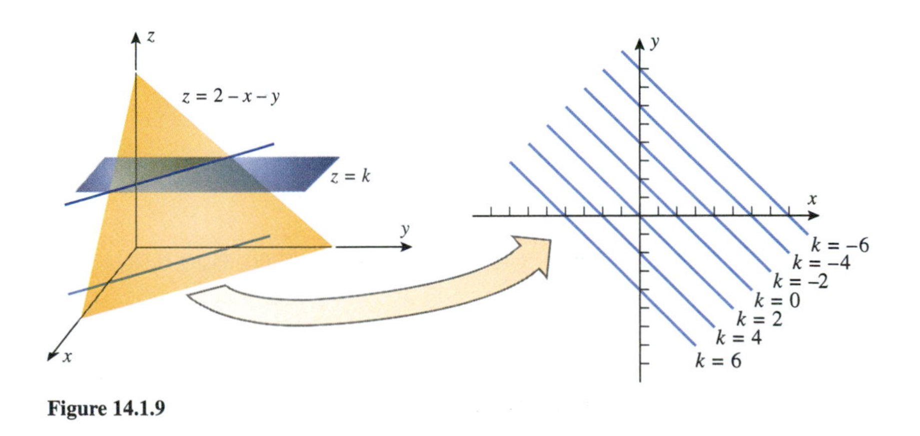

Figure 14.1.9

CONTOUR PLOTS USING TECHNOLOGY

Except in the simplest cases, contour plots can be difficult to produce without the help of a graphing utility. Figure 14.1.10 illustrates how graphing technology can be used to display level curves. The table shows two graphical representations of the level curves of the function $f(x, y) = |\sin x \sin y|$ produced with a CAS over the domain $0 \le x \le 2\pi$, $0 \le y \le 2\pi$.

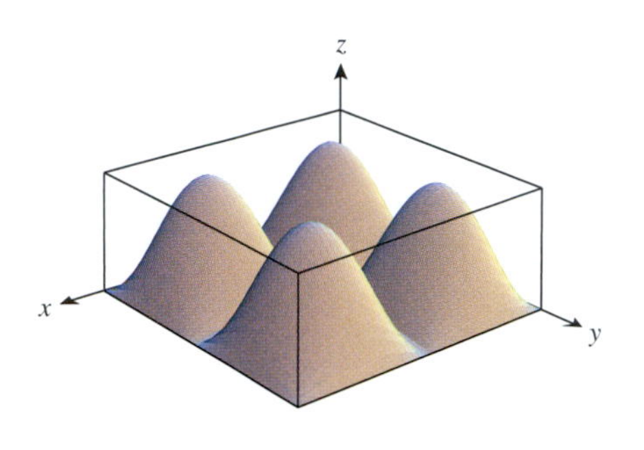

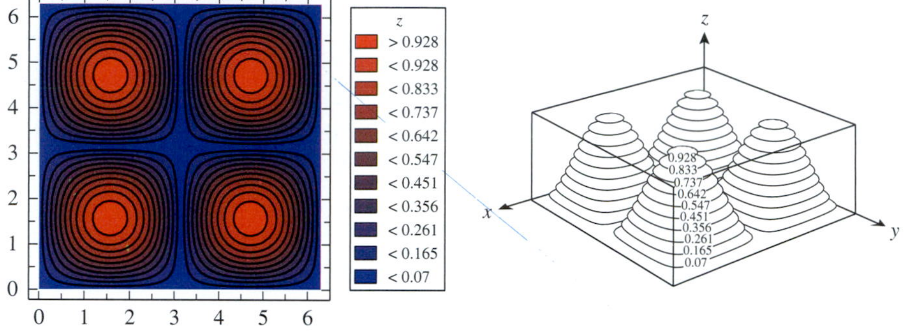

Figure 14.1.10

LEVEL SURFACES

The term "level surface" is standard but confusing, since a level surface need *not* be level in the sense of being horizontal—it is simply a surface on which all values of f are the same.

Observe that the graph of $y = f(x)$ is a curve in 2-space, and the graph of $z = f(x, y)$ is a surface in 3-space, so the number of dimensions required for these graphs is one greater than the number of independent variables. Accordingly, there is no "direct" way to graph a function of three variables since four dimensions are required. However, if k is a constant, then the graph of the equation $f(x, y, z) = k$ will generally be a surface in 3-space (e.g., the graph of $x^2 + y^2 + z^2 = 1$ is a sphere), which we call the ***level surface with constant k***. Some geometric insight into the behavior of the function f can sometimes be obtained by graphing these level surfaces for various values of k.

Example 8 Describe the level surfaces of

$$\text{(a)}\ f(x, y, z) = x^2 + y^2 + z^2 \qquad \text{(b)}\ f(x, y, z) = z^2 - x^2 - y^2$$

Solution (a). The level surfaces have equations of the form

$$x^2 + y^2 + z^2 = k$$

For $k > 0$ the graph of this equation is a sphere of radius $\sqrt{k}$, centered at the origin; for $k = 0$ the graph is the single point $(0, 0, 0)$; and for $k < 0$ there is no level surface (Figure 14.1.11).

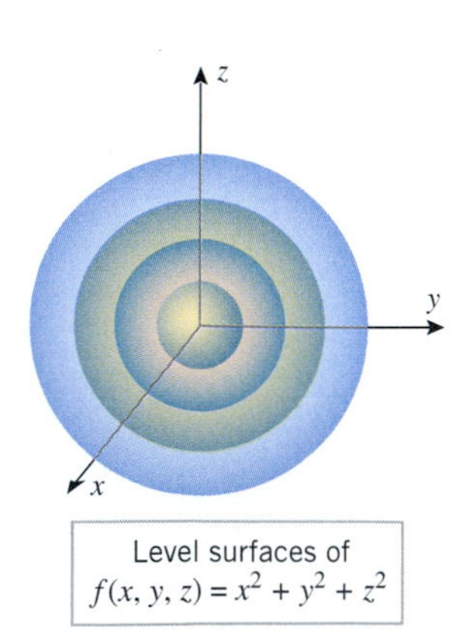

Figure 14.1.11

Solution (b). The level surfaces have equations of the form

$$z^2 - x^2 - y^2 = k$$

As discussed in Section 12.7, this equation represents a cone if $k = 0$, a hyperboloid of two sheets if $k > 0$, and a hyperboloid of one sheet if $k < 0$ (Figure 14.1.12). ◀

GRAPHING FUNCTIONS OF TWO VARIABLES USING TECHNOLOGY

Generating surfaces with a graphing utility is more complicated than generating plane curves because there are more factors that must be taken into account. We can only touch on the ideas here, so if you want to use a graphing utility, its documentation will be your main source of information.

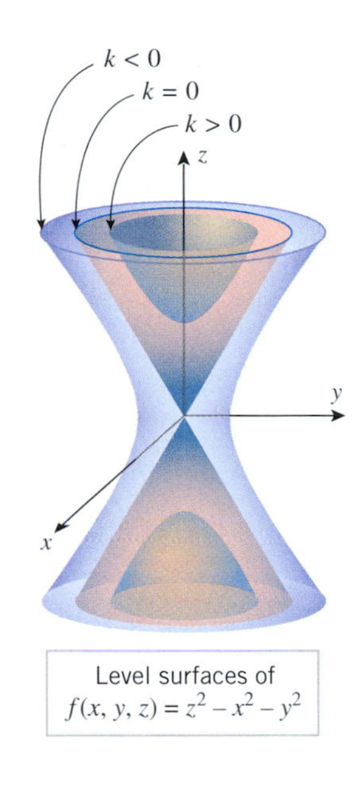

Figure 14.1.12

TECHNOLOGY MASTERY

If you have a graphing utility that can generate surfaces in 3-space, read the documentation and try to duplicate some of the surfaces in Figures 14.1.13 and 14.1.14 and Table 14.1.2.

Graphing utilities can only show a portion of xyz-space in a viewing screen, so the first step in graphing a surface is to determine which portion of xyz-space you want to display. This region is called the ***viewing box*** or ***viewing window***. For example, Figure 14.1.13 shows the effect of graphing the paraboloid $z = x^2 + y^2$ in three different viewing windows. However, within a fixed viewing box, the appearance of the surface is also affected by the ***viewpoint***, that is, the direction from which the surface is viewed, and the distance from the viewer to the surface. For example, Figure 14.1.14 shows the graph of the paraboloid $z = x^2 + y^2$ from three different viewpoints using the first viewing box in Figure 14.1.13.

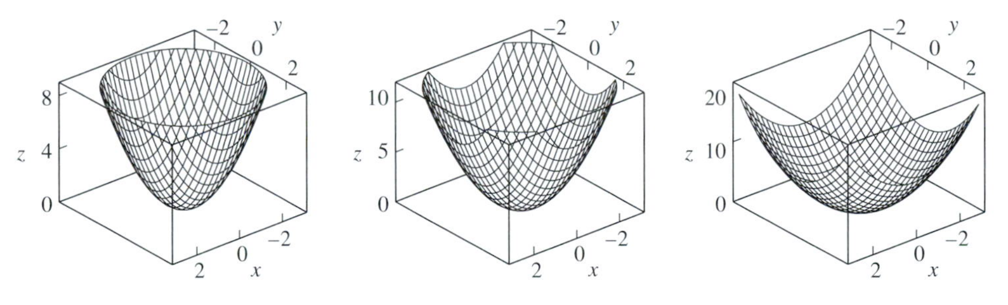

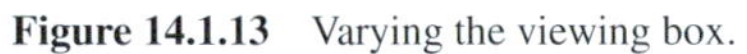

Figure 14.1.13 Varying the viewing box.

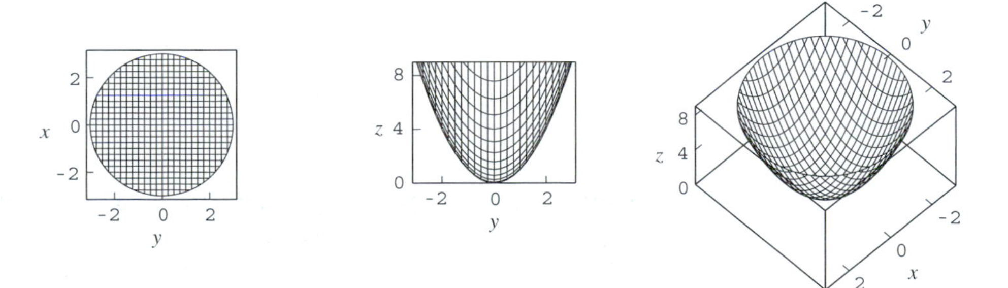

Figure 14.1.14 Viewing the viewpoint.

Table 14.1.2 shows six surfaces in 3-space along with their associated contour plots. Note that the mesh lines on the surface are traces in vertical planes, whereas the level curves correspond to traces in horizontal planes. In these contour plots the color gradation is from dark to light as z increases.

Table 14.1.2

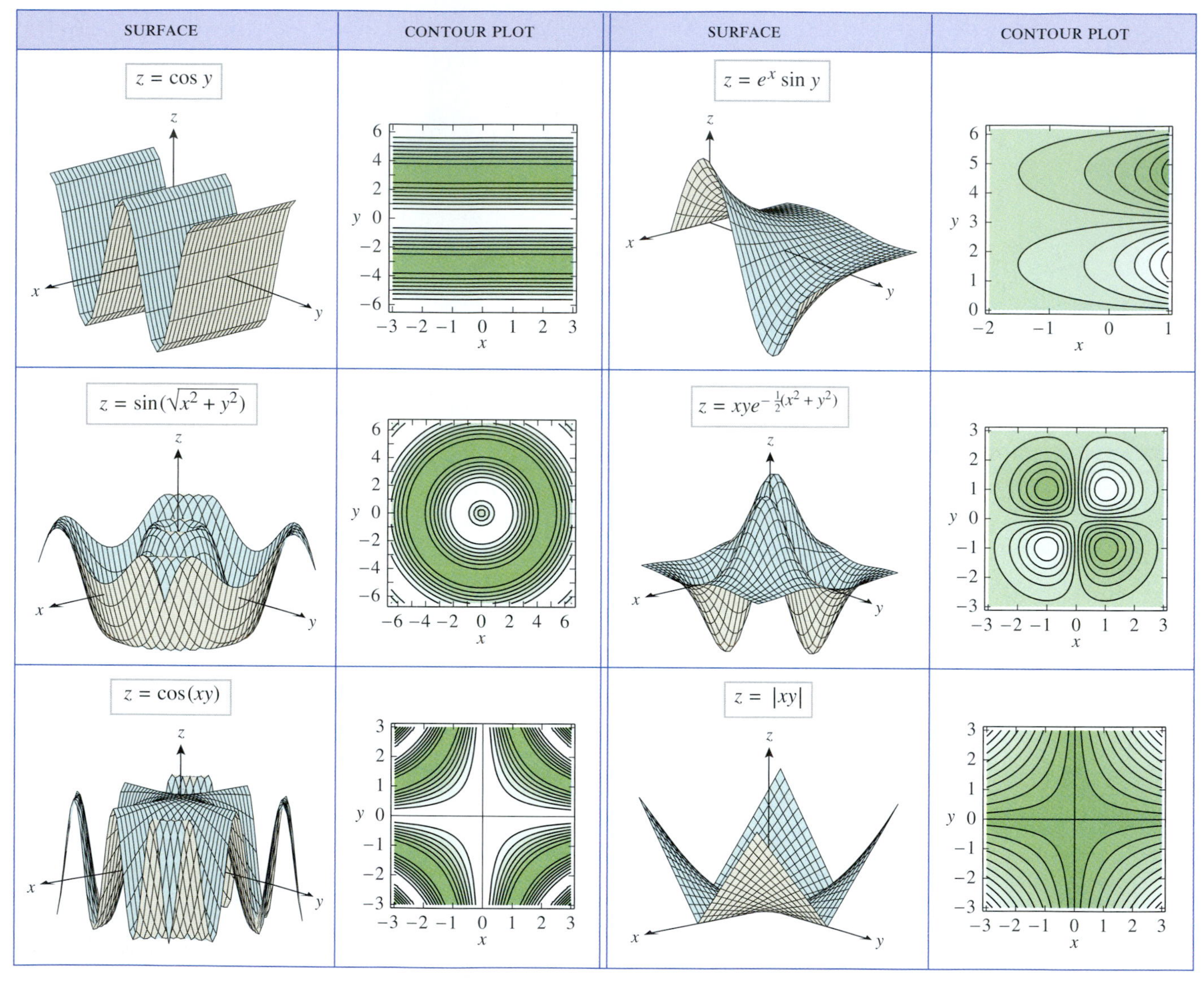

✔QUICK CHECK EXERCISES 14.1 (See page 936 for answers.)

1. The domain of $f(x, y) = \ln xy$ is ________ and the domain of $g(x, y) = \ln x + \ln y$ is ________.

2. Let $f(x, y) = \dfrac{x - y}{x + y + 1}$.
   (a) $f(2, 1) =$ ________ (b) $f(1, 2) =$ ________
   (c) $f(a, a) =$ ________ (d) $f(y + 1, y) =$ ________

3. Let $f(x, y) = e^{x+y}$.
   (a) for what values of k does the level curve $f(x, y) = k$ contain at least one point?
   (b) Describe the level curves $f(x, y) = k$ for the values of k obtained in part (a).

4. Let $f(x, y, z) = \dfrac{1}{x^2 + y^2 + z^2 + 1}$.
   (a) Determine all values of k such that the level surface $f(x, y, z) = k$ contains at least one point.
   (b) Describe the level surfaces $f(x, y, z) = k$ for the values of k obtained in part (a).

EXERCISE SET 14.1 Graphing Utility C CAS

1–8 These exercises are concerned with functions of two variables.

1. Let $f(x, y) = x^2y + 1$. Find
(a) $f(2, 1)$ (b) $f(1, 2)$ (c) $f(0, 0)$
(d) $f(1, -3)$ (e) $f(3a, a)$ (f) $f(ab, a - b)$.

2. Let $f(x, y) = x + \sqrt[3]{xy}$. Find
(a) $f(t, t^2)$ (b) $f(x, x^2)$ (c) $f(2y^2, 4y)$.

3. Let $f(x, y) = xy + 3$. Find
(a) $f(x + y, x - y)$ (b) $f(xy, 3x^2y^3)$.

4. Let $g(x) = x \sin x$. Find
(a) $g(x/y)$ (b) $g(xy)$ (c) $g(x - y)$.

5. Find $F(g(x), h(y))$ if $F(x, y) = xe^{xy}$, $g(x) = x^3$, and $h(y) = 3y + 1$.

6. Find $g(u(x, y), v(x, y))$ if $g(x, y) = y \sin(x^2y)$, $u(x, y) = x^2y^3$, and $v(x, y) = \pi xy$.

7. Let $f(x, y) = x + 3x^2y^2$, $x(t) = t^2$, and $y(t) = t^3$. Find
(a) $f(x(t), y(t))$ (b) $f(x(0), y(0))$
(c) $f(x(2), y(2))$.

8. Let $g(x, y) = ye^{-3x}$, $x(t) = \ln(t^2 + 1)$, and $y(t) = \sqrt{t}$. Find $g(x(t), y(t))$.

9. Refer to Table 14.1.1 to estimate the wind chill index when
(a) the temperature is 25°F and the wind speed is 7 mi/h
(b) the temperature is 28°F and the wind speed is 5 mi/h.

10. Refer to Table 14.1.1 to estimate the wind chill index when
(a) the temperature is 35°F and the wind speed is 14 mi/h
(b) the temperature is 32°F and the wind speed is 15 mi/h.

11. One method for determining relative humidity is to wet the bulb of a thermometer, whirl it through the air, and then compare the thermometer reading with the actual air temperature. If the relative humidity is less than 100%, the reading on the thermometer will be less than the temperature of the air. This difference in temperature is known as the *wet-bulb depression*. The accompanying table gives the relative humidity as a function of the air temperature and the wet-bulb depression. Use the table to complete parts (a)–(c).
(a) What is the relative humidity if the air temperature is 20°C and the wet-bulb thermometer reads 16°C?
(b) Estimate the relative humidity if the air temperature is 25°C and the wet-bulb depression is 3.5°C.
(c) Estimate the relative humidity if the air temperature is 22°C and the wet-bulb depression is 5°C.

AIR TEMPERATURE (°C)

WET-BULB DEPRESSION (°C)	15	20	25	30
3	71	74	77	79
4	62	66	70	73
5	53	59	63	67

Table Ex-11

12. Use the table in Exercise 11 to complete parts (a)–(c).
(a) What is the wet-bulb depression if the air temperature is 30°C and the relative humidity is 73%?
(b) Estimate the relative humidity if the air temperature is 15°C and the wet-bulb depression is 4.25°C.
(c) Estimate the relative humidity if the air temperature is 26°C and the wet-bulb depression is 3°C.

13–16 These exercises involve functions of three variables.

13. Let $f(x, y, z) = xy^2z^3 + 3$. Find
(a) $f(2, 1, 2)$ (b) $f(-3, 2, 1)$
(c) $f(0, 0, 0)$ (d) $f(a, a, a)$
(e) $f(t, t^2, -t)$ (f) $f(a + b, a - b, b)$.

14. Let $f(x, y, z) = zxy + x$. Find
(a) $f(x + y, x - y, x^2)$ (b) $f(xy, y/x, xz)$.

15. Find $F(f(x), g(y), h(z))$ if $F(x, y, z) = ye^{xyz}$, $f(x) = x^2$, $g(y) = y + 1$, and $h(z) = z^2$.

16. Find $g(u(x, y, z), v(x, y, z), w(x, y, z))$ if $g(x, y, z) = z \sin xy$, $u(x, y, z) = x^2z^3$, $v(x, y, z) = \pi xyz$, and $w(x, y, z) = xy/z$.

17–18 These exercises are concerned with functions of four or more variables.

17. (a) Let $f(x, y, z, t) = x^2y^3\sqrt{z + t}$. Find $f(\sqrt{5}, 2, \pi, 3\pi)$.
(b) Let $f(x_1, x_2, \ldots, x_n) = \displaystyle\sum_{k=1}^{n} kx_k$. Find $f(1, 1, \ldots, 1)$.

18. (a) Let $f(u, v, \lambda, \phi) = e^{u+v} \cos \lambda \tan \phi$. Find $f(-2, 2, 0, \pi/4)$.
(b) Let $f(x_1, x_2, \ldots, x_n) = x_1^2 + x_2^2 + \cdots + x_n^2$. Find $f(1, 2, \ldots, n)$.

19–22 Sketch the domain of f. Use solid lines for portions of the boundary included in the domain and dashed lines for portions not included.

19. $f(x, y) = \ln(1 - x^2 - y^2)$ **20.** $f(x, y) = \sqrt{x^2 + y^2 - 4}$

21. $f(x, y) = \dfrac{1}{x - y^2}$ **22.** $f(x, y) = \ln xy$

23–24 Describe the domain of f in words.

23. (a) $f(x, y) = xe^{-\sqrt{y+2}}$
(b) $f(x, y, z) = \sqrt{25 - x^2 - y^2 - z^2}$
(c) $f(x, y, z) = e^{xyz}$

24. (a) $f(x, y) = \dfrac{\sqrt{4 - x^2}}{y^2 + 3}$ (b) $f(x, y) = \ln(y - 2x)$

(c) $f(x, y, z) = \dfrac{xyz}{x + y + z}$

25–34 Sketch the graph of f.

25. $f(x, y) = 3$

26. $f(x, y) = \sqrt{9 - x^2 - y^2}$

27. $f(x, y) = \sqrt{x^2 + y^2}$

28. $f(x, y) = x^2 + y^2$

29. $f(x, y) = x^2 - y^2$

30. $f(x, y) = 4 - x^2 - y^2$

31. $f(x, y) = \sqrt{x^2 + y^2 + 1}$

32. $f(x, y) = \sqrt{x^2 + y^2 - 1}$

33. $f(x, y) = y + 1$

34. $f(x, y) = x^2$

FOCUS ON CONCEPTS

35. In each part, match the contour plot with one of the functions

$$f(x, y) = \sqrt{x^2 + y^2}, \quad f(x, y) = x^2 + y^2,$$
$$f(x, y) = 1 - x^2 - y^2$$

by inspection, and explain your reasoning. Larger values of z are indicated by lighter colors in the contour plot, and the concentric contours correspond to equally spaced values of z.

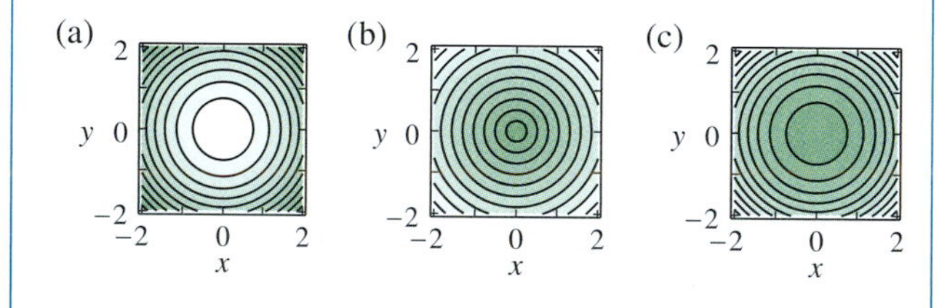

36. In each part, match the contour plot with one of the surfaces in the accompanying figure by inspection, and explain your reasoning. The larger the value of z, the lighter the color in the contour plot.

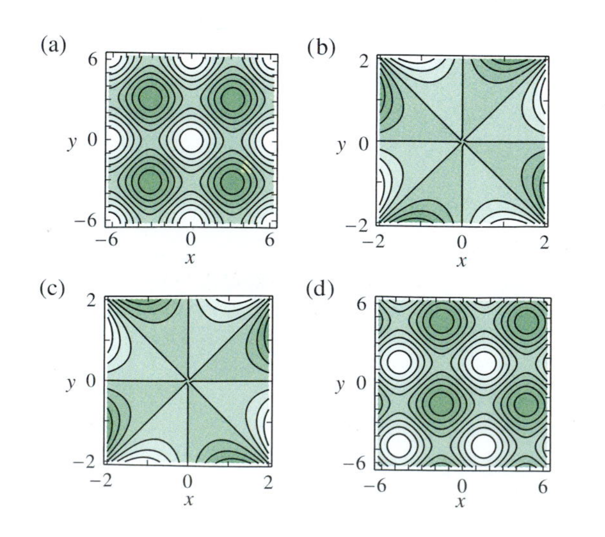

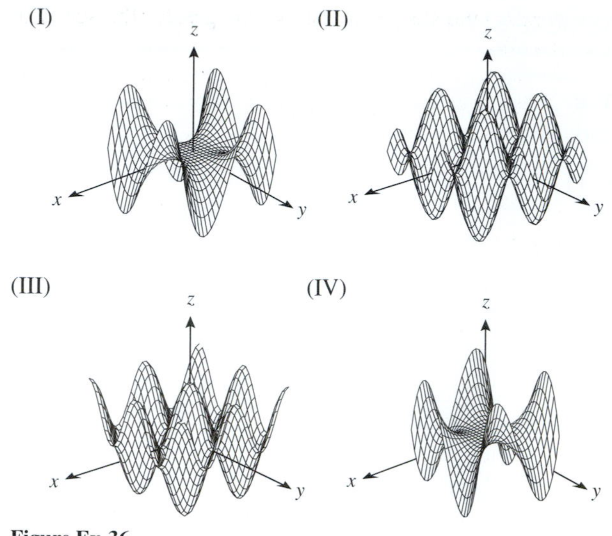

Figure Ex-36

37. In each part, the questions refer to the contour map in the accompanying figure.

(a) Is A or B the higher point? Explain your reasoning.

(b) Is the slope steeper at point A or at point B? Explain your reasoning.

(c) Starting at A and moving so that y remains constant and x increases, will the elevation begin to increase or decrease?

(d) Starting at B and moving so that y remains constant and x increases, will the elevation begin to increase or decrease?

(e) Starting at A and moving so that x remains constant and y decreases, will the elevation begin to increase or decrease?

(f) Starting at B and moving so that x remains constant and y decreases, will the elevation begin to increase or decrease?

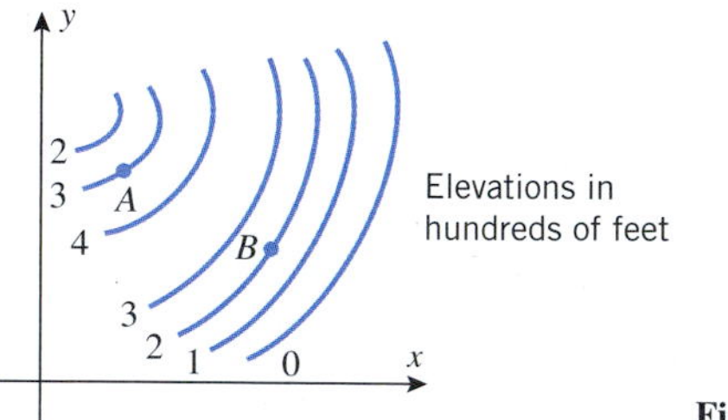

Figure Ex-37

38. A curve connecting points of equal atmospheric pressure on a weather map is called an ***isobar***. On a typical weather map the isobars refer to pressure at mean sea level and are given in units of ***millibars*** (mb). Mathematically, isobars are level curves for the pressure function $p(x, y)$ defined at the geographic points (x, y) represented on the map. Tightly packed isobars correspond to steep slopes on the graph of the pressure function, and these are usually associated with strong winds—the steeper the slope, the greater the speed of the wind.

(a) Referring to the accompanying weather map, is the wind speed greater in Medicine Hat, Alberta or in Chicago? Explain your reasoning.

(b) Estimate the average rate of change in atmospheric pressure (in mb/mi) from Medicine Hat to Chicago, given that the distance between the two cities is approximately 1400 mi.

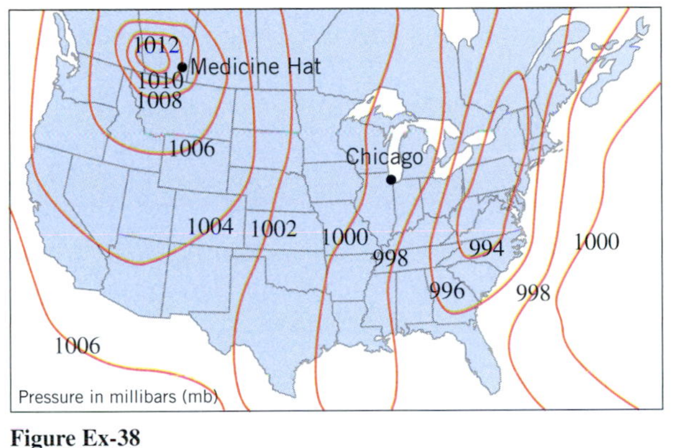

Figure Ex-38

39–44 Sketch the level curve $z = k$ for the specified values of k.

39. $z = x^2 + y^2;\ k = 0, 1, 2, 3, 4$

40. $z = y/x;\ k = -2, -1, 0, 1, 2$

41. $z = x^2 + y;\ k = -2, -1, 0, 1, 2$

42. $z = x^2 + 9y^2;\ k = 0, 1, 2, 3, 4$

43. $z = x^2 - y^2;\ k = -2, -1, 0, 1, 2$

44. $z = y \csc x;\ k = -2, -1, 0, 1, 2$

45–48 Sketch the level surface $f(x, y, z) = k$.

45. $f(x, y, z) = 4x^2 + y^2 + 4z^2;\ k = 16$

46. $f(x, y, z) = x^2 + y^2 - z^2;\ k = 0$

47. $f(x, y, z) = z - x^2 - y^2 + 4;\ k = 7$

48. $f(x, y, z) = 4x - 2y + z;\ k = 1$

49–52 Describe the level surfaces in words.

49. $f(x, y, z) = (x - 2)^2 + y^2 + z^2$

50. $f(x, y, z) = 3x - y + 2z$

51. $f(x, y, z) = x^2 + z^2$

52. $f(x, y, z) = z - x^2 - y^2$

53. Let $f(x, y) = x^2 - 2x^3 + 3xy$. Find an equation of the level curve that passes through the point
(a) $(-1, 1)$ (b) $(0, 0)$ (c) $(2, -1)$.

54. Let $f(x, y) = ye^x$. Find an equation of the level curve that passes through the point
(a) $(\ln 2, 1)$ (b) $(0, 3)$ (c) $(1, -2)$.

55. Let $f(x, y, z) = x^2 + y^2 - z$. Find an equation of the level surface that passes through the point
(a) $(1, -2, 0)$ (b) $(1, 0, 3)$ (c) $(0, 0, 0)$.

56. Let $f(x, y, z) = xyz + 3$. Find an equation of the level surface that passes through the point
(a) $(1, 0, 2)$ (b) $(-2, 4, 1)$ (c) $(0, 0, 0)$.

57. If $T(x, y)$ is the temperature at a point (x, y) on a thin metal plate in the xy-plane, then the level curves of T are called ***isothermal curves***. All points on such a curve are at the same temperature. Suppose that a plate occupies the first quadrant and $T(x, y) = xy$.

(a) Sketch the isothermal curves on which $T = 1$, $T = 2$, and $T = 3$.

(b) An ant, initially at $(1, 4)$, wants to walk on the plate so that the temperature along its path remains constant. What path should the ant take and what is the temperature along that path?

58. If $V(x, y)$ is the voltage or potential at a point (x, y) in the xy-plane, then the level curves of V are called ***equipotential curves***. Along such a curve, the voltage remains constant. Given that

$$V(x, y) = \frac{8}{\sqrt{16 + x^2 + y^2}}$$

sketch the equipotential curves at which $V = 2.0$, $V = 1.0$, and $V = 0.5$.

59. Let $f(x, y) = x^2 + y^3$.

(a) Use a graphing utility to generate the level curve that passes through the point $(2, -1)$.

(b) Generate the level curve of height 1.

60. Let $f(x, y) = 2\sqrt{xy}$.

(a) Use a graphing utility to generate the level curve that passes through the point $(2, 2)$.

(b) Generate the level curve of height 8.

C **61.** Let $f(x, y) = xe^{-(x^2+y^2)}$.

(a) Use a CAS to generate the graph of f for $-2 \le x \le 2$ and $-2 \le y \le 2$.

(b) Generate a contour plot for the surface, and confirm visually that it is consistent with the surface obtained in part (a).

(c) Read the appropriate documentation and explore the effect of generating the graph of f from various viewpoints.

C **62.** Let $f(x, y) = \frac{1}{10}e^x \sin y$.

(a) Use a CAS to generate the graph of f for $0 \le x \le 4$ and $0 \le y \le 2\pi$.

(b) Generate a contour plot for the surface, and confirm visually that it is consistent with the surface obtained in part (a).

(c) Read the appropriate documentation and explore the effect of generating the graph of f from various viewpoints.

63. In each part, describe in words how the graph of g is related to the graph of f.
(a) $g(x, y) = f(x - 1, y)$ (b) $g(x, y) = 1 + f(x, y)$
(c) $g(x, y) = -f(x, y + 1)$

64. (a) Sketch the graph of $f(x, y) = e^{-(x^2+y^2)}$.
(b) Describe in words how the graph of the function $g(x, y) = e^{-a(x^2+y^2)}$ is related to the graph of f for positive values of a.

✔ QUICK CHECK ANSWERS 14.1

1. points (x, y) in the first or third quadrants; points (x, y) in the first quadrant **2.** (a) $\frac{1}{4}$ (b) $-\frac{1}{4}$ (c) 0 (d) $1/(2y+2)$ **3.** (a) $k > 0$ (b) the lines $x + y = \ln k$ **4.** (a) $0 < k \leq 1$ (b) spheres of radius $\sqrt{(1-k)/k}$ for $0 < k < 1$, the single point $(0, 0, 0)$ for $k = 1$

14.2 LIMITS AND CONTINUITY

In this section we will introduce the notions of limit and continuity for functions of two or more variables. We will not go into great detail—our objective is to develop the basic concepts accurately and to obtain results needed in later sections. A more extensive study of these topics is usually given in advanced calculus.

LIMITS ALONG CURVES

For a function of one variable there are two one-sided limits at a point x_0, namely,

$$\lim_{x \to x_0^+} f(x) \quad \text{and} \quad \lim_{x \to x_0^-} f(x)$$

reflecting the fact that there are only two directions from which x can approach x_0, the right or the left. For functions of two or three variables the situation is more complicated because there are infinitely many different curves along which one point can approach another (Figure 14.2.1). Our first objective in this section is to define the limit of $f(x, y)$ as (x, y) approaches a point (x_0, y_0) along a curve C (and similarly for functions of three variables).

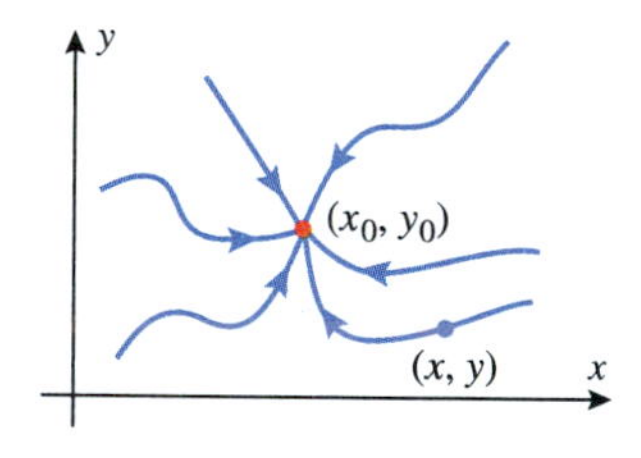

Figure 14.2.1

If C is a smooth parametric curve in 2-space or 3-space that is represented by the equations

$$x = x(t), \quad y = y(t) \qquad \text{or} \qquad x = x(t), \quad y = y(t), \quad z = z(t)$$

and if $x_0 = x(t_0)$, $y_0 = y(t_0)$, and $z_0 = z(t_0)$, then the limits

$$\lim_{\substack{(x, y) \to (x_0, y_0) \\ \text{(along } C)}} f(x, y) \quad \text{and} \quad \lim_{\substack{(x, y, z) \to (x_0, y_0, z_0) \\ \text{(along } C)}} f(x, y, z)$$

are defined by

$$\lim_{\substack{(x, y) \to (x_0, y_0) \\ \text{(along } C)}} f(x, y) = \lim_{t \to t_0} f(x(t), y(t)) \tag{1}$$

$$\lim_{\substack{(x, y, z) \to (x_0, y_0, z_0) \\ \text{(along } C)}} f(x, y, z) = \lim_{t \to t_0} f(x(t), y(t), z(t)) \tag{2}$$

In words, Formulas (1) and (2) state that a limit of a function f along a parametric curve can be obtained by substituting the parametric equations for the curve into the formula for the function and then computing the limit of the resulting function of one variable at the appropriate point.

In these formulas the limit of the function of t must be treated as a one-sided limit if (x_0, y_0) or (x_0, y_0, z_0) is an endpoint of C.

A geometric interpretation of the limit along a curve for a function of two variables is shown in Figure 14.2.2: As the point $(x(t), y(t))$ moves along the curve C in the xy-plane toward (x_0, y_0), the point $(x(t), y(t), f(x(t), y(t)))$ moves directly above it along the graph of $z = f(x, y)$ with $f(x(t), y(t))$ approaching the limiting value L. In the figure we followed a common practice of omitting the zero z-coordinate for points in the xy-plane.

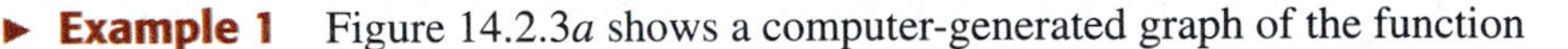

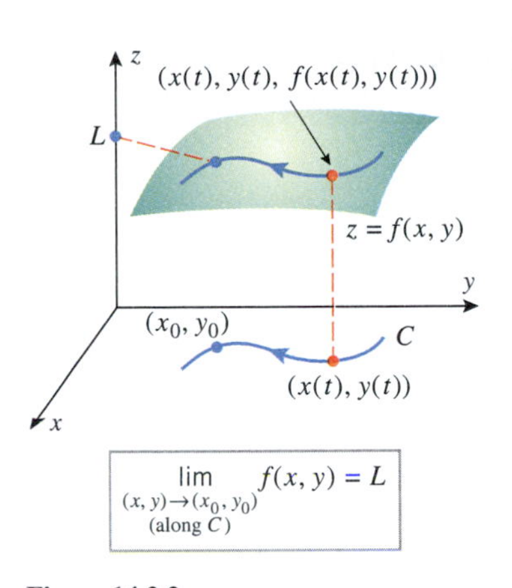

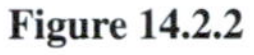

Figure 14.2.2

▶ **Example 1** Figure 14.2.3*a* shows a computer-generated graph of the function

$$f(x, y) = -\frac{xy}{x^2 + y^2}$$

The graph reveals that the surface has a ridge above the line $y = -x$, which is to be expected since $f(x, y)$ has a constant value of $\frac{1}{2}$ for $y = -x$, except at $(0, 0)$ where f is undefined (verify). Moreover, the graph suggests that the limit of $f(x, y)$ as $(x, y) \to (0, 0)$ along a line through the origin varies with the direction of the line. Find this limit along

(a) the x-axis
(b) the y-axis
(c) the line $y = x$
(d) the line $y = -x$
(e) the parabola $y = x^2$

Solution (a). The x-axis has parametric equations $x = t$, $y = 0$, with $(0, 0)$ corresponding to $t = 0$, so

$$\lim_{\substack{(x, y)\to(0, 0) \\ (\text{along } y = 0)}} f(x, y) = \lim_{t\to 0} f(t, 0) = \lim_{t\to 0} \left(-\frac{0}{t^2}\right) = \lim_{t\to 0} 0 = 0$$

which is consistent with Figure 14.2.3*b*.

Solution (b). The y-axis has parametric equations $x = 0$, $y = t$, with $(0, 0)$ corresponding to $t = 0$, so

$$\lim_{\substack{(x, y)\to(0, 0) \\ (\text{along } x = 0)}} f(x, y) = \lim_{t\to 0} f(0, t) = \lim_{t\to 0} \left(-\frac{0}{t^2}\right) = \lim_{t\to 0} 0 = 0$$

which is consistent with Figure 14.2.3*b*.

Solution (c). The line $y = x$ has parametric equations $x = t$, $y = t$, with $(0, 0)$ corresponding to $t = 0$, so

$$\lim_{\substack{(x, y)\to(0, 0) \\ (\text{along } y = x)}} f(x, y) = \lim_{t\to 0} f(t, t) = \lim_{t\to 0} \left(-\frac{t^2}{2t^2}\right) = \lim_{t\to 0} \left(-\frac{1}{2}\right) = -\frac{1}{2}$$

which is consistent with Figure 14.2.3*b*.

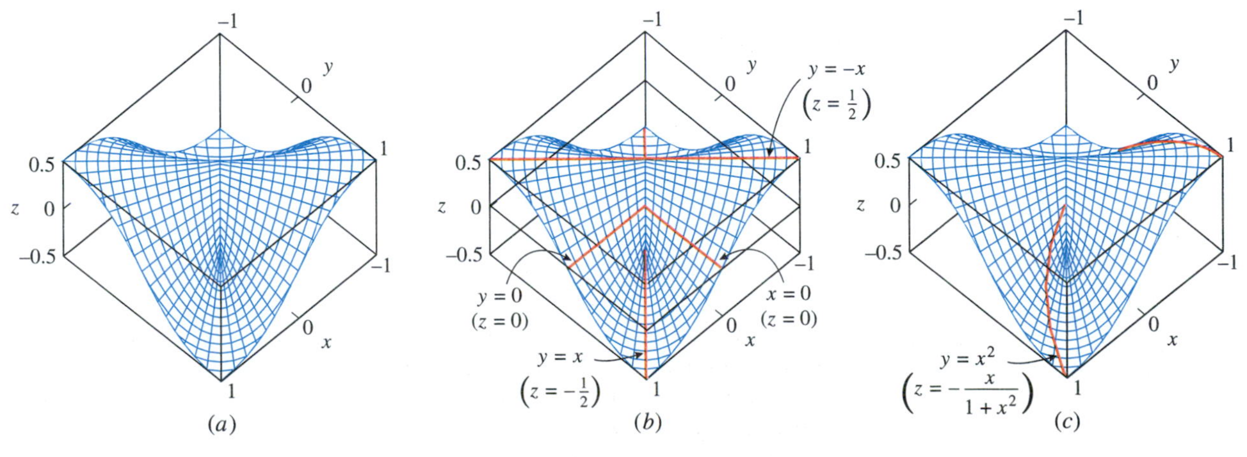

Figure 14.2.3

Solution (d). The line $y = -x$ has parametric equations $x = t$, $y = -t$, with $(0, 0)$ corresponding to $t = 0$, so

$$\lim_{\substack{(x,y)\to(0,0)\\ (\text{along } y=-x)}} f(x, y) = \lim_{t\to 0} f(t, -t) = \lim_{t\to 0} \frac{t^2}{2t^2} = \lim_{t\to 0} \frac{1}{2} = \frac{1}{2}$$

which is consistent with Figure 14.2.3*b*.

Solution (e). The parabola $y = x^2$ has parametric equations $x = t$, $y = t^2$, with $(0, 0)$ corresponding to $t = 0$, so

$$\lim_{\substack{(x,y)\to(0,0)\\ (\text{along } y=x^2)}} f(x, y) = \lim_{t\to 0} f(t, t^2) = \lim_{t\to 0} \left(-\frac{t^3}{t^2 + t^4}\right) = \lim_{t\to 0} \left(-\frac{t}{1 + t^2}\right) = 0$$

This is consistent with Figure 14.2.3*c*, which shows the parametric curve

$$x = t, \quad y = t^2, \quad z = -\frac{t}{1 + t^2}$$

superimposed on the surface. ◀

OPEN AND CLOSED SETS

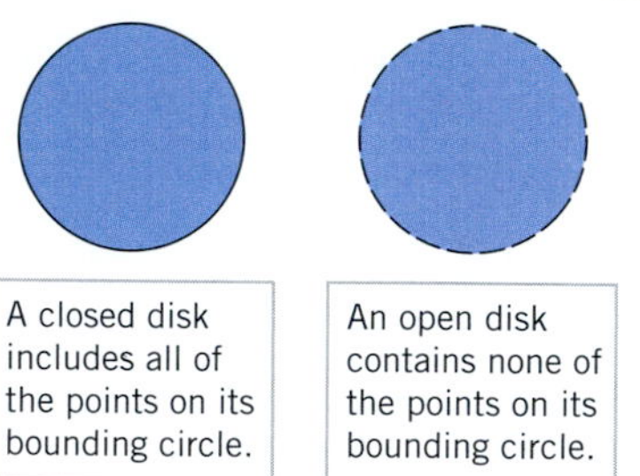

Figure 14.2.4

Although limits along specific curves are useful for many purposes, they do not always tell the complete story about the limiting behavior of a function at a point; what is required is a limit concept that accounts for the behavior of the function in an *entire vicinity* of a point, not just along smooth curves passing through the point. For this purpose, we start by introducing some terminology.

Let C be a circle in 2-space that is centered at (x_0, y_0) and has positive radius δ. The set of points that are enclosed by the circle, but do not lie on the circle, is called the ***open disk*** of radius δ centered at (x_0, y_0), and the set of points that lie on the circle together with those enclosed by the circle is called the ***closed disk*** of radius δ centered at (x_0, y_0) (Figure 14.2.4). Analogously, if S is a sphere in 3-space that is centered at (x_0, y_0, z_0) and has positive radius δ, then the set of points that are enclosed by the sphere, but do not lie on the sphere, is called the ***open ball*** of radius δ centered at (x_0, y_0, z_0), and the set of points that lie on the sphere together with those enclosed by the sphere is called the ***closed ball*** of radius δ centered at (x_0, y_0, z_0). Disks and balls are the two-dimensional and three-dimensional analogs of intervals on a line.

The notions of "open" and "closed" can be extended to more general sets in 2-space and 3-space. If D is a set of points in 2-space, then a point (x_0, y_0) is called an ***interior point*** of D if there is *some* open disk centered at (x_0, y_0) that contains only points of D, and (x_0, y_0) is called a ***boundary point*** of D if *every* open disk centered at (x_0, y_0) contains both points in D and points not in D. The same terminology applies to sets in 3-space, but in that case the definitions use balls rather than disks (Figure 14.2.5).

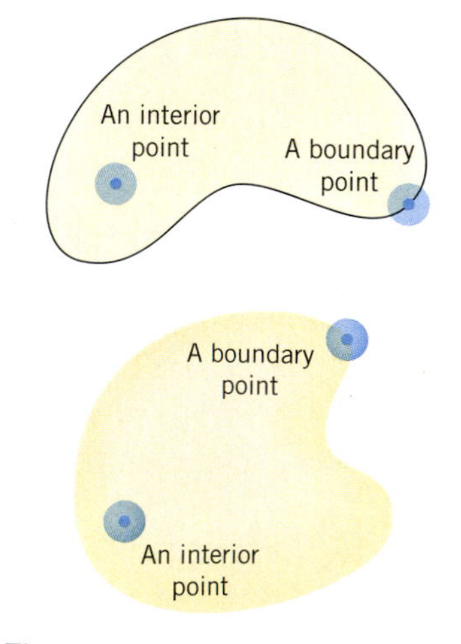

Figure 14.2.5

For a set D in either 2-space or 3-space, the set of all interior points is called the ***interior*** of D and the set of all boundary points is called the ***boundary*** of D. Moreover, just as for disks, we say that D is ***closed*** if it contains all of its boundary points and ***open*** if it contains *none* of its boundary points. The set of all points in 2-space and the set of all points in 3-space have no boundary points (why?), so by agreement they are regarded to be both open and closed.

GENERAL LIMITS OF FUNCTIONS OF TWO VARIABLES

The statement

$$\lim_{(x,y)\to(x_0,y_0)} f(x, y) = L$$

is intended to convey the idea that the value of $f(x, y)$ can be made as close as we like to the number L by restricting the point (x, y) to be sufficiently close to (but different from) the point (x_0, y_0). This idea has a formal expression in the following definition and is illustrated in Figure 14.2.6.

14.2.1 DEFINITION. Let f be a function of two variables, and assume that f is defined at all points of some open disk centered at (x_0, y_0), except possibly at (x_0, y_0). We will write

$$\lim_{(x,y)\to(x_0,y_0)} f(x, y) = L \tag{3}$$

if given any number $\epsilon > 0$, we can find a number $\delta > 0$ such that $f(x, y)$ satisfies

$$|f(x, y) - L| < \epsilon$$

whenever the distance between (x, y) and (x_0, y_0) satisfies

$$0 < \sqrt{(x - x_0)^2 + (y - y_0)^2} < \delta$$

When convenient, (3) can also be written as

$$\lim_{\substack{x\to x_0 \\ y\to y_0}} f(x, y) = L$$

or as

$f(x, y) \to L$ as $(x, y) \to (x_0, y_0)$

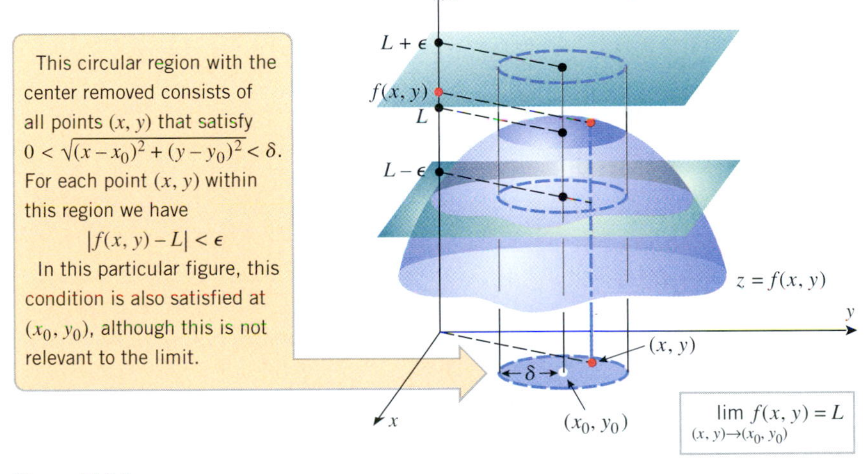

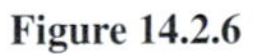

Figure 14.2.6

Another illustration of Definition 14.2.1 is shown in the "arrow diagram" of Figure 14.2.7. As in Figure 14.2.6, this figure is intended to convey the idea that the values of $f(x, y)$ can be forced within ϵ units of L on the z-axis by restricting (x, y) to lie within δ units of (x_0, y_0) in the xy-plane. We used a white dot at (x_0, y_0) to suggest that the epsilon condition need not hold at this point.

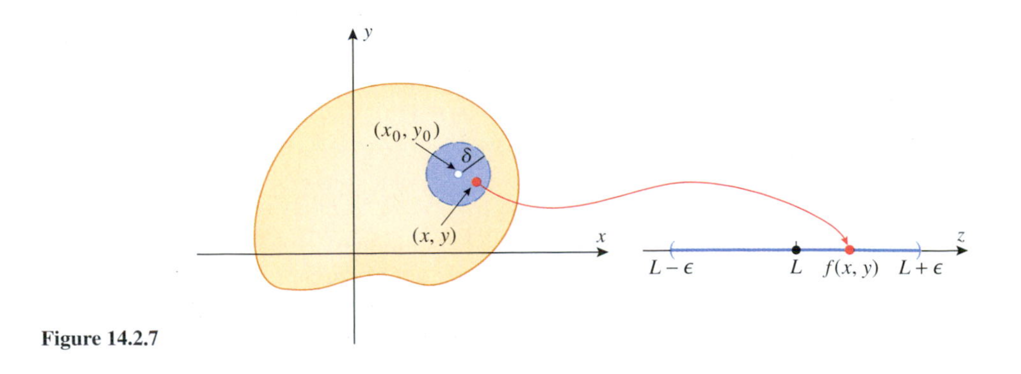

Figure 14.2.7

We note without proof that the standard properties of limits hold for limits along curves and for general limits of functions of two variables, so that computations involving such limits can be performed in the usual way.

Example 2

$$\lim_{(x,y)\to(1,4)} [5x^3y^2 - 9] = \lim_{(x,y)\to(1,4)} [5x^3y^2] - \lim_{(x,y)\to(1,4)} 9$$
$$= 5\left[\lim_{(x,y)\to(1,4)} x\right]^3\left[\lim_{(x,y)\to(1,4)} y\right]^2 - 9$$
$$= 5(1)^3(4)^2 - 9 = 71$$

RELATIONSHIPS BETWEEN GENERAL LIMITS AND LIMITS ALONG SMOOTH CURVES

Stated informally, if $f(x, y)$ has limit L as (x, y) approaches (x_0, y_0), then the value of $f(x, y)$ gets closer and closer to L as the distance between (x, y) and (x_0, y_0) approaches zero. Since this statement imposes no restrictions on the direction in which (x, y) approaches (x_0, y_0), it is plausible that the function $f(x, y)$ will also have the limit L as (x, y) approaches (x_0, y_0) along *any* smooth curve C. This is the implication of the following theorem, which we state without proof.

WARNING

In general, one cannot show that

$$\lim_{(x,y)\to(x_0,y_0)} f(x, y) = L$$

by showing that this limit holds along a specific curve, or even some specific family of curves. The problem is there may be some other curve along which the limit does not exist or has a value different from L (see Exercise 26, for example).

14.2.2 THEOREM.

(a) *If $f(x, y) \to L$ as $(x, y) \to (x_0, y_0)$, then $f(x, y) \to L$ as $(x, y) \to (x_0, y_0)$ along any smooth curve.*

(b) *If the limit of $f(x, y)$ fails to exist as $(x, y) \to (x_0, y_0)$ along some smooth curve, or if $f(x, y)$ has different limits as $(x, y) \to (x_0, y_0)$ along two different smooth curves, then the limit of $f(x, y)$ does not exist as $(x, y) \to (x_0, y_0)$.*

Example 3 The limit

$$\lim_{(x,y)\to(0,0)} -\frac{xy}{x^2 + y^2}$$

does not exist because in Example 1 we found two different smooth curves along which this limit had different values. Specifically,

$$\lim_{\substack{(x,y)\to(0,0)\\ (\text{along } x=0)}} -\frac{xy}{x^2 + y^2} = 0 \quad \text{and} \quad \lim_{\substack{(x,y)\to(0,0)\\ (\text{along } y=x)}} -\frac{xy}{x^2 + y^2} = -\frac{1}{2}$$

CONTINUITY

Stated informally, a function of one variable is continuous if its graph is an unbroken curve without jumps or holes. To extend this idea to functions of two variables, imagine that the graph of $z = f(x, y)$ is molded from a thin sheet of clay that has been hollowed or pinched into peaks and valleys. We will regard f as being continuous if the clay surface has no tears or holes. The functions graphed in Figure 14.2.8 fail to be continuous because of their behavior at $(0, 0)$.

The precise definition of continuity at a point for functions of two variables is similar to that for functions of one variable—we require the limit of the function and the value of the function to be the same at the point.

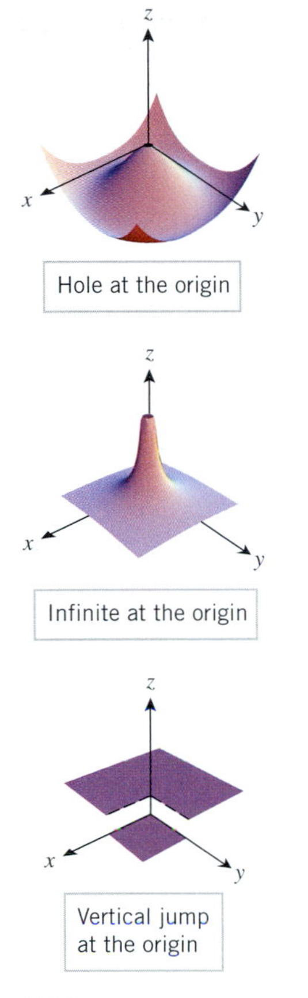

Figure 14.2.8

14.2.3 DEFINITION. A function $f(x, y)$ is said to be ***continuous at*** (x_0, y_0) if $f(x_0, y_0)$ is defined and if

$$\lim_{(x,y)\to(x_0,y_0)} f(x, y) = f(x_0, y_0)$$

In addition, if f is continuous at every point in an open set D, then we say that f is ***continuous on D***, and if f is continuous at every point in the xy-plane, then we say that f is ***continuous everywhere***.

The following theorem, which we state without proof, illustrates some of the ways in which continuous functions can be combined to produce new continuous functions.

14.2.4 THEOREM.

(*a*) *If $g(x)$ is continuous at x_0 and $h(y)$ is continuous at y_0, then $f(x, y) = g(x)h(y)$ is continuous at (x_0, y_0).*

(*b*) *If $h(x, y)$ is continuous at (x_0, y_0) and $g(u)$ is continuous at $u = h(x_0, y_0)$, then the composition $f(x, y) = g(h(x, y))$ is continuous at (x_0, y_0).*

(*c*) *If $f(x, y)$ is continuous at (x_0, y_0), and if $x(t)$ and $y(t)$ are continuous at t_0 with $x(t_0) = x_0$ and $y(t_0) = y_0$, then the composition $f(x(t), y(t))$ is continuous at t_0.*

▶ **Example 4** Use Theorem 14.2.4 to show that the functions $f(x, y) = 3x^2y^5$ and $f(x, y) = \sin(3x^2y^5)$ are continuous everywhere.

Solution. The polynomials $g(x) = 3x^2$ and $h(y) = y^5$ are continuous at every real number, and therefore by part (*a*) of Theorem 14.2.4, the function $f(x, y) = 3x^2y^5$ is continuous at every point (x, y) in the xy-plane. Since $3x^2y^5$ is continuous at every point in the xy-plane and $\sin u$ is continuous at every real number u, it follows from part (*b*) of Theorem 14.2.4 that the composition $f(x, y) = \sin(3x^2y^5)$ is continuous everywhere. ◀

Theorem 14.2.4 is one of a whole class of theorems about continuity of functions in two or more variables. The content of these theorems can be summarized informally with three basic principles:

Recognizing Continuous Functions

- A composition of continuous functions is continuous.
- A sum, difference, or product of continuous functions is continuous.
- A quotient of continuous functions is continuous, except where the denominator is zero.

By using these principles and Theorem 14.2.4, you should be able to confirm that the following functions are all continuous everywhere:

$$xe^{xy} + y^{2/3}, \quad \cosh(xy^3) - |xy|, \quad \frac{xy}{1 + x^2 + y^2}$$

▶ **Example 5** Evaluate $\lim_{(x,y)\to(-1,2)} \dfrac{xy}{x^2+y^2}$.

Solution. Since $f(x, y) = xy/(x^2 + y^2)$ is continuous at $(-1, 2)$ (why?), it follows from the definition of continuity for functions of two variables that

$$\lim_{(x,y)\to(-1,2)} \frac{xy}{x^2+y^2} = \frac{(-1)(2)}{(-1)^2+(2)^2} = -\frac{2}{5} \blacktriangleleft$$

▶ **Example 6** Since the function

$$f(x, y) = \frac{x^3y^2}{1-xy}$$

is a quotient of continuous functions, it is continuous except where $1 - xy = 0$. Thus, $f(x, y)$ is continuous everywhere except on the hyperbola $xy = 1$. ◀

LIMITS AT DISCONTINUITIES

Sometimes it is easy to recognize when a limit does not exist. For example, it is evident that

$$\lim_{(x,y)\to(0,0)} \frac{1}{x^2+y^2} = +\infty$$

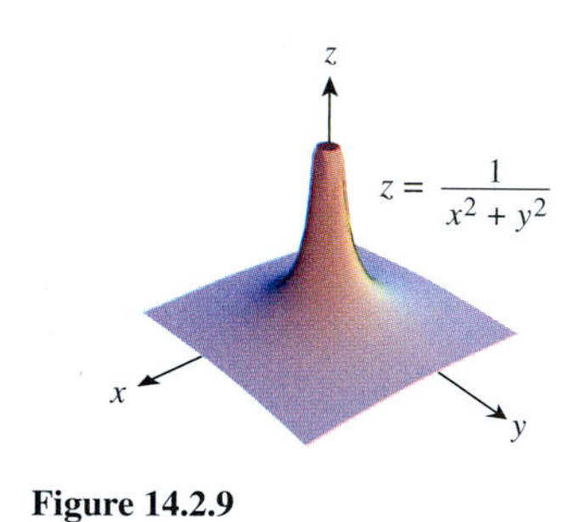

Figure 14.2.9

which implies that the values of the function approach $+\infty$ as $(x, y) \to (0, 0)$ along any smooth curve (Figure 14.2.9). However, it is not evident whether the limit

$$\lim_{(x,y)\to(0,0)} (x^2+y^2)\ln(x^2+y^2)$$

exists because it is an indeterminate form of type $0 \cdot \infty$. Although L'Hôpital's rule cannot be applied directly, the following example illustrates a method for finding this limit by converting to polar coordinates.

▶ **Example 7** Find $\lim_{(x,y)\to(0,0)} (x^2+y^2)\ln(x^2+y^2)$.

Solution. Let (r, θ) be polar coordinates of the point (x, y) with $r \geq 0$. Then we have

$$x = r\cos\theta, \quad y = r\sin\theta, \quad r^2 = x^2 + y^2$$

Moreover, since $r \geq 0$ we have $r = \sqrt{x^2+y^2}$, so that $r \to 0^+$ if and only if $(x, y) \to (0, 0)$. Thus, we can rewrite the given limit as

$$\begin{aligned}
\lim_{(x,y)\to(0,0)} (x^2+y^2)\ln(x^2+y^2) &= \lim_{r\to0^+} r^2 \ln r^2 \\
&= \lim_{r\to0^+} \frac{2\ln r}{1/r^2} && \text{This converts the limit to an indeterminate form of type } \infty/\infty. \\
&= \lim_{r\to0^+} \frac{2/r}{-2/r^3} && \text{L'Hôpital's rule} \\
&= \lim_{r\to0^+} (-r^2) = 0 \blacktriangleleft
\end{aligned}$$

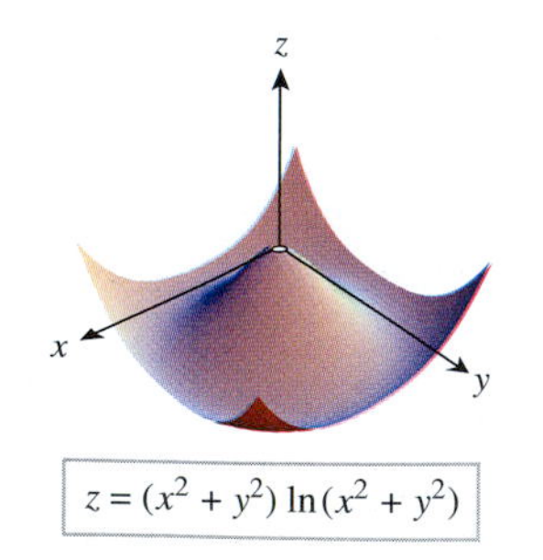

$z = (x^2+y^2)\ln(x^2+y^2)$

Figure 14.2.10

The graph of $f(x, y) = (x^2+y^2)\ln(x^2+y^2)$ in Example 7 is a surface with a hole (sometimes called a *puncture*) at the origin (Figure 14.2.10). We can remove this discontinuity by *defining* $f(0, 0)$ to be 0. (See Exercises 31 and 32, which also deal with the notion of a "removable" discontinuity.)

CONTINUITY AT BOUNDARY POINTS

Recall that in our study of continuity for functions of one variable, we first defined continuity at a point, then continuity on an open interval, and then, by using one-sided limits, we extended the notion of continuity to include the boundary points of the interval. Similarly, for functions of two variables one can extend the notion of continuity of $f(x, y)$ to the boundary of its domain by modifying Definition 14.2.1 appropriately so that (x, y) is restricted to approach (x_0, y_0) through points lying wholly in the domain of f. We will omit the details.

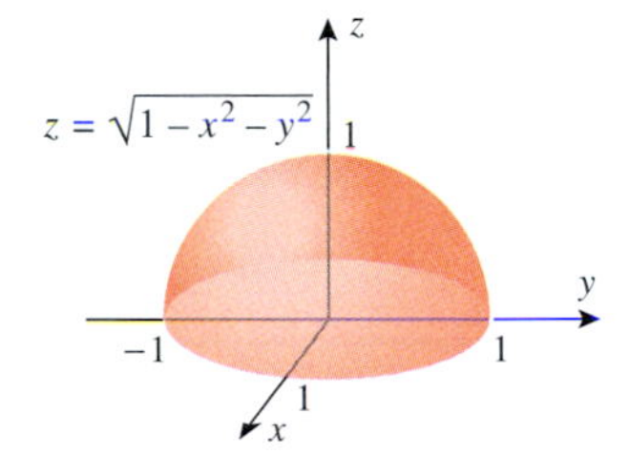

Figure 14.2.11

▸ **Example 8** The graph of the function $f(x, y) = \sqrt{1 - x^2 - y^2}$ is the upper hemisphere shown in Figure 14.2.11, and the natural domain of f is the closed unit disk

$$x^2 + y^2 \leq 1$$

The graph of f has no tears or holes, so it passes our "intuitive test" of continuity. In this case the continuity at a point (x_0, y_0) on the boundary reflects the fact that

$$\lim_{(x,y)\to(x_0,y_0)} \sqrt{1 - x^2 - y^2} = \sqrt{1 - x_0^2 - y_0^2} = 0$$

when (x, y) is restricted to points on the closed unit disk $x^2 + y^2 \leq 1$. It follows that f is continuous on its domain. ◂

EXTENSIONS TO THREE VARIABLES

All of the results in this section can be extended to functions of three or more variables. For example, the distance between the points (x, y, z) and (x_0, y_0, z_0) in 3-space is

$$\sqrt{(x - x_0)^2 + (y - y_0)^2 + (z - z_0)^2}$$

so the natural extension of Definition 14.2.1 to 3-space is as follows:

14.2.5 DEFINITION. Let f be a function of three variables, and assume that f is defined at all points within a ball centered at (x_0, y_0, z_0), except possibly at (x_0, y_0, z_0). We will write

$$\lim_{(x,y,z)\to(x_0,y_0,z_0)} f(x, y, z) = L \tag{4}$$

if given any number $\epsilon > 0$, we can find a number $\delta > 0$ such that $f(x, y, z)$ satisfies

$$|f(x, y, z) - L| < \epsilon$$

whenever the distance between (x, y, z) and (x_0, y_0, z_0) satisfies

$$0 < \sqrt{(x - x_0)^2 + (y - y_0)^2 + (z - z_0)^2} < \delta$$

As with functions of one and two variables, we define a function $f(x, y, z)$ of three variables to be continuous at a point (x_0, y_0, z_0) if the limit of the function and the value of the function are the same at this point; that is,

$$\lim_{(x,y,z)\to(x_0,y_0,z_0)} f(x, y, z) = f(x_0, y_0, z_0)$$

Although we will omit the details, the properties of limits and continuity that we discussed for functions of two variables, including the notion of continuity at boundary points, carry over to functions of three variables.

✔ QUICK CHECK EXERCISES 14.2 (See page 945 for answers.)

1. Let

$$f(x, y) = \frac{x^2 - y^2}{x^2 + y^2}$$

Determine the limit of $f(x, y)$ as (x, y) approaches $(0, 0)$ along the curve C.

(a) $C: x = 0$ (b) $C: y = 0$
(c) $C: y = x$ (d) $C: y = x^2$

2. (a) $\lim_{(x,y)\to(3,2)} x \cos \pi y =$ ______

(b) $\lim_{(x,y)\to(0,1)} e^{xy^2} =$ ______

(c) $\lim_{(x,y)\to(0,0)} (x^2 + y^2) \sin\left(\frac{1}{x^2 + y^2}\right) =$ ______

3. A function $f(x, y)$ is continuous at (x_0, y_0) provided $f(x_0, y_0)$ exists and provided $f(x, y)$ has limit ______ as (x, y) approaches ______.

4. Determine all values of the constant a such that the function $f(x, y) = \sqrt{x^2 - ay^2 + 1}$ is continuous everywhere.

EXERCISE SET 14.2

1–6 Use limit laws and continuity properties to evaluate the limit.

1. $\lim_{(x,y)\to(1,3)} (4xy^2 - x)$

2. $\lim_{(x,y)\to(1/2,\pi)} (xy^2 \sin xy)$

3. $\lim_{(x,y)\to(-1,2)} \frac{xy^3}{x + y}$

4. $\lim_{(x,y)\to(1,-3)} e^{2x-y^2}$

5. $\lim_{(x,y)\to(0,0)} \ln(1 + x^2y^3)$

6. $\lim_{(x,y)\to(4,-2)} x\sqrt[3]{y^3 + 2x}$

7–8 Show that the limit does not exist by considering the limits as $(x, y) \to (0, 0)$ along the coordinate axes.

7. (a) $\lim_{(x,y)\to(0,0)} \frac{3}{x^2 + 2y^2}$ (b) $\lim_{(x,y)\to(0,0)} \frac{x + y}{2x^2 + y^2}$

8. (a) $\lim_{(x,y)\to(0,0)} \frac{x - y}{x^2 + y^2}$ (b) $\lim_{(x,y)\to(0,0)} \frac{\cos xy}{x^2 + y^2}$

9–12 Evaluate the limit using the substitution $z = x^2 + y^2$ and observing that $z \to 0^+$ if and only if $(x, y) \to (0, 0)$.

9. $\lim_{(x,y)\to(0,0)} \frac{\sin(x^2 + y^2)}{x^2 + y^2}$

10. $\lim_{(x,y)\to(0,0)} \frac{1 - \cos(x^2 + y^2)}{x^2 + y^2}$

11. $\lim_{(x,y)\to(0,0)} e^{-1/(x^2+y^2)}$

12. $\lim_{(x,y)\to(0,0)} \frac{e^{-1/\sqrt{x^2+y^2}}}{\sqrt{x^2 + y^2}}$

13–20 Determine whether the limit exists. If so, find its value.

13. $\lim_{(x,y)\to(0,0)} \frac{x^4 - y^4}{x^2 + y^2}$

14. $\lim_{(x,y)\to(0,0)} \frac{x^4 - 16y^4}{x^2 + 4y^2}$

15. $\lim_{(x,y)\to(0,0)} \frac{xy}{3x^2 + 2y^2}$

16. $\lim_{(x,y)\to(0,0)} \frac{1 - x^2 - y^2}{x^2 + y^2}$

17. $\lim_{(x,y,z)\to(2,-1,2)} \frac{xz^2}{\sqrt{x^2 + y^2 + z^2}}$

18. $\lim_{(x,y,z)\to(2,0,-1)} \ln(2x + y - z)$

19. $\lim_{(x,y,z)\to(0,0,0)} \frac{\sin(x^2 + y^2 + z^2)}{\sqrt{x^2 + y^2 + z^2}}$

20. $\lim_{(x,y,z)\to(0,0,0)} \frac{\sin\sqrt{x^2 + y^2 + z^2}}{x^2 + y^2 + z^2}$

21–22 Evaluate the limit, if it exists, by converting to polar coordinates, as in Example 7.

21. $\lim_{(x,y)\to(0,0)} y \ln(x^2 + y^2)$

22. $\lim_{(x,y)\to(0,0)} \frac{x^2y^2}{\sqrt{x^2 + y^2}}$

23–24 Evaluate the limit, if it exists, by converting to spherical coordinates (ρ, θ, ϕ) and observe that $\rho \to 0^+$ if and only if $(x, y, z) \to (0, 0, 0)$, since $\rho = \sqrt{x^2 + y^2 + z^2}$.

23. $\lim_{(x,y,z)\to(0,0,0)} \frac{e^{\sqrt{x^2+y^2+z^2}}}{\sqrt{x^2 + y^2 + z^2}}$

24. $\lim_{(x,y,z)\to(0,0,0)} \tan^{-1}\left[\frac{1}{x^2 + y^2 + z^2}\right]$

FOCUS ON CONCEPTS

25. The accompanying figure shows a portion of the graph of

$$f(x, y) = \frac{x^2y}{x^4 + y^2}$$

(a) Based on the graph in the figure, does $f(x, y)$ have a limit as $(x, y) \to (0, 0)$? Explain your reasoning.

(b) Show that $f(x, y) \to 0$ as $(x, y) \to (0, 0)$ along any line $y = mx$. Does this imply that $f(x, y) \to 0$ as $(x, y) \to (0, 0)$? Explain.

(c) Show that $f(x, y) \to \frac{1}{2}$ as $(x, y) \to (0, 0)$ along the parabola $y = x^2$, and confirm visually that this is consistent with the graph of $f(x, y)$.

(d) Based on parts (b) and (c), does $f(x, y)$ have a limit as $(x, y) \to (0, 0)$? Is this consistent with your answer to part (a)?

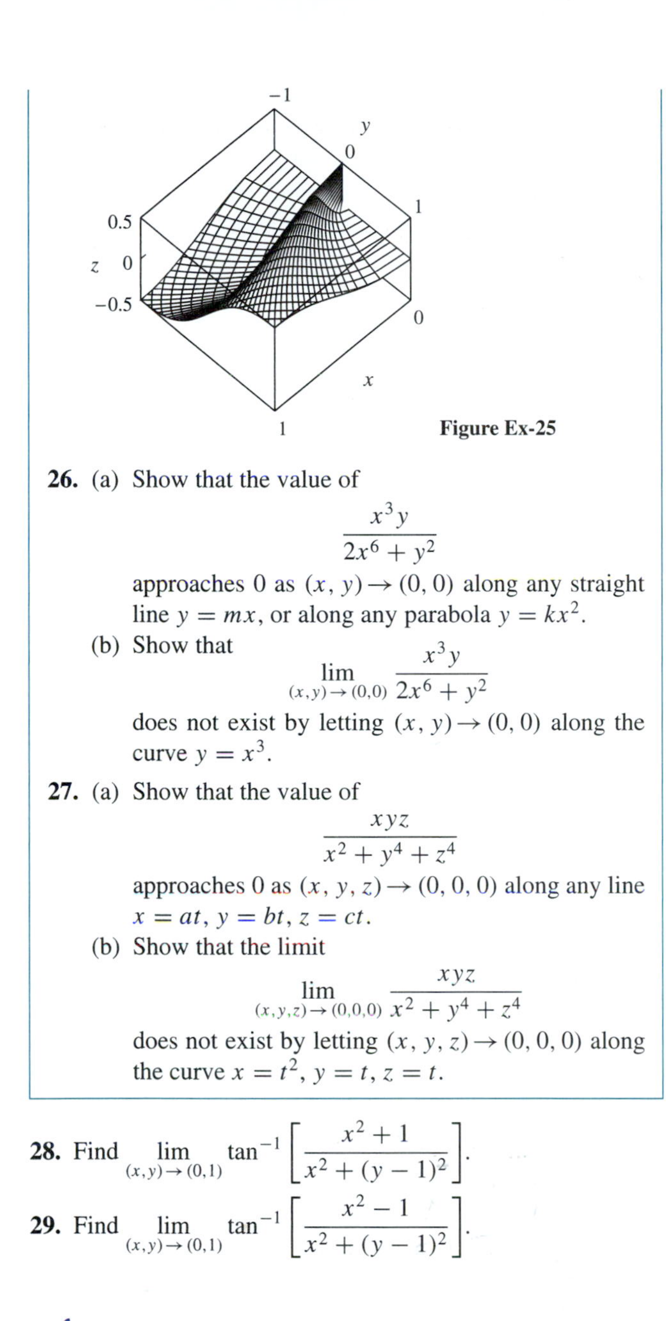

Figure Ex-25

26. (a) Show that the value of
$$\frac{x^3y}{2x^6+y^2}$$
approaches 0 as $(x, y) \to (0, 0)$ along any straight line $y = mx$, or along any parabola $y = kx^2$.

(b) Show that
$$\lim_{(x,y)\to(0,0)} \frac{x^3y}{2x^6+y^2}$$
does not exist by letting $(x, y) \to (0, 0)$ along the curve $y = x^3$.

27. (a) Show that the value of
$$\frac{xyz}{x^2+y^4+z^4}$$
approaches 0 as $(x, y, z) \to (0, 0, 0)$ along any line $x = at$, $y = bt$, $z = ct$.

(b) Show that the limit
$$\lim_{(x,y,z)\to(0,0,0)} \frac{xyz}{x^2+y^4+z^4}$$
does not exist by letting $(x, y, z) \to (0, 0, 0)$ along the curve $x = t^2$, $y = t$, $z = t$.

28. Find $\displaystyle\lim_{(x,y)\to(0,1)} \tan^{-1}\left[\frac{x^2+1}{x^2+(y-1)^2}\right]$.

29. Find $\displaystyle\lim_{(x,y)\to(0,1)} \tan^{-1}\left[\frac{x^2-1}{x^2+(y-1)^2}\right]$.

30. Let $f(x, y) = \begin{cases} \dfrac{\sin(x^2+y^2)}{x^2+y^2}, & (x, y) \neq (0, 0) \\ 1, & (x, y) = (0, 0). \end{cases}$

Show that f is continuous at $(0, 0)$.

31–32 A function $f(x, y)$ is said to have a ***removable discontinuity*** at (x_0, y_0) if $\lim_{(x,y)\to(x_0,y_0)} f(x, y)$ exists but f is not continuous at (x_0, y_0), either because f is not defined at (x_0, y_0) or because $f(x_0, y_0)$ differs from the value of the limit. Determine whether $f(x, y)$ has a removable discontinuity at $(0, 0)$.

31. $f(x, y) = \dfrac{x^2}{x^2+y^2}$

32. $f(x, y) = xy\ln(x^2+y^2)$

33–40 Sketch the largest region on which the function f is continuous.

33. $f(x, y) = y\ln(1+x)$

34. $f(x, y) = \sqrt{x-y}$

35. $f(x, y) = \dfrac{x^2y}{\sqrt{25-x^2-y^2}}$

36. $f(x, y) = \ln(2x - y + 1)$

37. $f(x, y) = \cos\left(\dfrac{xy}{1+x^2+y^2}\right)$

38. $f(x, y) = e^{1-xy}$

39. $f(x, y) = \sin^{-1}(xy)$

40. $f(x, y) = \tan^{-1}(y - x)$

41–44 Describe the largest region on which the function f is continuous.

41. $f(x, y, z) = 3x^2e^{yz}\cos(xyz)$

42. $f(x, y, z) = \ln(4 - x^2 - y^2 - z^2)$

43. $f(x, y, z) = \dfrac{y+1}{x^2+z^2-1}$

44. $f(x, y, z) = \sin\sqrt{x^2+y^2+3z^2}$

✔ QUICK CHECK ANSWERS 14.2

1. (a) −1 (b) 1 (c) 0 (d) 1 **2.** (a) 3 (b) 1 (c) 0 **3.** $f(x_0, y_0)$; (x_0, y_0) **4.** $a \leq 0$

14.3 PARTIAL DERIVATIVES

In this section we will develop the mathematical tools for studying rates of change that involve two or more independent variables.

PARTIAL DERIVATIVES OF FUNCTIONS OF TWO VARIABLES

If $z = f(x, y)$, then one can inquire how the value of z changes if y is held fixed and x is allowed to vary, or if x is held fixed and y is allowed to vary. For example, the ideal gas law in physics states that under appropriate conditions the pressure exerted by a gas is a function of the volume of the gas and its temperature. Thus, a physicist studying gases might be interested in the rate of change of the pressure if the volume is held fixed and the temperature is allowed to vary, or if the temperature is held fixed and the volume is allowed to vary. We now define a derivative that describes such rates of change.

Suppose that (x_0, y_0) is a point in the domain of a function $f(x, y)$. If we fix $y = y_0$, then $f(x, y_0)$ is a function of the variable x alone. The value of the derivative

$$\frac{d}{dx}[f(x, y_0)]$$

at x_0 then gives us a measure of the instantaneous rate of change of f with respect to x at the point (x_0, y_0). Similarly, the value of the derivative

$$\frac{d}{dy}[f(x_0, y)]$$

at y_0 gives us a measure of the instantaneous rate of change of f with respect to y at the point (x_0, y_0). These derivatives are so basic to the study of differential calculus of multivariable functions that they have their own name and notation.

14.3.1 DEFINITION. If $z = f(x, y)$ and (x_0, y_0) is a point in the domain of f, then the ***partial derivative of f with respect to x*** at (x_0, y_0) [also called the ***partial derivative of z with respect to x*** at (x_0, y_0)] is the derivative at x_0 of the function that results when $y = y_0$ is held fixed and x is allowed to vary. This partial derivative is denoted by $f_x(x_0, y_0)$ and is given by

$$f_x(x_0, y_0) = \frac{d}{dx}[f(x, y_0)]\bigg|_{x=x_0} = \lim_{\Delta x \to 0} \frac{f(x_0 + \Delta x, y_0) - f(x_0, y_0)}{\Delta x} \tag{1}$$

Similarly, the ***partial derivative of f with respect to y*** at (x_0, y_0) [also called the ***partial derivative of z with respect to y*** at (x_0, y_0)] is the derivative at y_0 of the function that results when $x = x_0$ is held fixed and y is allowed to vary. This partial derivative is denoted by $f_y(x_0, y_0)$ and is given by

$$f_y(x_0, y_0) = \frac{d}{dy}[f(x_0, y)]\bigg|_{y=y_0} = \lim_{\Delta y \to 0} \frac{f(x_0, y_0 + \Delta y) - f(x_0, y_0)}{\Delta y} \tag{2}$$

The limits in (1) and (2) show the relationship between partial derivatives and derivatives of functions of one variable. In practice, our usual method for computing partial derivatives is to hold one variable fixed and then differentiate the resulting function using the derivative rules for functions of one variable.

► **Example 1** Find $f_x(1, 3)$ and $f_y(1, 3)$ for the function $f(x, y) = 2x^3y^2 + 2y + 4x$.

Solution. Since

$$f_x(x, 3) = \frac{d}{dx}[f(x, 3)] = \frac{d}{dx}[18x^3 + 4x + 6] = 54x^2 + 4$$

we have $f_x(1, 3) = 54 + 4 = 58$. Also, since

$$f_y(1, y) = \frac{d}{dy}[f(1, y)] = \frac{d}{dy}[2y^2 + 2y + 4] = 4y + 2$$

we have $f_y(1, 3) = 4(3) + 2 = 14$. ◄

THE PARTIAL DERIVATIVE FUNCTIONS

Formulas (1) and (2) define the partial derivatives of a function at a specific point (x_0, y_0). However, often it will be desirable to omit the subscripts and think of the partial derivatives as functions of the variables x and y. These functions are

$$f_x(x, y) = \lim_{\Delta x \to 0} \frac{f(x + \Delta x, y) - f(x, y)}{\Delta x} \qquad f_y(x, y) = \lim_{\Delta y \to 0} \frac{f(x, y + \Delta y) - f(x, y)}{\Delta y}$$

The following example gives an alternative way of performing the computations in Example 1.

▶ **Example 2** Find $f_x(x, y)$ and $f_y(x, y)$ for $f(x, y) = 2x^3y^2 + 2y + 4x$, and use those partial derivatives to compute $f_x(1, 3)$ and $f_y(1, 3)$.

Solution. Keeping y fixed and differentiating with respect x yields

$$f_x(x, y) = \frac{d}{dx}[2x^3y^2 + 2y + 4x] = 6x^2y^2 + 4$$

and keeping x fixed and differentiating with respect to y yields

$$f_y(x, y) = \frac{d}{dy}[2x^3y^2 + 2y + 4x] = 4x^3y + 2$$

Thus,

$$f_x(1, 3) = 6(1^2)(3^2) + 4 = 58 \quad \text{and} \quad f_y(1, 3) = 4(1^3)3 + 2 = 14$$

which agree with the results in Example 1. ◀

TECHNOLOGY MASTERY

Computer algebra systems have specific commands for calculating partial derivatives. If you have a CAS, use it to find the partial derivatives $f_x(x, y)$ and $f_y(x, y)$ in Example 2.

PARTIAL DERIVATIVES VIEWED AS RATES OF CHANGE AND SLOPES

Recall that if $y = f(x)$, then the value of $f'(x_0)$ can be interpreted either as the rate of change of y with respect to x at x_0 or as the slope of the tangent line to the graph of f at x_0. Partial derivatives have analogous interpretations. To see that this is so, suppose that C_1 is the intersection of the surface $z = f(x, y)$ with the plane $y = y_0$ and that C_2 is its intersection with the plane $x = x_0$ (Figure 14.3.1). Thus, $f_x(x, y_0)$ can be interpreted as the rate of change of z with respect to x along the curve C_1, and $f_y(x_0, y)$ can be interpreted as the rate of change of z with respect to y along the curve C_2. In particular, $f_x(x_0, y_0)$ is the rate of change of z with respect to x along the curve C_1 at the point (x_0, y_0), and $f_y(x_0, y_0)$ is the rate of change of z with respect to y along the curve C_2 at the point (x_0, y_0).

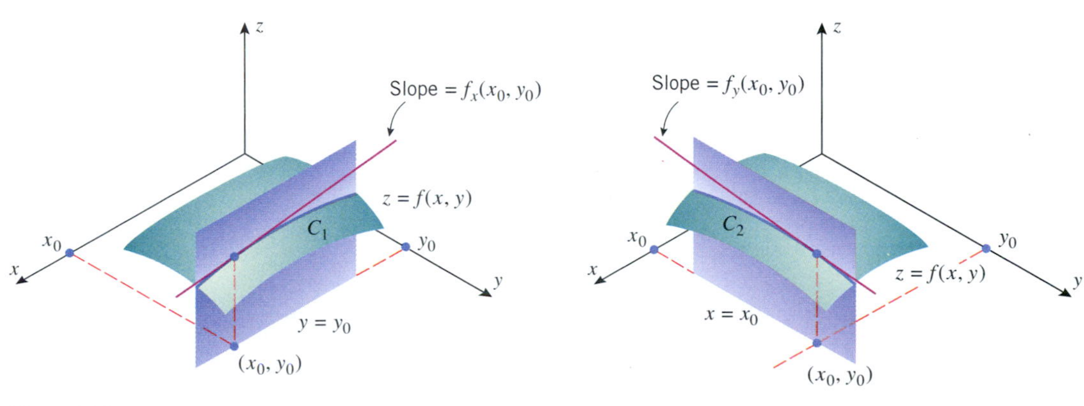

Figure 14.3.1

In an applied problem, the interpretations of $f_x(x_0, y_0)$ and $f_y(x_0, y_0)$ must be accompanied by the proper units. See Example 3.

▶ **Example 3** Recall that the wind chill temperature index is given by the formula

$$W = 35.74 + 0.6215T + (0.4275T - 35.75)v^{0.16}$$

Compute the partial derivative of W with respect to v at the point $(T, v) = (25, 10)$ and interpret this partial derivative as a rate of change.

Solution. Holding T fixed and differentiating with respect to v yields

$$\frac{\partial W}{\partial v}(T, v) = 0 + 0 + (0.4275T - 35.75)(0.16)v^{0.16-1} = (0.4275T - 35.75)(0.16)v^{-0.84}$$

Since W is in degrees Fahrenheit and v is in miles per hour, a rate of change of W with respect to v will have units °F/(mi/h) (which may also be written as °F·h/mi). Substituting $T = 25$ and $v = 10$ gives

$$\frac{\partial W}{\partial v}(25, 10) = (-4.01)10^{-0.84} \approx -0.58\frac{°\text{F}}{\text{mi/h}}$$

as the instantaneous rate of change of W with respect to v at $(T, v) = (25, 10)$. We conclude that if the air temperature is a constant 25°F and the wind speed changes by a small amount from an initial speed of 10 mi/h, then the ratio of the change in the wind chill index to the change in wind speed should be about −0.58°F/(mi/h). ◀

Confirm the conclusion of Example 3 by calculating

$$\frac{W(25, 10 + \Delta v) - W(25, 10)}{\Delta v}$$

for values of Δv near 0.

Geometrically, $f_x(x_0, y_0)$ can be viewed as the slope of the tangent line to the curve C_1 at the point (x_0, y_0), and $f_y(x_0, y_0)$ can be viewed as the slope of the tangent line to the curve C_2 at the point (x_0, y_0) (Figure 14.3.1). We will call $f_x(x_0, y_0)$ the ***slope of the surface in the x-direction*** at (x_0, y_0) and $f_y(x_0, y_0)$ the ***slope of the surface in the y-direction*** at (x_0, y_0).

▶ Example 4 Let $f(x, y) = x^2y + 5y^3$.

(a) Find the slope of the surface $z = f(x, y)$ in the x-direction at the point $(1, -2)$.

(b) Find the slope of the surface $z = f(x, y)$ in the y-direction at the point $(1, -2)$.

Solution (a). Differentiating f with respect to x with y held fixed yields

$$f_x(x, y) = 2xy$$

Thus, the slope in the x-direction is $f_x(1, -2) = -4$; that is, z is decreasing at the rate of 4 units per unit increase in x.

Solution (b). Differentiating f with respect to y with x held fixed yields

$$f_y(x, y) = x^2 + 15y^2$$

Thus, the slope in the y-direction is $f_y(1, -2) = 61$; that is, z is increasing at the rate of 61 units per unit increase in y. ◀

▶ Example 5 Let

$$f(x, y) = \begin{cases} -\dfrac{xy}{x^2 + y^2}, & (x, y) \neq (0, 0) \\ 0, & (x, y) = (0, 0) \end{cases} \tag{3}$$

(a) Show that $f_x(x, y)$ and $f_y(x, y)$ exist at all points (x, y).

(b) Explain why f is not continuous at $(0, 0)$.

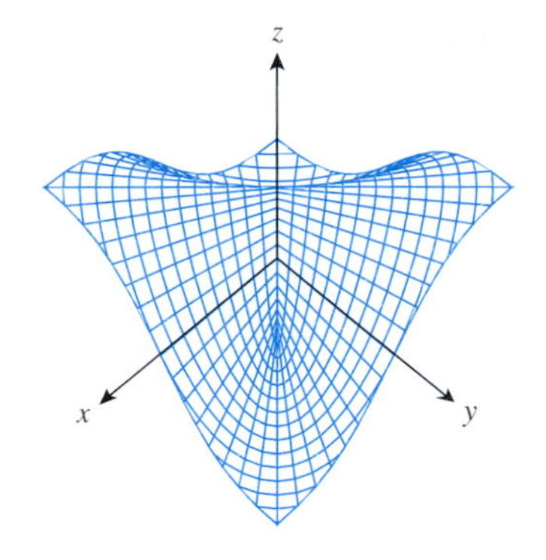

Figure 14.3.2

Solution (a). Figure 14.3.2 shows the graph of f. Note that f is similar to the function considered in Example 1 of Section 14.2, except that here we have assigned f a value of 0 at (0, 0). Except at this point, the partial derivatives of f are

$$f_x(x, y) = -\frac{(x^2 + y^2)y - xy(2x)}{(x^2 + y^2)^2} = \frac{x^2y - y^3}{(x^2 + y^2)^2} \tag{4}$$

$$f_y(x, y) = -\frac{(x^2 + y^2)x - xy(2y)}{(x^2 + y^2)^2} = \frac{xy^2 - x^3}{(x^2 + y^2)^2} \tag{5}$$

It is not evident from Formula (3) whether f has partial derivatives at (0, 0), and if so, what the values of those derivatives are. To answer that question we will have to use the definitions of the partial derivatives (Definition 14.3.1). Applying Formulas (1) and (2) to (3) we obtain

$$f_x(0, 0) = \lim_{\Delta x \to 0} \frac{f(\Delta x, 0) - f(0, 0)}{\Delta x} = \lim_{\Delta x \to 0} \frac{0 - 0}{\Delta x} = 0$$

$$f_y(0, 0) = \lim_{\Delta y \to 0} \frac{f(0, \Delta y) - f(0, 0)}{\Delta y} = \lim_{\Delta y \to 0} \frac{0 - 0}{\Delta y} = 0$$

This shows that f has partial derivatives at (0, 0) and the values of both partial derivatives are 0 at that point.

Solution (b). We saw in Example 3 of Section 14.2 that

$$\lim_{(x,y) \to (0,0)} -\frac{xy}{x^2 + y^2}$$

does not exist. Thus, f is not continuous at (0, 0). ◄

Example 5 shows that, in contrast to the case of functions of a single variable, the existence of partial derivatives for a multivariable function does not guarantee the continuity of the function. We will return to this issue in the next section.

PARTIAL DERIVATIVE NOTATION

The symbol ∂ is called a partial derivative sign. It is derived from the Cyrillic alphabet.

If $z = f(x, y)$, then the partial derivatives f_x and f_y are also denoted by the symbols

$$\frac{\partial f}{\partial x}, \quad \frac{\partial z}{\partial x} \quad \text{and} \quad \frac{\partial f}{\partial y}, \quad \frac{\partial z}{\partial y}$$

Some typical notations for the partial derivatives of $z = f(x, y)$ at a point (x_0, y_0) are

$$\left.\frac{\partial f}{\partial x}\right|_{x=x_0,\, y=y_0}, \quad \left.\frac{\partial z}{\partial x}\right|_{(x_0, y_0)}, \quad \left.\frac{\partial f}{\partial x}\right|_{(x_0, y_0)}, \quad \frac{\partial f}{\partial x}(x_0, y_0), \quad \frac{\partial z}{\partial x}(x_0, y_0)$$

► **Example 6** Find $\partial z/\partial x$ and $\partial z/\partial y$ if $z = x^4 \sin(xy^3)$.

Solution.

$$\begin{aligned}\frac{\partial z}{\partial x} &= \frac{\partial}{\partial x}[x^4 \sin(xy^3)] = x^4 \frac{\partial}{\partial x}[\sin(xy^3)] + \sin(xy^3) \cdot \frac{\partial}{\partial x}(x^4) \\ &= x^4 \cos(xy^3) \cdot y^3 + \sin(xy^3) \cdot 4x^3 = x^4y^3 \cos(xy^3) + 4x^3 \sin(xy^3) \\ \frac{\partial z}{\partial y} &= \frac{\partial}{\partial y}[x^4 \sin(xy^3)] = x^4 \frac{\partial}{\partial y}[\sin(xy^3)] + \sin(xy^3) \cdot \frac{\partial}{\partial y}(x^4) \\ &= x^4 \cos(xy^3) \cdot 3xy^2 + \sin(xy^3) \cdot 0 = 3x^5y^2 \cos(xy^3)\end{aligned}$$ ◄

For functions that are presented in tabular form, we can estimate partial derivatives by using adjacent entries within the table.

Table 14.3.1

TEMPERATURE T (°F)

WIND SPEED v (mi/h)	20	25	30	35
5	13	19	25	31
10	9	15	21	27
15	6	13	19	25
20	4	11	17	24

► **Example 7** Use the values of the wind chill index function $W(T, v)$ displayed in Table 14.3.1 to estimate the partial derivative of W with respect to v at $(T, v) = (25, 10)$. Compare this estimate with the value of the partial derivative obtained in Example 3.

Solution. Since

$$\frac{\partial W}{\partial v}(25, 10) = \lim_{\Delta v \to 0} \frac{W(25, 10 + \Delta v) - W(25, 10)}{\Delta v} = \lim_{\Delta v \to 0} \frac{W(25, 10 + \Delta v) - 15}{\Delta v}$$

we can approximate the partial derivative by

$$\frac{\partial W}{\partial v}(25, 10) \approx \frac{W(25, 10 + \Delta v) - 15}{\Delta v}$$

With $\Delta v = 5$ this approximation is

$$\frac{\partial W}{\partial v}(25, 10) \approx \frac{W(25, 10 + 5) - 15}{5} = \frac{W(25, 15) - 15}{5} = \frac{13 - 15}{5} = -\frac{2}{5} \frac{°\text{F}}{\text{mi/h}}$$

and with $\Delta v = -5$ this approximation is

$$\frac{\partial W}{\partial v}(25, 10) \approx \frac{W(25, 10 - 5) - 15}{-5} = \frac{W(25, 5) - 15}{-5} = \frac{19 - 15}{-5} = -\frac{4}{5} \frac{°\text{F}}{\text{mi/h}}$$

We will take the average, $-\frac{3}{5} = -0.6°\text{F/(mi/h)}$, of these two approximations as our estimate of $(\partial W/\partial v)(25, 10)$. This is close to the value

$$\frac{\partial W}{\partial v}(25, 10) = (-4.01)10^{-0.84} \approx -0.58 \frac{°\text{F}}{\text{mi/h}}$$

found in Example 3. ◄

IMPLICIT PARTIAL DIFFERENTIATION

► **Example 8** Find the slope of the sphere $x^2 + y^2 + z^2 = 1$ in the y-direction at the points $\left(\frac{2}{3}, \frac{1}{3}, \frac{2}{3}\right)$ and $\left(\frac{2}{3}, \frac{1}{3}, -\frac{2}{3}\right)$ (Figure 14.3.3).

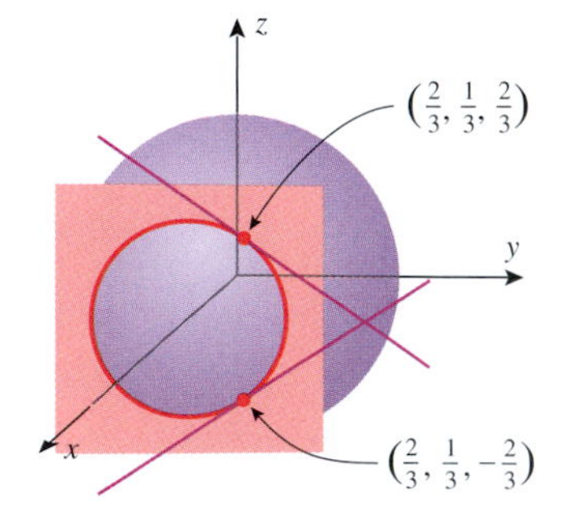

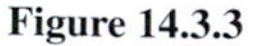

Figure 14.3.3

Solution. The point $\left(\frac{2}{3}, \frac{1}{3}, \frac{2}{3}\right)$ lies on the upper hemisphere $z = \sqrt{1 - x^2 - y^2}$, and the point $\left(\frac{2}{3}, \frac{1}{3}, -\frac{2}{3}\right)$ lies on the lower hemisphere $z = -\sqrt{1 - x^2 - y^2}$. We could find the slopes by differentiating each expression for z separately with respect to y and then evaluating the derivatives at $x = \frac{2}{3}$ and $y = \frac{1}{3}$. However, it is more efficient to differentiate the given equation

$$x^2 + y^2 + z^2 = 1$$

implicitly with respect to y, since this will give us both slopes with one differentiation. To perform the implicit differentiation, we view z as a function of x and y and differentiate both sides with respect to y, taking x to be fixed. The computations are as follows:

$$\frac{\partial}{\partial y}[x^2 + y^2 + z^2] = \frac{\partial}{\partial y}[1]$$

$$0 + 2y + 2z\frac{\partial z}{\partial y} = 0$$

$$\frac{\partial z}{\partial y} = -\frac{y}{z}$$

Check the results in Example 8 by differentiating the functions

$$z = \sqrt{1 - x^2 - y^2}$$

and

$$z = -\sqrt{1 - x^2 - y^2}$$

directly.

Substituting the y- and z-coordinates of the points $\left(\frac{2}{3}, \frac{1}{3}, \frac{2}{3}\right)$ and $\left(\frac{2}{3}, \frac{1}{3}, -\frac{2}{3}\right)$ in this expression, we find that the slope at the point $\left(\frac{2}{3}, \frac{1}{3}, \frac{2}{3}\right)$ is $-\frac{1}{2}$ and the slope at $\left(\frac{2}{3}, \frac{1}{3}, -\frac{2}{3}\right)$ is $\frac{1}{2}$. ◀

▶ **Example 9** Suppose that $D = \sqrt{x^2 + y^2}$ is the length of the diagonal of a rectangle whose sides have lengths x and y that are allowed to vary. Find a formula for the rate of change of D with respect to x if x varies with y held constant, and use this formula to find the rate of change of D with respect to x at the point where $x = 3$ and $y = 4$.

Solution. Differentiating both sides of the equation $D^2 = x^2 + y^2$ with respect to x yields

$$2D\frac{\partial D}{\partial x} = 2x \quad \text{and thus} \quad D\frac{\partial D}{\partial x} = x$$

Since $D = 5$ when $x = 3$ and $y = 4$, it follows that

$$5\left.\frac{\partial D}{\partial x}\right|_{x=3,\,y=4} = 3 \quad \text{or} \quad \left.\frac{\partial D}{\partial x}\right|_{x=3,\,y=4} = \frac{3}{5}$$

Thus, D is increasing at a rate of $\frac{3}{5}$ unit per unit increase in x at the point (3, 4). ◀

PARTIAL DERIVATIVES OF FUNCTIONS WITH MORE THAN TWO VARIABLES

For a function $f(x, y, z)$ of three variables, there are three ***partial derivatives***:

$$f_x(x, y, z), \quad f_y(x, y, z), \quad f_z(x, y, z)$$

The partial derivative f_x is calculated by holding y and z constant and differentiating with respect to x. For f_y the variables x and z are held constant, and for f_z the variables x and y are held constant. If a dependent variable

$$w = f(x, y, z)$$

is used, then the three partial derivatives of f can be denoted by

$$\frac{\partial w}{\partial x}, \quad \frac{\partial w}{\partial y}, \quad \text{and} \quad \frac{\partial w}{\partial z}$$

▶ **Example 10** If $f(x, y, z) = x^3y^2z^4 + 2xy + z$, then

$$\begin{aligned}
f_x(x, y, z) &= 3x^2y^2z^4 + 2y \\
f_y(x, y, z) &= 2x^3yz^4 + 2x \\
f_z(x, y, z) &= 4x^3y^2z^3 + 1 \\
f_z(-1, 1, 2) &= 4(-1)^3(1)^2(2)^3 + 1 = -31
\end{aligned}$$ ◀

▶ **Example 11** If $f(\rho, \theta, \phi) = \rho^2 \cos\phi \sin\theta$, then

$$\begin{aligned}
f_\rho(\rho, \theta, \phi) &= 2\rho \cos\phi \sin\theta \\
f_\theta(\rho, \theta, \phi) &= \rho^2 \cos\phi \cos\theta \\
f_\phi(\rho, \theta, \phi) &= -\rho^2 \sin\phi \sin\theta
\end{aligned}$$ ◀

In general, if $f(v_1, v_2, \ldots, v_n)$ is a function of n variables, there are n partial derivatives of f, each of which is obtained by holding $n - 1$ of the variables fixed and differentiating the function f with respect to the remaining variable. If $w = f(v_1, v_2, \ldots, v_n)$, then these partial derivatives are denoted by

$$\frac{\partial w}{\partial v_1}, \frac{\partial w}{\partial v_2}, \ldots, \frac{\partial w}{\partial v_n}$$

where $\partial w/\partial v_i$ is obtained by holding all variables except v_i fixed and differentiating with respect to v_i.

▶ **Example 12** Find

$$\frac{\partial}{\partial x_i}\left[\sqrt{x_1^2 + x_2^2 + \cdots + x_n^2}\right]$$

for $i = 1, 2, \ldots, n$.

Solution. For each $i = 1, 2, \ldots, n$ we obtain

$$\begin{aligned}
\frac{\partial}{\partial x_i}\left[\sqrt{x_1^2 + x_2^2 + \cdots + x_n^2}\right] &= \frac{1}{2\sqrt{x_1^2 + x_2^2 + \cdots + x_n^2}} \cdot \frac{\partial}{\partial x_i}[x_1^2 + x_2^2 + \cdots + x_n^2] \\
&= \frac{1}{2\sqrt{x_1^2 + x_2^2 + \cdots + x_n^2}}[2x_i] \\
&= \frac{x_i}{\sqrt{x_1^2 + x_2^2 + \cdots + x_n^2}}
\end{aligned}$$

All terms except x_i^2 are constant.

◀

HIGHER-ORDER PARTIAL DERIVATIVES

Suppose that f is a function of two variables x and y. Since the partial derivatives $\partial f/\partial x$ and $\partial f/\partial y$ are also functions of x and y, these functions may themselves have partial derivatives. This gives rise to four possible ***second-order*** partial derivatives of f, which are defined by

$$\frac{\partial^2 f}{\partial x^2} = \frac{\partial}{\partial x}\left(\frac{\partial f}{\partial x}\right) = f_{xx}$$

Differentiate twice with respect to x.

$$\frac{\partial^2 f}{\partial y^2} = \frac{\partial}{\partial y}\left(\frac{\partial f}{\partial y}\right) = f_{yy}$$

Differentiate twice with respect to y.

$$\frac{\partial^2 f}{\partial y \partial x} = \frac{\partial}{\partial y}\left(\frac{\partial f}{\partial x}\right) = f_{xy}$$

Differentiate first with respect to x and then with respect to y.

$$\frac{\partial^2 f}{\partial x \partial y} = \frac{\partial}{\partial x}\left(\frac{\partial f}{\partial y}\right) = f_{yx}$$

Differentiate first with respect to y and then with respect to x.

The last two cases are called the ***mixed second-order partial derivatives*** or the ***mixed second partials***. Also, the derivatives $\partial f/\partial x$ and $\partial f/\partial y$ are often called the ***first-order partial derivatives*** when it is necessary to distinguish them from higher-order partial derivatives. Similar conventions apply to the second-order partial derivatives of a function of three variables.

WARNING

Observe that the two notations for the mixed second partials have opposite conventions for the order of differentiation. In the "∂" notation the derivatives are taken right to left, and in the "subscript" notation they are taken left to right. The conventions are logical if you insert parentheses:

$$\frac{\partial^2 f}{\partial y \partial x} = \frac{\partial}{\partial y}\left(\frac{\partial f}{\partial x}\right)$$

Right to left. Differentiate inside the parentheses first.

$$f_{xy} = (f_x)_y$$

Left to right. Differentiate inside the parentheses first.

▶ **Example 13** Find the second-order partial derivatives of $f(x, y) = x^2y^3 + x^4y$.

Solution. We have

$$\frac{\partial f}{\partial x} = 2xy^3 + 4x^3y \quad \text{and} \quad \frac{\partial f}{\partial y} = 3x^2y^2 + x^4$$

so that

$$\frac{\partial^2 f}{\partial x^2} = \frac{\partial}{\partial x}\left(\frac{\partial f}{\partial x}\right) = \frac{\partial}{\partial x}(2xy^3 + 4x^3y) = 2y^3 + 12x^2y$$

$$\frac{\partial^2 f}{\partial y^2} = \frac{\partial}{\partial y}\left(\frac{\partial f}{\partial y}\right) = \frac{\partial}{\partial y}(3x^2y^2 + x^4) = 6x^2y$$

$$\frac{\partial^2 f}{\partial x \partial y} = \frac{\partial}{\partial x}\left(\frac{\partial f}{\partial y}\right) = \frac{\partial}{\partial x}(3x^2y^2 + x^4) = 6xy^2 + 4x^3$$

$$\frac{\partial^2 f}{\partial y \partial x} = \frac{\partial}{\partial y}\left(\frac{\partial f}{\partial x}\right) = \frac{\partial}{\partial y}(2xy^3 + 4x^3y) = 6xy^2 + 4x^3 \blacktriangleleft$$

Third-order, fourth-order, and higher-order partial derivatives can be obtained by successive differentiation. Some possibilities are

$$\frac{\partial^3 f}{\partial x^3} = \frac{\partial}{\partial x}\left(\frac{\partial^2 f}{\partial x^2}\right) = f_{xxx} \qquad \frac{\partial^4 f}{\partial y^4} = \frac{\partial}{\partial y}\left(\frac{\partial^3 f}{\partial y^3}\right) = f_{yyyy}$$

$$\frac{\partial^3 f}{\partial y^2 \partial x} = \frac{\partial}{\partial y}\left(\frac{\partial^2 f}{\partial y \partial x}\right) = f_{xyy} \qquad \frac{\partial^4 f}{\partial y^2 \partial x^2} = \frac{\partial}{\partial y}\left(\frac{\partial^3 f}{\partial y \partial x^2}\right) = f_{xxyy}$$

▶ **Example 14** Let $f(x, y) = y^2e^x + y$. Find f_{xyy}.

Solution.

$$f_{xyy} = \frac{\partial^3 f}{\partial y^2 \partial x} = \frac{\partial^2}{\partial y^2}\left(\frac{\partial f}{\partial x}\right) = \frac{\partial^2}{\partial y^2}(y^2e^x) = \frac{\partial}{\partial y}(2ye^x) = 2e^x \blacktriangleleft$$

EQUALITY OF MIXED PARTIALS

For a function $f(x, y)$ it might be expected that there would be four distinct second-order partial derivatives: f_{xx}, f_{xy}, f_{yx}, and f_{yy}. However, observe that the mixed second-order partial derivatives in Example 13 are equal. The following theorem (proved in advanced courses) explains why this is so.

If f is a function of three variables, then the analog of Theorem 14.3.2 holds for each pair of mixed second-order partials if we replace "open disk" by "open ball." How many second-order partials does $f(x, y, z)$ have?

14.3.2 THEOREM. *Let f be a function of two variables. If f_{xy} and f_{yx} are continuous on some open disk, then $f_{xy} = f_{yx}$ on that disk.*

It follows from this theorem that if $f_{xy}(x, y)$ and $f_{yx}(x, y)$ are continuous everywhere, then $f_{xy}(x, y) = f_{yx}(x, y)$ for all values of x and y. Since polynomials are continuous everywhere, this explains why the mixed second-order partials in Example 13 are equal.

THE WAVE EQUATION

Consider a string of length L that is stretched taut between $x = 0$ and $x = L$ on an x-axis, and suppose that the string is set into vibratory motion by "plucking" it at time $t = 0$ (Figure

14.3.4*a*). The displacement of a point on the string depends both on its coordinate x and the elapsed time t, and hence is described by a function $u(x, t)$ of two variables. For a fixed value t, the function $u(x, t)$ depends on x alone, and the graph of u versus x describes the shape of the string—think of it as a "snapshot" of the string at time t (Figure 14.3.4*b*). It follows that at a fixed time t, the partial derivative $\partial u/\partial x$ represents the slope of the string at x, and the sign of the second partial derivative $\partial^2 u/\partial x^2$ tells us whether the string is concave up or concave down at x (Figure 14.3.4*c*).

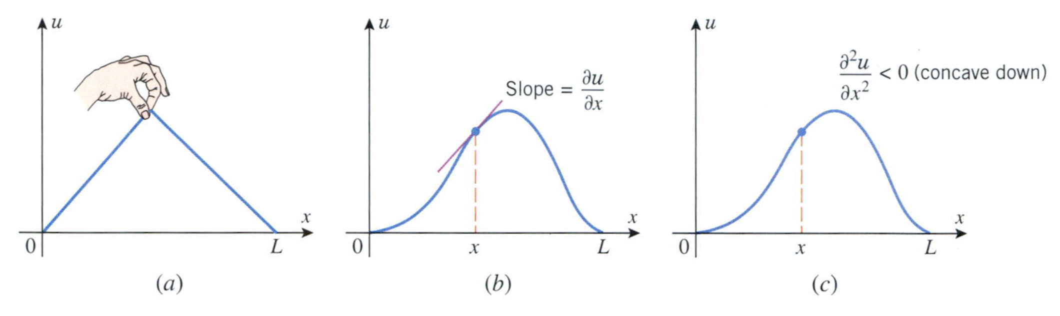

Figure 14.3.4

For a fixed value of x, the function $u(x, t)$ depends on t alone, and the graph of u versus t is the position versus time curve of the point on the string with coordinate x. Thus, for a fixed value of x, the partial derivative $\partial u/\partial t$ is the velocity of the point with coordinate x, and $\partial^2 u/\partial t^2$ is the acceleration of that point.

The vibration of a plucked string is governed by the wave equation.

It can be proved that under appropriate conditions the function $u(x, t)$ satisfies an equation of the form

$$\frac{\partial^2 u}{\partial t^2} = c^2 \frac{\partial^2 u}{\partial x^2} \tag{6}$$

where c is a positive constant that depends on the physical characteristics of the string. This equation, which is called the ***one-dimensional wave equation***, involves partial derivatives of the unknown function $u(x, t)$ and hence is classified as a ***partial differential equation***. Techniques for solving partial differential equations are studied in advanced courses and will not be discussed in this text.

▶ **Example 15** Show that the function $u(x, t) = \sin(x - ct)$ is a solution of Equation (6).

Solution. We have

$$\frac{\partial u}{\partial x} = \cos(x - ct), \qquad \frac{\partial^2 u}{\partial x^2} = -\sin(x - ct)$$

$$\frac{\partial u}{\partial t} = -c\cos(x - ct), \qquad \frac{\partial^2 u}{\partial t^2} = -c^2 \sin(x - ct)$$

Thus, $u(x, t)$ satisfies (6). ◀

✔QUICK CHECK EXERCISES 14.3 *(See page 959 for answers.)*

1. Let $f(x, y) = x \sin xy$. Then $f_x(x, y) =$ ______ and $f_y(x, y) =$ ______.

2. The slope of the surface $z = xy^2$ in the x-direction at the point $(2, 3)$ is ______, and the slope of this surface in the y-direction at the point $(2, 3)$ is ______.

3. The volume V of a right circular cone of radius r and height h is given by $V = \frac{1}{3}\pi r^2 h$.
   (a) Find a formula for the instantaneous rate of change of V with respect to r if r changes and h remains constant.
   (b) Find a formula for the instantaneous rate of change of V with respect to h if h changes and r remains constant.

4. Find all second-order partial derivatives for the function $f(x, y) = x^2y^3$.

EXERCISE SET 14.3 Graphing Utility

1. Let $f(x, y) = 3x^3y^2$. Find
   (a) $f_x(x, y)$ (b) $f_y(x, y)$ (c) $f_x(1, y)$
   (d) $f_x(x, 1)$ (e) $f_y(1, y)$ (f) $f_y(x, 1)$
   (g) $f_x(1, 2)$ (h) $f_y(1, 2)$.

2. Let $z = e^{2x} \sin y$. Find
   (a) $\partial z/\partial x$ (b) $\partial z/\partial y$ (c) $\partial z/\partial x|_{(0,y)}$
   (d) $\partial z/\partial x|_{(x,0)}$ (e) $\partial z/\partial y|_{(0,y)}$ (f) $\partial z/\partial y|_{(x,0)}$
   (g) $\partial z/\partial x|_{(\ln 2,0)}$ (h) $\partial z/\partial y|_{(\ln 2,0)}$.

3. Let $f(x, y) = \sqrt{3x + 2y}$.
   (a) Find the slope of the surface $z = f(x, y)$ in the x-direction at the point (4, 2).
   (b) Find the slope of the surface $z = f(x, y)$ in the y-direction at the point (4, 2).

4. Let $f(x, y) = xe^{-y} + 5y$.
   (a) Find the slope of the surface $z = f(x, y)$ in the x-direction at the point (3, 0).
   (b) Find the slope of the surface $z = f(x, y)$ in the y-direction at the point (3, 0).

5. Let $z = \sin(y^2 - 4x)$.
   (a) Find the rate of change of z with respect to x at the point (2, 1) with y held fixed.
   (b) Find the rate of change of z with respect to y at the point (2, 1) with x held fixed.

6. Let $z = (x + y)^{-1}$.
   (a) Find the rate of change of z with respect to x at the point (−2, 4) with y held fixed.
   (b) Find the rate of change of z with respect to y at the point (−2, 4) with x held fixed.

FOCUS ON CONCEPTS

7. Use the information in the accompanying figure to find the values of the first-order partial derivatives of f at the point (1, 2).

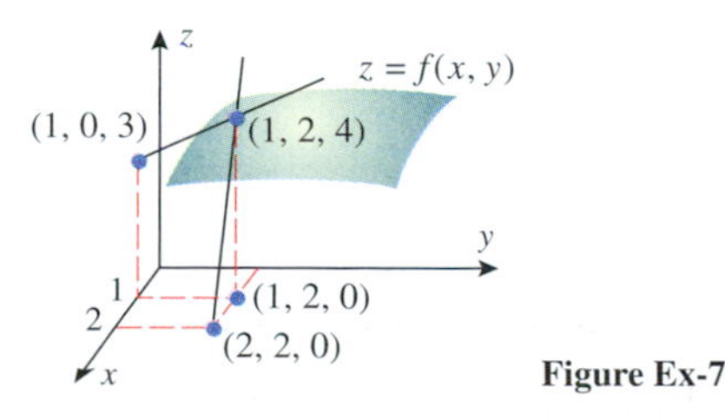

Figure Ex-7

8. The accompanying figure shows a contour plot for an unspecified function $f(x, y)$. Make a conjecture about the signs of the partial derivatives $f_x(x_0, y_0)$ and $f_y(x_0, y_0)$, and explain your reasoning.

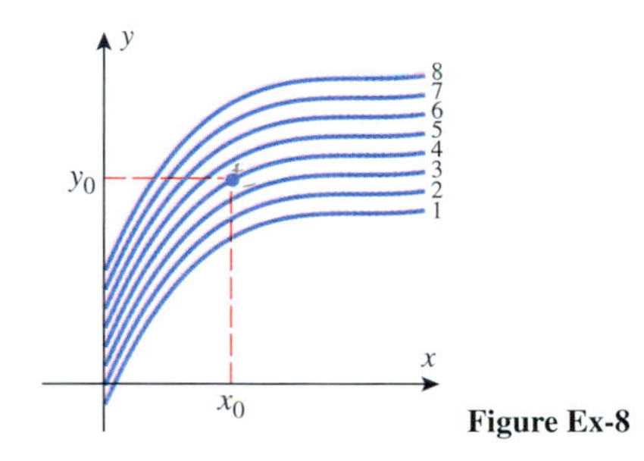

Figure Ex-8

9. Suppose that Nolan throws a baseball to Ryan and that the baseball leaves Nolan's hand at the same height at which it is caught by Ryan. It we ignore air resistance, the horizontal range r of the baseball is a function of the initial speed v of the ball when it leaves Nolan's hand and the angle θ above the horizontal at which it is thrown. Use the accompanying table and the method of Example 7 to estimate
   (a) the partial derivative of r with respect to v when $v = 80$ ft/s and $\theta = 40°$
   (b) the partial derivative of r with respect to θ when $v = 80$ ft/s and $\theta = 40°$.

ANGLE θ (degrees) \ SPEED v (ft/s)	75	80	85	90
35	165	188	212	238
40	173	197	222	249
45	176	200	226	253
50	173	197	222	249

Table Ex-9

10. Use the table in Exercise 9 and the method of Example 7 to estimate
   (a) the partial derivative of r with respect to v when $v = 85$ ft/s and $\theta = 45°$
   (b) the partial derivative of r with respect to θ when $v = 85$ ft/s and $\theta = 45°$.

11. The accompanying figure (next page) shows the graphs of an unspecified function $f(x, y)$ and its partial derivatives $f_x(x, y)$ and $f_y(x, y)$. Determine which is which, and explain your reasoning.

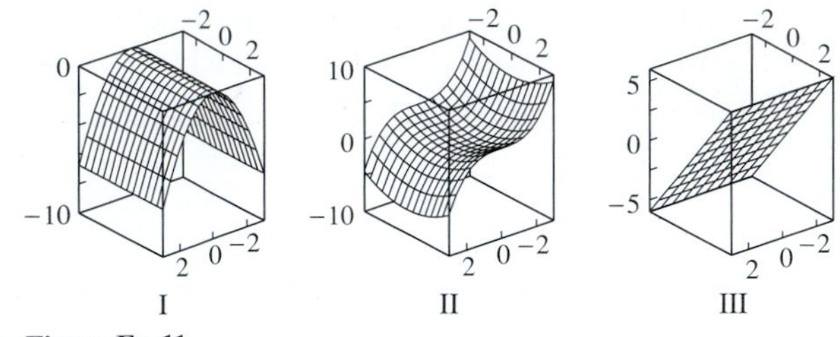

Figure Ex-11

12. What can you say about the signs of $\partial z/\partial x$, $\partial^2 z/\partial x^2$, $\partial z/\partial y$, and $\partial^2 z/\partial y^2$ at the point P in the accompanying figure? Explain your reasoning.

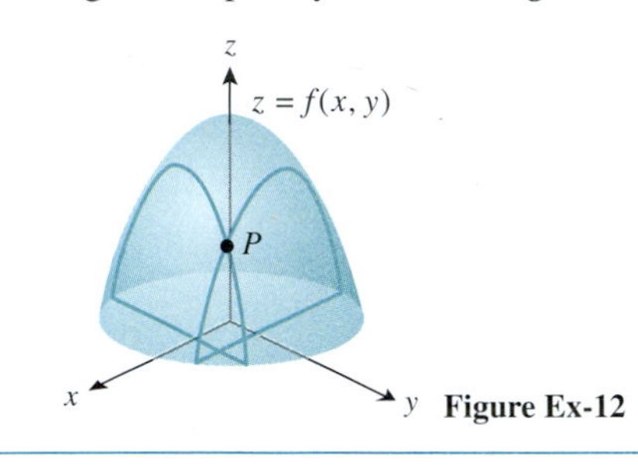

Figure Ex-12

13–18 Find $\partial z/\partial x$ and $\partial z/\partial y$.

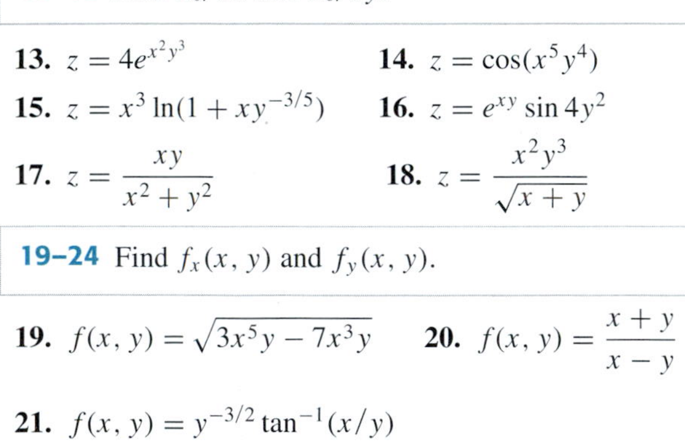

13. $z = 4e^{x^2y^3}$ **14.** $z = \cos(x^5y^4)$

15. $z = x^3 \ln(1 + xy^{-3/5})$ **16.** $z = e^{xy} \sin 4y^2$

17. $z = \dfrac{xy}{x^2 + y^2}$ **18.** $z = \dfrac{x^2y^3}{\sqrt{x + y}}$

19–24 Find $f_x(x, y)$ and $f_y(x, y)$.

19. $f(x, y) = \sqrt{3x^5y - 7x^3y}$ **20.** $f(x, y) = \dfrac{x + y}{x - y}$

21. $f(x, y) = y^{-3/2} \tan^{-1}(x/y)$

22. $f(x, y) = x^3e^{-y} + y^3 \sec \sqrt{x}$

23. $f(x, y) = (y^2 \tan x)^{-4/3}$

24. $f(x, y) = \cosh(\sqrt{x}) \sinh^2(xy^2)$

25–28 Evaluate the indicated partial derivatives.

25. $f(x, y) = 9 - x^2 - 7y^3$; $f_x(3, 1)$, $f_y(3, 1)$

26. $f(x, y) = x^2ye^{xy}$; $\partial f/\partial x(1, 1)$, $\partial f/\partial y(1, 1)$

27. $z = \sqrt{x^2 + 4y^2}$; $\partial z/\partial x(1, 2)$, $\partial z/\partial y(1, 2)$

28. $w = x^2 \cos xy$; $\partial w/\partial x\left(\frac{1}{2}, \pi\right)$, $\partial w/\partial y\left(\frac{1}{2}, \pi\right)$

29. Let $f(x, y, z) = x^2y^4z^3 + xy + z^2 + 1$. Find
(a) $f_x(x, y, z)$ (b) $f_y(x, y, z)$ (c) $f_z(x, y, z)$
(d) $f_x(1, y, z)$ (e) $f_y(1, 2, z)$ (f) $f_z(1, 2, 3)$.

30. Let $w = x^2y \cos z$. Find
(a) $\partial w/\partial x(x, y, z)$ (b) $\partial w/\partial y(x, y, z)$
(c) $\partial w/\partial z(x, y, z)$ (d) $\partial w/\partial x(2, y, z)$
(e) $\partial w/\partial y(2, 1, z)$ (f) $\partial w/\partial z(2, 1, 0)$.

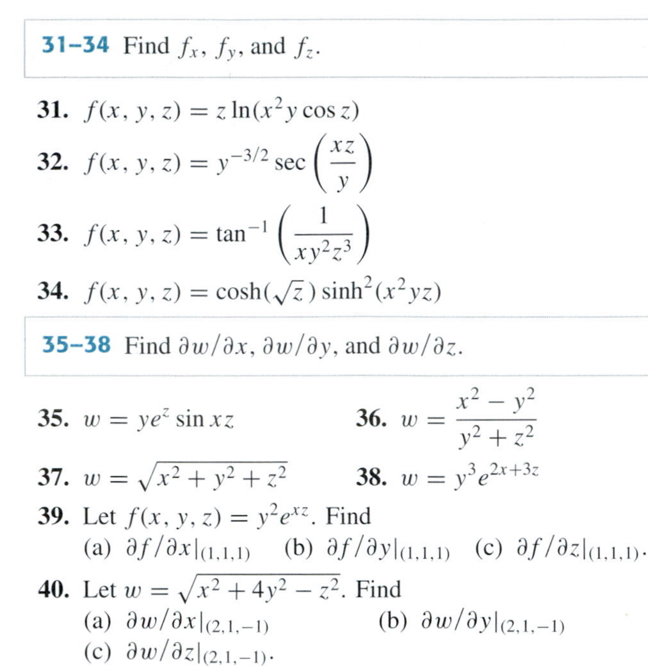

31–34 Find f_x, f_y, and f_z.

31. $f(x, y, z) = z \ln(x^2y \cos z)$

32. $f(x, y, z) = y^{-3/2} \sec\left(\dfrac{xz}{y}\right)$

33. $f(x, y, z) = \tan^{-1}\left(\dfrac{1}{xy^2z^3}\right)$

34. $f(x, y, z) = \cosh(\sqrt{z}) \sinh^2(x^2yz)$

35–38 Find $\partial w/\partial x$, $\partial w/\partial y$, and $\partial w/\partial z$.

35. $w = ye^z \sin xz$ **36.** $w = \dfrac{x^2 - y^2}{y^2 + z^2}$

37. $w = \sqrt{x^2 + y^2 + z^2}$ **38.** $w = y^3e^{2x+3z}$

39. Let $f(x, y, z) = y^2e^{xz}$. Find
(a) $\partial f/\partial x|_{(1,1,1)}$ (b) $\partial f/\partial y|_{(1,1,1)}$ (c) $\partial f/\partial z|_{(1,1,1)}$.

40. Let $w = \sqrt{x^2 + 4y^2 - z^2}$. Find
(a) $\partial w/\partial x|_{(2,1,-1)}$ (b) $\partial w/\partial y|_{(2,1,-1)}$
(c) $\partial w/\partial z|_{(2,1,-1)}$.

41. Let $f(x, y) = e^x \cos y$. Use a graphing utility to graph the functions $f_x(0, y)$ and $f_y(x, \pi/2)$.

42. Let $f(x, y) = e^x \sin y$. Use a graphing utility to graph the functions $f_x(0, y)$ and $f_y(x, 0)$.

43. A point moves along the intersection of the elliptic paraboloid $z = x^2 + 3y^2$ and the plane $y = 1$. At what rate is z changing with respect to x when the point is at $(2, 1, 7)$?

44. A point moves along the intersection of the elliptic paraboloid $z = x^2 + 3y^2$ and the plane $x = 2$. At what rate is z changing with respect to y when the point is at $(2, 1, 7)$?

45. A point moves along the intersection of the plane $y = 3$ and the surface $z = \sqrt{29 - x^2 - y^2}$. At what rate is z changing with respect to x when the point is at $(4, 3, 2)$?

46. Find the slope of the tangent line at $(-1, 1, 5)$ to the curve of intersection of the surface $z = x^2 + 4y^2$ and
(a) the plane $x = -1$ (b) the plane $y = 1$.

47. The volume V of a right circular cylinder is given by the formula $V = \pi r^2h$, where r is the radius and h is the height.
(a) Find a formula for the instantaneous rate of change of V with respect to r if r changes and h remains constant.
(b) Find a formula for the instantaneous rate of change of V with respect to h if h changes and r remains constant.
(c) Suppose that h has aconstant value of 4 in, but r varies. Find the rate of change of V with respect to r at the point where $r = 6$ in.
(d) Suppose that r has a constant value of 8 in, but h varies. Find the instantaneous rate of change of V with respect to h at the point where $h = 10$ in.

48. The volume V of a right circular cone is given by

$$V = \frac{\pi}{24}d^2\sqrt{4s^2 - d^2}$$

where s is the slant height and d is the diameter of the base.

(a) Find a formula for the instantaneous rate of change of V with respect to s if d remains constant.
(b) Find a formula for the instantaneous rate of change of V with respect to d if s remains constant.
(c) Suppose that d has a constant value of 16 cm, but s varies. Find the rate of change of V with respect to s when $s = 10$ cm.
(d) Suppose that s has a constant value of 10 cm, but d varies. Find the rate of change of V with respect to d when $d = 16$ cm.

49. According to the ideal gas law, the pressure, temperature, and volume of a gas are related by $P = kT/V$, where k is a constant of proportionality. Suppose that V is measured in cubic inches (in^3), T is measured in kelvins (K), and that for a certain gas the constant of proportionality is $k = 10$ in·lb/K.
(a) Find the instantaneous rate of change of pressure with respect to temperature if the temperature is 80 K and the volume remains fixed at 50 in^3.
(b) Find the instantaneous rate of change of volume with respect to pressure if the volume is 50 in^3 and the temperature remains fixed at 80 K.

50. The temperature at a point (x, y) on a metal plate in the xy-plane is $T(x, y) = x^3 + 2y^2 + x$ degrees Celsius. Assume that distance is measured in centimeters and find the rate at which temperature changes with respect to distance if we start at the point $(1, 2)$ and move
(a) to the right and parallel to the x-axis
(b) upward and parallel to the y-axis.

51. The length, width, and height of a rectangular box are $l = 5$, $w = 2$, and $h = 3$, respectively.
(a) Find the instantaneous rate of change of the volume of the box with respect to the length if w and h are held constant.
(b) Find the instantaneous rate of change of the volume of the box with respect to the width if l and h are held constant.
(c) Find the instantaneous rate of change of the volume of the box with respect to the height if l and w are held constant.

52. The area A of a triangle is given by $A = \frac{1}{2}ab\sin\theta$, where a and b are the lengths of two sides and θ is the angle between these sides. Suppose that $a = 5$, $b = 10$, and $\theta = \pi/3$.
(a) Find the rate at which A changes with respect to a if b and θ are held constant.
(b) Find the rate at which A changes with respect to θ if a and b are held constant.
(c) Find the rate at which b changes with respect to a if A and θ are held constant.

53. The volume of a right circular cone of radius r and height h is $V = \frac{1}{3}\pi r^2 h$. Show that if the height remains constant while the radius changes, then the volume satisfies

$$\frac{\partial V}{\partial r} = \frac{2V}{r}$$

54. Find parametric equations for the tangent line at $(1, 3, 3)$ to the curve of intersection of the surface $z = x^2 y$ and
(a) the plane $x = 1$ (b) the plane $y = 3$.

55. (a) By differentiating implicitly, find the slope of the hyperboloid $x^2 + y^2 - z^2 = 1$ in the x-direction at the points $(3, 4, 2\sqrt{6})$ and $(3, 4, -2\sqrt{6})$.
(b) Check the results in part (a) by solving for z and differentiating the resulting functions directly.

56. (a) By differentiating implicitly, find the slope of the hyperboloid $x^2 + y^2 - z^2 = 1$ in the y-direction at the points $(3, 4, 2\sqrt{6})$ and $(3, 4, -2\sqrt{6})$.
(b) Check the results in part (a) by solving for z and differentiating the resulting functions directly.

57–60 Calculate $\partial z/\partial x$ and $\partial z/\partial y$ using implicit differentiation. Leave your answers in terms of x, y, and z.

57. $(x^2 + y^2 + z^2)^{3/2} = 1$ **58.** $\ln(2x^2 + y - z^3) = x$

59. $x^2 + z\sin xyz = 0$ **60.** $e^{xy}\sinh z - z^2 x + 1 = 0$

61–64 Find $\partial w/\partial x$, $\partial w/\partial y$, and $\partial w/\partial z$ using implicit differentiation. Leave your answers in terms of x, y, z, and w.

61. $(x^2 + y^2 + z^2 + w^2)^{3/2} = 4$

62. $\ln(2x^2 + y - z^3 + 3w) = z$

63. $w^2 + w\sin xyz = 1$

64. $e^{xy}\sinh w - z^2 w + 1 = 0$

65–66 Find f_x and f_y.

65. $f(x, y) = \int_y^x e^{t^2}\,dt$ **66.** $f(x, y) = \int_1^{xy} e^{t^2}\,dt$

67. Let $z = \sqrt{x}\cos y$. Find
(a) $\partial^2 z/\partial x^2$ (b) $\partial^2 z/\partial y^2$
(c) $\partial^2 z/\partial x\partial y$ (d) $\partial^2 z/\partial y\partial x$.

68. Let $f(x, y) = 4x^2 - 2y + 7x^4 y^5$. Find
(a) f_{xx} (b) f_{yy} (c) f_{xy} (d) f_{yx}.

69–76 Confirm that the mixed second-order partial derivatives of f are the same.

69. $f(x, y) = 4x^2 - 8xy^4 + 7y^5 - 3$

70. $f(x, y) = \sqrt{x^2 + y^2}$ **71.** $f(x, y) = e^x\cos y$

72. $f(x, y) = e^{x-y^2}$ **73.** $f(x, y) = \ln(4x - 5y)$

74. $f(x, y) = \ln(x^2 + y^2)$

75. $f(x, y) = (x - y)/(x + y)$

76. $f(x, y) = (x^2 - y^2)/(x^2 + y^2)$

77. Express the following derivatives in "∂" notation.
(a) f_{xxx} (b) f_{xyy} (c) f_{yyxx} (d) f_{xyyy}

78. Express the derivatives in "subscript" notation.
(a) $\dfrac{\partial^3 f}{\partial y^2\partial x}$ (b) $\dfrac{\partial^4 f}{\partial x^4}$ (c) $\dfrac{\partial^4 f}{\partial y^2\partial x^2}$ (d) $\dfrac{\partial^5 f}{\partial x^2\partial y^3}$

79. Given $f(x, y) = x^3y^5 - 2x^2y + x$, find
(a) f_{xxy} (b) f_{yxy} (c) f_{yyy}.

80. Given $z = (2x - y)^5$, find
(a) $\dfrac{\partial^3 z}{\partial y \partial x \partial y}$ (b) $\dfrac{\partial^3 z}{\partial x^2 \partial y}$ (c) $\dfrac{\partial^4 z}{\partial x^2 \partial y^2}$.

81. Given $f(x, y) = y^3e^{-5x}$, find
(a) $f_{xyy}(0, 1)$ (b) $f_{xxx}(0, 1)$ (c) $f_{yyxx}(0, 1)$.

82. Given $w = e^y \cos x$, find
(a) $\left.\dfrac{\partial^3 w}{\partial y^2 \partial x}\right|_{(\pi/4,0)}$ (b) $\left.\dfrac{\partial^3 w}{\partial x^2 \partial y}\right|_{(\pi/4,0)}$

83. Let $f(x, y, z) = x^3y^5z^7 + xy^2 + y^3z$. Find
(a) f_{xy} (b) f_{yz} (c) f_{xz} (d) f_{zz}
(e) f_{zyy} (f) f_{xxy} (g) f_{zyx} (h) f_{xxyz}.

84. Let $w = (4x - 3y + 2z)^5$. Find
(a) $\dfrac{\partial^2 w}{\partial x \partial z}$ (b) $\dfrac{\partial^3 w}{\partial x \partial y \partial z}$ (c) $\dfrac{\partial^4 w}{\partial z^2 \partial y \partial x}$.

85. Show that the function satisfies ***Laplace's equation***

$$\frac{\partial^2 z}{\partial x^2} + \frac{\partial^2 z}{\partial y^2} = 0$$

(a) $z = x^2 - y^2 + 2xy$
(b) $z = e^x \sin y + e^y \cos x$
(c) $z = \ln(x^2 + y^2) + 2\tan^{-1}(y/x)$

86. Show that the function satisfies the ***heat equation***

$$\frac{\partial z}{\partial t} = c^2 \frac{\partial^2 z}{\partial x^2} \quad (c > 0,\ \text{constant})$$

(a) $z = e^{-t} \sin(x/c)$ (b) $z = e^{-t} \cos(x/c)$

87. Show that the function $u(x, t) = \sin c\omega t \sin \omega x$ satisfies the wave equation [Equation (6)] for all real values of ω.

88. In each part, show that $u(x, y)$ and $v(x, y)$ satisfy the ***Cauchy–Riemann equations***

$$\frac{\partial u}{\partial x} = \frac{\partial v}{\partial y} \quad \text{and} \quad \frac{\partial u}{\partial y} = -\frac{\partial v}{\partial x}$$

(a) $u = x^2 - y^2$, $v = 2xy$
(b) $u = e^x \cos y$, $v = e^x \sin y$
(c) $u = \ln(x^2 + y^2)$, $v = 2\tan^{-1}(y/x)$

89. Show that if $u(x, y)$ and $v(x, y)$ each have equal mixed second partials, and if u and v satisfy the Cauchy–Riemann equations (Exercise 88), then u, v, and $u + v$ satisfy Laplace's equation (Exercise 85).

90. When two resistors having resistances R_1 ohms and R_2 ohms are connected in parallel, their combined resistance R in ohms is $R = R_1R_2/(R_1 + R_2)$. Show that

$$\frac{\partial^2 R}{\partial R_1^2}\frac{\partial^2 R}{\partial R_2^2} = \frac{4R^2}{(R_1 + R_2)^4}$$

91–94 Find the indicated partial derivatives.

91. $f(v, w, x, y) = 4v^2w^3x^4y^5$;
$\partial f/\partial v$, $\partial f/\partial w$, $\partial f/\partial x$, $\partial f/\partial y$

92. $w = r \cos st + e^u \sin ur$;
$\partial w/\partial r$, $\partial w/\partial s$, $\partial w/\partial t$, $\partial w/\partial u$

93. $f(v_1, v_2, v_3, v_4) = \dfrac{v_1^2 - v_2^2}{v_3^2 + v_4^2}$;
$\partial f/\partial v_1$, $\partial f/\partial v_2$, $\partial f/\partial v_3$, $\partial f/\partial v_4$

94. $V = xe^{2x-y} + we^{zw} + yw$;
$\partial V/\partial x$, $\partial V/\partial y$, $\partial V/\partial z$, $\partial V/\partial w$

95. Let $u(w, x, y, z) = xe^{yw} \sin^2 z$. Find
(a) $\dfrac{\partial u}{\partial x}(0, 0, 1, \pi)$ (b) $\dfrac{\partial u}{\partial y}(0, 0, 1, \pi)$
(c) $\dfrac{\partial u}{\partial w}(0, 0, 1, \pi)$ (d) $\dfrac{\partial u}{\partial z}(0, 0, 1, \pi)$
(e) $\dfrac{\partial^4 u}{\partial x \partial y \partial w \partial z}$ (f) $\dfrac{\partial^4 u}{\partial w \partial z \partial y^2}$.

96. Let $f(v, w, x, y) = 2v^{1/2}w^4x^{1/2}y^{2/3}$. Find $f_v(1, -2, 4, 8)$, $f_w(1, -2, 4, 8)$, $f_x(1, -2, 4, 8)$, and $f_y(1, -2, 4, 8)$.

97–98 Find $\partial w/\partial x_i$ for $i = 1, 2, \ldots, n$.

97. $w = \cos(x_1 + 2x_2 + \cdots + nx_n)$

98. $w = \left(\displaystyle\sum_{k=1}^{n} x_k\right)^{1/n}$

99–100 Describe the largest set on which Theorem 14.3.2 may be used to prove that f_{xy} and f_{yx} are equal on that set. Then confirm by direct computation that $f_{xy} = f_{yx}$ on the given set.

99. (a) $f(x, y) = 4x^3y + 3x^2y$ (b) $f(x, y) = x^3/y$

100. (a) $f(x, y) = \sqrt{x^2 + y^2 - 1}$
(b) $f(x, y) = \sin(x^2 + y^3)$

101. Let $f(x, y) = 2x^2 - 3xy + y^2$. Find $f_x(2, -1)$ and $f_y(2, -1)$ by evaluating the limits in Definition 14.3.1. Then check your work by calculating the derivative in the usual way.

102. Let $f(x, y) = (x^2 + y^2)^{2/3}$. Show that

$$f_x(x, y) = \begin{cases} \dfrac{4x}{3(x^2 + y^2)^{1/3}}, & (x, y) \neq (0, 0) \\ 0, & (x, y) = (0, 0) \end{cases}$$

Source: This problem, due to Don Cohen, appeared in *Mathematics and Computer Education*, Vol. 25, No. 2, 1991, p. 179.

103. Let $f(x, y) = (x^3 + y^3)^{1/3}$.
(a) Show that $f_y(0, 0) = 1$.
(b) At what points, if any, does $f_y(x, y)$ fail to exist?

✔QUICK CHECK ANSWERS 14.3

1. $\sin xy + xy\cos xy$; $x^2\cos xy$ **2.** 9; 12 **3.** (a) $\frac{2}{3}\pi rh$ (b) $\frac{1}{3}\pi r^2 s$
4. $f_{xx}(x, y) = 2y^3$, $f_{yy}(x, y) = 6x^2 y$, $f_{xy}(x, y) = f_{yx}(x, y) = 6xy^2$

14.4 DIFFERENTIABILITY, DIFFERENTIALS, AND LOCAL LINEARITY

In this section we will extend the notion of differentiability to functions of two or three variables. Our definition of differentiability will be based on the idea that a function is differentiable at a point provided it can be very closely approximated by a linear function near that point. In the process, we will expand the concept of a "differential" to functions of more than one variable and define the "local linear approximation" of a function.

DIFFERENTIABILITY

Recall that a function f of one variable is called differentiable at x_0 if it has a derivative at x_0, that is, if the limit

$$f'(x_0) = \lim_{\Delta x \to 0} \frac{f(x_0 + \Delta x) - f(x_0)}{\Delta x} \tag{1}$$

exists. As a consequence of (1) a differentiable function enjoys a number of other important properties:

- The graph of $y = f(x)$ has a nonvertical tangent line at the point $(x_0, f(x_0))$;
- f may be closely approximated by a linear function near x_0 (Section 3.8);
- f is continuous at x_0.

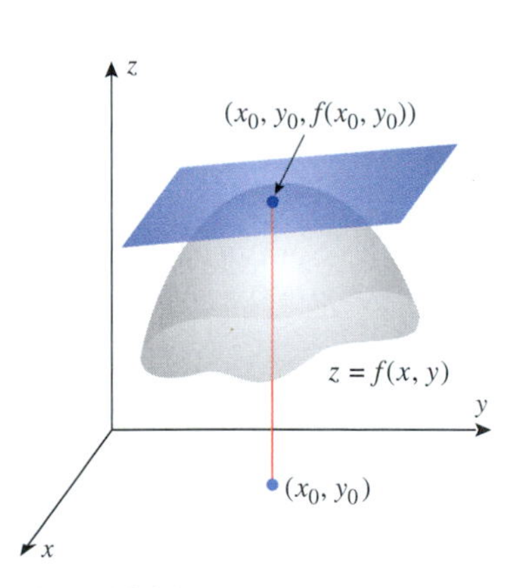

Figure 14.4.1

Our primary objective in this section is to extend the notion of differentiability to functions of two or three variables in such a way that the natural analogs of these properties hold. For example, if a function $f(x, y)$ of two variables is differentiable at a point (x_0, y_0), we want it to be the case that

- the surface $z = f(x, y)$ has a nonvertical tangent plane at the point $(x_0, y_0, f(x_0, y_0))$ (Figure 14.4.1);
- the values of f at points near (x_0, y_0) can be very closely approximated by the values of a linear function;
- f is continuous at (x_0, y_0).

One could reasonably conjecture that a function f of two or three variables should be called differentiable at a point if all the first-order partial derivatives of the function exist at that point. Unfortunately, this condition is not strong enough to guarantee that the properties above hold. For instance, we saw in Example 5 of Section 14.3 that the mere existence of both first-order partial derivatives for a function is not sufficient to guarantee the continuity of the function. To determine what else we should include in our definition, it will be helpful to reexamine one of the consequences of differentiability for a *single-variable* function $f(x)$. Suppose that $f(x)$ is differentiable at $x = x_0$ and let

$$\Delta f = f(x_0 + \Delta x) - f(x_0)$$

denote the change in f that corresponds to the change Δx in x from x_0 to $x_0 + \Delta x$. We saw in Section 3.8 that

$$\Delta f \approx f'(x_0)\Delta x$$

provide Δx is close to 0. In fact, for Δx close to 0 the error $\Delta f - f'(x_0)\Delta x$ in this approximation will have magnitude much smaller than that of Δx because

$$\lim_{\Delta x \to 0} \frac{\Delta f - f'(x_0)\Delta x}{\Delta x} = \lim_{\Delta x \to 0} \left(\frac{f(x_0 + \Delta x) - f(x_0)}{\Delta x} - f'(x_0) \right) = f'(x_0) - f'(x_0) = 0$$

Since the magnitude of Δx is just the distance between the points x_0 and $x_0 + \Delta x$, we see that when the two points are close together, the magnitude of the error in the approximation will be much smaller than the distance between the two points (Figure 14.4.2). The extension of this idea to functions of two or three variables is the "extra ingredient" needed in our definition of differentiability for multivariable functions.

For a function $f(x, y)$, the symbol Δf, called the ***increment*** of f, denotes the change in the value of $f(x, y)$ that results when (x, y) varies from some initial position (x_0, y_0) to some new position $(x_0 + \Delta x, y_0 + \Delta y)$; thus

$$\Delta f = f(x_0 + \Delta x, y_0 + \Delta y) - f(x_0, y_0) \tag{2}$$

(see Figure 14.4.3). [If a dependent variable $z = f(x, y)$ is used, then we will sometimes write Δz rather than Δf.] Let us assume that both $f_x(x_0, y_0)$ and $f_y(x_0, y_0)$ exist and (by analogy with the one-variable case) make the approximation

$$\Delta f \approx f_x(x_0, y_0)\Delta x + f_y(x_0, y_0)\Delta y \tag{3}$$

Show that if $f(x, y)$ is a linear function, then (3) becomes an equality.

For Δx and Δy close to 0, we would like the error

$$\Delta f - f_x(x_0, y_0)\Delta x - f_y(x_0, y_0)\Delta y$$

in this approximation to be much smaller than the distance $\sqrt{(\Delta x)^2 + (\Delta y)^2}$ between (x_0, y_0) and $(x_0 + \Delta x, y_0 + \Delta y)$. We can guarantee this by requiring that

$$\lim_{(\Delta x, \Delta y) \to (0,0)} \frac{\Delta f - f_x(x_0, y_0)\Delta x - f_y(x_0, y_0)\Delta y}{\sqrt{(\Delta x)^2 + (\Delta y)^2}} = 0$$

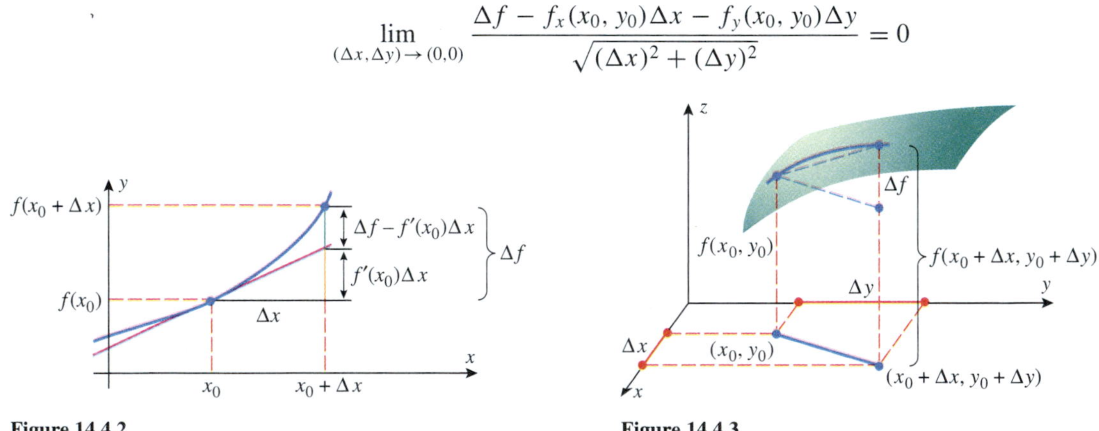

Figure 14.4.2 **Figure 14.4.3**

Based on these ideas, we can now give our definition of differentiability for functions of two variables.

14.4.1 DEFINITION. A function f of two variables is said to be ***differentiable*** at (x_0, y_0) provided $f_x(x_0, y_0)$ and $f_y(x_0, y_0)$ both exist and

$$\lim_{(\Delta x, \Delta y) \to (0,0)} \frac{\Delta f - f_x(x_0, y_0)\Delta x - f_y(x_0, y_0)\Delta y}{\sqrt{(\Delta x)^2 + (\Delta y)^2}} = 0 \tag{4}$$

As with the one-variable case, verification of differentiability using this definition involves the computation of a limit.

Example 1 Use Definition 14.4.1 to prove that $f(x, y) = x^2 + y^2$ is differentiable at $(0, 0)$.

Solution. The increment is

$$\Delta f = f(0 + \Delta x, 0 + \Delta y) - f(0, 0) = (\Delta x)^2 + (\Delta y)^2$$

Since $f_x(x, y) = 2x$ and $f_y(x, y) = 2y$, we have $f_x(0, 0) = f_y(0, 0) = 0$, and (4) becomes

$$\lim_{(\Delta x, \Delta y) \to (0,0)} \frac{(\Delta x)^2 + (\Delta y)^2}{\sqrt{(\Delta x)^2 + (\Delta y)^2}} = \lim_{(\Delta x, \Delta y) \to (0,0)} \sqrt{(\Delta x)^2 + (\Delta y)^2} = 0$$

Therefore, f is differentiable at $(0, 0)$. ◀

We now derive an important consequence of limit (4). Define a function

$$\epsilon = \epsilon(\Delta x, \Delta y) = \frac{\Delta f - f_x(x_0, y_0)\Delta x - f_y(x_0, y_0)\Delta y}{\sqrt{(\Delta x)^2 + (\Delta y)^2}} \quad \text{for } (\Delta x, \Delta y) \neq (0, 0)$$

and define $\epsilon(0, 0)$ to be 0. Equation (4) then implies that

$$\lim_{(\Delta x, \Delta y) \to (0,0)} \epsilon(\Delta x, \Delta y) = 0$$

Furthermore, it immediately follows from the definition of ϵ that

$$\Delta f = f_x(x_0, y_0)\Delta x + f_y(x_0, y_0)\Delta y + \epsilon\sqrt{(\Delta x)^2 + (\Delta y)^2} \tag{5}$$

In other words, if f is differentiable at (x_0, y_0), then Δf may be expressed as shown in (5), where $\epsilon \to 0$ as $(\Delta x, \Delta y) \to 0$ and where $\epsilon = 0$ if $(\Delta x, \Delta y) = (0, 0)$.

For functions of three variables we have an analogous definition of differentiability in terms of the increment

$$\Delta f = f(x_0 + \Delta x, y_0 + \Delta y, z_0 + \Delta z) - f(x_0, y_0, z_0)$$

14.4.2 DEFINITION. A function f of three variables is said to be ***differentiable*** at (x_0, y_0, z_0) provided $f_x(x_0, y_0, z_0)$, $f_y(x_0, y_0, z_0)$, and $f_z(x_0, y_0, z_0)$ exist and

$$\lim_{(\Delta x, \Delta y, \Delta z) \to (0,0,0)} \frac{\Delta f - f_x(x_0, y_0, z_0)\Delta x - f_y(x_0, y_0, z_0)\Delta y - f_z(x_0, y_0, z_0)\Delta z}{\sqrt{(\Delta x)^2 + (\Delta y)^2 + (\Delta z)^2}} = 0 \tag{6}$$

In a manner similar to the two-variable case, we can express the limit (6) in terms of a function $\epsilon(\Delta x, \Delta y, \Delta z)$ that vanishes at $(\Delta x, \Delta y, \Delta z) = (0, 0, 0)$ and is continuous there. The details are left as an exercise for the reader.

If a function f of two variables is differentiable at each point of a region R in the xy-plane, then we say that f is ***differentiable on R***; and if f is differentiable at every point in the xy-plane, then we say that f is ***differentiable everywhere***. For a function f of three variables we have corresponding conventions.

DIFFERENTIABILITY AND CONTINUITY

Recall that we want a function to be continuous at every point at which it is differentiable. The next result shows this to be the case.

14.4.3 THEOREM. *If a function is differentiable at a point, then it is continuous at that point.*

PROOF. We will give the proof for $f(x, y)$, a function of two variables, since that will reveal the essential ideas. Assume that f is differentiable at (x_0, y_0). To prove that f is continuous at (x_0, y_0) we must show that

$$\lim_{(x,y)\to(x_0,y_0)} f(x, y) = f(x_0, y_0)$$

which, on letting $x = x_0 + \Delta x$ and $y = y_0 + \Delta y$, is equivalent to

$$\lim_{(\Delta x,\Delta y)\to(0,0)} f(x_0 + \Delta x, y_0 + \Delta y) = f(x_0, y_0)$$

By Equation (2) this is equivalent to

$$\lim_{(\Delta x,\Delta y)\to(0,0)} \Delta f = 0$$

The converse of Theorem 14.4.3 is false. For example, explain why

$$f(x, y) = \sqrt{x^2 + y^2}$$

is continuous at $(0, 0)$ but not differentiable at $(0, 0)$.

However, from Equation (5)

$$\begin{aligned}\lim_{(\Delta x,\Delta y)\to(0,0)} \Delta f &= \lim_{(\Delta x,\Delta y)\to(0,0)} [f_x(x_0, y_0)\Delta x + f_y(x_0, y_0)\Delta y \\ &\qquad + \epsilon(\Delta x, \Delta y)\sqrt{(\Delta x)^2 + (\Delta y)^2}\,] \\ &= 0 + 0 + 0 \cdot 0 = 0\end{aligned}$$ ■

It can be difficult to verify that a function is differentiable at a point directly from the definition. The next theorem, whose proof is usually studied in more advanced courses, provides simple conditions for a function to be differentiable at a point.

14.4.4 THEOREM. *If all first-order partial derivatives of f exist and are continuous at a point, then f is differentiable at that point.*

For example, consider the function

$$f(x, y, z) = x + yz$$

Since $f_x(x, y, z) = 1$, $f_y(x, y, z) = z$, and $f_z(x, y, z) = y$ are defined and continuous everywhere, we conclude from Theorem 14.4.4 that f is differentiable everywhere.

DIFFERENTIALS

As with the one-variable case, the approximations

$$\Delta f \approx f_x(x_0, y_0)\Delta x + f_y(x_0, y_0)\Delta y$$

for a function of two variables and the approximation

$$\Delta f \approx f_x(x_0, y_0, z_0)\Delta x + f_y(x_0, y_0, z_0)\Delta y + f_z(x_0, y_0, z_0)\Delta z$$

for a function of three variables have a convenient formulation in the language of differentials. If $z = f(x, y)$ is differentiable at a point (x_0, y_0), we let

$$dz = f_x(x_0, y_0)\, dx + f_y(x_0, y_0)\, dy \tag{7}$$

denote a new function with dependent variable dz and independent variables dx and dy. We refer to this function (also denoted df) as the ***total differential of z*** at (x_0, y_0) or as the

total differential of f at (x_0, y_0). Similarly, for a function $w = f(x, y, z)$ of three variables we have the ***total differential of w*** at (x_0, y_0, z_0),

$$dw = f_x(x_0, y_0, z_0)\,dx + f_y(x_0, y_0, z_0)\,dy + f_z(x_0, y_0, z_0)\,dz \tag{8}$$

which is also referred to as the ***total differential of f*** at (x_0, y_0, z_0). It is common practice to omit the subscripts and write Equations (7) and (8) as

$$dz = f_x(x, y)\,dx + f_y(x, y)\,dy \tag{9}$$

and

$$dw = f_x(x, y, z)\,dx + f_y(x, y, z)\,dy + f_z(x, y, z)\,dz \tag{10}$$

In the two-variable case, the approximation

$$\Delta f \approx f_x(x_0, y_0)\Delta x + f_y(x_0, y_0)\Delta y$$

can be written in the form

$$\Delta f \approx df \tag{11}$$

for $dx = \Delta x$ and $dy = \Delta y$. Equivalently, we can write approximation (11) as

$$\Delta z \approx dz \tag{12}$$

In other words, we can estimate the change Δz in z by the value of the differential dz where dx is the change in x and dy is the change in y. Furthermore, it follows from (4) that if Δx and Δy are close to 0, then the magnitude of the error in approximation (12) will be much smaller than the distance $\sqrt{(\Delta x)^2 + (\Delta y)^2}$ between (x_0, y_0) and $(x_0 + \Delta x, y_0 + \Delta y)$.

▶ **Example 2** Use (12) to approximate the change in $z = xy^2$ from its value at (0.5, 1.0) to its value at (0.503, 1.004). Compare the magnitude of the error in this approximation with the distance between the points (0.5, 1.0) and (0.503, 1.004).

Solution. For $z = xy^2$ we have $dz = y^2\,dx + 2xy\,dy$. Evaluating this differential at $(x, y) = (0.5, 1.0)$, $dx = \Delta x = 0.503 - 0.5 = 0.003$, and $dy = \Delta y = 1.004 - 1.0 = 0.004$ yields

$$dz = 1.0^2(0.003) + 2(0.5)(1.0)(0.004) = 0.007$$

Since $z = 0.5$ at $(x, y) = (0.5, 1.0)$ and $z = 0.507032048$ at $(x, y) = (0.503, 1.004)$, we have

$$\Delta z = 0.507032048 - 0.5 = 0.007032048$$

and the error in approximating Δz by dz has magnitude

$$|dz - \Delta z| = |0.007 - 0.007032048| = 0.000032048$$

Since the distance between (0.5, 1.0) and (0.503, 1.004) $= (0.5 + \Delta x, 1.0 + \Delta y)$ is

$$\sqrt{(\Delta x)^2 + (\Delta y)^2} = \sqrt{(0.003)^2 + (0.004)^2} = \sqrt{0.000025} = 0.005$$

we have

$$\frac{|dz - \Delta z|}{\sqrt{(\Delta x)^2 + (\Delta y)^2}} = \frac{0.000032048}{0.005} = 0.0064096 < \frac{1}{150}$$

Thus, the magnitude of the error in our approximation is less than $\frac{1}{150}$ of the distance between the two points. ◀

With the appropriate changes in notation, the preceding analysis can be extended to functions of three or more variables.

▶ **Example 3** The length, width, and height of a rectangular box are measured with an error of at most 5%. Use a total differential to estimate the maximum percentage error that results if these quantities are used to calculate the diagonal of the box.

Solution. The diagonal D of a box with length x, width y, and height z is given by

$$D = \sqrt{x^2 + y^2 + z^2}$$

Let x_0, y_0, z_0, and $D_0 = \sqrt{x_0^2 + y_0^2 + z_0^2}$ denote the actual values of the length, width, height, and diagonal of the box. The total differential dD of D at (x_0, y_0, z_0) is given by

$$dD = \frac{x_0}{\sqrt{x_0^2 + y_0^2 + z_0^2}}\,dx + \frac{y_0}{\sqrt{x_0^2 + y_0^2 + z_0^2}}\,dy + \frac{z_0}{\sqrt{x_0^2 + y_0^2 + z_0^2}}\,dz$$

If x, y, z, and $D = \sqrt{x^2 + y^2 + z^2}$ are the measured and computed values of the length, width, height, and diagonal, respectively, then $\Delta x = x - x_0$, $\Delta y = y - y_0$, $\Delta z = z - z_0$, and

$$\left|\frac{\Delta x}{x_0}\right| \le 0.05, \quad \left|\frac{\Delta y}{y_0}\right| \le 0.05, \quad \left|\frac{\Delta z}{z_0}\right| \le 0.05$$

We are seeking an estimate for the maximum size of $\Delta D/D_0$. With the aid of Equation (10) we have

$$\begin{aligned}\frac{\Delta D}{D_0} \approx \frac{dD}{D_0} &= \frac{1}{x_0^2 + y_0^2 + z_0^2}[x_0\Delta x + y_0\Delta y + z_0\Delta z] \\ &= \frac{1}{x_0^2 + y_0^2 + z_0^2}\left[x_0^2\frac{\Delta x}{x_0} + y_0^2\frac{\Delta y}{y_0} + z_0^2\frac{\Delta z}{z_0}\right]\end{aligned}$$

Since

$$\begin{aligned}\left|\frac{dD}{D_0}\right| &= \frac{1}{x_0^2 + y_0^2 + z_0^2}\left|x_0^2\frac{\Delta x}{x_0} + y_0^2\frac{\Delta y}{y_0} + z_0^2\frac{\Delta z}{z_0}\right| \\ &\le \frac{1}{x_0^2 + y_0^2 + z_0^2}\left(x_0^2\left|\frac{\Delta x}{x_0}\right| + y_0^2\left|\frac{\Delta y}{y_0}\right| + z_0^2\left|\frac{\Delta z}{z_0}\right|\right) \\ &\le \frac{1}{x_0^2 + y_0^2 + z_0^2}\left(x_0^2(0.05) + y_0^2(0.05) + z_0^2(0.05)\right) = 0.05\end{aligned}$$

we estimate the maximum percentage error in D to be 5%. ◀

LOCAL LINEAR APPROXIMATIONS

We now show that if a function f is differentiable at a point, then it can be very closely approximated by a linear function near that point. For example, suppose that $f(x, y)$ is differentiable at the point (x_0, y_0). Then approximation (3) can be written in the form

$$f(x_0 + \Delta x, y_0 + \Delta y) \approx f(x_0, y_0) + f_x(x_0, y_0)\Delta x + f_y(x_0, y_0)\Delta y$$

Show that if $f(x, y)$ is a linear function, then (13) becomes an equality.

If we let $x = x_0 + \Delta x$ and $y = x_0 + \Delta y$, this approximation becomes

$$f(x, y) \approx f(x_0, y_0) + f_x(x_0, y_0)(x - x_0) + f_y(x_0, y_0)(y - y_0) \tag{13}$$

which yields a linear approximation of $f(x, y)$. Since the error in this approximation is equal to the error in the approximation (3), we conclude that for (x, y) close to (x_0, y_0), the error in (13) will be much smaller than the distance between these two points. When $f(x, y)$ is differentiable at (x_0, y_0) we get

Explain why the error in approximation (13) is the same as the error in approximation (3).

$$L(x, y) = f(x_0, y_0) + f_x(x_0, y_0)(x - x_0) + f_y(x_0, y_0)(y - y_0) \tag{14}$$

and refer to $L(x, y)$ as the ***local linear approximation to f at*** (x_0, y_0).

▸ Example 4 Let $L(x, y)$ denote the local linear approximation to $f(x, y) = \sqrt{x^2 + y^2}$ at the point $(3, 4)$. Compare the error in approximating

$$f(3.04, 3.98) = \sqrt{(3.04)^2 + (3.98)^2}$$

by $L(3.04, 3.98)$ with the distance between the points $(3, 4)$ and $(3.04, 3.98)$.

Solution. We have

$$f_x(x, y) = \frac{x}{\sqrt{x^2 + y^2}} \quad \text{and} \quad f_y(x, y) = \frac{y}{\sqrt{x^2 + y^2}}$$

with $f_x(3, 4) = \frac{3}{5}$ and $f_y(3, 4) = \frac{4}{5}$. Therefore, the local linear approximation to f at $(3, 4)$ is given by

$$L(x, y) = 5 + \tfrac{3}{5}(x - 3) + \tfrac{4}{5}(y - 4)$$

Consequently,

$$f(3.04, 3.98) \approx L(3.04, 3.98) = 5 + \tfrac{3}{5}(0.04) + \tfrac{4}{5}(-0.02) = 5.008$$

Since

$$f(3.04, 3.98) = \sqrt{(3.04)^2 + (3.98)^2} \approx 5.00819$$

the error in the approximation is about $5.00819 - 5.008 = 0.00019$. This is less than $\frac{1}{200}$ of the distance

$$\sqrt{(3.04 - 3)^2 + (3.98 - 4)^2} \approx 0.045$$

between the points $(3, 4)$ and $(3.04, 3.98)$. ◂

Similarly, for a function $f(x, y, z)$ that is differentiable at (x_0, y_0, z_0), the local linear approximation is

$$\begin{aligned} L(x, y, z) = f(x_0, y_0, z_0) &+ f_x(x_0, y_0, z_0)(x - x_0) \\ &+ f_y(x_0, y_0, z_0)(y - y_0) + f_z(x_0, y_0, z_0)(z - z_0) \end{aligned} \tag{15}$$

We have formulated our definitions in this section in such a way that continuity and local linearity are consequences of differentiability. In Section 14.7 we will show that if a function $f(x, y)$ is differentiable at a point (x_0, y_0), then the graph of $L(x, y)$ is a nonvertical tangent plane to the graph of f at the point $(x_0, y_0, f(x_0, y_0))$.

✔ QUICK CHECK EXERCISES 14.4

(See page 967 for answers.)

1. Assume that $f(x, y)$ is differentiable at (x_0, y_0) and let Δf denote the change in f from its value at (x_0, y_0) to its value at $(x_0 + \Delta x, y_0 + \Delta y)$.
   (a) $\Delta f \approx$ ________
   (b) The limit that guarantees the error in the approximation in part (a) is very small when both Δx and Δy are close to 0 is ________.
2. Compute the differential of each function.
   (a) $z = xe^{y^2}$ (b) $w = x \sin(yz)$
3. If f is differentiable at (x_0, y_0), then the local linear approximation to f at (x_0, y_0) is $L(x) =$ ________.
4. Assume that $f(1, -2) = 4$ and $f(x, y)$ is differentiable at $(1, -2)$ with $f_x(1, -2) = 2$ and $f_y(1, -2) = -3$. Estimate the value of $f(0.9, -1.950)$.

EXERCISE SET 14.4

FOCUS ON CONCEPTS

1. Suppose that a function $f(x, y)$ is differentiable at the point $(3, 4)$ with $f_x(3, 4) = 2$ and $f_y(3, 4) = -1$. If $f(3, 4) = 5$, estimate the value of $f(3.01, 3.98)$.
2. Suppose that a function $f(x, y)$ is differentiable at the point $(-1, 2)$ with $f_x(-1, 2) = 1$ and $f_y(-1, 2) = 3$. If $f(-1, 2) = 2$, estimate the value of $f(-0.99, 2.02)$.

3. Suppose that a function $f(x, y, z)$ is differentiable at the point $(1, 2, 3)$ with $f_x(1, 2, 3) = 1$, $f_y(1, 2, 3) = 2$, and $f_z(1, 2, 3) = 3$. If $f(1, 2, 3) = 4$, estimate the value of $f(1.01, 2.02, 3.03)$.

4. Suppose that a function $f(x, y, z)$ is differentiable at the point $(2, 1, -2)$, $f_x(2, 1, -2) = -1$, $f_y(2, 1, -2) = 1$, and $f_z(2, 1, -2) = -2$. If $f(2, 1, -2) = 0$, estimate the value of $f(1.98, 0.99, -1.97)$.

5. Use Definitions 14.4.1 and 14.4.2 to prove that a constant function of two or three variables is differentiable everywhere.

6. Use Definitions 14.4.1 and 14.4.2 to prove that a linear function of two or three variables is differentiable everywhere.

7. Use Definition 14.4.2 to prove that

$$f(x, y, z) = x^2 + y^2 + z^2$$

is differentiable at $(0, 0, 0)$.

8. Use Definition 14.4.2 to determine all values of r such that $f(x, y, z) = (x^2 + y^2 + z^2)^r$ is differentiable at $(0, 0, 0)$.

9–20 Compute the differential dz or dw of the specified function.

9. $z = 7x - 2y$ **10.** $z = e^{xy}$ **11.** $z = x^3y^2$

12. $z = 5x^2y^5 - 2x + 4y + 7$

13. $z = \tan^{-1} xy$ **14.** $z = \sec^2(x - 3y)$

15. $w = 8x - 3y + 4z$ **16.** $w = e^{xyz}$

17. $w = x^3y^2z$

18. $w = 4x^2y^3z^7 - 3xy + z + 5$

19. $w = \tan^{-1}(xyz)$ **20.** $w = \sqrt{x} + \sqrt{y} + \sqrt{z}$

21–26 Use a total differential to approximate the change in the values of f from P to Q. Compare your estimate with the actual change in f.

21. $f(x, y) = x^2 + 2xy - 4x$; $P(1, 2)$, $Q(1.01, 2.04)$

22. $f(x, y) = x^{1/3}y^{1/2}$; $P(8, 9)$, $Q(7.78, 9.03)$

23. $f(x, y) = \dfrac{x + y}{xy}$; $P(-1, -2)$, $Q(-1.02, -2.04)$

24. $f(x, y) = \ln \sqrt{1 + xy}$; $P(0, 2)$, $Q(-0.09, 1.98)$

25. $f(x, y, z) = 2xy^2z^3$; $P(1, -1, 2)$, $Q(0.99, -1.02, 2.02)$

26. $f(x, y, z) = \dfrac{xyz}{x + y + z}$; $P(-1, -2, 4)$, $Q(-1.04, -1.98, 3.97)$

27. In the accompanying figure a rectangle with initial length x_0 and initial width y_0 has been enlarged, resulting in a rectangle with length $x_0 + \Delta x$ and width $y_0 + \Delta y$. What portion of the figure represents the increase in the area of the rectangle? What portion of the figure represents an approximation of the increase in area by a total differential?

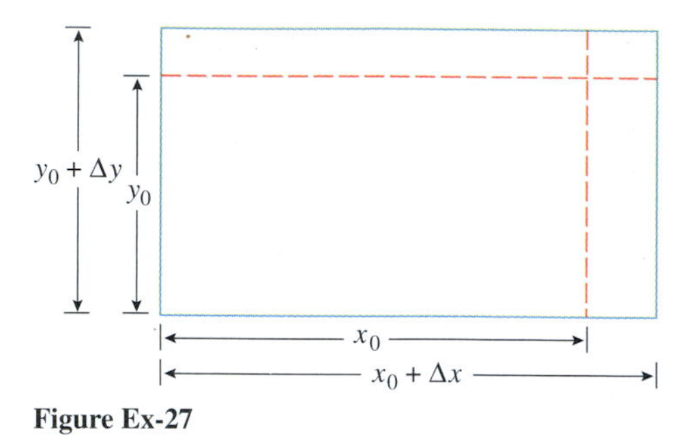

Figure Ex-27

28. The volume V of a right circular cone of radius r and height h is given by $V = \frac{1}{3}\pi r^2 h$. Suppose that the height decreases from 20 in to 19.95 in and the radius increases from 4 in to 4.05 in. Compare the change in volume of the cone with an approximation of this change using a total differential.

29–36 (a) Find the local linear approximation L to the specified function f at the designated point P. (b) Compare the error in approximating f by L at the specified point Q with the distance between P and Q.

29. $f(x, y) = \dfrac{1}{\sqrt{x^2 + y^2}}$; $P(4, 3)$, $Q(3.92, 3.01)$

30. $f(x, y) = x^{0.5}y^{0.3}$; $P(1, 1)$, $Q(1.05, 0.97)$

31. $f(x, y) = x \sin y$; $P(0, 0)$, $Q(0.003, 0.004)$

32. $f(x, y) = \ln xy$; $P(1, 2)$, $Q(1.01, 2.02)$

33. $f(x, y, z) = xyz$; $P(1, 2, 3)$, $Q(1.001, 2.002, 3.003)$

34. $f(x, y, z) = \dfrac{x + y}{y + z}$; $P(-1, 1, 1)$, $Q(-0.99, 0.99, 1.01)$

35. $f(x, y, z) = xe^{yz}$; $P(1, -1, -1)$, $Q(0.99, -1.01, -0.99)$

36. $f(x, y, z) = \ln(x + yz)$; $P(2, 1, -1)$, $Q(2.02, 0.97, -1.01)$

37. In each part, confirm that the stated formula is the local linear approximation at $(0, 0)$.

(a) $e^x \sin y \approx y$ (b) $\dfrac{2x + 1}{y + 1} \approx 1 + 2x - y$

38. Show that the local linear approximation of the function $f(x, y) = x^\alpha y^\beta$ at $(1, 1)$ is

$$x^\alpha y^\beta \approx 1 + \alpha(x - 1) + \beta(y - 1)$$

39. In each part, confirm that the stated formula is the local linear approximation at $(1, 1, 1)$.

(a) $xyz + 2 \approx x + y + z$ (b) $\dfrac{4x}{y + z} \approx 2x - y - z + 2$

40. Based on Exercise 38, what would you conjecture is the local linear approximation to $x^\alpha y^\beta z^\gamma$ at $(1, 1, 1)$? Verify your conjecture by finding this local linear approximation.

41. Suppose that a function $f(x, y)$ is differentiable at the point $(1, 1)$ with $f_x(1, 1) = 2$ and $f(1, 1) = 3$. Let $L(x, y)$ denote the local linear approximation of f at $(1, 1)$. If $L(1.1, 0.9) = 3.15$, find the value of $f_y(1, 1)$.

42. Suppose that a function $f(x, y)$ is differentiable at the point $(0, -1)$ with $f_y(0, -1) = -2$ and $f(0, -1) = 3$. Let

$L(x, y)$ denote the local linear approximation of f at $(0, -1)$. If $L(0.1, -1.1) = 3.3$, find the value of $f_x(0, -1)$.

43. Suppose that a function $f(x, y, z)$ is differentiable at the point $(3, 2, 1)$ and $L(x, y, z) = x - y + 2z - 2$ is the local linear approximation to f at $(3, 2, 1)$. Find $f(3, 2, 1)$, $f_x(3, 2, 1)$, $f_y(3, 2, 1)$, and $f_z(3, 2, 1)$.

44. Suppose that a function $f(x, y, z)$ is differentiable at the point $(0, -1, -2)$ and $L(x, y, z) = x + 2y + 3z + 4$ is the local linear approximation to f at $(0, -1, -2)$. Find $f(0, -1, -2)$, $f_x(0, -1, -2)$, $f_y(0, -1, -2)$, and $f_z(0, -1, -2)$.

45–48 A function f is given along with a local linear approximation L to f at a point P. Use the information given to determine point P.

45. $f(x, y) = x^2 + y^2$; $L(x, y) = 2y - 2x - 2$

46. $f(x, y) = x^2 y$; $L(x, y) = 4y - 4x + 8$

47. $f(x, y, z) = xy + z^2$; $L(x, y, z) = y + 2z - 1$

48. $f(x, y, z) = xyz$; $L(x, y, z) = x - y - z - 2$

49. The length and width of a rectangle are measured with errors of at most 3% and 5%, respectively. Use differentials to approximate the maximum percentage error in the calculated area.

50. The radius and height of a right circular cone are measured with errors of at most 1% and 4%, respectively. Use differentials to approximate the maximum percentage error in the calculated volume.

51. The length and width of a rectangle are measured with errors of at most r%, where r is small. Use differentials to approximate the maximum percentage error in the calculated length of the diagonal.

52. The legs of a right triangle are measured to be 3 cm and 4 cm, with a maximum error of 0.05 cm in each measurement. Use differentials to approximate the maximum possible error in the calculated value of (a) the hypotenuse and (b) the area of the triangle.

53. The period T of a simple pendulum with small oscillations is calculated from the formula $T = 2\pi\sqrt{L/g}$, where L is the length of the pendulum and g is the acceleration due to gravity. Suppose that measured values of L and g have errors of at most 0.5% and 0.1%, respectively. Use differentials to approximate the maximum percentage error in the calculated value of T.

54. According to the ideal gas law, the pressure, temperature, and volume of a confined gas are related by $P = kT/V$, where k is a constant. Use differentials to approximate the percentage change in pressure if the temperature of a gas is increased 3% and the volume is increased 5%.

55. Suppose that certain measured quantities x and y have errors of at most r% and s%, respectively. For each of the following formulas in x and y, use differentials to approximate the maximum possible error in the calculated result.
(a) xy (b) x/y (c) x^2y^3 (d) $x^3\sqrt{y}$

56. The total resistance R of three resistances R_1, R_2, and R_3, connected in parallel, is given by

$$\frac{1}{R} = \frac{1}{R_1} + \frac{1}{R_2} + \frac{1}{R_3}$$

Suppose that R_1, R_2, and R_3 are measured to be 100 ohms, 200 ohms, and 500 ohms, respectively, with a maximum error of 10% in each. Use differentials to approximate the maximum percentage error in the calculated value of R.

57. The area of a triangle is to be computed from the formula $A = \frac{1}{2}ab\sin\theta$, where a and b are the lengths of two sides and θ is the included angle. Suppose that a, b, and θ are measured to be 40 ft, 50 ft, and 30°, respectively. Use differentials to approximate the maximum error in the calculated value of A if the maximum errors in a, b, and θ are $\frac{1}{2}$ ft, $\frac{1}{4}$ ft, and 2°, respectively.

58. The length, width, and height of a rectangular box are measured with errors of at most r% (where r is small). Use differentials to approximate the maximum percentage error in the computed value of the volume.

59. Use Theorem 14.4.4 to prove that $f(x, y) = x^2 \sin y$ is differentiable everywhere.

60. Use Theorem 14.4.4 to prove that $f(x, y, z) = xy\sin z$ is differentiable everywhere.

61. Suppose that $f(x, y)$ is differentiable at the point (x_0, y_0) and let $z_0 = f(x_0, y_0)$. Prove that $g(x, y, z) = z - f(x, y)$ is differentiable at (x_0, y_0, z_0).

62. Suppose that Δf satisfies an equation in the form of (5), where $\epsilon(\Delta x, \Delta y)$ is continuous at $(\Delta x, \Delta y) = (0, 0)$ with $\epsilon(0, 0) = 0$. Prove that f is differentiable at (x_0, y_0).

✔ QUICK CHECK ANSWERS 14.4

1. (a) $f_x(x_0, y_0)\Delta x + f_y(x_0, y_0)\Delta y$ (b) $\lim\limits_{(\Delta x, \Delta y)\to(0,0)} \dfrac{\Delta f - f_x(x_0, y_0)\Delta x - f_y(x_0, y_0)\Delta y}{\sqrt{(\Delta x)^2 + (\Delta y)^2}} = 0$ **2.** (a) $dz = e^{y^2}dx + 2xye^{y^2}dy$ (b) $dw = \sin(yz)\,dx + xz\cos(yz)\,dy + xy\cos(yz)\,dz$ **3.** $f(x_0, y_0) + f_x(x_0, y_0)(x - x_0) + f_y(x_0, y_0)(y - y_0)$ **4.** 3.65

14.5 THE CHAIN RULE

In this section we will derive versions of the chain rule for functions of two or three variables. These new versions will allow us to generate useful relationships among the derivatives and partial derivatives of various functions.

THE CHAIN RULE FOR DERIVATIVES

If y is a differentiable function of x and x is a differentiable function of t, then the chain rule for functions of one variable states that, under composition, y becomes a differentiable function of t with

$$\frac{dy}{dt} = \frac{dy}{dx}\frac{dx}{dt}$$

We will now derive a version of the chain rule for functions of two variables.

Assume that $z = f(x, y)$ is a function of x and y, and suppose that x and y are in turn functions of a single variable t, say

$$x = x(t), \quad y = y(t)$$

The composition $z = f(x(t), y(t))$ then expresses z as a function of the single variable t. Thus, we can ask for the derivative dz/dt and we can inquire about its relationship to the derivatives $\partial z/\partial x$, $\partial z/\partial y$, dx/dt, and dy/dt. Letting Δx, Δy, and Δz denote the changes in x, y, and z, respectively, that correspond to a change of Δt in t, we have

$$\frac{dz}{dt} = \lim_{\Delta t \to 0} \frac{\Delta z}{\Delta t}, \quad \frac{dx}{dt} = \lim_{\Delta t \to 0} \frac{\Delta x}{\Delta t}, \quad \text{and} \quad \frac{dy}{dt} = \lim_{\Delta t \to 0} \frac{\Delta y}{\Delta t}$$

It follows from (3) of Section 14.4 that

$$\Delta z \approx \frac{\partial z}{\partial x}\Delta x + \frac{\partial z}{\partial y}\Delta y \tag{1}$$

where the partial derivatives are evaluated at $(x(t), y(t))$. Dividing both sides of (1) by Δt yields

$$\frac{\Delta z}{\Delta t} \approx \frac{\partial z}{\partial x}\frac{\Delta x}{\Delta t} + \frac{\partial z}{\partial y}\frac{\Delta y}{\Delta t} \tag{2}$$

Taking the limit as $\Delta t \to 0$ of both sides of (2) suggests the following result (whose complete proof can be found in Appendix C).

14.5.1 THEOREM (*Two-Variable Chain Rule*). *If $x = x(t)$ and $y = y(t)$ are differentiable at t, and if $z = f(x, y)$ is differentiable at the point $(x, y) = (x(t), y(t))$, then $z = f(x(t), y(t))$ is differentiable at t and*

$$\frac{dz}{dt} = \frac{\partial z}{\partial x}\frac{dx}{dt} + \frac{\partial z}{\partial y}\frac{dy}{dt} \tag{3}$$

where the ordinary derivatives are evaluated at t and the partial derivatives are evaluated at (x, y).

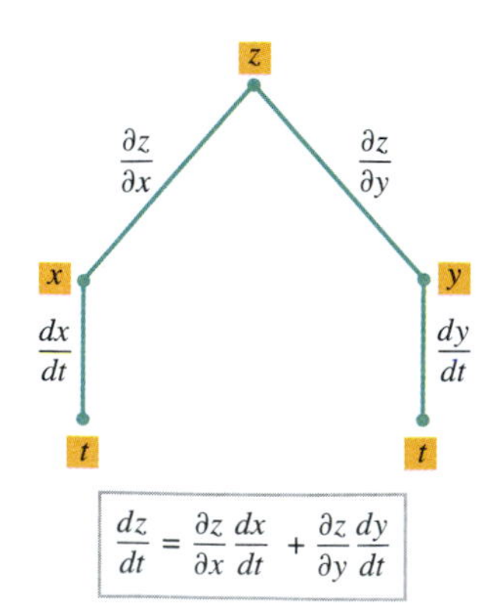

Figure 14.5.1

Formula (3) can be represented schematically by a "tree diagram" that is constructed as follows (Figure 14.5.1). Starting with z at the top of the tree and moving downward, join each variable by lines (or branches) to those variables on which it depends *directly*. Thus, z is joined to x and y and these in turn are joined to t. Next, label each branch with a derivative whose "numerator" contains the variable at the top end of that branch and whose "denominator" contains the variable at the bottom end of that branch. This completes the "tree." To find the formula for dz/dt, follow the two paths through the tree that start with z and end with t. Each such path corresponds to a term in Formula (3).

▶ Example 1 Suppose that

$$z = x^2y, \quad x = t^2, \quad y = t^3$$

Use the chain rule to find dz/dt, and check the result by expressing z as a function of t and differentiating directly.

Solution. By the chain rule

$$\begin{aligned}\frac{dz}{dt} &= \frac{\partial z}{\partial x}\frac{dx}{dt} + \frac{\partial z}{\partial y}\frac{dy}{dt} = (2xy)(2t) + (x^2)(3t^2) \\ &= (2t^5)(2t) + (t^4)(3t^2) = 7t^6\end{aligned}$$

Alternatively, we can express z directly as a function of t,

$$z = x^2y = (t^2)^2(t^3) = t^7$$

and then differentiate to obtain $dz/dt = 7t^6$. However, this procedure may not always be convenient. ◀

▶ Example 2 Suppose that

$$z = \sqrt{xy + y}, \quad x = \cos\theta, \quad y = \sin\theta$$

Use the chain rule to find $dz/d\theta$ when $\theta = \pi/2$.

Solution. From the chain rule with θ in place of t,

$$\frac{dz}{d\theta} = \frac{\partial z}{\partial x}\frac{dx}{d\theta} + \frac{\partial z}{\partial y}\frac{dy}{d\theta}$$

we obtain

$$\frac{dz}{d\theta} = \frac{1}{2}(xy + y)^{-1/2}(y)(-\sin\theta) + \frac{1}{2}(xy + y)^{-1/2}(x + 1)(\cos\theta)$$

When $\theta = \pi/2$, we have

$$x = \cos\frac{\pi}{2} = 0, \quad y = \sin\frac{\pi}{2} = 1$$

Substituting $x = 0$, $y = 1$, $\theta = \pi/2$ in the formula for $dz/d\theta$ yields

$$\left.\frac{dz}{d\theta}\right|_{\theta=\pi/2} = \frac{1}{2}(1)(1)(-1) + \frac{1}{2}(1)(1)(0) = -\frac{1}{2} \;◀$$

Confirm the result of Example 2 by expressing z directly as a function of θ.

There are many variations in derivative notations, each of which gives the chain rule a different look. If $z = f(x, y)$, where x and y are functions of t, then some possibilities are

$$\frac{dz}{dt} = f_x\frac{dx}{dt} + f_y\frac{dy}{dt}$$

$$\frac{df}{dt} = \frac{\partial f}{\partial x}\frac{dx}{dt} + \frac{\partial f}{\partial y}\frac{dy}{dt}$$

$$\frac{df}{dt} = f_x x'(t) + f_y y'(t)$$

Theorem 14.5.1 has a natural extension to functions $w = f(x, y, z)$ of three variables, which we state without proof.

14.5.2 THEOREM (*Three-Variable Chain Rule*). *If each of the functions $x = x(t)$, $y = y(t)$, and $z = z(t)$ is differentiable at t, and if $w = f(x, y, z)$ is differentiable at the point $(x, y, z) = (x(t), y(t), z(t))$, then $w = f(x(t), y(t), z(t))$ is differentiable at t and*

$$\frac{dw}{dt} = \frac{\partial w}{\partial x}\frac{dx}{dt} + \frac{\partial w}{\partial y}\frac{dy}{dt} + \frac{\partial w}{\partial z}\frac{dz}{dt} \tag{4}$$

where the ordinary derivatives are evaluated at t and the partial derivatives are evaluated at (x, y, z).

One of the principal uses of the chain rule for functions of a *single* variable was to compute formulas for the derivatives of compositions of functions. Theorems 14.5.1 and 14.5.2 are important not so much for the computation of formulas but because they allow us to express *relationships* among various derivatives. As illustrations, we revisit the topics of implicit differentiation and related rates problems.

IMPLICIT DIFFERENTIATION

Consider the special case where $z = f(x, y)$ is a function of x and y and y is a differentiable function of x. Equation (3) then becomes

$$\frac{dz}{dx} = \frac{\partial f}{\partial x}\frac{dx}{dx} + \frac{\partial f}{\partial y}\frac{dy}{dx} = \frac{\partial f}{\partial x} + \frac{\partial f}{\partial y}\frac{dy}{dx} \tag{5}$$

This result can be used to find derivatives of functions that are defined implicitly. For example, suppose that the equation

$$f(x, y) = c \tag{6}$$

defines y implicitly as a differentiable function of x and we are interested in finding dy/dx. Differentiating both sides of (6) with respect to x and applying (5) yields

$$\frac{\partial f}{\partial x} + \frac{\partial f}{\partial y}\frac{dy}{dx} = 0$$

Thus, if $\partial f/\partial y \neq 0$, we obtain

$$\frac{dy}{dx} = -\frac{\partial f/\partial x}{\partial f/\partial y}$$

In summary, we have the following result.

Show that the function $y = x$ is defined implicitly by the equation

$$x^2 - 2xy + y^2 = 0$$

but that Theorem 14.5.3 is not applicable for finding dy/dx.

14.5.3 THEOREM. *If the equation $f(x, y) = c$ defines y implicitly as a differentiable function of x, and if $\partial f/\partial y \neq 0$, then*

$$\frac{dy}{dx} = -\frac{\partial f/\partial x}{\partial f/\partial y} \tag{7}$$

Example 3 Given that

$$x^3 + y^2x - 3 = 0$$

find dy/dx using (7), and check the result using implicit differentiation.

Solution. By (7) with $f(x, y) = x^3 + y^2x - 3$,

$$\frac{dy}{dx} = -\frac{\partial f/\partial x}{\partial f/\partial y} = -\frac{3x^2 + y^2}{2yx}$$

Alternatively, differentiating the given equation implicitly yields

$$3x^2 + y^2 + x\left(2y\frac{dy}{dx}\right) - 0 = 0 \quad \text{or} \quad \frac{dy}{dx} = -\frac{3x^2 + y^2}{2yx}$$

which agrees with the result obtained by (7). ◀

RELATED RATES PROBLEMS

Theorems 14.5.1 and 14.5.2 provide us with additional perspective on related rates problems such as those in Section 3.7.

▶ **Example 4** At what rate is the volume of a box changing if its length is 8 ft and increasing at 3 ft/s, its width is 6 ft and increasing at 2 ft/s, and its height is 4 ft and increasing at 1 ft/s?

Solution. Let x, y, and z denote the length, width, and height of the box, respectively, and let t denote time in seconds. We can interpret the given rates to mean that

$$\frac{dx}{dt} = 3, \quad \frac{dy}{dt} = 2, \quad \text{and} \quad \frac{dz}{dt} = 1 \tag{8}$$

at the instant when

$$x = 8, \quad y = 6, \quad \text{and} \quad z = 4 \tag{9}$$

We want to find dV/dt at that instant. For this purpose we use the volume formula $V = xyz$ to obtain

$$\frac{dV}{dt} = \frac{\partial V}{\partial x}\frac{dx}{dt} + \frac{\partial V}{\partial y}\frac{dy}{dt} + \frac{\partial V}{\partial z}\frac{dz}{dt} = yz\frac{dx}{dt} + xz\frac{dy}{dt} + xy\frac{dz}{dt}$$

Substituting (8) and (9) into this equation yields

$$\frac{dV}{dt} = (6)(4)(3) + (8)(4)(2) + (8)(6)(1) = 184$$

Thus, the volume is increasing at a rate of 184 ft^3/s at the given instant. ◀

Redo Example 4 using the methods of Section 3.7. What derivative rule replaces the chain rule in your solution?

THE CHAIN RULE FOR PARTIAL DERIVATIVES

In Theorem 14.5.1 the variables x and y are each functions of a single variable t. We now consider the case where x and y are each functions of two variables. Let

$$z = f(x, y) \tag{10}$$

and suppose that x and y are functions of u and v, say

$$x = x(u, v), \quad y = y(u, v)$$

On substituting these functions of u and v into (10), we obtain the relationship

$$z = f(x(u, v), y(u, v))$$

which expresses z as a function of the two variables u and v. Thus, we can ask for the partial derivatives $\partial z/\partial u$ and $\partial z/\partial v$; and we can inquire about the relationship between these derivatives and the derivatives $\partial z/\partial x$, $\partial z/\partial y$, $\partial x/\partial u$, $\partial x/\partial v$, $\partial y/\partial u$, and $\partial y/\partial v$.

14.5.4 THEOREM (*Two-Variable Chain Rule*). *If $x = x(u, v)$ and $y = y(u, v)$ have first-order partial derivatives at the point (u, v), and if $z = f(x, y)$ is differentiable at the point $(x(u, v), y(u, v))$, then $z = f(x(u, v), y(u, v))$ has first-order partial derivatives at (u, v) given by*

$$\frac{\partial z}{\partial u} = \frac{\partial z}{\partial x}\frac{\partial x}{\partial u} + \frac{\partial z}{\partial y}\frac{\partial y}{\partial u} \quad \text{and} \quad \frac{\partial z}{\partial v} = \frac{\partial z}{\partial x}\frac{\partial x}{\partial v} + \frac{\partial z}{\partial y}\frac{\partial y}{\partial v}$$

PROOF. If v is held fixed, then $x = x(u, v)$ and $y = y(u, v)$ become functions of u alone. Thus, we are back to the case of Theorem 14.5.1. If we apply that theorem with u in place of t, and if we use ∂ rather than d to indicate that the variable v is fixed, we obtain

$$\frac{\partial z}{\partial u} = \frac{\partial z}{\partial x}\frac{\partial x}{\partial u} + \frac{\partial z}{\partial y}\frac{\partial y}{\partial u}$$

The formula for $\partial z/\partial v$ is derived similarly. ■

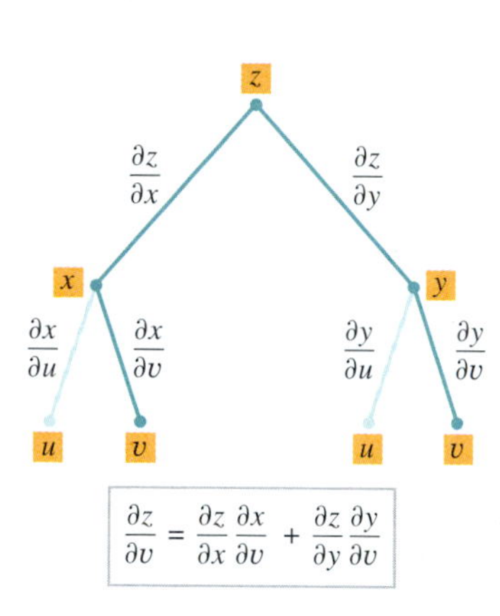

Figure 14.5.2

Figure 14.5.2 shows tree diagrams for the formulas in Theorem 14.5.4. The formula for $\partial z/\partial u$ can be obtained by tracing all paths through the tree that start with z and end with u, and the formula for $\partial z/\partial v$ can be obtained by tracing all paths through the tree that start with z and end with v.

▶ Example 5 Given that

$$z = e^{xy}, \quad x = 2u + v, \quad y = u/v$$

find $\partial z/\partial u$ and $\partial z/\partial v$ using the chain rule.

Solution.

$$\begin{aligned}
\frac{\partial z}{\partial u} &= \frac{\partial z}{\partial x}\frac{\partial x}{\partial u} + \frac{\partial z}{\partial y}\frac{\partial y}{\partial u} = (ye^{xy})(2) + (xe^{xy})\left(\frac{1}{v}\right) = \left[2y + \frac{x}{v}\right]e^{xy} \\
&= \left[\frac{2u}{v} + \frac{2u+v}{v}\right]e^{(2u+v)(u/v)} = \left[\frac{4u}{v} + 1\right]e^{(2u+v)(u/v)} \\
\frac{\partial z}{\partial v} &= \frac{\partial z}{\partial x}\frac{\partial x}{\partial v} + \frac{\partial z}{\partial y}\frac{\partial y}{\partial v} = (ye^{xy})(1) + (xe^{xy})\left(-\frac{u}{v^2}\right) \\
&= \left[y - x\left(\frac{u}{v^2}\right)\right]e^{xy} = \left[\frac{u}{v} - (2u+v)\left(\frac{u}{v^2}\right)\right]e^{(2u+v)(u/v)} \\
&= -\frac{2u^2}{v^2}e^{(2u+v)(u/v)} \quad \blacktriangleleft
\end{aligned}$$

Theorem 14.5.4 has a natural extension to functions $w = f(x, y, z)$ of three variables, which we state without proof.

14.5.5 THEOREM (*Three-Variable Chain Rule*). *If $x = x(u, v)$, $y = y(u, v)$, and $z = z(u, v)$ have first-order partial derivatives at the point (u, v), and if the function $w = f(x, y, z)$ is differentiable at the point $(x(u, v), y(u, v), z(u, v))$, then the function $w = f(x(u, v), y(u, v), z(u, v))$ has first-order partial derivatives at (u, v) given by*

$$\frac{\partial w}{\partial u} = \frac{\partial w}{\partial x}\frac{\partial x}{\partial u} + \frac{\partial w}{\partial y}\frac{\partial y}{\partial u} + \frac{\partial w}{\partial z}\frac{\partial z}{\partial u} \quad \text{and} \quad \frac{\partial w}{\partial v} = \frac{\partial w}{\partial x}\frac{\partial x}{\partial v} + \frac{\partial w}{\partial y}\frac{\partial y}{\partial v} + \frac{\partial w}{\partial z}\frac{\partial z}{\partial v}$$

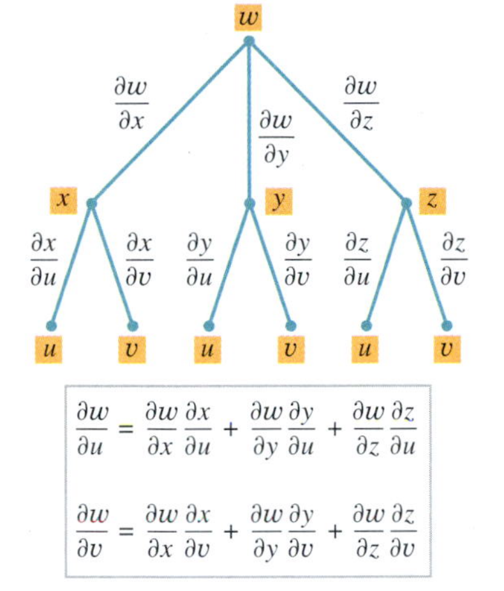

Figure 14.5.3

Example 6 Suppose that

$$w = e^{xyz}, \quad x = 3u + v, \quad y = 3u - v, \quad z = u^2 v$$

Use appropriate forms of the chain rule to find $\partial w/\partial u$ and $\partial w/\partial v$.

Solution. From the tree diagram and corresponding formulas in Figure 14.5.3 we obtain

$$\frac{\partial w}{\partial u} = yze^{xyz}(3) + xze^{xyz}(3) + xye^{xyz}(2uv) = e^{xyz}(3yz + 3xz + 2xyuv)$$

and

$$\frac{\partial w}{\partial v} = yze^{xyz}(1) + xze^{xyz}(-1) + xye^{xyz}(u^2) = e^{xyz}(yz - xz + xyu^2)$$

If desired, we can express $\partial w/\partial u$ and $\partial w/\partial v$ in terms of u and v alone by replacing x, y, and z by their expressions in terms of u and v. ◄

OTHER VERSIONS OF THE CHAIN RULE

Although we will not prove it, the chain rule extends to functions $w = f(v_1, v_2, \ldots, v_n)$ of n variables. For example, if each v_i is a function of t, $i = 1, 2, \ldots, n$, the relevant formula is

$$\frac{dw}{dt} = \frac{\partial w}{\partial v_1}\frac{dv_1}{dt} + \frac{\partial w}{\partial v_2}\frac{dv_2}{dt} + \cdots + \frac{\partial w}{\partial v_n}\frac{dv_n}{dt} \tag{11}$$

Note that (11) is a natural extension of Formula (3) in Theorem 14.5.1 and Formula (4) in Theorem 14.5.2.

There are infinitely many variations of the chain rule, depending on the number of variables and the choice of independent and dependent variables. A good working procedure is to use tree diagrams to derive new versions of the chain rule as needed. This approach will give correct results for the functions that we will usually encounter.

Example 7 Suppose that $w = x^2 + y^2 - z^2$ and

$$x = \rho \sin\phi \cos\theta, \quad y = \rho \sin\phi \sin\theta, \quad z = \rho\cos\phi$$

Use appropriate forms of the chain rule to find $\partial w/\partial \rho$ and $\partial w/\partial \theta$.

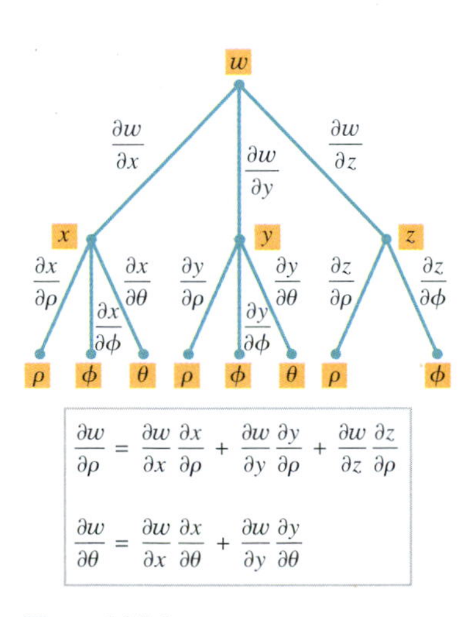

Figure 14.5.4

Solution. From the tree diagram and corresponding formulas in Figure 14.5.4 we obtain

$$\begin{aligned}
\frac{\partial w}{\partial \rho} &= 2x\sin\phi\cos\theta + 2y\sin\phi\sin\theta - 2z\cos\phi \\
&= 2\rho\sin^2\phi\cos^2\theta + 2\rho\sin^2\phi\sin^2\theta - 2\rho\cos^2\phi \\
&= 2\rho\sin^2\phi(\cos^2\theta + \sin^2\theta) - 2\rho\cos^2\phi \\
&= 2\rho(\sin^2\phi - \cos^2\phi) \\
&= -2\rho\cos 2\phi \\
\frac{\partial w}{\partial \theta} &= (2x)(-\rho\sin\phi\sin\theta) + (2y)\rho\sin\phi\cos\theta \\
&= -2\rho^2\sin^2\phi\sin\theta\cos\theta + 2\rho^2\sin^2\phi\sin\theta\cos\theta \\
&= 0
\end{aligned}$$

This result is explained by the fact that w does not vary with θ. You can see this directly by expressing the variables x, y, and z in terms of ρ, ϕ, and θ in the formula for w. (Verify that $w = -\rho^2\cos 2\phi$.) ◄

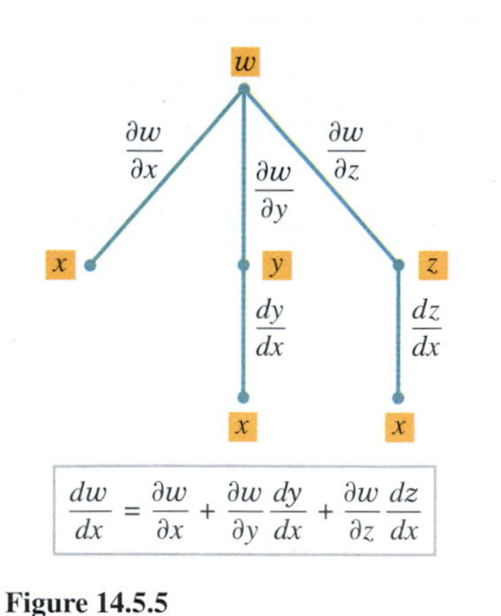

Figure 14.5.5

Example 8 Suppose that

$$w = xy + yz, \quad y = \sin x, \quad z = e^x$$

Use an appropriate form of the chain rule to find dw/dx.

Solution. From the tree diagram and corresponding formulas in Figure 14.5.5 we obtain

$$\begin{aligned} \frac{dw}{dx} &= y + (x+z)\cos x + ye^x \\ &= \sin x + (x + e^x)\cos x + e^x \sin x \end{aligned}$$

This result can also be obtained by first expressing w explicitly in terms of x as

$$w = x \sin x + e^x \sin x$$

and then differentiating with respect to x; however, such direct substitution is not always possible. ◄

WARNING The symbol ∂z, unlike the differential dz, has no meaning of its own. For example, if we were to "cancel" partial symbols in the chain-rule formula

$$\frac{\partial z}{\partial u} = \frac{\partial z}{\partial x}\frac{\partial x}{\partial u} + \frac{\partial z}{\partial y}\frac{\partial y}{\partial u}$$

we would obtain

$$\frac{\partial z}{\partial u} = \frac{\partial z}{\partial u} + \frac{\partial z}{\partial u}$$

which is false in cases where $\partial z/\partial u \neq 0$.

In each of the expressions

$$z = \sin xy, \quad z = \frac{xy}{1+xy}, \quad z = e^{xy}$$

the independent variables occur only in the combination xy, so the substitution $t = xy$ reduces the expression to a function of one variable:

$$z = \sin t, \quad z = \frac{t}{1+t}, \quad z = e^t$$

Conversely, if we begin with a function of one variable $z = f(t)$ and substitute $t = xy$, we obtain a function $z = f(xy)$ in which the variables appear only in the combination xy. Functions whose variables occur in fixed combinations arise frequently in applications.

Example 9 Show that when f is differentiable, a function of the form $z = f(xy)$ satisfies the equation

$$x\frac{\partial z}{\partial x} - y\frac{\partial z}{\partial y} = 0$$

Solution. Let $t = xy$, so that $z = f(t)$. From the tree diagram in Figure 14.5.6 we obtain the formulas

$$\frac{\partial z}{\partial x} = \frac{dz}{dt}\frac{\partial t}{\partial x} = y\frac{dz}{dt} \quad \text{and} \quad \frac{\partial z}{\partial y} = \frac{dz}{dt}\frac{\partial t}{\partial y} = x\frac{dz}{dt}$$

from which it follows that

$$x\frac{\partial z}{\partial x} - y\frac{\partial z}{\partial y} = xy\frac{dz}{dt} - yx\frac{dz}{dt} = 0 \text{ ◄}$$

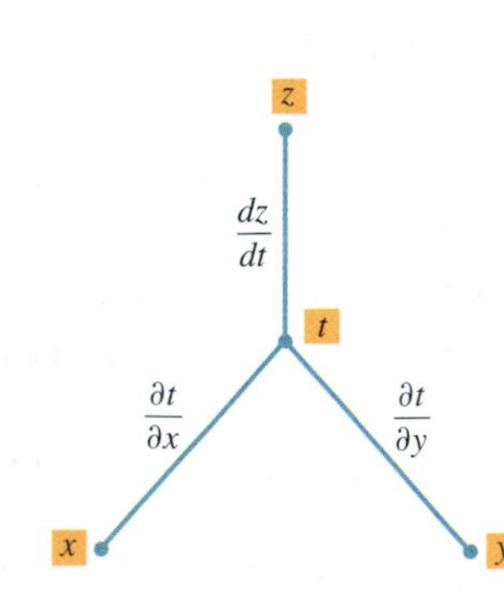

Figure 14.5.6

✔ QUICK CHECK EXERCISES 14.5

(See page 978 for answers.)

1. Suppose that $z = xy^2$ and x and y are differentiable functions of t with $x = 1$, $y = -1$, $dx/dt = -2$, and $dy/dt = 3$ when $t = -1$. Then $dz/dt =$ ________ when $t = -1$.
2. Suppose that C is the graph of the equation $f(x, y) = 1$ and that this equation defines y implicitly as a differentiable function of x. If the point $(2, 1)$ belongs to C with $f_x(2, 1) = 3$ and $f_y(2, 1) = -1$, then the tangent line to C at the point $(2, 1)$ has slope ________.
3. A rectangle is growing in such a way that when its length is 5 ft and its width is 2 ft, the length is increasing at a rate of 3 ft/s and its width is increasing at a rate of 4 ft/s. At this instant the area of the rectangle is growing at a rate of ________.
4. Suppose that $z = x/y$, where x and y are differentiable functions of u and v such that $x = 3$, $y = 1$, $\partial x/\partial u = 4$, $\partial x/\partial v = -2$, $\partial y/\partial u = 1$, and $\partial y/\partial v = -1$ when $u = 2$ and $v = 1$. When $u = 2$ and $v = 1$, $\partial z/\partial u =$ ________ and $\partial z/\partial v =$ ________.

14.5 #55, 65
Investigation

EXERCISE SET 14.5

1–6 Use an appropriate form of the chain rule to find dz/dt.

1. $z = 3x^2y^3$; $x = t^4$, $y = t^2$
2. $z = \ln(2x^2 + y)$; $x = \sqrt{t}$, $y = t^{2/3}$
3. $z = 3\cos x - \sin xy$; $x = 1/t$, $y = 3t$
4. $z = \sqrt{1 + x - 2xy^4}$; $x = \ln t$, $y = t$
5. $z = e^{1-xy}$; $x = t^{1/3}$, $y = t^3$
6. $z = \cosh^2 xy$; $x = t/2$, $y = e^t$

7–10 Use an appropriate form of the chain rule to find dw/dt.

7. $w = 5x^2y^3z^4$; $x = t^2$, $y = t^3$, $z = t^5$
8. $w = \ln(3x^2 - 2y + 4z^3)$; $x = t^{1/2}$, $y = t^{2/3}$, $z = t^{-2}$
9. $w = 5\cos xy - \sin xz$; $x = 1/t$, $y = t$, $z = t^3$
10. $w = \sqrt{1 + x - 2yz^4x}$; $x = \ln t$, $y = t$, $z = 4t$

FOCUS ON CONCEPTS

11. Suppose that
$$w = x^3y^2z^4; \quad x = t^2, \quad y = t + 2, \quad z = 2t^4$$
Find the rate of change of w with respect to t at $t = 1$ by using the chain rule, and then check your work by expressing w as a function of t and differentiating.
12. Suppose that
$$w = x\sin yz^2; \quad x = \cos t, \quad y = t^2, \quad z = e^t$$
Find the rate of change of w with respect to t at $t = 0$ by using the chain rule, and then check your work by expressing w as a function of t and differentiating.
13. Suppose that $z = f(x, y)$ is differentiable at the point $(4, 8)$ with $f_x(4, 8) = 3$ and $f_y(4, 8) = -1$. If $x = t^2$ and $y = t^3$, find dz/dt when $t = 2$.
14. Suppose that $w = f(x, y, z)$ is differentiable at the point $(1, 0, 2)$ with $f_x(1, 0, 2) = 1$, $f_y(1, 0, 2) = 2$, and $f_z(1, 0, 2) = 3$. If $x = t$, $y = \sin(\pi t)$, and $z = t^2 + 1$, find dw/dt when $t = 1$.
15. Explain how the product rule for functions of a single variable may be viewed as a consequence of the chain rule applied to a particular function of two variables.
16. A student attempts to differentiate the function x^x using the power rule, mistakenly getting $x \cdot x^{x-1}$. A second student attempts to differentiate x^x by treating it as an exponential function, mistakenly getting $(\ln x)x^x$. Use the chain rule to explain why the correct derivative is the sum of these two incorrect results.

17–22 Use appropriate forms of the chain rule to find $\partial z/\partial u$ and $\partial z/\partial v$.

17. $z = 8x^2y - 2x + 3y$; $x = uv$, $y = u - v$
18. $z = x^2 - y\tan x$; $x = u/v$, $y = u^2v^2$
19. $z = x/y$; $x = 2\cos u$, $y = 3\sin v$
20. $z = 3x - 2y$; $x = u + v\ln u$, $y = u^2 - v\ln v$
21. $z = e^{x^2y}$; $x = \sqrt{uv}$, $y = 1/v$
22. $z = \cos x \sin y$; $x = u - v$, $y = u^2 + v^2$

23–30 Use appropriate forms of the chain rule to find the derivatives.

23. Let $T = x^2y - xy^3 + 2$; $x = r\cos\theta$, $y = r\sin\theta$. Find $\partial T/\partial r$ and $\partial T/\partial\theta$.
24. Let $R = e^{2s-t^2}$; $s = 3\phi$, $t = \phi^{1/2}$. Find $dR/d\phi$.
25. Let $t = u/v$; $u = x^2 - y^2$, $v = 4xy^3$. Find $\partial t/\partial x$ and $\partial t/\partial y$.
26. Let $w = rs/(r^2 + s^2)$; $r = uv$, $s = u - 2v$. Find $\partial w/\partial u$ and $\partial w/\partial v$.
27. Let $z = \ln(x^2 + 1)$, where $x = r\cos\theta$. Find $\partial z/\partial r$ and $\partial z/\partial\theta$.

28. Let $u = rs^2 \ln t$, $r = x^2$, $s = 4y + 1$, $t = xy^3$. Find $\partial u/\partial x$ and $\partial u/\partial y$.

29. Let $w = 4x^2 + 4y^2 + z^2$, $x = \rho \sin\phi \cos\theta$, $y = \rho \sin\phi \sin\theta$, $z = \rho \cos\phi$. Find $\partial w/\partial\rho$, $\partial w/\partial\phi$, and $\partial w/\partial\theta$.

30. Let $w = 3xy^2z^3$, $y = 3x^2 + 2$, $z = \sqrt{x-1}$. Find dw/dx.

31. Use a chain rule to find the value of $\left.\dfrac{dw}{ds}\right|_{s=1/4}$ if $w = r^2 - r\tan\theta$; $r = \sqrt{s}$, $\theta = \pi s$.

32. Use a chain rule to find the values of

$$\left.\frac{\partial f}{\partial u}\right|_{u=1,v=-2} \quad \text{and} \quad \left.\frac{\partial f}{\partial v}\right|_{u=1,v=-2}$$

if $f(x, y) = x^2y^2 - x + 2y$; $x = \sqrt{u}$, $y = uv^3$.

33. Use a chain rule to find the values of

$$\left.\frac{\partial z}{\partial r}\right|_{r=2,\theta=\pi/6} \quad \text{and} \quad \left.\frac{\partial z}{\partial \theta}\right|_{r=2,\theta=\pi/6}$$

if $z = xye^{x/y}$; $x = r\cos\theta$, $y = r\sin\theta$.

34. Use a chain rule to find $\left.\dfrac{dz}{dt}\right|_{t=3}$ if $z = x^2y$; $x = t^2$, $y = t + 7$.

35–38 Use Theorem 14.5.3 to find dy/dx and check your result using implicit differentiation.

35. $x^2y^3 + \cos y = 0$

36. $x^3 - 3xy^2 + y^3 = 5$

37. $e^{xy} + ye^y = 1$

38. $x - \sqrt{xy} + 3y = 4$

39. Assume that $F(x, y, z) = 0$ defines z implicitly as a function of x and y. Show that if $\partial F/\partial z \neq 0$, then

$$\frac{\partial z}{\partial x} = -\frac{\partial F/\partial x}{\partial F/\partial z}$$

40. Assume that $F(x, y, z) = 0$ defines z implicitly as a function of x and y. Show that if $\partial F/\partial z \neq 0$, then

$$\frac{\partial z}{\partial y} = -\frac{\partial F/\partial y}{\partial F/\partial z}$$

41–44 Find $\partial z/\partial x$ and $\partial z/\partial y$ by implicit differentiation, and confirm that the results obtained agree with those predicted by the formulas in Exercises 39 and 40.

41. $x^2 - 3yz^2 + xyz - 2 = 0$

42. $\ln(1 + z) + xy^2 + z = 1$

43. $ye^x - 5\sin 3z = 3z$

44. $e^{xy}\cos yz - e^{yz}\sin xz + 2 = 0$

45. Two straight roads intersect at right angles. Car A, moving on one of the roads, approaches the intersection at 25 mi/h and car B, moving on the other road, approaches the intersection at 30 mi/h. At what rate is the distance between the cars changing when A is 0.3 mile from the intersection and B is 0.4 mile from the intersection?

46. Use the ideal gas law $P = kT/V$ with V in cubic inches (in^3), T in kelvins (K), and $k = 10$ in·lb/K to find the rate at which the temperature of a gas is changing when the volume is 200 in^3 and increasing at the rate of 4 in^3/s, while the pressure is 5 lb/in^2 and decreasing at the rate of 1 lb/in^2/s.

47. Two sides of a triangle have lengths $a = 4$ cm and $b = 3$ cm but are increasing at the rate of 1 cm/s. If the area of the triangle remains constant, at what rate is the angle θ between a and b changing when $\theta = \pi/6$?

48. Two sides of a triangle have lengths $a = 5$ cm and $b = 10$ cm, and the included angle is $\theta = \pi/3$. If a is increasing at a rate of 2 cm/s, b is increasing at a rate of 1 cm/s, and θ remains constant, at what rate is the third side changing? Is it increasing or decreasing? [*Hint:* Use the law of cosines.]

49. Suppose that the portion of a tree that is usable for lumber is a right circular cylinder. If the usable height of a tree increases 2 ft per year and the usable diameter increases 3 in per year, how fast is the volume of usable lumber increasing when the usable height of the tree is 20 ft and the usable diameter is 30 in?

50. Suppose that a particle moving along a metal plate in the xy-plane has velocity $\mathbf{v} = \mathbf{i} - 4\mathbf{j}$ (cm/s) at the point $(3, 2)$. Given that the temperature of the plate at points in the xy-plane is $T(x, y) = y^2 \ln x$, $x \geq 1$, in degrees Celsius, find dT/dt at the point $(3, 2)$.

51. The length, width, and height of a rectangular box are increasing at rates of 1 in/s, 2 in/s, and 3 in/s, respectively.
(a) At what rate is the volume increasing when the length is 2 in, the width is 3 in, and the height is 6 in?
(b) At what rate is the length of the diagonal increasing at that instant?

52. Consider the box in Exercise 51. At what rate is the surface area of the box increasing at the given instant?

53–54 A function $f(x, y)$ is said to be ***homogeneous of degree n*** if $f(tx, ty) = t^n f(x, y)$ for $t > 0$. This terminology is needed in these exercises.

53. In each part, show that the function is homogeneous, and find its degree.
(a) $f(x, y) = 3x^2 + y^2$
(b) $f(x, y) = \sqrt{x^2 + y^2}$
(c) $f(x, y) = x^2y - 2y^3$
(d) $f(x, y) = \dfrac{5}{(x^2 + 2y^2)^2}$

54. (a) Show that if $f(x, y)$ is a homogeneous function of degree n, then

$$x\frac{\partial f}{\partial x} + y\frac{\partial f}{\partial y} = nf$$

[*Hint:* Let $u = tx$ and $v = ty$ in $f(tx, ty)$, and differentiate both sides of $f(u, v) = t^n f(x, y)$ with respect to t.]
(b) Confirm that the functions in Exercise 53 satisfy the equation in part (a).

55. (a) Suppose that $z = f(u)$ and $u = g(x, y)$. Draw a tree diagram, and use it to construct chain rules that express $\partial z/\partial x$ and $\partial z/\partial y$ in terms of dz/du, $\partial u/\partial x$, and $\partial u/\partial y$.

(b) Show that

$$\frac{\partial^2 z}{\partial x^2} = \frac{dz}{du}\frac{\partial^2 u}{\partial x^2} + \frac{d^2 z}{du^2}\left(\frac{\partial u}{\partial x}\right)^2$$

$$\frac{\partial^2 z}{\partial y^2} = \frac{dz}{du}\frac{\partial^2 u}{\partial y^2} + \frac{d^2 z}{du^2}\left(\frac{\partial u}{\partial y}\right)^2$$

$$\frac{\partial^2 z}{\partial y \partial x} = \frac{dz}{du}\frac{\partial^2 u}{\partial y \partial x} + \frac{d^2 z}{du^2}\frac{\partial u}{\partial x}\frac{\partial u}{\partial y}$$

56. (a) Let $z = f(x^2 - y^2)$. Use the result in Exercise 55(a) to show that

$$y\frac{\partial z}{\partial x} + x\frac{\partial z}{\partial y} = 0$$

(b) Let $z = f(xy)$. Use the result in Exercise 55(a) to show that

$$x\frac{\partial z}{\partial x} - y\frac{\partial z}{\partial y} = 0$$

(c) Confirm the result in part (a) in the case where $z = \sin(x^2 - y^2)$.

(d) Confirm the result in part (b) in the case where $z = e^{xy}$.

57. Let f be a differentiable function of one variable, and let $z = f(x + 2y)$. Show that

$$2\frac{\partial z}{\partial x} - \frac{\partial z}{\partial y} = 0$$

58. Let f be a differentiable function of one variable, and let $z = f(x^2 + y^2)$. Show that

$$y\frac{\partial z}{\partial x} - x\frac{\partial z}{\partial y} = 0$$

59. Let f be a differentiable function of one variable, and let $w = f(u)$, where $u = x + 2y + 3z$. Show that

$$\frac{\partial w}{\partial x} + \frac{\partial w}{\partial y} + \frac{\partial w}{\partial z} = 6\frac{dw}{du}$$

60. Let f be a differentiable function of one variable, and let $w = f(\rho)$, where $\rho = (x^2 + y^2 + z^2)^{1/2}$. Show that

$$\left(\frac{\partial w}{\partial x}\right)^2 + \left(\frac{\partial w}{\partial y}\right)^2 + \left(\frac{\partial w}{\partial z}\right)^2 = \left(\frac{dw}{d\rho}\right)^2$$

61. Let $z = f(x - y, y - x)$. Show that $\partial z/\partial x + \partial z/\partial y = 0$.

62. Let f be a differentiable function of three variables and suppose that $w = f(x - y, y - z, z - x)$. Show that

$$\frac{\partial w}{\partial x} + \frac{\partial w}{\partial y} + \frac{\partial w}{\partial z} = 0$$

63. In parts (a)–(e), suppose that the equation $z = f(x, y)$ is expressed in the polar form $z = g(r, \theta)$ by making the substitution $x = r\cos\theta$ and $y = r\sin\theta$.

(a) View r and θ as functions of x and y and use implicit differentiation to show that

$$\frac{\partial r}{\partial x} = \cos\theta \quad \text{and} \quad \frac{\partial \theta}{\partial x} = -\frac{\sin\theta}{r}$$

(b) View r and θ as functions of x and y and use implicit differentiation to show that

$$\frac{\partial r}{\partial y} = \sin\theta \quad \text{and} \quad \frac{\partial \theta}{\partial y} = \frac{\cos\theta}{r}$$

(c) Use the results in parts (a) and (b) to show that

$$\frac{\partial z}{\partial x} = \frac{\partial z}{\partial r}\cos\theta - \frac{1}{r}\frac{\partial z}{\partial \theta}\sin\theta$$

$$\frac{\partial z}{\partial y} = \frac{\partial z}{\partial r}\sin\theta + \frac{1}{r}\frac{\partial z}{\partial \theta}\cos\theta$$

(d) Use the result in part (c) to show that

$$\left(\frac{\partial z}{\partial x}\right)^2 + \left(\frac{\partial z}{\partial y}\right)^2 = \left(\frac{\partial z}{\partial r}\right)^2 + \frac{1}{r^2}\left(\frac{\partial z}{\partial \theta}\right)^2$$

(e) Use the result in part (c) to show that if $z = f(x, y)$ satisfies Laplace's equation

$$\frac{\partial^2 z}{\partial x^2} + \frac{\partial^2 z}{\partial y^2} = 0$$

then $z = g(r, \theta)$ satisfies the equation

$$\frac{\partial^2 z}{\partial r^2} + \frac{1}{r^2}\frac{\partial^2 z}{\partial \theta^2} + \frac{1}{r}\frac{\partial z}{\partial r} = 0$$

and conversely. The latter equation is called the ***polar form of Laplace's equation***.

64. Show that the function

$$z = \tan^{-1}\frac{2xy}{x^2 - y^2}$$

satisfies Laplace's equation; then make the substitution $x = r\cos\theta$, $y = r\sin\theta$, and show that the resulting function of r and θ satisfies the polar form of Laplace's equation given in part (e) of Exercise 63.

65. (a) Show that if $u(x, y)$ and $v(x, y)$ satisfy the Cauchy–Riemann equations (Exercise 88, Section 14.3), and if $x = r\cos\theta$ and $y = r\sin\theta$, then

$$\frac{\partial u}{\partial r} = \frac{1}{r}\frac{\partial v}{\partial \theta} \quad \text{and} \quad \frac{\partial v}{\partial r} = -\frac{1}{r}\frac{\partial u}{\partial \theta}$$

This is called the ***polar form of the Cauchy–Riemann equations***.

(b) Show that the functions

$$u = \ln(x^2 + y^2), \quad v = 2\tan^{-1}(y/x)$$

satisfy the Cauchy–Riemann equations; then make the substitution $x = r\cos\theta$, $y = r\sin\theta$, and show that the resulting functions of r and θ satisfy the polar form of the Cauchy–Riemann equations.

66. In parts (a)–(d), recall from Formula (6) of Section 14.3 that under appropriate conditions a plucked string satisfies the wave equation

$$\frac{\partial^2 u}{\partial t^2} = c^2\frac{\partial^2 u}{\partial x^2}$$

where c is a positive constant.

(a) Show that a function of the form $u(x, t) = f(x + ct)$ satisfies the wave equation.

(b) Show that a function of the form $u(x, t) = g(x - ct)$ satisfies the wave equation.

(c) Show that a function of the form

$$u(x, t) = f(x + ct) + g(x - ct)$$

satisfies the wave equation.

(d) It can be proved that every solution of the wave equation is expressible in the form stated in part (c). Confirm that $u(x, t) = \sin t \sin x$ satisfies the wave equation in which $c = 1$, and then use appropriate trigonometric identities to express this function in the form $f(x + t) + g(x - t)$.

67. Let f be a differentiable function of three variables, and let $w = f(x, y, z)$, $x = \rho \sin\phi \cos\theta$, $y = \rho \sin\phi \sin\theta$, and $z = \rho\cos\phi$. Express $\partial w/\partial\rho$, $\partial w/\partial\phi$, and $\partial w/\partial\theta$ in terms of $\partial w/\partial x$, $\partial w/\partial y$, and $\partial w/\partial z$.

68. Let $w = f(x, y, z)$ be differentiable, where $z = g(x, y)$. Taking x and y as the independent variables, express each of the following in terms of $\partial f/\partial x$, $\partial f/\partial y$, $\partial f/\partial z$, $\partial z/\partial x$, and $\partial z/\partial y$.
(a) $\partial w/\partial x$ (b) $\partial w/\partial y$

69. Let $w = \ln(e^r + e^s + e^t + e^u)$. Show that

$$w_{rstu} = -6e^{r+s+t+u-4w}$$

[*Hint:* Take advantage of the relationship $e^w = e^r + e^s + e^t + e^u$.]

70. Suppose that w is a differentiable function of x_1, x_2, and x_3, and

$$x_1 = a_1y_1 + b_1y_2$$
$$x_2 = a_2y_1 + b_2y_2$$
$$x_3 = a_3y_1 + b_3y_2$$

where the a's and b's are constants. Express $\partial w/\partial y_1$ and $\partial w/\partial y_2$ in terms of $\partial w/\partial x_1$, $\partial w/\partial x_2$, and $\partial w/\partial x_3$.

71. (a) Let w be a differentiable function of x_1, x_2, x_3, and x_4, and let each x_i be a differentiable function of t. Find a chain-rule formula for dw/dt.
(b) Let w be a differentiable function of x_1, x_2, x_3, and x_4, and let each x_i be a differentiable function of v_1, v_2, and v_3. Find chain-rule formulas for $\partial w/\partial v_1$, $\partial w/\partial v_2$, and $\partial w/\partial v_3$.

72. Let $w = (x_1^2 + x_2^2 + \cdots + x_n^2)^k$, where $n \geq 2$. For what values of k does

$$\frac{\partial^2 w}{\partial x_1^2} + \frac{\partial^2 w}{\partial x_2^2} + \cdots + \frac{\partial^2 w}{\partial x_n^2} = 0$$

hold?

73. We showed in Exercise 24 of Section 6.9 that

$$\frac{d}{dx}\int_{h(x)}^{g(x)} f(t)\,dt = f(g(x))g'(x) - f(h(x))h'(x)$$

Derive this same result by letting $u = g(x)$ and $v = h(x)$ and then differentiating the function

$$F(u, v) = \int_v^u f(t)\,dt$$

with respect to x.

74. Prove: If f, f_x, and f_y are continuous on a circular region containing $A(x_0, y_0)$ and $B(x_1, y_1)$, then there is a point (x^*, y^*) on the line segment joining A and B such that

$$f(x_1, y_1) - f(x_0, y_0) = f_x(x^*, y^*)(x_1 - x_0) + f_y(x^*, y^*)(y_1 - y_0)$$

This result is the two-dimensional version of the Mean-Value Theorem. [*Hint:* Express the line segment joining A and B in parametric form and use the Mean-Value Theorem for functions of one variable.]

75. Prove: If $f_x(x, y) = 0$ and $f_y(x, y) = 0$ throughout a circular region, then $f(x, y)$ is constant on that region. [*Hint:* Use the result of Exercise 74.]

✔ QUICK CHECK ANSWERS 14.5

1. -8 **2.** 3 **3.** 26 ft^2/s **4.** 1; 1

14.6 DIRECTIONAL DERIVATIVES AND GRADIENTS

The partial derivatives $f_x(x, y)$ and $f_y(x, y)$ represent the rates of change of $f(x, y)$ in directions parallel to the x- and y-axes. In this section we will investigate rates of change of $f(x, y)$ in other directions.

DIRECTIONAL DERIVATIVES

In this section we extend the concept of a *partial* derivative to the more general notion of a *directional* derivative. We have seen that the partial derivatives of a function give the instantaneous rates of change of that function in directions parallel to the coordinate axes. Directional derivatives allow us to compute the rates of change of a function with respect to distance in *any* direction.

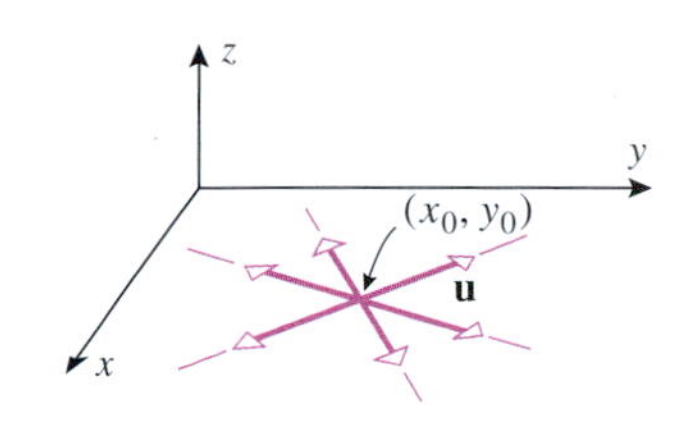

Figure 14.6.1

Suppose that we wish to compute the instantaneous rate of change of a function $f(x, y)$ with respect to distance from a point (x_0, y_0) in some direction. Since there are infinitely many different directions from (x_0, y_0) in which we could move, we need a convenient method for describing a specific direction starting at (x_0, y_0). One way to do this is to use a unit vector

$$\mathbf{u} = u_1\mathbf{i} + u_2\mathbf{j}$$

that has its initial point at (x_0, y_0) and points in the desired direction (Figure 14.6.1). This vector determines a line l in the xy-plane that can be expressed parametrically as

$$x = x_0 + su_1, \quad y = y_0 + su_2 \tag{1}$$

where s is the arc length parameter that has its reference point at (x_0, y_0) and has positive values in the direction of $\mathbf{u}$. For $s = 0$, the point (x, y) is at the reference point (x_0, y_0), and as s increases, the point (x, y) moves along l in the direction of $\mathbf{u}$. On the line l the variable $z = f(x_0 + su_1, y_0 + su_2)$ is a function of the parameter s. The value of the derivative dz/ds at $s = 0$ then gives an instantaneous rate of change of $f(x, y)$ with respect to distance from (x_0, y_0) in the direction of $\mathbf{u}$.

14.6.1 DEFINITION. If $f(x, y)$ is a function of x and y, and if $\mathbf{u} = u_1\mathbf{i} + u_2\mathbf{j}$ is a unit vector, then the ***directional derivative of f in the direction of*** $\mathbf{u}$ at (x_0, y_0) is denoted by $D_{\mathbf{u}}f(x_0, y_0)$ and is defined by

$$D_{\mathbf{u}}f(x_0, y_0) = \frac{d}{ds}[f(x_0 + su_1, y_0 + su_2)]_{s=0} \tag{2}$$

provided this derivative exists.

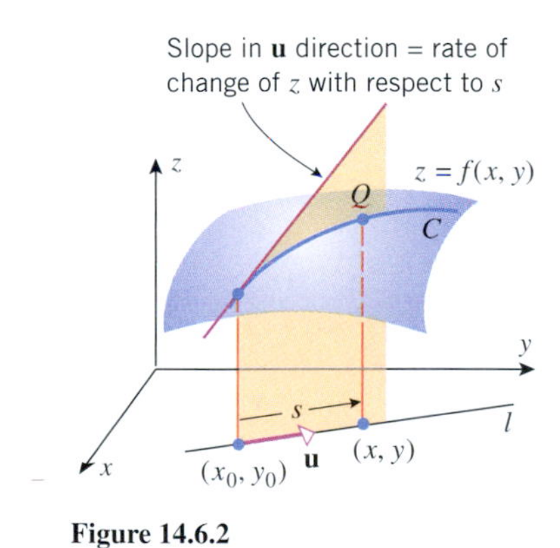

Figure 14.6.2

Geometrically, $D_{\mathbf{u}}f(x_0, y_0)$ can be interpreted as the ***slope of the surface z = f(x, y) in the direction of*** $\mathbf{u}$ at the point $(x_0, y_0, f(x_0, y_0))$ (Figure 14.6.2). Usually the value of $D_{\mathbf{u}}f(x_0, y_0)$ will depend on both the point (x_0, y_0) and the direction $\mathbf{u}$. Thus, at a fixed point the slope of the surface may vary with the direction (Figure 14.6.3). Analytically, the directional derivative represents the ***instantaneous rate of change of sfmar f(x, y) with respect to distance in the direction of*** $\mathbf{u}$ at the point (x_0, y_0).

The slope of the surface varies with the direction of **u**.

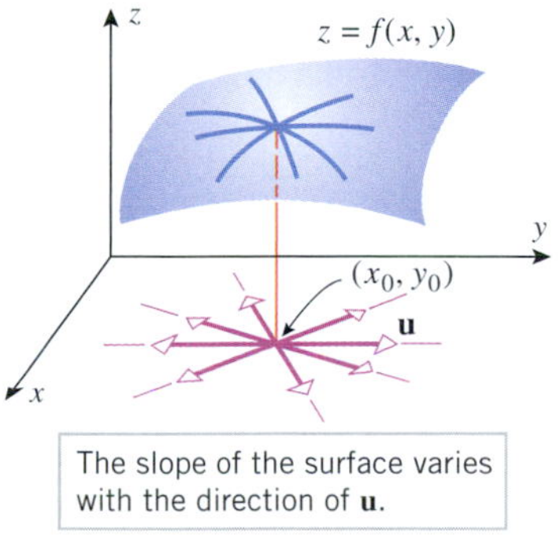

Figure 14.6.3

Example 1 Let $f(x, y) = xy$ and find $D_{\mathbf{u}}f(1, 2)$, where $\mathbf{u} = \frac{\sqrt{3}}{2}\mathbf{i} + \frac{1}{2}\mathbf{j}$.

Solution. It follows from Equation (2) that

$$D_{\mathbf{u}}f(1, 2) = \frac{d}{ds}\left[f\left(1 + \frac{\sqrt{3}s}{2}, 2 + \frac{s}{2}\right)\right]_{s=0}$$

Since

$$f\left(1 + \frac{\sqrt{3}s}{2}, 2 + \frac{s}{2}\right) = \left(1 + \frac{\sqrt{3}s}{2}\right)\left(2 + \frac{s}{2}\right) = \frac{\sqrt{3}}{4}s^2 + \left(\frac{1}{2} + \sqrt{3}\right)s + 2$$

we have

$$\frac{d}{ds}\left[f\left(1 + \frac{\sqrt{3}s}{2}, 2 + \frac{s}{2}\right)\right] = \frac{\sqrt{3}}{2}s + \frac{1}{2} + \sqrt{3}$$

and thus

$$D_{\mathbf{u}}f(1, 2) = \frac{d}{ds}\left[f\left(1 + \frac{\sqrt{3}s}{2}, 2 + \frac{s}{2}\right)\right]_{s=0} = \frac{1}{2} + \sqrt{3}$$

Since $\frac{1}{2} + \sqrt{3} \approx 2.23$, we conclude that if we move a small distance from the point $(1, 2)$ in the direction of $\mathbf{u}$, the function $f(x, y) = xy$ will increase by about 2.23 times the distance moved. ◀

The definition of a directional derivative for a function $f(x, y, z)$ of three variables is similar to Definition 14.6.1.

14.6.2 DEFINITION. If $\mathbf{u} = u_1\mathbf{i} + u_2\mathbf{j} + u_3\mathbf{k}$ is a unit vector, and if $f(x, y, z)$ is a function of x, y, and z, then the ***directional derivative of f in the direction of*** $\mathbf{u}$ at (x_0, y_0, z_0) is denoted by $D_{\mathbf{u}}f(x_0, y_0, z_0)$ and is defined by

$$D_{\mathbf{u}}f(x_0, y_0, z_0) = \frac{d}{ds}[f(x_0 + su_1, y_0 + su_2, z_0 + su_3)]_{s=0} \tag{3}$$

provided this derivative exists.

What are the difficulties in interpreting (3) as a "slope"?

Although Equation (3) does not have a convenient geometric interpretation, we can still interpret directional derivatives for functions of three variables in terms of instantaneous rates of change in a specified direction.

For a function that is differentiable at a point, directional derivatives exist in every direction from the point and can be computed directly in terms of the first-order partial derivatives of the function.

14.6.3 THEOREM.

(a) *If $f(x, y)$ is differentiable at (x_0, y_0), and if $\mathbf{u} = u_1\mathbf{i} + u_2\mathbf{j}$ is a unit vector, then the directional derivative $D_{\mathbf{u}}f(x_0, y_0)$ exists and is given by*

$$D_{\mathbf{u}}f(x_0, y_0) = f_x(x_0, y_0)u_1 + f_y(x_0, y_0)u_2 \tag{4}$$

(b) *If $f(x, y, z)$ is differentiable at (x_0, y_0, z_0), and if $\mathbf{u} = u_1\mathbf{i} + u_2\mathbf{j} + u_3\mathbf{k}$ is a unit vector, then the directional derivative $D_{\mathbf{u}}f(x_0, y_0, z_0)$ exists and is given by*

$$D_{\mathbf{u}}f(x_0, y_0, z_0) = f_x(x_0, y_0, z_0)u_1 + f_y(x_0, y_0, z_0)u_2 + f_z(x_0, y_0, z_0)u_3 \tag{5}$$

PROOF. We will give the proof of part (a); the proof of part (b) is similar and will be omitted. The function $z = f(x_0 + su_1, y_0 + su_2)$ is the composition of the function $z = f(x, y)$ with the functions

$$x = x(s) = x_0 + su_1 \quad \text{and} \quad y = y(s) = y_0 + su_2$$

As such, the chain rule in Theorem 14.5.1 immediately gives

$$\begin{aligned} D_{\mathbf{u}}f(x_0, y_0) &= \frac{d}{ds}[f(x_0 + su_1, y_0 + su_2)]_{s=0} \\ &= \frac{dz}{ds}(0) = f_x(x_0, y_0)u_1 + f_y(x_0, y_0)u_2 \end{aligned}$$ ■

We can use Theorem 14.6.3 to confirm the result of Example 1. For $f(x, y) = xy$ we have $f_x(1, 2) = 2$ and $f_y(1, 2) = 1$ (verify). With

$$\mathbf{u} = \frac{\sqrt{3}}{2}\mathbf{i} + \frac{1}{2}\mathbf{j}$$

Equation (4) becomes

$$D_{\mathbf{u}}f(1,2) = 2\left(\frac{\sqrt{3}}{2}\right) + \frac{1}{2} = \sqrt{3} + \frac{1}{2}$$

which agrees with our solution in Example 1.

Recall from Formula (13) of Section 12.2 that a unit vector $\mathbf{u}$ in the xy-plane can be expressed as

$$\mathbf{u} = \cos\phi\mathbf{i} + \sin\phi\mathbf{j} \tag{6}$$

where ϕ is the angle from the positive x-axis to $\mathbf{u}$. Thus, Formula (4) can also be expressed as

$$D_{\mathbf{u}}f(x_0, y_0) = f_x(x_0, y_0)\cos\phi + f_y(x_0, y_0)\sin\phi \tag{7}$$

▶ **Example 2** Find the directional derivative of $f(x, y) = e^{xy}$ at $(-2, 0)$ in the direction of the unit vector that makes an angle of $\pi/3$ with the positive x-axis.

Solution. The partial derivatives of f are

$$f_x(x, y) = ye^{xy}, \quad f_y(x, y) = xe^{xy}$$
$$f_x(-2, 0) = 0, \quad f_y(-2, 0) = -2$$

The unit vector $\mathbf{u}$ that makes an angle of $\pi/3$ with the positive x-axis is

$$\mathbf{u} = \cos(\pi/3)\mathbf{i} + \sin(\pi/3)\mathbf{j} = \frac{1}{2}\mathbf{i} + \frac{\sqrt{3}}{2}\mathbf{j}$$

Thus, from (7)

$$\begin{aligned} D_{\mathbf{u}}f(-2, 0) &= f_x(-2, 0)\cos(\pi/3) + f_y(-2, 0)\sin(\pi/3) \\ &= 0(1/2) + (-2)(\sqrt{3}/2) = -\sqrt{3} \end{aligned}$$ ◀

It is important that the direction of a directional derivative be specified by a *unit vector* when applying either Equation (4) or Equation (5).

▶ **Example 3** Find the directional derivative of $f(x, y, z) = x^2y - yz^3 + z$ at the point $(1, -2, 0)$ in the direction of the vector $\mathbf{a} = 2\mathbf{i} + \mathbf{j} - 2\mathbf{k}$.

Solution. The partial derivatives of f are

$$f_x(x, y, z) = 2xy, \quad f_y(x, y, z) = x^2 - z^3, \quad f_z(x, y, z) = -3yz^2 + 1$$
$$f_x(1, -2, 0) = -4, \quad f_y(1, -2, 0) = 1, \quad f_z(1, -2, 0) = 1$$

Since $\mathbf{a}$ is not a unit vector, we normalize it, getting

$$\mathbf{u} = \frac{\mathbf{a}}{\|\mathbf{a}\|} = \frac{1}{\sqrt{9}}(2\mathbf{i} + \mathbf{j} - 2\mathbf{k}) = \frac{2}{3}\mathbf{i} + \frac{1}{3}\mathbf{j} - \frac{2}{3}\mathbf{k}$$

Formula (5) then yields

$$D_{\mathbf{u}}f(1, -2, 0) = (-4)\left(\frac{2}{3}\right) + \frac{1}{3} - \frac{2}{3} = -3$$ ◀

THE GRADIENT

Formula (4) can be expressed in the form of a dot product as

$$\begin{aligned} D_{\mathbf{u}}f(x_0, y_0) &= (f_x(x_0, y_0)\mathbf{i} + f_y(x_0, y_0)\mathbf{j}) \cdot (u_1\mathbf{i} + u_2\mathbf{j}) \\ &= (f_x(x_0, y_0)\mathbf{i} + f_y(x_0, y_0)\mathbf{j}) \cdot \mathbf{u} \end{aligned}$$

Similarly, Formula (5) can be expressed as

$$D_{\mathbf{u}}f(x_0, y_0, z_0) = (f_x(x_0, y_0, z_0)\mathbf{i} + f_y(x_0, y_0, z_0)\mathbf{j} + f_z(x_0, y_0, z_0)\mathbf{k}) \cdot \mathbf{u}$$

In both cases the directional derivative is obtained by dotting the direction vector **u** with a new vector constructed from the first-order partial derivatives of f.

Remember that ∇f is not a product of ∇ and f. Think of ∇ as an "operator" that acts on a function f to produce the gradient ∇f.

14.6.4 DEFINITION.

(a) If f is a function of x and y, then the ***gradient of f*** is defined by

$$\nabla f(x, y) = f_x(x, y)\mathbf{i} + f_y(x, y)\mathbf{j} \tag{8}$$

(b) If f is a function of x, y, and z, then the ***gradient of f*** is defined by

$$\nabla f(x, y, z) = f_x(x, y, z)\mathbf{i} + f_y(x, y, z)\mathbf{j} + f_z(x, y, z)\mathbf{k} \tag{9}$$

The symbol ∇ (read "del") is an inverted delta. (It is sometimes called a "nabla" because of its similarity in form to an ancient Hebrew ten-stringed harp of that name.)

Formulas (4) and (5) can now be written as

$$D_{\mathbf{u}}f(x_0, y_0) = \nabla f(x_0, y_0) \cdot \mathbf{u} \tag{10}$$

and

$$D_{\mathbf{u}}f(x_0, y_0, z_0) = \nabla f(x_0, y_0, z_0) \cdot \mathbf{u} \tag{11}$$

respectively. For example, using Formula (11) our solution to Example 3 would take the form

$$\begin{aligned} D_{\mathbf{u}}f(1, -2, 0) = \nabla f(1, -2, 0) \cdot \mathbf{u} &= (-4\mathbf{i} + \mathbf{j} + \mathbf{k}) \cdot \left(\tfrac{2}{3}\mathbf{i} + \tfrac{1}{3}\mathbf{j} - \tfrac{2}{3}\mathbf{k}\right) \\ &= (-4)\left(\tfrac{2}{3}\right) + \tfrac{1}{3} - \tfrac{2}{3} = -3 \end{aligned}$$

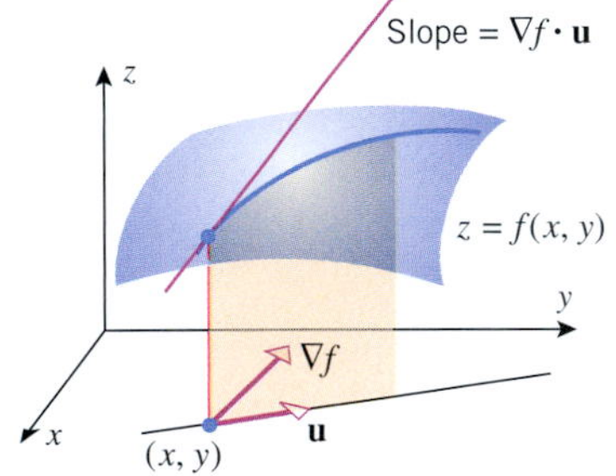

Figure 14.6.4

Formula (10) can be interpreted to mean that the slope of the surface $z = f(x, y)$ at the point (x_0, y_0) in the direction of **u** is the dot product of the gradient with **u** (Figure 14.6.4).

PROPERTIES OF THE GRADIENT

The gradient is not merely a notational device to simplify the formula for the directional derivative; we will see that the length and direction of the gradient ∇f provide important information about the function f and the surface $z = f(x, y)$. For example, suppose that $\nabla f(x, y) \neq \mathbf{0}$, and let us use Formula (4) of Section 12.3 to rewrite (10) as

$$D_{\mathbf{u}}f(x, y) = \nabla f(x, y) \cdot \mathbf{u} = \|\nabla f(x, y)\|\|\mathbf{u}\|\cos\theta = \|\nabla f(x, y)\|\cos\theta \tag{12}$$

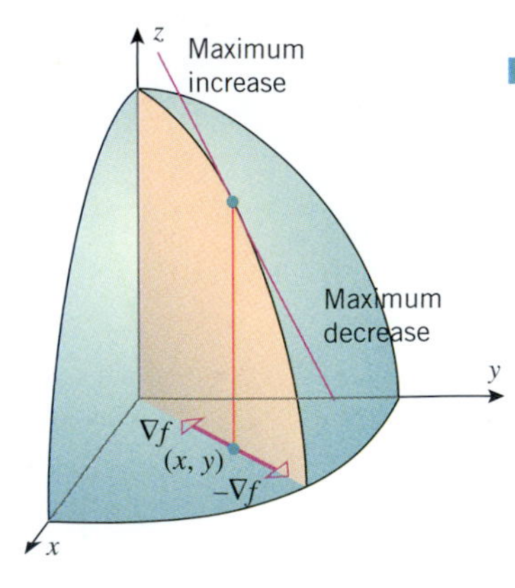

Figure 14.6.5

where θ is the angle between $\nabla f(x, y)$ and **u**. This equation tells us that the maximum value of $D_{\mathbf{u}}f(x, y)$ is $\|\nabla f(x, y)\|$, and this maximum occurs when $\theta = 0$, that is, when **u** is in the direction of $\nabla f(x, y)$. Geometrically, this means that *the surface $z = f(x, y)$ has its maximum slope at a point (x, y) in the direction of the gradient, and the maximum slope is* $\|\nabla f(x, y)\|$ (Figure 14.6.5). Similarly, (12) tells us that the minimum value of $D_{\mathbf{u}}f(x, y)$ is $-\|\nabla f(x, y)\|$, and this minimum occurs when $\theta = \pi$, that is, when **u** is oppositely directed to $\nabla f(x, y)$. Geometrically, this means that *the surface* $z = f(x, y)$ *has its minimum slope at a point* (x, y) *in the direction that is opposite to the gradient, and the minimum slope is* $-\|\nabla f(x, y)\|$ (Figure 14.6.5).

Finally, in the case where $\nabla f(x, y) = \mathbf{0}$, it follows from (12) that $D_{\mathbf{u}} f(x, y) = 0$ in all directions at the point (x, y). This typically occurs where the surface $z = f(x, y)$ has a "relative maximum," a "relative minimum," or a saddle point.

A similar analysis applies to functions of three variables. As a consequence, we have the following result.

14.6.5 THEOREM. *Let f be a function of either two variables or three variables, and let P denote the point $P(x_0, y_0)$ or $P(x_0, y_0, z_0)$, respectively. Assume that f is differentiable at P.*

(*a*) *If $\nabla f = \mathbf{0}$ at P, then all directional derivatives of f at P are zero.*

(*b*) *If $\nabla f \neq \mathbf{0}$ at P, then among all possible directional derivatives of f at P, the derivative in the direction of ∇f at P has the largest value. The value of this largest directional derivative is $\|\nabla f\|$ at P.*

(*c*) *If $\nabla f \neq \mathbf{0}$ at P, then among all possible directional derivatives of f at P, the derivative in the direction opposite to that of ∇f at P has the smallest value. The value of this smallest directional derivative is $-\|\nabla f\|$ at P.*

Example 4 Let $f(x, y) = x^2 e^y$. Find the maximum value of a directional derivative at $(-2, 0)$, and find the unit vector in the direction in which the maximum value occurs.

Solution. Since

$$\nabla f(x, y) = f_x(x, y)\mathbf{i} + f_y(x, y)\mathbf{j} = 2xe^y\mathbf{i} + x^2 e^y\mathbf{j}$$

the gradient of f at $(-2, 0)$ is

$$\nabla f(-2, 0) = -4\mathbf{i} + 4\mathbf{j}$$

By Theorem 14.6.5, the maximum value of the directional derivative is

$$\|\nabla f(-2, 0)\| = \sqrt{(-4)^2 + 4^2} = \sqrt{32} = 4\sqrt{2}$$

This maximum occurs in the direction of $\nabla f(-2, 0)$. The unit vector in this direction is

$$\mathbf{u} = \frac{\nabla f(-2, 0)}{\|\nabla f(-2, 0)\|} = \frac{1}{4\sqrt{2}}(-4\mathbf{i} + 4\mathbf{j}) = -\frac{1}{\sqrt{2}}\mathbf{i} + \frac{1}{\sqrt{2}}\mathbf{j}$$ ◀

What would be the minimum value of a directional derivative of

$$f(x, y) = x^2 e^y$$

at $(-2, 0)$?

GRADIENTS ARE NORMAL TO LEVEL CURVES

We have seen that the gradient points in the direction in which a function increases most rapidly. For a function $f(x, y)$ of two variables, we will now consider how this direction of maximum rate of increase can be determined from a contour map of the function. Suppose that (x_0, y_0) is a point on a level curve $f(x, y) = c$ of f, and assume that this curve can be smoothly parametrized as

$$x = x(s), \quad y = y(s) \tag{13}$$

where s is an arc length parameter. Recall from Formula (6) of Section 13.4 that the unit tangent vector to (13) is

$$\mathbf{T} = \mathbf{T}(s) = \left(\frac{dx}{ds}\right)\mathbf{i} + \left(\frac{dy}{ds}\right)\mathbf{j}$$

Since $\mathbf{T}$ gives a direction along which f is nearly constant, we would expect the instantaneous rate of change of f with respect to distance in the direction of $\mathbf{T}$ to be 0. That is, we would expect that

$$D_{\mathbf{T}} f(x, y) = \nabla f(x, y) \cdot \mathbf{T}(s) = 0$$

To show this to be the case, we differentiate both sides of the equation $f(x, y) = c$ with respect to s. Assuming that f is differentiable at (x, y), we can use the chain rule to obtain

$$\frac{\partial f}{\partial x}\frac{dx}{ds} + \frac{\partial f}{\partial y}\frac{dy}{ds} = 0$$

which we can rewrite as

$$\left(\frac{\partial f}{\partial x}\mathbf{i} + \frac{\partial f}{\partial y}\mathbf{j}\right) \cdot \left(\frac{dx}{ds}\mathbf{i} + \frac{dy}{ds}\mathbf{j}\right) = 0$$

or, alternatively, as

$$\nabla f(x, y) \cdot \mathbf{T} = 0$$

Therefore, if $\nabla f(x, y) \neq \mathbf{0}$, then $\nabla f(x, y)$ should be normal to the level curve $f(x, y) = c$ at any point (x, y) on the curve.

It is proved in advanced courses that if $f(x, y)$ has continuous first-order partial derivatives, and if $\nabla f(x_0, y_0) \neq \mathbf{0}$, then near (x_0, y_0) the graph of $f(x, y) = c$ is indeed a smooth curve through (x_0, y_0). Furthermore, we also know from Theorem 14.4.4 that f will be differentiable at (x_0, y_0). We therefore have the following result.

Verify Theorem 14.6.6 for

$$f(x, y) = x^2 + y^2$$

and $(x_0, y_0) = (3, 4)$.

14.6.6 THEOREM. *Assume that $f(x, y)$ has continuous first-order partial derivatives in an open disk centered at (x_0, y_0) and that $\nabla f(x_0, y_0) \neq \mathbf{0}$. Then $\nabla f(x_0, y_0)$ is normal to the level curve of f through (x_0, y_0).*

When we examine a contour map, we instinctively regard the distance between adjacent contours to be measured in a normal direction. If the contours correspond to equally spaced values of f, then the closer together the contours appear to be, the more rapidly the values of f will be changing in that normal direction. It follows from Theorems 14.6.5 and 14.6.6 that this rate of change of f is given by $\|\nabla f(x, y)\|$. Thus, the closer together the contours appear to be, the greater the length of the gradient of f.

Example 5 A contour plot of a function f is given in Figure 14.6.6*a*. Sketch the directions of the gradient of f at the points P, Q, and R. At which of these three points does the gradient vector have maximum length? Minimum length?

Solution. It follows from Theorems 14.6.5 and 14.6.6 that the directions of the gradient vectors will be as given in Figure 14.6.6*b*. Based on the density of the contour lines, we would guess that the gradient of f has maximum length at R and minimum length at P, with the length at Q somewhere in between. ◀

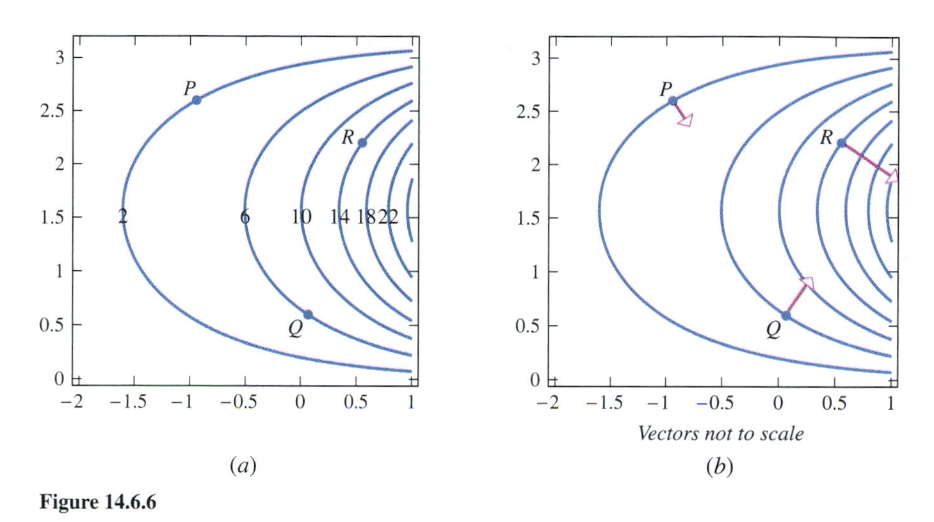

Figure 14.6.6

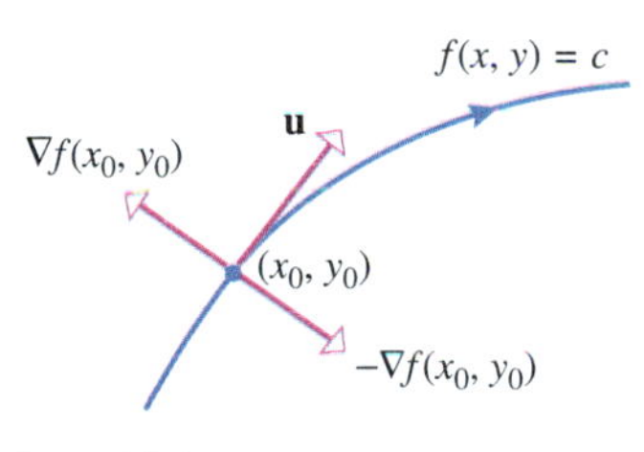

Figure 14.6.7

If (x_0, y_0) is a point on the level curve $f(x, y) = c$, then the slope of the surface $z = f(x, y)$ at that point in the direction of $\mathbf{u}$ is

$$D_{\mathbf{u}} f(x_0, y_0) = \nabla f(x_0, y_0) \cdot \mathbf{u}$$

If $\mathbf{u}$ is tangent to the level curve at (x_0, y_0), then $f(x, y)$ is neither increasing nor decreasing in that direction, so $D_{\mathbf{u}} f(x_0, y_0) = 0$. Thus, $\nabla f(x_0, y_0)$, $-\nabla f(x_0, y_0)$, and the tangent vector $\mathbf{u}$ mark the directions of maximum slope, minimum slope, and zero slope at a point (x_0, y_0) on a level curve (Figure 14.6.7). Good skiers use these facts intuitively to control their speed by zigzagging down ski slopes—they ski across the slope with their skis tangential to a level curve to stop their downhill motion, and they point their skis down the slope and normal to the level curve to obtain the most rapid descent.

AN APPLICATION OF GRADIENTS

There are numerous applications in which the motion of an object must be controlled so that it moves toward a heat source. For example, in medical applications the operation of certain diagnostic equipment is designed to locate heat sources generated by tumors or infections, and in military applications the trajectories of heat-seeking missiles are controlled to seek and destroy enemy aircraft. The following example illustrates how gradients are used to solve such problems.

► **Example 6** A heat-seeking particle is located at the point (2, 3) on a flat metal plate whose temperature at a point (x, y) is

$$T(x, y) = 10 - 8x^2 - 2y^2$$

Find an equation for the trajectory of the particle if it moves continuously in the direction of maximum temperature increase.

Solution. Assume that the trajectory is represented parametrically by the equations

$$x = x(t), \quad y = y(t)$$

where the particle is at the point (2, 3) at time $t = 0$. Because the particle moves in the direction of maximum temperature increase, its direction of motion at time t is in the direction of the gradient of $T(x, y)$, and hence its velocity vector $\mathbf{v}(t)$ at time t points in the direction of the gradient. Thus, there is a scalar k that depends on t such that

$$\mathbf{v}(t) = k\nabla T(x, y)$$

from which we obtain

$$\frac{dx}{dt}\mathbf{i} + \frac{dy}{dt}\mathbf{j} = k(-16x\mathbf{i} - 4y\mathbf{j})$$

Equating components yields

$$\frac{dx}{dt} = -16kx, \quad \frac{dy}{dt} = -4ky$$

and dividing to eliminate k yields

$$\frac{dy}{dx} = \frac{-4ky}{-16kx} = \frac{y}{4x}$$

Thus, we can obtain the trajectory by solving the initial-value problem

$$\frac{dy}{dx} - \frac{y}{4x} = 0, \quad y(2) = 3$$

The differential equation is a separable first-order linear equation and hence can be solved by separating the variables or by the method of integrating factors discussed in Section 9.1. We leave it for you to show that the solution of the initial-value problem is

$$y = \frac{3}{\sqrt[4]{2}}x^{1/4}$$

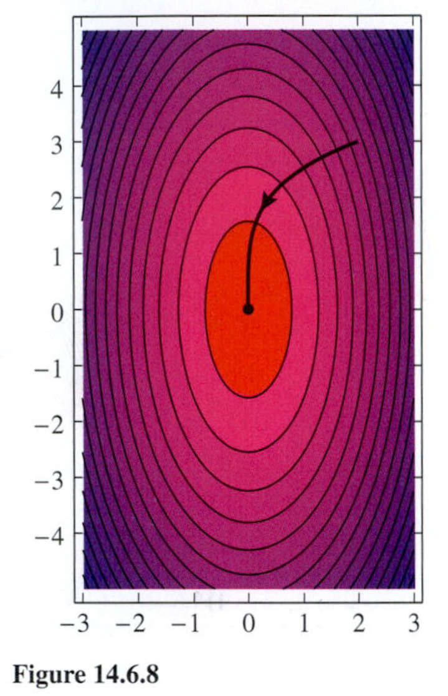

Figure 14.6.8

The graph of the trajectory and a contour plot of the temperature function are shown in Figure 14.6.8. ◀

✔QUICK CHECK EXERCISES 14.6

(See page 989 for answers.)

1. The gradient of $f(x, y, z) = xy^2z^3$ at the point $(1, 1, 1)$ is ________.

2. Suppose that the differentiable function $f(x, y)$ has the property that

$$f\left(2 + \frac{s\sqrt{3}}{2}, 1 + \frac{s}{2}\right) = 3se^s$$

The directional derivative of f in the direction of

$$\mathbf{u} = \frac{\sqrt{3}}{2}\mathbf{i} + \frac{1}{2}\mathbf{j}$$

at $(2, 1)$ is ________.

3. If the gradient of $f(x, y)$ at the origin is $6\mathbf{i} + 8\mathbf{j}$, then the directional derivative of f in the direction of $\mathbf{a} = 3\mathbf{i} + 4\mathbf{j}$ at the origin is ________. The slope of the tangent line to the level curve of f through the origin at $(0, 0)$ is ________.

4. If the gradient of $f(x, y, z)$ at $(1, 2, 3)$ is $2\mathbf{i} - 2\mathbf{j} + \mathbf{k}$, then the maximum value for a directional derivative of f at $(1, 2, 3)$ is ________ and the minimum value for a directional derivative at this point is ________.

EXERCISE SET 14.6 Graphing Utility C CAS

1–8 Find $D_{\mathbf{u}}f$ at P.

1. $f(x, y) = (1 + xy)^{3/2}$; $P(3, 1)$; $\mathbf{u} = \frac{1}{\sqrt{2}}\mathbf{i} + \frac{1}{\sqrt{2}}\mathbf{j}$

2. $f(x, y) = e^{2xy}$; $P(4, 0)$; $\mathbf{u} = -\frac{3}{5}\mathbf{i} + \frac{4}{5}\mathbf{j}$

3. $f(x, y) = \ln(1 + x^2 + y)$; $P(0, 0)$; $\mathbf{u} = -\frac{1}{\sqrt{10}}\mathbf{i} - \frac{3}{\sqrt{10}}\mathbf{j}$

4. $f(x, y) = \frac{cx + dy}{x - y}$; $P(3, 4)$; $\mathbf{u} = \frac{4}{5}\mathbf{i} + \frac{3}{5}\mathbf{j}$

5. $f(x, y, z) = 4x^5y^2z^3$; $P(2, -1, 1)$; $\mathbf{u} = \frac{1}{3}\mathbf{i} + \frac{2}{3}\mathbf{j} - \frac{2}{3}\mathbf{k}$

6. $f(x, y, z) = ye^{xz} + z^2$; $P(0, 2, 3)$; $\mathbf{u} = \frac{2}{7}\mathbf{i} - \frac{3}{7}\mathbf{j} + \frac{6}{7}\mathbf{k}$

7. $f(x, y, z) = \ln(x^2 + 2y^2 + 3z^2)$; $P(-1, 2, 4)$; $\mathbf{u} = -\frac{3}{13}\mathbf{i} - \frac{4}{13}\mathbf{j} - \frac{12}{13}\mathbf{k}$

8. $f(x, y, z) = \sin xyz$; $P\left(\frac{1}{2}, \frac{1}{3}, \pi\right)$;
$\mathbf{u} = \dfrac{1}{\sqrt{3}}\mathbf{i} - \dfrac{1}{\sqrt{3}}\mathbf{j} + \dfrac{1}{\sqrt{3}}\mathbf{k}$

9–18 Find the directional derivative of f at P in the direction of $\mathbf{a}$.

9. $f(x, y) = 4x^3y^2$; $P(2, 1)$; $\mathbf{a} = 4\mathbf{i} - 3\mathbf{j}$
10. $f(x, y) = x^2 - 3xy + 4y^3$; $P(-2, 0)$; $\mathbf{a} = \mathbf{i} + 2\mathbf{j}$
11. $f(x, y) = y^2 \ln x$; $P(1, 4)$; $\mathbf{a} = -3\mathbf{i} + 3\mathbf{j}$
12. $f(x, y) = e^x \cos y$; $P(0, \pi/4)$; $\mathbf{a} = 5\mathbf{i} - 2\mathbf{j}$
13. $f(x, y) = \tan^{-1}(y/x)$; $P(-2, 2)$; $\mathbf{a} = -\mathbf{i} - \mathbf{j}$
14. $f(x, y) = xe^y - ye^x$; $P(0, 0)$; $\mathbf{a} = 5\mathbf{i} - 2\mathbf{j}$
15. $f(x, y, z) = x^3z - yx^2 + z^2$; $P(2, -1, 1)$;
$\mathbf{a} = 3\mathbf{i} - \mathbf{j} + 2\mathbf{k}$
16. $f(x, y, z) = y - \sqrt{x^2 + z^2}$; $P(-3, 1, 4)$;
$\mathbf{a} = 2\mathbf{i} - 2\mathbf{j} - \mathbf{k}$
17. $f(x, y, z) = \dfrac{z - x}{z + y}$; $P(1, 0, -3)$; $\mathbf{a} = -6\mathbf{i} + 3\mathbf{j} - 2\mathbf{k}$
18. $f(x, y, z) = e^{x+y+3z}$; $P(-2, 2, -1)$; $\mathbf{a} = 20\mathbf{i} - 4\mathbf{j} + 5\mathbf{k}$

19–22 Find the directional derivative of f at P in the direction of a vector making the counterclockwise angle θ with the positive x-axis.

19. $f(x, y) = \sqrt{xy}$; $P(1, 4)$; $\theta = \pi/3$
20. $f(x, y) = \dfrac{x - y}{x + y}$; $P(-1, -2)$; $\theta = \pi/2$
21. $f(x, y) = \tan(2x + y)$; $P(\pi/6, \pi/3)$; $\theta = 7\pi/4$
22. $f(x, y) = \sinh x \cosh y$; $P(0, 0)$; $\theta = \pi$
23. Find the directional derivative of
$$f(x, y) = \frac{x}{x + y}$$
at $P(1, 0)$ in the direction of $Q(-1, -1)$.
24. Find the directional derivative of $f(x, y) = e^{-x} \sec y$ at $P(0, \pi/4)$ in the direction of the origin.
25. Find the directional derivative of $f(x, y) = \sqrt{xy}e^y$ at $P(1, 1)$ in the direction of the negative y-axis.
26. Let
$$f(x, y) = \frac{y}{x + y}$$
Find a unit vector $\mathbf{u}$ for which $D_{\mathbf{u}}f(2, 3) = 0$.
27. Find the directional derivative of
$$f(x, y, z) = \frac{y}{x + z}$$
at $P(2, 1, -1)$ in the direction from P to $Q(-1, 2, 0)$.
28. Find the directional derivative of the function
$$f(x, y, z) = x^3y^2z^5 - 2xz + yz + 3x$$
at $P(-1, -2, 1)$ in the direction of the negative z-axis.

FOCUS ON CONCEPTS

29. Suppose that $D_{\mathbf{u}}f(1, 2) = -5$ and $D_{\mathbf{v}}f(1, 2) = 10$, where $\mathbf{u} = \frac{3}{5}\mathbf{i} - \frac{4}{5}\mathbf{j}$ and $\mathbf{v} = \frac{4}{5}\mathbf{i} + \frac{3}{5}\mathbf{j}$. Find
(a) $f_x(1, 2)$
(b) $f_y(1, 2)$
(c) the directional derivative of f at $(1, 2)$ in the direction of the origin.
30. Given that $f_x(-5, 1) = -3$ and $f_y(-5, 1) = 2$, find the directional derivative of f at $P(-5, 1)$ in the direction of the vector from P to $Q(-4, 3)$.
31. The accompanying figure shows some level curves of an unspecified function $f(x, y)$. Which of the three vectors shown in the figure is most likely to be ∇f? Explain.
32. The accompanying figure shows some level curves of an unspecified function $f(x, y)$. Of the gradients at P and Q, which probably has the greater length? Explain.

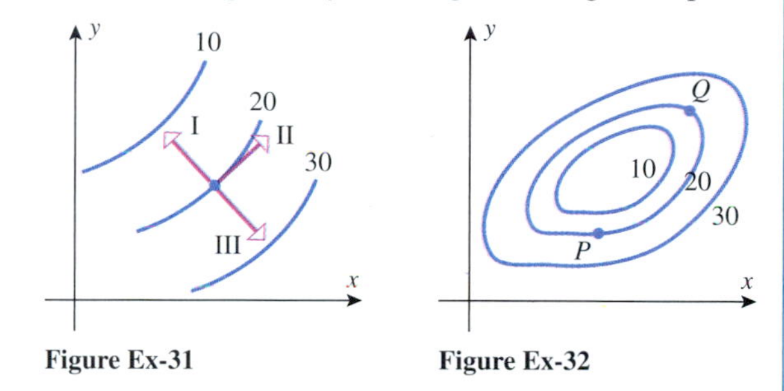

Figure Ex-31

Figure Ex-32

33–36 Find ∇z or ∇w.

33. $z = 4x - 8y$
34. $z = e^{-3y} \cos 4x$
35. $w = \ln \sqrt{x^2 + y^2 + z^2}$
36. $w = e^{-5x} \sec x^2yz$

37–40 Find the gradient of f at the indicated point.

37. $f(x, y) = (x^2 + xy)^3$; $(-1, -1)$
38. $f(x, y) = (x^2 + y^2)^{-1/2}$; $(3, 4)$
39. $f(x, y, z) = y \ln(x + y + z)$; $(-3, 4, 0)$
40. $f(x, y, z) = y^2z \tan^3 x$; $(\pi/4, -3, 1)$

41–44 Sketch the level curve of $f(x, y)$ that passes through P and draw the gradient vector at P.

41. $f(x, y) = 4x - 2y + 3$; $P(1, 2)$
42. $f(x, y) = y/x^2$; $P(-2, 2)$
43. $f(x, y) = x^2 + 4y^2$; $P(-2, 0)$
44. $f(x, y) = x^2 - y^2$; $P(2, -1)$
45. Find a unit vector $\mathbf{u}$ that is normal at $P(1, -2)$ to the level curve of $f(x, y) = 4x^2y$ through P.
46. Find a unit vector $\mathbf{u}$ that is normal at $P(2, -3)$ to the level curve of $f(x, y) = 3x^2y - xy$ through P.

47–54 Find a unit vector in the direction in which f increases most rapidly at P, and find the rate of change of f at P in that direction.

47. $f(x, y) = 4x^3y^2$; $P(-1, 1)$

48. $f(x, y) = 3x - \ln y$; $P(2, 4)$

49. $f(x, y) = \sqrt{x^2 + y^2}$; $P(4, -3)$

50. $f(x, y) = \dfrac{x}{x + y}$; $P(0, 2)$

51. $f(x, y, z) = x^3z^2 + y^3z + z - 1$; $P(1, 1, -1)$

52. $f(x, y, z) = \sqrt{x - 3y + 4z}$; $P(0, -3, 0)$

53. $f(x, y, z) = \dfrac{x}{z} + \dfrac{z}{y^2}$; $P(1, 2, -2)$

54. $f(x, y, z) = \tan^{-1}\left(\dfrac{x}{y + z}\right)$; $P(4, 2, 2)$

55–60 Find a unit vector in the direction in which f decreases most rapidly at P, and find the rate of change of f at P in that direction.

55. $f(x, y) = 20 - x^2 - y^2$; $P(-1, -3)$

56. $f(x, y) = e^{xy}$; $P(2, 3)$

57. $f(x, y) = \cos(3x - y)$; $P(\pi/6, \pi/4)$

58. $f(x, y) = \sqrt{\dfrac{x - y}{x + y}}$; $P(3, 1)$

59. $f(x, y, z) = \dfrac{x + z}{z - y}$; $P(5, 7, 6)$

60. $f(x, y, z) = 4e^{xy} \cos z$; $P(0, 1, \pi/4)$

FOCUS ON CONCEPTS

61. Given that $\nabla f(4, -5) = 2\mathbf{i} - \mathbf{j}$, find the directional derivative of the function f at the point $(4, -5)$ in the direction of $\mathbf{a} = 5\mathbf{i} + 2\mathbf{j}$.

62. Given that $\nabla f(x_0, y_0) = \mathbf{i} - 2\mathbf{j}$ and $D_{\mathbf{u}} f(x_0, y_0) = -2$, find $\mathbf{u}$ (two answers).

63. The accompanying figure shows some level curves of an unspecified function $f(x, y)$.

(a) Use the available information to approximate the length of the vector $\nabla f(1, 2)$, and sketch the approximation. Explain how you approximated the length and determined the direction of the vector.

(b) Sketch an approximation of the vector $-\nabla f(4, 4)$.

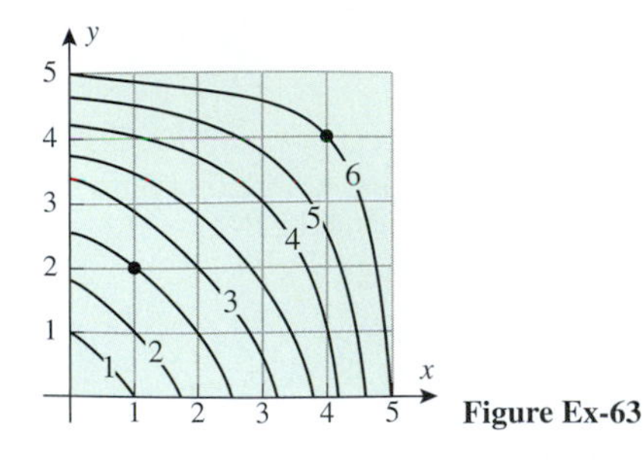

Figure Ex-63

64. The accompanying figure shows a topographic map of a hill and a point P at the bottom of the hill. Suppose that you want to climb from the point P toward the top of the hill in such a way that you are always ascending in the direction of steepest slope. Sketch the projection of your path on the contour map. This is called the ***path of steepest ascent***. Explain how you determined the path.

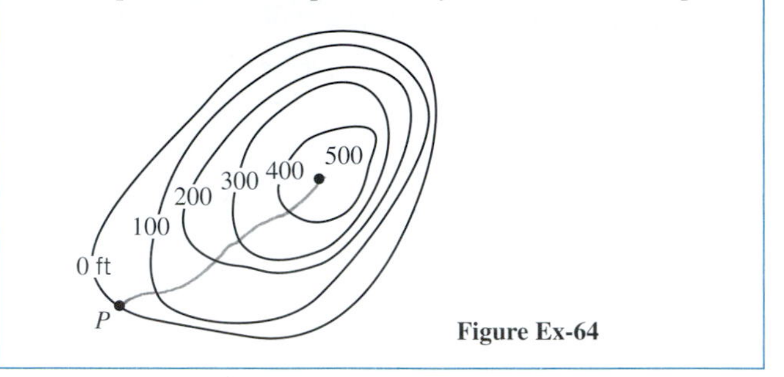

Figure Ex-64

65. Let $z = 3x^2 - y^2$. Find all points at which $\|\nabla z\| = 6$.

66. Given that $z = 3x + y^2$, find $\nabla\|\nabla z\|$ at the point $(5, 2)$.

67. A particle moves along a path C given by the equations $x = t$ and $y = -t^2$. If $z = x^2 + y^2$, find dz/ds along C at the instant when the particle is at the point $(2, -4)$.

68. The temperature (in degrees Celsius) at a point (x, y) on a metal plate in the xy-plane is

$$T(x, y) = \frac{xy}{1 + x^2 + y^2}$$

(a) Find the rate of change of temperature at $(1, 1)$ in the direction of $\mathbf{a} = 2\mathbf{i} - \mathbf{j}$.

(b) An ant at $(1, 1)$ wants to walk in the direction in which the temperature drops most rapidly. Find a unit vector in that direction.

69. If the electric potential at a point (x, y) in the xy-plane is $V(x, y)$, then the ***electric intensity vector*** at the point (x, y) is $\mathbf{E} = -\nabla V(x, y)$. Suppose that $V(x, y) = e^{-2x} \cos 2y$.

(a) Find the electric intensity vector at $(\pi/4, 0)$.

(b) Show that at each point in the plane, the electric potential decreases most rapidly in the direction of the vector $\mathbf{E}$.

70. On a certain mountain, the elevation z above a point (x, y) in an xy-plane at sea level is $z = 2000 - 0.02x^2 - 0.04y^2$, where x, y, and z are in meters. The positive x-axis points east, and the positive y-axis north. A climber is at the point $(-20, 5, 1991)$.

(a) If the climber uses a compass reading to walk due west, will she begin to ascend or descend?

(b) If the climber uses a compass reading to walk northeast, will she ascend or descend? At what rate?

(c) In what compass direction should the climber begin walking to travel a level path (two answers)?

71. Given that the directional derivative of $f(x, y, z)$ at the point $(3, -2, 1)$ in the direction of $\mathbf{a} = 2\mathbf{i} - \mathbf{j} - 2\mathbf{k}$ is -5 and that $\|\nabla f(3, -2, 1)\| = 5$, find $\nabla f(3, -2, 1)$.

72. The temperature (in degrees Celsius) at a point (x, y, z) in a metal solid is

$$T(x, y, z) = \frac{xyz}{1 + x^2 + y^2 + z^2}$$

(a) Find the rate of change of temperature with respect to distance at $(1, 1, 1)$ in the direction of the origin.
(b) Find the direction in which the temperature rises most rapidly at the point $(1, 1, 1)$. (Express your answer as a unit vector.)
(c) Find the rate at which the temperature rises moving from $(1, 1, 1)$ in the direction obtained in part (b).

73. Let $r = \sqrt{x^2 + y^2}$.
(a) Show that $\nabla r = \dfrac{\mathbf{r}}{r}$, where $\mathbf{r} = x\mathbf{i} + y\mathbf{j}$.
(b) Show that $\nabla f(r) = f'(r)\nabla r = \dfrac{f'(r)}{r}\mathbf{r}$.

74. Use the formula in part (b) of Exercise 73 to find
(a) $\nabla f(r)$ if $f(r) = re^{-3r}$
(b) $f(r)$ if $\nabla f(r) = 3r^2\mathbf{r}$ and $f(2) = 1$.

75. Let $\mathbf{u}_r$ be a unit vector whose counterclockwise angle from the positive x-axis is θ, and let $\mathbf{u}_\theta$ be a unit vector $90°$ counterclockwise from $\mathbf{u}_r$. Show that if $z = f(x, y)$, $x = r\cos\theta$, and $y = r\sin\theta$, then

$$\nabla z = \frac{\partial z}{\partial r}\mathbf{u}_r + \frac{1}{r}\frac{\partial z}{\partial \theta}\mathbf{u}_\theta$$

[*Hint:* Use part (c) of Exercise 63, Section 14.5.]

76. Prove: If f and g are differentiable, then
(a) $\nabla(f + g) = \nabla f + \nabla g$
(b) $\nabla(cf) = c\nabla f$ (c constant)
(c) $\nabla(fg) = f\nabla g + g\nabla f$
(d) $\nabla\left(\dfrac{f}{g}\right) = \dfrac{g\nabla f - f\nabla g}{g^2}$
(e) $\nabla(f^n) = nf^{n-1}\nabla f$.

77–78 A heat-seeking particle is located at the point P on a flat metal plate whose temperature at a point (x, y) is $T(x, y)$. Find parametric equations for the trajectory of the particle if it moves continuously in the direction of maximum temperature increase.

77. $T(x, y) = 5 - 4x^2 - y^2$; $P(1, 4)$

78. $T(x, y) = 100 - x^2 - 2y^2$; $P(5, 3)$

79. Use a graphing utility to generate the trajectory of the particle together with some representative level curves of the temperature function in Exercise 77.

80. Use a graphing utility to generate the trajectory of the particle together with some representative level curves of the temperature function in Exercise 78.

81. (a) Use a CAS to graph $f(x, y) = (x^2 + 3y^2)e^{-(x^2+y^2)}$.
(b) At how many points do you think it is true that $D_{\mathbf{u}}f(x, y) = 0$ for all unit vectors $\mathbf{u}$?
(c) Use a CAS to find ∇f.
(d) Use a CAS to solve the equation $\nabla f(x, y) = 0$ for x and y.
(e) Use the result in part (d) together with Theorem 14.6.5 to check your conjecture in part (b).

82. Prove: If $x = x(t)$ and $y = y(t)$ are differentiable at t, and if $z = f(x, y)$ is differentiable at the point $(x(t), y(t))$, then

$$\frac{dz}{dt} = \nabla z \cdot \mathbf{r}'(t)$$

where $\mathbf{r}(t) = x(t)\mathbf{i} + y(t)\mathbf{j}$.

83. Prove: If f, f_x, and f_y are continuous on a circular region, and if $\nabla f(x, y) = \mathbf{0}$ throughout the region, then $f(x, y)$ is constant on the region. [*Hint:* See Exercise 75, Section 14.5.]

84. Prove: If the function f is differentiable at the point (x, y) and if $D_{\mathbf{u}}f(x, y) = 0$ in two nonparallel directions, then $D_{\mathbf{u}}f(x, y) = 0$ in all directions.

85. Given that the functions $u = u(x, y, z)$, $v = v(x, y, z)$, $w = w(x, y, z)$, and $f(u, v, w)$ are all differentiable, show that

$$\nabla f(u, v, w) = \frac{\partial f}{\partial u}\nabla u + \frac{\partial f}{\partial v}\nabla v + \frac{\partial f}{\partial w}\nabla w$$

✔ QUICK CHECK ANSWERS 14.6

1. $\langle 1, 2, 3\rangle$ **2.** 3 **3.** 10; $-\frac{3}{4}$ **4.** 3; -3

14.7 TANGENT PLANES AND NORMAL VECTORS

In this section we will discuss tangent planes to surfaces in three-dimensional space. We will be concerned with three main questions: What is a tangent plane? When do tangent planes exist? How do we find equations of tangent planes?

TANGENT PLANES

Recall from Section 14.4 that if a function $f(x, y)$ is differentiable at a point (x_0, y_0), then we want it to be the case that the surface $z = f(x, y)$ has a nonvertical tangent plane at the point $P_0(x_0, y_0, f(x_0, y_0))$. We also saw in Section 14.4 that the linear function

$$L(x, y) = f(x_0, y_0) + f_x(x_0, y_0)(x - x_0) + f_y(x_0, y_0)(y - y_0)$$

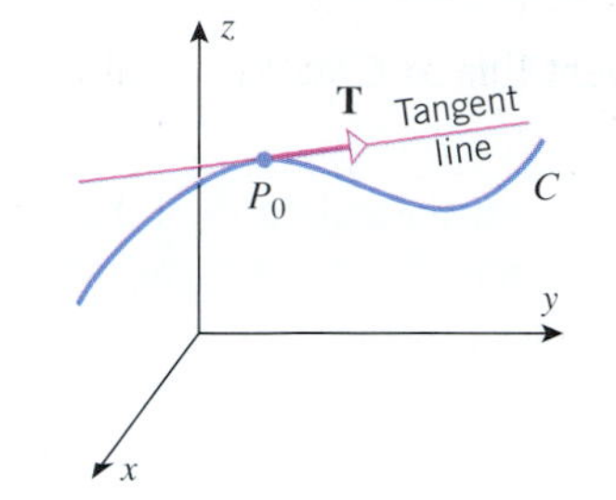

Figure 14.7.1

approximates $f(x, y)$ very closely near (x_0, y_0) and the graph of L is a nonvertical plane through the point P_0. This suggests that the graph of L is the tangent plane we seek. We can now provide some *geometric* justification for this conclusion.

We will base our concept of a tangent plane to a surface $S: z = f(x, y)$ on the more elementary notion of a tangent line to a curve C in 3-space (Figure 14.7.1). Intuitively, we would expect a tangent plane to S at a point P_0 to be composed of the tangent lines at P_0 of all curves on S that pass through P_0 (Figure 14.7.2). The following theorem shows that the graph of the local linear approximation is indeed tangent to the surface $z = f(x, y)$ in this geometric sense.

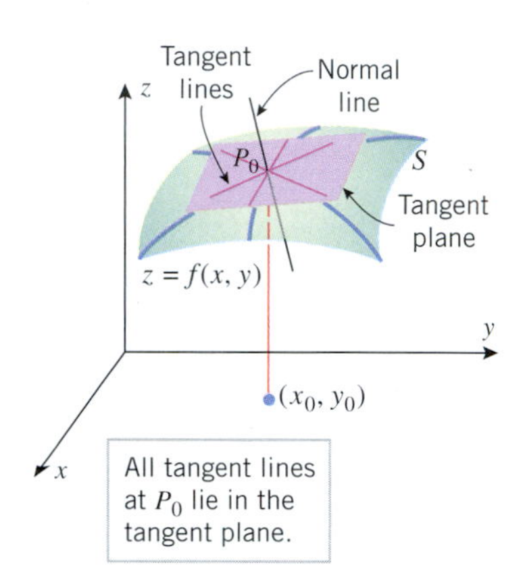

Figure 14.7.2

14.7.1 THEOREM. *Assume that the function $f(x, y)$ is differentiable at (x_0, y_0) and let $P_0(x_0, y_0, f(x_0, y_0))$ denote the corresponding point on the graph of f. Let T denote the graph of the local linear approximation*

$$L(x, y) = f(x_0, y_0) + f_x(x_0, y_0)(x - x_0) + f_y(x_0, y_0)(y - y_0) \tag{1}$$

to f at (x_0, y_0). Then a line is tangent at P_0 to a curve C on the surface $z = f(x, y)$ if and only if the line is contained in T.

PROOF. The graph T of (1) is the plane

$$z = f(x_0, y_0) + f_x(x_0, y_0)(x - x_0) + f_y(x_0, y_0)(y - y_0)$$

for which

$$\mathbf{n} = f_x(x_0, y_0)\mathbf{i} + f_y(x_0, y_0)\mathbf{j} - \mathbf{k}$$

is a normal vector (verify). Let C denote a curve on the surface $z = f(x, y)$ through P_0 and assume that C is parametrized by

$$x = x(t), \quad y = y(t), \quad z = z(t)$$

with

$$x_0 = x(t_0), \quad y_0 = y(t_0), \quad f(x_0, y_0) = z(t_0)$$

The tangent line l to C through P_0 is then parallel to the vector

$$\mathbf{r}' = x'(t_0)\mathbf{i} + y'(t_0)\mathbf{j} + z'(t_0)\mathbf{k}$$

where we assume that $\mathbf{r}' \neq \mathbf{0}$ (Definition 13.2.7). To prove that l is contained in T, it suffices to prove that $\mathbf{n} \cdot \mathbf{r}' = 0$. Since C lies on the graph of f, we have

$$z(t) = f(x(t), y(t))$$

Using the chain rule to compute the derivative of $z(t)$ at t_0 yields

$$z'(t_0) = f_x(x_0, y_0)x'(t_0) + f_y(x_0, y_0)y'(t_0)$$

or, equivalently, that

$$(f_x(x_0, y_0)\mathbf{i} + f_y(x_0, y_0)\mathbf{j} - \mathbf{k}) \cdot (x'(t_0)\mathbf{i} + y'(t_0)\mathbf{j} + z'(t_0)\mathbf{k}) = 0$$

But this is just the equation $\mathbf{n} \cdot \mathbf{r}' = 0$, which completes the proof that l is contained in T.

Conversely, let $\mathbf{a} = a_1\mathbf{i} + a_2\mathbf{j} + a_3\mathbf{k}$ denote the direction vector for a line l through P_0 contained in T. Then

$$0 = \mathbf{n} \cdot \mathbf{a} = a_1 f_x(x_0, y_0) + a_2 f_y(x_0, y_0) - a_3$$

which implies that

$$a_3 = a_1 f_x(x_0, y_0) + a_2 f_y(x_0, y_0)$$

Let C denote the curve with parametric equations

$$x = x(t) = x_0 + a_1 t, \quad y = y(t) = y_0 + a_2 t, \quad z = z(t) = f(x(t), y(t))$$

The curve C passes through P_0 when $t = 0$ and the tangent line to C at P_0 has direction vector

$$\mathbf{r}' = x'(0)\mathbf{i} + y'(0)\mathbf{j} + z'(0)\mathbf{k} = a_1\mathbf{i} + a_2\mathbf{j} + z'(0)\mathbf{k}$$

It follows from the chain rule that

$$z'(0) = a_1 f_x(x_0, y_0) + a_2 f_y(x_0, y_0) = a_3$$

and therefore

$$\mathbf{r}' = x'(0)\mathbf{i} + y'(0)\mathbf{j} + z'(0)\mathbf{k} = a_1\mathbf{i} + a_2\mathbf{j} + a_3\mathbf{k} = \mathbf{a}$$

Thus, the vector $\mathbf{a} = a_1\mathbf{i} + a_2\mathbf{j} + a_3\mathbf{k}$ is the direction vector $\mathbf{r}'$ for the line through P_0 tangent to C. Therefore, this line is l, which completes the proof that l is tangent at P_0 to a curve C on the surface $z = f(x, y)$. ■

Use Theorem 14.7.1 to show that if a curve is contained in a plane, then so is any tangent line to the curve.

Based on Theorem 14.7.1 we make the following definition.

14.7.2 DEFINITION. If $f(x, y)$ is differentiable at the point (x_0, y_0), then the ***tangent plane*** to the surface $z = f(x, y)$ at the point $P_0(x_0, y_0, f(x_0, y_0))$ [or (x_0, y_0)] is the plane

$$z = f(x_0, y_0) + f_x(x_0, y_0)(x - x_0) + f_y(x_0, y_0)(y - y_0) \tag{2}$$

The line through the point P_0 parallel to the vector $\mathbf{n}$ is perpendicular to the tangent plane (2). We will refer to this line as the ***normal line*** to the surface $z = f(x, y)$ at P_0. It follows that this normal line can be expressed parametrically as

$$x = x_0 + f_x(x_0, y_0)t, \quad y = y_0 + f_y(x_0, y_0)t, \quad z = f(x_0, y_0) - t \tag{3}$$

► **Example 1** Find an equation for the tangent plane and parametric equations for the normal line to the surface $z = x^2y$ at the point $(2, 1, 4)$.

Solution. The partial derivatives of f are

$$f_x(x, y) = 2xy, \quad f_y(x, y) = x^2$$
$$f_x(2, 1) = 4, \quad f_y(2, 1) = 4$$

Therefore, the tangent plane has equation

$$z = 4 + 4(x - 2) + 4(y - 1) = 4x + 4y - 8$$

and the normal line has equations

$$x = 2 + 4t, \quad y = 1 + 4t, \quad z = 4 - t$$ ◄

TANGENT PLANES AND TOTAL DIFFERENTIALS

Recall that for a function $z = f(x, y)$ of two variables, the approximation by differentials is

$$\Delta z = \Delta f = f(x, y) - f(x_0, y_0) \approx dz = f_x(x_0, y_0)(x - x_0) + f_y(x_0, y_0)(y - y_0)$$

The tangent plane provides a geometric interpretation of this approximation. We see in Figure 14.7.3 that Δz is the change in z *along the surface* $z = f(x, y)$ from the point $P_0(x_0, y_0, f(x_0, y_0))$ to the point $P(x, y, f(x, y))$, and dz is the change in z *along the tangent plane* from P_0 to $Q(x, y, L(x, y))$. The small vertical displacement at (x, y) between the surface and the plane represents the error in the local linear approximation to

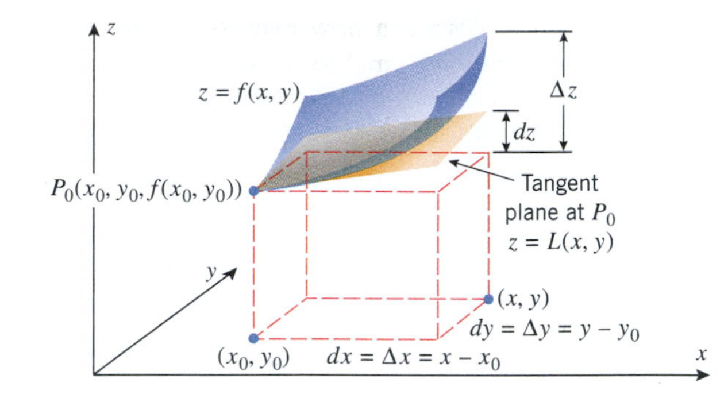

Figure 14.7.3

Note that the tangent plane in Figure 14.7.3 is analogous to the tangent line in Figure 14.7.2.

f at (x_0, y_0). We have seen that near (x_0, y_0) this error term has magnitude much smaller than the distance between (x, y) and (x_0, y_0).

TANGENT PLANES TO LEVEL SURFACES

We now consider the problem of finding tangent planes to surfaces that can be represented implicitly by equations of the form $F(x, y, z) = c$. We will assume that F has continuous first-order partial derivatives. This assumption poses no real restriction on the functions we will routinely encounter and has an important geometric consequence. In advanced courses it is proved that if F has continuous first-order partial derivatives, and if $\nabla F(x_0, y_0, z_0) \neq \mathbf{0}$, then near $P_0(x_0, y_0, z_0)$ the graph of $F(x, y, z) = c$ is actually the graph of an implicitly defined differentiable function of (at least) one of the following forms:

$$z = f(x, y), \quad y = g(x, z), \quad x = h(y, z) \tag{4}$$

This guarantees that near P_0 the graph of $F(x, y, z) = c$ is indeed a "surface" (rather than some possibly exotic-looking set of points in 3-space), and it follows from Theorem 14.7.1 that there is a tangent plane to the surface at the point P_0.

Fortunately, we do not need to solve the equation $F(x, y, z) = c$ for one of the functions in (4) in order to find the tangent plane at P_0. (In practice, this may be impossible.) We know from Theorem 14.7.1 that a line through P_0 will belong to this tangent plane if and only if it is a tangent line at P_0 of a curve C on the surface $F(x, y, z) = c$. Suppose that C is parametrized by

$$x = x(t), \quad y = y(t), \quad z = z(t)$$

with

$$x_0 = x(t_0), \quad y_0 = y(t_0), \quad z_0 = z(t_0)$$

The tangent line l to C through P_0 is then parallel to the vector

$$\mathbf{r}' = x'(t_0)\mathbf{i} + y'(t_0)\mathbf{j} + z'(t_0)\mathbf{k}$$

where we assume that $\mathbf{r}' \neq \mathbf{0}$ (Definition 13.2.7). Since C is on the surface $F(x, y, z) = c$, we have

$$c = F(x(t), y(t), z(t)) \tag{5}$$

Computing the derivative at t_0 of both sides of (5), we have by the chain rule that

$$0 = F_x(x_0, y_0, z_0)x'(t_0) + F_y(x_0, y_0, z_0)y'(t_0) + F_z(x_0, y_0, z_0)z'(t_0)$$

We can write this equation in vector form as

$$0 = (F_x(x_0, y_0, z_0)\mathbf{i} + F_y(x_0, y_0, z_0)\mathbf{j} + F_z(x_0, y_0, z_0)\mathbf{k}) \cdot (x'(t_0)\mathbf{i} + y'(t_0)\mathbf{j} + z'(t_0)\mathbf{k})$$

or

$$0 = \nabla F(x_0, y_0, z_0) \cdot \mathbf{r}'$$

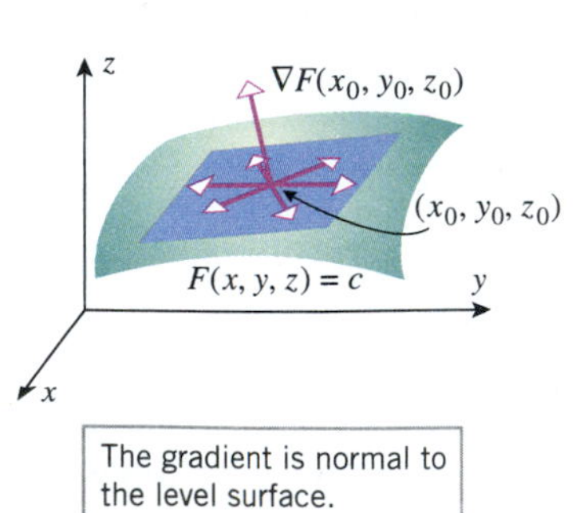

The gradient is normal to the level surface.

Figure 14.7.4

It follows that if $\nabla F(x_0, y_0, z_0) \neq \mathbf{0}$, then $\nabla F(x_0, y_0, z_0)$ is normal to line l. We conclude that if $\nabla F(x_0, y_0, z_0) \neq \mathbf{0}$, then $\nabla F(x_0, y_0, z_0)$ is normal to any line through P_0 that is contained in the tangent plane to the surface $F(x, y, z) = c$ at P_0. It follows that $\nabla F(x_0, y_0, z_0)$ is a normal vector to this plane and hence is normal to the level surface (Figure 14.7.4).

We can now express the equation of the tangent plane to the level surface at P_0 in point-normal form as

$$F_x(x_0, y_0, z_0)(x - x_0) + F_y(x_0, y_0, z_0)(y - y_0) + F_z(x_0, y_0, z_0)(z - z_0) = 0$$

[see Formula (3) of Section 12.6]. Based on this analysis we have the following theorem.

> **14.7.3 THEOREM.** *Assume that $F(x, y, z)$ has continuous first-order partial derivatives and let $c = F(x_0, y_0, z_0)$. If $\nabla F(x_0, y_0, z_0) \neq \mathbf{0}$, then $\nabla F(x_0, y_0, z_0)$ is a **normal vector** to the surface $F(x, y, z) = c$ at the point $P_0(x_0, y_0, z_0)$, and the **tangent plane** to this surface at P_0 is the plane with equation*
>
> $$F_x(x_0, y_0, z_0)(x - x_0) + F_y(x_0, y_0, z_0)(y - y_0) + F_z(x_0, y_0, z_0)(z - z_0) = 0 \quad (6)$$

Theorem 14.7.3 can be viewed as an extension of Theorem 14.6.6 from curves to surfaces.

▶ **Example 2** Find an equation of the tangent plane to the ellipsoid $x^2 + 4y^2 + z^2 = 18$ at the point $(1, 2, 1)$, and determine the acute angle that this plane makes with the xy-plane.

Solution. The ellipsoid is a level surface of the function $F(x, y, z) = x^2 + 4y^2 + z^2$, so we begin by finding the gradient of this function at the point $(1, 2, 1)$. The computations are

$$\nabla F(x, y, z) = \frac{\partial F}{\partial x}\mathbf{i} + \frac{\partial F}{\partial y}\mathbf{j} + \frac{\partial F}{\partial z}\mathbf{k} = 2x\mathbf{i} + 8y\mathbf{j} + 2z\mathbf{k}$$

$$\nabla F(1, 2, 1) = 2\mathbf{i} + 16\mathbf{j} + 2\mathbf{k}$$

Thus,

$$F_x(1, 2, 1) = 2, \quad F_y(1, 2, 1) = 16, \quad F_z(1, 2, 1) = 2$$

and hence from (6) the equation of the tangent plane is

$$2(x - 1) + 16(y - 2) + 2(z - 1) = 0 \quad \text{or} \quad x + 8y + z = 18$$

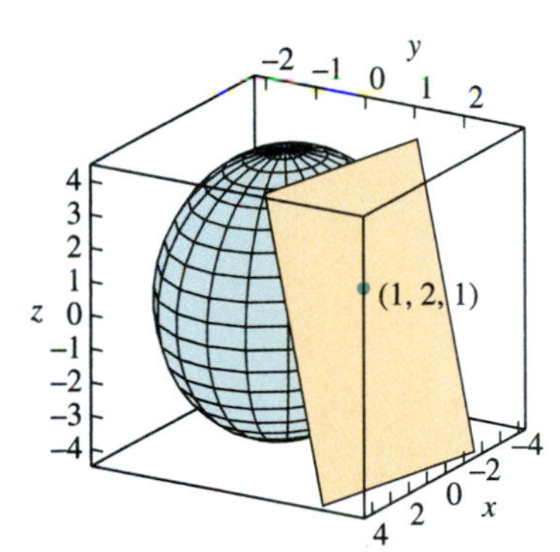

Figure 14.7.5

To find the acute angle θ between the tangent plane and the xy-plane, we will apply Formula (9) of Section 12.6 with $\mathbf{n}_1 = \nabla F(1, 2, 1) = 2\mathbf{i} + 16\mathbf{j} + 2\mathbf{k}$ and $\mathbf{n}_2 = \mathbf{k}$. This yields

$$\cos\theta = \frac{|\nabla F(1, 2, 1) \cdot \mathbf{k}|}{\|\nabla F(1, 2, 1)\|\|\mathbf{k}\|} = \frac{2}{(2\sqrt{66})(1)} = \frac{1}{\sqrt{66}}$$

Thus,

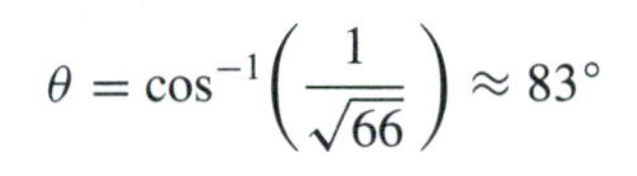

$$\theta = \cos^{-1}\left(\frac{1}{\sqrt{66}}\right) \approx 83^\circ$$

(Figure 14.7.5). ◀

USING GRADIENTS TO FIND TANGENT LINES TO INTERSECTIONS OF SURFACES

In general, the intersection of two surfaces $F(x, y, z) = 0$ and $G(x, y, z) = 0$ will be a curve in 3-space. If (x_0, y_0, z_0) is a point on this curve, then $\nabla F(x_0, y_0, z_0)$ will be normal to the surface $F(x, y, z) = 0$ at (x_0, y_0, z_0) and $\nabla G(x_0, y_0, z_0)$ will be normal to the surface $G(x, y, z) = 0$ at (x_0, y_0, z_0). Thus, if the curve of intersection can be smoothly parametrized, then its unit tangent vector $\mathbf{T}$ at (x_0, y_0, z_0) will be orthogonal to both $\nabla F(x_0, y_0, z_0)$ and $\nabla G(x_0, y_0, z_0)$ (Figure 14.7.6). Consequently, if

$$\nabla F(x_0, y_0, z_0) \times \nabla G(x_0, y_0, z_0) \neq \mathbf{0}$$

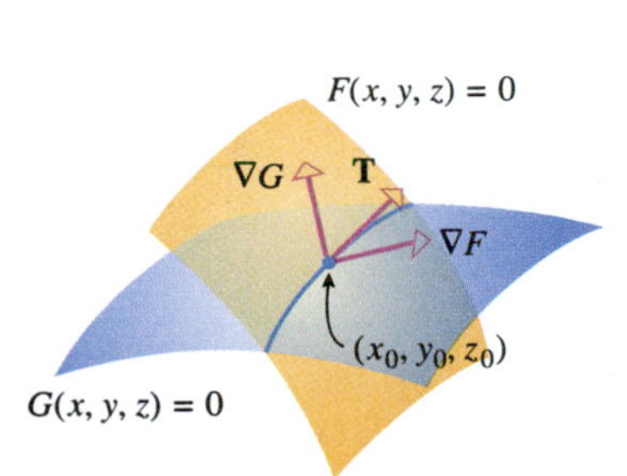

Figure 14.7.6

then this cross product will be parallel to $\mathbf{T}$ and hence will be tangent to the curve of intersection. This tangent vector can be used to determine the direction of the tangent line to the curve of intersection at the point (x_0, y_0, z_0).

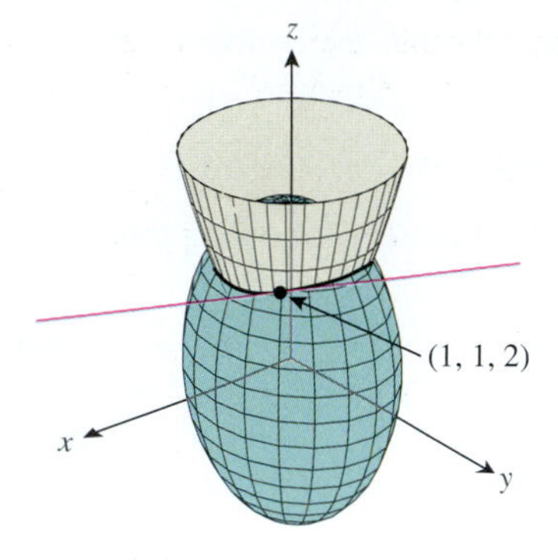

Figure 14.7.7

► **Example 3** Find parametric equations of the tangent line to the curve of intersection of the paraboloid $z = x^2 + y^2$ and the ellipsoid $3x^2 + 2y^2 + z^2 = 9$ at the point $(1, 1, 2)$ (Figure 14.7.7).

Solution. We begin by rewriting the equations of the surfaces as

$$x^2 + y^2 - z = 0 \quad \text{and} \quad 3x^2 + 2y^2 + z^2 - 9 = 0$$

and we take

$$F(x, y, z) = x^2 + y^2 - z \quad \text{and} \quad G(x, y, z) = 3x^2 + 2y^2 + z^2 - 9$$

We will need the gradients of these functions at the point $(1, 1, 2)$. The computations are

$$\begin{aligned} \nabla F(x, y, z) &= 2x\mathbf{i} + 2y\mathbf{j} - \mathbf{k}, & \nabla G(x, y, z) &= 6x\mathbf{i} + 4y\mathbf{j} + 2z\mathbf{k} \\ \nabla F(1, 1, 2) &= 2\mathbf{i} + 2\mathbf{j} - \mathbf{k}, & \nabla G(1, 1, 2) &= 6\mathbf{i} + 4\mathbf{j} + 4\mathbf{k} \end{aligned}$$

Thus, a tangent vector at $(1, 1, 2)$ to the curve of intersection is

$$\nabla F(1, 1, 2) \times \nabla G(1, 1, 2) = \begin{vmatrix} \mathbf{i} & \mathbf{j} & \mathbf{k} \\ 2 & 2 & -1 \\ 6 & 4 & 4 \end{vmatrix} = 12\mathbf{i} - 14\mathbf{j} - 4\mathbf{k}$$

Since any scalar multiple of this vector will do just as well, we can multiply by $\frac{1}{2}$ to reduce the size of the coefficients and use the vector of $6\mathbf{i} - 7\mathbf{j} - 2\mathbf{k}$ to determine the direction of the tangent line. This vector and the point $(1, 1, 2)$ yield the parametric equations

$$x = 1 + 6t, \quad y = 1 - 7t, \quad z = 2 - 2t \blacktriangleleft$$

✔ QUICK CHECK EXERCISES 14.7 *(See page 996 for answers.)*

1. Suppose that $f(x, y)$ is differentiable at the point $(3, 1)$ with $f(3, 1) = 4$, $f_x(3, 1) = 2$, and $f_y(3, 1) = -3$. An equation for the tangent plane to the graph of f at the point $(3, 1, 4)$ is ________, and parametric equations for the normal line to the graph of f through the point $(3, 1, 4)$ are

   $x =$ ________, $y =$ ________, $z =$ ________

2. An equation for the tangent plane to the graph of $z = x^2\sqrt{y}$ at the point $(2, 4, 8)$ is ________, and parametric equations for the normal line to the graph of $z = x^2\sqrt{y}$ through the point $(2, 4, 8)$ are

   $x =$ ________, $y =$ ________, $z =$ ________

3. Suppose that $f(1, 0, -1) = 2$, and $f(x, y, z)$ is differentiable at $(1, 0, -1)$ with $\nabla f(1, 0, -1) = \langle 2, 1, 1 \rangle$. An equation for the tangent plane to the level surface $f(x, y, z) = 2$ at the point $(1, 0, -1)$ is ________, and parametric equations for the normal line to the level surface through the point $(1, 0, -1)$ are

   $x =$ ________, $y =$ ________, $z =$ ________

4. The sphere $x^2 + y^2 + z^2 = 9$ and the plane $x + y + z = 5$ intersect in a circle that passes through the point $(2, 1, 2)$. Parametric equations for the tangent line to this circle at $(2, 1, 2)$ are

   $x =$ ________, $y =$ ________, $z =$ ________

EXERCISE SET 14.7 [C] CAS

1–8 Find an equation for the tangent plane and parametric equations for the normal line to the surface at the point P.

1. $z = 4x^3y^2 + 2y$; $P(1, -2, 12)$
2. $z = \frac{1}{2}x^7y^{-2}$; $P(2, 4, 4)$
3. $z = xe^{-y}$; $P(1, 0, 1)$
4. $z = \ln\sqrt{x^2 + y^2}$; $P(-1, 0, 0)$
5. $z = e^{3y}\sin 3x$; $P(\pi/6, 0, 1)$
6. $z = x^{1/2} + y^{1/2}$; $P(4, 9, 5)$
7. $x^2 + y^2 + z^2 = 25$; $P(-3, 0, 4)$
8. $x^2y - 4z^2 = -7$; $P(-3, 1, -2)$

FOCUS ON CONCEPTS

9. Find all points on the surface at which the tangent plane is horizontal.
(a) $z = x^3y^2$
(b) $z = x^2 - xy + y^2 - 2x + 4y$

10. Find a point on the surface $z = 3x^2 - y^2$ at which the tangent plane is parallel to the plane $6x + 4y - z = 5$.

11. Find a point on the surface $z = 8 - 3x^2 - 2y^2$ at which the tangent plane is perpendicular to the line $x = 2 - 3t$, $y = 7 + 8t$, $z = 5 - t$.

12. Show that the surfaces

$$z = \sqrt{x^2 + y^2} \quad \text{and} \quad z = \tfrac{1}{10}(x^2 + y^2) + \tfrac{5}{2}$$

intersect at $(3, 4, 5)$ and have a common tangent plane at that point.

13. (a) Find all points of intersection of the line

$$x = -1 + t, \quad y = 2 + t, \quad z = 2t + 7$$

and the surface

$$z = x^2 + y^2$$

(b) At each point of intersection, find the cosine of the acute angle between the given line and the line normal to the surface.

14. Show that if f is differentiable and $z = xf(x/y)$, then all tangent planes to the graph of this equation pass through the origin.

15. Consider the ellipsoid $x^2 + y^2 + 4z^2 = 12$.
(a) Use the method of Example 2 to find an equation of the tangent plane to the ellipsoid at the point $(2, 2, 1)$.
(b) Find parametric equations of the line that is normal to the ellipsoid at the point $(2, 2, 1)$.
(c) Find the acute angle that the tangent plane at the point $(2, 2, 1)$ makes with the xy-plane.

16. Consider the surface $xz - yz^3 + yz^2 = 2$.
(a) Use the method of Example 2 to find an equation of the tangent plane to the surface at the point $(2, -1, 1)$.
(b) Find parametric equations of the line that is normal to the surface at the point $(2, -1, 1)$.
(c) Find the acute angle that the tangent plane at the point $(2, -1, 1)$ makes with the xy-plane.

17–18 Find two unit vectors that are normal to the given surface at the point P.

17. $\sqrt{\dfrac{z + x}{y - 1}} = z^2$; $P(3, 5, 1)$

18. $\sin xz - 4\cos yz = 4$; $P(\pi, \pi, 1)$

19. Show that every line that is normal to the sphere

$$x^2 + y^2 + z^2 = 1$$

passes through the origin.

20. Find all points on the ellipsoid $2x^2 + 3y^2 + 4z^2 = 9$ at which the plane tangent to the ellipsoid is parallel to the plane $x - 2y + 3z = 5$.

21. Find all points on the surface $x^2 + y^2 - z^2 = 1$ at which the normal line is parallel to the line through $P(1, -2, 1)$ and $Q(4, 0, -1)$.

22. Show that the ellipsoid $2x^2 + 3y^2 + z^2 = 9$ and the sphere

$$x^2 + y^2 + z^2 - 6x - 8y - 8z + 24 = 0$$

have a common tangent plane at the point $(1, 1, 2)$.

23. Find parametric equations for the tangent line to the curve of intersection of the paraboloid $z = x^2 + y^2$ and the ellipsoid $x^2 + 4y^2 + z^2 = 9$ at the point $(1, -1, 2)$.

24. Find parametric equations for the tangent line to the curve of intersection of the cone $z = \sqrt{x^2 + y^2}$ and the plane $x + 2y + 2z = 20$ at the point $(4, 3, 5)$.

25. Find parametric equations for the tangent line to the curve of intersection of the cylinders $x^2 + z^2 = 25$ and $y^2 + z^2 = 25$ at the point $(3, -3, 4)$.

C **26.** The accompanying figure shows the intersection of the surfaces $z = 8 - x^2 - y^2$ and $4x + 2y - z = 0$.
(a) Find parametric equations for the tangent line to the curve of intersection at the point $(0, 2, 4)$.
(b) Use a CAS to generate a reasonable facsimile of the figure. You need not generate the colors, but try to obtain a similar viewpoint.

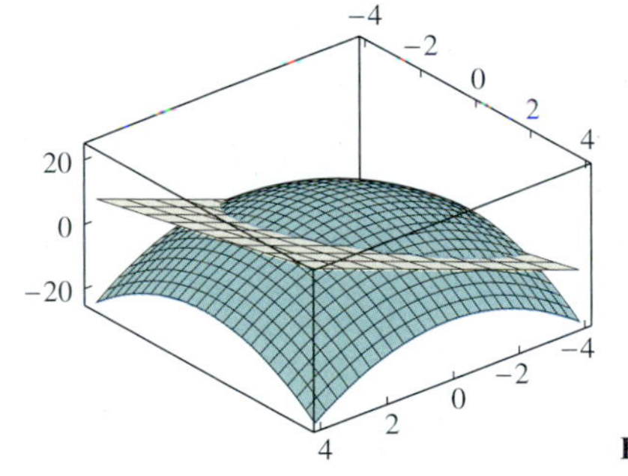

Figure Ex-26

27. Show that the equation of the plane that is tangent to the ellipsoid

$$\frac{x^2}{a^2} + \frac{y^2}{b^2} + \frac{z^2}{c^2} = 1$$

at (x_0, y_0, z_0) can be written in the form

$$\frac{x_0x}{a^2} + \frac{y_0y}{b^2} + \frac{z_0z}{c^2} = 1$$

28. Show that the equation of the plane that is tangent to the paraboloid

$$z = \frac{x^2}{a^2} + \frac{y^2}{b^2}$$

at (x_0, y_0, z_0) can be written in the form

$$z + z_0 = \frac{2x_0x}{a^2} + \frac{2y_0y}{b^2}$$

29. Prove: If the surfaces $z = f(x, y)$ and $z = g(x, y)$ intersect at $P(x_0, y_0, z_0)$, and if f and g are differentiable at (x_0, y_0), then the normal lines at P are perpendicular if and only if

$$f_x(x_0, y_0)g_x(x_0, y_0) + f_y(x_0, y_0)g_y(x_0, y_0) = -1$$

30. Use the result in Exercise 29 to show that the normal lines to the cones $z = \sqrt{x^2 + y^2}$ and $z = -\sqrt{x^2 + y^2}$ are perpendicular to the normal lines to the sphere $x^2 + y^2 + z^2 = a^2$ at every point of intersection (see Figure Ex-32).

31. Two surfaces $f(x, y, z) = 0$ and $g(x, y, z) = 0$ are said to be ***orthogonal*** at a point P of intersection if ∇f and ∇g are nonzero at P and the normal lines to the surfaces are perpendicular at P. Show that if $\nabla f(x_0, y_0, z_0) \neq \mathbf{0}$ and $\nabla g(x_0, y_0, z_0) \neq \mathbf{0}$, then the surfaces $f(x, y, z) = 0$ and $g(x, y, z) = 0$ are orthogonal at the point (x_0, y_0, z_0) if and only if

$$f_x g_x + f_y g_y + f_z g_z = 0$$

at this point. [*Note:* This is a more general version of the result in Exercise 29.]

32. Use the result of Exercise 31 to show that the sphere $x^2 + y^2 + z^2 = a^2$ and the cone $z^2 = x^2 + y^2$ are orthogonal at every point of intersection (see the accompanying figure).

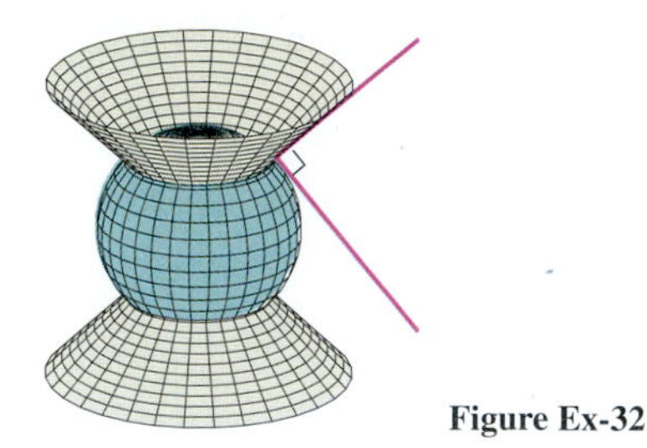

Figure Ex-32

33. Show that the volume of the solid bounded by the coordinate planes and a plane tangent to the portion of the surface $xyz = k$, $k > 0$, in the first octant does not depend on the point of tangency.

✔ QUICK CHECK ANSWERS 14.7

1. $z = 4 + 2(x-3) - 3(y-1)$; $x = 3 + 2t$; $y = 1 - 3t$; $z = 4 - t$
2. $z = 8 + 8(x-2) + (y-4)$; $x = 2 + 8t$; $y = 4 + t$; $z = 8 - t$
3. $2(x-1) + y + (z+1) = 0$; $x = 1 + 2t$; $y = t$; $z = -1 + t$ **4.** $x = 2 + t$; $y = 1$; $z = 2 - t$

14.8 MAXIMA AND MINIMA OF FUNCTIONS OF TWO VARIABLES

Earlier in this text we learned how to find maximum and minimum values of a function of one variable. In this section we will develop similar techniques for functions of two variables.

EXTREMA

If we imagine the graph of a function f of two variables to be a mountain range (Figure 14.8.1), then the mountaintops, which are the high points in their immediate vicinity, are called *relative maxima* of f, and the valley bottoms, which are the low points in their immediate vicinity, are called *relative minima* of f.

Just as a geologist might be interested in finding the highest mountain and deepest valley in an entire mountain range, so a mathematician might be interested in finding the largest and smallest values of $f(x, y)$ over the *entire* domain of f. These are called the *absolute maximum* and *absolute minimum values* of f. The following definitions make these informal ideas precise.

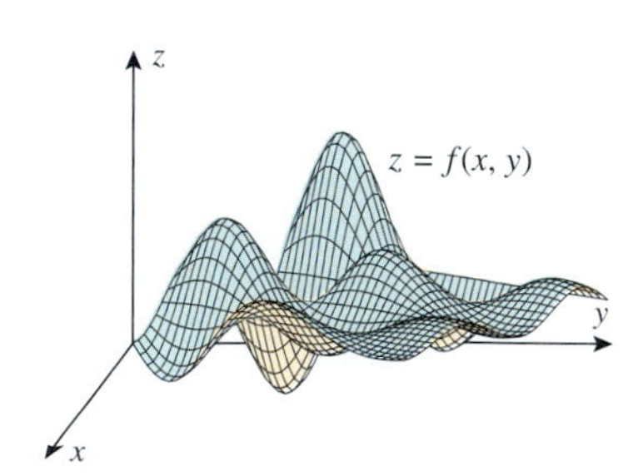

Figure 14.8.1

> **14.8.1 DEFINITION.** A function f of two variables is said to have a ***relative maximum*** at a point (x_0, y_0) if there is a disk centered at (x_0, y_0) such that $f(x_0, y_0) \geq f(x, y)$ for all points (x, y) that lie inside the disk, and f is said to have an ***absolute maximum*** at (x_0, y_0) if $f(x_0, y_0) \geq f(x, y)$ for all points (x, y) in the domain of f.

14.8.2 DEFINITION. A function f of two variables is said to have a ***relative minimum*** at a point (x_0, y_0) if there is a disk centered at (x_0, y_0) such that $f(x_0, y_0) \leq f(x, y)$ for all points (x, y) that lie inside the disk, and f is said to have an ***absolute minimum*** at (x_0, y_0) if $f(x_0, y_0) \leq f(x, y)$ for all points (x, y) in the domain of f.

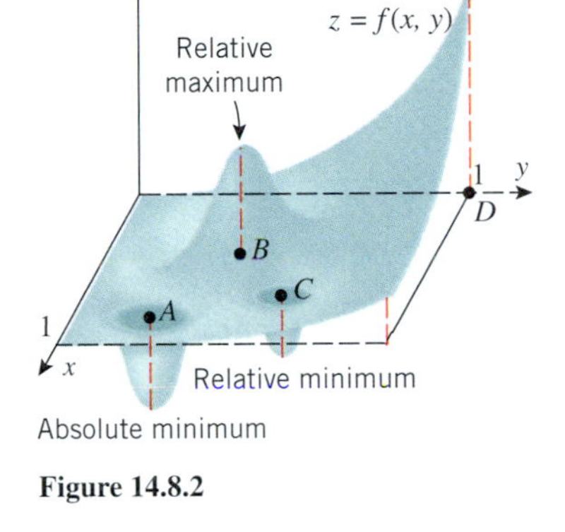

Figure 14.8.2

If f has a relative maximum or a relative minimum at (x_0, y_0), then we say that f has a ***relative extremum*** at (x_0, y_0), and if f has an absolute maximum or absolute minimum at (x_0, y_0), then we say that f has an ***absolute extremum*** at (x_0, y_0).

Figure 14.8.2 shows the graph of a function f whose domain is the square region in the xy-plane whose points satisfy the inequalities $0 \leq x \leq 1, 0 \leq y \leq 1$. The function f has relative minima at the points A and C and a relative maximum at B. There is an absolute minimum at A and an absolute maximum at D.

For functions of two variables we will be concerned with two important questions:

- Are there any relative or absolute extrema?
- If so, where are they located?

BOUNDED SETS

Just as we distinguished between finite intervals and infinite intervals on the real line, so we will want to distinguish between regions of "finite extent" and regions of "infinite extent" in 2-space and 3-space. A set of points in 2-space is called ***bounded*** if the entire set can be contained within some rectangle, and is called ***unbounded*** if there is no rectangle that contains all the points of the set. Similarly, a set of points in 3-space is ***bounded*** if the entire set can be contained within some box, and is unbounded otherwise (Figure 14.8.3).

Explain why any subset of a bounded set is also bounded.

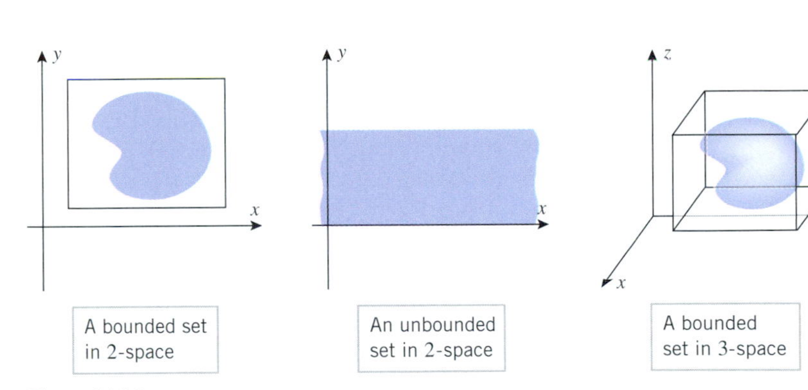

Figure 14.8.3

THE EXTREME-VALUE THEOREM

For functions of one variable that are continuous on a closed interval, the Extreme-Value Theorem (Theorem 5.4.2) answered the existence question for absolute extrema. The following theorem, which we state without proof, is the corresponding result for functions of two variables.

14.8.3 THEOREM (*Extreme-Value Theorem*). *If $f(x, y)$ is continuous on a closed and bounded set R, then f has both an absolute maximum and an absolute minimum on R.*

Example 1 The square region R whose points satisfy the inequalities

$$0 \le x \le 1 \quad \text{and} \quad 0 \le y \le 1$$

is a closed and bounded set in the xy-plane. The function f whose graph is shown in Figure 14.8.2 is continuous on R; thus, it is guaranteed to have an absolute maximum and minimum on R by the last theorem. These occur at points D and A that are shown in the figure. ◀

If any of the conditions in the Extreme-Value Theorem fail to hold, then there is no guarantee that an absolute maximum or absolute minimum exists on the region R. Thus, a discontinuous function on a closed and bounded set need not have any absolute extrema, and a continuous function on a set that is not closed and bounded also need not have any absolute extrema.

FINDING RELATIVE EXTREMA

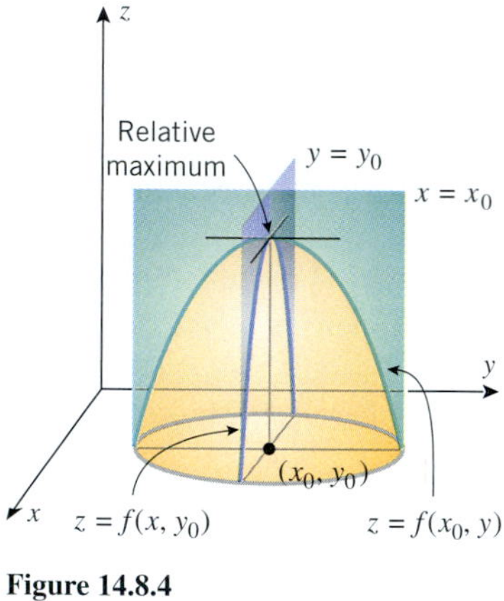

Figure 14.8.4

Recall that if a function g of one variable has a relative extremum at a point x_0 where g is differentiable, then $g'(x_0) = 0$. To obtain the analog of this result for functions of two variables, suppose that $f(x, y)$ has a relative maximum at a point (x_0, y_0) and that the partial derivatives of f exist at (x_0, y_0). It seems plausible geometrically that the traces of the surface $z = f(x, y)$ on the planes $x = x_0$ and $y = y_0$ have horizontal tangent lines at (x_0, y_0) (Figure 14.8.4), so

$$f_x(x_0, y_0) = 0 \quad \text{and} \quad f_y(x_0, y_0) = 0$$

The same conclusion holds if f has a relative minimum at (x_0, y_0), all of which suggests the following result, which we state without formal proof.

14.8.4 THEOREM. *If f has a relative extremum at a point (x_0, y_0), and if the first-order partial derivatives of f exist at this point, then*

$$f_x(x_0, y_0) = 0 \quad \textit{and} \quad f_y(x_0, y_0) = 0$$

Recall that the *critical points* of a function f of one variable are those values of x in the domain of f at which $f'(x) = 0$ or f is not differentiable. The following definition is the analog for functions of two variables.

Explain why

$$D_{\mathbf{u}} f(x_0, y_0) = 0$$

for all $\mathbf{u}$ if (x_0, y_0) is a critical point of f and f is differentiable at (x_0, y_0).

14.8.5 DEFINITION. A point (x_0, y_0) in the domain of a function $f(x, y)$ is called a ***critical point*** of the function if $f_x(x_0, y_0) = 0$ and $f_y(x_0, y_0) = 0$ or if one or both partial derivatives do not exist at (x_0, y_0).

It follows from this definition and Theorem 14.8.4 that relative extrema occur at critical points, just as for a function of one variable. However, recall that for a function of one variable a relative extremum need not occur at *every* critical point. For example, the function might have an inflection point with a horizontal tangent line at the critical point (see Figure 5.2.6). Similarly, a function of two variables need not have a relative extremum at every critical point. For example, consider the function

$$f(x, y) = y^2 - x^2$$

This function, whose graph is the hyperbolic paraboloid shown in Figure 14.8.5, has a

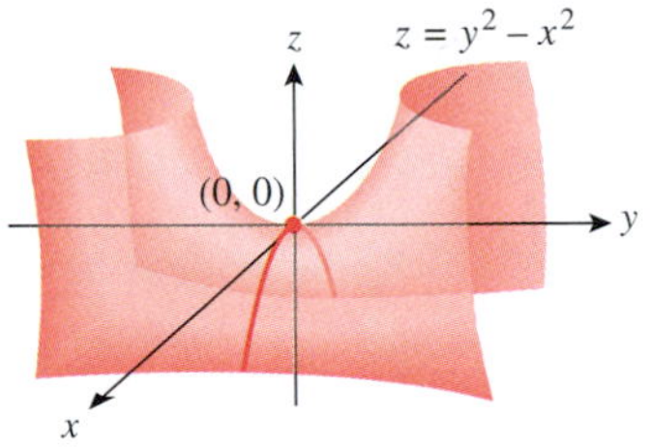

The function $f(x, y) = y^2 - x^2$ has neither a relative maximum nor a relative minimum at the critical point $(0, 0)$.

Figure 14.8.5

critical point at $(0, 0)$, since

$$f_x(x, y) = -2x \quad \text{and} \quad f_y(x, y) = 2y$$

from which it follows that

$$f_x(0, 0) = 0 \quad \text{and} \quad f_y(0, 0) = 0$$

However, the function f has neither a relative maximum nor a relative minimum at $(0, 0)$. For obvious reasons, the point $(0, 0)$ is called a *saddle point* of f. In general, we will say that a surface $z = f(x, y)$ has a ***saddle point*** at (x_0, y_0) if there are two distinct vertical planes through this point such that the trace of the surface in one of the planes has a relative maximum at (x_0, y_0) and the trace in the other has a relative minimum at (x_0, y_0).

Example 2 The three functions graphed in Figure 14.8.6 all have critical points at $(0, 0)$. For the paraboloids, the partial derivatives at the origin are zero. You can check this algebraically by evaluating the partial derivatives at $(0, 0)$, but you can see it geometrically by observing that the traces in the xz-plane and yz-plane have horizontal tangent lines at $(0, 0)$. For the cone neither partial derivative exists at the origin because the traces in the xz-plane and the yz-plane have corners there. The paraboloid in part (a) and the cone in part (c) have a relative minimum and absolute minimum at the origin, and the paraboloid in part (b) has a relative maximum and an absolute maximum at the origin. ◀

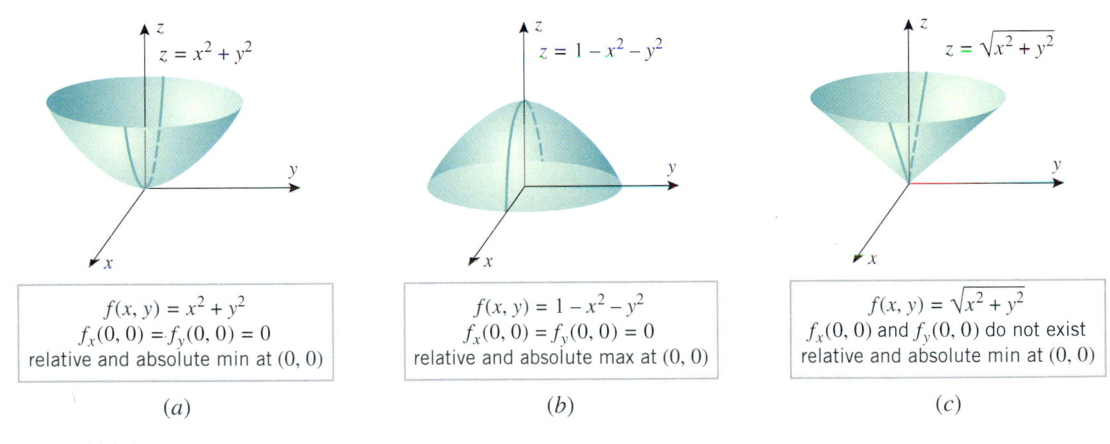

$f(x, y) = x^2 + y^2$
$f_x(0, 0) = f_y(0, 0) = 0$
relative and absolute min at $(0, 0)$
(a)

$f(x, y) = 1 - x^2 - y^2$
$f_x(0, 0) = f_y(0, 0) = 0$
relative and absolute max at $(0, 0)$
(b)

$f(x, y) = \sqrt{x^2 + y^2}$
$f_x(0, 0)$ and $f_y(0, 0)$ do not exist
relative and absolute min at $(0, 0)$
(c)

Figure 14.8.6

THE SECOND PARTIALS TEST

For functions of one variable the second derivative test (Theorem 5.2.4) was used to determine the behavior of a function at a critical point. The following theorem, which is usually proved in advanced calculus, is the analog of that theorem for functions of two variables.

14.8.6 THEOREM (*The Second Partials Test*). *Let f be a function of two variables with continuous second-order partial derivatives in some disk centered at a critical point (x_0, y_0), and let*

$$D = f_{xx}(x_0, y_0) f_{yy}(x_0, y_0) - f_{xy}^2(x_0, y_0)$$

(a) *If $D > 0$ and $f_{xx}(x_0, y_0) > 0$, then f has a relative minimum at (x_0, y_0).*

(b) *If $D > 0$ and $f_{xx}(x_0, y_0) < 0$, then f has a relative maximum at (x_0, y_0).*

(c) *If $D < 0$, then f has a saddle point at (x_0, y_0).*

(d) *If $D = 0$, then no conclusion can be drawn.*

With the notation of Theorem 14.8.6, show that if $D > 0$, then $f_{xx}(x_0, y_0)$ and $f_{yy}(x_0, y_0)$ have the same sign. Thus, we can replace $f_{xx}(x_0, y_0)$ by $f_{yy}(x_0, y_0)$ in parts (*a*) and (*b*) of the theorem.

▸ Example 3 Locate all relative extrema and saddle points of

$$f(x, y) = 3x^2 - 2xy + y^2 - 8y$$

Solution. Since $f_x(x, y) = 6x - 2y$ and $f_y(x, y) = -2x + 2y - 8$, the critical points of f satisfy the equations

$$6x - 2y = 0$$
$$-2x + 2y - 8 = 0$$

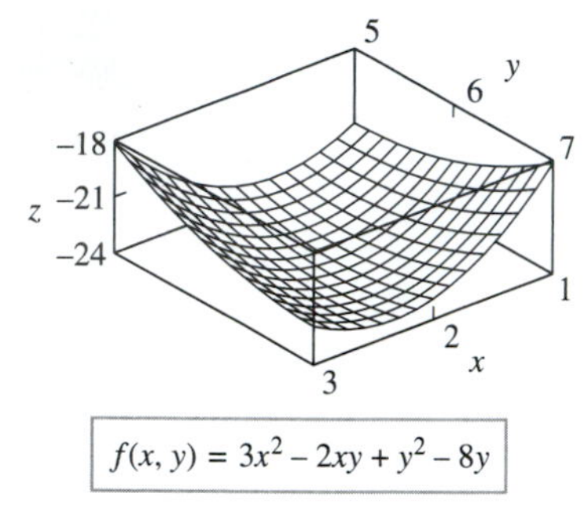

Figure 14.8.7

Solving these for x and y yields $x = 2$, $y = 6$ (verify), so $(2, 6)$ is the only critical point. To apply Theorem 14.8.6 we need the second-order partial derivatives

$$f_{xx}(x, y) = 6, \quad f_{yy}(x, y) = 2, \quad f_{xy}(x, y) = -2$$

At the point $(2, 6)$ we have

$$D = f_{xx}(2, 6)f_{yy}(2, 6) - f_{xy}^2(2, 6) = (6)(2) - (-2)^2 = 8 > 0$$

and

$$f_{xx}(2, 6) = 6 > 0$$

so f has a relative minimum at $(2, 6)$ by part (*a*) of the second partials test. Figure 14.8.7 shows a graph of f in the vicinity of the relative minimum. ◂

▸ Example 4 Locate all relative extrema and saddle points of

$$f(x, y) = 4xy - x^4 - y^4$$

Solution. Since

$$f_x(x, y) = 4y - 4x^3$$
$$f_y(x, y) = 4x - 4y^3 \tag{1}$$

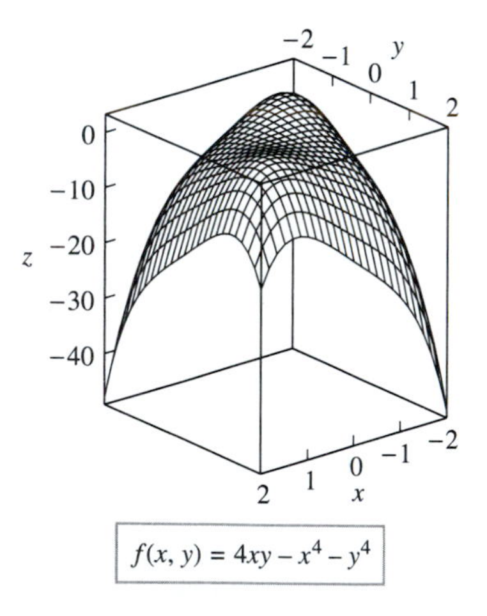

the critical points of f have coordinates satisfying the equations

$$4y - 4x^3 = 0 \quad \text{or} \quad y = x^3$$
$$4x - 4y^3 = 0 \quad\quad\quad x = y^3 \tag{2}$$

Substituting the top equation in the bottom yields $x = (x^3)^3$ or equivalently $x^9 - x = 0$ or $x(x^8 - 1) = 0$, which has solutions $x = 0$, $x = 1$, $x = -1$. Substituting these values in the top equation of (2), we obtain the corresponding y-values $y = 0$, $y = 1$, $y = -1$. Thus, the critical points of f are $(0, 0)$, $(1, 1)$, and $(-1, -1)$.

From (1),

$$f_{xx}(x, y) = -12x^2, \quad f_{yy}(x, y) = -12y^2, \quad f_{xy}(x, y) = 4$$

which yields the following table:

CRITICAL POINT (x_0, y_0)	$f_{xx}(x_0, y_0)$	$f_{yy}(x_0, y_0)$	$f_{xy}(x_0, y_0)$	$D = f_{xx}f_{yy} - f_{xy}^2$
(0, 0)	0	0	4	−16
(1, 1)	−12	−12	4	128
(−1, −1)	−12	−12	4	128

Figure 14.8.8

At the points $(1, 1)$ and $(-1, -1)$, we have $D > 0$ and $f_{xx} < 0$, so relative maxima occur at these critical points. At $(0, 0)$ there is a saddle point since $D < 0$. The surface and a contour plot are shown in Figure 14.8.8. ◂

> The "figure eight" pattern at $(0, 0)$ in the contour plot for the surface in Figure 14.8.8 is typical for level curves that pass through a saddle point. If a bug starts at the point $(0, 0, 0)$ on the surface, in how many directions can it walk and remain in the xy-plane?

The following theorem, which is the analog for functions of two variables of Theorem 5.4.3, will lead to an important method for finding absolute extrema.

14.8.7 THEOREM. *If a function f of two variables has an absolute extremum (either an absolute maximum or an absolute minimum) at an interior point of its domain, then this extremum occurs at a critical point.*

PROOF. If f has an absolute maximum at the point (x_0, y_0) in the interior of the domain of f, then f has a relative maximum at (x_0, y_0). If both partial derivatives exist at (x_0, y_0), then

$$f_x(x_0, y_0) = 0 \quad \text{and} \quad f_y(x_0, y_0) = 0$$

by Theorem 14.8.4, so (x_0, y_0) is a critical point of f. If either partial derivative does not exist, then again (x_0, y_0) is a critical point, so (x_0, y_0) is a critical point in all cases. The proof for an absolute minimum is similar. ■

FINDING ABSOLUTE EXTREMA ON CLOSED AND BOUNDED SETS

If $f(x, y)$ is continuous on a closed and bounded set R, then the Extreme-Value Theorem (Theorem 14.8.3) guarantees the existence of an absolute maximum and an absolute minimum of f on R. These absolute extrema can occur either on the boundary of R or in the interior of R, but if an absolute extremum occurs in the interior, then it occurs at a critical point by Theorem 14.8.7. Thus, we are led to the following procedure for finding absolute extrema:

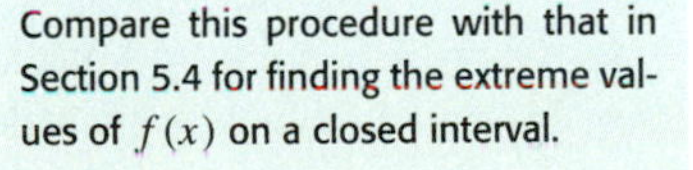

Compare this procedure with that in Section 5.4 for finding the extreme values of $f(x)$ on a closed interval.

How to Find the Absolute Extrema of a Continuous Function f of Two Variables on a Closed and Bounded Set R

Step 1. Find the critical points of f that lie in the interior of R.

Step 2. Find all boundary points at which the absolute extrema can occur.

Step 3. Evaluate $f(x, y)$ at the points obtained in the preceding steps. The largest of these values is the absolute maximum and the smallest the absolute minimum.

▶ Example 5 Find the absolute maximum and minimum values of

$$f(x, y) = 3xy - 6x - 3y + 7 \tag{3}$$

on the closed triangular region R with vertices $(0, 0)$, $(3, 0)$, and $(0, 5)$.

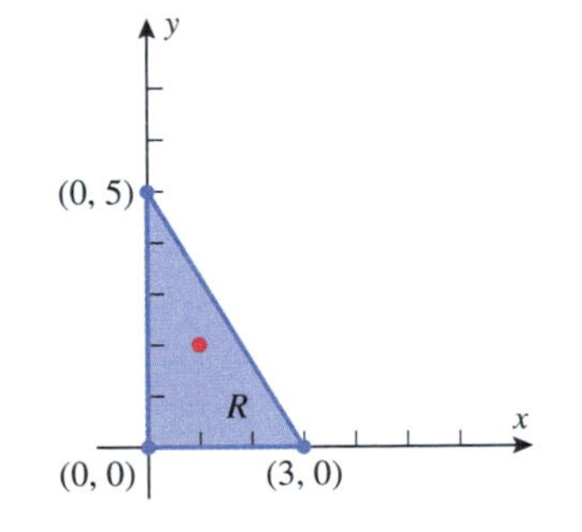

Figure 14.8.9

Solution. The region R is shown in Figure 14.8.9. We have

$$\frac{\partial f}{\partial x} = 3y - 6 \quad \text{and} \quad \frac{\partial f}{\partial y} = 3x - 3$$

so all critical points occur where

$$3y - 6 = 0 \quad \text{and} \quad 3x - 3 = 0$$

Solving these equations yields $x = 1$ and $y = 2$, so $(1, 2)$ is the only critical point. As shown in Figure 14.8.9, this critical point is in the interior of R.

Next we want to determine the locations of the points on the boundary of R at which the absolute extrema might occur. The boundary of R consists of three line segments, each of which we will treat separately:

The line segment between $(0, 0)$ *and* $(3, 0)$: On this line segment we have $y = 0$, so (3) simplifies to a function of the single variable x,

$$u(x) = f(x, 0) = -6x + 7, \quad 0 \le x \le 3$$

This function has no critical points because $u'(x) = -6$ is nonzero for all x. Thus the extreme values of $u(x)$ occur at the endpoints $x = 0$ and $x = 3$, which correspond to the points $(0, 0)$ and $(3, 0)$ of R.

The line segment between $(0, 0)$ *and* $(0, 5)$: On this line segment we have $x = 0$, so (3) simplifies to a function of the single variable y,

$$v(y) = f(0, y) = -3y + 7, \quad 0 \le y \le 5$$

This function has no critical points because $v'(y) = -3$ is nonzero for all y. Thus, the extreme values of $v(y)$ occur at the endpoints $y = 0$ and $y = 5$, which correspond to the points $(0, 0)$ and $(0, 5)$ of R.

The line segment between $(3, 0)$ *and* $(0, 5)$: In the xy-plane, an equation for this line segment is

$$y = -\tfrac{5}{3}x + 5, \quad 0 \le x \le 3 \tag{4}$$

so (3) simplifies to a function of the single variable x,

$$\begin{aligned} w(x) = f\left(x, -\tfrac{5}{3}x + 5\right) &= 3x\left(-\tfrac{5}{3}x + 5\right) - 6x - 3\left(-\tfrac{5}{3}x + 5\right) + 7 \\ &= -5x^2 + 14x - 8, \quad 0 \le x \le 3 \end{aligned}$$

Since $w'(x) = -10x + 14$, the equation $w'(x) = 0$ yields $x = \frac{7}{5}$ as the only critical point of w. Thus, the extreme values of w occur either at the critical point $x = \frac{7}{5}$ or at the endpoints $x = 0$ and $x = 3$. The endpoints correspond to the points $(0, 5)$ and $(3, 0)$ of R, and from (4) the critical point corresponds to $\left(\frac{7}{5}, \frac{8}{3}\right)$.

Finally, Table 14.8.1 lists the values of $f(x, y)$ at the interior critical point and at the points on the boundary where an absolute extremum can occur. From the table we conclude that the absolute maximum value of f is $f(0, 0) = 7$ and the absolute minimum value is $f(3, 0) = -11$. ◄

Table 14.8.1

(x, y)	$(0, 0)$	$(3, 0)$	$(0, 5)$	$\left(\frac{7}{5}, \frac{8}{3}\right)$	$(1, 2)$
$f(x, y)$	7	−11	−8	$\frac{9}{5}$	1

► Example 6 Determine the dimensions of a rectangular box, open at the top, having a volume of 32 ft^3, and requiring the least amount of material for its construction.

Solution. Let

x = length of the box (in feet)

y = width of the box (in feet)

z = height of the box (in feet)

S = surface area of the box (in square feet)

We may reasonably assume that the box with least surface area requires the least amount of material, so our objective is to minimize the surface area

$$S = xy + 2xz + 2yz \tag{5}$$

Two sides each have area xz.
Two sides each have area yz.
The base has area xy.

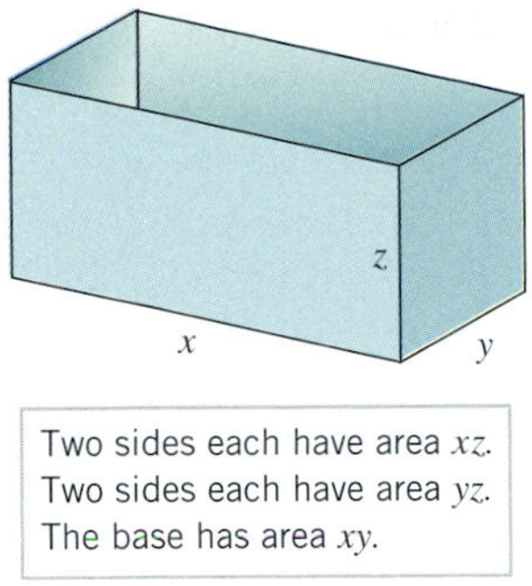

Figure 14.8.10

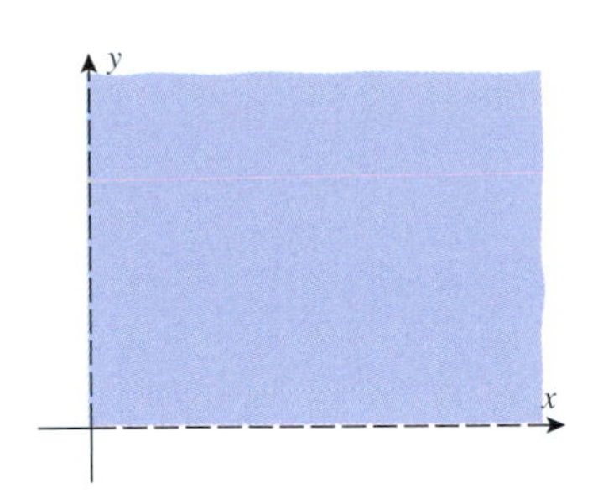

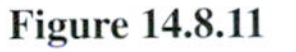
Figure 14.8.11

(Figure 14.8.10) subject to the volume requirement

$$xyz = 32 \tag{6}$$

From (6) we obtain $z = 32/xy$, so (5) can be rewritten as

$$S = xy + \frac{64}{y} + \frac{64}{x} \tag{7}$$

which expresses S as a function of two variables. The dimensions x and y in this formula must be positive, but otherwise have no limitation, so our problem reduces to finding the absolute minimum value of S over the first quadrant: $x > 0$, $y > 0$ (Figure 14.8.11). Because this region is neither closed nor bounded, we have no mathematical guarantee at this stage that an absolute minimum exists. However, note that S will have a large value at any point (x, y) in the first quadrant for which the product xy is large or for which either x or y is close to 0. We can use this observation to prove the existence of an absolute minimum value of S.

Let R denote the region in the first quadrant defined by the inequalities

$$\tfrac{1}{2} \le x, \quad \tfrac{1}{2} \le y, \quad \text{and} \quad xy \le 128$$

This region is both closed and bounded (verify) and the function S is continuous on R. It follows from Theorem 14.8.3 that S has an absolute minimum on R. Furthermore, note that $S > 128$ at any point (x, y) not in R and that the point $(8, 8)$ belongs to R with $S = 80 < 128$ at this point. We conclude that the minimum value of S on R is also the minimum value of S on the entire first quadrant.

Since S has an absolute minimum value in the first quadrant, it must occur at a critical point of S. Differentiating (7) we obtain

$$\frac{\partial S}{\partial x} = y - \frac{64}{x^2}, \quad \frac{\partial S}{\partial y} = x - \frac{64}{y^2} \tag{8}$$

so the coordinates of the critical points of S satisfy

$$y - \frac{64}{x^2} = 0, \quad x - \frac{64}{y^2} = 0$$

Solving the first equation for y yields

$$y = \frac{64}{x^2} \tag{9}$$

and substituting this expression in the second equation yields

$$x - \frac{64}{(64/x^2)^2} = 0$$

which can be rewritten as

$$x\left(1 - \frac{x^3}{64}\right) = 0$$

The solutions of this equation are $x = 0$ and $x = 4$. Since we require $x > 0$, the only solution of significance is $x = 4$. Substituting this value into (9) yields $y = 4$. Substituting $x = 4$ and $y = 4$ into (6) yields $z = 2$, so the box using least material has a height of 2 ft and a square base whose edges are 4 ft long. ◂

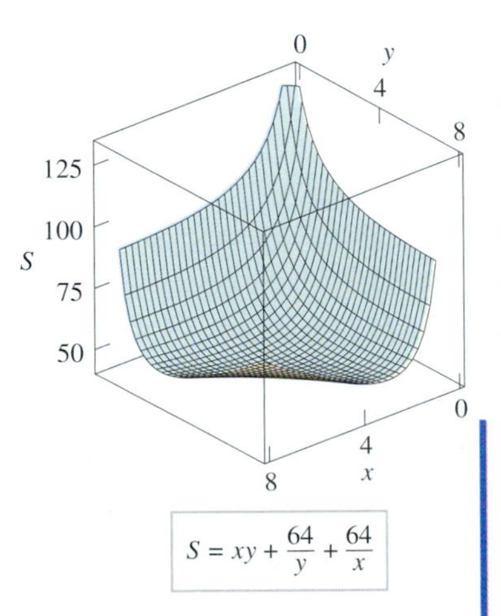

$S = xy + \frac{64}{y} + \frac{64}{x}$

Figure 14.8.12

Fortunately, in our solution to Example 6 we were able to prove the existence of an absolute minimum of S in the first quadrant. The general problem of finding the absolute extrema of a function on an unbounded region, or on a region that is not closed, can be difficult and will not be considered in this text. However, in applied problems we can sometimes use physical considerations to deduce that an absolute extremum has been found. For example, the graph of Equation (7) in Figure 14.8.12 strongly suggests that the relative minimum at $x = 4$ and $y = 4$ is also an absolute minimum.

✔ QUICK CHECK EXERCISES 14.8 (See page 1007 for answers.)

1. The critical points of the function $f(x, y) = x^3 + xy + y^2$ are ________.

2. Suppose that $f(x, y)$ has continuous second-order partial derivatives everywhere and that the origin is a critical point for f. State what information (if any) is provided by the second partials test if
   (a) $f_{xx}(0,0) = 2,\ f_{xy}(0,0) = 2,\ f_{yy}(0,0) = 2$
   (b) $f_{xx}(0,0) = -2,\ f_{xy}(0,0) = 2,\ f_{yy}(0,0) = 2$
   (c) $f_{xx}(0,0) = 3,\ f_{xy}(0,0) = 2,\ f_{yy}(0,0) = 2$
   (d) $f_{xx}(0,0) = -3,\ f_{xy}(0,0) = 2,\ f_{yy}(0,0) = -2$.

3. For the function $f(x, y) = x^3 - 3xy + y^3$, state what information (if any) is provided by the second partials test at the point
   (a) $(0, 0)$ (b) $(-1, -1)$ (c) $(1, 1)$.

4. A rectangular box has total surface area of 2 ft^2. Express the volume of the box as a function of the dimensions x and y of the base of the box.

EXERCISE SET 14.8 Graphing Utility C CAS

1–2 Locate all absolute maxima and minima, if any, by inspection. Then check your answers using calculus.

1. (a) $f(x, y) = (x-2)^2 + (y+1)^2$
   (b) $f(x, y) = 1 - x^2 - y^2$
   (c) $f(x, y) = x + 2y - 5$

2. (a) $f(x, y) = 1 - (x+1)^2 - (y-5)^2$
   (b) $f(x, y) = e^{xy}$
   (c) $f(x, y) = x^2 - y^2$

3–4 Complete the squares and locate all absolute maxima and minima, if any, by inspection. Then check your answers using calculus.

3. $f(x, y) = 13 - 6x + x^2 + 4y + y^2$

4. $f(x, y) = 1 - 2x - x^2 + 4y - 2y^2$

FOCUS ON CONCEPTS

5–8 The contour plots show all significant features of the function. Make a conjecture about the number and the location of all relative extrema and saddle points, and then use calculus to check your conjecture.

5. 6.

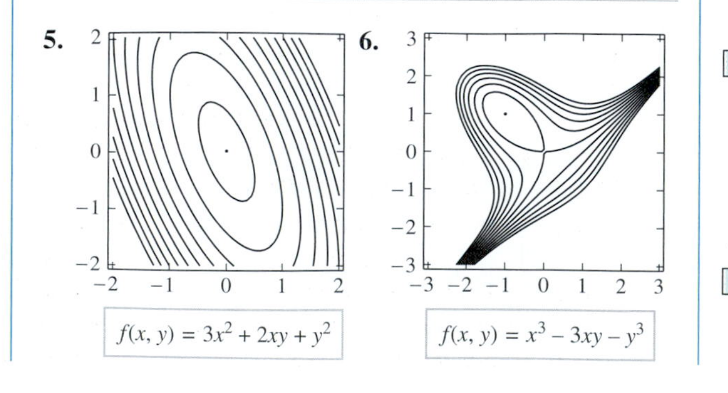

$f(x, y) = 3x^2 + 2xy + y^2$ $f(x, y) = x^3 - 3xy - y^3$

7. 8.

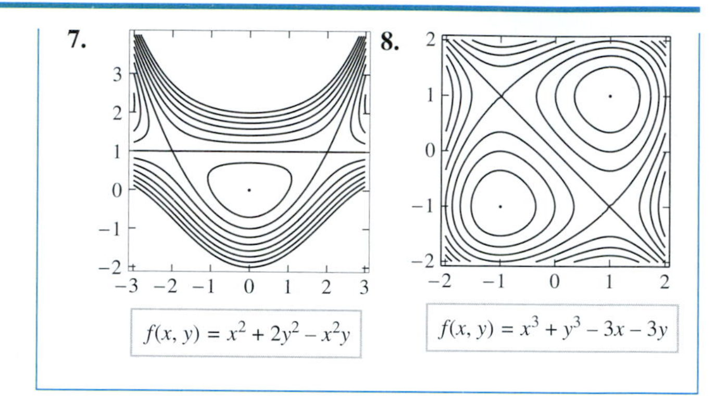

$f(x, y) = x^2 + 2y^2 - x^2y$ $f(x, y) = x^3 + y^3 - 3x - 3y$

9–20 Locate all relative maxima, relative minima, and saddle points, if any.

9. $f(x, y) = y^2 + xy + 3y + 2x + 3$

10. $f(x, y) = x^2 + xy - 2y - 2x + 1$

11. $f(x, y) = x^2 + xy + y^2 - 3x$

12. $f(x, y) = xy - x^3 - y^2$

13. $f(x, y) = x^2 + y^2 + \dfrac{2}{xy}$

14. $f(x, y) = xe^y$

15. $f(x, y) = x^2 + y - e^y$

16. $f(x, y) = xy + \dfrac{2}{x} + \dfrac{4}{y}$

17. $f(x, y) = e^x \sin y$

18. $f(x, y) = y \sin x$

19. $f(x, y) = e^{-(x^2+y^2+2x)}$

20. $f(x, y) = xy + \dfrac{a^3}{x} + \dfrac{b^3}{y}\quad (a \neq 0, b \neq 0)$

C 21. Use a CAS to generate a contour plot of
$$f(x, y) = 2x^2 - 4xy + y^4 + 2$$
for $-2 \le x \le 2$ and $-2 \le y \le 2$, and use the plot to approximate the locations of all relative extrema and saddle points in the region. Check your answer using calculus, and identify the relative extrema as relative maxima or minima.

C 22. Use a CAS to generate a contour plot of
$$f(x, y) = 2y^2x - yx^2 + 4xy$$
for $-5 \le x \le 5$ and $-5 \le y \le 5$, and use the plot to ap-

proximate the locations of all relative extrema and saddle points in the region. Check your answer using calculus, and identify the relative extrema as relative maxima or minima.

FOCUS ON CONCEPTS

23. (a) Show that the second partials test provides no information about the critical points of the function $f(x, y) = x^4 + y^4$.
(b) Classify all critical points of f as relative maxima, relative minima, or saddle points.

24. (a) Show that the second partials test provides no information about the critical points of the function $f(x, y) = x^4 - y^4$.
(b) Classify all critical points of f as relative maxima, relative minima, or saddle points.

25. Recall from Theorem 5.4.4 that if a continuous function of one variable has exactly one relative extremum on an interval, then that relative extremum is an absolute extremum on the interval. This exercise shows that this result does not extend to functions of two variables.
(a) Show that $f(x, y) = 3xe^y - x^3 - e^{3y}$ has only one critical point and that a relative maximum occurs there. (See the accompanying figure.)
(b) Show that f does not have an absolute maximum.

Source: This exercise is based on the article "The Only Critical Point in Town Test" by Ira Rosenholtz and Lowell Smylie, *Mathematics Magazine*, Vol. 58, No. 3, May 1985, pp. 149–150.

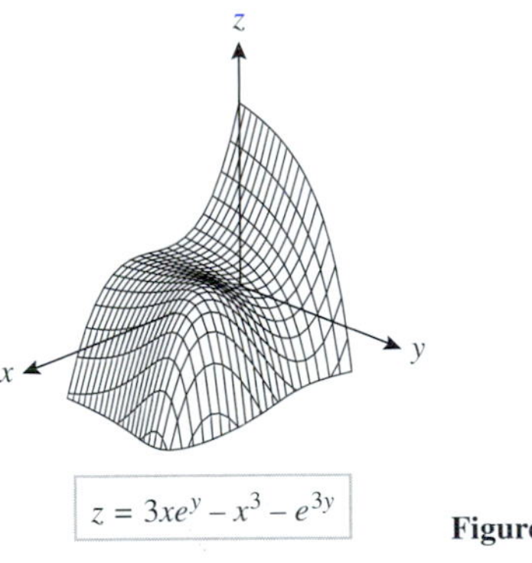

Figure Ex-25

26. If f is a continuous function of one variable with two relative maxima on an interval, then there must be a relative minimum between the relative maxima. (Convince yourself of this by drawing some pictures.) The purpose of this exercise is to show that this result does not extend to functions of two variables. Show that $f(x, y) = 4x^2e^y - 2x^4 - e^{4y}$ has two relative maxima but no other critical points (see Figure Ex-26).

Source: This exercise is based on the problem "Two Mountains Without a Valley" proposed and solved by Ira Rosenholtz, *Mathematics Magazine*, Vol. 60, No. 1, February 1987, p. 48.

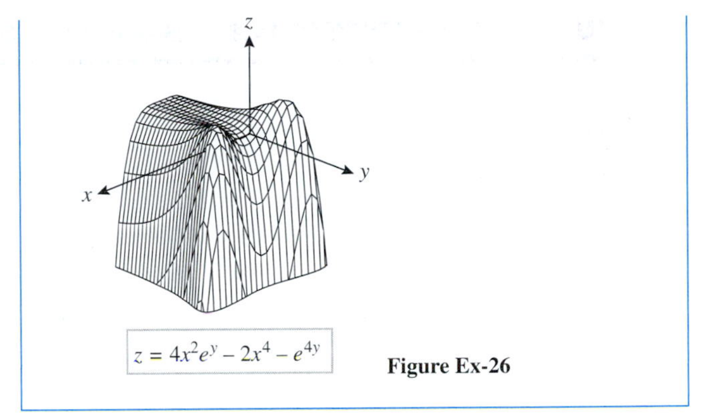

Figure Ex-26

27–32 Find the absolute extrema of the given function on the indicated closed and bounded set R.

27. $f(x, y) = xy - x - 3y$; R is the triangular region with vertices $(0, 0)$, $(0, 4)$, and $(5, 0)$.

28. $f(x, y) = xy - 2x$; R is the triangular region with vertices $(0, 0)$, $(0, 4)$, and $(4, 0)$.

29. $f(x, y) = x^2 - 3y^2 - 2x + 6y$; R is the region bounded by the square with vertices $(0, 0)$, $(0, 2)$, $(2, 2)$, and $(2, 0)$.

30. $f(x, y) = xe^y - x^2 - e^y$; R is the rectangular region with vertices $(0, 0)$, $(0, 1)$, $(2, 1)$, and $(2, 0)$.

31. $f(x, y) = x^2 + 2y^2 - x$; R is the disk $x^2 + y^2 \le 4$.

32. $f(x, y) = xy^2$; R is the region that satisfies the inequalities $x \ge 0$, $y \ge 0$, and $x^2 + y^2 \le 1$.

33. Find three positive numbers whose sum is 48 and such that their product is as large as possible.

34. Find three positive numbers whose sum is 27 and such that the sum of their squares is as small as possible.

35. Find all points on the portion of the plane $x + y + z = 5$ in the first octant at which $f(x, y, z) = xy^2z^2$ has a maximum value.

36. Find the points on the surface $x^2 - yz = 5$ that are closest to the origin.

37. Find the dimensions of the rectangular box of maximum volume that can be inscribed in a sphere of radius a.

38. Find the maximum volume of a rectangular box with three faces in the coordinate planes and a vertex in the first octant on the plane $x + y + z = 1$.

39. A closed rectangular box with a volume of 16 ft^3 is made from two kinds of materials. The top and bottom are made of material costing 10¢ per square foot and the sides from material costing 5¢ per square foot. Find the dimensions of the box so that the cost of materials is minimized.

40. A manufacturer makes two models of an item, standard and deluxe. It costs \$40 to manufacture the standard model and \$60 for the deluxe. A market research firm estimates that if the standard model is priced at x dollars and the deluxe at y dollars, then the manufacturer will sell $500(y - x)$ of the standard items and $45{,}000 + 500(x - 2y)$ of the deluxe each year. How should the items be priced to maximize the profit?

41. Consider the function

$$f(x, y) = 4x^2 - 3y^2 + 2xy$$

over the unit square $0 \le x \le 1, 0 \le y \le 1$.

(a) Find the maximum and minimum values of f on each edge of the square.

(b) Find the maximum and minimum values of f on each diagonal of the square.

(c) Find the maximum and minimum values of f on the entire square.

42. Show that among all parallelograms with perimeter l, a square with sides of length $l/4$ has maximum area. [*Hint:* The area of a parallelogram is given by the formula $A = ab \sin \alpha$, where a and b are the lengths of two adjacent sides and α is the angle between them.]

43. Determine the dimensions of a rectangular box, open at the top, having volume V, and requiring the least amount of material for its construction.

44. A length of sheet metal 27 inches wide is to be made into a water trough by bending up two sides as shown in the accompanying figure. Find x and ϕ so that the trapezoid-shaped cross section has a maximum area.

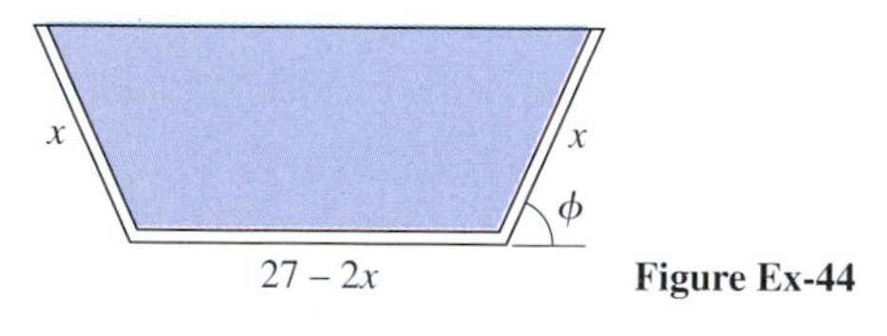

Figure Ex-44

45–46 A common problem in experimental work is to obtain a mathematical relationship $y = f(x)$ between two variables x and y by "fitting" a curve to points in the plane that correspond to experimentally determined values of x and y, say

$$(x_1, y_1), (x_2, y_2), \ldots, (x_n, y_n)$$

The curve $y = f(x)$ is called a ***mathematical model*** of the data. The general form of the function f is commonly determined by some underlying physical principle, but sometimes it is just determined by the pattern of the data. We are concerned with fitting a straight line $y = mx + b$ to data. Usually, the data will not lie on a line (possibly due to experimental error or variations in experimental conditions), so the problem is to find a line that fits the data "best" according to some criterion. One criterion for selecting the line of best fit is to choose m and b to minimize the function

$$g(m, b) = \sum_{i=1}^{n} (mx_i + b - y_i)^2$$

This is called the ***method of least squares***, and the resulting line is called the ***regression line*** or the ***least squares line of best fit***. Geometrically, $|mx_i + b - y_i|$ is the vertical distance between the data point (x_i, y_i) and the line $y = mx + b$.

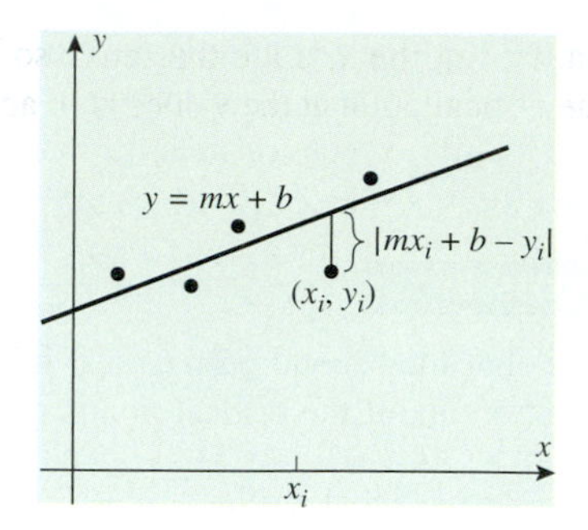

These vertical distances are called the ***residuals*** of the data points, so the effect of minimizing $g(m, b)$ is to minimize the sum of the squares of the residuals. In these exercises, we will derive a formula for the regression line. More on this topic can be found in Section 1.7.

45. The purpose of this exercise is to find the values of m and b that produce the regression line.

(a) To minimize $g(m, b)$, we start by finding values of m and b such that $\partial g/\partial m = 0$ and $\partial g/\partial b = 0$. Show that these equations are satisfied if m and b satisfy the conditions

$$\left(\sum_{i=1}^{n} x_i^2\right) m + \left(\sum_{i=1}^{n} x_i\right) b = \sum_{i=1}^{n} x_i y_i$$

$$\left(\sum_{i=1}^{n} x_i\right) m + nb = \sum_{i=1}^{n} y_i$$

(b) Let $\bar{x} = (x_1 + x_2 + \cdots + x_n)/n$ denote the arithmetic average of $x_1, x_2, \ldots, x_n$. Use the fact that

$$\sum_{i=1}^{n} (x_i - \bar{x})^2 \ge 0$$

to show that

$$n\left(\sum_{i=1}^{n} x_i^2\right) - \left(\sum_{i=1}^{n} x_i\right)^2 \ge 0$$

with equality if and only if all the x_i's are the same.

(c) Assuming that not all the x_i's are the same, prove that the equations in part (a) have the unique solution

$$m = \frac{n\sum_{i=1}^{n} x_i y_i - \sum_{i=1}^{n} x_i \sum_{i=1}^{n} y_i}{n\sum_{i=1}^{n} x_i^2 - \left(\sum_{i=1}^{n} x_i\right)^2}$$

$$b = \frac{1}{n}\left(\sum_{i=1}^{n} y_i - m\sum_{i=1}^{n} x_i\right)$$

[*Note:* We have shown that g has a critical point at these values of m and b. In the next exercise we will show that g has an absolute minimum at this critical point. Accepting this to be so, we have shown that the line $y = mx + b$ is the regression line for these values of m and b.]

46. Assume that not all the x_i's are the same, so that $g(m, b)$ has a unique critical point at the values of m and b obtained in Exercise 45(c). The purpose of this exercise is to show that g has an absolute minimum value at this point.

(a) Find the partial derivatives $g_{mm}(m, b)$, $g_{bb}(m, b)$, and $g_{mb}(m, b)$, and then apply the second partials test to show that g has a relative minimum at the critical point obtained in Exercise 45.

(b) Show that the graph of the equation $z = g(m, b)$ is a quadric surface. [*Hint:* See Formula (4) of Section 12.7.]

(c) It can be proved that the graph of $z = g(m, b)$ is an elliptic paraboloid. Accepting this to be so, show that this paraboloid opens in the positive z-direction, and explain how this shows that g has an absolute minimum at the critical point obtained in Exercise 45.

47–50 Use the formulas obtained in Exercise 45 to find and draw the regression line. If you have a calculating utility that can calculate regression lines, use it to check your work.

47. **48.**

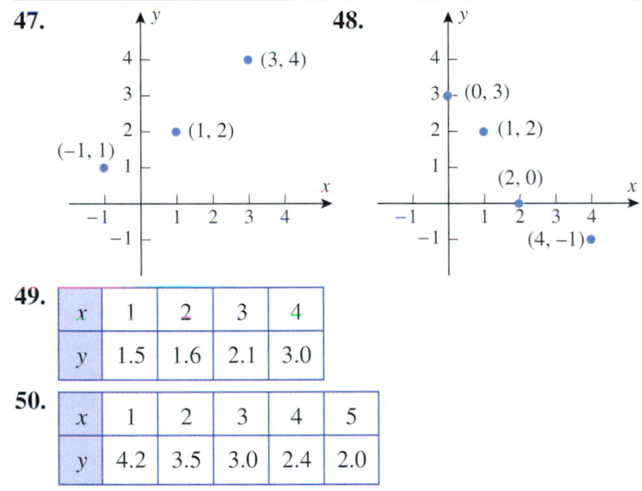

49.

x	1	2	3	4
y	1.5	1.6	2.1	3.0

50.

x	1	2	3	4	5
y	4.2	3.5	3.0	2.4	2.0

51. The following table shows the life expectancy by year of birth of females in the United States:

YEAR OF BIRTH	1930	1940	1950	1960	1970	1980	1990	2000
LIFE EXPECTANCY	61.6	65.2	71.1	73.1	74.7	77.5	78.8	79.7

(a) Take $t = 0$ to be the year 1930, and let y be the life expectancy for birth year t. Use the regression capability of a calculating utility to find the regression line of y as a function of t.

(b) Use a graphing utility to make a graph that shows the data points and the regression line.

(c) Use the regression line to make a conjecture about the life expectancy of females born in the year 2010.

52. A company manager wants to establish a relationship between the sales of a certain product and the price. The company research department provides the following data:

PRICE (x) IN DOLLARS	\$35.00	\$40.00	\$45.00	\$48.00	\$50.00
DAILY SALES VOLUME (y) IN UNITS	80	75	68	66	63

(a) Use a calculating utility to find the regression line of y as a function of x.

(b) Use a graphing utility to make a graph that shows the data points and the regression line.

(c) Use the regression line to make a conjecture about the number of units that would be sold at a price of \$60.00.

53. If a gas is cooled with its volume held constant, then it follows from the ***ideal gas law*** in physics that its pressure drops proportionally to the drop in temperature. The temperature that, in theory, corresponds to a pressure of zero is called ***absolute zero***. Suppose that an experiment produces the following data for pressure P versus temperature T with the volume held constant:

P (KILOPASCALS)	134	142	155	160	171	184
T (°CELSIUS)	0	20	40	60	80	100

(a) Use a calculating utility to find the regression line of P as a function of T.

(b) Use a graphing utility to make a graph that shows the data points and the regression line.

(c) Use the regression line to estimate the value of absolute zero in degrees Celsius.

54. Find

(a) a continuous function $f(x, y)$ that is defined on the entire xy-plane and has no absolute extrema on the xy-plane;

(b) a function $f(x, y)$ that is defined everywhere on the rectangle $0 \le x \le 1, 0 \le y \le 1$ and has no absolute extrema on the rectangle.

55. Show that if f has a relative maximum at (x_0, y_0), then $G(x) = f(x, y_0)$ has a relative maximum at $x = x_0$ and $H(y) = f(x_0, y)$ has a relative maximum at $y = y_0$.

✔QUICK CHECK ANSWERS 14.8

1. $(0, 0)$ and $\left(\frac{1}{6}, -\frac{1}{12}\right)$ **2.** (a) no information (b) a saddle point at $(0, 0)$ (c) a relative minimum at $(0, 0)$ (d) a relative maximum at $(0, 0)$ **3.** (a) a saddle point at $(0, 0)$ (b) no information, since $(-1, -1)$ is not a critical point (c) a relative minimum at $(1, 1)$ **4.** $V = \dfrac{xy(1 - xy)}{x + y}$

14.9 LAGRANGE MULTIPLIERS

In this section we will study a powerful new method for maximizing or minimizing a function subject to constraints on the variables. This method will help us to solve certain optimization problems that are difficult or impossible to solve using the methods studied in the last section.

EXTREMUM PROBLEMS WITH CONSTRAINTS

In Example 6 of the last section, we solved the problem of minimizing

$$S = xy + 2xz + 2yz \tag{1}$$

subject to the constraint

$$xyz - 32 = 0 \tag{2}$$

This is a special case of the following general problem:

14.9.1 ***Three-Variable Extremum Problem with One Constraint***
Maximize or minimize the function $f(x, y, z)$ subject to the constraint $g(x, y, z) = 0$.

We will also be interested in the following two-variable version of this problem:

14.9.2 ***Two-Variable Extremum Problem with One Constraint***
Maximize or minimize the function $f(x, y)$ subject to the constraint $g(x, y) = 0$.

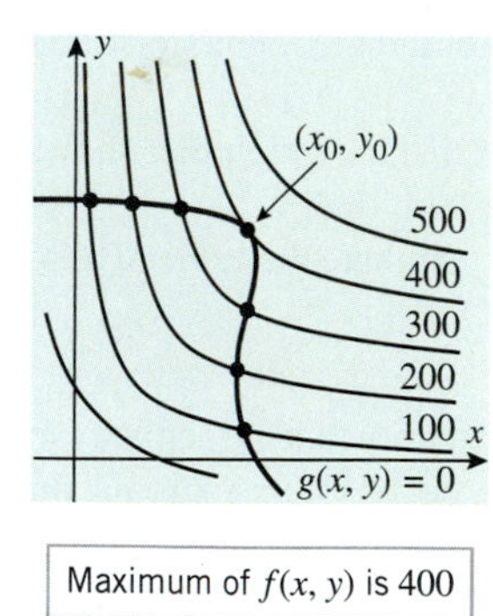

Maximum of $f(x, y)$ is 400

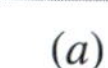

(*a*)

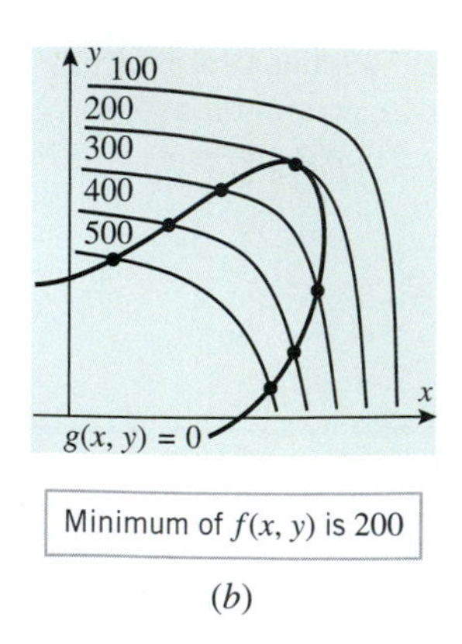

Minimum of $f(x, y)$ is 200

(*b*)

Figure 14.9.1

LAGRANGE MULTIPLIERS

One way to attack problems of these types is to solve the constraint equation for one of the variables in terms of the others and substitute the result into f. This produces a new function of one or two variables that incorporates the constraint and can be maximized or minimized by applying standard methods. For example, to solve the problem in Example 6 of the last section we substituted (2) into (1) to obtain

$$S = xy + \frac{64}{y} + \frac{64}{x}$$

which we then minimized by finding the critical points and applying the second partials test. However, this approach hinges on our ability to solve the constraint equation for one of the variables in terms of the others. If this cannot be done, then other methods must be used. One such method, called the *method of Lagrange multipliers*, will be discussed in this section.

To motivate the method of Lagrange multipliers, suppose that we are trying to maximize a function $f(x, y)$ subject to the constraint $g(x, y) = 0$. Geometrically, this means that we are looking for a point (x_0, y_0) on the graph of the constraint curve at which $f(x, y)$ is as large as possible. To help locate such a point, let us construct a contour plot of $f(x, y)$ in the same coordinate system as the graph of $g(x, y) = 0$. For example, Figure 14.9.1*a* shows some typical level curves of $f(x, y) = c$, which we have labeled $c = 100, 200, 300, 400$, and 500 for purposes of illustration. In this figure, each point of intersection of $g(x, y) = 0$ with a level curve is a candidate for a solution, since these points lie on the constraint curve. Among the seven such intersections shown in the figure, the maximum value of $f(x, y)$ occurs at the intersection (x_0, y_0), where $f(x, y)$ has a value of 400. Note that at (x_0, y_0) the constraint curve and the level curve just touch and thus have a *common* tangent line at this point. Since $\nabla f(x_0, y_0)$ is normal to the level curve $f(x, y) = 400$ at (x_0, y_0), and since $\nabla g(x_0, y_0)$ is normal to the constraint curve $g(x, y) = 0$ at (x_0, y_0), we conclude that the vectors $\nabla f(x_0, y_0)$ and $\nabla g(x_0, y_0)$ must be parallel. That is,

$$\nabla f(x_0, y_0) = \lambda \nabla g(x_0, y_0) \tag{3}$$

for some scalar λ. The same condition holds at points on the constraint curve where $f(x, y)$ has a minimum. For example, if the level curves are as shown in Figure 14.9.1*b*, then the minimum value of $f(x, y)$ occurs where the constraint curve just touches a level curve. Thus, to find the maximum or minimum of $f(x, y)$ subject to the constraint $g(x, y) = 0$, we look for points at which (3) holds—this is the method of Lagrange multipliers.

Our next objective in this section is to make the preceding intuitive argument more precise. For this purpose it will help to begin with some terminology about the problem of maximizing or minimizing a function $f(x, y)$ subject to a constraint $g(x, y) = 0$. As with other kinds of maximization and minimization problems, we need to distinguish between relative and absolute extrema. We will say that f has a ***constrained absolute maximum (minimum)*** at (x_0, y_0) if $f(x_0, y_0)$ is the largest (smallest) value of f on the constraint curve, and we will say that f has a ***constrained relative maximum (minimum)*** at (x_0, y_0) if $f(x_0, y_0)$ is the largest (smallest) value of f on some segment of the constraint curve that extends on both sides of the point (x_0, y_0) (Figure 14.9.2).

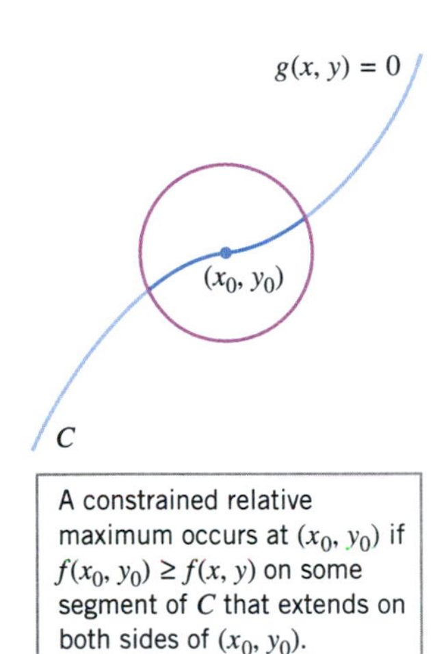

A constrained relative maximum occurs at (x_0, y_0) if $f(x_0, y_0) \geq f(x, y)$ on some segment of C that extends on both sides of (x_0, y_0).

Figure 14.9.2

Let us assume that a constrained relative maximum or minimum occurs at the point (x_0, y_0) and for simplicity, let us further assume that the equation $g(x, y) = 0$ can be smoothly parametrized as

$$x = x(s), \quad y = y(s)$$

where s is an arc length parameter with reference point (x_0, y_0) at $s = 0$. Thus, the quantity

$$z = f(x(s), y(s))$$

has a relative maximum or minimum at $s = 0$, and this implies that $dz/ds = 0$ at that point. From the chain rule, this equation can be expressed as

$$\frac{dz}{ds} = \frac{\partial f}{\partial x}\frac{dx}{ds} + \frac{\partial f}{\partial y}\frac{dy}{ds} = \left(\frac{\partial f}{\partial x}\mathbf{i} + \frac{\partial f}{\partial y}\mathbf{j}\right) \cdot \left(\frac{dx}{ds}\mathbf{i} + \frac{dy}{ds}\mathbf{j}\right) = 0$$

where the derivatives are all evaluated at $s = 0$. However, the first factor in the dot product is the gradient of f, and the second factor is the unit tangent vector to the constraint curve.

Joseph Louis Lagrange (1736–1813) French–Italian mathematician and astronomer. Lagrange, the son of a public official, was born in Turin, Italy. (Baptismal records list his name as Giuseppe Lodovico Lagrangia.) Although his father wanted him to be a lawyer, Lagrange was attracted to mathematics and astronomy after reading a memoir by the astronomer Halley. At age 16 he began to study mathematics on his own and by age 19 was appointed to a professorship at the Royal Artillery School in Turin. The following year Lagrange sent Euler solutions to some famous problems using new methods that eventually blossomed into a branch of mathematics called calculus of variations. These methods and Lagrange's applications of them to problems in celestial mechanics were so monumental that by age 25 he was regarded by many of his contemporaries as the greatest living mathematician.

In 1776, on the recommendations of Euler, he was chosen to succeed Euler as the director of the Berlin Academy. During his stay in Berlin, Lagrange distinguished himself not only in celestial mechanics, but also in algebraic equations and the theory of numbers. After twenty years in Berlin, he moved to Paris at the invitation of Louis XVI. He was given apartments in the Louvre and treated with great honor, even during the revolution.

Napoleon was a great admirer of Lagrange and showered him with honors—count, senator, and Legion of Honor. The years Lagrange spent in Paris were devoted primarily to didactic treatises summarizing his mathematical conceptions. One of Lagrange's most famous works is a memoir, *Mécanique Analytique*, in which he reduced the theory of mechanics to a few general formulas from which all other necessary equations could be derived.

It is an interesting historical fact that Lagrange's father speculated unsuccessfully in several financial ventures, so his family was forced to live quite modestly. Lagrange himself stated that if his family had money, he would not have made mathematics his vocation. In spite of his fame, Lagrange was always a shy and modest man. On his death, he was buried with honor in the Pantheon.

Since the point (x_0, y_0) corresponds to $s = 0$, it follows from this equation that

$$\nabla f(x_0, y_0) \cdot \mathbf{T}(0) = 0$$

which implies that the gradient is either $\mathbf{0}$ or is normal to the constraint curve at a constrained relative extremum. However, the constraint curve $g(x, y) = 0$ is a level curve for the function $g(x, y)$, so that if $\nabla g(x_0, y_0) \neq \mathbf{0}$, then $\nabla g(x_0, y_0)$ is normal to this curve at (x_0, y_0). It then follows that there is some scalar λ such that

$$\nabla f(x_0, y_0) = \lambda \nabla g(x_0, y_0) \tag{4}$$

This scalar is called a ***Lagrange multiplier***. Thus, the ***method of Lagrange multipliers*** for finding constrained relative extrema is to look for points on the constraint curve $g(x, y) = 0$ at which Equation (4) is satisfied for some scalar λ.

14.9.3 THEOREM ***(Constrained-Extremum Principle for Two Variables and One Constraint).*** *Let f and g be functions of two variables with continuous first partial derivatives on some open set containing the constraint curve $g(x, y) = 0$, and assume that $\nabla g \neq \mathbf{0}$ at any point on this curve. If f has a constrained relative extremum, then this extremum occurs at a point (x_0, y_0) on the constraint curve at which the gradient vectors $\nabla f(x_0, y_0)$ and $\nabla g(x_0, y_0)$ are parallel; that is, there is some number λ such that*

$$\nabla f(x_0, y_0) = \lambda \nabla g(x_0, y_0)$$

► **Example 1** At what point or points on the circle $x^2 + y^2 = 1$ does $f(x, y) = xy$ have an absolute maximum, and what is that maximum?

Solution. The circle $x^2 + y^2 = 1$ is a closed and bounded set and $f(x, y) = xy$ is a continuous function, so it follows from the Extreme-Value Theorem (Theorem 14.8.3) that f has an absolute maximum and an absolute minimum on the circle. To find these extrema, we will use Lagrange multipliers to find the constrained relative extrema, and then we will evaluate f at those relative extrema to find the absolute extrema.

We want to maximize $f(x, y) = xy$ subject to the constraint

$$g(x, y) = x^2 + y^2 - 1 = 0 \tag{5}$$

First we will look for constrained *relative* extrema. For this purpose we will need the gradients

$$\nabla f = y\mathbf{i} + x\mathbf{j} \quad \text{and} \quad \nabla g = 2x\mathbf{i} + 2y\mathbf{j}$$

From the formula for ∇g we see that $\nabla g = \mathbf{0}$ if and only if $x = 0$ and $y = 0$, so $\nabla g \neq \mathbf{0}$ at any point on the circle $x^2 + y^2 = 1$. Thus, at a constrained relative extremum we must have

$$\nabla f = \lambda \nabla g \quad \text{or} \quad y\mathbf{i} + x\mathbf{j} = \lambda(2x\mathbf{i} + 2y\mathbf{j})$$

which is equivalent to the pair of equations

$$y = 2x\lambda \quad \text{and} \quad x = 2y\lambda$$

It follows from these equations that if $x = 0$, then $y = 0$, and if $y = 0$, then $x = 0$. In either case we have $x^2 + y^2 = 0$, so the constraint equation $x^2 + y^2 = 1$ is not satisfied. Thus, we can assume that x and y are nonzero, and we can rewrite the equations as

$$\lambda = \frac{y}{2x} \quad \text{and} \quad \lambda = \frac{x}{2y}$$

from which we obtain

$$\frac{y}{2x} = \frac{x}{2y}$$

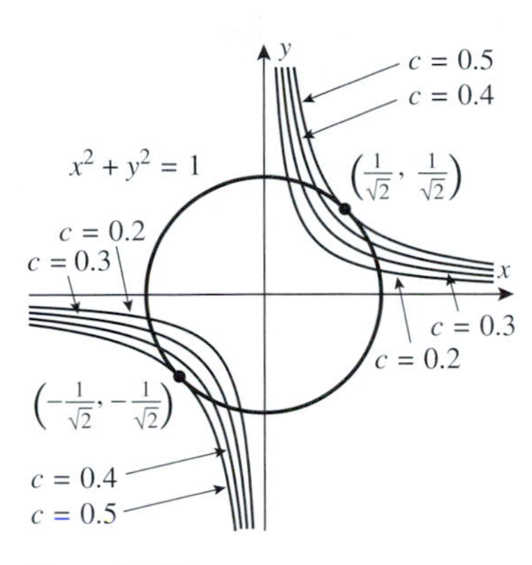

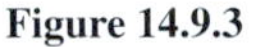

Figure 14.9.3

or

$$y^2 = x^2 \tag{6}$$

Substituting this in (5) yields

$$2x^2 - 1 = 0$$

from which we obtain $x = \pm 1/\sqrt{2}$. Each of these values, when substituted in Equation (6), produces y-values of $y = \pm 1/\sqrt{2}$. Thus, constrained relative extrema occur at the points $(1/\sqrt{2}, 1/\sqrt{2})$, $(1/\sqrt{2}, -1/\sqrt{2})$, $(-1/\sqrt{2}, 1/\sqrt{2})$, and $(-1/\sqrt{2}, -1/\sqrt{2})$. The values of xy at these points are as follows:

(x, y)	$(1/\sqrt{2}, 1/\sqrt{2})$	$(1/\sqrt{2}, -1/\sqrt{2})$	$(-1/\sqrt{2}, 1/\sqrt{2})$	$(-1/\sqrt{2}, -1/\sqrt{2})$
xy	1/2	−1/2	−1/2	1/2

Give another solution to Example 1 using the parametrization

$$x = \cos\theta, \quad y = \sin\theta$$

and the identity

$$\sin 2\theta = 2\sin\theta\cos\theta$$

Thus, the function $f(x, y) = xy$ has an absolute maximum of $\frac{1}{2}$ occurring at the two points $(1/\sqrt{2}, 1/\sqrt{2})$ and $(-1/\sqrt{2}, -1/\sqrt{2})$. Although it was not asked for, we can also see that f has an absolute minimum of $-\frac{1}{2}$ occurring at the points $(1/\sqrt{2}, -1/\sqrt{2})$ and $(-1/\sqrt{2}, 1/\sqrt{2})$. Figure 14.9.3 shows some level curves $xy = c$ and the constraint curve in the vicinity of the maxima. A similar figure for the minima can be obtained using negative values of c for the level curves $xy = c$. ◂

If c is a constant, then the functions $g(x, y)$ and $g(x, y) - c$ have the same gradient since the constant c drops out when we differentiate. Consequently, it is *not* essential to rewrite a constraint of the form $g(x, y) = c$ as $g(x, y) - c = 0$ in order to apply the constrained-extremum principle. Thus, in the last example, we could have kept the constraint in the form $x^2 + y^2 = 1$ and then taken $g(x, y) = x^2 + y^2$ rather than $g(x, y) = x^2 + y^2 - 1$.

▸ **Example 2** Use the method of Lagrange multipliers to find the dimensions of a rectangle with perimeter p and maximum area.

Solution. Let

x = length of the rectangle, y = width of the rectangle, A = area of the rectangle

We want to maximize $A = xy$ on the line segment

$$2x + 2y = p, \quad 0 \le x, y \tag{7}$$

that corresponds to the perimeter constraint. This segment is a closed and bounded set, and since $f(x, y) = xy$ is a continuous function, it follows from the Extreme-Value Theorem (Theorem 14.8.3) that f has an absolute maximum on this segment. This absolute maximum must also be a constrained relative maximum since f is 0 at the endpoints of the segment and positive elsewhere on the segment. If $g(x, y) = 2x + 2y$, then we have

$$\nabla f = y\mathbf{i} + x\mathbf{j} \quad \text{and} \quad \nabla g = 2\mathbf{i} + 2\mathbf{j}$$

Noting that $\nabla g \neq \mathbf{0}$, it follows from (4) that

$$y\mathbf{i} + x\mathbf{j} = \lambda(2\mathbf{i} + 2\mathbf{j})$$

at a constrained relative maximum. This is equivalent to the two equations

$$y = 2\lambda \quad \text{and} \quad x = 2\lambda$$

Eliminating λ from these equations we obtain $x = y$, which shows that the rectangle is actually a square. Using this condition and constraint (7), we obtain $x = p/4$, $y = p/4$. ◂

THREE VARIABLES AND ONE CONSTRAINT

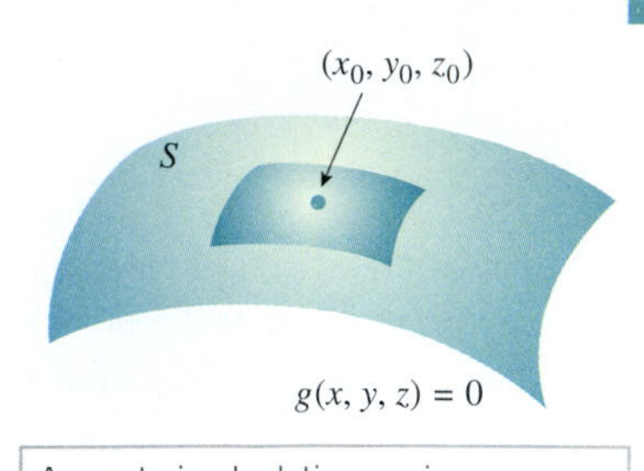

A constrained relative maximum occurs at (x_0, y_0, z_0) if $f(x_0, y_0, z_0) \geq f(x, y, z)$ at all points of S near (x_0, y_0, z_0).

Figure 14.9.4

The method of Lagrange multipliers can also be used to maximize or minimize a function of three variables $f(x, y, z)$ subject to a constraint $g(x, y, z) = 0$. As a rule, the graph of $g(x, y, z) = 0$ will be some surface S in 3-space. Thus, from a geometric viewpoint, the problem is to maximize or minimize $f(x, y, z)$ as (x, y, z) varies over the surface S (Figure 14.9.4). As usual, we distinguish between relative and absolute extrema. We will say that f has a ***constrained absolute maximum (minimum)*** at (x_0, y_0, z_0) if $f(x_0, y_0, z_0)$ is the largest (smallest) value of $f(x, y, z)$ on S, and we will say that f has a ***constrained relative maximum (minimum)*** at (x_0, y_0, z_0) if $f(x_0, y_0, z_0)$ is the largest (smallest) value of $f(x, y, z)$ at all points of S "near" (x_0, y_0, z_0).

The following theorem, which we state without proof, is the three-variable analog of Theorem 14.9.3.

14.9.4 THEOREM *(Constrained-Extremum Principle for Three Variables and One Constraint).* *Let f and g be functions of three variables with continuous first partial derivatives on some open set containing the constraint surface $g(x, y, z) = 0$, and assume that $\nabla g \neq \mathbf{0}$ at any point on this surface. If f has a constrained relative extremum, then this extremum occurs at a point (x_0, y_0, z_0) on the constraint surface at which the gradient vectors $\nabla f(x_0, y_0, z_0)$ and $\nabla g(x_0, y_0, z_0)$ are parallel; that is, there is some number λ such that*

$$\nabla f(x_0, y_0, z_0) = \lambda \nabla g(x_0, y_0, z_0)$$

▶ **Example 3** Find the points on the sphere $x^2 + y^2 + z^2 = 36$ that are closest to and farthest from the point $(1, 2, 2)$.

Solution. To avoid radicals, we will find points on the sphere that minimize and maximize the *square* of the distance to $(1, 2, 2)$. Thus, we want to find the relative extrema of

$$f(x, y, z) = (x-1)^2 + (y-2)^2 + (z-2)^2$$

subject to the constraint

$$x^2 + y^2 + z^2 = 36 \tag{8}$$

If we let $g(x, y, z) = x^2 + y^2 + z^2$, then $\nabla g = 2x\mathbf{i} + 2y\mathbf{j} + 2z\mathbf{k}$. Thus, $\nabla g = \mathbf{0}$ if and only if $x = y = z = 0$. It follows that $\nabla g \neq \mathbf{0}$ at any point of the sphere (8), and hence the constrained relative extrema must occur at points where

$$\nabla f(x, y, z) = \lambda \nabla g(x, y, z)$$

That is,

$$2(x-1)\mathbf{i} + 2(y-2)\mathbf{j} + 2(z-2)\mathbf{k} = \lambda(2x\mathbf{i} + 2y\mathbf{j} + 2z\mathbf{k})$$

which leads to the equations

$$2(x-1) = 2x\lambda, \quad 2(y-2) = 2y\lambda, \quad 2(z-2) = 2z\lambda \tag{9}$$

We may assume that x, y, and z are nonzero since $x = 0$ does not satisfy the first equation, $y = 0$ does not satisfy the second, and $z = 0$ does not satisfy the third. Thus, we can rewrite (9) as

$$\frac{x-1}{x} = \lambda, \quad \frac{y-2}{y} = \lambda, \quad \frac{z-2}{z} = \lambda$$

The first two equations imply that

$$\frac{x-1}{x} = \frac{y-2}{y}$$

from which it follows that

$$y = 2x \tag{10}$$

Similarly, the first and third equations imply that

$$z = 2x \tag{11}$$

Substituting (10) and (11) in the constraint equation (8), we obtain

$$9x^2 = 36 \quad \text{or} \quad x = \pm 2$$

Substituting these values in (10) and (11) yields two points:

$$(2, 4, 4) \quad \text{and} \quad (-2, -4, -4)$$

Since $f(2, 4, 4) = 9$ and $f(-2, -4, -4) = 81$, it follows that $(2, 4, 4)$ is the point on the sphere closest to $(1, 2, 2)$, and $(-2, -4, -4)$ is the point that is farthest (Figure 14.9.5). ◄

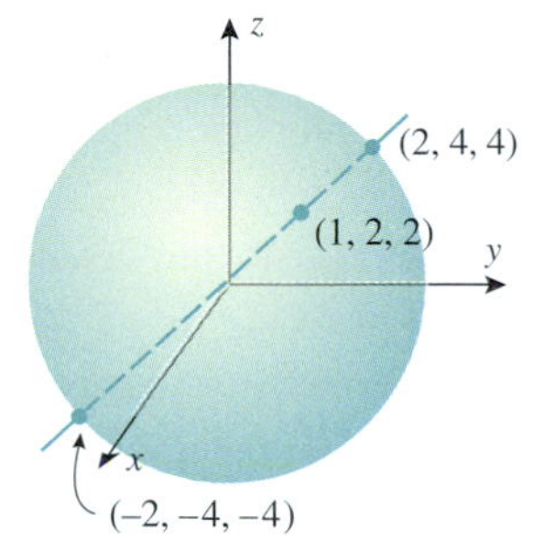

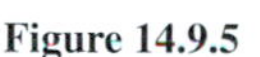
Figure 14.9.5

Next we will use Lagrange multipliers to solve the problem of Example 6 in the last section.

► Example 4 Use Lagrange multipliers to determine the dimensions of a rectangular box, open at the top, having a volume of 32 ft^3, and requiring the least amount of material for its construction.

Solution. With the notation introduced in Example 6 of the last section, the problem is to minimize the surface area

$$S = xy + 2xz + 2yz$$

subject to the volume constraint

$$xyz = 32 \tag{12}$$

If we let $f(x, y, z) = xy + 2xz + 2yz$ and $g(x, y, z) = xyz$, then

$$\nabla f = (y + 2z)\mathbf{i} + (x + 2z)\mathbf{j} + (2x + 2y)\mathbf{k} \quad \text{and} \quad \nabla g = yz\mathbf{i} + xz\mathbf{j} + xy\mathbf{k}$$

It follows that $\nabla g \neq \mathbf{0}$ at any point on the surface $xyz = 32$, since x, y, and z are all nonzero on this surface. Thus, at a constrained relative extremum we must have $\nabla f = \lambda \nabla g$, that is,

$$(y + 2z)\mathbf{i} + (x + 2z)\mathbf{j} + (2x + 2y)\mathbf{k} = \lambda(yz\mathbf{i} + xz\mathbf{j} + xy\mathbf{k})$$

This condition yields the three equations

$$y + 2z = \lambda yz, \quad x + 2z = \lambda xz, \quad 2x + 2y = \lambda xy$$

Because x, y, and z are nonzero, these equations can be rewritten as

$$\frac{1}{z} + \frac{2}{y} = \lambda, \quad \frac{1}{z} + \frac{2}{x} = \lambda, \quad \frac{2}{y} + \frac{2}{x} = \lambda$$

From the first two equations,

$$y = x \tag{13}$$

and from the first and third equations,

$$z = \tfrac{1}{2}x \tag{14}$$

Substituting (13) and (14) in the volume constraint (12) yields

$$\tfrac{1}{2}x^3 = 32$$

This equation, together with (13) and (14), yields

$$x = 4, \quad y = 4, \quad z = 2$$

which agrees with the result that was obtained in Example 6 of the last section. ◄

There are variations in the method of Lagrange multipliers that can be used to solve problems with two or more constraints. However, we will not discuss that topic here.

✔ QUICK CHECK EXERCISES 14.9 (See page 1015 for answers.)

1. (a) Suppose that $f(x, y)$ and $g(x, y)$ are differentiable at the origin and have nonzero gradients there, and that $g(0, 0) = 0$. If the maximum value of $f(x, y)$ subject to the constraint $g(x, y) = 0$ occurs at the origin, how is the tangent line to the graph of $g(x, y) = 0$ related to the tangent line at the origin to the level curve of f through $(0, 0)$?
   (b) Suppose that $f(x, y, z)$ and $g(x, y, z)$ are differentiable at the origin and have nonzero gradients there, and that $g(0, 0, 0) = 0$. If the maximum value of $f(x, y, z)$ subject to the constraint $g(x, y, z) = 0$ occurs at the origin, how is the tangent plane to the graph of the constraint $g(x, y, z) = 0$ related to the tangent plane at the origin to the level surface of f through $(0, 0, 0)$?
2. The maximum value of $x + y$ subject to the constraint $x^2 + y^2 = 1$ is ________.
3. The maximum value of $x + y + z$ subject to the constraint $x^2 + y^2 + z^2 = 1$ is ________.
4. The maximum and minimum values of $2x + 3y$ subject to the constraint $x + y = 1$, where $0 \le x, 0 \le y$, are ________ and ________, respectively.

EXERCISE SET 14.9 Graphing Utility C CAS

FOCUS ON CONCEPTS

1. The accompanying figure shows graphs of the line $x + y = 4$ and the level curves of height $c = 2, 4, 6,$ and 8 for the function $f(x, y) = xy$.
   (a) Use the figure to find the maximum value of the function $f(x, y) = xy$ subject to $x + y = 4$, and explain your reasoning.
   (b) How can you tell from the figure that your answer to part (a) is not the minimum value of f subject to the constraint?
   (c) Use Lagrange multipliers to check your work.
2. The accompanying figure shows the graphs of the line $3x + 4y = 25$ and the level curves of height $c = 9, 16, 25, 36,$ and 49 for the function $f(x, y) = x^2 + y^2$.
   (a) Use the accompanying figure to find the minimum value of the function $f(x, y) = x^2 + y^2$ subject to $3x + 4y = 25$, and explain your reasoning.
   (b) How can you tell from the accompanying figure that your answer to part (a) is not the maximum value of f subject to the constraint?
   (c) Use Lagrange multipliers to check your work.

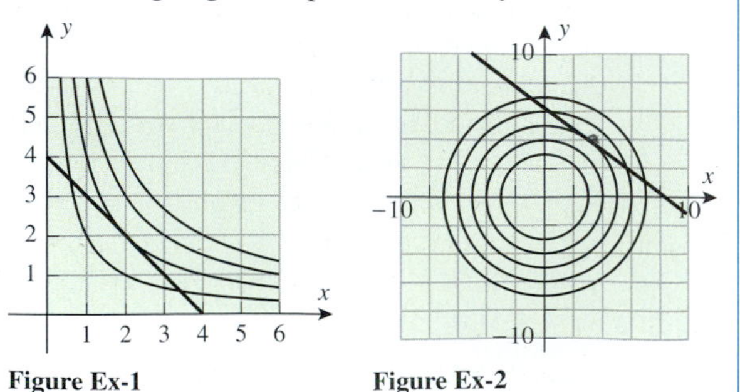

Figure Ex-1 **Figure Ex-2**

3. (a) On a graphing utility, graph the circle $x^2 + y^2 = 25$ and two distinct level curves of $f(x, y) = x^2 - y$ that just touch the circle in a single point.
   (b) Use the results you obtained in part (a) to approximate the maximum and minimum values of f subject to the constraint $x^2 + y^2 = 25$.
   (c) Check your approximations in part (b) using Lagrange multipliers.
4. C (a) If you have a CAS with implicit plotting capability, use it to graph the circle $(x - 4)^2 + (y - 4)^2 = 4$ and two level curves of $f(x, y) = x^3 + y^3 - 3xy$ that just touch the circle.
   (b) Use the result you obtained in part (a) to approximate the minimum value of f subject to the constraint $(x - 4)^2 + (y - 4)^2 = 4$.
   (c) Confirm graphically that you have found a minimum and not a maximum.
   (d) Check your approximation using Lagrange multipliers and solving the required equations numerically.

5–12 Use Lagrange multipliers to find the maximum and minimum values of f subject to the given constraint. Also, find the points at which these extreme values occur.

5. $f(x, y) = xy;\ 4x^2 + 8y^2 = 16$
6. $f(x, y) = x^2 - y^2;\ x^2 + y^2 = 25$
7. $f(x, y) = 4x^3 + y^2;\ 2x^2 + y^2 = 1$
8. $f(x, y) = x - 3y - 1;\ x^2 + 3y^2 = 16$
9. $f(x, y, z) = 2x + y - 2z;\ x^2 + y^2 + z^2 = 4$
10. $f(x, y, z) = 3x + 6y + 2z;\ 2x^2 + 4y^2 + z^2 = 70$
11. $f(x, y, z) = xyz;\ x^2 + y^2 + z^2 = 1$
12. $f(x, y, z) = x^4 + y^4 + z^4;\ x^2 + y^2 + z^2 = 1$

13–20 Solve using Lagrange multipliers.

13. Find the point on the line $2x - 4y = 3$ that is closest to the origin.
14. Find the point on the line $y = 2x + 3$ that is closest to $(4, 2)$.
15. Find the point on the plane $x + 2y + z = 1$ that is closest to the origin.
16. Find the point on the plane $4x + 3y + z = 2$ that is closest to $(1, -1, 1)$.

17. Find the points on the circle $x^2 + y^2 = 45$ that are closest to and farthest from $(1, 2)$.

18. Find the points on the surface $xy - z^2 = 1$ that are closest to the origin.

19. Find a vector in 3-space whose length is 5 and whose components have the largest possible sum.

20. Suppose that the temperature at a point (x, y) on a metal plate is $T(x, y) = 4x^2 - 4xy + y^2$. An ant, walking on the plate, traverses a circle of radius 5 centered at the origin. What are the highest and lowest temperatures encountered by the ant?

21–28 Use Lagrange multipliers to solve the indicated problems from Section 14.8.

21. Exercise 34 **22.** Exercise 35

23. Exercise 36 **24.** Exercise 37

25. Exercise 39 **26.** Exercises 41(a) and (b)

27. Exercise 42 **28.** Exercise 43

C **29.** Let α, β, and γ be the angles of a triangle.

(a) Use Lagrange multipliers to find the maximum value of $f(\alpha, \beta, \gamma) = \cos\alpha \cos\beta \cos\gamma$, and determine the angles for which the maximum occurs.

(b) Express $f(\alpha, \beta, \gamma)$ as a function of α and β alone, and use a CAS to graph this function of two variables. Confirm that the result obtained in part (a) is consistent with the graph.

30. The accompanying figure shows the intersection of the elliptic paraboloid $z = x^2 + 4y^2$ and the right circular cylinder $x^2 + y^2 = 1$. Use Lagrange multipliers to find the highest and lowest points on the curve of intersection.

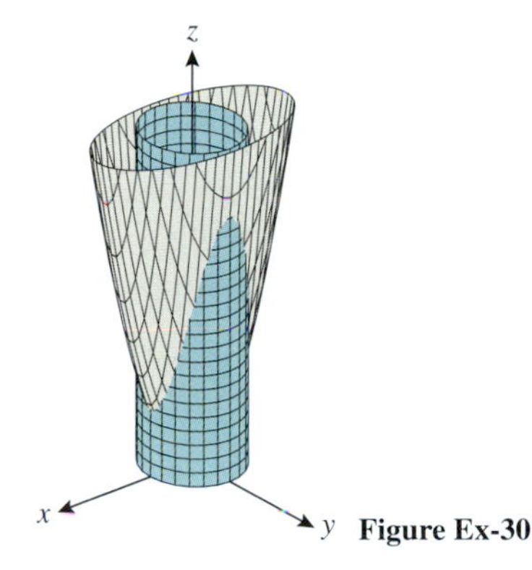

Figure Ex-30

✔ QUICK CHECK ANSWERS 14.9

1. (a) They are the same line. (b) They are the same plane. **2.** $\sqrt{2}$ **3.** $\sqrt{3}$ **4.** 3; 2

CHAPTER REVIEW EXERCISES

Graphing Utility

1. Let $f(x, y) = e^x \ln y$. Find

(a) $f(\ln y, e^x)$ (b) $f(r + s, rs)$.

2. Sketch the domain of f using solid lines for portions of the boundary included in the domain and dashed lines for portions not included.

(a) $f(x, y) = \ln(xy - 1)$ (b) $f(x, y) = (\sin^{-1} x)/e^y$

3. Show that the level curves of the cone $z = \sqrt{x^2 + y^2}$ and the paraboloid $z = x^2 + y^2$ are circles, and make a sketch that illustrates the difference between the contour plots of the two functions.

4. (a) In words, describe the level surfaces of the function $f(x, y, z) = a^2x^2 + a^2y^2 + z^2$, where $a > 0$.

(b) Find a function $f(x, y, z)$ whose level surfaces form a family of circular paraboloids that open in the positive z-direction.

5–6 (a) Find the limit of $f(x, y)$ as $(x, y) \to (0, 0)$ if it exists, and (b) determine whether f is continuous at $(0, 0)$.

5. $f(x, y) = \dfrac{x^4 - x + y - x^3y}{x - y}$

6. $f(x, y) = \begin{cases} \dfrac{x^4 - y^4}{x^2 + y^2} & \text{if } (x, y) \neq (0, 0) \\ 0 & \text{if } (x, y) = (0, 0) \end{cases}$

7. (a) A company manufactures two types of computer monitors: standard monitors and high resolution monitors. Suppose that $P(x, y)$ is the profit that results from producing and selling x standard monitors and y high-resolution monitors. What do the two partial derivatives $\partial P/\partial x$ and $\partial P/\partial y$ represent?

(b) Suppose that the temperature at time t at a point (x, y) on the surface of a lake is $T(x, y, t)$. What do the partial derivatives $\partial T/\partial x$, $\partial T/\partial y$, and $\partial T/\partial t$ represent?

8. Let $z = f(x, y)$.

(a) Express $\partial z/\partial x$ and $\partial z/\partial y$ as limits.

(b) In words, what do the derivatives $f_x(x_0, y_0)$ and $f_y(x_0, y_0)$ tell you about the surface $z = f(x, y)$?

(c) In words, what do the derivatives $\partial z/\partial x(x_0, y_0)$ and $\partial z/\partial y(x_0, y_0)$ tell you about the rates of change of z with respect to x and y?

9. The pressure in newtons per square meter (N/m^2) of a gas in a cylinder is given by $P = 10T/V$ with T in kelvins (K) and V in cubic meters (m^3).
   (a) If T is increasing at a rate of 3 K/min with V held fixed at 2.5 m^3, find the rate at which the pressure is changing when $T = 50$ K.
   (b) If T is held fixed at 50 K while V is decreasing at the rate of 3 m^3/min, find the rate at which the pressure is changing when $V = 2.5\ m^3$.

10. Find the slope of the tangent line at the point $(1, -2, -3)$ on the curve of intersection of the surface $z = 5 - 4x^2 - y^2$ with
   (a) the plane $x = 1$ (b) the plane $y = -2$.

11–14 Verify the assertion.

11. If $w = \tan(x^2 + y^2) + x\sqrt{y}$, then $w_{xy} = w_{yx}$.

12. If $w = \ln(3x - 3y) + \cos(x + y)$, then $\partial^2 w/\partial x^2 = \partial^2 w/\partial y^2$.

13. If $F(x, y, z) = 2z^3 - 3(x^2 + y^2)z$, then F satisfies the equation $F_{xx} + F_{yy} + F_{zz} = 0$.

14. If $f(x, y, z) = xyz + x^2 + \ln(y/z)$, then $f_{xyzx} = f_{zxxy}$.

15. What do Δf and df represent, and how are they related?

16. If $w = x^2y - 2xy + y^2x$, find the increment Δw and the differential dw if (x, y) varies from $(1, 0)$ to $(1.1, -0.1)$.

17. Use differentials to estimate the change in the volume $V = \frac{1}{3}x^2h$ of a pyramid with a square base when its height h is increased from 2 to 2.2 m and its base dimension x is decreased from 1 to 0.9 m. Compare this to ΔV.

18. Find the local linear approximation of $f(x, y) = \sin(xy)$ at $\left(\frac{1}{3}, \pi\right)$.

19. Suppose that z is a differentiable function of x and y with

$$\frac{\partial z}{\partial x}(1, 2) = 4 \quad \text{and} \quad \frac{\partial z}{\partial y}(1, 2) = 2$$

If $x = x(t)$ and $y = y(t)$ are differentiable functions of t with $x(0) = 1$, $y(0) = 2$, $x'(0) = -\frac{1}{2}$, and (under composition) $z'(0) = 2$, find $y'(0)$.

20. In each part, use Theorem 14.5.3 to find dy/dx.
   (a) $3x^2 - 5xy + \tan xy = 0$
   (b) $x \ln y + \sin(x - y) = \pi$

21. Given that $f(x, y) = 0$, use Theorem 14.5.3 to express d^2y/dx^2 in terms of partial derivatives of f.

22. Let $z = f(x, y)$, where $x = g(t)$ and $y = h(t)$.
   (a) Show that

$$\frac{d}{dt}\left(\frac{\partial z}{\partial x}\right) = \frac{\partial^2 z}{\partial x^2}\frac{dx}{dt} + \frac{\partial^2 z}{\partial y \partial x}\frac{dy}{dt}$$

and

$$\frac{d}{dt}\left(\frac{\partial z}{\partial y}\right) = \frac{\partial^2 z}{\partial x \partial y}\frac{dx}{dt} + \frac{\partial^2 z}{\partial y^2}\frac{dy}{dt}$$

   (b) Use the formulas in part (a) to help find a formula for d^2z/dt^2.

23. (a) How are the directional derivative and the gradient of a function related?
   (b) Under what conditions is the directional derivative of a differentiable function 0?
   (c) In what direction does the directional derivative of a differentiable function have its maximum value? Its minimum value?

24. In words, what does the derivative $D_{\mathbf{u}}f(x_0, y_0)$ tell you about the surface $z = f(x, y)$?

25. Find $D_{\mathbf{u}}f(-3, 5)$ for $f(x, y) = y\ln(x + y)$ if $\mathbf{u} = \frac{3}{5}\mathbf{i} + \frac{4}{5}\mathbf{j}$.

26. Suppose that $\nabla f(0, 0) = 2\mathbf{i} + \frac{3}{2}\mathbf{j}$.
   (a) Find a unit vector $\mathbf{u}$ such that $D_{\mathbf{u}}f(0, 0)$ is a maximum. What is this maximum value?
   (b) Find a unit vector $\mathbf{u}$ such that $D_{\mathbf{u}}f(0, 0)$ is a minimum. What is this minimum value?

27. At the point $(1, 2)$, the directional derivative $D_{\mathbf{u}}f$ is $2\sqrt{2}$ toward $P_1(2, 3)$ and -3 toward $P_2(1, 0)$. Find $D_{\mathbf{u}}f(1, 2)$ toward the origin.

28. Find equations for the tangent plane and normal line to the given surface at P_0.
   (a) $z = x^2e^{2y}$; $P_0(1, \ln 2, 4)$
   (b) $x^2y^3z^4 + xyz = 2$; $P_0(2, 1, -1)$

29. Find all points P_0 on the surface $z = 2 - xy$ at which the normal line passes through the origin.

30. Show that for all tangent planes to the surface

$$x^{2/3} + y^{2/3} + z^{2/3} = 1$$

the sum of the squares of the x-, y-, and z-intercepts is 1.

31. Find all points on the paraboloid $z = 9x^2 + 4y^2$ at which the normal line is parallel to the line through the points $P(4, -2, 5)$ and $Q(-2, -6, 4)$.

32. Suppose the equations of motion of a particle are $x = t - 1$, $y = 4e^{-t}$, $z = 2 - \sqrt{t}$, where $t > 0$. Find, to the nearest tenth of a degree, the acute angle between the velocity vector and the normal line to the surface $(x^2/4) + y^2 + z^2 = 1$ at the points where the particle collides with the surface. Use a calculating utility with a root-finding capability where needed.

33–36 Locate all relative minima, relative maxima, and saddle points.

33. $f(x, y) = x^2 + 3xy + 3y^2 - 6x + 3y$

34. $f(x, y) = x^2y - 6y^2 - 3x^2$

35. $f(x, y) = x^3 - 3xy + \frac{1}{2}y^2$

36. $f(x, y) = 4x^2 - 12xy + 9y^2$

37–39 Solve these exercises two ways:
(a) Use the constraint to eliminate a variable.
(b) Use Lagrange multipliers.

37. Find all relative extrema of x^2y^2 subject to the constraint $4x^2 + y^2 = 8$.

38. Find the dimensions of the rectangular box of maximum volume that can be inscribed in the ellipsoid

$$(x/a)^2 + (y/b)^2 + (z/c)^2 = 1$$

39. As illustrated in the accompanying figure, suppose that a current I branches into currents I_1, I_2, and I_3 through resistors R_1, R_2, and R_3 in such a way that the total power dissipated in the three resistors is a minimum. Find the ratios $I_1 : I_2 : I_3$ if the power dissipated in R_i is $I_i^2 R_i (i = 1, 2, 3)$ and $I_1 + I_2 + I_3 = I$.

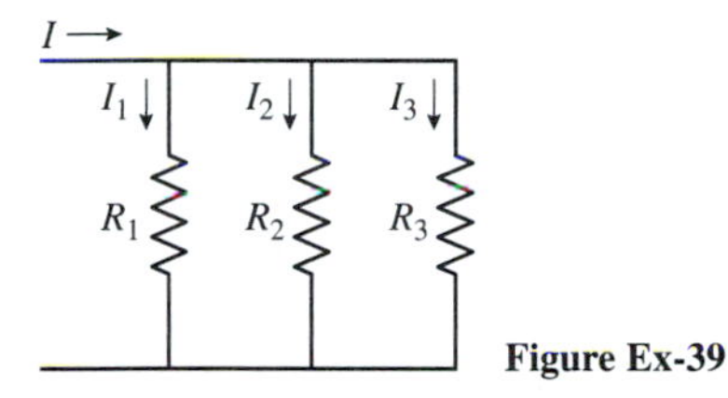

Figure Ex-39

40–42 In economics, a ***production model*** is a mathematical relationship between the output of a company or a country and the labor and capital equipment required to produce that output. Much of the pioneering work in the field of production models occurred in the 1920s when Paul Douglas of the University of Chicago and his collaborator Charles Cobb proposed that the output P can be expressed in terms of the labor L and the capital equipment K by an equation of the form

$$P = cL^{\alpha}K^{\beta}$$

where c is a constant of proportionality and α and β are constants such that $0 < \alpha < 1$ and $0 < \beta < 1$. This is called the ***Cobb–Douglas production model***. Typically, P, L, and K are all expressed in terms of their equivalent monetary values. These exercises explore properties of this model.

40. (a) Consider the Cobb–Douglas production model given by the formula $P = L^{0.75}K^{0.25}$. Sketch the level curves $P(L, K) = 1$, $P(L, K) = 2$, and $P(L, K) = 3$ in an LK-coordinate system (L horizontal and K vertical). Your sketch need not be accurate numerically, but it should show the general shape of the curves and their relative positions.

(b) Use a graphing utility to make a more extensive contour plot of the model.

41. (a) Find $\partial P/\partial L$ and $\partial P/\partial K$ for the Cobb–Douglas production model $P = cL^{\alpha}K^{\beta}$.

(b) The derivative $\partial P/\partial L$ is called the ***marginal productivity of labor***, and the derivative $\partial P/\partial K$ is called the ***marginal productivity of capital***. Explain what these quantities mean in practical terms.

(c) Show that if $\beta = 1 - \alpha$, then P satisfies the partial differential equation

$$K\frac{\partial P}{\partial K} + L\frac{\partial P}{\partial L} = P$$

42. Consider the Cobb–Douglas production model

$$P = 1000\, L^{0.6}K^{0.4}$$

(a) Find the maximum output value of P if labor costs \$50.00 per unit, capital costs \$100.00 per unit, and the total cost of labor and capital is set at \$200,000.

(b) How should the \$200,000 be allocated between labor and capital to achieve the maximum?

MULTIPLE INTEGRALS

For the things of this world cannot be made known without a knowledge of mathematics.
—Roger Bacon
Mathematician and Scientist

In this chapter we will extend the concept of a definite integral to functions of two and three variables. Whereas functions of one variable are usually integrated over intervals, functions of two variables are usually integrated over regions in 2-space and functions of three variables over regions in 3-space. Calculating such integrals will require some new techniques that will be a central focus in this chapter. Once we have developed the basic methods for integrating functions of two and three variables, we will show how such integrals can be used to calculate surface areas and volumes of solids; and we will also show how they can be used to find masses and centers of gravity of flat plates and three-dimensional solids. In addition to our study of integration, we will generalize the concept of a parametic curve in 2-space to a parametric surface in 3-space. This will allow us to work with a wider variety of surfaces than previously possible and will provide a powerful tool for generating surfaces using computers and other graphing utilities.

Photo: *Finding the area of complex surfaces such as those used in the design of the Denver International Airport require integration methods studied in this chapter.*

15.1 DOUBLE INTEGRALS

The notion of a definite integral can be extended to functions of two or more variables. In this section we will discuss the double integral, which is the extension to functions of two variables.

VOLUME

Recall that the definite integral of a function of one variable

$$\int_a^b f(x)\,dx = \lim_{\max \Delta x_k \to 0} \sum_{k=1}^{n} f(x_k^*)\Delta x_k = \lim_{n\to+\infty} \sum_{k=1}^{n} f(x_k^*)\Delta x_k \tag{1}$$

arose from the problem of finding areas under curves. [In the rightmost expression in (1), we use the "limit as $n \to +\infty$" to encapsulate the process by which we increase the number of subintervals of $[a, b]$ in such a way that the lengths of the subintervals approach zero.] Integrals of functions of two variables arise from the problem of finding volumes under surfaces.

> **15.1.1 THE VOLUME PROBLEM.** Given a function f of two variables that is continuous and nonnegative on a region R in the xy-plane, find the volume of the solid enclosed between the surface $z = f(x, y)$ and the region R (Figure 15.1.1).

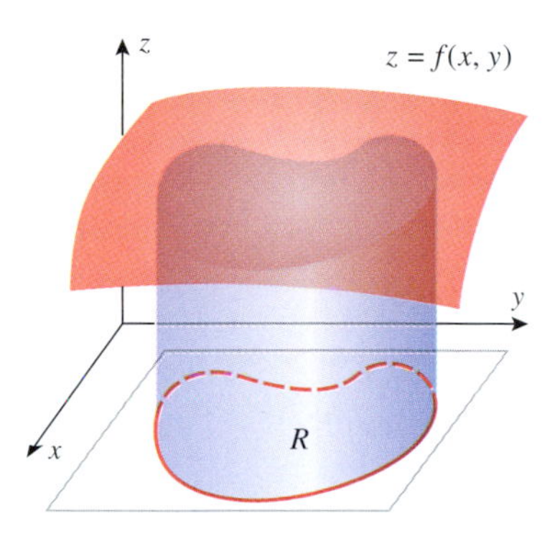

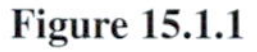
Figure 15.1.1

Later, we will place more restrictions on the region R, but for now we will just assume that the entire region can be enclosed within some suitably large rectangle with sides parallel to the coordinate axes. This ensures that R does not extend indefinitely in any direction.

The procedure for finding the volume V of the solid in Figure 15.1.1 will be similar to the limiting process used for finding areas, except that now the approximating elements will be rectangular parallelepipeds rather than rectangles. We proceed as follows:

- Using lines parallel to the coordinate axes, divide the rectangle enclosing the region R into subrectangles, and exclude from consideration all those subrectangles that contain any points outside of R. This leaves only rectangles that are subsets of R (Figure 15.1.2). Assume that there are n such rectangles, and denote the area of the kth such rectangle by ΔA_k.
- Choose any arbitrary point in each subrectangle, and denote the point in the kth subrectangle by (x_k^*, y_k^*). As shown in Figure 15.1.3, the product $f(x_k^*, y_k^*)\Delta A_k$ is the volume of a rectangular parallelepiped with base area ΔA_k and height $f(x_k^*, y_k^*)$, so the sum

$$\sum_{k=1}^{n} f(x_k^*, y_k^*)\Delta A_k$$

can be viewed as an approximation to the volume V of the entire solid.
- There are two sources of error in the approximation: first, the parallelepipeds have flat tops, whereas the surface $z = f(x, y)$ may be curved; second, the rectangles that form the bases of the parallelepipeds may not completely cover the region R. However, if we repeat the above process with more and more subdivisions in such a way that both the lengths and the widths of the subrectangles approach zero, then it is plausible that the errors of both types approach zero, and the exact volume of the solid will be

$$V = \lim_{n\to+\infty} \sum_{k=1}^{n} f(x_k^*, y_k^*)\Delta A_k$$

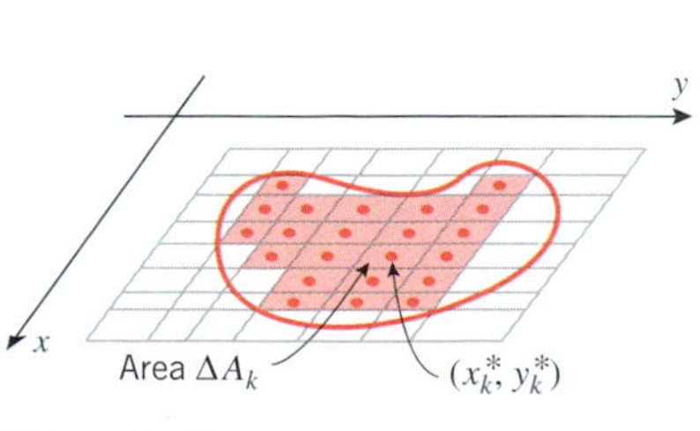

Figure 15.1.2

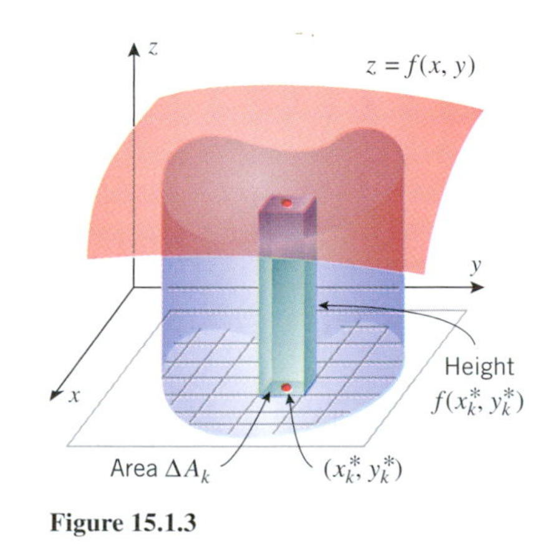

Figure 15.1.3

This suggests the following definition.

Definition 15.1.2 is satisfactory for our present purposes, but some issues would have to be resolved before it could be regarded as rigorous. For example, we would have to prove that the limit actually exists and that its value does not depend on how the points $(x_1^*, y_1^*), (x_2^*, y_2^*), \dots, (x_n^*, y_n^*)$ are chosen. This can be verified if the region R is not too "complicated" and if f is continuous on R. The details are beyond the scope of this text.

15.1.2 DEFINITION (*Volume Under a Surface*). If f is a function of two variables that is continuous and nonnegative on a region R in the xy-plane, then the volume of the solid enclosed between the surface $z = f(x, y)$ and the region R is defined by

$$V = \lim_{n \to +\infty} \sum_{k=1}^{n} f(x_k^*, y_k^*)\Delta A_k \tag{2}$$

Here, $n \to +\infty$ indicates the process of increasing the number of subrectangles of the rectangle enclosing R in such a way that both the lengths and the widths of the subrectangles approach zero.

It is assumed in Definition 15.1.2 that f is nonnegative on the region R. If f is continuous on R and has both positive and negative values, then the limit

$$\lim_{n \to +\infty} \sum_{k=1}^{n} f(x_k^*, y_k^*)\Delta A_k \tag{3}$$

no longer represents the volume between R and the surface $z = f(x, y)$; rather, it represents a *difference* of volumes—the volume between R and the portion of the surface that is above the xy-plane minus the volume between R and the portion of the surface below the xy-plane. We call this the ***net signed volume*** between the region R and the surface $z = f(x, y)$.

DEFINITION OF A DOUBLE INTEGRAL

As in Definition 15.1.2, the notation $n \to +\infty$ in (3) encapsulates a process in which the enclosing rectangle for R is repeatedly subdivided in such a way that both the lengths and the widths of the subrectangles approach zero. Note that subdividing so that the subrectangle lengths approach zero forces the mesh of the partition of the length of the enclosing rectangle for R to approach zero. Similarly, subdividing so that the subrectangle widths approach zero forces the mesh of the partition of the width of the enclosing rectangle for R to approach zero. Thus, we have extended the notion conveyed by Formula (1) where the definite integral of a one-variable function is expressed as a limit of Riemann sums. By extension, the sums in (3) are also called ***Riemann sums***, and the limit of the Riemann sums is denoted by

$$\iint_R f(x, y)\, dA = \lim_{n \to +\infty} \sum_{k=1}^{n} f(x_k^*, y_k^*)\Delta A_k \tag{4}$$

which is called the ***double integral*** of $f(x, y)$ over R.

If f is continuous and nonnegative on the region R, then the volume formula in (2) can be expressed as

$$V = \iint_R f(x, y)\, dA \tag{5}$$

If f has both positive and negative values on R, then a positive value for the double integral of f over R means that there is more volume above R than below, a negative value for the double integral means that there is more volume below R than above, and a value of zero means that the volume above R is the same as the volume below R.

EVALUATING DOUBLE INTEGRALS

Except in the simplest cases, it is impractical to obtain the value of a double integral from the limit in (4). However, we will now show how to evaluate double integrals by calculating two successive single integrals. For the rest of this section we will limit our discussion to the case where R is a rectangle; in the next section we will consider double integrals over more complicated regions.

The partial derivatives of a function $f(x, y)$ are calculated by holding one of the variables fixed and differentiating with respect to the other variable. Let us consider the reverse of this process, ***partial integration***. The symbols

$$\int_a^b f(x, y)\, dx \quad \text{and} \quad \int_c^d f(x, y)\, dy$$

denote ***partial definite integrals***; the first integral, called the ***partial definite integral with respect to x***, is evaluated by holding y fixed and integrating with respect to x, and the second integral, called the ***partial definite integral with respect to y***, is evaluated by holding x fixed and integrating with respect to y. As the following example shows, the partial definite integral with respect to x is a function of y, and the partial definite integral with respect to y is a function of x.

▶ Example 1

$$\int_0^1 xy^2\, dx = y^2 \int_0^1 x\, dx = \left.\frac{y^2x^2}{2}\right]_{x=0}^1 = \frac{y^2}{2}$$

$$\int_0^1 xy^2\, dy = x \int_0^1 y^2\, dy = \left.\frac{xy^3}{3}\right]_{y=0}^1 = \frac{x}{3} \blacktriangleleft$$

A partial definite integral with respect to x is a function of y and hence can be integrated with respect to y; similarly, a partial definite integral with respect to y can be integrated with respect to x. This two-stage integration process is called ***iterated*** (or ***repeated***) ***integration***. We introduce the following notation:

$$\int_c^d \int_a^b f(x, y)\, dx\, dy = \int_c^d \left[\int_a^b f(x, y)\, dx\right] dy \tag{6}$$

$$\int_a^b \int_c^d f(x, y)\, dy\, dx = \int_a^b \left[\int_c^d f(x, y)\, dy\right] dx \tag{7}$$

These integrals are called ***iterated integrals***.

▶ Example 2 Evaluate

(a) $\displaystyle\int_1^3 \int_2^4 (40 - 2xy)\, dy\, dx$ (b) $\displaystyle\int_2^4 \int_1^3 (40 - 2xy)\, dx\, dy$

Solution (a).

$$\begin{aligned}\int_1^3 \int_2^4 (40 - 2xy)\, dy\, dx &= \int_1^3 \left[\int_2^4 (40 - 2xy)\, dy\right] dx \\ &= \int_1^3 (40y - xy^2)\Big]_{y=2}^4 dx \\ &= \int_1^3 [(160 - 16x) - (80 - 4x)]\, dx \\ &= \int_1^3 (80 - 12x)\, dx \\ &= (80x - 6x^2)\Big]_1^3 = 112\end{aligned}$$

Solution (b).

$$\int_2^4 \int_1^3 (40 - 2xy)\,dx\,dy = \int_2^4 \left[\int_1^3 (40 - 2xy)\,dx\right] dy$$
$$= \int_2^4 (40x - x^2 y)\Big]_{x=1}^{3}\,dy$$
$$= \int_2^4 [(120 - 9y) - (40 - y)]\,dy$$
$$= \int_2^4 (80 - 8y)\,dy$$
$$= (80y - 4y^2)\Big]_2^4 = 112 \blacktriangleleft$$

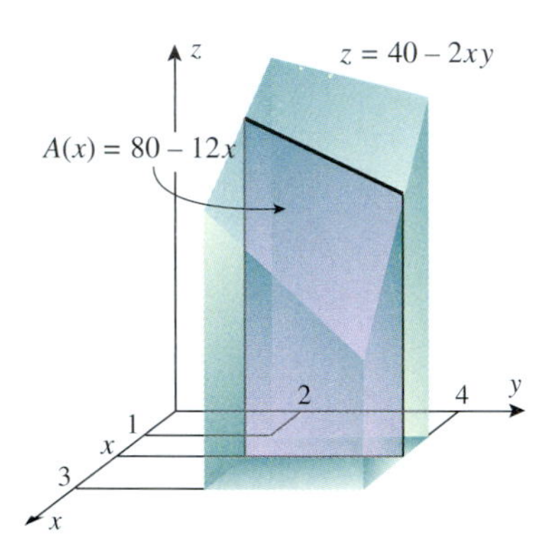

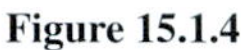
Figure 15.1.4

It is no accident that both parts of Example 2 produced the same answer. Consider the solid S bounded above by the surface $z = 40 - 2xy$ and below by the rectangle R defined by $1 \le x \le 3$ and $2 \le y \le 4$. By the method of slicing discussed in Section 7.2, the volume of S is given by

$$V = \int_1^3 A(x)\,dx$$

where $A(x)$ is the area of a vertical cross section of S taken perpendicular to the x-axis (Figure 15.1.4). For a fixed value of x, $1 \le x \le 3$, $z = 40 - 2xy$ is a function of y, so the integral

$$A(x) = \int_2^4 (40 - 2xy)\,dy$$

represents the area under the graph of this function of y. Thus,

$$V = \int_1^3 \left[\int_2^4 (40 - 2xy)\,dy\right] dx = \int_1^3 \int_2^4 (40 - 2xy)\,dy\,dx$$

is the volume of S. Similarly, by the method of slicing with cross sections of S taken perpendicular to the y-axis, the volume of S is given by

$$V = \int_2^4 A(y)\,dy = \int_2^4 \left[\int_1^3 (40 - 2xy)\,dx\right] dy = \int_2^4 \int_1^3 (40 - 2xy)\,dx\,dy$$

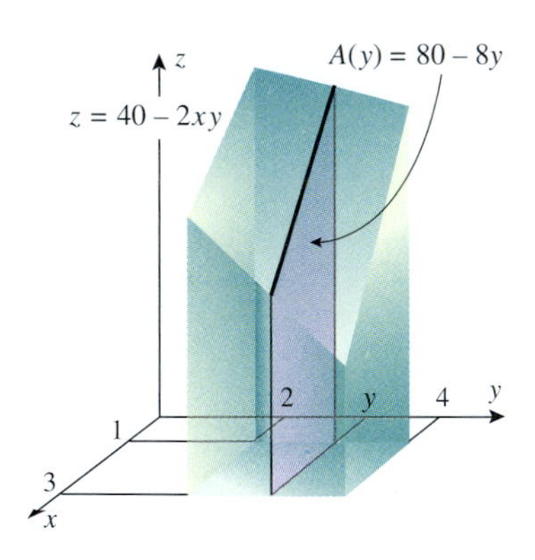

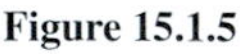
Figure 15.1.5

(Figure 15.1.5). Thus, the iterated integrals in parts (a) and (b) of Example 2 both measure the volume of S, which by Formula (5) is the double integral of $z = 40 - 2xy$ over R. That is,

$$\int_1^3 \int_2^4 (40 - 2xy)\,dy\,dx = \iint_R (40 - 2xy)\,dA = \int_2^4 \int_1^3 (40 - 2xy)\,dx\,dy$$

We will often denote the rectangle

$$\{(x, y) : a \le x \le b, c \le y \le d\}$$

as $[a, b] \times [c, d]$ for simplicity.

The geometric argument above applies to any continuous function $f(x, y)$ that is nonnegative on a rectangle $R = [a, b] \times [c, d]$, as is the case for $f(x, y) = 40 - 2xy$ on $[1, 3] \times [2, 4]$. The conclusion that the double integral of $f(x, y)$ over R has the same value as either of the two possible iterated integrals is true even when f is negative at some points in R. We state this result in the following theorem and omit a formal proof.

15.1.3 THEOREM. *Let R be the rectangle defined by the inequalities*

$$a \le x \le b, \quad c \le y \le d$$

If $f(x, y)$ is continuous on this rectangle, then

$$\iint_R f(x, y)\,dA = \int_c^d \int_a^b f(x, y)\,dx\,dy = \int_a^b \int_c^d f(x, y)\,dy\,dx$$

Theorem 15.1.3 allows us to evaluate a double integral over a rectangle by converting it to an iterated integral. This can be done in two ways, both of which produce the value of the double integral.

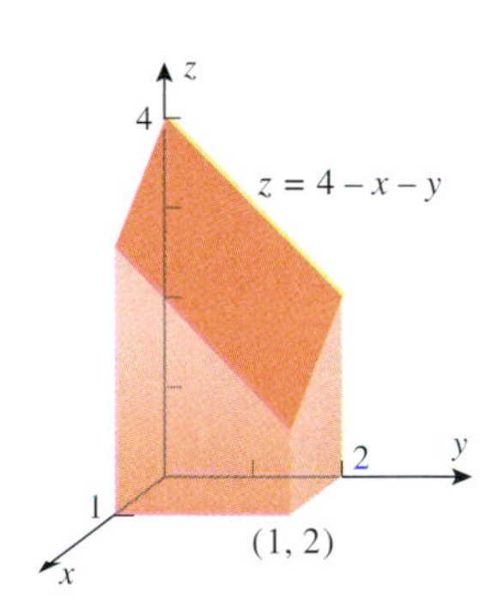

Figure 15.1.6

Example 3 Use a double integral to find the volume of the solid that is bounded above by the plane $z = 4 - x - y$ and below by the rectangle $R = [0, 1] \times [0, 2]$ (Figure 15.1.6).

Solution. The volume is the double integral of $z = 4 - x - y$ over R. Using Theorem 15.1.3, this can be obtained from either of the iterated integrals

$$\int_0^2 \int_0^1 (4 - x - y)\, dx\, dy \quad \text{or} \quad \int_0^1 \int_0^2 (4 - x - y)\, dy\, dx \tag{8}$$

Using the first of these, we obtain

$$\begin{aligned} V = \iint_R (4 - x - y)\, dA &= \int_0^2 \int_0^1 (4 - x - y)\, dx\, dy \\ &= \int_0^2 \left[4x - \frac{x^2}{2} - xy\right]_{x=0}^{1} dy = \int_0^2 \left(\frac{7}{2} - y\right) dy \\ &= \left[\frac{7}{2}y - \frac{y^2}{2}\right]_0^2 = 5 \end{aligned}$$

You can check this result by evaluating the second integral in (8). ◀

TECHNOLOGY MASTERY

If you have a CAS with a built-in capability for computing iterated double integrals, use it to check Example 3.

Theorem 15.1.3 guarantees that the double integral in Example 4 can also be evaluated by integrating first with respect to y and then with respect to x. Verify this.

Example 4 Evaluate the double integral

$$\iint_R y^2 x\, dA$$

over the rectangle $R = \{(x, y) : -3 \le x \le 2, 0 \le y \le 1\}$.

Solution. In view of Theorem 15.1.3, the value of the double integral can be obtained by evaluating one of two possible iterated double integrals. We choose to integrate first with respect to x and then with respect to y.

$$\begin{aligned} \iint_R y^2 x\, dA &= \int_0^1 \int_{-3}^2 y^2 x\, dx\, dy = \int_0^1 \left[\frac{1}{2}y^2 x^2\right]_{x=-3}^{2} dy \\ &= \int_0^1 \left(-\frac{5}{2}y^2\right) dy = -\frac{5}{6}y^3\Big]_0^1 = -\frac{5}{6} \end{aligned}$$ ◀

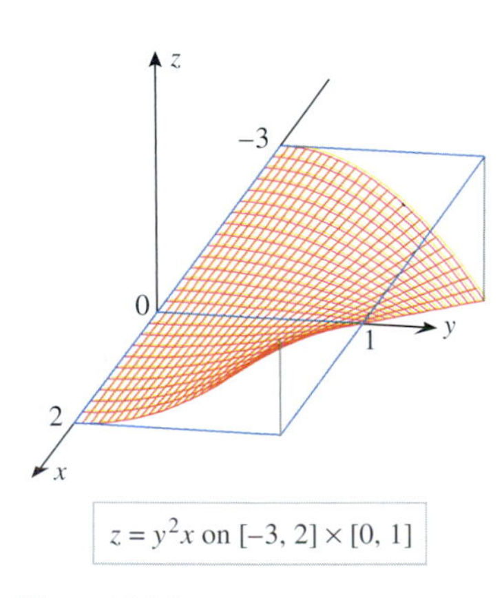

$z = y^2 x$ on $[-3, 2] \times [0, 1]$

Figure 15.1.7

The integral in Example 4 can be interpreted as the net signed volume between the rectangle $[-3, 2] \times [0, 1]$ and the surface $z = y^2 x$. That is, it is the volume below $z = y^2 x$ and above $[0, 2] \times [0, 1]$ minus the volume above $z = y^2 x$ and below $[-3, 0] \times [0, 1]$ (Figure 15.1.7).

PROPERTIES OF DOUBLE INTEGRALS

To distinguish between double integrals of functions of two variables and definite integrals of functions of one variable, we will refer to the latter as ***single integrals***. Because double integrals, like single integrals, are defined as limits, they inherit many of the properties of

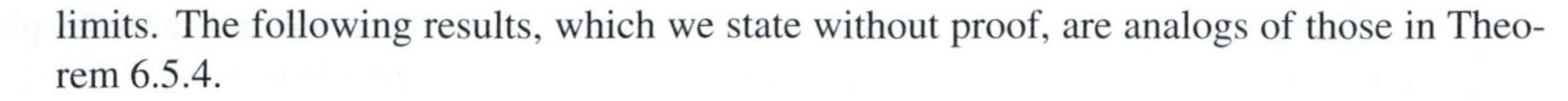

limits. The following results, which we state without proof, are analogs of those in Theorem 6.5.4.

$$\iint_R cf(x, y)\, dA = c \iint_R f(x, y)\, dA \quad (c \text{ a constant}) \tag{9}$$

$$\iint_R [f(x, y) + g(x, y)]\, dA = \iint_R f(x, y)\, dA + \iint_R g(x, y)\, dA \tag{10}$$

$$\iint_R [f(x, y) - g(x, y)]\, dA = \iint_R f(x, y)\, dA - \iint_R g(x, y)\, dA \tag{11}$$

It is evident intuitively that if $f(x, y)$ is nonnegative on a region R, then subdividing R into two regions R_1 and R_2 has the effect of subdividing the solid between R and $z = f(x, y)$ into two solids, the sum of whose volumes is the volume of the entire solid (Figure 15.1.8). This suggests the following result, which holds even if f has negative values:

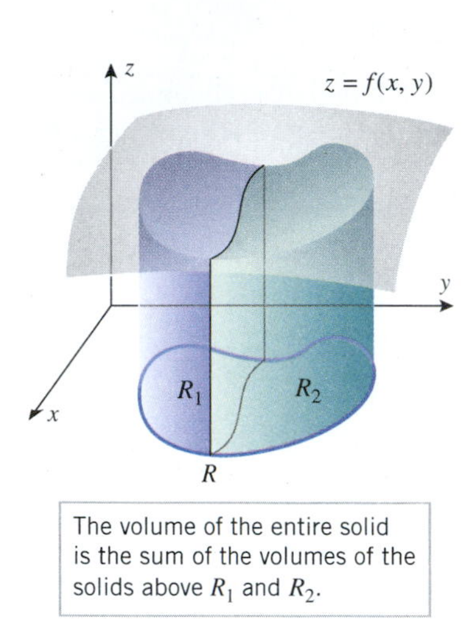

The volume of the entire solid is the sum of the volumes of the solids above R_1 and R_2.

Figure 15.1.8

$$\iint_R f(x, y)\, dA = \iint_{R_1} f(x, y)\, dA + \iint_{R_2} f(x, y)\, dA \tag{12}$$

The proof of this result will be omitted.

✔ QUICK CHECK EXERCISES 15.1 *(See page 1026 for answers.)*

1. The double integral is defined as a limit of Riemann sums by

$$\iint_R f(x, y)\, dA = \_\_\_\_\_\_$$

2. The iterated integral

$$\int_1^5 \int_2^4 f(x, y)\, dx\, dy$$

integrates f over the rectangle defined by

$\_\_\_\_ \le x \le \_\_\_\_, \quad \_\_\_\_ \le y \le \_\_\_\_$

3. Supply the missing integrand and limits of integration.

$$\int_1^5 \int_2^4 (3x^2 - 2xy + y^2)\, dx\, dy = \int_{\square}^{\square} \_\_\_\_\_\_ \, dy$$

4. The volume of the solid enclosed by the surface $z = x/y$ and the rectangle $0 \le x \le 4$, $1 \le y \le e^2$ in the xy-plane is $\_\_\_\_$.

EXERCISE SET 15.1 [C] CAS

1–12 Evaluate the iterated integrals.

1. $\displaystyle\int_0^1 \int_0^2 (x + 3)\, dy\, dx$

2. $\displaystyle\int_1^3 \int_{-1}^1 (2x - 4y)\, dy\, dx$

3. $\displaystyle\int_2^4 \int_0^1 x^2 y\, dx\, dy$

4. $\displaystyle\int_{-2}^0 \int_{-1}^2 (x^2 + y^2)\, dx\, dy$

5. $\displaystyle\int_0^{\ln 3} \int_0^{\ln 2} e^{x+y}\, dy\, dx$

6. $\displaystyle\int_0^2 \int_0^1 y \sin x\, dy\, dx$

7. $\displaystyle\int_{-1}^0 \int_2^5 dx\, dy$

8. $\displaystyle\int_4^6 \int_{-3}^7 dy\, dx$

9. $\displaystyle\int_0^1 \int_0^1 \frac{x}{(xy + 1)^2}\, dy\, dx$

10. $\displaystyle\int_{\pi/2}^{\pi} \int_1^2 x \cos xy\, dy\, dx$

11. $\displaystyle\int_0^{\ln 2} \int_0^1 xy e^{y^2 x}\, dy\, dx$

12. $\displaystyle\int_3^4 \int_1^2 \frac{1}{(x + y)^2}\, dy\, dx$

13–16 Evaluate the double integral over the rectangular region R.

13. $\displaystyle\iint_R 4xy^3\, dA$; $R = \{(x, y) : -1 \le x \le 1, -2 \le y \le 2\}$

14. $\displaystyle\iint_R \frac{xy}{\sqrt{x^2+y^2+1}}\,dA;$
$R = \{(x, y) : 0 \le x \le 1, 0 \le y \le 1\}$

15. $\displaystyle\iint_R x\sqrt{1-x^2}\,dA;\ R = \{(x, y) : 0 \le x \le 1, 2 \le y \le 3\}$

16. $\displaystyle\iint_R (x \sin y - y \sin x)\,dA;$
$R = \{(x, y) : 0 \le x \le \pi/2, 0 \le y \le \pi/3\}$

FOCUS ON CONCEPTS

17. (a) Let $f(x, y) = x^2 + y$, and as shown in the accompanying figure, let the rectangle $R = [0, 2] \times [0, 2]$ be subdivided into 16 subrectangles. Take (x_k^*, y_k^*) to be the center of the kth rectangle, and approximate the double integral of f over R by the resulting Riemann sum.
(b) Compare the result in part (a) to the exact value of the integral.

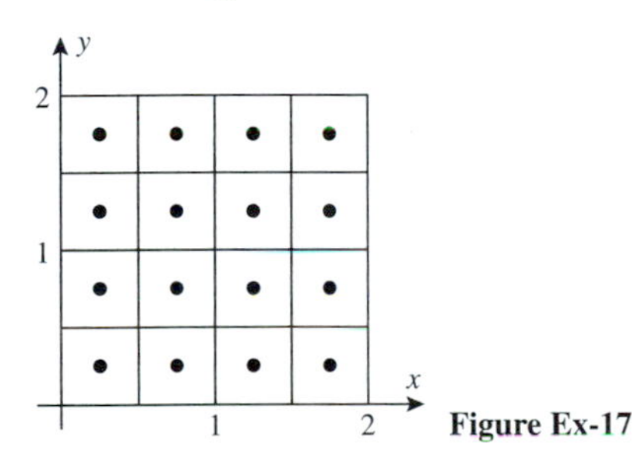

Figure Ex-17

18. (a) Let $f(x, y) = x - 2y$, and as shown in Exercise 17, let the rectangle $R = [0, 2] \times [0, 2]$ be subdivided into 16 subrectangles. Take (x_k^*, y_k^*) to be the center of the kth rectangle, and approximate the double integral of f over R by the resulting Riemann sum.
(b) Compare the result in part (a) to the exact value of the integral.

19–20 Each iterated integral represents the volume of a solid. Make a sketch of the solid. (You do *not* have to find the volume.)

19. (a) $\displaystyle\int_0^5\int_1^2 4\,dx\,dy$
(b) $\displaystyle\int_0^3\int_0^4 \sqrt{25-x^2-y^2}\,dy\,dx$

20. (a) $\displaystyle\int_0^1\int_0^1 (2-x-y)\,dy\,dx$
(b) $\displaystyle\int_{-2}^2\int_{-2}^2 (x^2+y^2)\,dx\,dy$

21–24 Use a double integral to find the volume.

21. The volume under the plane $z = 2x + y$ and over the rectangle $R = \{(x, y) : 3 \le x \le 5, 1 \le y \le 2\}$.

22. The volume under the surface $z = 3x^3 + 3x^2y$ and over the rectangle $R = \{(x, y) : 1 \le x \le 3, 0 \le y \le 2\}$.

23. The volume of the solid enclosed by the surface $z = x^2$ and the planes $x = 0$, $x = 2$, $y = 3$, $y = 0$, and $z = 0$.

24. The volume in the first octant bounded by the coordinate planes, the plane $y = 4$, and the plane $(x/3) + (z/5) = 1$.

25. Evaluate the integral by choosing a convenient order of integration:
$$\iint_R x\cos(xy)\cos^2 \pi x\,dA;\ R = \left[0, \tfrac{1}{2}\right] \times [0, \pi]$$

26. (a) Sketch the solid in the first octant that is enclosed by the planes $x = 0$, $z = 0$, $x = 5$, $z - y = 0$, and $z = -2y + 6$.
(b) Find the volume of the solid by breaking it into two parts.

27–30 The ***average value*** or ***mean value*** of a continuous function $f(x, y)$ over a rectangle $R = [a, b] \times [c, d]$ is defined as
$$f_{\text{ave}} = \frac{1}{A(R)}\iint_R f(x, y)\,dA$$
where $A(R) = (b - a)(d - c)$ is the area of the rectangle R (compare to Definition 7.6.1). Use this definition in these exercises.

27. Find the average value of $f(x, y) = y \sin xy$ over the rectangle $[0, 1] \times [0, \pi/2]$.

28. Find the average value of $f(x, y) = x(x^2 + y)^{1/2}$ over the interval $[0, 1] \times [0, 3]$.

29. Suppose that the temperature in degrees Celsius at a point (x, y) on a flat metal plate is $T(x, y) = 10 - 8x^2 - 2y^2$, where x and y are in meters. Find the average temperature of the rectangular portion of the plate for which $0 \le x \le 1$ and $0 \le y \le 2$.

30. Show that if $f(x, y)$ is constant on the rectangle $R = [a, b] \times [c, d]$, say $f(x, y) = k$, then $f_{\text{ave}} = k$ over R.

31–32 Most computer algebra systems have commands for approximating double integrals numerically. Read the relevant documentation and use a CAS to find a numerical approximation of the double integral in these exercises.

C 31. $\displaystyle\int_0^2\int_0^1 \sin\sqrt{x^3+y^3}\,dx\,dy$

C 32. $\displaystyle\int_{-1}^1\int_{-1}^1 e^{-(x^2+y^2)}\,dx\,dy$

33. In this exercise, suppose that $f(x, y) = g(x)h(y)$ and $R = \{(x, y) : a \le x \le b, c \le y \le d\}$. Show that
$$\iint_R f(x, y)\,dA = \left[\int_a^b g(x)\,dx\right]\left[\int_c^d h(y)\,dy\right]$$

34. Use the result in Exercise 33 to evaluate the integral

$$\int_0^{\ln 2} \int_{-1}^{1} \sqrt{e^y + 1} \tan x \, dx \, dy$$

by inspection. Explain your reasoning.

C **35.** Use a CAS to evaluate the iterated integrals

$$\int_0^1 \int_0^1 \frac{y - x}{(x + y)^3} \, dx \, dy \quad \text{and} \quad \int_0^1 \int_0^1 \frac{y - x}{(x + y)^3} \, dy \, dx$$

Does this violate Theorem 15.1.3? Explain.

C **36.** Use a CAS to show that the volume V under the surface $z = xy^3 \sin xy$ over the rectangle shown in the accompanying figure is $V = 3/\pi$.

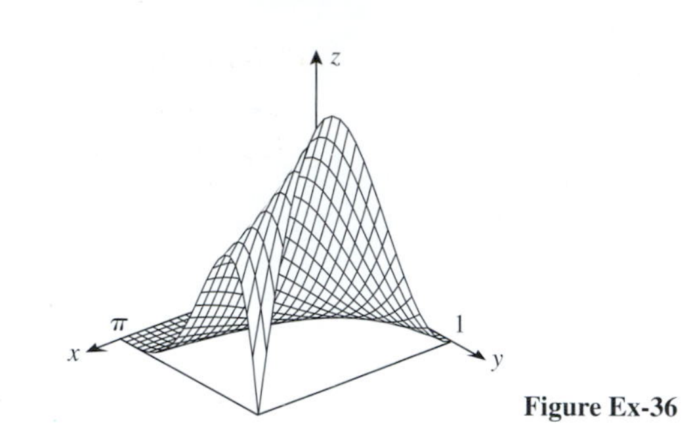

Figure Ex-36

✔QUICK CHECK ANSWERS 15.1

1. $\lim_{n \to +\infty} \sum_{k=1}^{n} f(x_k^*, y_k^*) \Delta A_k$ **2.** $2 \le x \le 4$; $1 \le y \le 5$ **3.** $\int_1^5 (56 - 12y + 2y^2) \, dy$ **4.** 16

15.2 DOUBLE INTEGRALS OVER NONRECTANGULAR REGIONS

In this section we will show how to evaluate double integrals over regions other than rectangles.

ITERATED INTEGRALS WITH NONCONSTANT LIMITS OF INTEGRATION

Later in this section we will see that double integrals over nonrectangular regions can often be evaluated as iterated integrals of the following types:

$$\int_a^b \int_{g_1(x)}^{g_2(x)} f(x, y) \, dy \, dx = \int_a^b \left[\int_{g_1(x)}^{g_2(x)} f(x, y) \, dy \right] dx \tag{1}$$

$$\int_c^d \int_{h_1(y)}^{h_2(y)} f(x, y) \, dx \, dy = \int_c^d \left[\int_{h_1(y)}^{h_2(y)} f(x, y) \, dx \right] dy \tag{2}$$

We begin with an example that illustrates how to evaluate such integrals.

► **Example 1** Evaluate

$$\text{(a)} \int_0^1 \int_{-x}^{x^2} y^2 x \, dy \, dx \qquad \text{(b)} \int_0^{\pi/3} \int_0^{\cos y} x \sin y \, dx \, dy$$

Solution (a).

$$\int_0^1 \int_{-x}^{x^2} y^2 x \, dy \, dx = \int_0^1 \left[\int_{-x}^{x^2} y^2 x \, dy \right] dx = \int_0^1 \frac{y^3 x}{3} \bigg]_{y=-x}^{x^2} dx$$

$$= \int_0^1 \left[\frac{x^7}{3} + \frac{x^4}{3} \right] dx = \left(\frac{x^8}{24} + \frac{x^5}{15} \right) \bigg]_0^1 = \frac{13}{120}$$

Solution (b).

$$\int_0^{\pi/3}\int_0^{\cos y} x\sin y\,dx\,dy = \int_0^{\pi/3}\left[\int_0^{\cos y} x\sin y\,dx\right]dy = \int_0^{\pi/3}\left.\frac{x^2}{2}\sin y\right]_{x=0}^{\cos y} dy$$

$$= \int_0^{\pi/3}\left[\frac{1}{2}\cos^2 y\sin y\right]dy = \left.-\frac{1}{6}\cos^3 y\right]_0^{\pi/3} = \frac{7}{48} \blacktriangleleft$$

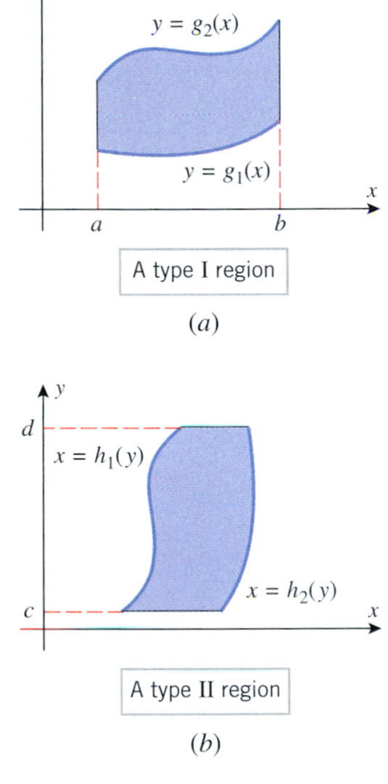

Figure 15.2.1

DOUBLE INTEGRALS OVER NONRECTANGULAR REGIONS

Plane regions can be extremely complex, and the theory of double integrals over very general regions is a topic for advanced courses in mathematics. We will limit our study of double integrals to two basic types of regions, which we will call *type I* and *type II*; they are defined as follows.

15.2.1 DEFINITION.

(a) A ***type I region*** is bounded on the left and right by vertical lines $x = a$ and $x = b$ and is bounded below and above by continuous curves $y = g_1(x)$ and $y = g_2(x)$, where $g_1(x) \le g_2(x)$ for $a \le x \le b$ (Figure 15.2.1*a*).

(b) A ***type II region*** is bounded below and above by horizontal lines $y = c$ and $y = d$ and is bounded on the left and right by continuous curves $x = h_1(y)$ and $x = h_2(y)$ satisfying $h_1(y) \le h_2(y)$ for $c \le y \le d$ (Figure 15.2.1*b*).

The following theorem will enable us to evaluate double integrals over type I and type II regions using iterated integrals.

15.2.2 THEOREM.

(a) If R is a type I region on which $f(x, y)$ is continuous, then

$$\iint\limits_R f(x,y)\,dA = \int_a^b \int_{g_1(x)}^{g_2(x)} f(x,y)\,dy\,dx \tag{3}$$

(b) If R is a type II region on which $f(x, y)$ is continuous, then

$$\iint\limits_R f(x,y)\,dA = \int_c^d \int_{h_1(y)}^{h_2(y)} f(x,y)\,dx\,dy \tag{4}$$

Using Theorem 15.2.2, the integral in Example 1(a) is the double integral of the function $f(x, y) = y^2x$ over the type I region bounded on the left and right by the vertical lines $x = 0$ and $x = 1$ and bounded below and above by the curves $y = -x$ and $y = x^2$ (Figure 15.2.2). Also, the integral in Example 1(b) is the double integral of $f(x, y) = x\sin y$ over the type II region bounded below and above by the horizontal lines $y = 0$ and $y = \pi/3$ and bounded on the left and right by the curves $x = 0$ and $x = \cos y$ (Figure 15.2.3).

We will not prove Theorem 15.2.2, but for the case where $f(x, y)$ is nonnegative on the region R, it can be made plausible by a geometric argument that is similar to that given for Theorem 15.1.3. Since $f(x, y)$ is nonnegative, the double integral can be interpreted as the volume of the solid S that is bounded above by the surface $z = f(x, y)$ and below

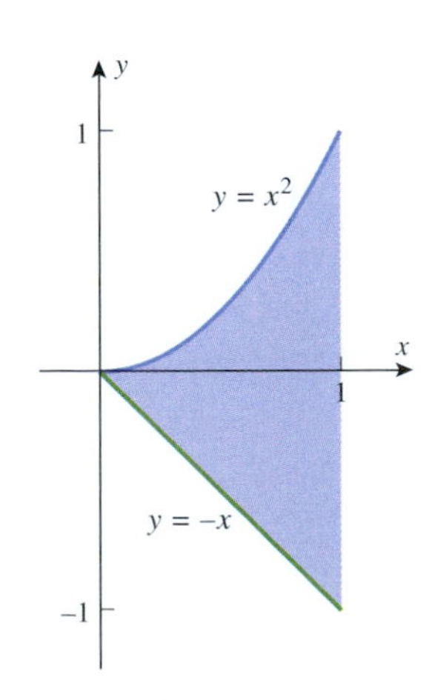

Figure 15.2.2

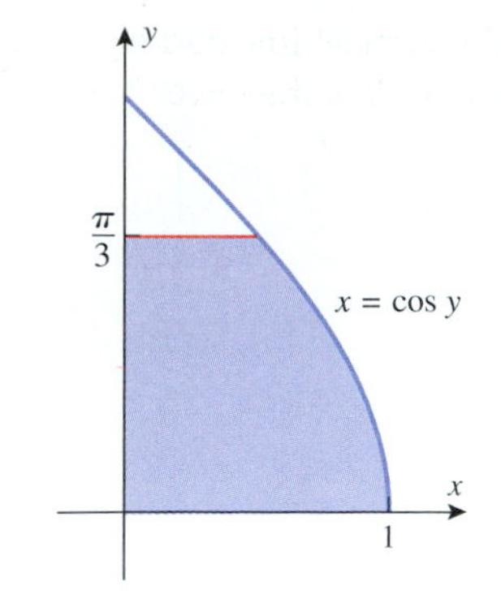

Figure 15.2.3

by the region R, so it suffices to show that the iterated integrals also represent this volume. Consider the iterated integral in (3), for example. For a fixed value of x, the function $f(x, y)$ is a function of y, and hence the integral

$$A(x) = \int_{g_1(x)}^{g_2(x)} f(x, y)\, dy$$

represents the area under the graph of this function of y between $y = g_1(x)$ and $y = g_2(x)$. This area, shown in yellow in Figure 15.2.4, is the cross-sectional area at x of the solid S, and hence by the method of slicing, the volume V of the solid S is

$$V = \int_a^b \int_{g_1(x)}^{g_2(x)} f(x, y)\, dy\, dx$$

which shows that in (3) the iterated integral is equal to the double integral. Similarly for (4).

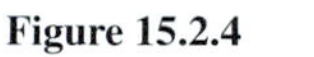

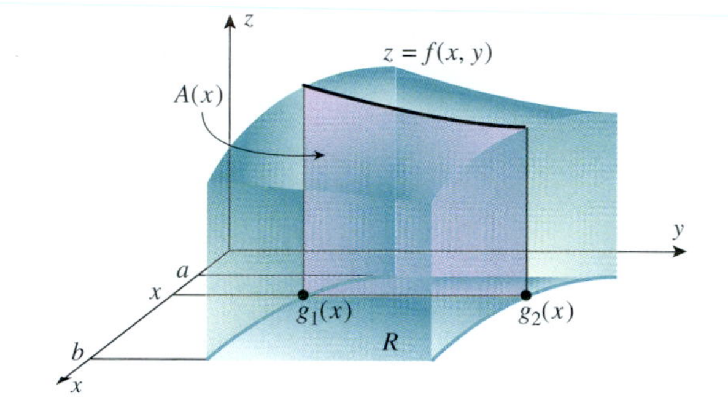

Figure 15.2.4

SETTING UP LIMITS OF INTEGRATION FOR EVALUATING DOUBLE INTEGRALS

To apply Theorem 15.2.2, it is helpful to start with a two-dimensional sketch of the region R. [It is not necessary to graph $f(x, y)$.] For a type I region, the limits of integration in Formula (3) can be obtained as follows:

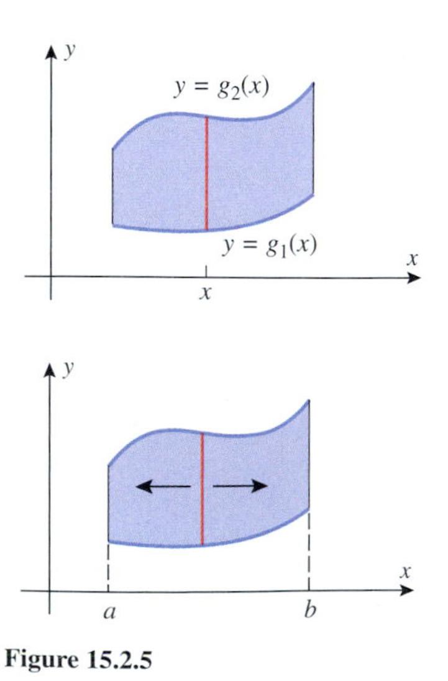

Figure 15.2.5

Determining Limits of Integration: Type I Region

Step 1. Since x is held fixed for the first integration, we draw a vertical line through the region R at an arbitrary fixed value x (Figure 15.2.5). This line crosses the boundary of R twice. The lower point of intersection is on the curve $y = g_1(x)$ and the higher point is on the curve $y = g_2(x)$. These two intersections determine the lower and upper y-limits of integration in Formula (3).

Step 2. Imagine moving the line drawn in Step 1 first to the left and then to the right (Figure 15.2.5). The leftmost position where the line intersects the region R is $x = a$, and the rightmost position where the line intersects the region R is $x = b$. This yields the limits for the x-integration in Formula (3).

▶ **Example 2** Evaluate

$$\iint_R xy\, dA$$

over the region R enclosed between $y = \frac{1}{2}x$, $y = \sqrt{x}$, $x = 2$, and $x = 4$.

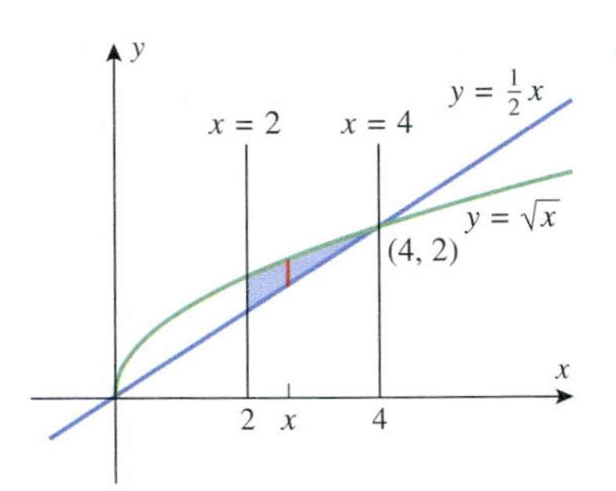

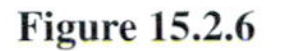
Figure 15.2.6

Solution. We view R as a type I region. The region R and a vertical line corresponding to a fixed x are shown in Figure 15.2.6. This line meets the region R at the lower boundary $y = \frac{1}{2}x$ and the upper boundary $y = \sqrt{x}$. These are the y-limits of integration. Moving this line first left and then right yields the x-limits of integration, $x = 2$ and $x = 4$. Thus,

$$\iint_R xy\,dA = \int_2^4 \int_{x/2}^{\sqrt{x}} xy\,dy\,dx = \int_2^4 \left[\frac{xy^2}{2}\right]_{y=x/2}^{\sqrt{x}} dx = \int_2^4 \left(\frac{x^2}{2} - \frac{x^3}{8}\right) dx$$

$$= \left[\frac{x^3}{6} - \frac{x^4}{32}\right]_2^4 = \left(\frac{64}{6} - \frac{256}{32}\right) - \left(\frac{8}{6} - \frac{16}{32}\right) = \frac{11}{6} \blacktriangleleft$$

If R is a type II region, then the limits of integration in Formula (4) can be obtained as follows:

Determining Limits of Integration: Type II Region

Step 1. Since y is held fixed for the first integration, we draw a horizontal line through the region R at a fixed value y (Figure 15.2.7). This line crosses the boundary of R twice. The leftmost point of intersection is on the curve $x = h_1(y)$ and the rightmost point is on the curve $x = h_2(y)$. These intersections determine the x-limits of integration in (4).

Step 2. Imagine moving the line drawn in Step 1 first down and then up (Figure 15.2.7). The lowest position where the line intersects the region R is $y = c$, and the highest position where the line intersects the region R is $y = d$. This yields the y-limits of integration in (4).

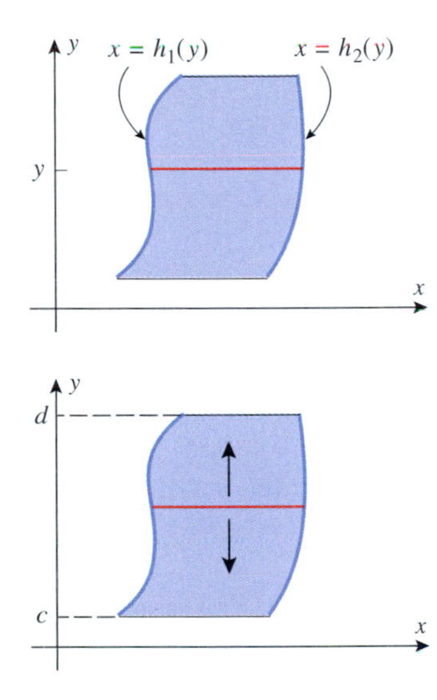

Figure 15.2.7

▶ **Example 3** Evaluate

$$\iint_R (2x - y^2)\,dA$$

over the triangular region R enclosed between the lines $y = -x + 1$, $y = x + 1$, and $y = 3$.

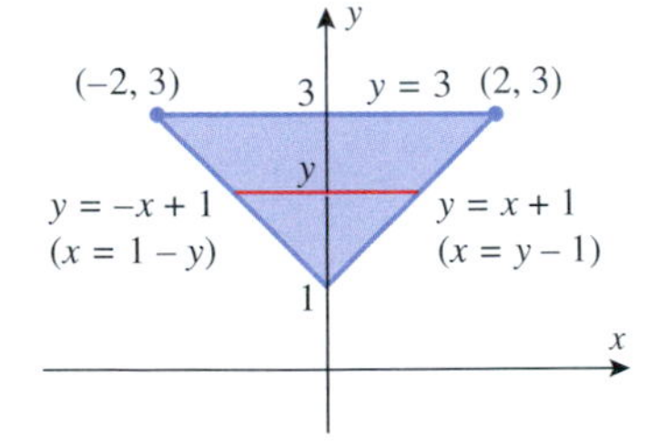

Figure 15.2.8

Solution. We view R as a type II region. The region R and a horizontal line corresponding to a fixed y are shown in Figure 15.2.8. This line meets the region R at its left-hand boundary $x = 1 - y$ and its right-hand boundary $x = y - 1$. These are the x-limits of integration. Moving this line first down and then up yields the y-limits, $y = 1$ and $y = 3$. Thus,

$$\iint_R (2x - y^2)\,dA = \int_1^3 \int_{1-y}^{y-1} (2x - y^2)\,dx\,dy = \int_1^3 \left[x^2 - y^2x\right]_{x=1-y}^{y-1} dy$$

$$= \int_1^3 [(1 - 2y + 2y^2 - y^3) - (1 - 2y + y^3)]\,dy$$

$$= \int_1^3 (2y^2 - 2y^3)\,dy = \left[\frac{2y^3}{3} - \frac{y^4}{2}\right]_1^3 = -\frac{68}{3} \blacktriangleleft$$

To integrate over a type II region, the left- and right-hand boundaries must be expressed in the form $x = h_1(y)$ and $x = h_2(y)$. This is why we rewrote the boundary equations

$$y = -x + 1 \quad \text{and} \quad y = x + 1$$

as

$$x = 1 - y \quad \text{and} \quad x = y - 1$$

in Example 3.

In Example 3 we could have treated R as a type I region, but with an added complication. Viewed as a type I region, the upper boundary of R is the line $y = 3$ (Figure 15.2.9) and the lower boundary consists of two parts, the line $y = -x + 1$ to the left of the y-axis and

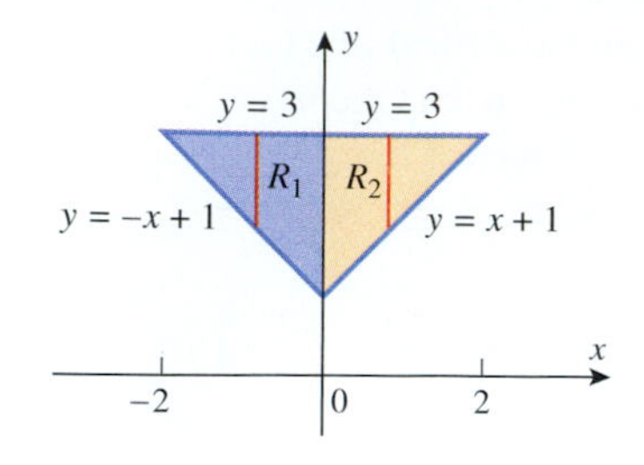

Figure 15.2.9

the line $y = x + 1$ to the right of the y-axis. To carry out the integration it is necessary to decompose the region R into two parts, R_1 and R_2, as shown in Figure 15.2.9, and write

$$\iint_R (2x - y^2)\,dA = \iint_{R_1} (2x - y^2)\,dA + \iint_{R_2} (2x - y^2)\,dA$$

$$= \int_{-2}^{0} \int_{-x+1}^{3} (2x - y^2)\,dy\,dx + \int_{0}^{2} \int_{x+1}^{3} (2x - y^2)\,dy\,dx$$

This will yield the same result that was obtained in Example 3. (Verify.)

▶ Example 4 Use a double integral to find the volume of the tetrahedron bounded by the coordinate planes and the plane $z = 4 - 4x - 2y$.

Solution. The tetrahedron in question is bounded above by the plane

$$z = 4 - 4x - 2y \tag{5}$$

and below by the triangular region R shown in Figure 15.2.10. Thus, the volume is given by

$$V = \iint_R (4 - 4x - 2y)\,dA$$

The region R is bounded by the x-axis, the y-axis, and the line $y = 2 - 2x$ [set $z = 0$ in (5)], so that treating R as a type I region yields

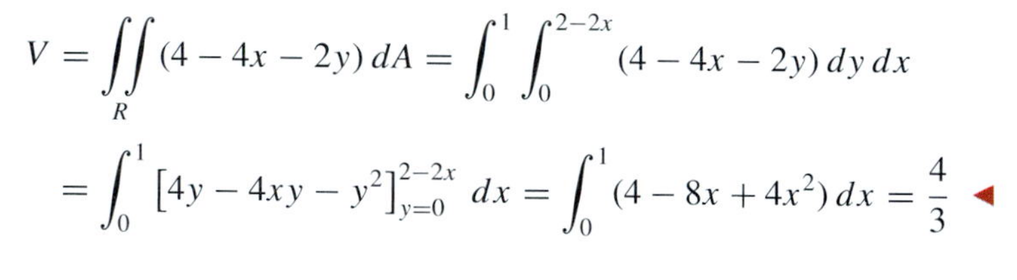

$$V = \iint_R (4 - 4x - 2y)\,dA = \int_0^1 \int_0^{2-2x} (4 - 4x - 2y)\,dy\,dx$$

$$= \int_0^1 \left[4y - 4xy - y^2\right]_{y=0}^{2-2x} dx = \int_0^1 (4 - 8x + 4x^2)\,dx = \frac{4}{3} \blacktriangleleft$$

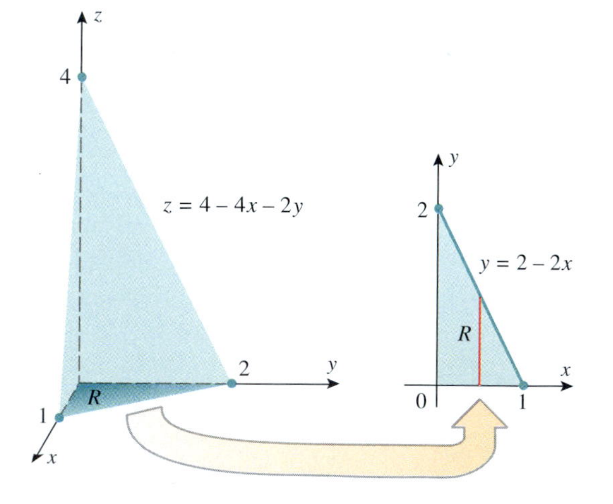

Figure 15.2.10

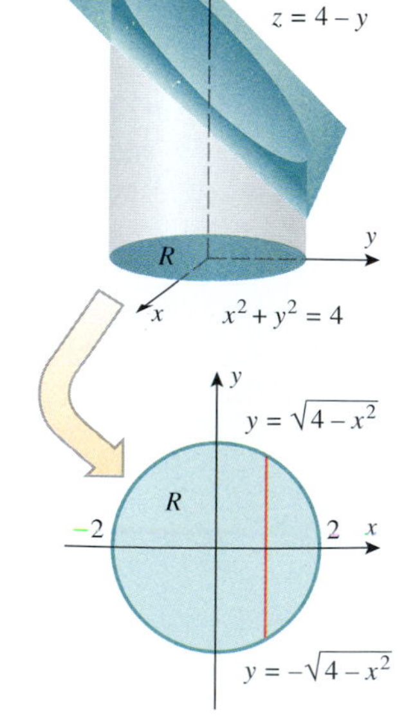

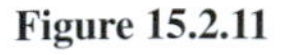
Figure 15.2.11

▶ Example 5 Find the volume of the solid bounded by the cylinder $x^2 + y^2 = 4$ and the planes $y + z = 4$ and $z = 0$.

Solution. The solid shown in Figure 15.2.11 is bounded above by the plane $z = 4 - y$ and below by the region R within the circle $x^2 + y^2 = 4$. The volume is given by

$$V = \iint_R (4 - y)\,dA$$

Treating R as a type I region we obtain

$$V = \int_{-2}^{2}\int_{-\sqrt{4-x^2}}^{\sqrt{4-x^2}} (4-y)\,dy\,dx = \int_{-2}^{2}\left[4y - \frac{1}{2}y^2\right]_{y=-\sqrt{4-x^2}}^{\sqrt{4-x^2}} dx$$

$$= \int_{-2}^{2} 8\sqrt{4-x^2}\,dx = 8(2\pi) = 16\pi$$

See Formula (3) of Section 8.4. ◀

REVERSING THE ORDER OF INTEGRATION

Sometimes the evaluation of an iterated integral can be simplified by reversing the order of integration. The next example illustrates how this is done.

Example 6 Since there is no elementary antiderivative of e^{x^2}, the integral

$$\int_0^2 \int_{y/2}^1 e^{x^2}\,dx\,dy$$

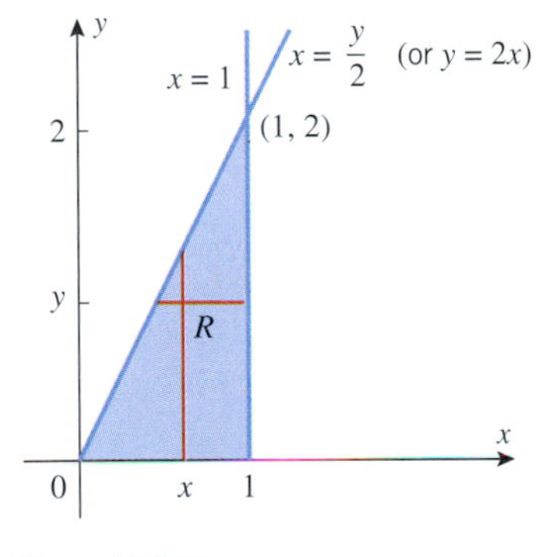

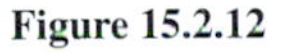
Figure 15.2.12

cannot be evaluated by performing the x-integration first. Evaluate this integral by expressing it as an equivalent iterated integral with the order of integration reversed.

Solution. For the inside integration, y is fixed and x varies from the line $x = y/2$ to the line $x = 1$ (Figure 15.2.12). For the outside integration, y varies from 0 to 2, so the given iterated integral is equal to a double integral over the triangular region R in Figure 15.2.12.

To reverse the order of integration, we treat R as a type I region, which enables us to write the given integral as

$$\int_0^2 \int_{y/2}^1 e^{x^2}\,dx\,dy = \iint_R e^{x^2}\,dA = \int_0^1 \int_0^{2x} e^{x^2}\,dy\,dx = \int_0^1 \left[e^{x^2}y\right]_{y=0}^{2x} dx$$

$$= \int_0^1 2xe^{x^2}\,dx = e^{x^2}\Big]_0^1 = e - 1 \blacktriangleleft$$

AREA CALCULATED AS A DOUBLE INTEGRAL

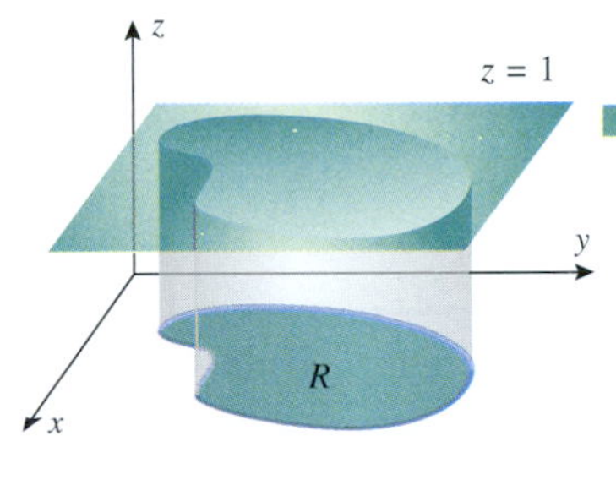

Cylinder with base R and height 1

Figure 15.2.13

Although double integrals arose in the context of calculating volumes, they can also be used to calculate areas. To see why this is so, recall that a *right cylinder* is a solid that is generated when a plane region is translated along a line that is perpendicular to the region. In Formula (2) of Section 7.2 we stated that the volume V of a right cylinder with cross-sectional area A and height h is

$$V = A \cdot h \tag{6}$$

Now suppose that we are interested in finding the area A of a region R in the xy-plane. If we translate the region R upward 1 unit, then the resulting solid will be a right cylinder that has cross-sectional area A, base R, and the plane $z = 1$ as its top (Figure 15.2.13). Thus, it follows from (6) that

$$\iint_R 1\,dA = (\text{area of } R) \cdot 1$$

which we can rewrite as

$$\text{area of } R = \iint_R 1\,dA = \iint_R dA \tag{7}$$

Formula (7) can be confusing because it equates an area and a volume; the formula is intended to equate only the *numerical values* of the area and volume and not the units, which must, of course, be different.

▶ Example 7 Use a double integral to find the area of the region R enclosed between the parabola $y = \frac{1}{2}x^2$ and the line $y = 2x$.

Solution. The region R may be treated equally well as type I (Figure 15.2.14a) or type II (Figure 15.2.14b). Treating R as type I yields

$$\text{area of } R = \iint_R dA = \int_0^4 \int_{x^2/2}^{2x} dy\,dx = \int_0^4 \left[y\right]_{y=x^2/2}^{2x} dx$$
$$= \int_0^4 \left(2x - \frac{1}{2}x^2\right) dx = \left[x^2 - \frac{x^3}{6}\right]_0^4 = \frac{16}{3}$$

Treating R as type II yields

$$\text{area of } R = \iint_R dA = \int_0^8 \int_{y/2}^{\sqrt{2y}} dx\,dy = \int_0^8 \left[x\right]_{x=y/2}^{\sqrt{2y}} dy$$
$$= \int_0^8 \left(\sqrt{2y} - \frac{1}{2}y\right) dy = \left[\frac{2\sqrt{2}}{3}y^{3/2} - \frac{y^2}{4}\right]_0^8 = \frac{16}{3} \blacktriangleleft$$

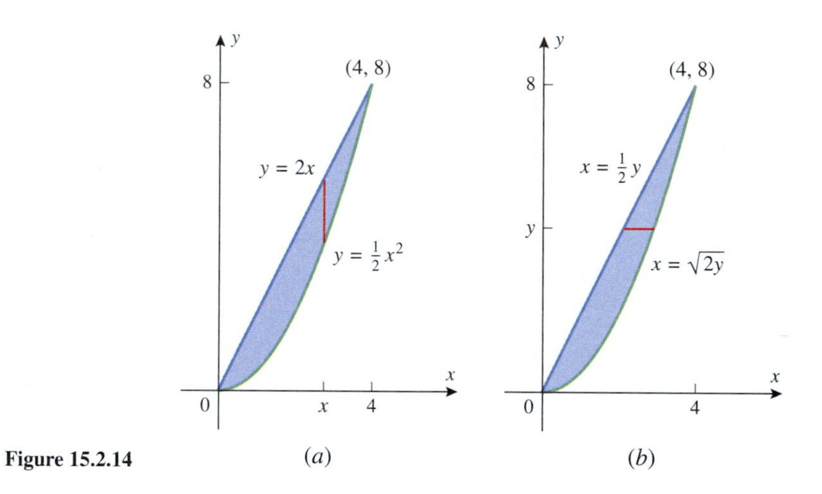

Figure 15.2.14

✔ QUICK CHECK EXERCISES 15.2 (See page 1035 for answers.)

1. Supply the missing integrand and limits of integration.
   (a) $\int_1^5 \int_2^{y/2} 6x^2y\,dx\,dy = \int_\square^\square$ ________ dy
   (b) $\int_1^5 \int_2^{x/2} 6x^2y\,dy\,dx = \int_\square^\square$ ________ dx

2. Let R be the triangular region in the xy-plane with vertices $(0, 0)$, $(3, 0)$, and $(0, 4)$. Supply the missing portions of the integrals.
   (a) Treating R as a type I region,
   $$\iint_R f(x, y)\,dA = \int_\square^\square \int_\square^\square f(x, y) \text{\_\_\_\_\_\_\_\_}$$
   (b) Treating R as a type II region,
   $$\iint_R f(x, y)\,dA = \int_\square^\square \int_\square^\square f(x, y) \text{\_\_\_\_\_\_\_\_}$$

3. Let R be the triangular region in the xy-plane with vertices $(0, 0)$, $(3, 3)$, and $(0, 4)$. Expressed as an iterated double integral, the area of R is $A(R) =$ ________.

4. The line $y = 2 - x$ and the parabola $y = x^2$ intersect at the points $(-2, 4)$ and $(1, 1)$. If R is the region enclosed by $y = 2 - x$ and $y = x^2$, then
   $$\iint_R (1 + 2y)\,dA = \text{\_\_\_\_\_\_\_\_}$$

EXERCISE SET 15.2 Graphing Utility C CAS

1–10 Evaluate the iterated integral.

1. $\int_0^1 \int_{x^2}^{x} xy^2\,dy\,dx$

2. $\int_1^{3/2} \int_y^{3-y} y\,dx\,dy$

3. $\int_0^3 \int_0^{\sqrt{9-y^2}} y\,dx\,dy$

4. $\int_{1/4}^1 \int_{x^2}^{x} \sqrt{\frac{x}{y}}\,dy\,dx$

5. $\int_{\sqrt{\pi}}^{\sqrt{2\pi}} \int_0^{x^3} \sin\frac{y}{x}\,dy\,dx$

6. $\int_{-1}^1 \int_{-x^2}^{x^2} (x^2 - y)\,dy\,dx$

7. $\int_{\pi/2}^{\pi} \int_0^{x^2} \frac{1}{x}\cos\frac{y}{x}\,dy\,dx$

8. $\int_0^1 \int_0^x e^{x^2}\,dy\,dx$

9. $\int_0^1 \int_0^x y\sqrt{x^2 - y^2}\,dy\,dx$

10. $\int_1^2 \int_0^{y^2} e^{x/y^2}\,dx\,dy$

FOCUS ON CONCEPTS

11. Let R be the region shown in the accompanying figure. Fill in the missing limits of integration.

(a) $\iint_R f(x, y)\,dA = \int_{\square}^{\square} \int_{\square}^{\square} f(x, y)\,dy\,dx$

(b) $\iint_R f(x, y)\,dA = \int_{\square}^{\square} \int_{\square}^{\square} f(x, y)\,dx\,dy$

12. Let R be the region shown in the accompanying figure. Fill in the missing limits of integration.

(a) $\iint_R f(x, y)\,dA = \int_{\square}^{\square} \int_{\square}^{\square} f(x, y)\,dy\,dx$

(b) $\iint_R f(x, y)\,dA = \int_{\square}^{\square} \int_{\square}^{\square} f(x, y)\,dx\,dy$

Figure Ex-11

Figure Ex-12

13. Let R be the region shown in the accompanying figure. Fill in the missing limits of integration.

(a) $\iint_R f(x, y)\,dA = \int_1^2 \int_{\square}^{\square} f(x, y)\,dy\,dx + \int_2^4 \int_{\square}^{\square} f(x, y)\,dy\,dx + \int_4^5 \int_{\square}^{\square} f(x, y)\,dy\,dx$

(b) $\iint_R f(x, y)\,dA = \int_{\square}^{\square} \int_{\square}^{\square} f(x, y)\,dx\,dy$

14. Let R be the region shown in the accompanying figure. Fill in the missing limits of integration.

(a) $\iint_R f(x, y)\,dA = \int_{\square}^{\square} \int_{\square}^{\square} f(x, y)\,dy\,dx$

(b) $\iint_R f(x, y)\,dA = \int_{\square}^{\square} \int_{\square}^{\square} f(x, y)\,dx\,dy$

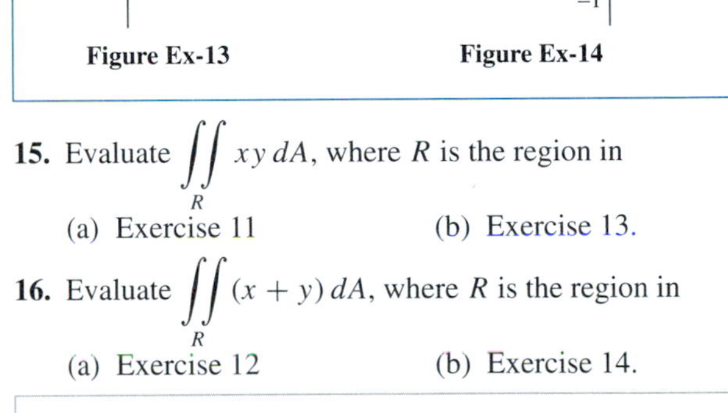

Figure Ex-13

Figure Ex-14

15. Evaluate $\iint_R xy\,dA$, where R is the region in

(a) Exercise 11 (b) Exercise 13.

16. Evaluate $\iint_R (x + y)\,dA$, where R is the region in

(a) Exercise 12 (b) Exercise 14.

17–20 Evaluate the double integral in two ways using iterated integrals: (a) viewing R as a type I region, and (b) viewing R as a type II region.

17. $\iint_R x^2\,dA$; R is the region bounded by $y = 16/x$, $y = x$, and $x = 8$.

18. $\iint_R xy^2\,dA$; R is the region enclosed by $y = 1$, $y = 2$, $x = 0$, and $y = x$.

19. $\iint_R (3x - 2y)\,dA$; R is the region enclosed by the circle $x^2 + y^2 = 1$.

20. $\iint_R y\,dA$; R is the region in the first quadrant enclosed between the circle $x^2 + y^2 = 25$ and the line $x + y = 5$.

21–26 Evaluate the double integral.

21. $\iint_R x(1 + y^2)^{-1/2}\,dA$; R is the region in the first quadrant enclosed by $y = x^2$, $y = 4$, and $x = 0$.

22. $\iint_R x \cos y \, dA$; R is the triangular region bounded by the lines $y = x$, $y = 0$, and $x = \pi$.

23. $\iint_R xy \, dA$; R is the region enclosed by $y = \sqrt{x}$, $y = 6 - x$, and $y = 0$.

24. $\iint_R x \, dA$; R is the region enclosed by $y = \sin^{-1} x$, $x = 1/\sqrt{2}$, and $y = 0$.

25. $\iint_R (x - 1) \, dA$; R is the region in the first quadrant enclosed between $y = x$ and $y = x^3$.

26. $\iint_R x^2 \, dA$; R is the region in the first quadrant enclosed by $xy = 1$, $y = x$, and $y = 2x$.

27. (a) By hand or with the help of a graphing utility, make a sketch of the region R enclosed between the curves $y = x + 2$ and $y = e^x$.
(b) Estimate the intersections of the curves in part (a).
(c) Viewing R as a type I region, estimate $\iint_R x \, dA$.
(d) Viewing R as a type II region, estimate $\iint_R x \, dA$.

28. (a) By hand or with the help of a graphing utility, make a sketch of the region R enclosed between the curves $y = 4x^3 - x^4$ and $y = 3 - 4x + 4x^2$.
(b) Find the intersections of the curves in part (a).
(c) Find $\iint_R x \, dA$.

29–32 Use double integration to find the area of the plane region enclosed by the given curves.

29. $y = \sin x$ and $y = \cos x$, for $0 \le x \le \pi/4$.

30. $y^2 = -x$ and $3y - x = 4$.

31. $y^2 = 9 - x$ and $y^2 = 9 - 9x$.

32. $y = \cosh x$, $y = \sinh x$, $x = 0$, and $x = 1$.

33–34 Use double integration to find the volume of the solid.

33. 34.

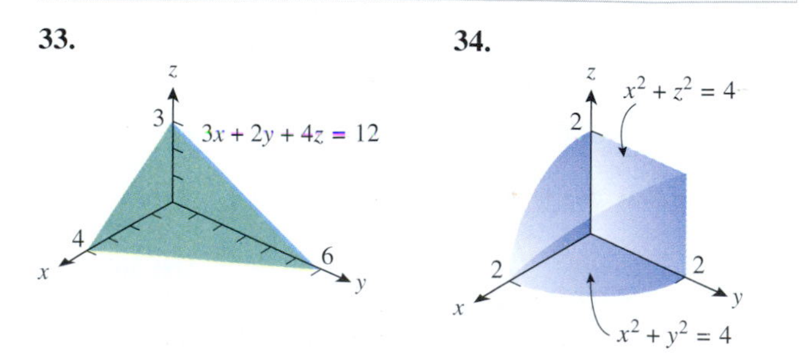

35–42 Use double integration to find the volume of each solid.

35. The solid bounded by the cylinder $x^2 + y^2 = 9$ and the planes $z = 0$ and $z = 3 - x$.

36. The solid in the first octant bounded above by the paraboloid $z = x^2 + 3y^2$, below by the plane $z = 0$, and laterally by $y = x^2$ and $y = x$.

37. The solid bounded above by the paraboloid $z = 9x^2 + y^2$, below by the plane $z = 0$, and laterally by the planes $x = 0$, $y = 0$, $x = 3$, and $y = 2$.

38. The solid enclosed by $y^2 = x$, $z = 0$, and $x + z = 1$.

39. The wedge cut from the cylinder $4x^2 + y^2 = 9$ by the planes $z = 0$ and $z = y + 3$.

40. The solid in the first octant bounded above by $z = 9 - x^2$, below by $z = 0$, and laterally by $y^2 = 3x$.

41. The solid that is common to the cylinders $x^2 + y^2 = 25$ and $x^2 + z^2 = 25$.

42. The solid bounded above by the paraboloid $z = x^2 + y^2$, below by the xy-plane, and laterally by the circular cylinder $x^2 + (y - 1)^2 = 1$.

43–44 Use a double integral and a CAS to find the volume of the solid.

C 43. The solid bounded above by the paraboloid $z = 1 - x^2 - y^2$ and below by the xy-plane.

C 44. The solid in the first octant that is bounded by the paraboloid $z = x^2 + y^2$, the cylinder $x^2 + y^2 = 4$ and the coordinate planes.

45–50 Express the integral as an equivalent integral with the order of integration reversed.

45. $\int_0^2 \int_0^{\sqrt{x}} f(x, y) \, dy \, dx$

46. $\int_0^4 \int_{2y}^8 f(x, y) \, dx \, dy$

47. $\int_0^2 \int_1^{e^y} f(x, y) \, dx \, dy$

48. $\int_1^e \int_0^{\ln x} f(x, y) \, dy \, dx$

49. $\int_0^1 \int_{\sin^{-1} y}^{\pi/2} f(x, y) \, dx \, dy$

50. $\int_0^1 \int_{y^2}^{\sqrt{y}} f(x, y) \, dx \, dy$

51–54 Evaluate the integral by first reversing the order of integration.

51. $\int_0^1 \int_{4x}^4 e^{-y^2} \, dy \, dx$

52. $\int_0^2 \int_{y/2}^1 \cos(x^2) \, dx \, dy$

53. $\int_0^4 \int_{\sqrt{y}}^2 e^{x^3} \, dx \, dy$

54. $\int_1^3 \int_0^{\ln x} x \, dy \, dx$

55. Evaluate $\iint_R \sin(y^3) \, dA$, where R is the region bounded by $y = \sqrt{x}$, $y = 2$, and $x = 0$. [*Hint:* Choose the order of integration carefully.]

56. Evaluate $\iint_R x\,dA$, where R is the region bounded by $x = \ln y$, $x = 0$, and $y = e$.

C **57.** Try to evaluate the integral with a CAS using the stated order of integration, and then by reversing the order of integration.

(a) $\displaystyle\int_0^4 \int_{\sqrt{x}}^2 \sin \pi y^3\, dy\, dx$

(b) $\displaystyle\int_0^1 \int_{\sin^{-1} y}^{\pi/2} \sec^2(\cos x)\, dx\, dy$

58. Use the appropriate Wallis formula (see Exercise Set 8.3) to find the volume of the solid enclosed between the circular paraboloid $z = x^2 + y^2$, the right circular cylinder $x^2 + y^2 = 4$, and the xy-plane (see the accompanying figure for cut view).

59. Evaluate $\iint_R xy^2\, dA$ over the region R shown in the accompanying figure.

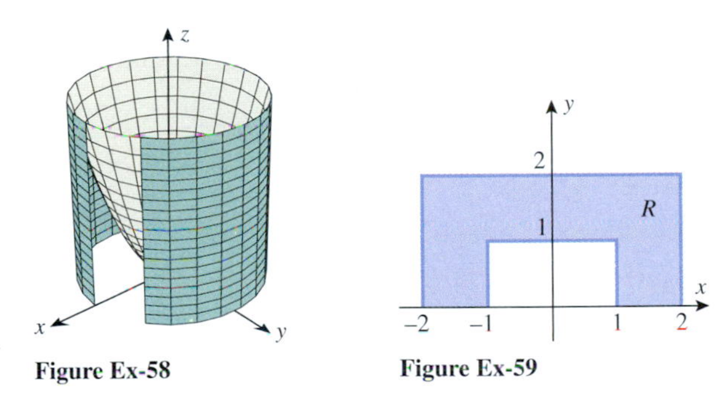

Figure Ex-58 **Figure Ex-59**

60. Give a geometric argument to show that

$$\int_0^1 \int_0^{\sqrt{1-y^2}} \sqrt{1 - x^2 - y^2}\, dx\, dy = \frac{\pi}{6}$$

61–62 The ***average value*** or ***mean value*** of a continuous function $f(x, y)$ over a region R in the xy-plane is defined as

$$f_{\text{ave}} = \frac{1}{A(R)} \iint_R f(x, y)\, dA$$

where $A(R)$ is the area of the region R (compare to the definition preceding Exercise 27 in Section 15.1). Use this definition in these exercises.

61. Find the average value of $1/(1 + x^2)$ over the triangular region with vertices $(0, 0)$, $(1, 1)$, and $(0, 1)$.

62. Find the average value of $f(x, y) = x^2 - xy$ over the region enclosed by $y = x$ and $y = 3x - x^2$.

63. Suppose that the temperature in degrees Celsius at a point (x, y) on a flat metal plate is $T(x, y) = 5xy + x^2$, where x and y are in meters. Find the average temperature of the diamond-shaped portion of the plate for which $|2x + y| \le 4$ and $|2x - y| \le 4$.

64. A circular lens of radius 2 inches has thickness $1 - (r^2/4)$ inches at all points r inches from the center of the lens. Find the average thickness of the lens.

C **65.** Use a CAS to approximate the intersections of the curves $y = \sin x$ and $y = x/2$, and then approximate the volume of the solid in the first octant that is below the surface $z = \sqrt{1 + x + y}$ and above the region in the xy-plane that is enclosed by the curves.

✔ QUICK CHECK ANSWERS 15.2

1. (a) $\displaystyle\int_1^5 \left(\frac{1}{4}y^4 - 16y\right) dy$ (b) $\displaystyle\int_1^5 \left(\frac{3}{4}x^4 - 12x^2\right) dx$ **2.** (a) $\displaystyle\int_0^3 \int_0^{-\frac{4}{3}x+4} f(x, y)\, dy\, dx$ (b) $\displaystyle\int_0^4 \int_0^{-\frac{3}{4}y+3} f(x, y)\, dx\, dy$ **3.** $\displaystyle\int_0^3 \int_x^{-\frac{1}{3}x+4} dy\, dx$ **4.** $\displaystyle\int_{-2}^1 \int_{x^2}^{2-x} (1 + 2y)\, dy\, dx = 18.9$

15.3 DOUBLE INTEGRALS IN POLAR COORDINATES

In this section we will study double integrals in which the integrand and the region of integration are expressed in polar coordinates. Such integrals are important for two reasons: first, they arise naturally in many applications, and second, many double integrals in rectangular coordinates can be evaluated more easily if they are converted to polar coordinates.

SIMPLE POLAR REGIONS

Some double integrals are easier to evaluate if the region of integration is expressed in polar coordinates. This is usually true if the region is bounded by a cardioid, a rose curve, a spiral, or, more generally, by any curve whose equation is simpler in polar coordinates than in rectangular coordinates. Moreover, double integrals whose integrands involve $x^2 + y^2$ also tend to be easier to evaluate in polar coordinates because this sum simplifies to r^2 when the conversion formulas $x = r\cos\theta$ and $y = r\sin\theta$ are applied.

An overview of polar coordinates can be found in Section 11.1.

Figure 15.3.1a shows a region R in a polar coordinate system that is enclosed between two rays, $\theta = \alpha$ and $\theta = \beta$, and two polar curves, $r = r_1(\theta)$ and $r = r_2(\theta)$. If, as shown in the figure, the functions $r_1(\theta)$ and $r_2(\theta)$ are continuous and their graphs do not cross, then the region R is called a *simple polar region*. If $r_1(\theta)$ is identically zero, then the boundary $r = r_1(\theta)$ reduces to a point (the origin), and the region has the general shape shown in Figure 15.3.1b. If, in addition, $\beta = \alpha + 2\pi$, then the rays coincide, and the region has the general shape shown in Figure 15.3.1c. The following definition expresses these geometric ideas algebraically.

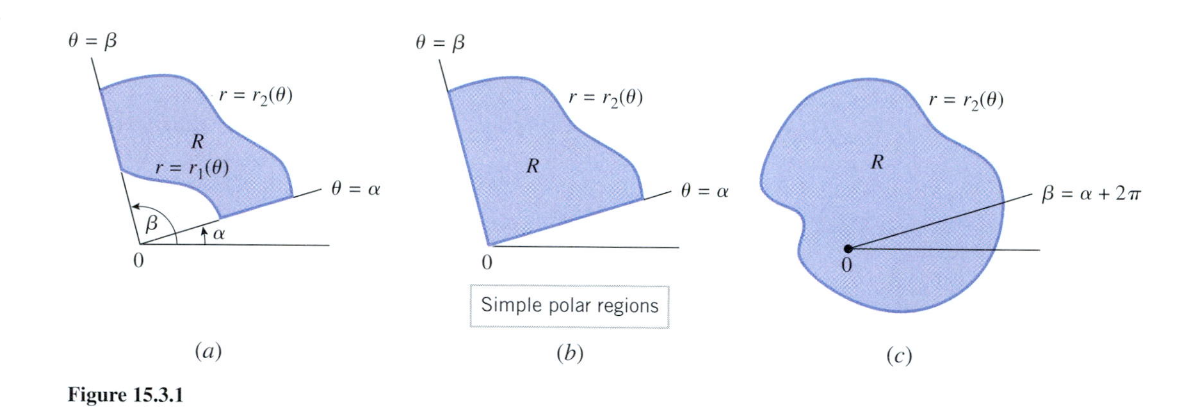

Figure 15.3.1

15.3.1 DEFINITION. A ***simple polar region*** in a polar coordinate system is a region that is enclosed between two rays, $\theta = \alpha$ and $\theta = \beta$, and two continuous polar curves, $r = r_1(\theta)$ and $r = r_2(\theta)$, where the equations of the rays and the polar curves satisfy the following conditions:

(i) $\alpha \le \beta$ (ii) $\beta - \alpha \le 2\pi$ (iii) $0 \le r_1(\theta) \le r_2(\theta)$

Conditions (i) and (ii) together imply that the ray $\theta = \beta$ can be obtained by rotating the ray $\theta = \alpha$ counterclockwise through an angle that is at most 2π radians. This is consistent with Figure 15.3.1. Condition (iii) implies that the boundary curves $r = r_1(\theta)$ and $r = r_2(\theta)$ can touch but cannot actually cross over one another (why?). Thus, in keeping with Figure 15.3.1, it is appropriate to describe $r = r_1(\theta)$ as the ***inner boundary*** of the region and $r = r_2(\theta)$ as the ***outer boundary***.

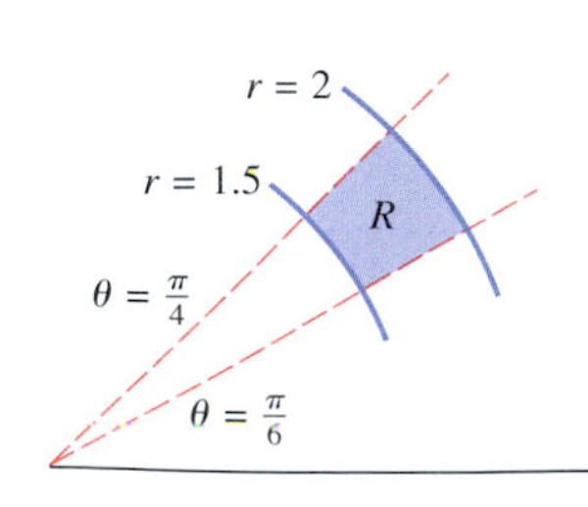

Figure 15.3.2

A ***polar rectangle*** is a simple polar region for which the bounding polar curves are circular arcs. For example, Figure 15.3.2 shows the polar rectangle R given by

$$1.5 \le r \le 2, \quad \frac{\pi}{6} \le \theta \le \frac{\pi}{4}$$

DOUBLE INTEGRALS IN POLAR COORDINATES

Next we will consider the polar version of Problem 15.1.1.

> **15.3.2 THE VOLUME PROBLEM IN POLAR COORDINATES.** Given a function $f(r, \theta)$ that is continuous and nonnegative on a simple polar region R, find the volume of the solid that is enclosed between the region R and the surface whose equation in cylindrical coordinates is $z = f(r, \theta)$ (Figure 15.3.3).

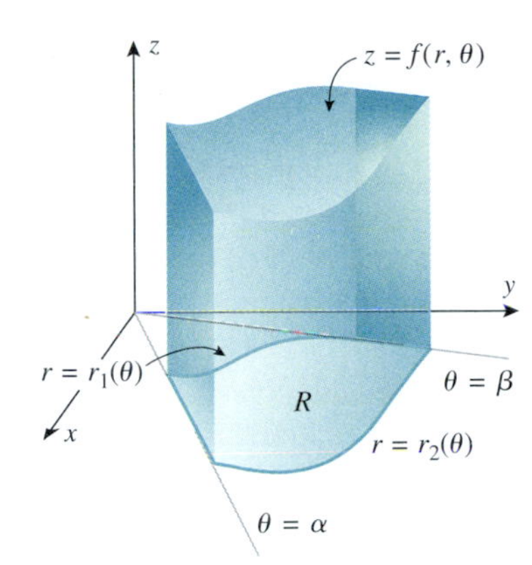

Figure 15.3.3

To motivate a formula for the volume V of the solid in Figure 15.3.3, we will use a limit process similar to that used to obtain Formula (2) of Section 15.1, except that here we will use circular arcs and rays to subdivide the region R into polar rectangles. As shown in Figure 15.3.4, we will exclude from consideration all polar rectangles that contain any points outside of R, leaving only polar rectangles that are subsets of R. Assume that there are n such polar rectangles, and denote the area of the kth polar rectangle by ΔA_k. Let (r_k^*, θ_k^*) be any point in this polar rectangle. As shown in Figure 15.3.5, the product $f(r_k^*, \theta_k^*)\Delta A_k$ is the volume of a solid with base area ΔA_k and height $f(r_k^*, \theta_k^*)$, so the sum

$$\sum_{k=1}^{n} f(r_k^*, \theta_k^*)\Delta A_k$$

can be viewed as an approximation to the volume V of the entire solid.

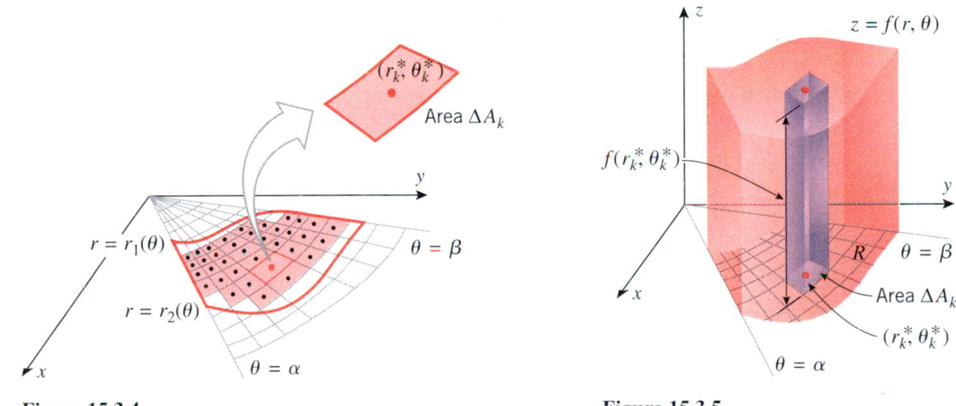

Figure 15.3.4

Figure 15.3.5

If we now increase the number of subdivisions in such a way that the dimensions of the polar rectangles approach zero, then it seems plausible that the errors in the approximations approach zero, and the exact volume of the solid is

$$V = \lim_{n\to+\infty} \sum_{k=1}^{n} f(r_k^*, \theta_k^*)\Delta A_k \tag{1}$$

If $f(r, \theta)$ is continuous on R and has both positive and negative values, then the limit

$$\lim_{n\to+\infty} \sum_{k=1}^{n} f(r_k^*, \theta_k^*)\Delta A_k \tag{2}$$

represents the net signed volume between the region R and the surface $z = f(r, \theta)$ (as with double integrals in rectangular coordinates). The sums in (2) are called ***polar Riemann sums***, and the limit of the polar Riemann sums is denoted by

$$\iint_R f(r, \theta)\, dA = \lim_{n\to+\infty} \sum_{k=1}^{n} f(r_k^*, \theta_k^*)\Delta A_k \tag{3}$$

Polar double integrals are also called ***double integrals in polar coordinates*** to distinguish them from double integrals over regions in the xy-plane; the latter are called ***double integrals in rectangular coordinates***. Double integrals in polar coordinates have the usual integral properties, such as those stated in Formulas (9), (10), and (11) of Section 15.1.

which is called the ***polar double integral*** of $f(r,\theta)$ over R. If $f(r,\theta)$ is continuous and nonnegative on R, then the volume formula (1) can be expressed as

$$V = \iint_R f(r,\theta)\,dA \tag{4}$$

EVALUATING POLAR DOUBLE INTEGRALS

In Sections 15.1 and 15.2 we evaluated double integrals in rectangular coordinates by expressing them as iterated integrals. Polar double integrals are evaluated the same way. To motivate the formula that expresses a double polar integral as an iterated integral, we will assume that $f(r,\theta)$ is nonnegative so that we can interpret (3) as a volume. However, the results that we will obtain will also be applicable if f has negative values. To begin, let us choose the arbitrary point (r_k^*, θ_k^*) in (3) to be at the "center" of the kth polar rectangle as shown in Figure 15.3.6. Suppose also that this polar rectangle has a central angle $\Delta\theta_k$ and a "radial thickness" Δr_k. Thus, the inner radius of this polar rectangle is $r_k^* - \frac{1}{2}\Delta r_k$ and the outer radius is $r_k^* + \frac{1}{2}\Delta r_k$. Treating the area ΔA_k of this polar rectangle as the difference in area of two sectors, we obtain

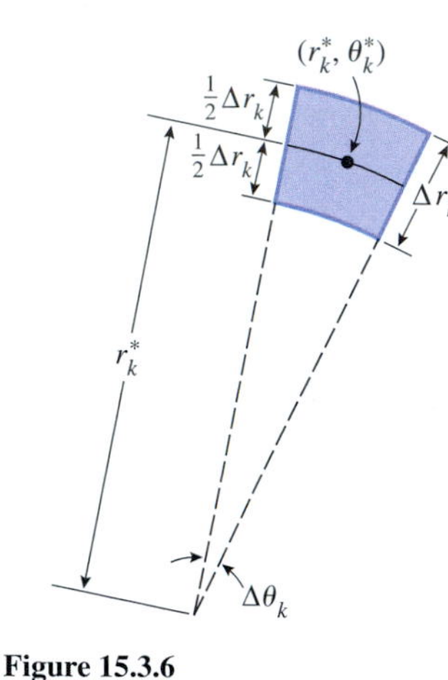

Figure 15.3.6

$$\Delta A_k = \tfrac{1}{2}\left(r_k^* + \tfrac{1}{2}\Delta r_k\right)^2 \Delta\theta_k - \tfrac{1}{2}\left(r_k^* - \tfrac{1}{2}\Delta r_k\right)^2 \Delta\theta_k$$

which simplifies to

$$\Delta A_k = r_k^* \Delta r_k \Delta\theta_k \tag{5}$$

Thus, from (3) and (4)

$$V = \iint_R f(r,\theta)\,dA = \lim_{n\to+\infty} \sum_{k=1}^{n} f(r_k^*, \theta_k^*) r_k^* \Delta r_k \Delta\theta_k$$

which suggests that the volume V can be expressed as the iterated integral

$$V = \iint_R f(r,\theta)\,dA = \int_\alpha^\beta \int_{r_1(\theta)}^{r_2(\theta)} f(r,\theta) r\,dr\,d\theta \tag{6}$$

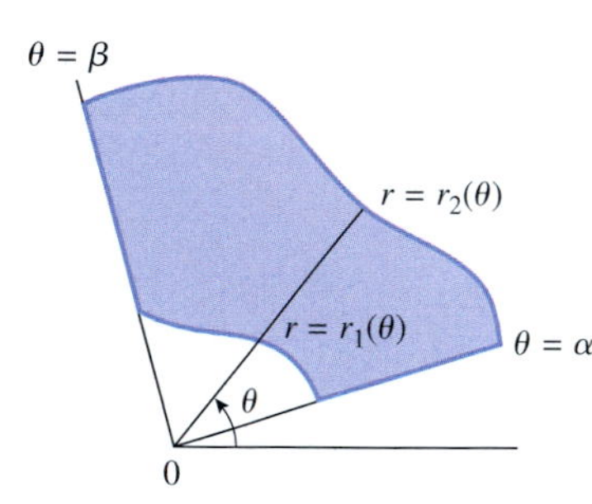

Figure 15.3.7

in which the limits of integration are chosen to cover the region R; that is, with θ fixed between α and β, the value of r varies from $r_1(\theta)$ to $r_2(\theta)$ (Figure 15.3.7).

Although we assumed $f(r,\theta)$ to be nonnegative in deriving Formula (6), it can be proved that the relationship between the polar double integral and the iterated integral in this formula also holds if f has negative values. Accepting this to be so, we obtain the following theorem, which we state without formal proof.

15.3.3 THEOREM. *If R is a simple polar region whose boundaries are the rays $\theta = \alpha$ and $\theta = \beta$ and the curves $r = r_1(\theta)$ and $r = r_2(\theta)$ shown in Figure 15.3.7, and if $f(r,\theta)$ is continuous on R, then*

$$\iint_R f(r,\theta)\,dA = \int_\alpha^\beta \int_{r_1(\theta)}^{r_2(\theta)} f(r,\theta) r\,dr\,d\theta \tag{7}$$

Note the extra factor of r that appears in the integrand when expressing a polar double integral as an iterated integral in polar coordinates.

To apply this theorem you will need to be able to find the rays and the curves that form the boundary of the region R, since these determine the limits of integration in the iterated integral. This can be done as follows:

Determining Limits of Integration for a Polar Double Integral: Simple Polar Region

Step 1. Since θ is held fixed for the first integration, draw a radial line from the origin through the region R at a fixed angle θ (Figure 15.3.8a). This line crosses the boundary of R at most twice. The innermost point of intersection is on the inner boundary curve $r = r_1(\theta)$ and the outermost point is on the outer boundary curve $r = r_2(\theta)$. These intersections determine the r-limits of integration in (7).

Step 2. Imagine rotating a ray along the polar x-axis one revolution counterclockwise about the origin. The smallest angle at which this ray intersects the region R is $\theta = \alpha$ and the largest angle is $\theta = \beta$ (Figure 15.3.8b). This determines the θ-limits of integration.

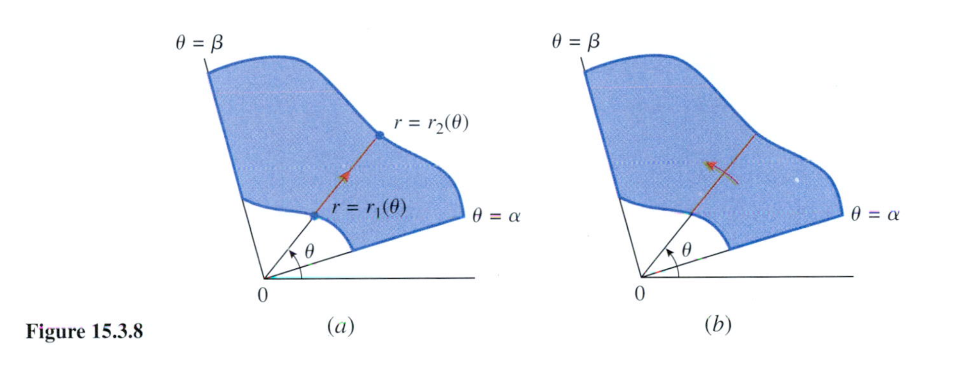

Figure 15.3.8

► Example 1 Evaluate

$$\iint_R \sin\theta \, dA$$

where R is the region in the first quadrant that is outside the circle $r = 2$ and inside the cardioid $r = 2(1 + \cos\theta)$.

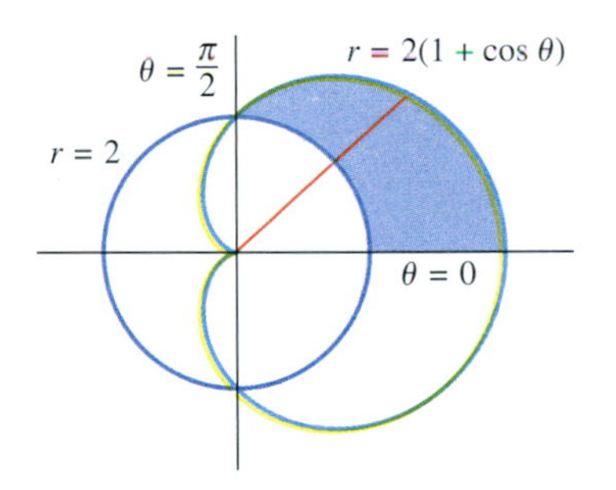

Figure 15.3.9

Solution. The region R is sketched in Figure 15.3.9. Following the two steps outlined above we obtain

$$\begin{aligned}
\iint_R \sin\theta \, dA &= \int_0^{\pi/2} \int_2^{2(1+\cos\theta)} (\sin\theta) r \, dr \, d\theta \\
&= \int_0^{\pi/2} \left[\frac{1}{2} r^2 \sin\theta\right]_{r=2}^{2(1+\cos\theta)} d\theta \\
&= 2\int_0^{\pi/2} [(1+\cos\theta)^2 \sin\theta - \sin\theta] \, d\theta \\
&= 2\left[-\frac{1}{3}(1+\cos\theta)^3 + \cos\theta\right]_0^{\pi/2} \\
&= 2\left[-\frac{1}{3} - \left(-\frac{5}{3}\right)\right] = \frac{8}{3} \quad ◄
\end{aligned}$$

► Example 2 The sphere of radius a centered at the origin is expressed in rectangular coordinates as $x^2 + y^2 + z^2 = a^2$, and hence its equation in cylindrical coordinates is $r^2 + z^2 = a^2$. Use this equation and a polar double integral to find the volume of the sphere.

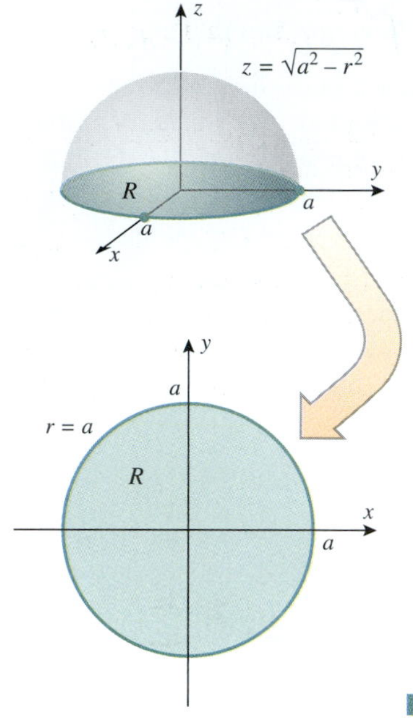

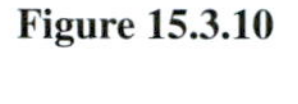
Figure 15.3.10

Solution. In cylindrical coordinates the upper hemisphere is given by the equation

$$z = \sqrt{a^2 - r^2}$$

so the volume enclosed by the entire sphere is

$$V = 2 \iint_R \sqrt{a^2 - r^2}\, dA$$

where R is the circular region shown in Figure 15.3.10. Thus,

$$\begin{aligned} V = 2 \iint_R \sqrt{a^2 - r^2}\, dA &= \int_0^{2\pi} \int_0^a \sqrt{a^2 - r^2}(2r)\, dr\, d\theta \\ &= \int_0^{2\pi} \left[-\frac{2}{3}(a^2 - r^2)^{3/2} \right]_{r=0}^a d\theta = \int_0^{2\pi} \frac{2}{3} a^3\, d\theta \\ &= \left[\frac{2}{3} a^3 \theta \right]_0^{2\pi} = \frac{4}{3}\pi a^3 \end{aligned}$$ ◀

FINDING AREAS USING POLAR DOUBLE INTEGRALS

Recall from Formula (7) of Section 15.2 that the area of a region R in the xy-plane can be expressed as

$$\text{area of } R = \iint_R 1\, dA = \iint_R dA \tag{8}$$

The argument used to derive this result can also be used to show that the formula applies to polar double integrals over regions in polar coordinates.

▶ **Example 3** Use a polar double integral to find the area enclosed by the three-petaled rose $r = \sin 3\theta$.

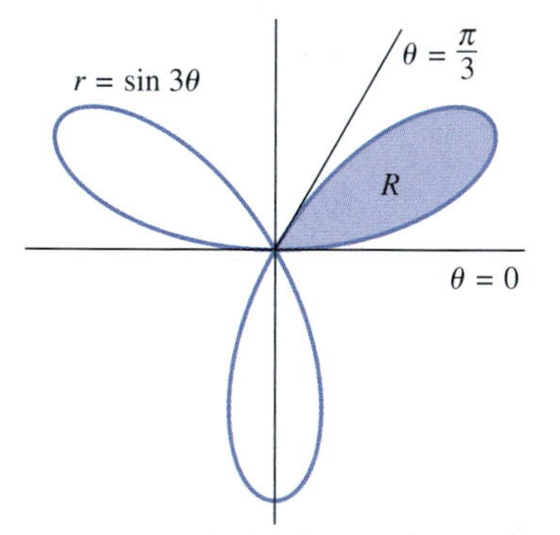

Figure 15.3.11

Solution. The rose is sketched in Figure 15.3.11. We will use Formula (8) to calculate the area of the petal R in the first quadrant and multiply by three.

$$\begin{aligned} A = 3 \iint_R dA &= 3 \int_0^{\pi/3} \int_0^{\sin 3\theta} r\, dr\, d\theta \\ &= \frac{3}{2} \int_0^{\pi/3} \sin^2 3\theta\, d\theta = \frac{3}{4} \int_0^{\pi/3} (1 - \cos 6\theta)\, d\theta \\ &= \frac{3}{4} \left[\theta - \frac{\sin 6\theta}{6} \right]_0^{\pi/3} = \frac{1}{4}\pi \end{aligned}$$ ◀

CONVERTING DOUBLE INTEGRALS FROM RECTANGULAR TO POLAR COORDINATES

Sometimes a double integral that is difficult to evaluate in rectangular coordinates can be evaluated more easily in polar coordinates by making the substitution $x = r\cos\theta$, $y = r\sin\theta$ and expressing the region of integration in polar form; that is, we rewrite the double integral in rectangular coordinates as

$$\iint_R f(x, y)\, dA = \iint_R f(r\cos\theta, r\sin\theta)\, dA = \iint_{\substack{\text{appropriate} \\ \text{limits}}} f(r\cos\theta, r\sin\theta) r\, dr\, d\theta \tag{9}$$

▶ Example 4 Use polar coordinates to evaluate $\displaystyle\int_{-1}^{1}\int_{0}^{\sqrt{1-x^2}}(x^2+y^2)^{3/2}\,dy\,dx$.

Solution. In this problem we are starting with an iterated integral in rectangular coordinates rather than a double integral, so before we can make the conversion to polar coordinates we will have to identify the region of integration. To do this, we observe that for fixed x the y-integration runs from $y=0$ to $y=\sqrt{1-x^2}$, which tells us that the lower boundary of the region is the x-axis and the upper boundary is a semicircle of radius 1 centered at the origin. From the x-integration we see that x varies from -1 to 1, so we conclude that the region of integration is as shown in Figure 15.3.12. In polar coordinates, this is the region swept out as r varies between 0 and 1 and θ varies between 0 and π. Thus,

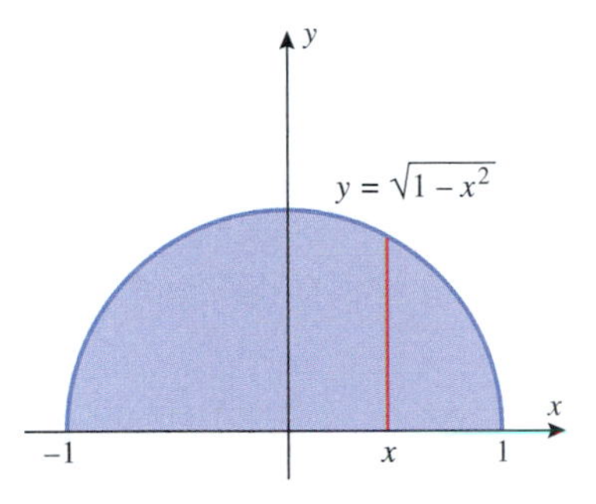

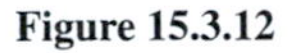
Figure 15.3.12

$$\int_{-1}^{1}\int_{0}^{\sqrt{1-x^2}}(x^2+y^2)^{3/2}\,dy\,dx=\iint_R (x^2+y^2)^{3/2}\,dA$$

$$=\int_0^{\pi}\int_0^1 (r^3)r\,dr\,d\theta=\int_0^{\pi}\frac{1}{5}\,d\theta=\frac{\pi}{5}\ ◀$$

The conversion to polar coordinates worked so nicely in Example 4 because the substitution $x=r\cos\theta$, $y=r\sin\theta$ collapsed the sum x^2+y^2 into the single term r^2, thereby simplifying the integrand. Whenever you see an expression involving x^2+y^2 in the integrand, you should consider the possibility of converting to polar coordinates.

✔QUICK CHECK EXERCISES 15.3

(See page 1043 for answers.)

1. The polar region inside the circle $r=2\sin\theta$ and outside the circle $r=1$ is a simple polar region given by the inequalities

   $\_\_\_\_\_ \le r \le \_\_\_\_\_,\quad \_\_\_\_\_ \le \theta \le \_\_\_\_\_$

2. Let R be the region in the first quadrant enclosed between the circles $x^2+y^2=9$ and $x^2+y^2=100$. Supply the missing limits of integration.

   $$\iint_R f(r,\theta)\,dA=\int_{\square}^{\square}\int_{\square}^{\square} f(r,\theta)r\,dr\,d\theta$$

3. Let V be the volume of the solid bounded above by the hemisphere $z=\sqrt{1-r^2}$ and bounded below by the disk enclosed within the circle $r=\sin\theta$. Expressed as a double integral in polar coordinates, $V=$ _____.

4. Express the iterated integral as a double integral in polar coordinates.

   $$\int_{1/\sqrt{2}}^{1}\int_{\sqrt{1-x^2}}^{x}\left(\frac{1}{x^2+y^2}\right)dy\,dx=\_\_\_\_\_$$

EXERCISE SET 15.3 [C] CAS

1–6 Evaluate the iterated integral.

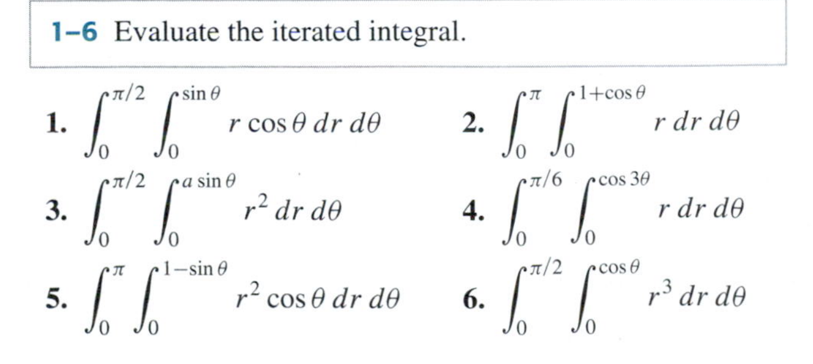

1. $\displaystyle\int_0^{\pi/2}\int_0^{\sin\theta} r\cos\theta\,dr\,d\theta$

2. $\displaystyle\int_0^{\pi}\int_0^{1+\cos\theta} r\,dr\,d\theta$

3. $\displaystyle\int_0^{\pi/2}\int_0^{a\sin\theta} r^2\,dr\,d\theta$

4. $\displaystyle\int_0^{\pi/6}\int_0^{\cos 3\theta} r\,dr\,d\theta$

5. $\displaystyle\int_0^{\pi}\int_0^{1-\sin\theta} r^2\cos\theta\,dr\,d\theta$

6. $\displaystyle\int_0^{\pi/2}\int_0^{\cos\theta} r^3\,dr\,d\theta$

7–10 Use a double integral in polar coordinates to find the area of the region described.

7. The region enclosed by the cardioid $r=1-\cos\theta$.
8. The region enclosed by the rose $r=\sin 2\theta$.
9. The region in the first quadrant bounded by $r=1$ and $r=\sin 2\theta$, with $\pi/4\le\theta\le\pi/2$.
10. The region inside the circle $x^2+y^2=4$ and to the right of the line $x=1$.

FOCUS ON CONCEPTS

11–12 Let R be the region described. Sketch the region R and fill in the missing limits of integration.

$$\iint_R f(r,\theta)\,dA = \int_{\square}^{\square}\int_{\square}^{\square} f(r,\theta)r\,dr\,d\theta$$

11. The region inside the circle $r = 4\sin\theta$ and outside the circle $r = 2$.

12. The region inside the circle $r = 1$ and outside the cardioid $r = 1 + \cos\theta$.

13–16 Express the volume of the solid described as a double integral in polar coordinates.

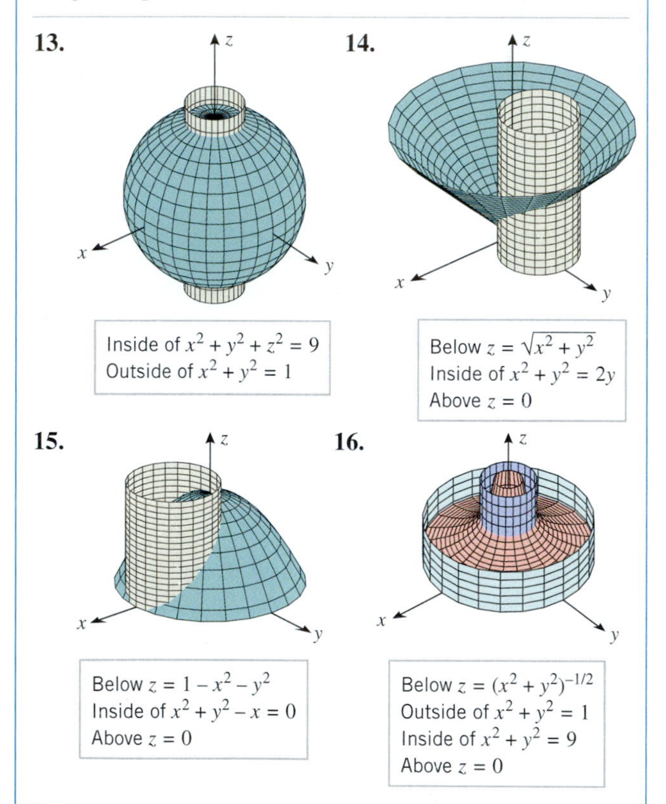

17. Find the volume of the solid described in Exercise 13.

18. Find the volume of the solid described in Exercise 14.

19. Find the volume of the solid described in Exercise 15.

20. Find the volume of the solid described in Exercise 16.

21. Find the volume of the solid in the first octant bounded above by the surface $z = r\sin\theta$, below by the xy-plane, and laterally by the plane $x = 0$ and the surface $r = 3\sin\theta$.

22. Find the volume of the solid inside the surface $r^2 + z^2 = 4$ and outside the surface $r = 2\cos\theta$.

23–26 Use polar coordinates to evaluate the double integral.

23. $\displaystyle\iint_R e^{-(x^2+y^2)}\,dA$, where R is the region enclosed by the circle $x^2 + y^2 = 1$.

24. $\displaystyle\iint_R \sqrt{9 - x^2 - y^2}\,dA$, where R is the region in the first quadrant within the circle $x^2 + y^2 = 9$.

25. $\displaystyle\iint_R \frac{1}{1 + x^2 + y^2}\,dA$, where R is the sector in the first quadrant bounded by $y = 0$, $y = x$, and $x^2 + y^2 = 4$.

26. $\displaystyle\iint_R 2y\,dA$, where R is the region in the first quadrant bounded above by the circle $(x - 1)^2 + y^2 = 1$ and below by the line $y = x$.

27–34 Evaluate the iterated integral by converting to polar coordinates.

27. $\displaystyle\int_0^1\int_0^{\sqrt{1-x^2}} (x^2 + y^2)\,dy\,dx$

28. $\displaystyle\int_{-2}^2\int_{-\sqrt{4-y^2}}^{\sqrt{4-y^2}} e^{-(x^2+y^2)}\,dx\,dy$

29. $\displaystyle\int_0^2\int_0^{\sqrt{2x-x^2}} \sqrt{x^2 + y^2}\,dy\,dx$

30. $\displaystyle\int_0^1\int_0^{\sqrt{1-y^2}} \cos(x^2 + y^2)\,dx\,dy$

31. $\displaystyle\int_0^a\int_0^{\sqrt{a^2-x^2}} \frac{dy\,dx}{(1 + x^2 + y^2)^{3/2}} \quad (a > 0)$

32. $\displaystyle\int_0^1\int_y^{\sqrt{y}} \sqrt{x^2 + y^2}\,dx\,dy$

33. $\displaystyle\int_0^{\sqrt{2}}\int_y^{\sqrt{4-y^2}} \frac{1}{\sqrt{1 + x^2 + y^2}}\,dx\,dy$

34. $\displaystyle\int_0^4\int_3^{\sqrt{25-x^2}} dy\,dx$

35. Use a double integral in polar coordinates to find the volume of a cylinder of radius a and height h.

36. (a) Use a double integral in polar coordinates to find the volume of the oblate spheroid

$$\frac{x^2}{a^2} + \frac{y^2}{a^2} + \frac{z^2}{c^2} = 1 \quad (0 < c < a)$$

(b) Use the result in part (a) and the World Geodetic System of 1984 (WGS-84) discussed in Exercise 50 of Section 12.7 to find the volume of the Earth in cubic meters.

37. Use polar coordinates to find the volume of the solid that is above the xy-plane, inside the cylinder $x^2 + y^2 - ay = 0$, and inside the ellipsoid

$$\frac{x^2}{a^2} + \frac{y^2}{a^2} + \frac{z^2}{c^2} = 1$$

38. Find the area of the region enclosed by the lemniscate $r^2 = 2a^2\cos 2\theta$.

39. Find the area in the first quadrant that is inside the circle $r = 4\sin\theta$ and outside the lemniscate $r^2 = 8\cos 2\theta$.

40. Show that the shaded area in the accompanying figure is $a^2\phi - \frac{1}{2}a^2 \sin 2\phi$.

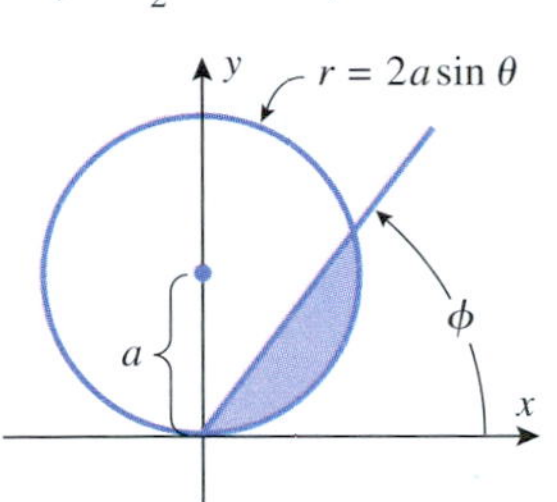

Figure Ex-40

41. The integral $\int_0^{+\infty} e^{-x^2}\,dx$, which arises in probability theory, can be evaluated using the following method. Let the value of the integral be I. Thus,

$$I = \int_0^{+\infty} e^{-x^2}\,dx = \int_0^{+\infty} e^{-y^2}\,dy$$

since the letter used for the variable of integration in a definite integral does not matter.

(a) Give a reasonable argument to show that

$$I^2 = \int_0^{+\infty}\int_0^{+\infty} e^{-(x^2+y^2)}\,dx\,dy$$

(b) Evaluate the iterated integral in part (a) by converting to polar coordinates.

(c) Use the result in part (b) to show that $I = \sqrt{\pi}/2$.

42. Show that

$$\int_0^{+\infty}\int_0^{+\infty} \frac{1}{(1+x^2+y^2)^2}\,dx\,dy = \frac{\pi}{4}$$

[*Hint:* See Exercise 41.]

C **43.** (a) Use the numerical integration capability of a CAS to approximate the value of the double integral

$$\int_{-1}^{1}\int_0^{\sqrt{1-x^2}} e^{-(x^2+y^2)^2}\,dy\,dx$$

(b) Compare the approximation obtained in part (a) to the approximation that results if the integral is first converted to polar coordinates.

44. Suppose that a geyser, centered at the origin of a polar coordinate system, sprays water in a circular pattern in such a way that the depth D of water that reaches a point at a distance of r feet from the origin in 1 hour is $D = ke^{-r}$. Find the total volume of water that the geyser sprays inside a circle of radius R centered at the origin.

45. Evaluate $\iint_R x^2\,dA$ over the region R shown in the accompanying figure.

Figure Ex-45

✔ QUICK CHECK ANSWERS 15.3

1. $1 \le r \le 2\sin\theta,\ \pi/6 \le \theta \le 5\pi/6$ **2.** $\int_0^{\pi/2}\int_3^{10} f(r,\theta)r\,dr\,d\theta$ **3.** $\int_0^{\pi}\int_0^{\sin\theta} r\sqrt{1-r^2}\,dr\,d\theta$ **4.** $\int_0^{\pi/4}\int_1^{\sec\theta} \frac{1}{r}\,dr\,d\theta$

15.4 PARAMETRIC SURFACES; SURFACE AREA

In previous sections we considered parametric curves in 2-space and 3-space. In this section we will discuss parametric surfaces in 3-space. As we will see, parametric representations of surfaces are not only important in computer graphics but also allow us to study more general kinds of surfaces than those encountered so far. In Section 7.5 we showed how to find the surface area of a surface of revolution. Our work on parametric surfaces will enable us to derive area formulas for more general kinds of surfaces.

PARAMETRIC REPRESENTATION OF SURFACES

We have seen that curves in 3-space can be represented by three equations involving one parameter, say

$$x = x(t), \quad y = y(t), \quad z = z(t)$$

Surfaces in 3-space can be represented parametrically by three equations involving two parameters, say

$$x = x(u, v), \quad y = y(u, v), \quad z = z(u, v) \tag{1}$$

To visualize why such equations represent a surface, think of (u, v) as a point that varies over some region in a uv-plane. If u is held constant, then v is the only varying parameter in (1), and hence these equations represent a curve in 3-space. We call this a ***constant u-curve*** (Figure 15.4.1). Similarly, if v is held constant, then u is the only varying parameter in (1), so again these equations represent a curve in 3-space. We call this a ***constant v-curve***. By varying the constants we generate a family of u-curves and a family of v-curves that together form a surface.

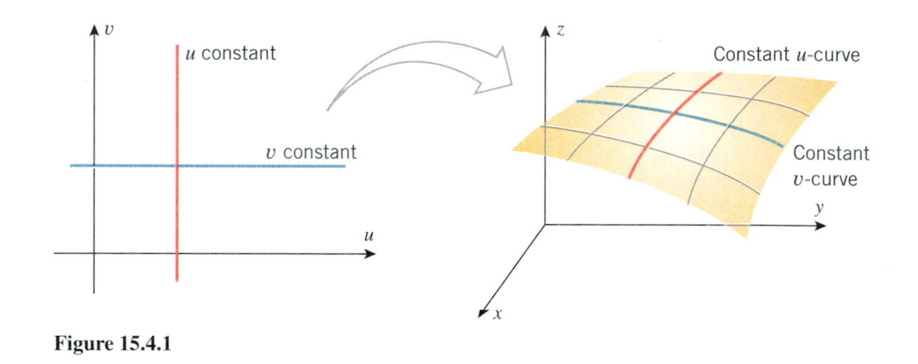

Figure 15.4.1

▶ Example 1 Consider the paraboloid $z = 4 - x^2 - y^2$. One way to parametrize this surface is to take $x = u$ and $y = v$ as the parameters, in which case the surface is represented by the parametric equations

$$x = u, \quad y = v, \quad z = 4 - u^2 - v^2 \tag{2}$$

Figure 15.4.2a shows a computer-generated graph of this surface. The constant u-curves correspond to constant x-values and hence appear on the surface as traces parallel to the yz-plane. Similarly, the constant v-curves correspond to constant y-values and hence appear on the surface as traces parallel to the xz-plane. ◀

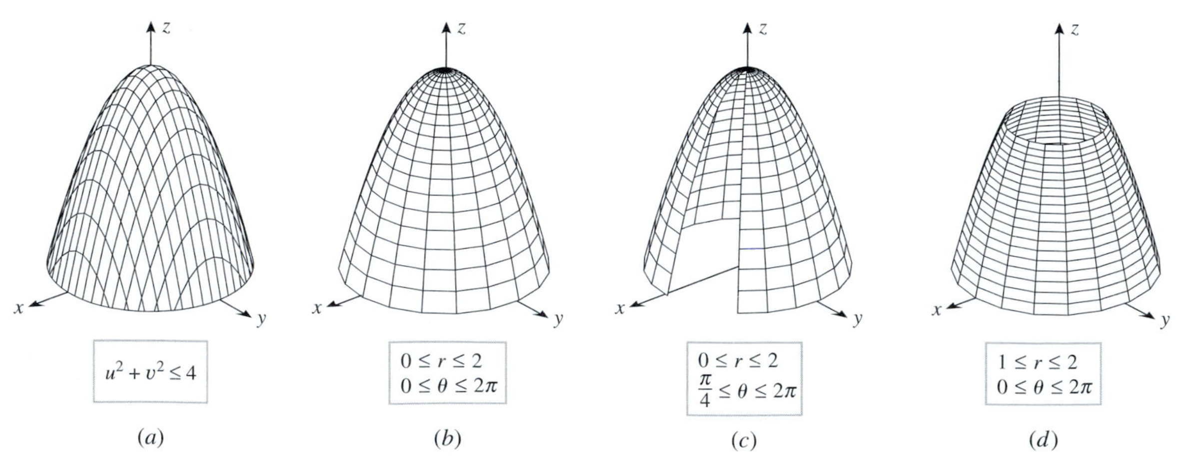

Figure 15.4.2

▶ Example 2 The paraboloid $z = 4 - x^2 - y^2$ that was considered in Example 1 can also be parametrized by first expressing the equation in cylindrical coordinates. For this

TECHNOLOGY MASTERY

If you have a graphing utility that can generate parametric surfaces, consult the relevant documentation and then try to generate the surfaces in Figure 15.4.2.

purpose, we make the substitution $x = r\cos\theta$, $y = r\sin\theta$, which yields $z = 4 - r^2$. Thus, the paraboloid can be represented parametrically in terms of r and θ as

$$x = r\cos\theta, \quad y = r\sin\theta, \quad z = 4 - r^2 \tag{3}$$

A computer-generated graph of this surface for $0 \le r \le 2$ and $0 \le \theta \le 2\pi$ is shown in Figure 15.4.2*b*. The constant r-curves correspond to constant z-values and hence appear on the surface as traces parallel to the xy-plane. The constant θ-curves appear on the surface as traces from vertical planes through the origin at varying angles with the x-axis. Parts (*c*) and (*d*) of Figure 15.4.2 show the effect of restrictions on the parameters r and θ. ◂

▸ **Example 3** One way to generate the sphere $x^2 + y^2 + z^2 = 1$ with a graphing utility is to graph the upper and lower hemispheres

$$z = \sqrt{1 - x^2 - y^2} \quad \text{and} \quad z = -\sqrt{1 - x^2 - y^2}$$

on the same screen. However, this usually produces a fragmented sphere (Figure 15.4.3*a*) because roundoff error sporadically produces negative values inside the radical when $1 - x^2 - y^2$ is near zero. A better graph can be generated by first expressing the sphere in spherical coordinates as $\rho = 1$ and then using the spherical-to-rectangular conversion formulas in Table 12.8.1 to obtain the parametric equations

$$x = \sin\phi\cos\theta, \quad y = \sin\phi\sin\theta, \quad z = \cos\phi$$

with parameters θ and ϕ. Figure 15.4.3*b* shows the graph of this parametric surface for $0 \le \theta \le 2\pi$ and $0 \le \phi \le \pi$. In the language of cartographers, the constant ϕ-curves are the ***lines of latitude*** and the constant θ-curves are the ***lines of longitude***. ◂

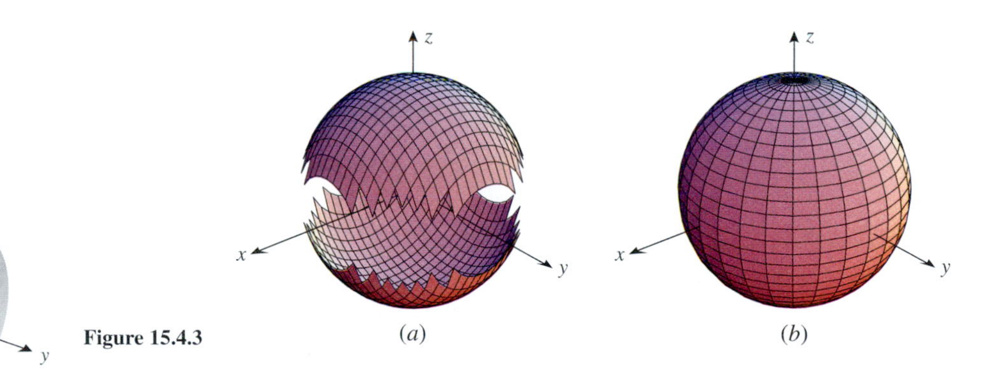

Figure 15.4.3

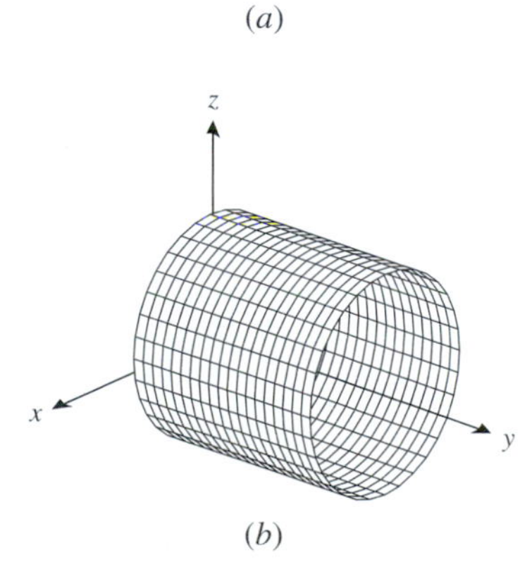

Figure 15.4.4

▸ **Example 4** Find parametric equations for the portion of the right circular cylinder

$$x^2 + z^2 = 9 \quad \text{for which} \quad 0 \le y \le 5$$

in terms of the parameters u and v shown in Figure 15.4.4*a*. The parameter u is the y-coordinate of a point $P(x, y, z)$ on the surface, and v is the angle shown in the figure.

Solution. The radius of the cylinder is 3, so it is evident from the figure that $y = u$, $x = 3\cos v$, and $z = 3\sin v$. Thus, the surface can be represented parametrically as

$$x = 3\cos v, \quad y = u, \quad z = 3\sin v$$

To obtain the portion of the surface from $y = 0$ to $y = 5$, we let the parameter u vary over the interval $0 \le u \le 5$, and to ensure that the entire lateral surface is covered, we let the parameter v vary over the interval $0 \le v \le 2\pi$. Figure 15.4.4*b* shows a computer-generated graph of the surface in which u and v vary over these intervals. Constant u-curves appear as circular traces parallel to the xz-plane, and constant v-curves appear as lines parallel to the y-axis. ◀

REPRESENTING SURFACES OF REVOLUTION PARAMETRICALLY

The basic idea of Example 4 can be adapted to obtain parametric equations for surfaces of revolution. For example, suppose that we want to find parametric equations for the surface generated by revolving the plane curve $y = f(x)$ about the x-axis. Figure 15.4.5 suggests that the surface can be represented parametrically as

$$x = u, \quad y = f(u)\cos v, \quad z = f(u)\sin v \tag{4}$$

where v is the angle shown. In the exercises we will discuss analogous formulas for surfaces of revolution about other axes.

▶ Example 5 Find parametric equations for the surface generated by revolving the curve $y = 1/x$ about the x-axis.

Solution. From (4) this surface can be represented parametrically as

$$x = u, \quad y = \frac{1}{u}\cos v, \quad z = \frac{1}{u}\sin v$$

Figure 15.4.6 shows a computer-generated graph of the surface in which $0.7 \le u \le 5$ and $0 \le v \le 2\pi$. This surface is a portion of Gabriel's horn, which was discussed in Exercise 49 of Section 8.8. ◀

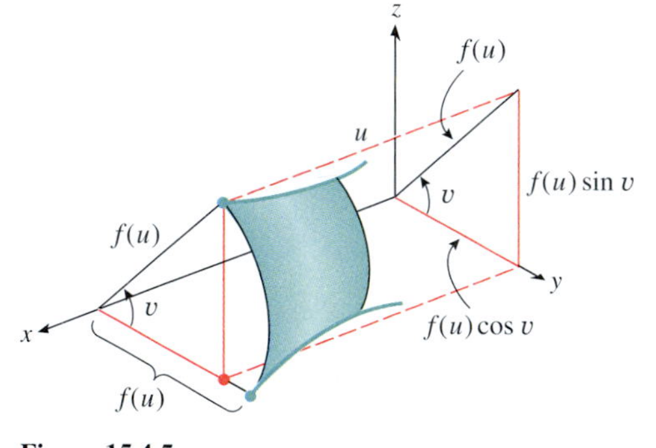

Figure 15.4.5

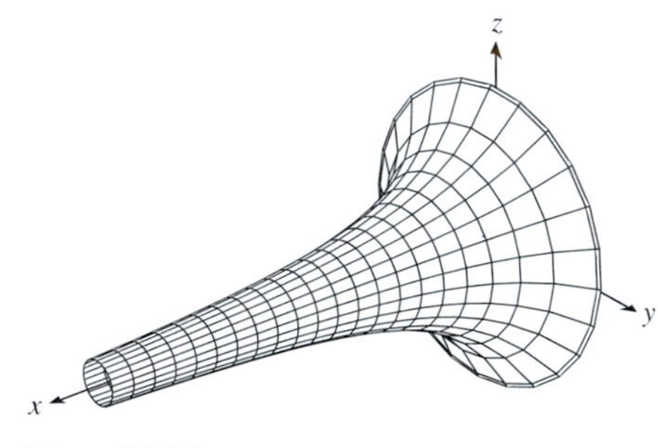

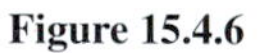

Figure 15.4.6

VECTOR-VALUED FUNCTIONS OF TWO VARIABLES

Recall that the parametric equations

$$x = x(t), \quad y = y(t), \quad z = z(t)$$

can be expressed in vector form as

$$\mathbf{r} = x(t)\mathbf{i} + y(t)\mathbf{j} + z(t)\mathbf{k}$$

where $\mathbf{r} = x\mathbf{i} + y\mathbf{j} + z\mathbf{k}$ is the radius vector and $\mathbf{r}(t) = x(t)\mathbf{i} + y(t)\mathbf{j} + z(t)\mathbf{k}$ is a vector-valued function of one variable. Similarly, the parametric equations

$$x = x(u, v), \quad y = y(u, v), \quad z = z(u, v)$$

can be expressed in vector form as

$$\mathbf{r} = x(u, v)\mathbf{i} + y(u, v)\mathbf{j} + z(u, v)\mathbf{k}$$

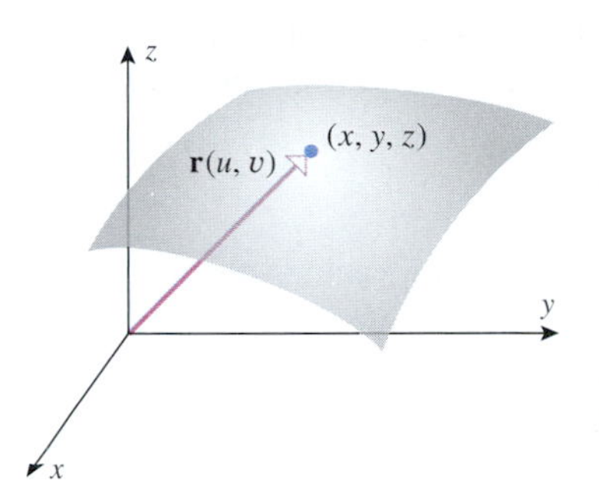

Figure 15.4.7

Here the function $\mathbf{r}(u, v) = x(u, v)\mathbf{i} + y(u, v)\mathbf{j} + z(u, v)\mathbf{k}$ is a ***vector-valued function of two variables***. We define the ***graph*** of $\mathbf{r}(u, v)$ to be the graph of the corresponding parametric equations. Geometrically, we can view $\mathbf{r}$ as a vector from the origin to a point (x, y, z) that moves over the surface $\mathbf{r} = \mathbf{r}(u, v)$ as u and v vary (Figure 15.4.7). As with vector-valued functions of one variable, we say that $\mathbf{r}(u, v)$ is ***continuous*** if each component is continuous.

▶ Example 6 The paraboloid in Example 1 was expressed parametrically as

$$x = u, \quad y = v, \quad z = 4 - u^2 - v^2$$

These equations can be expressed in vector form as

$$\mathbf{r} = u\mathbf{i} + v\mathbf{j} + (4 - u^2 - v^2)\mathbf{k} \;◀$$

PARTIAL DERIVATIVES OF VECTOR-VALUED FUNCTIONS

Partial derivatives of vector-valued functions of two variables are obtained by taking partial derivatives of the components. For example, if

$$\mathbf{r}(u, v) = x(u, v)\mathbf{i} + y(u, v)\mathbf{j} + z(u, v)\mathbf{k}$$

then

$$\frac{\partial \mathbf{r}}{\partial u} = \frac{\partial x}{\partial u}\mathbf{i} + \frac{\partial y}{\partial u}\mathbf{j} + \frac{\partial z}{\partial u}\mathbf{k}$$

$$\frac{\partial \mathbf{r}}{\partial v} = \frac{\partial x}{\partial v}\mathbf{i} + \frac{\partial y}{\partial v}\mathbf{j} + \frac{\partial z}{\partial v}\mathbf{k}$$

These derivatives can also be written as $\mathbf{r}_u$ and $\mathbf{r}_v$ or $\mathbf{r}_u(u, v)$ and $\mathbf{r}_v(u, v)$ and can be expressed as the limits

$$\frac{\partial \mathbf{r}}{\partial u} = \lim_{\Delta u \to 0} \frac{\mathbf{r}(u + \Delta u, v) - \mathbf{r}(u, v)}{\Delta u} = \lim_{w \to u} \frac{\mathbf{r}(w, v) - \mathbf{r}(u, v)}{w - u} \tag{5}$$

$$\frac{\partial \mathbf{r}}{\partial v} = \lim_{\Delta v \to 0} \frac{\mathbf{r}(u, v + \Delta v) - \mathbf{r}(u, v)}{\Delta v} = \lim_{w \to v} \frac{\mathbf{r}(u, w) - \mathbf{r}(u, v)}{w - v} \tag{6}$$

▶ Example 7 Find the partial derivatives of the vector-valued function $\mathbf{r}$ in Example 6.

Solution.

$$\frac{\partial \mathbf{r}}{\partial u} = \frac{\partial}{\partial u}[u\mathbf{i} + v\mathbf{j} + (4 - u^2 - v^2)\mathbf{k}] = \mathbf{i} - 2u\mathbf{k}$$

$$\frac{\partial \mathbf{r}}{\partial v} = \frac{\partial}{\partial v}[u\mathbf{i} + v\mathbf{j} + (4 - u^2 - v^2)\mathbf{k}] = \mathbf{j} - 2v\mathbf{k} \;◀$$

TANGENT PLANES TO PARAMETRIC SURFACES

Our next objective is to show how to find tangent planes to parametric surfaces. Let σ denote a parametric surface in 3-space, with P_0 a point on σ. We will say that a plane is ***tangent*** to σ at P_0 provided a line through P_0 lies in the plane if and only if it is a tangent line at P_0 to a curve on σ. We showed in Section 14.7 that if $z = f(x, y)$, then the graph of f has a tangent plane at a point if f is differentiable at that point. It is beyond the scope of this text to obtain precise conditions under which a parametric surface has a tangent plane at a point, so we will simply assume the existence of tangent planes at points of interest and focus on finding their equations.

Suppose that the parametric surface σ is the graph of the vector-valued function $\mathbf{r}(u, v)$ and that we are interested in the tangent plane at the point (x_0, y_0, z_0) on the surface that corresponds to the parameter values $u = u_0$ and $v = v_0$; that is,

$$\mathbf{r}(u_0, v_0) = x_0\mathbf{i} + y_0\mathbf{j} + z_0\mathbf{k}$$

If $v = v_0$ is kept fixed and u is allowed to vary, then $\mathbf{r}(u, v_0)$ is a vector-valued function of one variable whose graph is the constant v-curve through the point (u_0, v_0); similarly, if $u = u_0$ is kept fixed and v is allowed to vary, then $\mathbf{r}(u_0, v)$ is a vector-valued function of one variable whose graph is the constant u-curve through the point (u_0, v_0). Moreover, it follows from the geometric interpretation of the derivative developed in Section 13.2 that if $\partial\mathbf{r}/\partial u \neq \mathbf{0}$ at (u_0, v_0), then this vector is tangent to the constant v-curve through (u_0, v_0); and if $\partial\mathbf{r}/\partial v \neq \mathbf{0}$ at (u_0, v_0), then this vector is tangent to the constant u-curve through (u_0, v_0) (Figure 15.4.8). Thus, if $\partial\mathbf{r}/\partial u \times \partial\mathbf{r}/\partial v \neq \mathbf{0}$ at (u_0, v_0), then the vector

$$\frac{\partial\mathbf{r}}{\partial u} \times \frac{\partial\mathbf{r}}{\partial v} = \begin{vmatrix} \mathbf{i} & \mathbf{j} & \mathbf{k} \\ \dfrac{\partial x}{\partial u} & \dfrac{\partial y}{\partial u} & \dfrac{\partial z}{\partial u} \\ \dfrac{\partial x}{\partial v} & \dfrac{\partial y}{\partial v} & \dfrac{\partial z}{\partial v} \end{vmatrix} \tag{7}$$

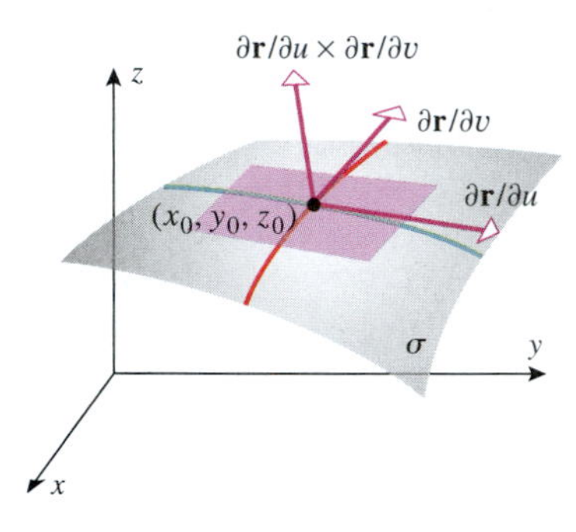

Figure 15.4.8

is orthogonal to both tangent vectors at the point (u_0, v_0) and hence is normal to the tangent plane and the surface at this point (Figure 15.4.8). Accordingly, we make the following definition.

15.4.1 DEFINITION. If a parametric surface σ is the graph of $\mathbf{r} = \mathbf{r}(u, v)$, and if $\partial\mathbf{r}/\partial u \times \partial\mathbf{r}/\partial v \neq \mathbf{0}$ at a point on the surface, then the ***principal unit normal vector*** to the surface at that point is denoted by $\mathbf{n}$ or $\mathbf{n}(u, v)$ and is defined as

$$\mathbf{n} = \frac{\dfrac{\partial\mathbf{r}}{\partial u} \times \dfrac{\partial\mathbf{r}}{\partial v}}{\left\| \dfrac{\partial\mathbf{r}}{\partial u} \times \dfrac{\partial\mathbf{r}}{\partial v} \right\|} \tag{8}$$

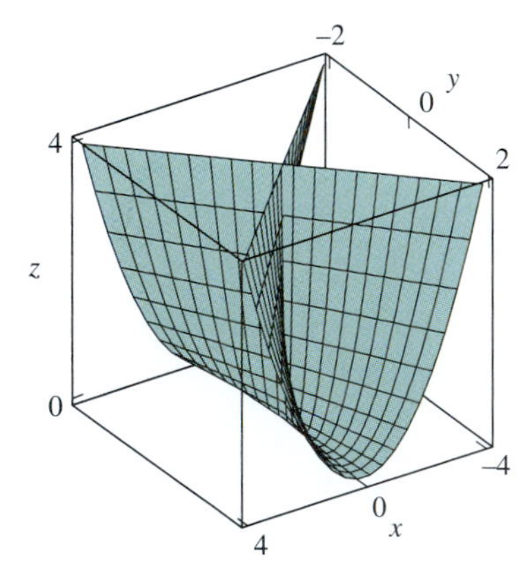

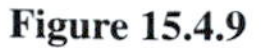
Figure 15.4.9

Example 8 Find an equation of the tangent plane to the parametric surface

$$x = uv, \quad y = u, \quad z = v^2$$

at the point where $u = 2$ and $v = -1$. This surface, called *Whitney's umbrella*, is an example of a self-intersecting parametric surface (Figure 15.4.9).

Solution. We start by writing the equations in the vector form

$$\mathbf{r} = uv\mathbf{i} + u\mathbf{j} + v^2\mathbf{k}$$

The partial derivatives of $\mathbf{r}$ are

$$\frac{\partial\mathbf{r}}{\partial u}(u, v) = v\mathbf{i} + \mathbf{j}$$

$$\frac{\partial\mathbf{r}}{\partial v}(u, v) = u\mathbf{i} + 2v\mathbf{k}$$

and at $u = 2$ and $v = -1$ these partial derivatives are

$$\frac{\partial\mathbf{r}}{\partial u}(2, -1) = -\mathbf{i} + \mathbf{j}$$

$$\frac{\partial\mathbf{r}}{\partial v}(2, -1) = 2\mathbf{i} - 2\mathbf{k}$$

Thus, from (7) and (8) a normal to the surface at this point is

$$\frac{\partial \mathbf{r}}{\partial u}(2,-1) \times \frac{\partial \mathbf{r}}{\partial v}(2,-1) = \begin{vmatrix} \mathbf{i} & \mathbf{j} & \mathbf{k} \\ -1 & 1 & 0 \\ 2 & 0 & -2 \end{vmatrix} = -2\mathbf{i} - 2\mathbf{j} - 2\mathbf{k}$$

Since any normal will suffice to find the tangent plane, it makes sense to multiply this vector by $-\frac{1}{2}$ and use the simpler normal $\mathbf{i}+\mathbf{j}+\mathbf{k}$. It follows from the given parametric equations that the point on the surface corresponding to $u = 2$ and $v = -1$ is $(-2, 2, 1)$, so the tangent plane at this point can be expressed in point-normal form as

$$(x+2)+(y-2)+(z-1)=0 \quad \text{or} \quad x+y+z=1 \blacktriangleleft$$

Convince yourself that the result obtained in Example 8 is consistent with Figure 15.4.9.

Example 9 The sphere $x^2+y^2+z^2=a^2$ can be expressed in spherical coordinates as $\rho = a$, and the spherical-to-rectangular conversion formulas in Table 12.8.1 can then be used to express the sphere as the graph of the vector-valued function

$$\mathbf{r}(\phi, \theta) = a \sin\phi \cos\theta \mathbf{i} + a \sin\phi \sin\theta \mathbf{j} + a \cos\phi \mathbf{k}$$

where $0 \le \phi \le \pi$ and $0 \le \theta \le 2\pi$ (verify). Use this function to show that the radius vector is normal to the tangent plane at each point on the sphere.

Solution. We will show that at each point of the sphere the unit normal vector $\mathbf{n}$ is a scalar multiple of $\mathbf{r}$ (and hence is parallel to $\mathbf{r}$). We have

$$\frac{\partial \mathbf{r}}{\partial \phi} \times \frac{\partial \mathbf{r}}{\partial \theta} = \begin{vmatrix} \mathbf{i} & \mathbf{j} & \mathbf{k} \\ \frac{\partial x}{\partial \phi} & \frac{\partial y}{\partial \phi} & \frac{\partial z}{\partial \phi} \\ \frac{\partial x}{\partial \theta} & \frac{\partial y}{\partial \theta} & \frac{\partial z}{\partial \theta} \end{vmatrix} = \begin{vmatrix} \mathbf{i} & \mathbf{j} & \mathbf{k} \\ a\cos\phi\cos\theta & a\cos\phi\sin\theta & -a\sin\phi \\ -a\sin\phi\sin\theta & a\sin\phi\cos\theta & 0 \end{vmatrix}$$

$$= a^2\sin^2\phi\cos\theta\mathbf{i} + a^2\sin^2\phi\sin\theta\mathbf{j} + a^2\sin\phi\cos\phi\mathbf{k}$$

and hence

$$\begin{aligned} \left\| \frac{\partial \mathbf{r}}{\partial \phi} \times \frac{\partial \mathbf{r}}{\partial \theta} \right\| &= \sqrt{a^4\sin^4\phi\cos^2\theta + a^4\sin^4\phi\sin^2\theta + a^4\sin^2\phi\cos^2\phi} \\ &= \sqrt{a^4\sin^4\phi + a^4\sin^2\phi\cos^2\phi} \\ &= a^2\sqrt{\sin^2\phi} = a^2|\sin\phi| = a^2\sin\phi \end{aligned}$$

For $\phi \ne 0$ or π, it follows from (8) that

$$\mathbf{n} = \sin\phi\cos\theta\mathbf{i} + \sin\phi\sin\theta\mathbf{j} + \cos\phi\mathbf{k} = \frac{1}{a}\mathbf{r}$$

Furthermore, the tangent planes at $\phi \ne 0$ or π are horizontal, to which $\mathbf{r} = \pm a\mathbf{k}$ is clearly normal. $\blacktriangleleft$

SURFACE AREA OF PARAMETRIC SURFACES

In Section 7.5 we obtained formulas for the surface area of a surface of revolution [see Formulas (4) and (5) and Exercise 36 in that section]. We will now obtain a formula for the surface area S of a parametric surface σ and from that formula we will then derive a formula for the surface area of a surface of the form $z = f(x, y)$.

Let σ be a parametric surface whose vector equation is

$$\mathbf{r} = x(u,v)\mathbf{i} + y(u,v)\mathbf{j} + z(u,v)\mathbf{k}$$

We will say that σ is a ***smooth parametric surface*** on a region R of the uv-plane if $\partial\mathbf{r}/\partial u$ and $\partial\mathbf{r}/\partial v$ are continuous on R and $\partial\mathbf{r}/\partial u \times \partial\mathbf{r}/\partial v \neq \mathbf{0}$ on R. Geometrically, this means that σ has a principal unit normal vector (and hence a tangent plane) for all (u, v) in R and $\mathbf{n} = \mathbf{n}(u, v)$ is a continuous function on R. Thus, on a smooth parametric surface the unit normal vector $\mathbf{n}$ varies continuously and has no abrupt changes in direction. We will derive a surface area formula for parametric surfaces that have no self-intersections and are smooth on a region R, with the possible exception that $\partial\mathbf{r}/\partial u \times \partial\mathbf{r}/\partial v$ may equal $\mathbf{0}$ on the boundary of R.

We begin by subdividing R into rectangular regions by lines parallel to the u- and v-axes and discarding any nonrectangular portions that contain points of the boundary. Assume that there are n rectangles, and let R_k denote the kth rectangle. Let (u_k, v_k) be the lower left corner of R_k, and assume that R_k has area $\Delta A_k = \Delta u_k \Delta v_k$, where Δu_k and Δv_k are the dimensions of R_k (Figure 15.4.10*a*). The image of R_k will be some *curvilinear patch* σ_k on the surface σ that has a corner at $\mathbf{r}(u_k, v_k)$; denote the area of this patch by ΔS_k (Figure 15.4.10*b*).

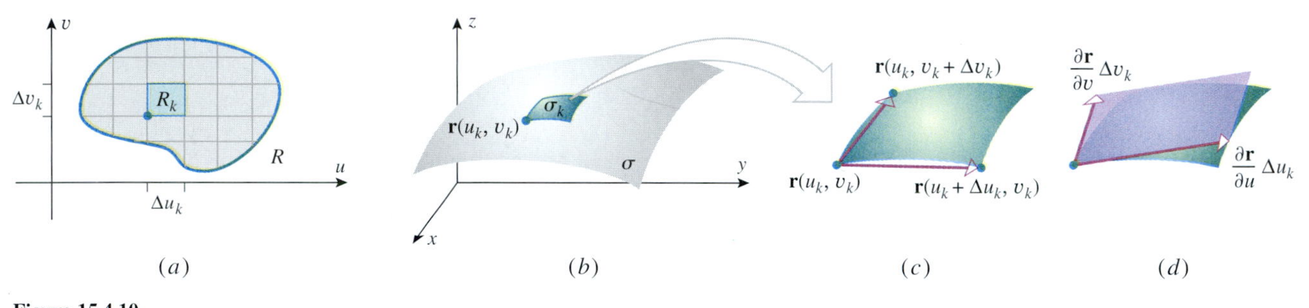

Figure 15.4.10

As suggested by Figure 15.4.10*c*, the two edges of the patch that meet at $\mathbf{r}(u_k, v_k)$ can be approximated by the "secant" vectors

$$\mathbf{r}(u_k + \Delta u_k, v_k) - \mathbf{r}(u_k, v_k)$$
$$\mathbf{r}(u_k, v_k + \Delta v_k) - \mathbf{r}(u_k, v_k)$$

and hence the area of σ_k can be approximated by the area of the parallelogram determined by these vectors. However, it follows from Formulas (5) and (6) that if Δu_k and Δv_k are small, then these secant vectors can in turn be approximated by the tangent vectors

$$\frac{\partial\mathbf{r}}{\partial u}\Delta u_k \quad \text{and} \quad \frac{\partial\mathbf{r}}{\partial v}\Delta v_k$$

where the partial derivatives are evaluated at (u_k, v_k). Thus, the area of the patch σ_k can be approximated by the area of the parallelogram determined by these vectors (Figure 15.4.10*d*); that is,

$$\Delta S_k \approx \left\| \frac{\partial\mathbf{r}}{\partial u}\Delta u_k \times \frac{\partial\mathbf{r}}{\partial v}\Delta v_k \right\| = \left\| \frac{\partial\mathbf{r}}{\partial u} \times \frac{\partial\mathbf{r}}{\partial v} \right\| \Delta u_k \Delta v_k = \left\| \frac{\partial\mathbf{r}}{\partial u} \times \frac{\partial\mathbf{r}}{\partial v} \right\| \Delta A_k \tag{9}$$

It follows that the surface area S of the entire surface σ can be approximated as

$$S \approx \sum_{k=1}^{n} \left\| \frac{\partial\mathbf{r}}{\partial u} \times \frac{\partial\mathbf{r}}{\partial v} \right\| \Delta A_k$$

Thus, if we assume that the errors in the approximations approach zero as n increases in such a way that the dimensions of the rectangles approach zero, then it is plausible that the exact value of S is

$$S = \lim_{n \to +\infty} \sum_{k=1}^{n} \left\| \frac{\partial\mathbf{r}}{\partial u} \times \frac{\partial\mathbf{r}}{\partial v} \right\| \Delta A_k$$

or, equivalently,

$$S = \iint_R \left\| \frac{\partial \mathbf{r}}{\partial u} \times \frac{\partial \mathbf{r}}{\partial v} \right\| dA \tag{10}$$

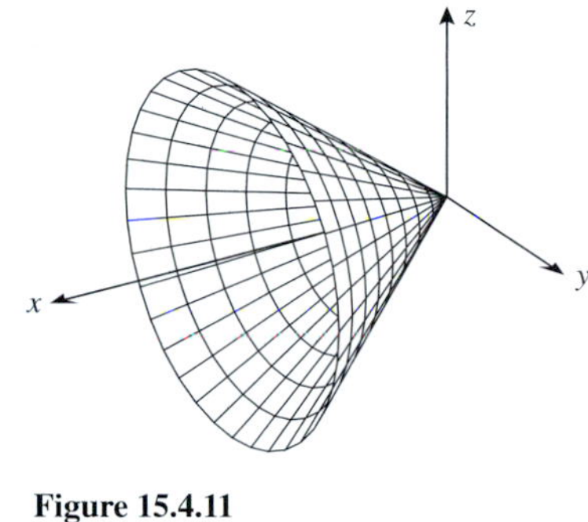

Figure 15.4.11

► **Example 10** It follows from (4) that the parametric equations

$$x = u, \quad y = u\cos v, \quad z = u\sin v$$

represent the cone that results when the line $y = x$ in the xy-plane is revolved about the x-axis. Use Formula (10) to find the surface area of that portion of the cone for which $0 \le u \le 2$ and $0 \le v \le 2\pi$ (Figure 15.4.11).

Solution. The surface can be expressed in vector form as

$$\mathbf{r} = u\mathbf{i} + u\cos v\mathbf{j} + u\sin v\mathbf{k} \quad (0 \le u \le 2,\ \ 0 \le v \le 2\pi)$$

Thus,

$$\frac{\partial \mathbf{r}}{\partial u} = \mathbf{i} + \cos v\mathbf{j} + \sin v\mathbf{k}$$

$$\frac{\partial \mathbf{r}}{\partial v} = -u\sin v\mathbf{j} + u\cos v\mathbf{k}$$

$$\frac{\partial \mathbf{r}}{\partial u} \times \frac{\partial \mathbf{r}}{\partial v} = \begin{vmatrix} \mathbf{i} & \mathbf{j} & \mathbf{k} \\ 1 & \cos v & \sin v \\ 0 & -u\sin v & u\cos v \end{vmatrix} = u\mathbf{i} - u\cos v\mathbf{j} - u\sin v\mathbf{k}$$

$$\left\| \frac{\partial \mathbf{r}}{\partial u} \times \frac{\partial \mathbf{r}}{\partial v} \right\| = \sqrt{u^2 + (-u\cos v)^2 + (-u\sin v)^2} = |u|\sqrt{2} = u\sqrt{2}$$

Thus, from (10)

$$S = \iint_R \left\| \frac{\partial \mathbf{r}}{\partial u} \times \frac{\partial \mathbf{r}}{\partial v} \right\| dA = \int_0^{2\pi} \int_0^2 \sqrt{2}u\, du\, dv = 2\sqrt{2} \int_0^{2\pi} dv = 4\pi\sqrt{2}$$ ◄

SURFACE AREA OF SURFACES OF THE FORM $z = f(x, y)$

In the case where σ is a surface of the form $z = f(x, y)$, we can take $x = u$ and $y = v$ as parameters and express the surface parametrically as

$$x = u, \quad y = v, \quad z = f(u, v)$$

or in vector form as

$$\mathbf{r} = u\mathbf{i} + v\mathbf{j} + f(u, v)\mathbf{k}$$

Thus,

$$\frac{\partial \mathbf{r}}{\partial u} = \mathbf{i} + \frac{\partial f}{\partial u}\mathbf{k} = \mathbf{i} + \frac{\partial z}{\partial x}\mathbf{k}$$

$$\frac{\partial \mathbf{r}}{\partial v} = \mathbf{j} + \frac{\partial f}{\partial v}\mathbf{k} = \mathbf{j} + \frac{\partial z}{\partial y}\mathbf{k}$$

$$\frac{\partial \mathbf{r}}{\partial u} \times \frac{\partial \mathbf{r}}{\partial v} = \begin{vmatrix} \mathbf{i} & \mathbf{j} & \mathbf{k} \\ 1 & 0 & \dfrac{\partial z}{\partial x} \\ 0 & 1 & \dfrac{\partial z}{\partial y} \end{vmatrix} = -\frac{\partial z}{\partial x}\mathbf{i} - \frac{\partial z}{\partial y}\mathbf{j} + \mathbf{k}$$

$$\left\| \frac{\partial \mathbf{r}}{\partial u} \times \frac{\partial \mathbf{r}}{\partial v} \right\| = \sqrt{\left(\frac{\partial z}{\partial x}\right)^2 + \left(\frac{\partial z}{\partial y}\right)^2 + 1}$$

In Formula (11) the region R lies in the xy-plane because the parameters are x and y. Geometrically, this region is the projection on the xy-plane of that portion of the surface $z = f(x, y)$ whose area is being determined by the formula (Figure 15.4.12).

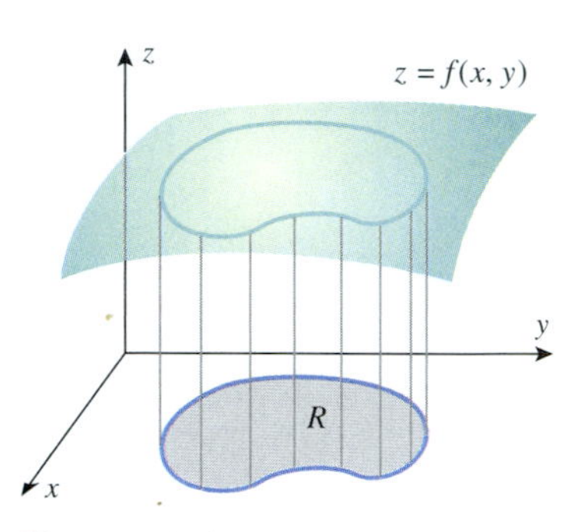

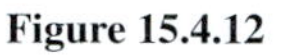

Figure 15.4.12

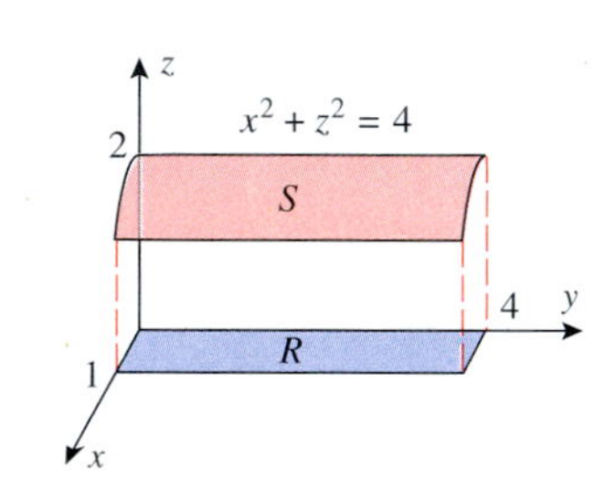

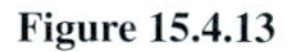

Figure 15.4.13

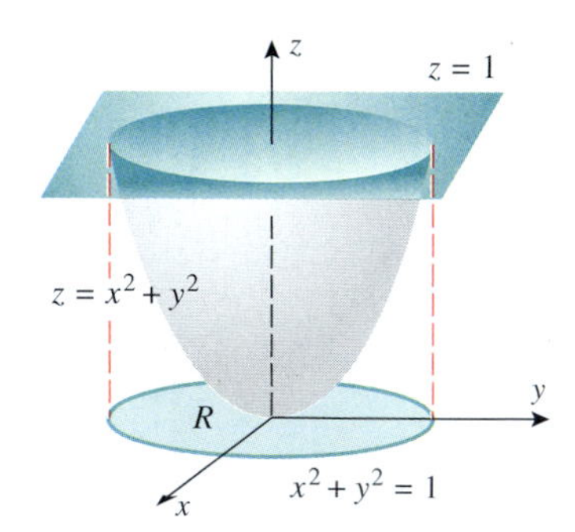

Figure 15.4.14

Thus, it follows from (10) that

$$S = \iint\limits_R \sqrt{\left(\frac{\partial z}{\partial x}\right)^2 + \left(\frac{\partial z}{\partial y}\right)^2 + 1}\, dA \tag{11}$$

▶ **Example 11** Find the surface area of that portion of the surface $z = \sqrt{4 - x^2}$ that lies above the rectangle R in the xy-plane whose coordinates satisfy $0 \le x \le 1$ and $0 \le y \le 4$.

Solution. As shown in Figure 15.4.13, the surface is a portion of the cylinder $x^2 + z^2 = 4$. It follows from (11) that the surface area is

$$\begin{aligned} S &= \iint\limits_R \sqrt{\left(\frac{\partial z}{\partial x}\right)^2 + \left(\frac{\partial z}{\partial y}\right)^2 + 1}\, dA \\ &= \iint\limits_R \sqrt{\left(-\frac{x}{\sqrt{4 - x^2}}\right)^2 + 0 + 1}\, dA = \int_0^4 \int_0^1 \frac{2}{\sqrt{4 - x^2}}\, dx\, dy \\ &= 2\int_0^4 \left[\sin^{-1}\left(\frac{1}{2}x\right)\right]_{x=0}^1 dy = 2\int_0^4 \frac{\pi}{6}\, dy = \frac{4}{3}\pi \end{aligned}$$ ◀

Formula 21 of Section 8.1

▶ **Example 12** Find the surface area of the portion of the paraboloid $z = x^2 + y^2$ below the plane $z = 1$.

Solution. The surface $z = x^2 + y^2$ is the circular paraboloid shown in Figure 15.4.14. The trace of the paraboloid in the plane $z = 1$ projects onto the circle $x^2 + y^2 = 1$ in the xy-plane, and the portion of the paraboloid that lies below the plane $z = 1$ projects onto the region R that is enclosed by this circle. Thus, it follows from (11) that the surface area is

$$S = \iint\limits_R \sqrt{4x^2 + 4y^2 + 1}\, dA$$

The expression $4x^2 + 4y^2 + 1 = 4(x^2 + y^2) + 1$ in the integrand suggests that we evaluate the integral in polar coordinates. In accordance with Formula (9) of Section 15.3, we substitute $x = r\cos\theta$ and $y = r\sin\theta$ in the integrand, replace dA by $r\,dr\,d\theta$, and find the limits of integration by expressing the region R in polar coordinates. This yields

$$\begin{aligned} S &= \int_0^{2\pi} \int_0^1 \sqrt{4r^2 + 1}\, r\, dr\, d\theta = \int_0^{2\pi} \left[\frac{1}{12}(4r^2 + 1)^{3/2}\right]_{r=0}^1 d\theta \\ &= \int_0^{2\pi} \frac{1}{12}(5\sqrt{5} - 1)\, d\theta = \frac{1}{6}\pi(5\sqrt{5} - 1) \end{aligned}$$ ◀

✔ QUICK CHECK EXERCISES 15.4 *(See page 1056 for answers.)*

1. Consider the surface represented parametrically by

$$x = 1 - u$$
$$y = (1 - u)\cos v \qquad (0 \le u \le 1,\ 0 \le v \le 2\pi)$$
$$z = (1 - u)\sin v$$

(a) Describe the constant u-curves.
(b) Describe the constant v-curves.

2. If

$$\mathbf{r}(u, v) = (1-u)\mathbf{i} + [(1-u)\cos v]\mathbf{j} + [(1-u)\sin v]\mathbf{k}$$

then

$$\frac{\partial \mathbf{r}}{\partial u} = \_\_\_\_\_\_ \quad \text{and} \quad \frac{\partial \mathbf{r}}{\partial v} = \_\_\_\_\_\_$$

3. If

$$\mathbf{r}(u, v) = (1-u)\mathbf{i} + [(1-u)\cos v]\mathbf{j} + [(1-u)\sin v]\mathbf{k}$$

the principal unit normal to the graph of $\mathbf{r}$ at the point where $u = 1/2$ and $v = \pi/6$ is given by ______.

4. Suppose σ is a parametric surface with vector equation

$$\mathbf{r}(u, v) = x(u, v)\mathbf{i} + y(u, v)\mathbf{j} + z(u, v)\mathbf{k}$$

If σ has no self-intersections and σ is smooth on a region R in the uv-plane, then the surface area of σ is given by

$$S = \iint_R \_\_\_\_\_\_ \, dA$$

5. The surface area of a surface of the form $z = f(x, y)$ over a region R in the xy-plane is given by

$$S = \iint_R \_\_\_\_\_\_ \, dA$$

EXERCISE SET 15.4

Graphing Utility · C CAS

1–2 Sketch the parametric surface.

1. (a) $x = u,\ y = v,\ z = \sqrt{u^2 + v^2}$
   (b) $x = u,\ y = \sqrt{u^2 + v^2},\ z = v$
   (c) $x = \sqrt{u^2 + v^2},\ y = u,\ z = v$
2. (a) $x = u,\ y = v,\ z = u^2 + v^2$
   (b) $x = u,\ y = u^2 + v^2,\ z = v$
   (c) $x = u^2 + v^2,\ y = u,\ z = v$

3–4 Find a parametric representation of the surface in terms of the parameters $u = x$ and $v = y$.

3. (a) $2z - 3x + 4y = 5$ (b) $z = x^2$
4. (a) $z + zx^2 - y = 0$ (b) $y^2 - 3z = 5$
5. (a) Find parametric equations for the portion of the cylinder $x^2 + y^2 = 5$ that extends between the planes $z = 0$ and $z = 1$.
   (b) Find parametric equations for the portion of the cylinder $x^2 + z^2 = 4$ that extends between the planes $y = 1$ and $y = 3$.
6. (a) Find parametric equations for the portion of the plane $x + y = 1$ that extends between the planes $z = -1$ and $z = 1$.
   (b) Find parametric equations for the portion of the plane $y - 2z = 5$ that extends between the planes $x = 0$ and $x = 3$.
7. Find parametric equations for the surface generated by revolving the curve $y = \sin x$ about the x-axis.
8. Find parametric equations for the surface generated by revolving the curve $y - e^x = 0$ about the x-axis.

9–14 Find a parametric representation of the surface in terms of the parameters r and θ, where (r, θ, z) are the cylindrical coordinates of a point on the surface.

9. $z = \dfrac{1}{1 + x^2 + y^2}$
10. $z = e^{-(x^2+y^2)}$
11. $z = 2xy$
12. $z = x^2 - y^2$
13. The portion of the sphere $x^2 + y^2 + z^2 = 9$ on or above the plane $z = 2$.
14. The portion of the cone $z = \sqrt{x^2 + y^2}$ on or below the plane $z = 3$.
15. Find a parametric representation of the cone

$$z = \sqrt{3x^2 + 3y^2}$$

in terms of parameters ρ and θ, where (ρ, θ, ϕ) are spherical coordinates of a point on the surface.

16. Describe the cylinder $x^2 + y^2 = 9$ in terms of parameters θ and ϕ, where (ρ, θ, ϕ) are spherical coordinates of a point on the surface.

FOCUS ON CONCEPTS

17–22 Eliminate the parameters to obtain an equation in rectangular coordinates, and describe the surface.

17. $x = 2u + v,\ y = u - v,\ z = 3v$ for $-\infty < u < +\infty$ and $-\infty < v < +\infty$.
18. $x = u \cos v,\ y = u^2,\ z = u \sin v$ for $0 \le u \le 2$ and $0 \le v < 2\pi$.
19. $x = 3 \sin u,\ y = 2 \cos u,\ z = 2v$ for $0 \le u < 2\pi$ and $1 \le v \le 2$.
20. $x = \sqrt{u} \cos v,\ y = \sqrt{u} \sin v,\ z = u$ for $0 \le u \le 4$ and $0 \le v < 2\pi$.
21. $\mathbf{r}(u, v) = 3u \cos v\mathbf{i} + 4u \sin v\mathbf{j} + u\mathbf{k}$ for $0 \le u \le 1$ and $0 \le v < 2\pi$.
22. $r(u, v) = \sin u \cos v\mathbf{i} + 2 \sin u \sin v\mathbf{j} + 3 \cos u\mathbf{k}$ for $0 \le u \le \pi$ and $0 \le v < 2\pi$.
23. The accompanying figure shows the graphs of two parametric representations of the cone $z = \sqrt{x^2 + y^2}$ for $0 \le z \le 2$.

(a) Find parametric equations that produce reasonable facsimiles of these surfaces.
(b) Use a graphing utility to check your answer in part (a).

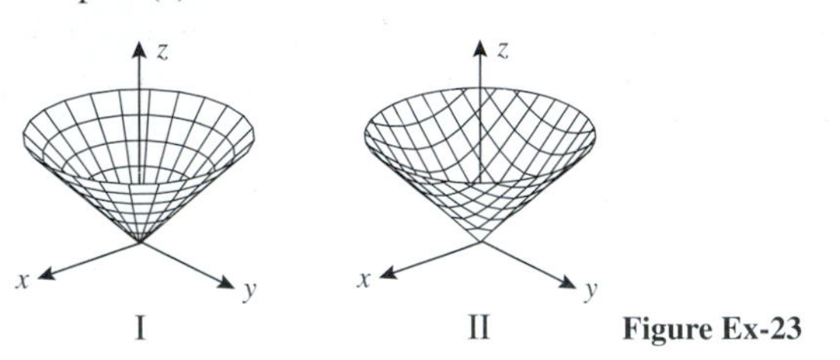

Figure Ex-23

24. The accompanying figure shows the graphs of two parametric representations of the paraboloid $z = x^2 + y^2$ for $0 \leq z \leq 2$.
(a) Find parametric equations that produce reasonable facsimiles of these surfaces.
(b) Use a graphing utility to check your answer in part (a).

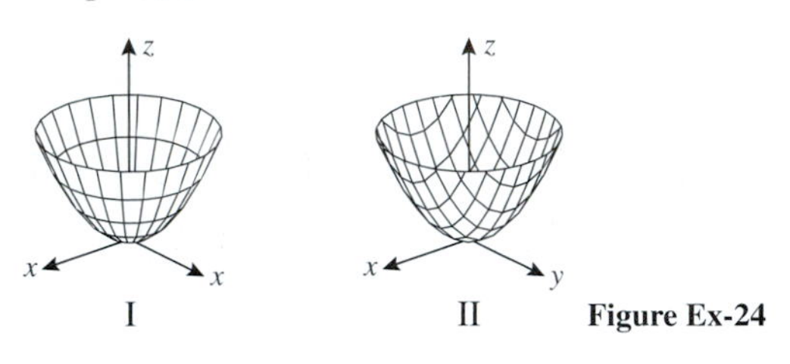

Figure Ex-24

25. In each part, the figure shows a portion of the parametric surface $x = 3\cos v$, $y = u$, $z = 3\sin v$. Find restrictions on u and v that produce the surface, and check your answer with a graphing utility.

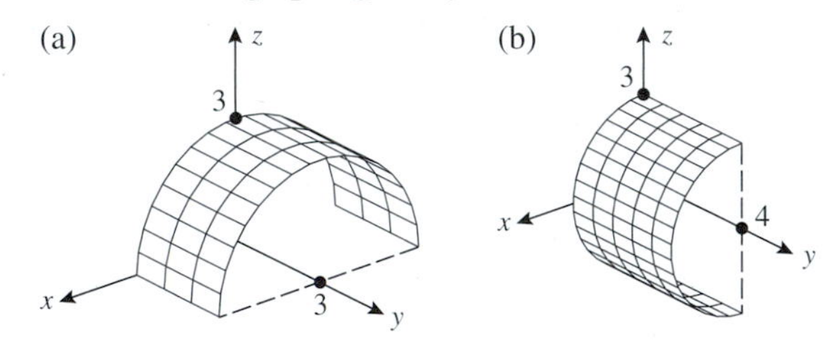

26. In each part, the figure shows a portion of the parametric surface $x = 3\cos v$, $y = 3\sin v$, $z = u$. Find restrictions on u and v that produce the surface, and check your answer with a graphing utility.

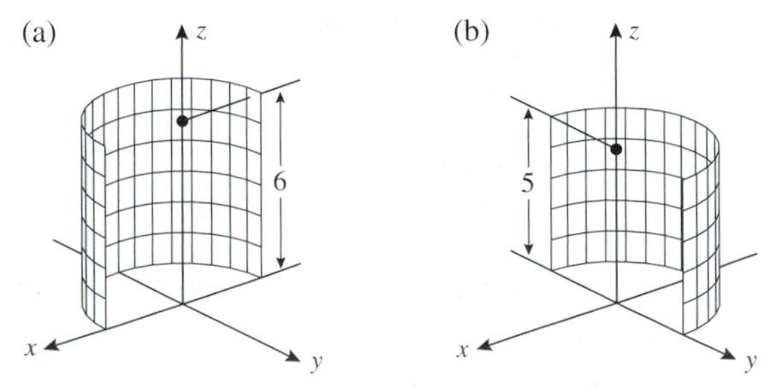

27. In each part, the figure shows a hemisphere that is a portion of the sphere $x = \sin\phi\cos\theta$, $y = \sin\phi\sin\theta$, $z = \cos\phi$. Find restrictions on ϕ and θ that produce the hemisphere, and check your answer with a graphing utility.

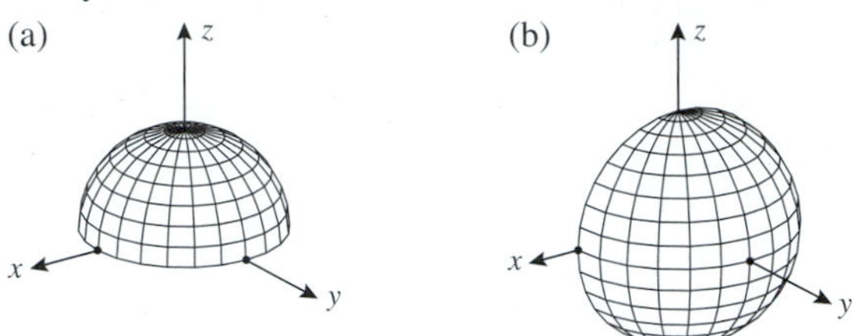

28. In each part, the figure shows a portion of the sphere $x = \sin\phi\cos\theta$, $y = \sin\phi\sin\theta$, $z = \cos\phi$. Find restrictions on ϕ and θ that produce the surface, and check your answer with a graphing utility.

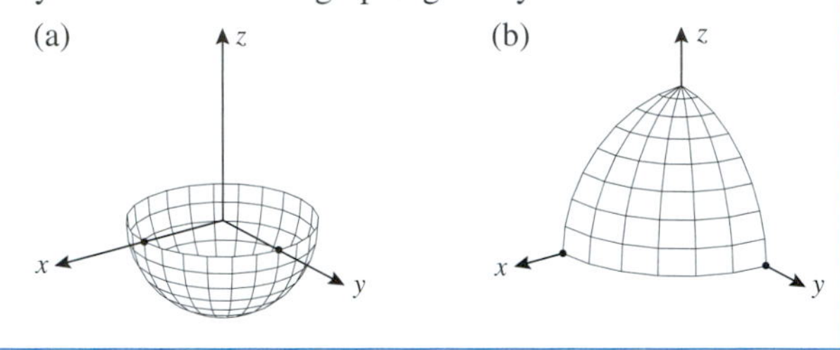

29–34 Find an equation of the tangent plane to the parametric surface at the stated point.

29. $x = u,\ y = v,\ z = u^2 + v^2$; $(1, 2, 5)$

30. $x = u^2,\ y = v^2,\ z = u + v$; $(1, 4, 3)$

31. $x = 3v\sin u,\ y = 2v\cos u,\ z = u^2$; $(0, 2, 0)$

32. $\mathbf{r} = uv\mathbf{i} + (u - v)\mathbf{j} + (u + v)\mathbf{k}$; $u = 1, v = 2$

33. $\mathbf{r} = u\cos v\mathbf{i} + u\sin v\mathbf{j} + v\mathbf{k}$; $u = 1/2, v = \pi/4$

34. $\mathbf{r} = uv\mathbf{i} + ue^v\mathbf{j} + ve^u\mathbf{k}$; $u = \ln 2, v = 0$

35–46 Find the area of the given surface.

35. The portion of the cylinder $y^2 + z^2 = 9$ that is above the rectangle $R = \{(x, y) : 0 \leq x \leq 2, -3 \leq y \leq 3\}$.

36. The portion of the plane $2x + 2y + z = 8$ in the first octant.

37. The portion of the cone $z^2 = 4x^2 + 4y^2$ that is above the region in the first quadrant bounded by the line $y = x$ and the parabola $y = x^2$.

38. The portion of the cone $z = \sqrt{x^2 + y^2}$ that lies inside the cylinder $x^2 + y^2 = 2x$.

39. The portion of the paraboloid $z = 1 - x^2 - y^2$ that is above the xy-plane.

40. The portion of the surface $z = 2x + y^2$ that is above the triangular region with vertices $(0, 0)$, $(0, 1)$, and $(1, 1)$.

41. The portion of the paraboloid

$$\mathbf{r}(u, v) = u\cos v\mathbf{i} + u\sin v\mathbf{j} + u^2\mathbf{k}$$

for which $1 \leq u \leq 2$, $0 \leq v \leq 2\pi$.

42. The portion of the cone

$$\mathbf{r}(u, v) = u\cos v\mathbf{i} + u\sin v\mathbf{j} + u\mathbf{k}$$

for which $0 \leq u \leq 2v$, $0 \leq v \leq \pi/2$.

43. The portion of the surface $z = xy$ that is above the sector in the first quadrant bounded by the lines $y = x/\sqrt{3}$, $y = 0$, and the circle $x^2 + y^2 = 9$.

44. The portion of the paraboloid $2z = x^2 + y^2$ that is inside the cylinder $x^2 + y^2 = 8$.

45. The portion of the sphere $x^2 + y^2 + z^2 = 16$ between the planes $z = 1$ and $z = 2$.

46. The portion of the sphere $x^2 + y^2 + z^2 = 8$ that is inside the cone $z = \sqrt{x^2 + y^2}$.

47. Use parametric equations to derive the formula for the surface area of a sphere of radius a.

48. Use parametric equations to derive the formula for the lateral surface area of a right circular cylinder of radius r and height h.

49. The portion of the surface

$$z = \frac{h}{a}\sqrt{x^2 + y^2} \quad (a, h > 0)$$

between the xy-plane and the plane $z = h$ is a right circular cone of height h and radius a. Use a double integral to show that the lateral surface area of this cone is $S = \pi a\sqrt{a^2 + h^2}$.

50. The accompanying figure shows the ***torus*** that is generated by revolving the circle

$$(x - a)^2 + z^2 = b^2 \quad (0 < b < a)$$

in the xz-plane about the z-axis.

(a) Show that this torus can be expressed parametrically as

$$x = (a + b\cos v)\cos u$$
$$y = (a + b\cos v)\sin u$$
$$z = b\sin v$$

where u and v are the parameters shown in the figure and $0 \le u \le 2\pi$, $0 \le v \le 2\pi$.

(b) Use a graphing utility to generate a torus.

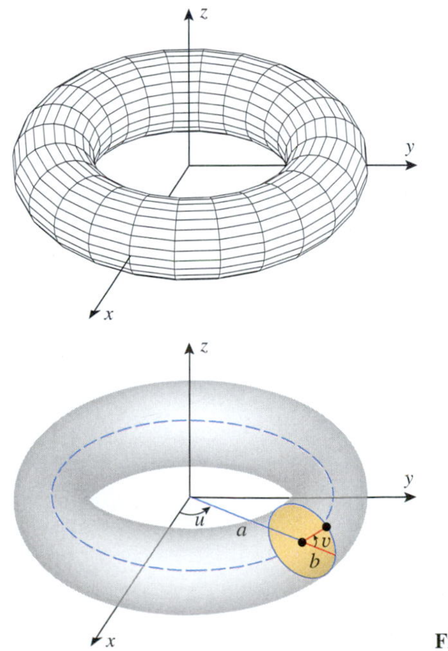

Figure Ex-50

51. Find the surface area of the torus in Exercise 50(a).

52. Use a CAS to graph the ***helicoid***

$$x = u\cos v, \quad y = u\sin v, \quad z = v$$

for $0 \le u \le 5$ and $0 \le v \le 4\pi$ (see the accompanying figure), and then use the numerical double integration operation of the CAS to approximate the surface area.

53. Use a CAS to graph the ***pseudosphere***

$$x = \cos u \sin v$$
$$y = \sin u \sin v$$
$$z = \cos v + \ln\left(\tan\frac{v}{2}\right)$$

for $0 \le u \le 2\pi$, $0 < v < \pi$ (see the accompanying figure), and then use the numerical double integration operation of the CAS to approximate the surface area between the planes $z = -1$ and $z = 1$.

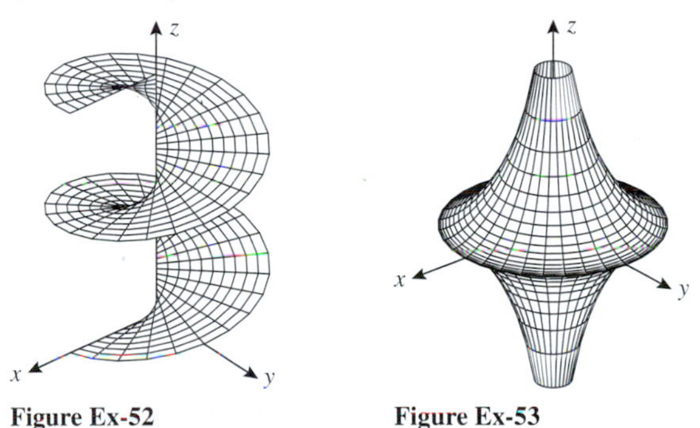

Figure Ex-52 Figure Ex-53

54. The accompanying figure shows the graph of an ***astroidal sphere***

$$x^{2/3} + y^{2/3} + z^{2/3} = a^{2/3}$$

(a) Show that this surface can be represented parametrically as

$$x = a(\sin u \cos v)^3$$
$$y = a(\sin u \sin v)^3 \qquad (0 \le u \le \pi,\ 0 \le v \le 2\pi)$$
$$z = a(\cos u)^3$$

(b) Use a CAS to approximate the surface area in the case where $a = 1$.

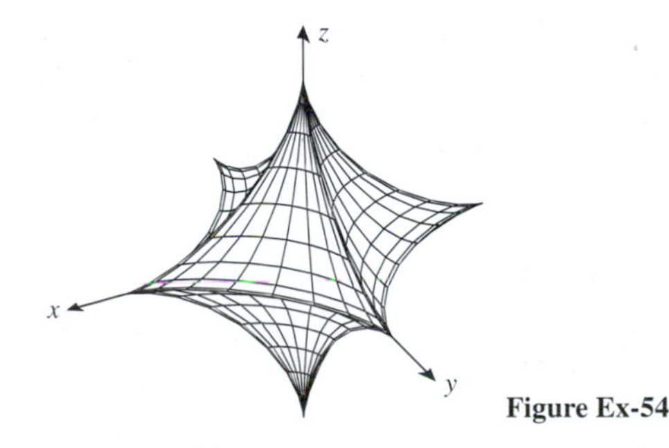

Figure Ex-54

55. (a) Describe the surface that is represented by the parametric equations

$$x = a\sin\phi\cos\theta$$
$$y = b\sin\phi\sin\theta \qquad (0 \le \phi \le \pi,\ 0 \le \theta \le 2\pi)$$
$$z = c\cos\phi$$

where $a > 0$, $b > 0$, and $c > 0$.

(b) Use a CAS to approximate the area of the surface for $a = 2, b = 3, c = 4$.

56. (a) Find parametric equations for the surface of revolution that is generated by revolving the curve $z = f(x)$ in the xz-plane about the z-axis.
(b) Use the result obtained in part (a) to find parametric equations for the surface of revolution that is generated by revolving the curve $z = 1/x^2$ in the xz-plane about the z-axis.
(c) Use a graphing utility to check your work by graphing the parametric surface.

57–59 The parametric equations in these exercises represent a quadric surface for positive values of a, b, and c. Identify the type of surface by eliminating the parameters u and v. Check your conclusion by choosing specific values for the constants and generating the surface with a graphing utility.

57. $x = a\cos u\cos v,\ \ y = b\sin u\cos v,\ \ z = c\sin v$

58. $x = a\cos u\cosh v,\ \ y = b\sin u\cosh v,\ \ z = c\sinh v$

59. $x = a\sinh v,\ \ y = b\sinh u\cosh v,\ \ z = c\cosh u\cosh v$

✔QUICK CHECK ANSWERS 15.4

1. (a) The constant u-curves are circles of radius $1 - u$ centered at $(1 - u, 0, 0)$ and parallel to the yz-plane. (b) The constant v-curves are line segments joining the points $(1, \cos v, \sin v)$ and $(0, 0, 0)$. **2.** $\dfrac{\partial \mathbf{r}}{\partial u} = -\mathbf{i} - (\cos v)\mathbf{j} - (\sin v)\mathbf{k}$; $\dfrac{\partial \mathbf{r}}{\partial v} = -[(1-u)\sin v]\mathbf{j} + [(1-u)\cos v]\mathbf{k}$ **3.** $\dfrac{1}{\sqrt{8}}(-2\mathbf{i} + \sqrt{3}\mathbf{j} + \mathbf{k})$ **4.** $\left\| \dfrac{\partial \mathbf{r}}{\partial u} \times \dfrac{\partial \mathbf{r}}{\partial v} \right\|$ **5.** $\sqrt{\left(\dfrac{\partial z}{\partial x}\right)^2 + \left(\dfrac{\partial z}{\partial y}\right)^2 + 1}$

15.5 TRIPLE INTEGRALS

In the preceding sections we defined and discussed properties of double integrals for functions of two variables. In this section we will define triple integrals for functions of three variables.

DEFINITION OF A TRIPLE INTEGRAL

A single integral of a function $f(x)$ is defined over a finite closed interval on the x-axis, and a double integral of a function $f(x, y)$ is defined over a finite closed region R in the xy-plane. Our first goal in this section is to define what is meant by a *triple integral* of $f(x, y, z)$ over a closed solid region G in an xyz-coordinate system. To ensure that G does not extend indefinitely in some direction, we will assume that it can be enclosed in a suitably large box whose sides are parallel to the coordinate planes (Figure 15.5.1). In this case we say that G is a ***finite solid***.

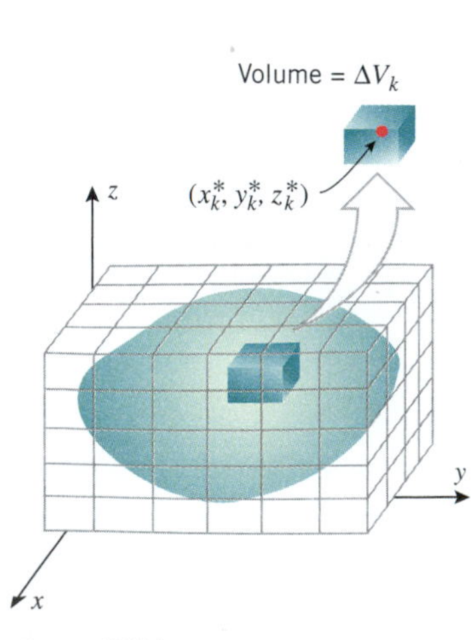

Figure 15.5.1

To define the triple integral of $f(x, y, z)$ over G, we first divide the box into n "subboxes" by planes parallel to the coordinate planes. We then discard those subboxes that contain any points outside of G and choose an arbitrary point in each of the remaining subboxes. As shown in Figure 15.5.1, we denote the volume of the kth remaining subbox by ΔV_k and the point selected in the kth subbox by (x_k^*, y_k^*, z_k^*). Next we form the product

$$f(x_k^*, y_k^*, z_k^*)\Delta V_k$$

for each subbox, then add the products for all of the subboxes to obtain the ***Riemann sum***

$$\sum_{k=1}^{n} f(x_k^*, y_k^*, z_k^*)\Delta V_k$$

Finally, we repeat this process with more and more subdivisions in such a way that the length, width, and height of each subbox approach zero, and n approaches $+\infty$. The limit

$$\iiint_G f(x, y, z)\,dV = \lim_{n\to+\infty} \sum_{k=1}^{n} f(x_k^*, y_k^*, z_k^*)\Delta V_k \tag{1}$$

is called the ***triple integral*** of $f(x, y, z)$ over the region G. Conditions under which the triple integral exists are studied in advanced calculus. However, for our purposes it suffices to say that existence is ensured when f is continuous on G and the region G is not too "complicated."

PROPERTIES OF TRIPLE INTEGRALS

Triple integrals enjoy many properties of single and double integrals:

$$\iiint_G cf(x, y, z)\,dV = c\iiint_G f(x, y, z)\,dV \quad (c \text{ a constant})$$

$$\iiint_G [f(x, y, z) + g(x, y, z)]\,dV = \iiint_G f(x, y, z)\,dV + \iiint_G g(x, y, z)\,dV$$

$$\iiint_G [f(x, y, z) - g(x, y, z)]\,dV = \iiint_G f(x, y, z)\,dV - \iiint_G g(x, y, z)\,dV$$

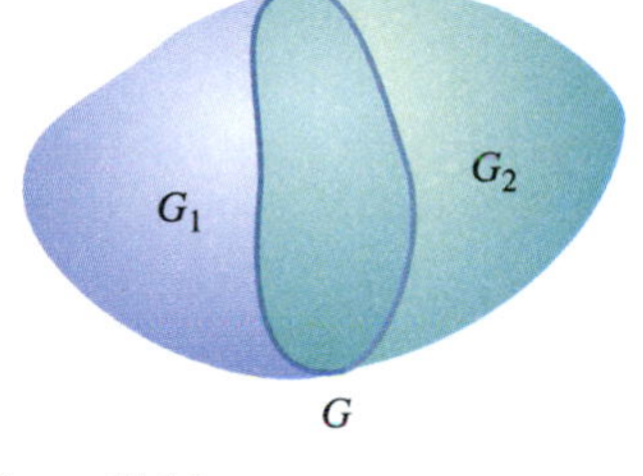

Figure 15.5.2

Moreover, if the region G is subdivided into two subregions G_1 and G_2 (Figure 15.5.2), then

$$\iiint_G f(x, y, z)\,dV = \iiint_{G_1} f(x, y, z)\,dV + \iiint_{G_2} f(x, y, z)\,dV$$

We omit the proofs.

EVALUATING TRIPLE INTEGRALS OVER RECTANGULAR BOXES

Just as a double integral can be evaluated by two successive single integrations, so a triple integral can be evaluated by three successive integrations. The following theorem, which we state without proof, is the analog of Theorem 15.1.3.

There are two possible orders of integration for the iterated integrals in Theorem 15.1.3:

$dx\,dy, \quad dy\,dx$

Six orders of integration are possible for the iterated integral in Theorem 15.5.1:

$dx\,dy\,dz, \quad dy\,dz\,dx, \quad dz\,dx\,dy$
$dx\,dz\,dy, \quad dz\,dy\,dx, \quad dy\,dx\,dz$

15.5.1 THEOREM. *Let G be the rectangular box defined by the inequalities*

$$a \le x \le b, \quad c \le y \le d, \quad k \le z \le l$$

If f is continuous on the region G, then

$$\iiint_G f(x, y, z)\,dV = \int_a^b \int_c^d \int_k^l f(x, y, z)\,dz\,dy\,dx \tag{2}$$

Moreover, the iterated integral on the right can be replaced with any of the five other iterated integrals that result by altering the order of integration.

▶ Example 1 Evaluate the triple integral

$$\iiint_G 12xy^2z^3\,dV$$

over the rectangular box G defined by the inequalities $-1 \le x \le 2$, $0 \le y \le 3$, $0 \le z \le 2$.

Solution. Of the six possible iterated integrals we might use, we will choose the one in (2). Thus, we will first integrate with respect to z, holding x and y fixed, then with respect to y, holding x fixed, and finally with respect to x.

$$\begin{aligned}\iiint_G 12xy^2z^3\,dV &= \int_{-1}^{2}\int_0^3\int_0^2 12xy^2z^3\,dz\,dy\,dx \\ &= \int_{-1}^{2}\int_0^3 \left[3xy^2z^4\right]_{z=0}^2\,dy\,dx = \int_{-1}^{2}\int_0^3 48xy^2\,dy\,dx \\ &= \int_{-1}^{2}\left[16xy^3\right]_{y=0}^3\,dx = \int_{-1}^{2} 432x\,dx \\ &= 216x^2\Big]_{-1}^{2} = 648 \end{aligned}$$

EVALUATING TRIPLE INTEGRALS OVER MORE GENERAL REGIONS

Next we will consider how triple integrals can be evaluated over solids that are not rectangular boxes. For the moment we will limit our discussion to solids of the type shown in Figure 15.5.3. Specifically, we will assume that the solid G is bounded above by a surface $z = g_2(x, y)$ and below by a surface $z = g_1(x, y)$ and that the projection of the solid on the xy-plane is a type I or type II region R (see Definition 15.2.1). In addition, we will assume that $g_1(x, y)$ and $g_2(x, y)$ are continuous on R and that $g_1(x, y) \le g_2(x, y)$ on R. Geometrically, this means that the surfaces may touch but cannot cross. We call a solid of this type a ***simple xy-solid***.

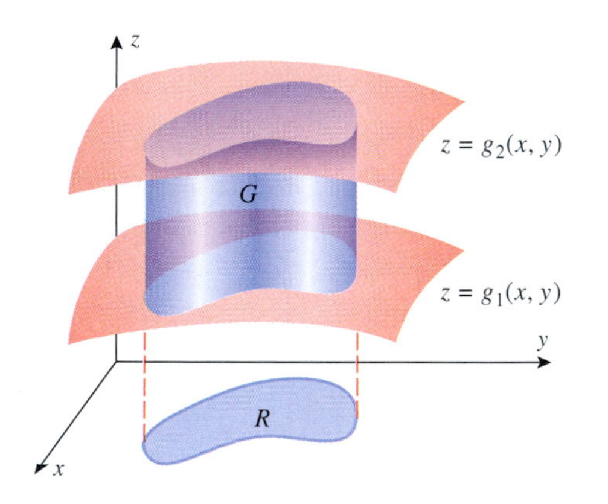

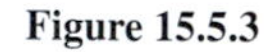

Figure 15.5.3

The following theorem, which we state without proof, will enable us to evaluate triple integrals over simple xy-solids.

15.5.2 THEOREM. *Let G be a simple xy-solid with upper surface $z = g_2(x, y)$ and lower surface $z = g_1(x, y)$, and let R be the projection of G on the xy-plane. If $f(x, y, z)$ is continuous on G, then*

$$\iiint_G f(x, y, z)\,dV = \iint_R \left[\int_{g_1(x,y)}^{g_2(x,y)} f(x, y, z)\,dz\right] dA \tag{3}$$

In (3), the first integration is with respect to z, after which a function of x and y remains. This function of x and y is then integrated over the region R in the xy-plane. To apply (3), it is helpful to begin with a three-dimensional sketch of the solid G. The limits of integration can be obtained from the sketch as follows:

Determining Limits of Integration: Simple xy-Solid

Step 1. Find an equation $z = g_2(x, y)$ for the upper surface and an equation $z = g_1(x, y)$ for the lower surface of G. The functions $g_1(x, y)$ and $g_2(x, y)$ determine the lower and upper z-limits of integration.

Step 2. Make a two-dimensional sketch of the projection R of the solid on the xy-plane. From this sketch determine the limits of integration for the double integral over R in (3).

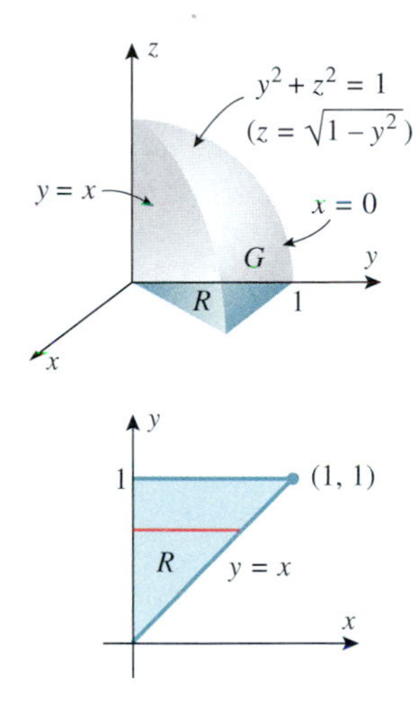

Figure 15.5.4

Example 2 Let G be the wedge in the first octant that is cut from the cylindrical solid $y^2 + z^2 \leq 1$ by the planes $y = x$ and $x = 0$. Evaluate

$$\iiint_G z\, dV$$

Solution. The solid G and its projection R on the xy-plane are shown in Figure 15.5.4. The upper surface of the solid is formed by the cylinder and the lower surface by the xy-plane. Since the portion of the cylinder $y^2 + z^2 = 1$ that lies above the xy-plane has the equation $z = \sqrt{1 - y^2}$, and the xy-plane has the equation $z = 0$, it follows from (3) that

$$\iiint_G z\, dV = \iint_R \left[\int_0^{\sqrt{1-y^2}} z\, dz \right] dA \tag{4}$$

For the double integral over R, the x- and y-integrations can be performed in either order, since R is both a type I and type II region. We will integrate with respect to x first. With this choice, (4) yields

$$\iiint_G z\, dV = \int_0^1 \int_0^y \int_0^{\sqrt{1-y^2}} z\, dz\, dx\, dy = \int_0^1 \int_0^y \frac{1}{2} z^2 \Big]_{z=0}^{\sqrt{1-y^2}} dx\, dy$$

$$= \int_0^1 \int_0^y \frac{1}{2}(1 - y^2)\, dx\, dy = \frac{1}{2} \int_0^1 (1 - y^2)x \Big]_{x=0}^{y} dy$$

$$= \frac{1}{2} \int_0^1 (y - y^3)\, dy = \frac{1}{2} \left[\frac{1}{2} y^2 - \frac{1}{4} y^4 \right]_0^1 = \frac{1}{8} \blacktriangleleft$$

TECHNOLOGY MASTERY

Most computer algebra systems have a built-in capability for computing iterated triple integrals. If you have a CAS, consult the relevant documentation and use the CAS to check Examples 1 and 2.

VOLUME CALCULATED AS A TRIPLE INTEGRAL

Triple integrals have many physical interpretations, some of which we will consider in the next section. However, in the special case where $f(x, y, z) = 1$, Formula (1) yields

$$\iiint_G dV = \lim_{n \to +\infty} \sum_{k=1}^{n} \Delta V_k$$

which Figure 15.5.1 suggests is the volume of G; that is,

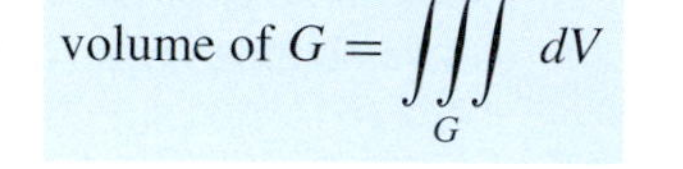

$$\text{volume of } G = \iiint_G dV \tag{5}$$

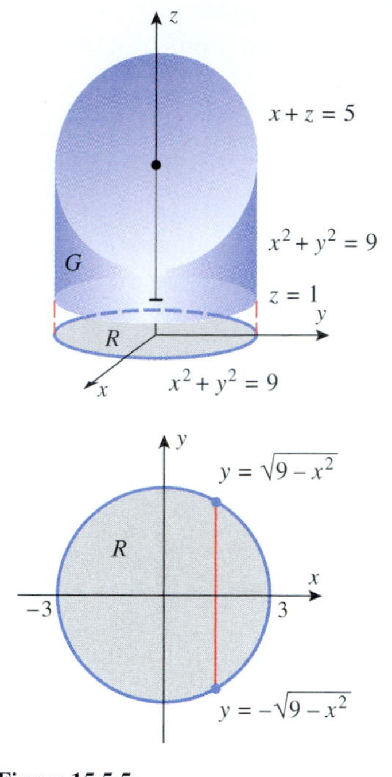

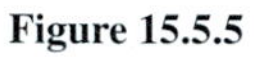
Figure 15.5.5

▶ **Example 3** Use a triple integral to find the volume of the solid within the cylinder $x^2 + y^2 = 9$ and between the planes $z = 1$ and $x + z = 5$.

Solution. The solid G and its projection R on the xy-plane are shown in Figure 15.5.5. The lower surface of the solid is the plane $z = 1$ and the upper surface is the plane $x + z = 5$ or, equivalently, $z = 5 - x$. Thus, from (3) and (5)

$$\text{volume of } G = \iiint_G dV = \iint_R \left[\int_1^{5-x} dz\right] dA \tag{6}$$

For the double integral over R, we will integrate with respect to y first. Thus, (6) yields

$$\begin{aligned}
\text{volume of } G &= \int_{-3}^{3}\int_{-\sqrt{9-x^2}}^{\sqrt{9-x^2}}\int_1^{5-x} dz\,dy\,dx = \int_{-3}^{3}\int_{-\sqrt{9-x^2}}^{\sqrt{9-x^2}} z\Big]_{z=1}^{5-x} dy\,dx \\
&= \int_{-3}^{3}\int_{-\sqrt{9-x^2}}^{\sqrt{9-x^2}} (4-x)\,dy\,dx = \int_{-3}^{3}(8-2x)\sqrt{9-x^2}\,dx \\
&= 8\int_{-3}^{3}\sqrt{9-x^2}\,dx - \int_{-3}^{3}2x\sqrt{9-x^2}\,dx \\
&= 8\left(\frac{9}{2}\pi\right) - \int_{-3}^{3}2x\sqrt{9-x^2}\,dx \\
&= 8\left(\frac{9}{2}\pi\right) - 0 = 36\pi \blacktriangleleft
\end{aligned}$$

For the first integral, see Formula (3) of Section 8.4.

The second integral is 0 because the integrand is an odd function.

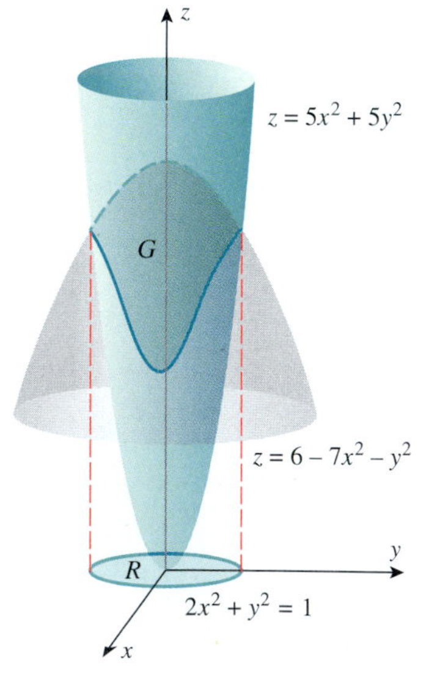

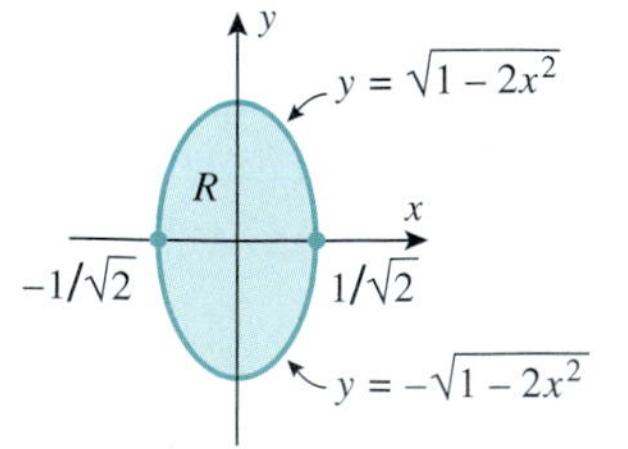

Figure 15.5.6

▶ **Example 4** Find the volume of the solid enclosed between the paraboloids

$$z = 5x^2 + 5y^2 \quad \text{and} \quad z = 6 - 7x^2 - y^2$$

Solution. The solid G and its projection R on the xy-plane are shown in Figure 15.5.6. The projection R is obtained by solving the given equations simultaneously to determine where the paraboloids intersect. We obtain

$$5x^2 + 5y^2 = 6 - 7x^2 - y^2$$

or

$$2x^2 + y^2 = 1 \tag{7}$$

which tells us that the paraboloids intersect in a curve on the elliptic cylinder given by (7).

The projection of this intersection on the xy-plane is an ellipse with this same equation. Therefore,

$$\text{volume of } G = \iiint\limits_G dV = \iint\limits_R \left[\int_{5x^2+5y^2}^{6-7x^2-y^2} dz\right] dA$$

$$= \int_{-1/\sqrt{2}}^{1/\sqrt{2}} \int_{-\sqrt{1-2x^2}}^{\sqrt{1-2x^2}} \int_{5x^2+5y^2}^{6-7x^2-y^2} dz\, dy\, dx$$

$$= \int_{-1/\sqrt{2}}^{1/\sqrt{2}} \int_{-\sqrt{1-2x^2}}^{\sqrt{1-2x^2}} (6 - 12x^2 - 6y^2)\, dy\, dx$$

$$= \int_{-1/\sqrt{2}}^{1/\sqrt{2}} \left[6(1-2x^2)y - 2y^3\right]_{y=-\sqrt{1-2x^2}}^{\sqrt{1-2x^2}} dx$$

$$= 8\int_{-1/\sqrt{2}}^{1/\sqrt{2}} (1-2x^2)^{3/2}\, dx = \frac{8}{\sqrt{2}} \int_{-\pi/2}^{\pi/2} \cos^4\theta\, d\theta = \frac{3\pi}{\sqrt{2}} \blacktriangleleft$$

Let $x = \frac{1}{\sqrt{2}}\sin\theta$.

Use the Wallis cosine formula in Exercise 66 of Section 8.3.

INTEGRATION IN OTHER ORDERS

In Formula (3) for integrating over a simple xy-solid, the z-integration was performed first. However, there are situations in which it is preferable to integrate in a different order. For example, Figure 15.5.7*a* shows a ***simple xz-solid***, and Figure 15.5.7*b* shows a ***simple yz-solid***. For a simple xz-solid it is usually best to integrate with respect to y first, and for a simple yz-solid it is usually best to integrate with respect to x first:

$$\underset{\substack{G \\ \text{simple } xz\text{-solid}}}{\iiint} f(x,y,z)\, dV = \iint\limits_R \left[\int_{g_1(x,z)}^{g_2(x,z)} f(x,y,z)\, dy\right] dA \qquad (8)$$

$$\underset{\substack{G \\ \text{simple } yz\text{-solid}}}{\iiint} f(x,y,z)\, dV = \iint\limits_R \left[\int_{g_1(y,z)}^{g_2(y,z)} f(x,y,z)\, dx\right] dA \qquad (9)$$

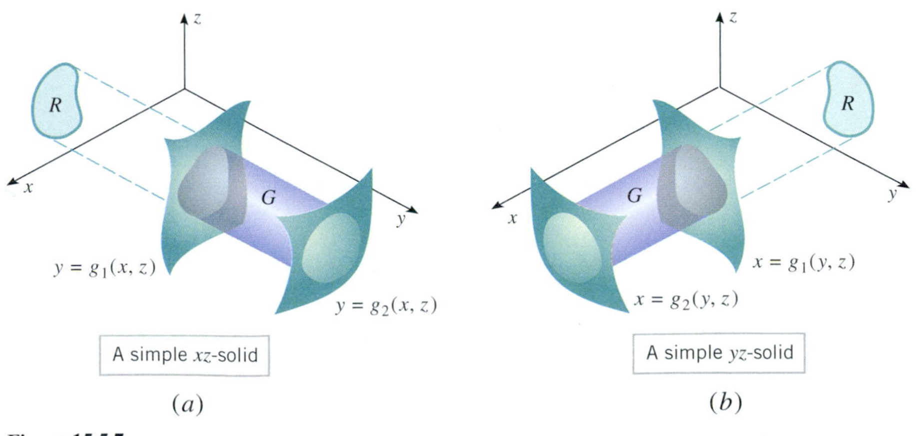

Figure 15.5.7

Sometimes a solid G can be viewed as a simple xy-solid, a simple xz-solid, and a simple yz-solid, in which case the order of integration can be chosen to simplify the computations.

Example 5 In Example 2, we evaluated

$$\iiint_G z\,dV$$

over the wedge in Figure 15.5.4 by integrating first with respect to z. Evaluate this integral by integrating first with respect to x.

Solution. The solid is bounded in the back by the plane $x = 0$ and in the front by the plane $x = y$, so

$$\iiint_G z\,dV = \iint_R \left[\int_0^y z\,dx\right] dA$$

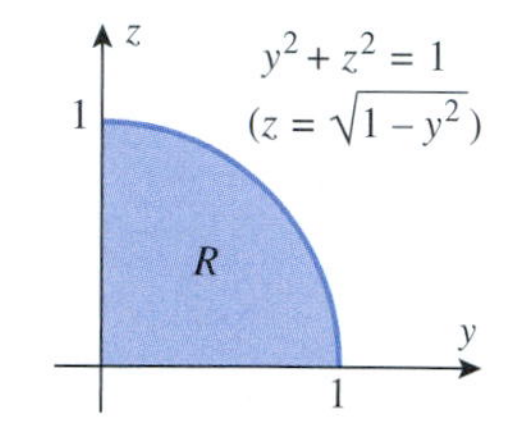

Figure 15.5.8

where R is the projection of G on the yz-plane (Figure 15.5.8). The integration over R can be performed first with respect to z and then y or vice versa. Performing the z-integration first yields

$$\iiint_G z\,dV = \int_0^1 \int_0^{\sqrt{1-y^2}} \int_0^y z\,dx\,dz\,dy = \int_0^1 \int_0^{\sqrt{1-y^2}} zx\Big]_{x=0}^{y} dz\,dy$$

$$= \int_0^1 \int_0^{\sqrt{1-y^2}} zy\,dz\,dy = \int_0^1 \frac{1}{2}z^2y\Big]_{z=0}^{\sqrt{1-y^2}} dy = \int_0^1 \frac{1}{2}(1-y^2)y\,dy = \frac{1}{8}$$

which agrees with the result in Example 2. ◀

✔QUICK CHECK EXERCISES 15.5 (See page 1064 for answers.)

1. The iterated integral
$$\int_1^5 \int_2^4 \int_3^6 f(x,y,z)\,dx\,dz\,dy$$
integrates f over the rectangular box defined by
____ $\le x \le$ ____, ____ $\le y \le$ ____,
____ $\le z \le$ ____

2. Let G be the solid in the first octant bounded below by the surface $z = y + x^2$ and bounded above by the plane $z = 4$. Supply the missing limits of integration.
(a) $\iiint_G f(x,y,z)\,dA = \int_\square^\square \int_\square^\square \int_{y+x^2}^4 f(x,y,z)\,dz\,dx\,dy$
(b) $\iiint_G f(x,y,z)\,dA = \int_\square^\square \int_\square^\square \int_{y+x^2}^4 f(x,y,z)\,dz\,dy\,dx$
(c) $\iiint_G f(x,y,z)\,dA = \int_\square^\square \int_\square^\square \int_\square^\square f(x,y,z)\,dy\,dz\,dx$

3. The volume of the solid G in Quick Check Exercises 2 is ____.

EXERCISE SET 15.5 [C] CAS

1–8 Evaluate the iterated integral.

1. $\int_{-1}^1 \int_0^2 \int_0^1 (x^2+y^2+z^2)\,dx\,dy\,dz$

2. $\int_{1/3}^{1/2} \int_0^\pi \int_0^1 zx \sin xy\,dz\,dy\,dx$

3. $\int_0^2 \int_{-1}^{y^2} \int_{-1}^z yz\,dx\,dz\,dy$

4. $\int_0^{\pi/4} \int_0^1 \int_0^{x^2} x\cos y\,dz\,dx\,dy$

5. $\int_0^3 \int_0^{\sqrt{9-z^2}} \int_0^x xy\,dy\,dx\,dz$

6. $\int_1^3 \int_x^{x^2} \int_0^{\ln z} xe^y\,dy\,dz\,dx$

7. $\int_0^2 \int_0^{\sqrt{4-x^2}} \int_{-5+x^2+y^2}^{3-x^2-y^2} x\,dz\,dy\,dx$

8. $\int_1^2 \int_z^2 \int_0^{\sqrt{3}y} \frac{y}{x^2+y^2}\,dx\,dy\,dz$

9–12 Evaluate the triple integral.

9. $\iiint_G xy \sin yz\, dV$, where G is the rectangular box defined by the inequalities $0 \le x \le \pi$, $0 \le y \le 1$, $0 \le z \le \pi/6$.

10. $\iiint_G y\, dV$, where G is the solid enclosed by the plane $z = y$, the xy-plane, and the parabolic cylinder $y = 1 - x^2$.

11. $\iiint_G xyz\, dV$, where G is the solid in the first octant that is bounded by the parabolic cylinder $z = 2 - x^2$ and the planes $z = 0$, $y = x$, and $y = 0$.

12. $\iiint_G \cos(z/y)\, dV$, where G is the solid defined by the inequalities $\pi/6 \le y \le \pi/2$, $y \le x \le \pi/2$, $0 \le z \le xy$.

C **13.** Use the numerical triple integral operation of a CAS to approximate

$$\iiint_G \frac{\sqrt{x+z^2}}{y}\, dV$$

where G is the rectangular box defined by the inequalities $0 \le x \le 3$, $1 \le y \le 2$, $-2 \le z \le 1$.

C **14.** Use the numerical triple integral operation of a CAS to approximate

$$\iiint_G e^{-x^2-y^2-z^2}\, dV$$

where G is the spherical region $x^2 + y^2 + z^2 \le 1$.

15–18 Use a triple integral to find the volume of the solid.

15. The solid in the first octant bounded by the coordinate planes and the plane $3x + 6y + 4z = 12$.

16. The solid bounded by the surface $z = \sqrt{y}$ and the planes $x + y = 1$, $x = 0$, and $z = 0$.

17. The solid bounded by the surface $y = x^2$ and the planes $y + z = 4$ and $z = 0$.

18. The wedge in the first octant that is cut from the solid cylinder $y^2 + z^2 \le 1$ by the planes $y = x$ and $x = 0$.

FOCUS ON CONCEPTS

19. Let G be the solid enclosed by the surfaces in the accompanying figure. Fill in the missing limits of integration.

(a) $$\iiint_G f(x,y,z)\, dV = \int_{\square}^{\square}\int_{\square}^{\square}\int_{\square}^{\square} f(x,y,z)\, dz\, dy\, dx$$

(b) $$\iiint_G f(x,y,z)\, dV = \int_{\square}^{\square}\int_{\square}^{\square}\int_{\square}^{\square} f(x,y,z)\, dz\, dx\, dy$$

20. Let G be the solid enclosed by the surfaces in the accompanying figure. Fill in the missing limits of integration.

(a) $$\iiint_G f(x,y,z)\, dV = \int_{\square}^{\square}\int_{\square}^{\square}\int_{\square}^{\square} f(x,y,z)\, dz\, dy\, dx$$

(b) $$\iiint_G f(x,y,z)\, dV = \int_{\square}^{\square}\int_{\square}^{\square}\int_{\square}^{\square} f(x,y,z)\, dz\, dx\, dy$$

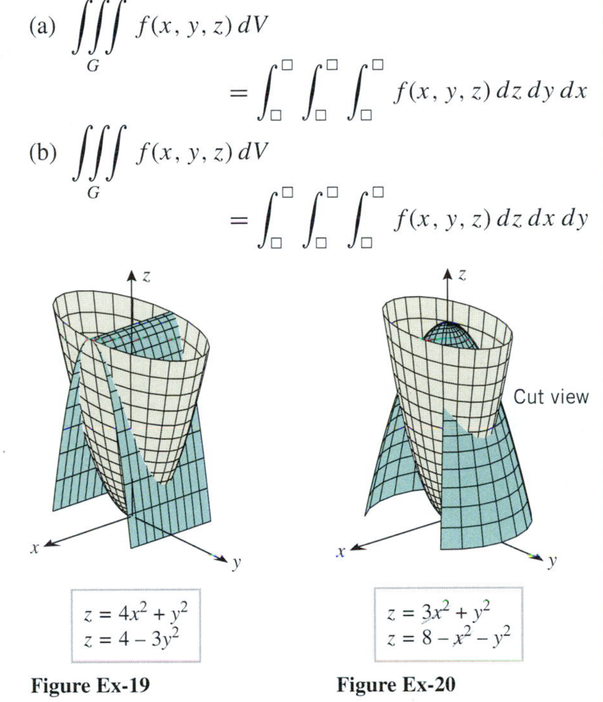

Figure Ex-19

Figure Ex-20

21–24 Set up (but do not evaluate) an iterated triple integral for the volume of the solid enclosed between the given surfaces.

21. The surfaces in Exercise 19.

22. The surfaces in Exercise 20.

23. The elliptic cylinder $x^2 + 9y^2 = 9$ and the planes $z = 0$ and $z = x + 3$.

24. The cylinders $x^2 + y^2 = 1$ and $x^2 + z^2 = 1$.

25–26 In each part, sketch the solid whose volume is given by the integral.

25. (a) $$\int_{-1}^{1}\int_{-\sqrt{1-x^2}}^{\sqrt{1-x^2}}\int_{0}^{y+1} dz\, dy\, dx$$

(b) $$\int_{0}^{9}\int_{0}^{y/3}\int_{0}^{\sqrt{y^2-9x^2}} dz\, dx\, dy$$

(c) $$\int_{0}^{1}\int_{0}^{\sqrt{1-x^2}}\int_{0}^{2} dy\, dz\, dx$$

26. (a) $$\int_{0}^{3}\int_{x^2}^{9}\int_{0}^{2} dz\, dy\, dx$$

(b) $$\int_{0}^{2}\int_{0}^{2-y}\int_{0}^{2-x-y} dz\, dx\, dy$$

(c) $$\int_{-2}^{2}\int_{0}^{4-y^2}\int_{0}^{2} dx\, dz\, dy$$

27–30 The ***average value*** or ***mean value*** of a continuous function $f(x, y, z)$ over a solid G is defined as

$$f_{\text{ave}} = \frac{1}{V(G)} \iiint_G f(x, y, z)\, dV$$

where $V(G)$ is the volume of the solid G (compare to the definition preceding Exercise 61 of Section 15.2). Use this definition in these exercises.

27. Find the average value of $f(x, y, z) = x + y + z$ over the tetrahedron shown in the accompanying figure.

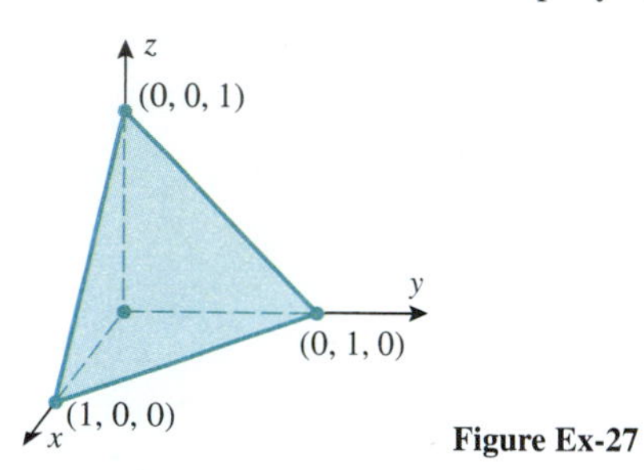

Figure Ex-27

28. Find the average value of $f(x, y, z) = xyz$ over the spherical region $x^2 + y^2 + z^2 \le 1$.

C **29.** Use the numerical triple integral operation of a CAS to approximate the average distance from the origin to a point in the solid of Example 4.

C **30.** Let $d(x, y, z)$ be the distance from the point (z, z, z) to the point $(x, y, 0)$. Use the numerical triple integral operation of a CAS to approximate the average value of d for $0 \le x \le 1$, $0 \le y \le 1$, and $0 \le z \le 1$. Write a short explanation as to why this value may be considered to be the average distance between a point on the diagonal from $(0, 0, 0)$ to $(1, 1, 1)$ and a point on the face in the xy-plane for the unit cube $0 \le x \le 1$, $0 \le y \le 1$, and $0 \le z \le 1$.

31. Let G be the tetrahedron in the first octant bounded by the coordinate planes and the plane

$$\frac{x}{a} + \frac{y}{b} + \frac{z}{c} = 1 \quad (a > 0, b > 0, c > 0)$$

(a) List six different iterated integrals that represent the volume of G.

(b) Evaluate any one of the six to show that the volume of G is $\frac{1}{6}abc$.

32. Use a triple integral to derive the formula for the volume of the ellipsoid

$$\frac{x^2}{a^2} + \frac{y^2}{b^2} + \frac{z^2}{c^2} = 1$$

FOCUS ON CONCEPTS

33–34 Express each integral as an equivalent integral in which the z-integration is performed first, the y-integration second, and the x-integration last.

33. (a) $\displaystyle\int_0^5 \int_0^2 \int_0^{\sqrt{4-y^2}} f(x, y, z)\, dx\, dy\, dz$

(b) $\displaystyle\int_0^9 \int_0^{3-\sqrt{x}} \int_0^{z} f(x, y, z)\, dy\, dz\, dx$

(c) $\displaystyle\int_0^4 \int_y^{8-y} \int_0^{\sqrt{4-y}} f(x, y, z)\, dx\, dz\, dy$

34. (a) $\displaystyle\int_0^3 \int_0^{\sqrt{9-z^2}} \int_0^{\sqrt{9-y^2-z^2}} f(x, y, z)\, dx\, dy\, dz$

(b) $\displaystyle\int_0^4 \int_0^2 \int_0^{x/2} f(x, y, z)\, dy\, dz\, dx$

(c) $\displaystyle\int_0^4 \int_0^{4-y} \int_0^{\sqrt{z}} f(x, y, z)\, dx\, dz\, dy$

C **35.** (a) Find the region G over which the triple integral

$$\iiint_G (1 - x^2 - y^2 - z^2)\, dV$$

has its maximum value.

(b) Use the numerical triple integral operation of a CAS to approximate the maximum value.

(c) Find the exact maximum value.

36. Let G be the rectangular box defined by the inequalities $a \le x \le b$, $c \le y \le d$, $k \le z \le l$. Show that

$$\iiint_G f(x)g(y)h(z)\, dV = \left[\int_a^b f(x)\, dx\right]\left[\int_c^d g(y)\, dy\right]\left[\int_k^l h(z)\, dz\right]$$

37. Use the result of Exercise 36 to evaluate

(a) $\displaystyle\iiint_G xy^2 \sin z\, dV$, where G is the set of points satisfying $-1 \le x \le 1$, $0 \le y \le 1$, $0 \le z \le \pi/2$;

(b) $\displaystyle\iiint_G e^{2x+y-z}\, dV$, where G is the set of points satisfying $0 \le x \le 1$, $0 \le y \le \ln 3$, $0 \le z \le \ln 2$.

✔ QUICK CHECK ANSWERS 15.5

1. $3 \le x \le 6,\ 1 \le y \le 5,\ 2 \le z \le 4$ **2.** (a) $\displaystyle\int_0^4 \int_0^{\sqrt{4-y}} \int_{y+x^2}^{4} f(x, y, z)\, dz\, dx\, dy$ (b) $\displaystyle\int_0^2 \int_0^{4-x^2} \int_{y+x^2}^{4} f(x, y, z)\, dz\, dy\, dx$ (c) $\displaystyle\int_0^2 \int_{x^2}^{4} \int_0^{z-x^2} f(x, y, z)\, dy\, dz\, dx$ **3.** $\dfrac{128}{15}$

15.6 CENTROID, CENTER OF GRAVITY, THEOREM OF PAPPUS

Suppose that a rigid physical body is acted on by a gravitational field. Because the body is composed of many particles, each of which is affected by gravity, the action of a constant gravitational field on the body consists of a large number of forces distributed over the entire body. However, these individual forces can be replaced by a single force acting at a point called the ***center of gravity*** *of the body. In this section we will show how double and triple integrals can be used to locate centers of gravity.*

DENSITY OF A LAMINA

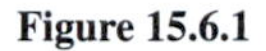
Figure 15.6.1

Let us consider an idealized flat object that is thin enough to be viewed as a two-dimensional plane region (Figure 15.6.1). Such an object is called a ***lamina***. A lamina is called ***homogeneous*** if its composition is uniform throughout and ***inhomogeneous*** otherwise. The ***density*** of a *homogeneous* lamina is defined to be its mass per unit area. Thus, the density δ of a homogeneous lamina of mass M and area A is given by $\delta = M/A$.

For an inhomogeneous lamina the composition may vary from point to point, and hence an appropriate definition of "density" must reflect this. To motivate such a definition, suppose that the lamina is placed in an xy-plane. The density at a point (x, y) can be specified by a function $\delta(x, y)$, called the ***density function***, which can be interpreted as follows. Construct a small rectangle centered at (x, y) and let ΔM and ΔA be the mass and area of the portion of the lamina enclosed by this rectangle (Figure 15.6.2). If the ratio $\Delta M/\Delta A$ approaches a limiting value as the dimensions (and hence the area) of the rectangle approach zero, then this limit is considered to be the density of the lamina at (x, y). Symbolically,

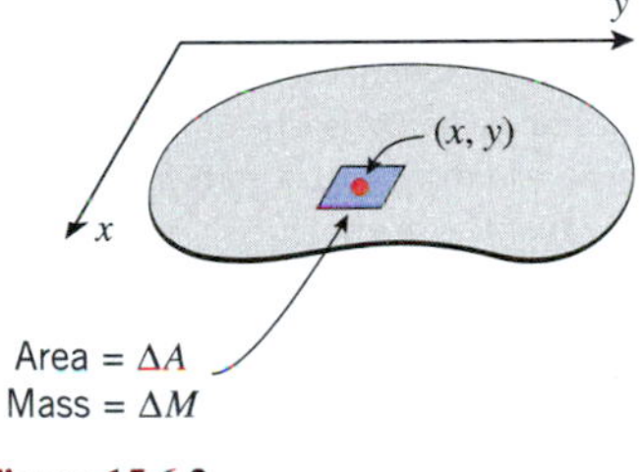

Figure 15.6.2

$$\delta(x, y) = \lim_{\Delta A \to 0} \frac{\Delta M}{\Delta A} \tag{1}$$

From this relationship we obtain the approximation

$$\Delta M \approx \delta(x, y)\Delta A \tag{2}$$

which relates the mass and area of a small rectangular portion of the lamina centered at (x, y). It is assumed that as the dimensions of the rectangle tend to zero, the error in this approximation also tends to zero.

MASS OF A LAMINA

The following result shows how to find the mass of a lamina from its density function.

> **15.6.1 MASS OF A LAMINA.** If a lamina with a continuous density function $\delta(x, y)$ occupies a region R in the xy-plane, then its total mass M is given by
>
> $$M = \iint_R \delta(x, y)\, dA \tag{3}$$

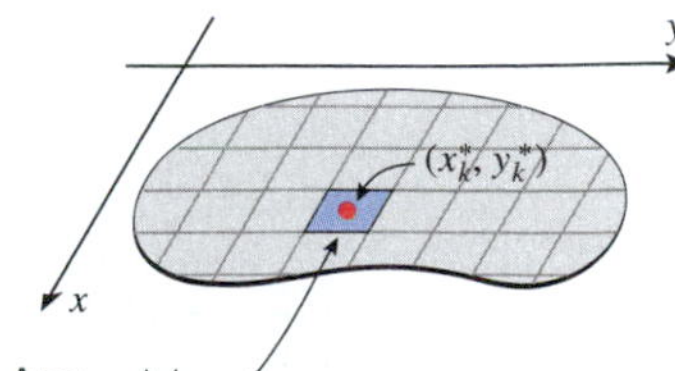

Figure 15.6.3

This formula can be motivated by a familiar limiting process that can be outlined as follows: Imagine the lamina to be subdivided into rectangular pieces using lines parallel to the coordinate axes and excluding from consideration any nonrectangular parts at the boundary (Figure 15.6.3). Assume that there are n such rectangular pieces, and suppose that the kth piece has area ΔA_k. If we let (x_k^*, y_k^*) denote the center of the kth piece, then from Formula (2), the mass ΔM_k of this piece can be approximated by

$$\Delta M_k \approx \delta(x_k^*, y_k^*)\Delta A_k \tag{4}$$

and hence the mass M of the entire lamina can be approximated by

$$M \approx \sum_{k=1}^{n} \delta(x_k^*, y_k^*)\Delta A_k$$

If we now increase n in such a way that the dimensions of the rectangles tend to zero, then it is plausible that the errors in our approximations will approach zero, so

$$M = \lim_{n \to +\infty} \sum_{k=1}^{n} \delta(x_k^*, y_k^*)\Delta A_k = \iint_R \delta(x, y)\, dA$$

▶ **Example 1** A triangular lamina with vertices (0, 0), (0, 1), and (1, 0) has density function $\delta(x, y) = xy$. Find its total mass.

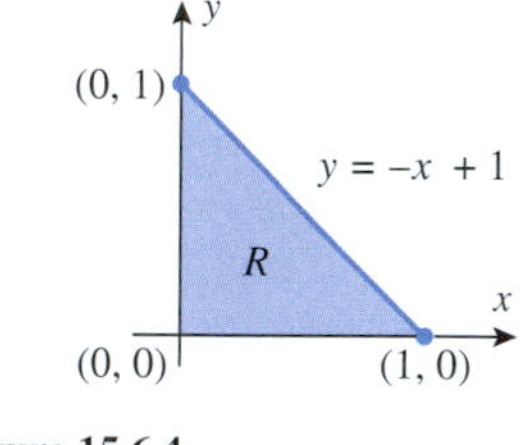

Figure 15.6.4

Solution. Referring to (3) and Figure 15.6.4, the mass M of the lamina is

$$M = \iint_R \delta(x, y)\, dA = \iint_R xy\, dA = \int_0^1 \int_0^{-x+1} xy\, dy\, dx$$

$$= \int_0^1 \left[\frac{1}{2}xy^2\right]_{y=0}^{-x+1} dx = \int_0^1 \left[\frac{1}{2}x^3 - x^2 + \frac{1}{2}x\right] dx = \frac{1}{24} \text{ (unit of mass)} ◀$$

CENTER OF GRAVITY OF A LAMINA

Assume that the acceleration due to the force of gravity is constant and acts downward, and suppose that a lamina occupies a region R in a horizontal xy-plane. It can be shown that there exists a unique point $(\bar{x}, \bar{y})$ (which may or may not belong to R) such that the effect of gravity on the lamina is "equivalent" to that of a single force acting at the point $(\bar{x}, \bar{y})$. This point is called the ***center of gravity*** of the lamina, and if it is in R, then the lamina will balance horizontally on the point of a support placed at $(\bar{x}, \bar{y})$. For example, the center of gravity of a disk of uniform density is at the center of the disk, and the center of gravity of a rectangular region of uniform density is at the center of the rectangle. For an irregularly shaped lamina or for a lamina in which the density varies from point to point, locating the center of gravity requires calculus.

> **15.6.2 PROBLEM.** Suppose that a lamina with a continuous density function $\delta(x, y)$ occupies a region R in a horizontal xy-plane. Find the coordinates $(\bar{x}, \bar{y})$ of the center of gravity.

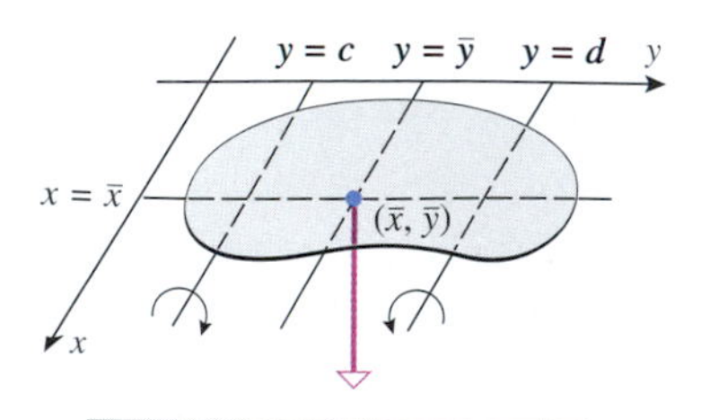

Force of gravity acting on the center of gravity of the lamina

Figure 15.6.5

To motivate the solution, consider what happens if we try to balance the lamina on a knife-edge parallel to the x-axis. Suppose the lamina in Figure 15.6.5 is placed on a knife-edge along a line $y = c$ that does not pass through the center of gravity. Because the lamina behaves as if its entire mass is concentrated at the center of gravity $(\bar{x}, \bar{y})$, the lamina will be rotationally unstable and the force of gravity will cause a rotation about $y = c$. Similarly, the lamina will undergo a rotation if placed on a knife-edge along $y = d$. However, if the knife-edge runs along the line $y = \bar{y}$ through the center of gravity, the lamina will be in perfect balance. Similarly, the lamina will be in perfect balance on a knife-edge along the line $x = \bar{x}$ through the center of gravity. This suggests that the center of gravity of a lamina can be determined as the intersection of two lines of balance, one parallel to the x-axis and the other parallel to the y-axis. In order to find these lines of balance, we will need some preliminary results about rotations.

Children on a seesaw learn by experience that a lighter child can balance a heavier one by sitting farther from the fulcrum or pivot point. This is because the tendency for an object to produce rotation is proportional not only to its mass but also to the distance between the object and the fulcrum. To make this more precise, consider an x-axis, which we view as a weightless beam. If a point-mass m is located on the axis at x, then the tendency for that mass to produce a rotation of the beam about a point a on the axis is measured by the following quantity, called the ***moment of m about x = a***:

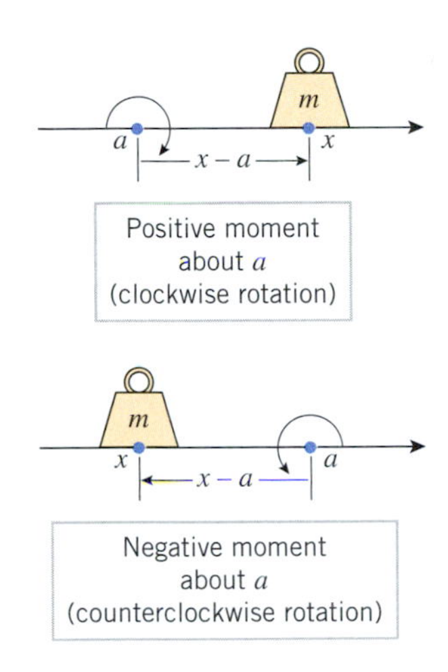

Figure 15.6.6

$$\begin{bmatrix}\text{moment of } m \\ \text{about } a\end{bmatrix} = m(x-a)$$

The number $x - a$ is called the ***lever arm***. Depending on whether the mass is to the right or left of a, the lever arm is either the distance between x and a or the negative of this distance (Figure 15.6.6). Positive lever arms result in positive moments and clockwise rotations, and negative lever arms result in negative moments and counterclockwise rotations.

Suppose that masses $m_1, m_2, \ldots, m_n$ are located at $x_1, x_2, \ldots, x_n$ on a coordinate axis and a fulcrum is positioned at the point a (Figure 15.6.7). Depending on whether the sum of the moments about a,

$$\sum_{k=1}^{n} m_k(x_k - a) = m_1(x_1 - a) + m_2(x_2 - a) + \cdots + m_n(x_n - a)$$

is positive, negative, or zero, a weightless beam along the axis will rotate clockwise about a, rotate counterclockwise about a, or balance perfectly. In the last case, the system of masses is said to be in ***equilibrium***.

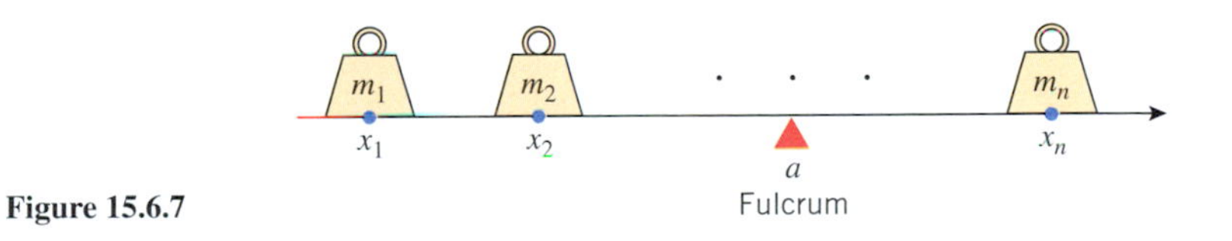

Figure 15.6.7

The preceding ideas can be extended to masses distributed in two-dimensional space. If we imagine the xy-plane to be a weightless sheet supporting a point-mass m located at a point (x, y), then the tendency for the mass to produce a rotation of the sheet about the line $x = a$ is $m(x - a)$, called the ***moment of m about x = a***, and the tendency for the mass to produce a rotation about the line $y = c$ is $m(y - c)$, called the ***moment of m about y = c*** (Figure 15.6.8). In summary,

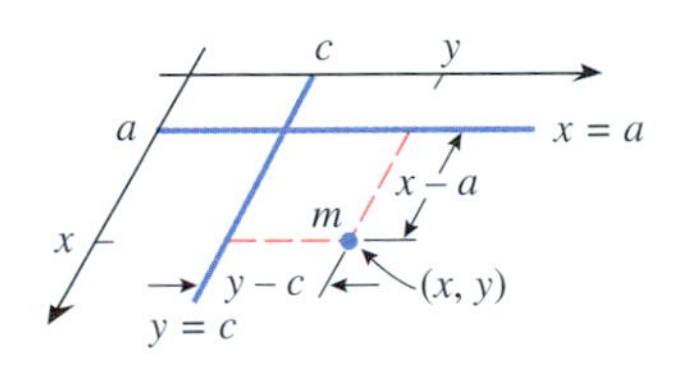

Figure 15.6.8

$$\begin{bmatrix}\text{moment of } m \\ \text{about the} \\ \text{line } x = a\end{bmatrix} = m(x-a) \quad \text{and} \quad \begin{bmatrix}\text{moment of } m \\ \text{about the} \\ \text{line } y = c\end{bmatrix} = m(y-c) \tag{5–6}$$

If a number of masses are distributed throughout the xy-plane, then the plane (viewed as a weightless sheet) will balance on a knife-edge along the line $x = a$ if the sum of the moments about the line is zero. Similarly for the line $y = c$.

We are now ready to solve Problem 15.6.2. We imagine the lamina to be subdivided into rectangular pieces using lines parallel to the coordinate axes and excluding from consideration any nonrectangular pieces at the boundary (Figure 15.6.3). We assume that there are n such rectangular pieces and that the kth piece has area ΔA_k and mass ΔM_k. We will let (x_k^*, y_k^*) be the center of the kth piece, and we will assume that the entire mass of the kth piece is concentrated at its center. From (4), the mass of the kth piece can be approximated by

$$\Delta M_k \approx \delta(x_k^*, y_k^*)\Delta A_k$$

Since the lamina balances on the lines $x = \bar{x}$ and $y = \bar{y}$, the sum of the moments of the rectangular pieces about those lines should be close to zero; that is,

$$\sum_{k=1}^{n}(x_k^* - \bar{x})\Delta M_k = \sum_{k=1}^{n}(x_k^* - \bar{x})\delta(x_k^*, y_k^*)\Delta A_k \approx 0$$

$$\sum_{k=1}^{n}(y_k^* - \bar{y})\Delta M_k = \sum_{k=1}^{n}(y_k^* - \bar{y})\delta(x_k^*, y_k^*)\Delta A_k \approx 0$$

If we now increase n in such a way that the dimensions of the rectangles tend to zero, then it is plausible that the errors in our approximations will approach zero, so that

$$\lim_{n\to+\infty}\sum_{k=1}^{n}(x_k^* - \bar{x})\delta(x_k^*, y_k^*)\Delta A_k = 0$$

$$\lim_{n\to+\infty}\sum_{k=1}^{n}(y_k^* - \bar{y})\delta(x_k^*, y_k^*)\Delta A_k = 0$$

from which we obtain

$$\iint_R (x - \bar{x})\delta(x, y)\,dA = 0$$

$$\iint_R (y - \bar{y})\delta(x, y)\,dA = 0$$

Since $\bar{x}$ and $\bar{y}$ are constant, these equations can be rewritten as

$$\iint_R x\delta(x, y)\,dA = \bar{x}\iint_R \delta(x, y)\,dA$$

$$\iint_R y\delta(x, y)\,dA = \bar{y}\iint_R \delta(x, y)\,dA$$

from which we obtain the following formulas for the center of gravity of the lamina:

Center of Gravity $(\bar{x}, \bar{y})$ ***of a Lamina***

$$\bar{x} = \frac{\displaystyle\iint_R x\delta(x, y)\,dA}{\displaystyle\iint_R \delta(x, y)\,dA}, \qquad \bar{y} = \frac{\displaystyle\iint_R y\delta(x, y)\,dA}{\displaystyle\iint_R \delta(x, y)\,dA} \tag{7–8}$$

Observe that in both formulas the denominator is the mass M of the lamina [see (3)]. The numerator in the formula for $\bar{x}$ is denoted by M_y and is called the ***first moment of the lamina about the y-axis***; the numerator of the formula for $\bar{y}$ is denoted by M_x and is called the ***first moment of the lamina about the x-axis***. Thus, Formulas (7) and (8) can be expressed as

$$\bar{x} = \frac{M_y}{M} = \frac{1}{\text{mass of } R}\iint_R x\delta(x, y)\,dA \tag{9}$$

$$\bar{y} = \frac{M_x}{M} = \frac{1}{\text{mass of } R}\iint_R y\delta(x, y)\,dA \tag{10}$$

Example 2 Find the center of gravity of the triangular lamina with vertices (0, 0), (0, 1), and (1, 0) and density function $\delta(x, y) = xy$.

Solution. The lamina is shown in Figure 15.6.4. In Example 1 we found the mass of the lamina to be

$$M = \iint_R \delta(x, y)\, dA = \iint_R xy\, dA = \frac{1}{24}$$

The moment of the lamina about the y-axis is

$$M_y = \iint_R x\delta(x, y)\, dA = \iint_R x^2 y\, dA = \int_0^1 \int_0^{-x+1} x^2 y\, dy\, dx$$

$$= \int_0^1 \left[\frac{1}{2}x^2 y^2\right]_{y=0}^{-x+1} dx = \int_0^1 \left(\frac{1}{2}x^4 - x^3 + \frac{1}{2}x^2\right) dx = \frac{1}{60}$$

and the moment about the x-axis is

$$M_x = \iint_R y\delta(x, y)\, dA = \iint_R xy^2\, dA = \int_0^1 \int_0^{-x+1} xy^2\, dy\, dx$$

$$= \int_0^1 \left[\frac{1}{3}xy^3\right]_{y=0}^{-x+1} dx = \int_0^1 \left(-\frac{1}{3}x^4 + x^3 - x^2 + \frac{1}{3}x\right) dx = \frac{1}{60}$$

From (9) and (10),

$$\bar{x} = \frac{M_y}{M} = \frac{1/60}{1/24} = \frac{2}{5}, \quad \bar{y} = \frac{M_x}{M} = \frac{1/60}{1/24} = \frac{2}{5}$$

so the center of gravity is $\left(\frac{2}{5}, \frac{2}{5}\right)$. ◀

CENTROIDS

In the special case of a *homogeneous* lamina, the center of gravity is called the ***centroid of the lamina*** or sometimes the ***centroid of the region R***. Because the density function δ is constant for a homogeneous lamina, the factor δ may be moved through the integral signs in (7) and (8) and canceled. Thus, the centroid $(\bar{x}, \bar{y})$ is a geometric property of the region R and is given by the following formulas:

Centroid of a Region R

$$\bar{x} = \frac{\displaystyle\iint_R x\, dA}{\displaystyle\iint_R dA} = \frac{1}{\text{area of } R} \iint_R x\, dA \tag{11}$$

$$\bar{y} = \frac{\displaystyle\iint_R y\, dA}{\displaystyle\iint_R dA} = \frac{1}{\text{area of } R} \iint_R y\, dA \tag{12}$$

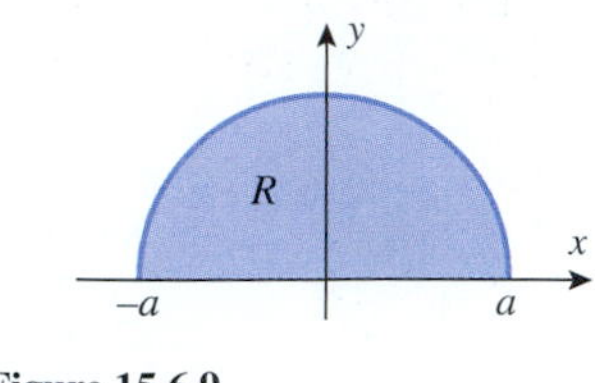

Figure 15.6.9

▶ **Example 3** Find the centroid of the semicircular region in Figure 15.6.9.

Solution. By symmetry, $\bar{x} = 0$ since the y-axis is obviously a line of balance. From (12),

$$\bar{y} = \frac{1}{\text{area of } R} \iint_R y\,dA = \frac{1}{\frac{1}{2}\pi a^2} \iint_R y\,dA$$

$$= \frac{1}{\frac{1}{2}\pi a^2} \int_0^{\pi} \int_0^{a} (r \sin\theta) r\,dr\,d\theta \qquad \text{Evaluating in polar coordinates}$$

$$= \frac{1}{\frac{1}{2}\pi a^2} \int_0^{\pi} \left[\frac{1}{3} r^3 \sin\theta\right]_{r=0}^{a} d\theta$$

$$= \frac{1}{\frac{1}{2}\pi a^2} \left(\frac{1}{3}a^3\right) \int_0^{\pi} \sin\theta\,d\theta = \frac{1}{\frac{1}{2}\pi a^2}\left(\frac{2}{3}a^3\right) = \frac{4a}{3\pi}$$

so the centroid is $\left(0, \dfrac{4a}{3\pi}\right)$. ◀

CENTER OF GRAVITY AND CENTROID OF A SOLID

For a three-dimensional solid G, the formulas for moments, center of gravity, and centroid are similar to those for laminas. If G is *homogeneous*, then its ***density*** is defined to be its mass per unit volume. Thus, if G is a homogeneous solid of mass M and volume V, then its density δ is given by $\delta = M/V$. If G is inhomogeneous and is in an xyz-coordinate system, then its density at a general point (x, y, z) is specified by a ***density function*** $\delta(x, y, z)$ whose value at a point can be viewed as a limit:

$$\delta(x, y, z) = \lim_{\Delta V \to 0} \frac{\Delta M}{\Delta V}$$

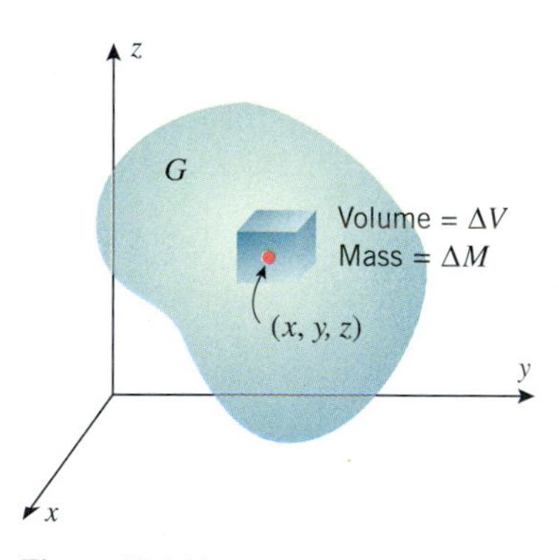

Figure 15.6.10

where ΔM and ΔV represent the mass and volume of a rectangular parallelepiped, centered at (x, y, z), whose dimensions tend to zero (Figure 15.6.10).

Using the discussion of laminas as a model, you should be able to show that the mass M of a solid with a continuous density function $\delta(x, y, z)$ is

$$M = \text{mass of } G = \iiint_G \delta(x, y, z)\,dV \tag{13}$$

The formulas for center of gravity and centroid are as follows:

Center of Gravity $(\bar{x}, \bar{y}, \bar{z})$ of a Solid G

$$\bar{x} = \frac{1}{M}\iiint_G x\delta(x, y, z)\,dV$$
$$\bar{y} = \frac{1}{M}\iiint_G y\delta(x, y, z)\,dV$$
$$\bar{z} = \frac{1}{M}\iiint_G z\delta(x, y, z)\,dV$$

Centroid $(\bar{x}, \bar{y}, \bar{z})$ of a Solid G

$$\bar{x} = \frac{1}{V}\iiint_G x\,dV$$
$$\bar{y} = \frac{1}{V}\iiint_G y\,dV$$
$$\bar{z} = \frac{1}{V}\iiint_G z\,dV$$

(14–15)

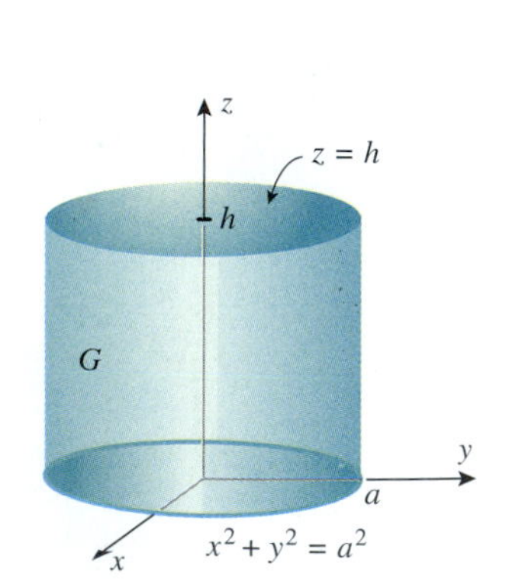

Figure 15.6.11

▶ **Example 4** Find the mass and the center of gravity of a cylindrical solid of height h and radius a (Figure 15.6.11), assuming that the density at each point is proportional to the distance between the point and the base of the solid.

Solution. Since the density is proportional to the distance z from the base, the density function has the form $\delta(x, y, z) = kz$, where k is some (unknown) positive constant of proportionality. From (13) the mass of the solid is

$$M = \iiint_G \delta(x, y, z)\, dV = \int_{-a}^{a} \int_{-\sqrt{a^2-x^2}}^{\sqrt{a^2-x^2}} \int_0^h kz\, dz\, dy\, dx$$

$$= k \int_{-a}^{a} \int_{-\sqrt{a^2-x^2}}^{\sqrt{a^2-x^2}} \frac{1}{2}h^2\, dy\, dx$$

$$= kh^2 \int_{-a}^{a} \sqrt{a^2 - x^2}\, dx$$

$$= \tfrac{1}{2}kh^2\pi a^2$$

Interpret the integral as the area of a semicircle.

Without additional information, the constant k cannot be determined. However, as we will now see, the value of k does not affect the center of gravity.

From (14),

$$\bar{z} = \frac{1}{M} \iiint_G z\delta(x, y, z)\, dV = \frac{1}{\frac{1}{2}kh^2\pi a^2} \iiint_G z\delta(x, y, z)\, dV$$

$$= \frac{1}{\frac{1}{2}kh^2\pi a^2} \int_{-a}^{a} \int_{-\sqrt{a^2-x^2}}^{\sqrt{a^2-x^2}} \int_0^h z(kz)\, dz\, dy\, dx$$

$$= \frac{k}{\frac{1}{2}kh^2\pi a^2} \int_{-a}^{a} \int_{-\sqrt{a^2-x^2}}^{\sqrt{a^2-x^2}} \frac{1}{3}h^3\, dy\, dx$$

$$= \frac{\frac{1}{3}kh^3}{\frac{1}{2}kh^2\pi a^2} \int_{-a}^{a} 2\sqrt{a^2 - x^2}\, dx$$

$$= \frac{\frac{1}{3}kh^3\pi a^2}{\frac{1}{2}kh^2\pi a^2} = \frac{2}{3}h$$

Similar calculations using (14) will yield $\bar{x} = \bar{y} = 0$. However, this is evident by inspection, since it follows from the symmetry of the solid and the form of its density function that the center of gravity is on the z-axis. Thus, the center of gravity is $\left(0, 0, \frac{2}{3}h\right)$. ◄

THEOREM OF PAPPUS

The following theorem, due to the Greek mathematician Pappus, gives an important relationship between the centroid of a plane region R and the volume of the solid generated when the region is revolved about a line.

Pappus of Alexandria (4th century A.D.) Greek mathematician. Pappus lived during the early Christian era when mathematical activity was in a period of decline. His main contributions to mathematics appeared in a series of eight books called *The Collection* (written about 340 A.D.). This work, which survives only partially, contained some original results but was devoted mostly to statements, refinements, and proofs of results by earlier mathematicians. Pappus' Theorem, stated without proof in Book VII of *The Collection*, was probably known and proved in earlier times. This result is sometimes called Guldin's Theorem in recognition of the Swiss mathematician, Paul Guldin (1577–1643), who rediscovered it independently.

15.6.3 THEOREM (*Theorem of Pappus*). *If R is a bounded plane region and L is a line that lies in the plane of R such that R is entirely on one side of L, then the volume of the solid formed by revolving R about L is given by*

$$\text{volume} = (\text{area of } R) \cdot \begin{pmatrix} \text{distance traveled} \\ \text{by the centroid} \end{pmatrix}$$

PROOF. Introduce an xy-coordinate system so that L is along the y-axis and the region R is in the first quadrant (Figure 15.6.12). Let R be partitioned into subregions in the usual way and let R_k be a typical rectangle interior to R. If (x_k^*, y_k^*) is the center of R_k, and if the area of R_k is $\Delta A_k = \Delta x_k \Delta y_k$, then from Formula (1) of Section 7.3 the volume generated by R_k as it revolves about L is

$$2\pi x_k^* \Delta x_k \Delta y_k = 2\pi x_k^* \Delta A_k$$

Therefore, the total volume of the solid is approximately

$$V \approx \sum_{k=1}^{n} 2\pi x_k^* \Delta A_k$$

from which it follows that the exact volume is

$$V = \iint_R 2\pi x \, dA = 2\pi \iint_R x \, dA$$

Thus, it follows from (11) that

$$V = 2\pi \cdot \bar{x} \cdot [\text{area of } R]$$

This completes the proof since $2\pi\bar{x}$ is the distance traveled by the centroid when R is revolved about the y-axis. ■

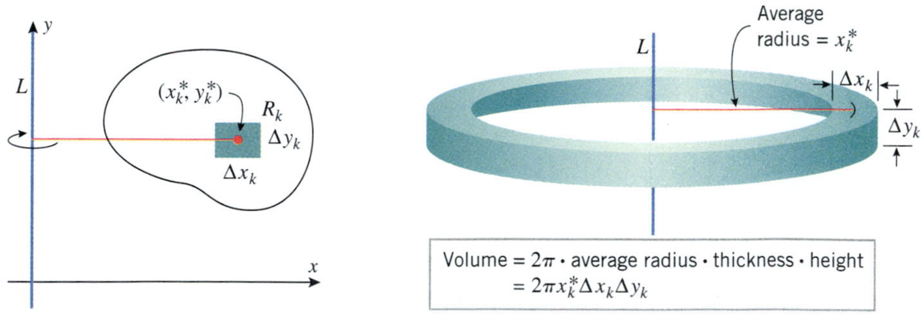

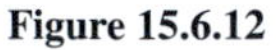
Figure 15.6.12

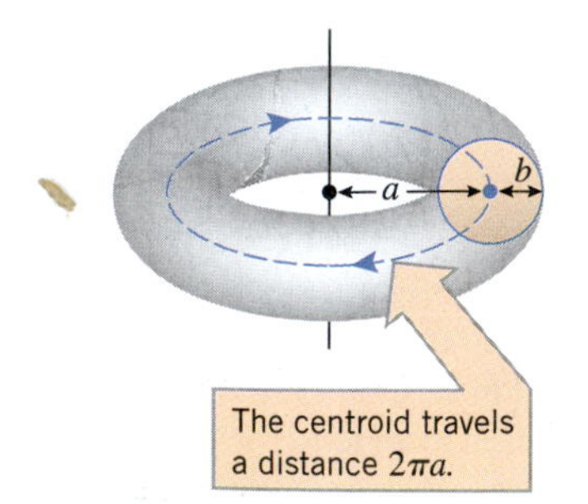

Figure 15.6.13

▸ Example 5 Use Pappus' Theorem to find the volume V of the torus generated by revolving a circular region of radius b about a line at a distance a (greater than b) from the center of the circle (Figure 15.6.13).

Solution. By symmetry, the centroid of a circular region is its center. Thus, the distance traveled by the centroid is $2\pi a$. Since the area of a circle of radius b is πb^2, it follows from Pappus' Theorem that the volume of the torus is

$$V = (2\pi a)(\pi b^2) = 2\pi^2 ab^2 \blacktriangleleft$$

✔QUICK CHECK EXERCISES 15.6 (See page 1075 for answers.)

1. The total mass of a lamina with continuous density function $\delta(x, y)$ that occupies a region R in the xy-plane is given by $M =$ ________.

2. Consider a lamina with mass M and continuous density function $\delta(x, y)$ that occupies a region R in the xy-plane. The x-coordinate of the center of gravity of the lamina is M_y/M, where M_y is called the ________ and is given by the double integral ________.

3. Let R be the region between the graphs of $y = x^2$ and $y = 2 - x$ for $0 \le x \le 1$. The area of R is $\frac{7}{6}$ and the centroid of R is ________.

4. If the region R in Quick Check Exercise 3 is used to generate a solid G by rotating R about a horizontal line 6 units above its centroid, then the volume of G is ________.

EXERCISE SET 15.6 Graphing Utility C CAS

FOCUS ON CONCEPTS

1. Masses $m_1 = 5$, $m_2 = 10$, and $m_3 = 20$ are positioned on a weightless beam as shown in the accompanying figure.
   (a) Suppose that the fulcrum is positioned at $x = 5$. Without computing the sum of moments about 5, determine whether the sum is positive, zero, or negative. Explain.
   (b) Where should the fulcrum be placed so that the beam is in equilibrium?

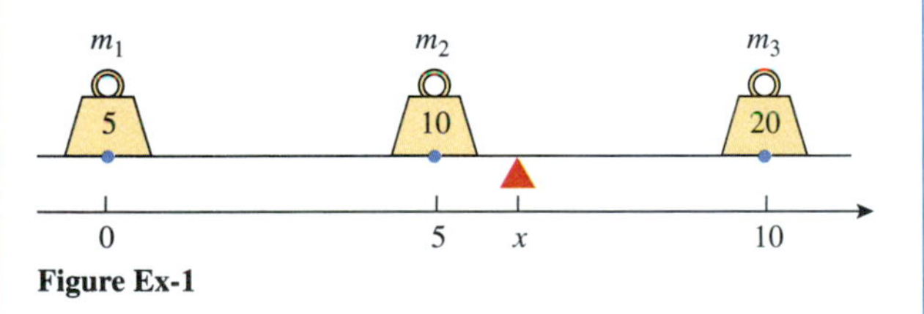

Figure Ex-1

2. Masses $m_1 = 10$, $m_2 = 3$, $m_3 = 4$, and m are positioned on a weightless beam, with the fulcrum positioned at point 4, as shown in the accompanying figure.
   (a) Suppose that $m = 14$. Without computing the sum of the moments about 4, determine whether the sum is positive, zero, or negative. Explain.
   (b) For what value of m is the beam in equilibrium?

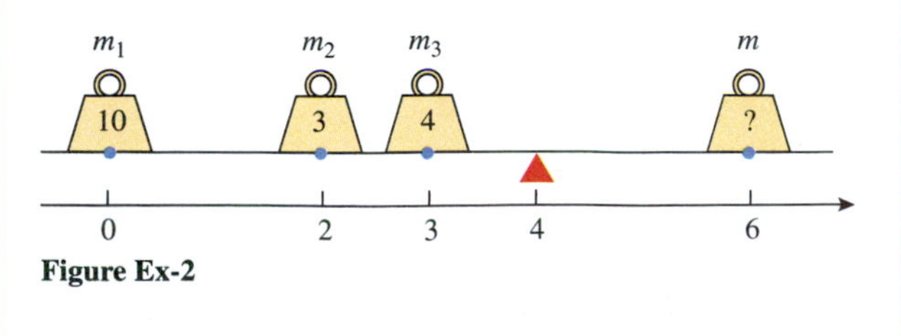

Figure Ex-2

3–4 Make a conjecture about the coordinates of the centroid of the region and confirm your conjecture by integrating.

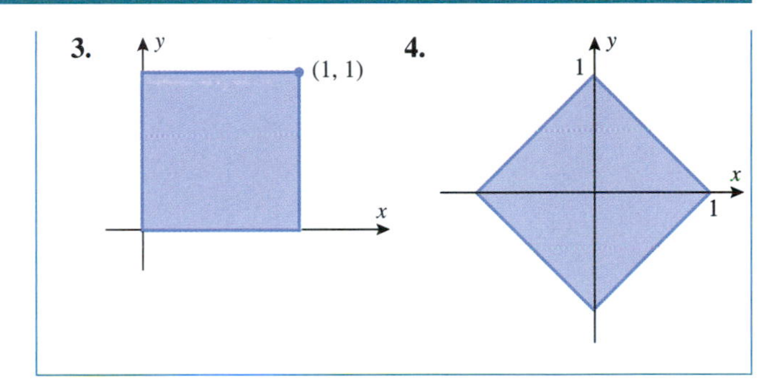

5–10 Find the centroid of the region.

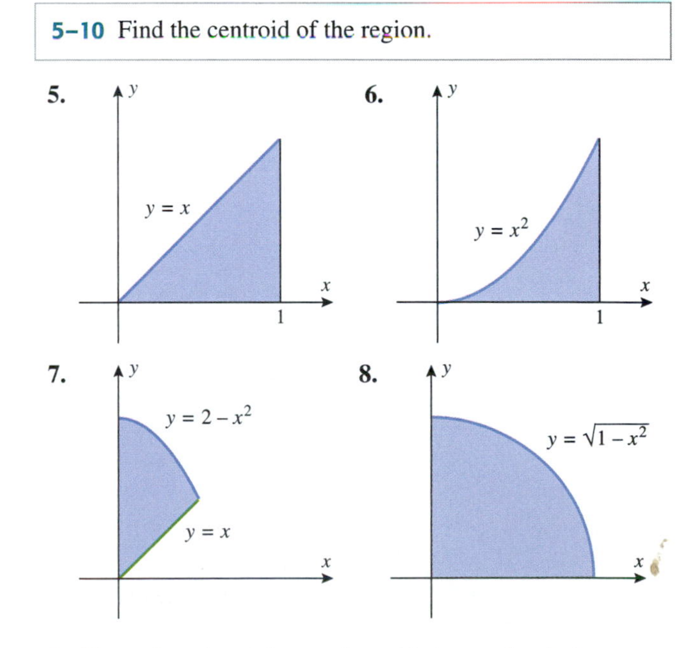

9. The region above the x-axis and between the circles $x^2 + y^2 = a^2$ and $x^2 + y^2 = b^2$ $(a < b)$.

10. The region enclosed between the y-axis and the right half of the circle $x^2 + y^2 = a^2$.

FOCUS ON CONCEPTS

11–12 Make a conjecture about the coordinates of the center of gravity and confirm your conjecture by integrating.

11. The lamina of Exercise 3 with density function $\delta(x, y) = |x + y - 1|$.

12. The lamina of Exercise 4 with density function $\delta(x, y) = 1 + x^2 + y^2$.

13–16 Find the mass and center of gravity of the lamina.

13. A lamina with density $\delta(x, y) = x + y$ is bounded by the x-axis, the line $x = 1$, and the curve $y = \sqrt{x}$.

14. A lamina with density $\delta(x, y) = y$ is bounded by $y = \sin x$, $y = 0$, $x = 0$, and $x = \pi$.

15. A lamina with density $\delta(x, y) = xy$ is in the first quadrant and is bounded by the circle $x^2 + y^2 = a^2$ and the coordinate axes.

16. A lamina with density $\delta(x, y) = x^2 + y^2$ is bounded by the x-axis and the upper half of the circle $x^2 + y^2 = 1$.

FOCUS ON CONCEPTS

17–18 Make a conjecture about the coordinates of the centroid and confirm your conjecture by integrating.

17. **18.**

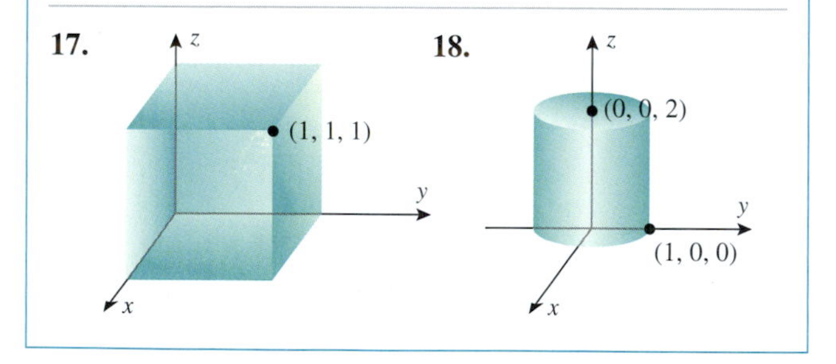

19–24 Find the centroid of the solid.

19. The tetrahedron in the first octant enclosed by the coordinate planes and the plane $x + y + z = 1$.

20. The solid bounded by the parabolic cylinder $z = 1 - y^2$ and the planes $x + z = 1$, $x = 0$, and $z = 0$.

21. The solid bounded by the surface $z = y^2$ and the planes $x = 0$, $x = 1$, and $z = 1$.

22. The solid in the first octant bounded by the surface $z = xy$ and the planes $z = 0$, $x = 2$, and $y = 2$.

23. The solid in the first octant that is bounded by the sphere $x^2 + y^2 + z^2 = a^2$ and the coordinate planes.

24. The solid enclosed by the xy-plane and the hemisphere $z = \sqrt{a^2 - x^2 - y^2}$.

25–28 Find the mass and center of gravity of the solid.

25. The cube that has density $\delta(x, y, z) = a - x$ and is defined by the inequalities $0 \le x \le a$, $0 \le y \le a$, and $0 \le z \le a$.

26. The cylindrical solid that has density $\delta(x, y, z) = h - z$ and is enclosed by $x^2 + y^2 = a^2$, $z = 0$, and $z = h$.

27. The solid that has density $\delta(x, y, z) = yz$ and is enclosed by $z = 1 - y^2$ (for $y \ge 0$), $z = 0$, $y = 0$, $x = -1$, and $x = 1$.

28. The solid that has density $\delta(x, y, z) = xz$ and is enclosed by $y = 9 - x^2$ (for $x \ge 0$), $x = 0$, $y = 0$, $z = 0$, and $z = 1$.

29. Find the center of gravity of the square lamina with vertices $(0, 0)$, $(1, 0)$, $(0, 1)$, and $(1, 1)$ if
(a) the density is proportional to the square of the distance from the origin;
(b) the density is proportional to the distance from the y-axis.

30. Find the center of gravity of the cube that is determined by the inequalities $0 \le x \le 1$, $0 \le y \le 1$, $0 \le z \le 1$ if
(a) the density is proportional to the square of the distance to the origin;
(b) the density is proportional to the sum of the distances to the faces that lie in the coordinate planes.

C **31.** Use the numerical triple integral capability of a CAS to approximate the location of the centroid of the solid that is bounded above by the surface $z = 1/(1 + x^2 + y^2)$, below by the xy-plane, and laterally by the plane $y = 0$ and the surface $y = \sin x$ for $0 \le x \le \pi$ (see the accompanying figure).

32. The accompanying figure shows the solid that is bounded above by the surface $z = 1/(x^2 + y^2 + 1)$, below by the xy-plane, and laterally by the surface $x^2 + y^2 = a^2$.
(a) By symmetry, the centroid of the solid lies on the z-axis. Make a conjecture about the behavior of the z-coordinate of the centroid as $a \to 0^+$ and as $a \to +\infty$.
(b) Find the z-coordinate of the centroid, and check your conjecture by calculating the appropriate limits.
(c) Use a graphing utility to plot the z-coordinate of the centroid versus a, and use the graph to estimate the value of a for which the centroid is $(0, 0, 0.25)$.

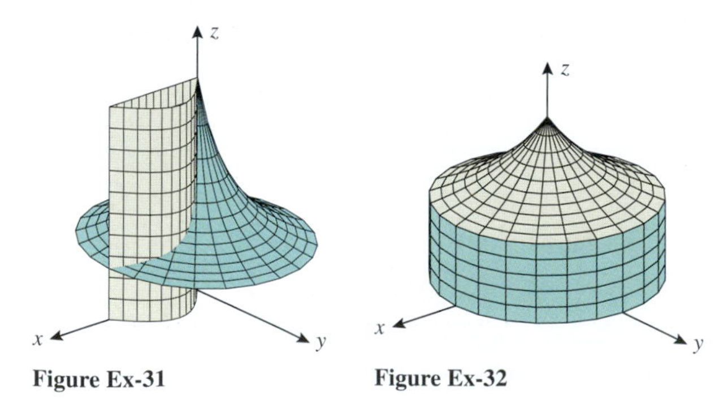

Figure Ex-31 Figure Ex-32

33. Show that in polar coordinates the formulas for the centroid $(\bar{x}, \bar{y})$ of a region R are

$$\bar{x} = \frac{1}{\text{area of } R} \iint_R r^2 \cos\theta \, dr \, d\theta$$

$$\bar{y} = \frac{1}{\text{area of } R} \iint_R r^2 \sin\theta \, dr \, d\theta$$

34. Use the result of Exercise 33 to find the centroid $(\bar{x}, \bar{y})$ of the region enclosed by the cardioid $r = a(1 + \sin\theta)$.

35. Use the result of Exercise 33 to find the centroid $(\bar{x}, \bar{y})$ of the petal of the rose $r = \sin 2\theta$ in the first quadrant.

36. Let R be the rectangle bounded by the lines $x = 0$, $x = 3$, $y = 0$, and $y = 2$. By inspection, find the centroid of R and use it to evaluate

$$\iint_R x \, dA \quad \text{and} \quad \iint_R y \, dA$$

37. Use the Theorem of Pappus and the fact that the volume of a sphere of radius a is $V = \frac{4}{3}\pi a^3$ to show that the centroid of the lamina that is bounded by the x-axis and the semicircle $y = \sqrt{a^2 - x^2}$ is $(0, 4a/(3\pi))$. (This problem was solved directly in Example 3.)

38. Use the Theorem of Pappus and the result of Exercise 37 to find the volume of the solid generated when the region bounded by the x-axis and the semicircle $y = \sqrt{a^2 - x^2}$ is revolved about

(a) the line $y = -a$ (b) the line $y = x - a$.

39. Use the Theorem of Pappus and the fact that the area of an ellipse with semiaxes a and b is πab to find the volume of the elliptical torus generated by revolving the ellipse

$$\frac{(x-k)^2}{a^2} + \frac{y^2}{b^2} = 1$$

about the y-axis. Assume that $k > a$.

40. Use the Theorem of Pappus to find the volume of the solid that is generated when the region enclosed by $y = x^2$ and $y = 8 - x^2$ is revolved about the x-axis.

41. Use the Theorem of Pappus to find the centroid of the triangular region with vertices $(0, 0)$, $(a, 0)$, and $(0, b)$, where $a > 0$ and $b > 0$. [*Hint:* Revolve the region about the x-axis to obtain $\bar{y}$ and about the y-axis to obtain $\bar{x}$.]

42. It can be proved that if a bounded plane region slides along a helix in such a way that the region is always orthogonal to the helix (i.e., orthogonal to the unit tangent vector to the helix), then the volume swept out by the region is equal to the area of the region times the distance traveled by its centroid. Use this result to find the volume of the "tube" in the accompanying figure that is swept out by sliding a circle of radius $\frac{1}{2}$ along the helix

$$x = \cos t, \quad y = \sin t, \quad z = \frac{t}{4} \qquad (0 \le t \le 4\pi)$$

in such a way that the circle is always centered on the helix and lies in the plane perpendicular to the helix.

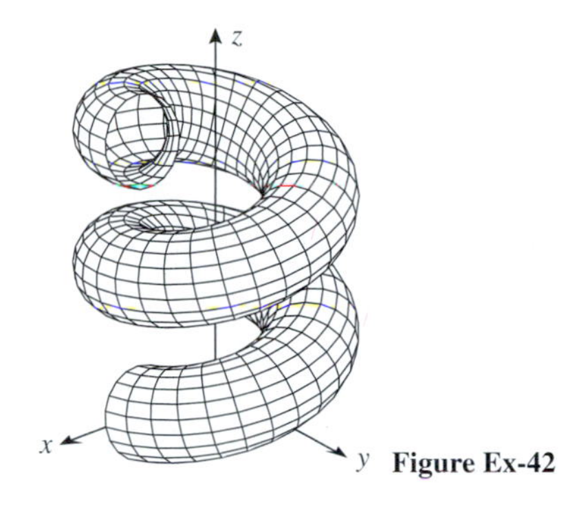

Figure Ex-42

43–44 The tendency of a lamina to resist a change in rotational motion about an axis is measured by its ***moment of inertia*** about that axis. If a lamina occupies a region R of the xy-plane, and if its density function $\delta(x, y)$ is continuous on R, then the moments of inertia about the x-axis, the y-axis, and the z-axis are denoted by I_x, I_y, and I_z, respectively, and are defined by

$$I_x = \iint_R y^2 \, \delta(x, y) \, dA, \quad I_y = \iint_R x^2 \, \delta(x, y) \, dA,$$

$$I_z = \iint_R (x^2 + y^2) \, \delta(x, y) \, dA$$

Use these definitions in Exercises 43 and 44.

43. Consider the rectangular lamina that occupies the region described by the inequalities $0 \le x \le a$ and $0 \le y \le b$. Assuming that the lamina has constant density δ, show that

$$I_x = \frac{\delta ab^3}{3}, \quad I_y = \frac{\delta a^3 b}{3}, \quad I_z = \frac{\delta ab(a^2 + b^2)}{3}$$

44. Consider the circular lamina that occupies the region described by the inequalities $0 \le x^2 + y^2 \le a^2$. Assuming that the lamina has constant density δ, show that

$$I_x = I_y = \frac{\delta\pi a^4}{4}, \quad I_z = \frac{\delta\pi a^4}{2}$$

✔ QUICK CHECK ANSWERS 15.6

1. $\iint_R \delta(x, y) \, dA$ **2.** first moment about the y-axis; $\iint_R x\delta(x, y) \, dA$ **3.** $\left(\frac{5}{14}, \frac{32}{35}\right)$ **4.** 14π

15.7 TRIPLE INTEGRALS IN CYLINDRICAL AND SPHERICAL COORDINATES

Earlier we saw that some double integrals are easier to evaluate in polar coordinates than in rectangular coordinates. Similarly, some triple integrals are easier to evaluate in cylindrical or spherical coordinates than in rectangular coordinates. In this section we will study triple integrals in these coordinate systems.

TRIPLE INTEGRALS IN CYLINDRICAL COORDINATES

Recall that in rectangular coordinates the triple integral of a continuous function f over a solid region G is defined as

$$\iiint_G f(x, y, z)\, dV = \lim_{n \to +\infty} \sum_{k=1}^{n} f(x_k^*, y_k^*, z_k^*)\Delta V_k$$

where ΔV_k denotes the volume of a rectangular parallelepiped interior to G and (x_k^*, y_k^*, z_k^*) is a point in this parallelepiped (see Figure 15.5.1). Triple integrals in cylindrical and spherical coordinates are defined similarly, except that the region G is divided not into rectangular parallelepipeds but into regions more appropriate to these coordinate systems.

In cylindrical coordinates, the simplest equations are of the form

$$r = \text{constant}, \quad \theta = \text{constant}, \quad z = \text{constant}$$

The first equation represents a right circular cylinder centered on the z-axis, the second a vertical half-plane hinged on the z-axis, and the third a horizontal plane. (See Figure 12.8.3.) These surfaces can be paired up to determine solids called ***cylindrical wedges*** or ***cylindrical elements of volume***. To be precise, a cylindrical wedge is a solid enclosed between six surfaces of the following form:

$$\begin{array}{llll} \text{two cylinders} & r = r_1, & r = r_2 & (r_1 < r_2) \\ \text{two half-planes} & \theta = \theta_1, & \theta = \theta_2 & (\theta_1 < \theta_2) \\ \text{two planes} & z = z_1, & z = z_2 & (z_1 < z_2) \end{array}$$

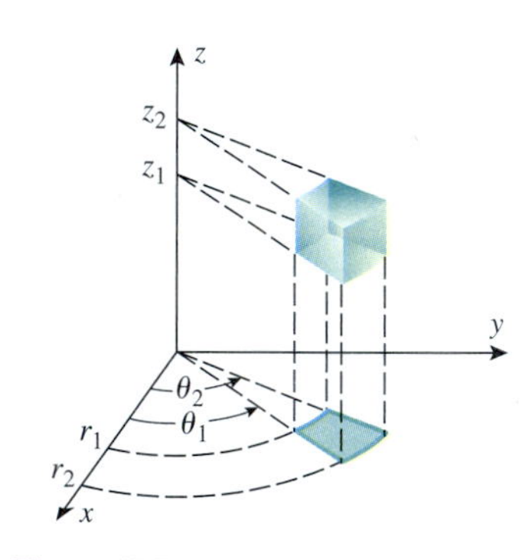

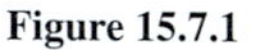
Figure 15.7.1

(Figure 15.7.1). The dimensions $\theta_2 - \theta_1$, $r_2 - r_1$, and $z_2 - z_1$ are called the ***central angle***, ***thickness***, and ***height*** of the wedge.

To define the triple integral over G of a function $f(r, \theta, z)$ in cylindrical coordinates we proceed as follows:

- Subdivide G into pieces by a three-dimensional grid consisting of concentric circular cylinders centered on the z-axis, half-planes hinged on the z-axis, and horizontal planes. Exclude from consideration all pieces that contain any points outside of G, thereby leaving only cylindrical wedges that are subsets of G.
- Assume that there are n such cylindrical wedges, and denote the volume of the kth cylindrical wedge by ΔV_k. As indicated in Figure 15.7.2, let $(r_k^*, \theta_k^*, z_k^*)$ be any point in the kth cylindrical wedge.
- Repeat this process with more and more subdivisions so that as n increases, the height, thickness, and central angle of the cylindrical wedges approach zero. Define

$$\iiint_G f(r, \theta, z)\, dV = \lim_{n \to +\infty} \sum_{k=1}^{n} f(r_k^*, \theta_k^*, z_k^*)\Delta V_k \tag{1}$$

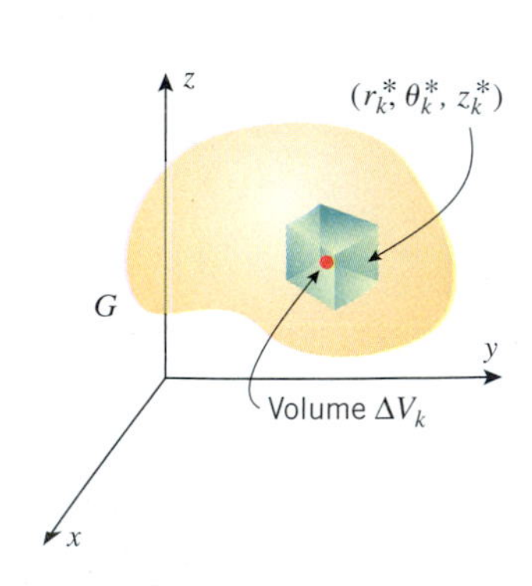

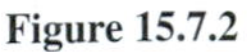
Figure 15.7.2

For computational purposes, it will be helpful to express (1) as an iterated integral. Toward this end we note that the volume ΔV_k of the kth cylindrical wedge can be expressed as

$$\Delta V_k = [\text{area of base}] \cdot [\text{height}] \tag{2}$$

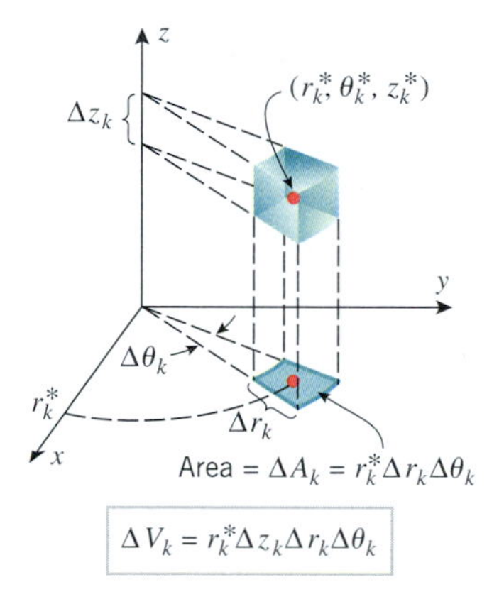

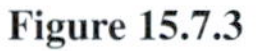

Figure 15.7.3

Note the extra factor of r that appears in the integrand on converting a triple integral to an iterated integral in cylindrical coordinates.

If we denote the thickness, central angle, and height of this wedge by Δr_k, $\Delta\theta_k$, and Δz_k, and if we choose the arbitrary point $(r_k^*, \theta_k^*, z_k^*)$ to lie above the "center" of the base (Figures 15.3.5 and 15.7.3), then it follows from (5) of Section 15.3 that the base has area $\Delta A_k = r_k^* \Delta r_k \Delta\theta_k$. Thus, (2) can be written as

$$\Delta V_k = r_k^* \Delta r_k \Delta\theta_k \Delta z_k = r_k^* \Delta z_k \Delta r_k \Delta\theta_k$$

Substituting this expression in (1) yields

$$\iiint_G f(r, \theta, z)\, dV = \lim_{n \to +\infty} \sum_{k=1}^{n} f(r_k^*, \theta_k^*, z_k^*) r_k^* \Delta z_k \Delta r_k \Delta\theta_k$$

which suggests that a triple integral in cylindrical coordinates can be evaluated as an iterated integral of the form

$$\iiint_G f(r, \theta, z)\, dV = \iiint_{\substack{\text{appropriate} \\ \text{limits}}} f(r, \theta, z) r\, dz\, dr\, d\theta \tag{3}$$

In this formula the integration with respect to z is done first, then with respect to r, and then with respect to θ, but any order of integration is allowable.

The following theorem, which we state without proof, makes the preceding ideas more precise.

15.7.1 THEOREM. *Let G be a solid region whose upper surface has the equation $z = g_2(r, \theta)$ and whose lower surface has the equation $z = g_1(r, \theta)$ in cylindrical coordinates. If the projection of the solid on the xy-plane is a simple polar region R, and if $f(r, \theta, z)$ is continuous on G, then*

$$\iiint_G f(r, \theta, z)\, dV = \iint_R \left[\int_{g_1(r,\theta)}^{g_2(r,\theta)} f(r, \theta, z)\, dz \right] dA \tag{4}$$

where the double integral over R is evaluated in polar coordinates. In particular, if the projection R is as shown in Figure 15.7.4, then (4) can be written as

$$\iiint_G f(r, \theta, z)\, dV = \int_{\theta_1}^{\theta_2} \int_{r_1(\theta)}^{r_2(\theta)} \int_{g_1(r,\theta)}^{g_2(r,\theta)} f(r, \theta, z) r\, dz\, dr\, d\theta \tag{5}$$

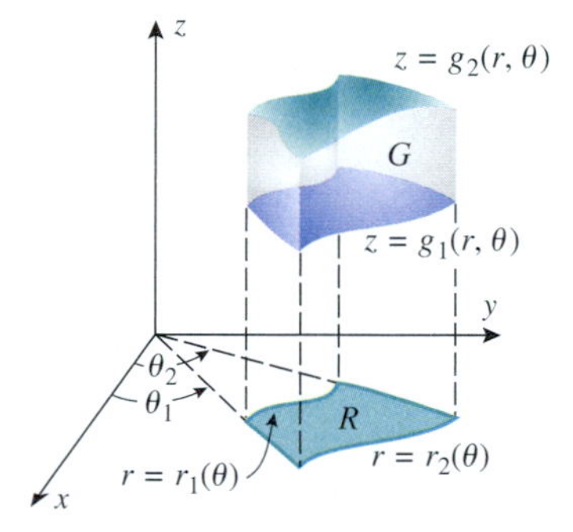

Figure 15.7.4

The type of solid to which Formula (5) applies is illustrated in Figure 15.7.4. To apply (4) and (5) it is best to begin with a three-dimensional sketch of the solid G, from which the limits of integration can be obtained as follows:

Determining Limits of Integration: Cylindrical Coordinates

Step 1. Identify the upper surface $z = g_2(r, \theta)$ and the lower surface $z = g_1(r, \theta)$ of the solid. The functions $g_1(r, \theta)$ and $g_2(r, \theta)$ determine the z-limits of integration. (If the upper and lower surfaces are given in rectangular coordinates, convert them to cylindrical coordinates.)

Step 2. Make a two-dimensional sketch of the projection R of the solid on the xy-plane. From this sketch the r- and θ-limits of integration may be obtained exactly as with double integrals in polar coordinates.

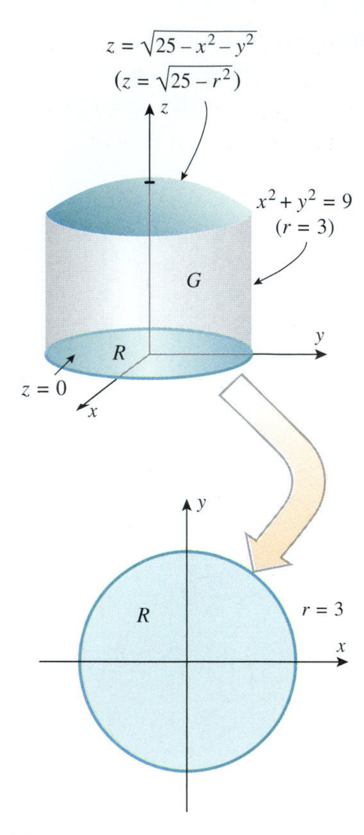

Figure 15.7.5

▶ Example 1 Use triple integration in cylindrical coordinates to find the volume and the centroid of the solid G that is bounded above by the hemisphere $z = \sqrt{25 - x^2 - y^2}$, below by the xy-plane, and laterally by the cylinder $x^2 + y^2 = 9$.

Solution. The solid G and its projection R on the xy-plane are shown in Figure 15.7.5. In cylindrical coordinates, the upper surface of G is the hemisphere $z = \sqrt{25 - r^2}$ and the lower surface is the plane $z = 0$. Thus, from (4), the volume of G is

$$V = \iiint_G dV = \iint_R \left[\int_0^{\sqrt{25-r^2}} dz\right] dA$$

For the double integral over R, we use polar coordinates:

$$\begin{aligned}
V &= \int_0^{2\pi}\int_0^3\int_0^{\sqrt{25-r^2}} r\,dz\,dr\,d\theta = \int_0^{2\pi}\int_0^3 \Big[rz\Big]_{z=0}^{\sqrt{25-r^2}} dr\,d\theta \\
&= \int_0^{2\pi}\int_0^3 r\sqrt{25-r^2}\,dr\,d\theta = \int_0^{2\pi}\left[-\frac{1}{3}(25-r^2)^{3/2}\right]_{r=0}^3 d\theta \\
&= \int_0^{2\pi}\frac{61}{3}\,d\theta = \frac{122}{3}\pi
\end{aligned}$$

$u = 25 - r^2$
$du = -2r\,dr$

From this result and (15) of Section 15.6,

$$\begin{aligned}
\bar{z} &= \frac{1}{V}\iiint_G z\,dV = \frac{3}{122\pi}\iiint_G z\,dV = \frac{3}{122\pi}\iint_R\left[\int_0^{\sqrt{25-r^2}} z\,dz\right] dA \\
&= \frac{3}{122\pi}\int_0^{2\pi}\int_0^3\int_0^{\sqrt{25-r^2}} zr\,dz\,dr\,d\theta = \frac{3}{122\pi}\int_0^{2\pi}\int_0^3\left[\frac{1}{2}rz^2\right]_{z=0}^{\sqrt{25-r^2}} dr\,d\theta \\
&= \frac{3}{244\pi}\int_0^{2\pi}\int_0^3 (25r - r^3)\,dr\,d\theta = \frac{3}{244\pi}\int_0^{2\pi}\frac{369}{4}\,d\theta = \frac{1107}{488}
\end{aligned}$$

By symmetry, the centroid $(\bar{x}, \bar{y}, \bar{z})$ of G lies on the z-axis, so $\bar{x} = \bar{y} = 0$. Thus, the centroid is at the point $(0, 0, 1107/488)$. ◀

CONVERTING TRIPLE INTEGRALS FROM RECTANGULAR TO CYLINDRICAL COORDINATES

Sometimes a triple integral that is difficult to integrate in rectangular coordinates can be evaluated more easily by making the substitution $x = r\cos\theta$, $y = r\sin\theta$, $z = z$ to convert it to an integral in cylindrical coordinates. Under such a substitution, a rectangular triple integral can be expressed as an iterated integral in cylindrical coordinates as

$$\iiint_G f(x, y, z)\,dV = \iiint_{\substack{\text{appropriate}\\ \text{limits}}} f(r\cos\theta, r\sin\theta, z)\,r\,dz\,dr\,d\theta \qquad (6)$$

The order of integration on the right side of (6) can be changed, provided the limits of integration are adjusted accordingly.

▶ Example 2 Use cylindrical coordinates to evaluate

$$\int_{-3}^{3}\int_{-\sqrt{9-x^2}}^{\sqrt{9-x^2}}\int_0^{9-x^2-y^2} x^2\,dz\,dy\,dx$$

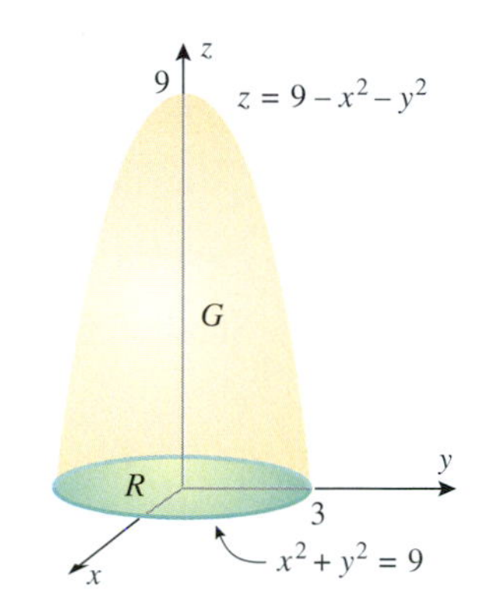

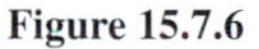
Figure 15.7.6

Solution. In problems of this type, it is helpful to sketch the region of integration G and its projection R on the xy-plane. From the z-limits of integration, the upper surface of G is the paraboloid $z = 9 - x^2 - y^2$ and the lower surface is the xy-plane $z = 0$. From the x- and y-limits of integration, the projection R is the region in the xy-plane enclosed by the circle $x^2 + y^2 = 9$ (Figure 15.7.6). Thus,

$$\int_{-3}^{3}\int_{-\sqrt{9-x^2}}^{\sqrt{9-x^2}}\int_{0}^{9-x^2-y^2} x^2\,dz\,dy\,dx = \iiint_G x^2\,dV$$

$$= \iint_R \left[\int_0^{9-r^2} r^2\cos^2\theta\,dz\right] dA = \int_0^{2\pi}\int_0^3\int_0^{9-r^2} (r^2\cos^2\theta)r\,dz\,dr\,d\theta$$

$$= \int_0^{2\pi}\int_0^3\int_0^{9-r^2} r^3\cos^2\theta\,dz\,dr\,d\theta = \int_0^{2\pi}\int_0^3 \left[zr^3\cos^2\theta\right]_{z=0}^{9-r^2} dr\,d\theta$$

$$= \int_0^{2\pi}\int_0^3 (9r^3 - r^5)\cos^2\theta\,dr\,d\theta = \int_0^{2\pi}\left[\left(\frac{9r^4}{4} - \frac{r^6}{6}\right)\cos^2\theta\right]_{r=0}^{3} d\theta$$

$$= \frac{243}{4}\int_0^{2\pi}\cos^2\theta\,d\theta = \frac{243}{4}\int_0^{2\pi}\frac{1}{2}(1+\cos 2\theta)\,d\theta = \frac{243\pi}{4} \blacktriangleleft$$

TRIPLE INTEGRALS IN SPHERICAL COORDINATES

In spherical coordinates, the simplest equations are of the form

$$\rho = \text{constant}, \quad \theta = \text{constant}, \quad \phi = \text{constant}$$

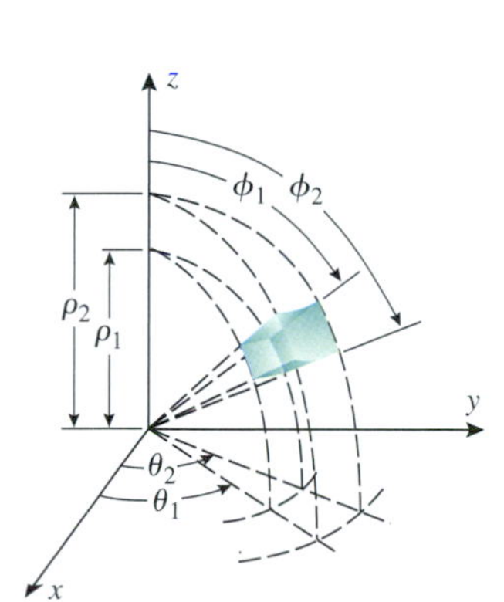

Figure 15.7.7

As indicated in Figure 12.8.4, the first equation represents a sphere centered at the origin and the second a half-plane hinged on the z-axis. The graph of the third equation is a right circular cone nappe with its vertex at the origin and its line of symmetry along the z-axis for $\phi \neq \pi/2$, and is the xy-plane if $\phi = \pi/2$. By a ***spherical wedge*** or ***spherical element of volume*** we mean a solid enclosed between six surfaces of the following form:

two spheres	$\rho = \rho_1$,	$\rho = \rho_2$	$(\rho_1 < \rho_2)$
two half-planes	$\theta = \theta_1$,	$\theta = \theta_2$	$(\theta_1 < \theta_2)$
nappes of two right circular cones	$\phi = \phi_1$,	$\phi = \phi_2$	$(\phi_1 < \phi_2)$

(Figure 15.7.7). We will refer to the numbers $\rho_2 - \rho_1$, $\theta_2 - \theta_1$, and $\phi_2 - \phi_1$ as the ***dimensions*** of a spherical wedge.

If G is a solid region in three-dimensional space, then the triple integral over G of a continuous function $f(\rho, \theta, \phi)$ in spherical coordinates is similar in definition to the triple integral in cylindrical coordinates, except that the solid G is partitioned into *spherical* wedges by a three-dimensional grid consisting of spheres centered at the origin, half-planes hinged on the z-axis, and nappes of right circular cones with vertices at the origin and lines of symmetry along the z-axis (Figure 15.7.8).

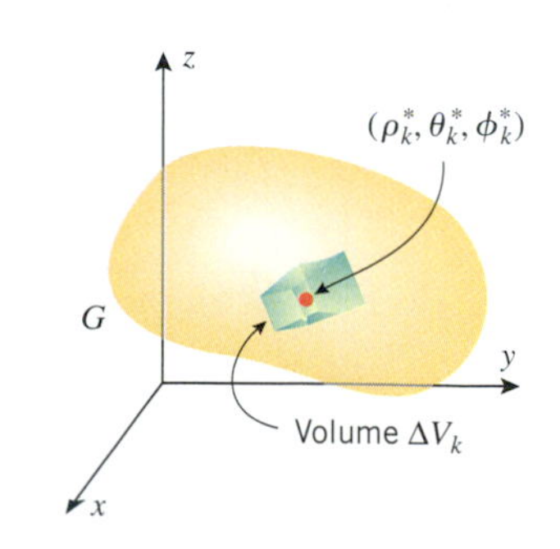

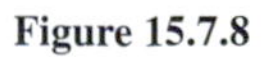
Figure 15.7.8

The defining equation of a triple integral in spherical coordinates is

$$\iiint_G f(\rho, \theta, \phi)\,dV = \lim_{n\to+\infty}\sum_{k=1}^{n} f(\rho_k^*, \theta_k^*, \phi_k^*)\Delta V_k \tag{7}$$

where ΔV_k is the volume of the kth spherical wedge that is interior to G, $(\rho_k^*, \theta_k^*, \phi_k^*)$ is an arbitrary point in this wedge, and n increases in such a way that the dimensions of each interior spherical wedge tend to zero.

For computational purposes, it will be desirable to express (7) as an iterated integral. In the exercises we will help you to show that if the point $(\rho_k^*, \theta_k^*, \phi_k^*)$ is suitably chosen, then the volume ΔV_k in (7) can be written as

$$\Delta V_k = \rho_k^{*2} \sin \phi_k^* \Delta\rho_k \Delta\phi_k \Delta\theta_k \tag{8}$$

where $\Delta\rho_k$, $\Delta\phi_k$, and $\Delta\theta_k$ are the dimensions of the wedge (Exercise 42). Substituting this in (7) we obtain

$$\iiint_G f(\rho, \theta, \phi)\, dV = \lim_{n \to +\infty} \sum_{k=1}^{n} f(\rho_k^*, \theta_k^*, \phi_k^*)\rho_k^{*2} \sin \phi_k^* \Delta\rho_k \Delta\phi_k \Delta\theta_k$$

which suggests that a triple integral in spherical coordinates can be evaluated as an iterated integral of the form

Note the extra factor of $\rho^2 \sin\phi$ that appears in the integrand on converting a triple integral to an iterated integral in spherical coordinates. This is analogous to the extra factor of r that appears in an iterated integral in cylindrical coordinates.

$$\iiint_G f(\rho, \theta, \phi)\, dV = \iiint_{\substack{\text{appropriate} \\ \text{limits}}} f(\rho, \theta, \phi)\rho^2 \sin\phi \, d\rho\, d\phi\, d\theta \tag{9}$$

The analog of Theorem 15.7.1 for triple integrals in spherical coordinates is tedious to state, so instead we will give some examples that illustrate techniques for obtaining the limits of integration. In all of our examples we will use the same order of integration—first with respect to ρ, then ϕ, and then θ. Once you have mastered the basic ideas, there should be no trouble using other orders of integration.

Suppose that we want to integrate $f(\rho, \theta, \phi)$ over the spherical solid G enclosed by the sphere $\rho = \rho_0$. The basic idea is to choose the limits of integration so that every point of the solid is accounted for in the integration process. Figure 15.7.9 illustrates one way of doing this. Holding θ and ϕ fixed for the first integration, we let ρ vary from 0 to ρ_0. This covers a radial line from the origin to the surface of the sphere. Next, keeping θ fixed, we let ϕ vary from 0 to π so that the radial line sweeps out a fan-shaped region. Finally, we let θ vary from 0 to 2π so that the fan-shaped region makes a complete revolution, thereby sweeping out the entire sphere. Thus, the triple integral of $f(\rho, \theta, \phi)$ over the spherical solid G may be evaluated by writing

$$\iiint_G f(\rho, \theta, \phi)\, dV = \int_0^{2\pi} \int_0^{\pi} \int_0^{\rho_0} f(\rho, \theta, \phi)\rho^2 \sin\phi \, d\rho\, d\phi\, d\theta$$

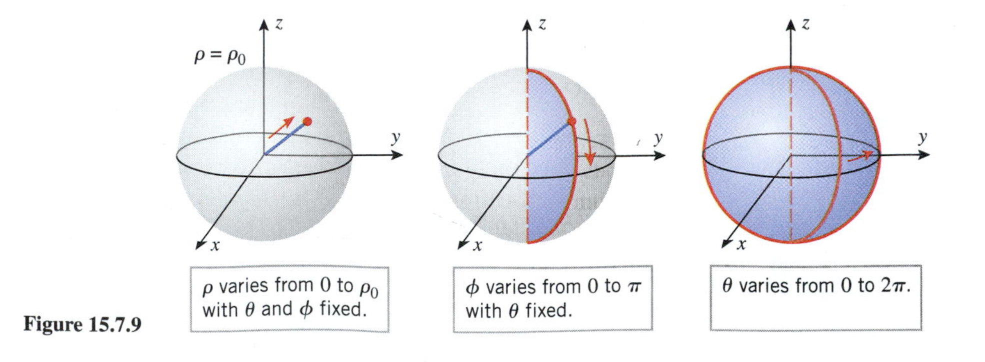

Figure 15.7.9

Table 15.7.1 suggests how the limits of integration in spherical coordinates can be obtained for some other common solids.

Table 15.7.1

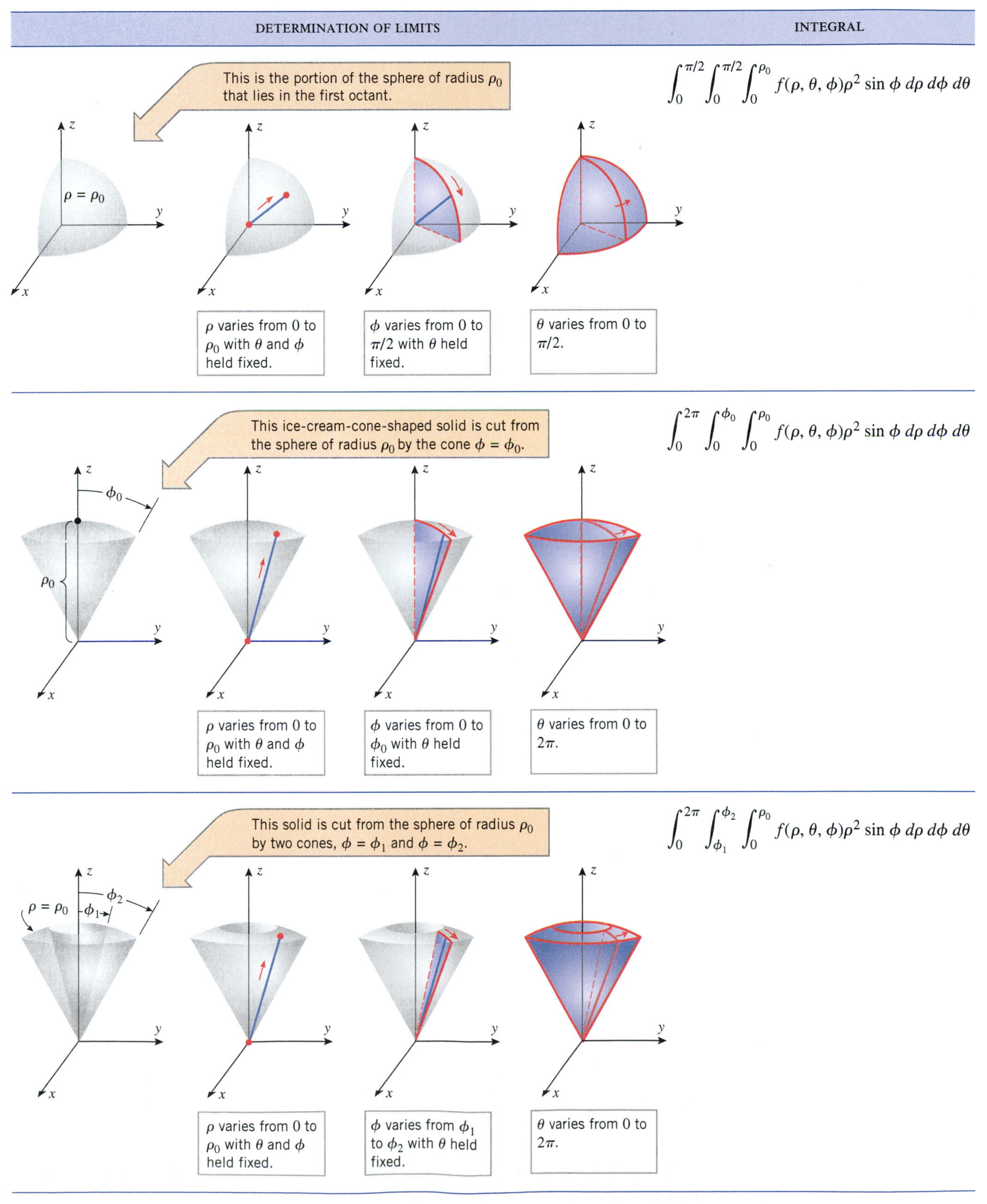
DETERMINATION OF LIMITS
INTEGRAL
This is the portion of the sphere of radius ρ_0 that lies in the first octant.
$\rho = \rho_0$
ρ varies from 0 to ρ_0 with θ and ϕ held fixed.
ϕ varies from 0 to $\pi/2$ with θ held fixed.
θ varies from 0 to $\pi/2$.
$\int_0^{\pi/2}\int_0^{\pi/2}\int_0^{\rho_0} f(\rho, \theta, \phi)\rho^2 \sin\phi \, d\rho \, d\phi \, d\theta$
This ice-cream-cone-shaped solid is cut from the sphere of radius ρ_0 by the cone $\phi = \phi_0$.
ϕ_0
ρ_0
ρ varies from 0 to ρ_0 with θ and ϕ held fixed.
ϕ varies from 0 to ϕ_0 with θ held fixed.
θ varies from 0 to 2π.
$\int_0^{2\pi}\int_0^{\phi_0}\int_0^{\rho_0} f(\rho, \theta, \phi)\rho^2 \sin\phi \, d\rho \, d\phi \, d\theta$
This solid is cut from the sphere of radius ρ_0 by two cones, $\phi = \phi_1$ and $\phi = \phi_2$.
$\rho = \rho_0$
ϕ_1
ϕ_2
ρ varies from 0 to ρ_0 with θ and ϕ held fixed.
ϕ varies from ϕ_1 to ϕ_2 with θ held fixed.
θ varies from 0 to 2π.
$\int_0^{2\pi}\int_{\phi_1}^{\phi_2}\int_0^{\rho_0} f(\rho, \theta, \phi)\rho^2 \sin\phi \, d\rho \, d\phi \, d\theta$

Table 15.7.1 *(continued)*

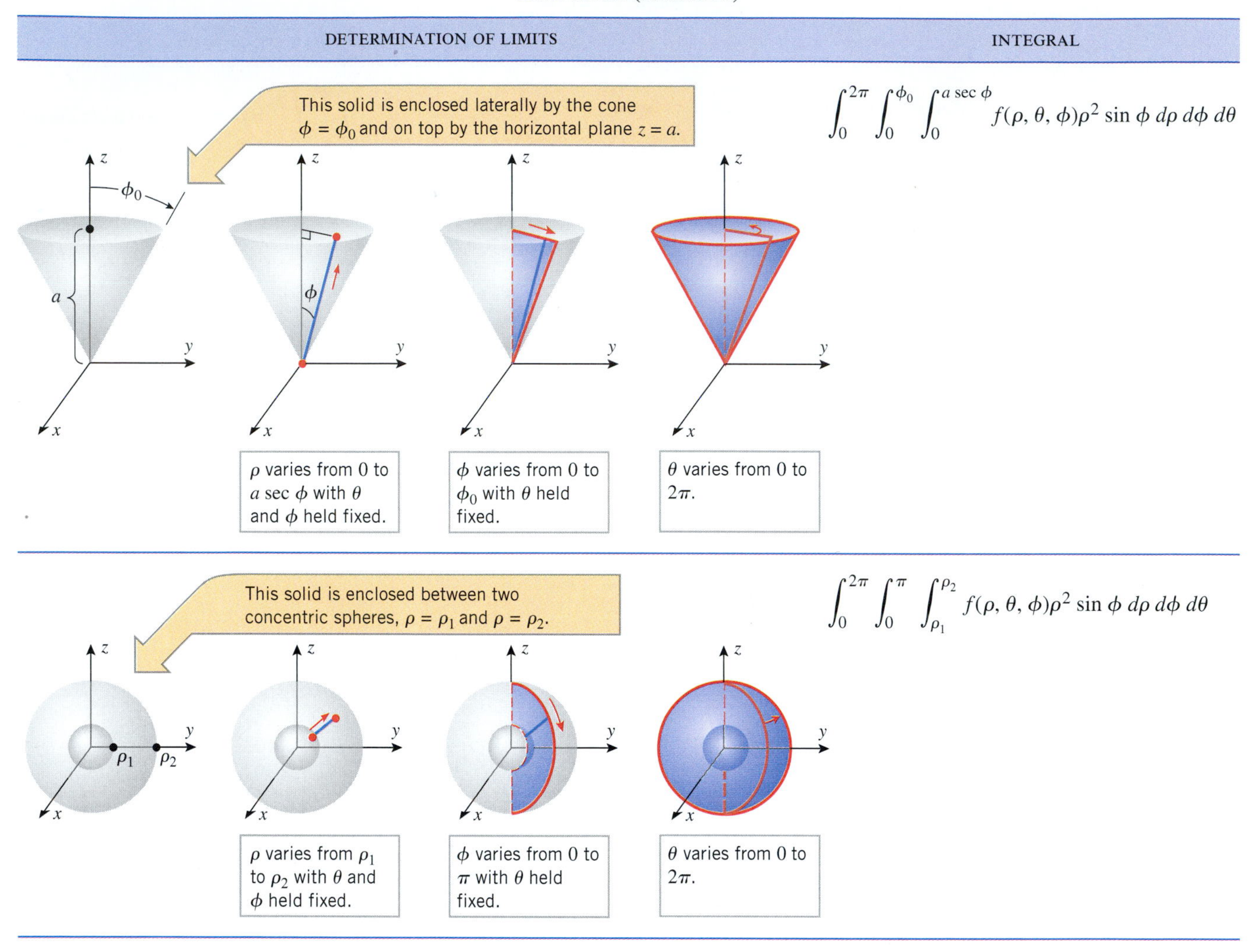

▶ **Example 3** Use spherical coordinates to find the volume and the centroid of the solid G bounded above by the sphere $x^2 + y^2 + z^2 = 16$ and below by the cone $z = \sqrt{x^2 + y^2}$.

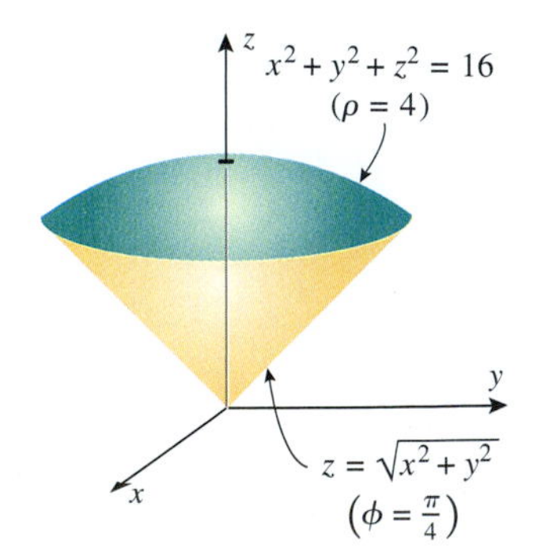

Figure 15.7.10

Solution. The solid G is sketched in Figure 15.7.10.

In spherical coordinates, the equation of the sphere $x^2 + y^2 + z^2 = 16$ is $\rho = 4$ and the equation of the cone $z = \sqrt{x^2 + y^2}$ is

$$\rho \cos \phi = \sqrt{\rho^2 \sin^2 \phi \cos^2 \theta + \rho^2 \sin^2 \phi \sin^2 \theta}$$

which simplifies to

$$\rho \cos \phi = \rho \sin \phi$$

or, on dividing both sides by $\rho \cos \phi$,

$$\tan \phi = 1$$

Thus $\phi = \pi/4$, and using the second entry in Table 15.7.1, the volume of G is

$$
\begin{aligned}
V = \iiint\limits_G dV &= \int_0^{2\pi}\int_0^{\pi/4}\int_0^4 \rho^2 \sin\phi \, d\rho \, d\phi \, d\theta \\
&= \int_0^{2\pi}\int_0^{\pi/4} \left[\frac{\rho^3}{3}\sin\phi\right]_{\rho=0}^{4} d\phi \, d\theta \\
&= \int_0^{2\pi}\int_0^{\pi/4} \frac{64}{3}\sin\phi \, d\phi \, d\theta \\
&= \frac{64}{3}\int_0^{2\pi} \left[-\cos\phi\right]_{\phi=0}^{\pi/4} d\theta = \frac{64}{3}\int_0^{2\pi}\left(1 - \frac{\sqrt{2}}{2}\right) d\theta \\
&= \frac{64\pi}{3}(2 - \sqrt{2})
\end{aligned}
$$

By symmetry, the centroid $(\bar{x}, \bar{y}, \bar{z})$ is on the z-axis, so $\bar{x} = \bar{y} = 0$. From (15) of Section 15.6 and the volume calculated above,

$$
\begin{aligned}
\bar{z} &= \frac{1}{V}\iiint\limits_G z \, dV = \frac{1}{V}\int_0^{2\pi}\int_0^{\pi/4}\int_0^4 (\rho\cos\phi)\rho^2 \sin\phi \, d\rho \, d\phi \, d\theta \\
&= \frac{1}{V}\int_0^{2\pi}\int_0^{\pi/4}\left[\frac{\rho^4}{4}\cos\phi\sin\phi\right]_{\rho=0}^{4} d\phi \, d\theta \\
&= \frac{64}{V}\int_0^{2\pi}\int_0^{\pi/4} \sin\phi\cos\phi \, d\phi \, d\theta = \frac{64}{V}\int_0^{2\pi}\left[\frac{1}{2}\sin^2\phi\right]_{\phi=0}^{\pi/4} d\theta \\
&= \frac{16}{V}\int_0^{2\pi} d\theta = \frac{32\pi}{V} = \frac{3}{2(2-\sqrt{2})}
\end{aligned}
$$

With the help of a calculator, $\bar{z} \approx 2.56$ (to two decimal places), so the approximate location of the centroid in the xyz-coordinate system is $(0, 0, 2.56)$. ◀

CONVERTING TRIPLE INTEGRALS FROM RECTANGULAR TO SPHERICAL COORDINATES

Referring to Table 12.8.1, triple integrals can be converted from rectangular coordinates to spherical coordinates by making the substitution $x = \rho\sin\phi\cos\theta$, $y = \rho\sin\phi\sin\theta$, $z = \rho\cos\phi$. The two integrals are related by the equation

$$
\iiint\limits_G f(x, y, z) \, dV = \iiint\limits_{\substack{\text{appropriate}\\ \text{limits}}} f(\rho\sin\phi\cos\theta, \rho\sin\phi\sin\theta, \rho\cos\phi)\rho^2\sin\phi \, d\rho \, d\phi \, d\theta \tag{10}
$$

▶ **Example 4** Use spherical coordinates to evaluate

$$
\int_{-2}^{2}\int_{-\sqrt{4-x^2}}^{\sqrt{4-x^2}}\int_0^{\sqrt{4-x^2-y^2}} z^2\sqrt{x^2+y^2+z^2} \, dz \, dy \, dx
$$

Solution. In problems like this, it is helpful to begin (when possible) with a sketch of the region G of integration. From the z-limits of integration, the upper surface of G is the hemisphere $z = \sqrt{4 - x^2 - y^2}$ and the lower surface is the xy-plane $z = 0$. From the x- and y-limits of integration, the projection of the solid G on the xy-plane is the region

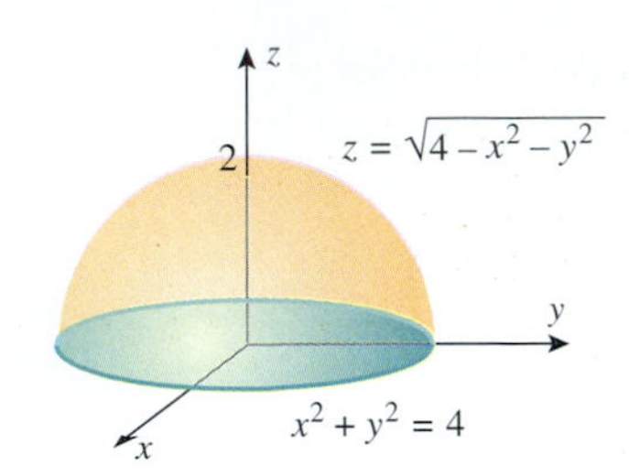

Figure 15.7.11

enclosed by the circle $x^2 + y^2 = 4$. From this information we obtain the sketch of G in Figure 15.7.11. Thus,

$$\int_{-2}^{2}\int_{-\sqrt{4-x^2}}^{\sqrt{4-x^2}}\int_{0}^{\sqrt{4-x^2-y^2}} z^2\sqrt{x^2+y^2+z^2}\,dz\,dy\,dx$$

$$= \iiint_G z^2\sqrt{x^2+y^2+z^2}\,dV$$

$$= \int_0^{2\pi}\int_0^{\pi/2}\int_0^{2} \rho^5\cos^2\phi\sin\phi\,d\rho\,d\phi\,d\theta$$

$$= \int_0^{2\pi}\int_0^{\pi/2} \frac{32}{3}\cos^2\phi\sin\phi\,d\phi\,d\theta$$

$$= \frac{32}{3}\int_0^{2\pi}\left[-\frac{1}{3}\cos^3\phi\right]_{\phi=0}^{\pi/2} d\theta = \frac{32}{9}\int_0^{2\pi} d\theta = \frac{64}{9}\pi \blacktriangleleft$$

✔ QUICK CHECK EXERCISES 15.7 (See page 1086 for answers.)

1. (a) The cylindrical wedge $1 \le r \le 3$, $\pi/6 \le \theta \le \pi/2$, $0 \le z \le 5$ has volume $V =$ ______.
(b) The spherical wedge $1 \le \rho \le 3$, $\pi/6 \le \theta \le \pi/2$, $0 \le \phi \le \pi/3$ has volume $V =$ ______.

2. Let G be the solid region inside the sphere of radius 2 centered at the origin and above the plane $z = 1$. In each part, supply the missing integrand and limits of integration for the iterated integral in cylindrical coordinates.
(a) The volume of G is

$$\iiint_G dV = \int_{\square}^{\square}\int_{\square}^{\square}\int_{\square}^{\square} \underline{\qquad\qquad}\,dz\,dr\,d\theta$$

(b) $$\iiint_G \frac{z}{x^2+y^2+z^2}\,dV = \int_{\square}^{\square}\int_{\square}^{\square}\int_{\square}^{\square} \underline{\qquad\qquad}\,dz\,dr\,d\theta$$

3. Let G be the solid region described in Quick Check Exercise 2. In each part, supply the missing integrand and limits of integration for the iterated integral in spherical coordinates.
(a) The volume of G is

$$\iiint_G dV = \int_{\square}^{\square}\int_{\square}^{\square}\int_{\square}^{\square} \underline{\qquad\qquad}\,d\rho\,d\phi\,d\theta$$

(b) $$\iiint_G \frac{z}{x^2+y^2+z^2}\,dV = \int_{\square}^{\square}\int_{\square}^{\square}\int_{\square}^{\square} \underline{\qquad\qquad}\,d\rho\,d\phi\,d\theta$$

EXERCISE SET 15.7 [C] CAS

1–4 Evaluate the iterated integral.

1. $\int_0^{2\pi}\int_0^{1}\int_0^{\sqrt{1-r^2}} zr\,dz\,dr\,d\theta$

2. $\int_0^{\pi/2}\int_0^{\cos\theta}\int_0^{r^2} r\sin\theta\,dz\,dr\,d\theta$

3. $\int_0^{\pi/2}\int_0^{\pi/2}\int_0^{1} \rho^3\sin\phi\cos\phi\,d\rho\,d\phi\,d\theta$

4. $\int_0^{2\pi}\int_0^{\pi/4}\int_0^{a\sec\phi} \rho^2\sin\phi\,d\rho\,d\phi\,d\theta \quad (a > 0)$

FOCUS ON CONCEPTS

5. Sketch the region G and identify the function f so that

$$\iiint_G f(r,\theta,z)\,dV$$

corresponds to the iterated integral in Exercise 1.

6. Sketch the region G and identify the function f so that

$$\iiint_G f(r, \theta, z)\, dV$$

corresponds to the iterated integral in Exercise 2.

7. Sketch the region G and identify the function f so that

$$\iiint_G f(\rho, \theta, \phi)\, dV$$

corresponds to the iterated integral in Exercise 3.

8. Sketch the region G and identify the function f so that

$$\iiint_G f(\rho, \theta, \phi)\, dV$$

corresponds to the iterated integral in Exercise 4.

9–12 Use cylindrical coordinates to find the volume of the solid.

9. The solid enclosed by the paraboloid $z = x^2 + y^2$ and the plane $z = 9$.

10. The solid that is bounded above and below by the sphere $x^2 + y^2 + z^2 = 9$ and inside the cylinder $x^2 + y^2 = 4$.

11. The solid that is inside the surface $r^2 + z^2 = 20$ but not above the surface $z = r^2$.

12. The solid enclosed between the cone $z = (hr)/a$ and the plane $z = h$.

13–16 Use spherical coordinates to find the volume of the solid.

13. The solid bounded above by the sphere $\rho = 4$ and below by the cone $\phi = \pi/3$.

14. The solid within the cone $\phi = \pi/4$ and between the spheres $\rho = 1$ and $\rho = 2$.

15. The solid enclosed by the sphere $x^2 + y^2 + z^2 = 4a^2$ and the planes $z = 0$ and $z = a$.

16. The solid within the sphere $x^2 + y^2 + z^2 = 9$, outside the cone $z = \sqrt{x^2 + y^2}$, and above the xy-plane.

17–20 Use cylindrical or spherical coordinates to evaluate the integral.

17. $\displaystyle\int_0^a \int_0^{\sqrt{a^2-x^2}} \int_0^{a^2-x^2-y^2} x^2\, dz\, dy\, dx \quad (a > 0)$

18. $\displaystyle\int_{-1}^1 \int_0^{\sqrt{1-x^2}} \int_0^{\sqrt{1-x^2-y^2}} e^{-(x^2+y^2+z^2)^{3/2}}\, dz\, dy\, dx$

19. $\displaystyle\int_0^2 \int_0^{\sqrt{4-y^2}} \int_{\sqrt{x^2+y^2}}^{\sqrt{8-x^2-y^2}} z^2\, dz\, dx\, dy$

20. $\displaystyle\int_{-3}^3 \int_{-\sqrt{9-y^2}}^{\sqrt{9-y^2}} \int_{-\sqrt{9-x^2-y^2}}^{\sqrt{9-x^2-y^2}} \sqrt{x^2 + y^2 + z^2}\, dz\, dx\, dy$

C **21.** (a) Use a CAS to evaluate

$$\int_{-2}^2 \int_1^4 \int_{\pi/6}^{\pi/3} \frac{r \tan^3 \theta}{\sqrt{1 + z^2}}\, d\theta\, dr\, dz$$

(b) Find a function $f(x, y, z)$ and sketch a region G in 3-space so that the triple integral in rectangular coordinates

$$\iiint_G f(x, y, z)\, dV$$

matches the iterated integral in cylindrical coordinates given in part (a).

C **22.** Use a CAS to evaluate

$$\int_0^{\pi/2} \int_0^{\pi/4} \int_0^{\cos\theta} \rho^{17} \cos\phi \cos^{19}\theta\, d\rho\, d\phi\, d\theta$$

23. Find the volume enclosed by $x^2 + y^2 + z^2 = a^2$ using
(a) cylindrical coordinates
(b) spherical coordinates.

24. Let G be the solid in the first octant bounded by the sphere $x^2 + y^2 + z^2 = 4$ and the coordinate planes. Evaluate

$$\iiint_G xyz\, dV$$

(a) using rectangular coordinates
(b) using cylindrical coordinates
(c) using spherical coordinates.

25–26 Use cylindrical coordinates.

25. Find the mass of the solid with density $\delta(x, y, z) = 3 - z$ that is bounded by the cone $z = \sqrt{x^2 + y^2}$ and the plane $z = 3$.

26. Find the mass of a right circular cylinder of radius a and height h if the density is proportional to the distance from the base. (Let k be the constant of proportionality.)

27–28 Use spherical coordinates.

27. Find the mass of a spherical solid of radius a if the density is proportional to the distance from the center. (Let k be the constant of proportionality.)

28. Find the mass of the solid enclosed between the spheres $x^2 + y^2 + z^2 = 1$ and $x^2 + y^2 + z^2 = 4$ if the density is $\delta(x, y, z) = (x^2 + y^2 + z^2)^{-1/2}$.

29–30 Use cylindrical coordinates to find the centroid of the solid.

29. The solid that is bounded above by the sphere

$$x^2 + y^2 + z^2 = 2$$

and below by the paraboloid $z = x^2 + y^2$.

30. The solid that is bounded by the cone $z = \sqrt{x^2 + y^2}$ and the plane $z = 2$.

31–32 Use spherical coordinates to find the centroid of the solid.

31. The solid in the first octant bounded by the coordinate planes and the sphere $x^2 + y^2 + z^2 = a^2$.

32. The solid bounded above by the sphere $\rho = 4$ and below by the cone $\phi = \pi/3$.

33–34 Use the Wallis formulas in Exercises 64 and 66 of Section 8.3.

33. Find the centroid of the solid bounded above by the paraboloid $z = x^2 + y^2$, below by the plane $z = 0$, and laterally by the cylinder $(x-1)^2 + y^2 = 1$.

34. Find the mass of the solid in the first octant bounded above by the paraboloid $z = 4 - x^2 - y^2$, below by the plane $z = 0$, and laterally by the cylinder $x^2 + y^2 = 2x$ and the plane $y = 0$, assuming the density to be $\delta(x, y, z) = z$.

35–40 Solve the problem using either cylindrical or spherical coordinates (whichever seems appropriate).

35. Find the volume of the solid in the first octant bounded by the sphere $\rho = 2$, the coordinate planes, and the cones $\phi = \pi/6$ and $\phi = \pi/3$.

36. Find the mass of the solid that is enclosed by the sphere $x^2 + y^2 + z^2 = 1$ and lies within the cone $z = \sqrt{x^2 + y^2}$ if the density is $\delta(x, y, z) = \sqrt{x^2 + y^2 + z^2}$.

37. Find the center of gravity of the solid bounded by the paraboloid $z = 1 - x^2 - y^2$ and the xy-plane, assuming the density to be $\delta(x, y, z) = x^2 + y^2 + z^2$.

38. Find the center of gravity of the solid that is bounded by the cylinder $x^2 + y^2 = 1$, the cone $z = \sqrt{x^2 + y^2}$, and the xy-plane if the density is $\delta(x, y, z) = z$.

39. Find the center of gravity of the solid hemisphere bounded by $z = \sqrt{a^2 - x^2 - y^2}$ and $z = 0$ if the density is proportional to the distance from the origin.

40. Find the centroid of the solid that is enclosed by the hemispheres $y = \sqrt{9 - x^2 - z^2}$, $y = \sqrt{4 - x^2 - z^2}$, and the plane $y = 0$.

41. Suppose that the density at a point in a gaseous spherical star is modeled by the formula

$$\delta = \delta_0 e^{-(\rho/R)^3}$$

where δ_0 is a positive constant, R is the radius of the star, and ρ is the distance from the point to the star's center. Find the mass of the star.

42. In this exercise we will obtain a formula for the volume of the spherical wedge in Figure 15.7.7.

(a) Use a triple integral in cylindrical coordinates to show that the volume of the solid bounded above by a sphere $\rho = \rho_0$, below by a cone $\phi = \phi_0$, and on the sides by $\theta = \theta_1$ and $\theta = \theta_2$ $(\theta_1 < \theta_2)$ is

$$V = \tfrac{1}{3}\rho_0^3(1 - \cos\phi_0)(\theta_2 - \theta_1)$$

[*Hint:* In cylindrical coordinates, the sphere has the equation $r^2 + z^2 = \rho_0^2$ and the cone has the equation $z = r \cot\phi_0$. For simplicity, consider only the case $0 < \phi_0 < \pi/2$.]

(b) Subtract appropriate volumes and use the result in part (a) to deduce that the volume ΔV of the spherical wedge is

$$\Delta V = \frac{\rho_2^3 - \rho_1^3}{3}(\cos\phi_1 - \cos\phi_2)(\theta_2 - \theta_1)$$

(c) Apply the Mean-Value Theorem to the functions $\cos\phi$ and ρ^3 to deduce that the formula in part (b) can be written as

$$\Delta V = \rho^{*2} \sin\phi^* \,\Delta\rho\,\Delta\phi\,\Delta\theta$$

where ρ^* is between ρ_1 and ρ_2, ϕ^* is between ϕ_1 and ϕ_2, and $\Delta\rho = \rho_2 - \rho_1$, $\Delta\phi = \phi_2 - \phi_1$, $\Delta\theta = \theta_2 - \theta_1$.

43–46 The tendency of a solid to resist a change in rotational motion about an axis is measured by its ***moment of inertia*** about that axis. If the solid occupies a region G in an xyz-coordinate system, and if its density function $\delta(x, y, z)$ is continuous on G, then the moments of inertia about the x-axis, the y-axis, and the z-axis are denoted by I_x, I_y, and I_z, respectively, and are defined by

$$I_x = \iiint_G (y^2 + z^2)\,\delta(x, y, z)\,dV$$

$$I_y = \iiint_G (x^2 + z^2)\,\delta(x, y, z)\,dV$$

$$I_z = \iiint_G (x^2 + y^2)\,\delta(x, y, z)\,dV$$

(compare with the discussion preceding Exercise 43 in Section 15.6). In these exercises, find the indicated moments of inertia of the solid, assuming that it has constant density δ.

43. I_z for the solid cylinder $x^2 + y^2 \le a^2$, $0 \le z \le h$.

44. I_y for the solid cylinder $x^2 + y^2 \le a^2$, $0 \le z \le h$.

45. I_z for the hollow cylinder $a_1^2 \le x^2 + y^2 \le a_2^2$, $0 \le z \le h$.

46. I_z for the solid sphere $x^2 + y^2 + z^2 \le a^2$.

✔QUICK CHECK ANSWERS 15.7

1. (a) $\frac{20}{3}\pi$ (b) $\frac{13}{9}\pi$ **2.** (a) $\int_0^{2\pi}\int_0^{\sqrt{3}}\int_1^{\sqrt{4-r^2}} r\,dz\,dr\,d\theta$ (b) $\int_0^{2\pi}\int_0^{\sqrt{3}}\int_1^{\sqrt{4-r^2}} \frac{rz}{r^2+z^2}\,dz\,dr\,d\theta$

3. (a) $\int_0^{2\pi}\int_0^{\pi/3}\int_{\sec\phi}^{2} \rho^2\sin\phi\,d\rho\,d\phi\,d\theta$ (b) $\int_0^{2\pi}\int_0^{\pi/3}\int_{\sec\phi}^{2} \rho\cos\phi\sin\phi\,d\rho\,d\phi\,d\theta$

15.8 CHANGE OF VARIABLES IN MULTIPLE INTEGRALS; JACOBIANS

In this section we will discuss a general method for evaluating double and triple integrals by substitution. Most of the results in this section are very difficult to prove, so our approach will be informal and motivational. Our goal is to provide a geometric understanding of the basic principles and an exposure to computational techniques.

CHANGE OF VARIABLE IN A SINGLE INTEGRAL

To motivate techniques for evaluating double and triple integrals by substitution, it will be helpful to consider the effect of a substitution $x = g(u)$ on a single integral over an interval $[a, b]$. If g is differentiable and either increasing or decreasing, then g is one-to-one and

$$\int_a^b f(x)\,dx = \int_{g^{-1}(a)}^{g^{-1}(b)} f(g(u))g'(u)\,du$$

In this relationship $f(x)$ and dx are expressed in terms of u, and the u-limits of integration result from solving the equations

$$a = g(u) \quad \text{and} \quad b = g(u)$$

In the case where g is decreasing we have $g^{-1}(b) < g^{-1}(a)$, which is contrary to our usual convention of writing definite integrals with the larger limit of integration at the top. We can remedy this by reversing the limits of integration and writing

$$\int_a^b f(x)\,dx = -\int_{g^{-1}(b)}^{g^{-1}(a)} f(g(u))g'(u)\,du = \int_{g^{-1}(b)}^{g^{-1}(a)} f(g(u))|g'(u)|\,du$$

where the absolute value results from the fact that $g'(u)$ is negative. Thus, regardless of whether g is increasing or decreasing we can write

$$\int_a^b f(x)\,dx = \int_\alpha^\beta f(g(u))|g'(u)|\,du \tag{1}$$

where α and β are the u-limits of integration and $\alpha < \beta$.

The expression $g'(u)$ that appears in (1) is called the ***Jacobian*** of the change of variable $x = g(u)$ in honor of C. G. J. Jacobi, who made the first serious study of change of variables in multiple integrals in the mid-1800s. Formula (1) reveals three effects of the change of variable $x = g(u)$:

- The new integrand becomes $f(g(u))$ times the absolute value of the Jacobian.
- dx becomes du.
- The x-interval of integration is transformed into a u-interval of integration.

Our goal in this section is to show that analogous results hold for changing variables in double and triple integrals.

TRANSFORMATIONS OF THE PLANE

In earlier sections we considered parametric equations of three kinds:

$$x = x(t), \quad y = y(t)$$ A curve in the plane

$$x = x(t), \quad y = y(t), \quad z = z(t)$$ A curve in 3-space

$$x = x(u, v), \quad y = y(u, v), \quad z = z(u, v)$$ A surface in 3-space

Now we will consider parametric equations of the form

$$x = x(u, v), \quad y = y(u, v) \tag{2}$$

Parametric equations of this type associate points in the xy-plane with points in the uv-plane. These equations can be written in vector form as

$$\mathbf{r} = \mathbf{r}(u, v) = x(u, v)\mathbf{i} + y(u, v)\mathbf{j}$$

where $\mathbf{r} = x\mathbf{i} + y\mathbf{j}$ is a position vector in the xy-plane and $\mathbf{r}(u, v)$ is a vector-valued function of the variables u and v.

It will also be useful in this section to think of the parametric equations in (2) in terms of inputs and outputs. If we think of the pair of numbers (u, v) as an input, then the two equations, in combination, produce a unique output (x, y), and hence define a function T that associates points in the xy-plane with points in the uv-plane. This function is described by the formula

$$T(u, v) = (x(u, v), y(u, v))$$

We call T a ***transformation*** from the uv-plane to the xy-plane and (x, y) the ***image*** of (u, v) under the transformation T. We also say that T ***maps*** (u, v) into (x, y). The set R of all images in the xy-plane of a set S in the uv-plane is called the ***image of S under T***. If distinct points in the uv-plane have distinct images in the xy-plane, then T is said to be ***one-to-one***. In this case the equations in (2) define u and v as functions of x and y, say

$$u = u(x, y), \quad v = v(x, y)$$

These equations, which can often be obtained by solving (2) for u and v in terms of x and y, define a transformation from the xy-plane to the uv-plane that maps the image of (u, v) under T back into (u, v). This transformation is denoted by T^{-1} and is called the ***inverse of T*** (Figure 15.8.1).

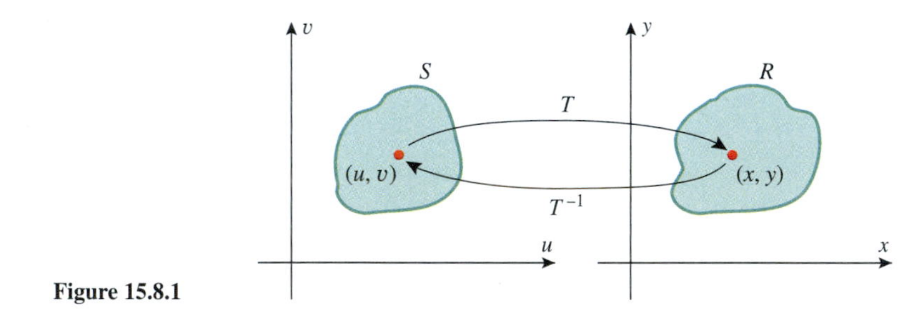

Figure 15.8.1

Carl Gustav Jacob Jacobi (1804–1851) German mathematician. Jacobi, the son of a banker, grew up in a background of wealth and culture and showed brilliance in mathematics early. He resisted studying mathematics by rote, preferring instead to learn general principles from the works of the masters, Euler and Lagrange. He entered the University of Berlin at age 16 as a student of mathematics and classical studies. However, he soon realized that he could not do both and turned fully to mathematics with a blazing intensity that he would maintain throughout his life. He received his Ph.D. in 1825 and was able to secure a position as a lecturer at the University of Berlin by giving up Judaism and becoming a Christian. However, his promotion opportunities remained limited and he moved on to the University of Königsberg. Jacobi was born to teach—he had a dynamic personality and delivered his lectures with a clarity and enthusiasm that frequently left his audience spellbound. In spite of extensive teaching commitments, he was able to publish volumes of revolutionary mathematical research that eventually made him the leading European mathematician after Gauss. His main body of research was in the area of elliptic functions, a branch of mathematics with important applications in astronomy and physics as well as in other fields of mathematics. Because of his family wealth, Jacobi was not dependent on his teaching salary in his early years. However, his comfortable world eventually collapsed. In 1840 his family went bankrupt and he was wiped out financially. In 1842 he had a nervous breakdown from overwork. In 1843 he became seriously ill with diabetes and moved to Berlin with the help of a government grant to defray his medical expenses. In 1848 he made an injudicious political speech that caused the government to withdraw the grant, eventually resulting in the loss of his home. His health continued to decline and in 1851 he finally succumbed to successive bouts of influenza and smallpox. In spite of all his problems, Jacobi was a tireless worker to the end. When a friend expressed concern about the effect of the hard work on his health, Jacobi replied, "Certainly, I have sometimes endangered my health by overwork, but what of it? Only cabbages have no nerves, no worries. And what do they get out of their perfect well-being?"

One way to visualize the geometric effect of a transformation T is to determine the images in the xy-plane of the vertical and horizontal lines in the uv-plane. Following the discussion on page 1044 in Section 15.4, sets of points in the xy-plane that are images of horizontal lines (v constant) are called ***constant v-curves***, and sets of points that are images of vertical lines (u constant) are called ***constant u-curves*** (Figure 15.8.2).

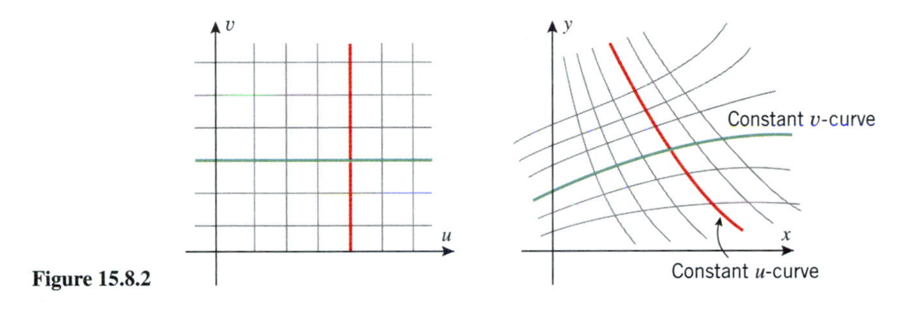

Figure 15.8.2

▶ Example 1 Let T be the transformation from the uv-plane to the xy-plane defined by the equations

$$x = \tfrac{1}{4}(u + v), \quad y = \tfrac{1}{2}(u - v) \tag{3}$$

(a) Find $T(1, 3)$.

(b) Sketch the constant v-curves corresponding to $v = -2, -1, 0, 1, 2$.

(c) Sketch the constant u-curves corresponding to $u = -2, -1, 0, 1, 2$.

(d) Sketch the image under T of the square region in the uv-plane bounded by the lines $u = -2$, $u = 2$, $v = -2$, and $v = 2$.

Solution (a). Substituting $u = 1$ and $v = 3$ in (3) yields $T(1, 3) = (1, -1)$.

Solutions (b and c). In these parts it will be convenient to express the transformation equations with u and v as functions of x and y. We leave it for you to show that

$$u = 2x + y, \quad v = 2x - y$$

Thus, the constant v-curves corresponding to $v = -2, -1, 0, 1$, and 2 are

$$2x - y = -2, \quad 2x - y = -1, \quad 2x - y = 0, \quad 2x - y = 1, \quad 2x - y = 2$$

and the constant u-curves corresponding to $u = -2, -1, 0, 1$, and 2 are

$$2x + y = -2, \quad 2x + y = -1, \quad 2x + y = 0, \quad 2x + y = 1, \quad 2x + y = 2$$

In Figure 15.8.3 the constant v-curves are shown in green and the constant u-curves in red.

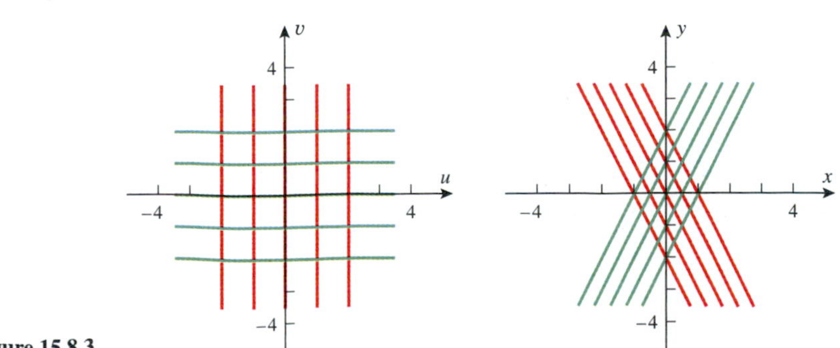

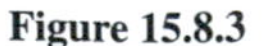

Figure 15.8.3

Solution (d). The image of a region can often be found by finding the image of its boundary. In this case the images of the boundary lines $u = -2$, $u = 2$, $v = -2$, and $v = 2$ enclose the diamond-shaped region in the xy-plane shown in Figure 15.8.4. ◀

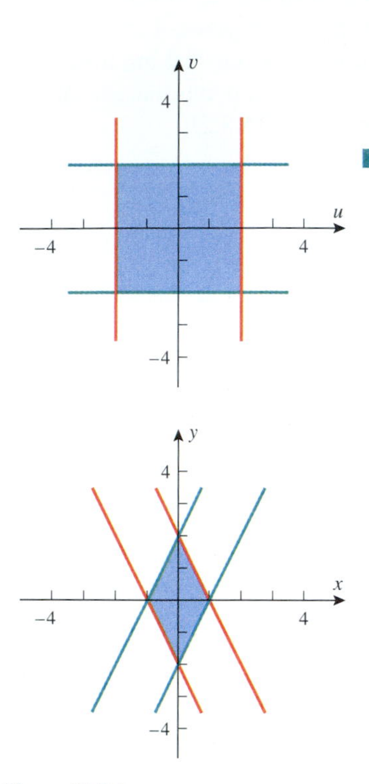

Figure 15.8.4

JACOBIANS IN TWO VARIABLES

To derive the change of variables formula for double integrals, we will need to understand the relationship between the area of a *small* rectangular region in the uv-plane and the area of its image in the xy-plane under a transformation T given by the equations

$$x = x(u, v), \quad y = y(u, v)$$

For this purpose, suppose that Δu and Δv are positive, and consider a rectangular region S in the uv-plane enclosed by the lines

$$u = u_0, \quad u = u_0 + \Delta u, \quad v = v_0, \quad v = v_0 + \Delta v$$

If the functions $x(u, v)$ and $y(u, v)$ are continuous, and if Δu and Δv are not too large, then the image of S in the xy-plane will be a region R that looks like a slightly distorted parallelogram (Figure 15.8.5). The sides of R are the constant u-curves and v-curves that correspond to the sides of S.

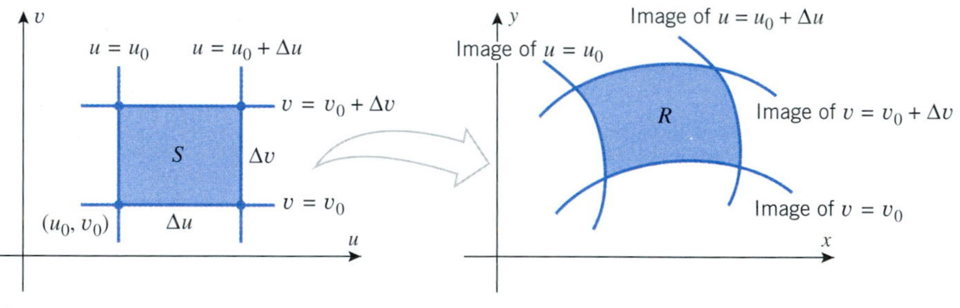

Figure 15.8.5

If we let

$$\mathbf{r} = \mathbf{r}(u, v) = x(u, v)\mathbf{i} + y(u, v)\mathbf{j}$$

be the position vector of the point in the xy-plane that corresponds to the point (u, v) in the uv-plane, then the constant v-curve corresponding to $v = v_0$ and the constant u-curve corresponding to $u = u_0$ can be represented in vector form as

$$\mathbf{r}(u, v_0) = x(u, v_0)\mathbf{i} + y(u, v_0)\mathbf{j} \quad \text{Constant } v\text{-curve}$$

$$\mathbf{r}(u_0, v) = x(u_0, v)\mathbf{i} + y(u_0, v)\mathbf{j} \quad \text{Constant } u\text{-curve}$$

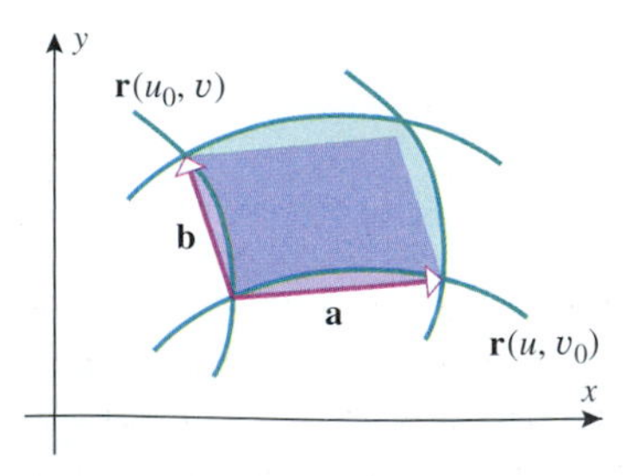

Figure 15.8.6

Since we are assuming Δu and Δv to be small, the region R can be approximated by a parallelogram determined by the "secant vectors"

$$\mathbf{a} = \mathbf{r}(u_0 + \Delta u, v_0) - \mathbf{r}(u_0, v_0) \tag{4}$$

$$\mathbf{b} = \mathbf{r}(u_0, v_0 + \Delta v) - \mathbf{r}(u_0, v_0) \tag{5}$$

shown in Figure 15.8.6. A more useful approximation of R can be obtained by using Formulas (5) and (6) of Section 15.4 to approximate these secant vectors by tangent vectors as follows:

$$\mathbf{a} = \frac{\mathbf{r}(u_0 + \Delta u, v_0) - \mathbf{r}(u_0, v_0)}{\Delta u}\Delta u$$

$$\approx \frac{\partial \mathbf{r}}{\partial u}\Delta u = \left(\frac{\partial x}{\partial u}\mathbf{i} + \frac{\partial y}{\partial u}\mathbf{j}\right)\Delta u$$

$$\mathbf{b} = \frac{\mathbf{r}(u_0, v_0 + \Delta v) - \mathbf{r}(u_0, v_0)}{\Delta v}\Delta v$$

$$\approx \frac{\partial \mathbf{r}}{\partial v}\Delta v = \left(\frac{\partial x}{\partial v}\mathbf{i} + \frac{\partial y}{\partial v}\mathbf{j}\right)\Delta v$$

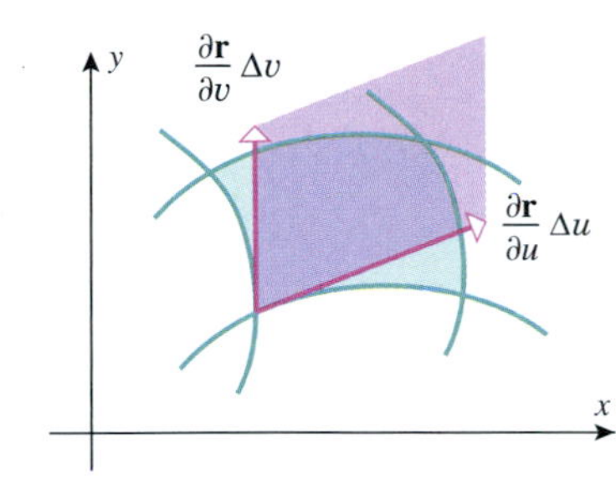

Figure 15.8.7

where the partial derivatives are evaluated at (u_0, v_0) (Figure 15.8.7). Hence, it follows that the area of the region R, which we will denote by ΔA, can be approximated by the area of the parallelogram determined by these vectors. Thus, from Formula (8) of Section 12.4 we have

$$\Delta A \approx \left\|\frac{\partial \mathbf{r}}{\partial u}\Delta u \times \frac{\partial \mathbf{r}}{\partial v}\Delta v\right\| = \left\|\frac{\partial \mathbf{r}}{\partial u} \times \frac{\partial \mathbf{r}}{\partial v}\right\|\Delta u\,\Delta v \tag{6}$$

where the derivatives are evaluated at (u_0, v_0). Computing the cross product, we obtain

$$\frac{\partial \mathbf{r}}{\partial u} \times \frac{\partial \mathbf{r}}{\partial v} = \begin{vmatrix} \mathbf{i} & \mathbf{j} & \mathbf{k} \\ \frac{\partial x}{\partial u} & \frac{\partial y}{\partial u} & 0 \\ \frac{\partial x}{\partial v} & \frac{\partial y}{\partial v} & 0 \end{vmatrix} = \begin{vmatrix} \frac{\partial x}{\partial u} & \frac{\partial y}{\partial u} \\ \frac{\partial x}{\partial v} & \frac{\partial y}{\partial v} \end{vmatrix}\mathbf{k} = \begin{vmatrix} \frac{\partial x}{\partial u} & \frac{\partial x}{\partial v} \\ \frac{\partial y}{\partial u} & \frac{\partial y}{\partial v} \end{vmatrix}\mathbf{k} \tag{7}$$

The determinant in (7) is sufficiently important that it has its own terminology and notation.

15.8.1 DEFINITION. If T is the transformation from the uv-plane to the xy-plane defined by the equations $x = x(u, v)$, $y = y(u, v)$, then the ***Jacobian of T*** is denoted by $J(u, v)$ or by $\partial(x, y)/\partial(u, v)$ and is defined by

$$J(u, v) = \frac{\partial(x, y)}{\partial(u, v)} = \begin{vmatrix} \frac{\partial x}{\partial u} & \frac{\partial x}{\partial v} \\ \frac{\partial y}{\partial u} & \frac{\partial y}{\partial v} \end{vmatrix} = \frac{\partial x}{\partial u}\frac{\partial y}{\partial v} - \frac{\partial y}{\partial u}\frac{\partial x}{\partial v}$$

Using the notation in this definition, it follows from (6) and (7) that

$$\Delta A \approx \left\|\frac{\partial(x, y)}{\partial(u, v)}\mathbf{k}\right\|\Delta u\,\Delta v$$

or, since $\mathbf{k}$ is a unit vector,

$$\Delta A \approx \left|\frac{\partial(x, y)}{\partial(u, v)}\right|\Delta u\,\Delta v \tag{8}$$

At the point (u_0, v_0) this important formula relates the areas of the regions R and S in Figure 15.8.5; it tells us that *for small values of Δu and Δv, the area of R is approximately the absolute value of the Jacobian times the area of S.* Moreover, it is proved in advanced calculus courses that the relative error in the approximation approaches zero as $\Delta u \to 0$ and $\Delta v \to 0$.

CHANGE OF VARIABLES IN DOUBLE INTEGRALS

Our next objective is to provide a geometric motivation for the following result.

A precise statement of conditions under which Formula (9) holds is beyond the scope of this course. Suffice it to say that the formula holds if T is a one-to-one transformation, $f(x, y)$ is continuous on R, the partial derivatives of $x(u, v)$ and $y(u, v)$ exist and are continuous on S, and the regions R and S are not complicated.

15.8.2 CHANGE OF VARIABLES FORMULA FOR DOUBLE INTEGRALS. If the transformation $x = x(u, v)$, $y = y(u, v)$ maps the region S in the uv-plane into the region R in the xy-plane, and if the Jacobian $\partial(x, y)/\partial(u, v)$ is nonzero and does not change sign on S, then with appropriate restrictions on the transformation and the regions it follows that

$$\iint_R f(x, y)\, dA_{xy} = \iint_S f(x(u, v), y(u, v)) \left| \frac{\partial(x, y)}{\partial(u, v)} \right| dA_{uv} \tag{9}$$

where we have attached subscripts to the dA's to help identify the associated variables.

To motivate Formula (9), we proceed as follows:

- Subdivide the region S in the uv-plane into pieces by lines parallel to the coordinate axes, and exclude from consideration any pieces that contain points outside of S. This leaves only rectangular regions that are subsets of S. Assume that there are n such regions and denote the kth such region by S_k. Assume that S_k has dimensions Δu_k by Δv_k and, as shown in Figure 15.8.8a, let (u_k^*, v_k^*) be its "lower left corner."
- As shown in Figure 15.8.8b, the transformation T defined by the equations $x = x(u, v)$, $y = y(u, v)$ maps S_k into a curvilinear parallelogram R_k in the xy-plane and maps the point (u_k^*, v_k^*) into the point $(x_k^*, y_k^*) = (x(u_k^*, v_k^*), y(u_k^*, v_k^*))$ in R_k. Denote the area of R_k by ΔA_k.
- In rectangular coordinates the double integral of $f(x, y)$ over a region R is defined as a limit of Riemann sums in which R is subdivided into *rectangular* subregions. It is proved in advanced calculus courses that under appropriate conditions subdivisions into *curvilinear* parallelograms can be used instead. Accepting this to be so, we can approximate the double integral of $f(x, y)$ over R as

$$\begin{aligned}\iint_R f(x, y)\, dA_{xy} &\approx \sum_{k=1}^{n} f(x_k^*, y_k^*)\, \Delta A_k \\ &\approx \sum_{k=1}^{n} f(x(u_k^*, v_k^*), y(u_k^*, v_k^*)) \left| \frac{\partial(x, y)}{\partial(u, v)} \right| \Delta u_k\, \Delta v_k\end{aligned}$$

where the Jacobian is evaluated at (u_k^*, v_k^*). But the last expression is a Riemann sum for the integral

$$\iint_S f(x(u, v), y(u, v)) \left| \frac{\partial(x, y)}{\partial(u, v)} \right| dA_{uv}$$

so Formula (9) follows if we assume that the errors in the approximations approach zero as $n \to +\infty$.

Example 2 Evaluate

$$\iint_R \frac{x - y}{x + y}\, dA$$

where R is the region enclosed by $x - y = 0$, $x - y = 1$, $x + y = 1$, and $x + y = 3$ (Figure 15.8.9a).

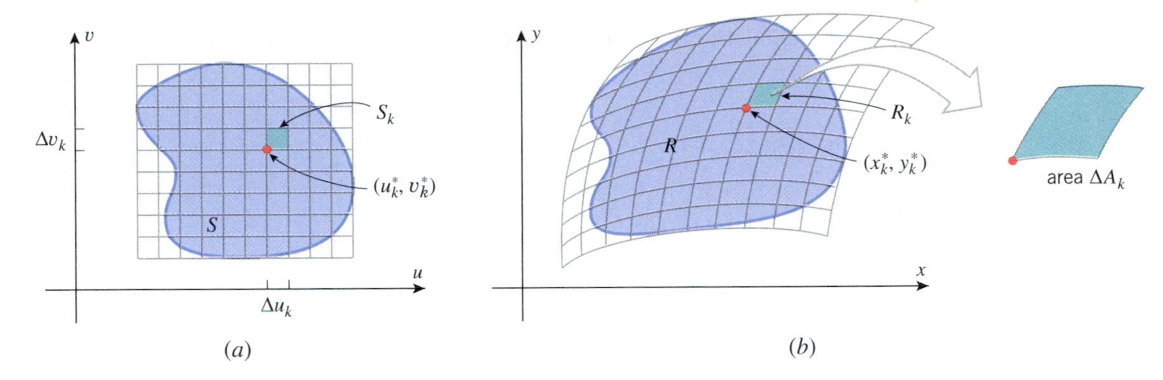

Figure 15.8.8

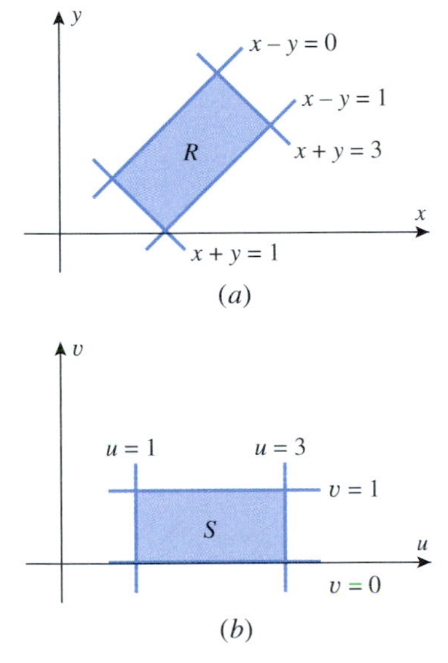

Figure 15.8.9

Solution. This integral would be tedious to evaluate directly because the region R is oriented in such a way that we would have to subdivide it and integrate over each part separately. However, the occurrence of the expressions $x - y$ and $x + y$ in the equations of the boundary suggests that the transformation

$$u = x + y, \quad v = x - y \tag{10}$$

would be helpful, since with this transformation the boundary lines

$$x + y = 1, \quad x + y = 3, \quad x - y = 0, \quad x - y = 1$$

are constant u-curves and constant v-curves corresponding to the lines

$$u = 1, \quad u = 3, \quad v = 0, \quad v = 1$$

in the uv-plane. These lines enclose the rectangular region S shown in Figure 15.8.9*b*. To find the Jacobian $\partial(x, y)/\partial(u, v)$ of this transformation, we first solve (10) for x and y in terms of u and v. This yields

$$x = \tfrac{1}{2}(u + v), \quad y = \tfrac{1}{2}(u - v)$$

from which we obtain

$$\frac{\partial(x, y)}{\partial(u, v)} = \begin{vmatrix} \dfrac{\partial x}{\partial u} & \dfrac{\partial x}{\partial v} \\[2mm] \dfrac{\partial y}{\partial u} & \dfrac{\partial y}{\partial v} \end{vmatrix} = \begin{vmatrix} \frac{1}{2} & \frac{1}{2} \\ \frac{1}{2} & -\frac{1}{2} \end{vmatrix} = -\tfrac{1}{4} - \tfrac{1}{4} = -\tfrac{1}{2}$$

Thus, from Formula (9), but with the notation dA rather than dA_{xy},

$$\begin{aligned} \iint_R \frac{x - y}{x + y}\, dA &= \iint_S \frac{v}{u} \left|\frac{\partial(x, y)}{\partial(u, v)}\right| dA_{uv} \\ &= \iint_S \frac{v}{u} \left|-\frac{1}{2}\right| dA_{uv} = \frac{1}{2} \int_0^1 \int_1^3 \frac{v}{u}\, du\, dv \\ &= \frac{1}{2} \int_0^1 v \ln|u| \Big]_{u=1}^{3} dv \\ &= \frac{1}{2} \ln 3 \int_0^1 v\, dv = \frac{1}{4} \ln 3 \end{aligned}$$

The underlying idea illustrated in Example 2 is to find a one-to-one transformation that maps a rectangle S in the uv-plane into the region R of integration, and then use that transformation as a substitution in the integral to produce an equivalent integral over S.

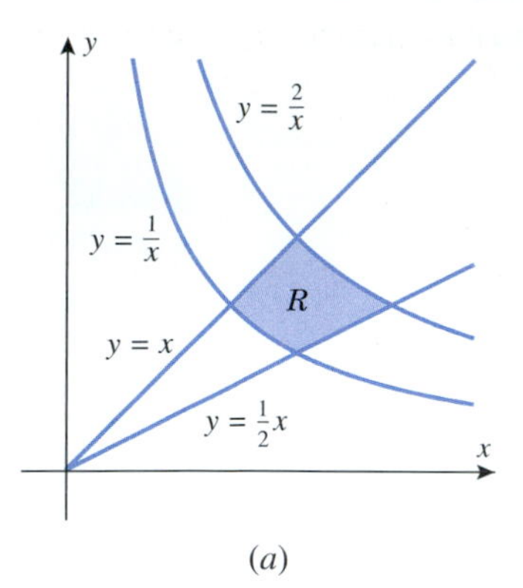

(a)

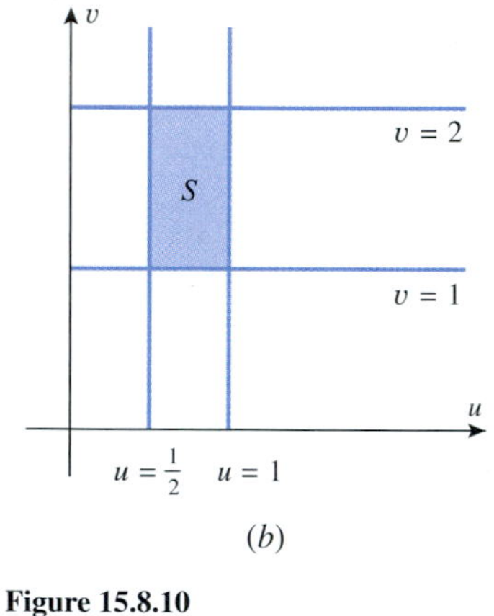

(b)

Figure 15.8.10

▶ **Example 3** Evaluate

$$\iint_R e^{xy}\, dA$$

where R is the region enclosed by the lines $y = \frac{1}{2}x$ and $y = x$ and the hyperbolas $y = 1/x$ and $y = 2/x$ (Figure 15.8.10a).

Solution. As in the last example, we look for a transformation in which the boundary curves in the xy-plane become constant v-curves and constant u-curves. For this purpose we rewrite the four boundary curves as

$$\frac{y}{x} = \frac{1}{2}, \quad \frac{y}{x} = 1, \quad xy = 1, \quad xy = 2$$

which suggests the transformation

$$u = \frac{y}{x}, \quad v = xy \tag{11}$$

With this transformation the boundary curves in the xy-plane are constant u-curves and constant v-curves corresponding to the lines

$$u = \tfrac{1}{2}, \quad u = 1, \quad v = 1, \quad v = 2$$

in the uv-plane. These lines enclose the region S shown in Figure 15.8.10b. To find the Jacobian $\partial(x, y)/\partial(u, v)$ of this transformation, we first solve (11) for x and y in terms of u and v. This yields

$$x = \sqrt{v/u}, \quad y = \sqrt{uv}$$

from which we obtain

$$\frac{\partial(x, y)}{\partial(u, v)} = \begin{vmatrix} \dfrac{\partial x}{\partial u} & \dfrac{\partial x}{\partial v} \\ \dfrac{\partial y}{\partial u} & \dfrac{\partial y}{\partial v} \end{vmatrix} = \begin{vmatrix} -\dfrac{1}{2u}\sqrt{\dfrac{v}{u}} & \dfrac{1}{2\sqrt{uv}} \\ \dfrac{1}{2}\sqrt{\dfrac{v}{u}} & \dfrac{1}{2}\sqrt{\dfrac{u}{v}} \end{vmatrix} = -\frac{1}{4u} - \frac{1}{4u} = -\frac{1}{2u}$$

Thus, from Formula (9), but with the notation dA rather than dA_{xy},

$$\begin{aligned} \iint_R e^{xy}\, dA &= \iint_S e^v \left| -\frac{1}{2u} \right| dA_{uv} = \frac{1}{2} \iint_S \frac{1}{u} e^v\, dA_{uv} \\ &= \frac{1}{2} \int_1^2 \int_{1/2}^1 \frac{1}{u} e^v\, du\, dv = \frac{1}{2} \int_1^2 e^v \ln|u| \Big]_{u=1/2}^1 dv \\ &= \frac{1}{2} \ln 2 \int_1^2 e^v\, dv = \frac{1}{2}(e^2 - e) \ln 2 \end{aligned}$$ ◀

■ CHANGE OF VARIABLES IN TRIPLE INTEGRALS

Equations of the form

$$x = x(u, v, w), \quad y = y(u, v, w), \quad z = z(u, v, w) \tag{12}$$

define a ***transformation*** T from uvw-space to xyz-space. Just as a transformation $x = x(u, v)$, $y = y(u, v)$ in two variables maps small rectangles in the uv-plane into curvilinear parallelograms in the xy-plane, so (12) maps small rectangular parallelepipeds in uvw-space into curvilinear parallelepipeds in xyz-space (Figure 15.8.11). The definition of the Jacobian of (12) is similar to Definition 15.8.1.

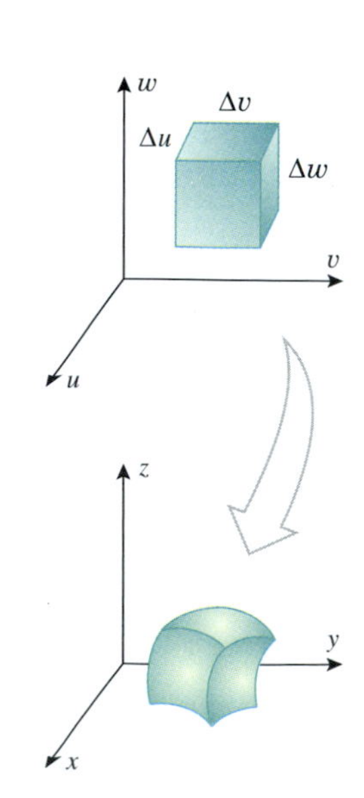

Figure 15.8.11

15.8.3 DEFINITION. If T is the transformation from uvw-space to xyz-space defined by the equations $x = x(u, v, w)$, $y = y(u, v, w)$, $z = z(u, v, w)$, then the ***Jacobian of T*** is denoted by $J(u, v, w)$ or $\partial(x, y, z)/\partial(u, v, w)$ and is defined by

$$J(u, v, w) = \frac{\partial(x, y, z)}{\partial(u, v, w)} = \begin{vmatrix} \dfrac{\partial x}{\partial u} & \dfrac{\partial x}{\partial v} & \dfrac{\partial x}{\partial w} \\ \dfrac{\partial y}{\partial u} & \dfrac{\partial y}{\partial v} & \dfrac{\partial y}{\partial w} \\ \dfrac{\partial z}{\partial u} & \dfrac{\partial z}{\partial v} & \dfrac{\partial z}{\partial w} \end{vmatrix}$$

For small values of Δu, Δv, and Δw, the volume ΔV of the curvilinear parallelepiped in Figure 15.8.11 is related to the volume $\Delta u\, \Delta v\, \Delta w$ of the rectangular parallelepiped by

$$\Delta V \approx \left| \frac{\partial(x, y, z)}{\partial(u, v, w)} \right| \Delta u\, \Delta v\, \Delta w \tag{13}$$

which is the analog of Formula (8). Using this relationship and an argument similar to the one that led to Formula (9), we can obtain the following result.

15.8.4 CHANGE OF VARIABLES FORMULA FOR TRIPLE INTEGRALS. If the transformation $x = x(u, v, w)$, $y = y(u, v, w)$, $z = z(u, v, w)$ maps the region S in uvw-space into the region R in xyz-space, and if the Jacobian $\partial(x, y, z)/\partial(u, v, w)$ is nonzero and does not change sign on S, then with appropriate restrictions on the transformation and the regions it follows that

$$\iiint_R f(x, y, z)\, dV_{xyz} = \iiint_S f(x(u, v, w), y(u, v, w), z(u, v, w)) \left| \frac{\partial(x, y, z)}{\partial(u, v, w)} \right| dV_{uvw} \tag{14}$$

▶ Example 4 Find the volume of the region G enclosed by the ellipsoid

$$\frac{x^2}{a^2} + \frac{y^2}{b^2} + \frac{z^2}{c^2} = 1$$

Solution. The volume V is given by the triple integral

$$V = \iiint_G dV$$

To evaluate this integral, we make the change of variables

$$x = au, \quad y = bv, \quad z = cw \tag{15}$$

which maps the region S in uvw-space enclosed by a sphere of radius 1 into the region G in xyz-space. This can be seen from (15) by noting that

$$\frac{x^2}{a^2} + \frac{y^2}{b^2} + \frac{z^2}{c^2} = 1 \quad \text{becomes} \quad u^2 + v^2 + w^2 = 1$$

The Jacobian of (15) is

$$\frac{\partial(x, y, z)}{\partial(u, v, w)} = \begin{vmatrix} \dfrac{\partial x}{\partial u} & \dfrac{\partial x}{\partial v} & \dfrac{\partial x}{\partial w} \\ \dfrac{\partial y}{\partial u} & \dfrac{\partial y}{\partial v} & \dfrac{\partial y}{\partial w} \\ \dfrac{\partial z}{\partial u} & \dfrac{\partial z}{\partial v} & \dfrac{\partial z}{\partial w} \end{vmatrix} = \begin{vmatrix} a & 0 & 0 \\ 0 & b & 0 \\ 0 & 0 & c \end{vmatrix} = abc$$

Thus, from Formula (14), but with the notation dV rather than dV_{xyz},

$$V = \iiint_G dV = \iiint_S \left| \frac{\partial(x, y, z)}{\partial(u, v, w)} \right| dV_{uvw} = abc \iiint_S dV_{uvw}$$

The last integral is the volume enclosed by a sphere of radius 1, which we know to be $\frac{4}{3}\pi$. Thus, the volume enclosed by the ellipsoid is $V = \frac{4}{3}\pi abc$. ◀

Jacobians also arise in converting triple integrals in rectangular coordinates to iterated integrals in cylindrical and spherical coordinates. For example, we will ask you to show in Exercise 47 that the Jacobian of the transformation

$$x = r\cos\theta, \quad y = r\sin\theta, \quad z = z$$

is

$$\frac{\partial(x, y, z)}{\partial(r, \theta, z)} = r$$

and the Jacobian of the transformation

$$x = \rho\sin\phi\cos\theta, \quad y = \rho\sin\phi\sin\theta, \quad z = \rho\cos\phi$$

is

$$\frac{\partial(x, y, z)}{\partial(\rho, \phi, \theta)} = \rho^2\sin\phi$$

Thus, Formulas (6) and (10) of Section 15.7 can be expressed in terms of Jacobians as

$$\iiint_G f(x, y, z)\,dV = \iiint_{\substack{\text{appropriate}\\\text{limits}}} f(r\cos\theta, r\sin\theta, z)\frac{\partial(x, y, z)}{\partial(r, \theta, z)}\,dz\,dr\,d\theta \tag{16}$$

$$\iiint_G f(x, y, z)\,dV = \iiint_{\substack{\text{appropriate}\\\text{limits}}} f(\rho\sin\phi\cos\theta, \rho\sin\phi\sin\theta, \rho\cos\phi)\frac{\partial(x, y, z)}{\partial(\rho, \phi, \theta)}\,d\rho\,d\phi\,d\theta \tag{17}$$

The absolute-value signs are omitted from Formulas (16) and (17) because the Jacobians are nonnegative (see the restrictions in Table 12.8.1).

✔QUICK CHECK EXERCISES 15.8 *(See page 1099 for answers.)*

1. Let T be the transformation from the uv-plane to the xy-plane defined by the equations

$$x = u - 2v, \quad y = 3u + v$$

   (a) Sketch the image under T of the rectangle $1 \le u \le 3$, $0 \le v \le 2$.

   (b) Solve for u and v in terms of x and y:

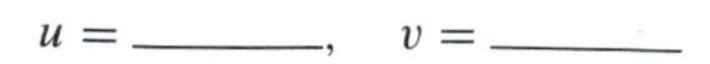

$u =$ ________, $v =$ ________

2. State the relationship between R and S in the change of variables formula

$$\iint_R f(x, y)\,dA_{xy} = \iint_S f(x(u, v), y(u, v)) \left| \frac{\partial(x, y)}{\partial(u, v)} \right| dA_{uv}$$

3. Let T be the transformation in Quick Check Exercise 1.

   (a) The Jacobian $\partial(x, y)/\partial(u, v)$ of T is ________.

(b) Let R be the region in Quick Check Exercise 1(a). Fill in the missing integrand and limits of integration for the change of variables given by T.

$$\iint_R e^{x+2y}\, dA = \int_{\square}^{\square}\int_{\square}^{\square} \_\_\_\_\_\_\_\_\, du\, dv$$

4. The Jacobian of the transformation

$$x = uv, \quad y = vw, \quad z = 2w$$

is

$$\frac{\partial(x, y, z)}{\partial(u, v, w)} = \_\_\_\_\_\_\_\_$$

EXERCISE SET 15.8

1–4 Find the Jacobian $\partial(x, y)/\partial(u, v)$.

1. $x = u + 4v, \quad y = 3u - 5v$
2. $x = u + 2v^2, \quad y = 2u^2 - v$
3. $x = \sin u + \cos v, \quad y = -\cos u + \sin v$
4. $x = \dfrac{2u}{u^2 + v^2}, \quad y = -\dfrac{2v}{u^2 + v^2}$

5–8 Solve for x and y in terms of u and v, and then find the Jacobian $\partial(x, y)/\partial(u, v)$.

5. $u = 2x - 5y, \quad v = x + 2y$
6. $u = e^x, \quad v = ye^{-x}$
7. $u = x^2 - y^2, \quad v = x^2 + y^2 \quad (x > 0, y > 0)$
8. $u = xy, \quad v = xy^3 \quad (x > 0, y > 0)$

9–12 Find the Jacobian $\partial(x, y, z)/\partial(u, v, w)$.

9. $x = 3u + v, \quad y = u - 2w, \quad z = v + w$
10. $x = u - uv, \quad y = uv - uvw, \quad z = uvw$
11. $u = xy, \quad v = y, \quad w = x + z$
12. $u = x + y + z, \quad v = x + y - z, \quad w = x - y + z$

FOCUS ON CONCEPTS

13–16 Sketch the image in the xy-plane of the set S under the given transformation.

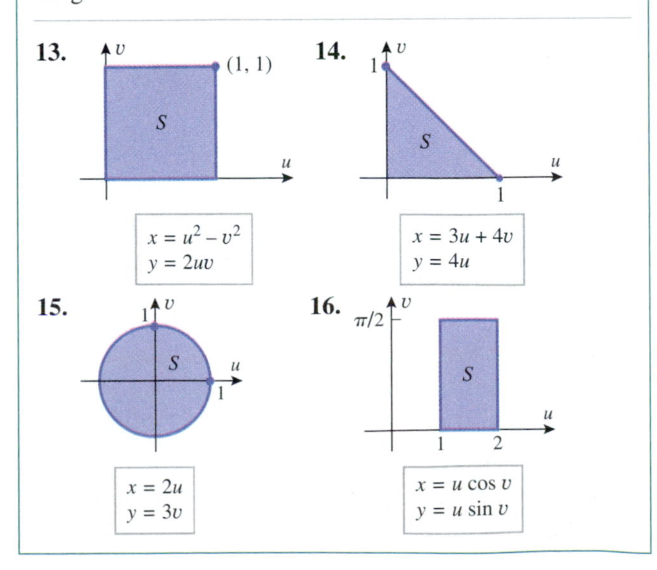

17. Use the transformation $u = x - 2y$, $v = 2x + y$ to find

$$\iint_R \frac{x - 2y}{2x + y}\, dA$$

where R is the rectangular region enclosed by the lines $x - 2y = 1$, $x - 2y = 4$, $2x + y = 1$, $2x + y = 3$.

18. Use the transformation $u = x + y$, $v = x - y$ to find

$$\iint_R (x - y)e^{x^2 - y^2}\, dA$$

over the rectangular region R enclosed by the lines $x + y = 0$, $x + y = 1$, $x - y = 1$, $x - y = 4$.

19. Use the transformation $u = \frac{1}{2}(x + y)$, $v = \frac{1}{2}(x - y)$ to find

$$\iint_R \sin \tfrac{1}{2}(x + y) \cos \tfrac{1}{2}(x - y)\, dA$$

over the triangular region R with vertices $(0, 0)$, $(2, 0)$, $(1, 1)$.

20. Use the transformation $u = y/x$, $v = xy$ to find

$$\iint_R xy^3\, dA$$

over the region R in the first quadrant enclosed by $y = x$, $y = 3x$, $xy = 1$, $xy = 4$.

21–24 The transformation $x = au$, $y = bv$ $(a > 0, b > 0)$ can be rewritten as $x/a = u$, $y/b = v$, and hence it maps the circular region

$$u^2 + v^2 \le 1$$

into the elliptical region

$$\frac{x^2}{a^2} + \frac{y^2}{b^2} \le 1$$

In these exercises, perform the integration by transforming the elliptical region of integration into a circular region of integration and then evaluating the transformed integral in polar coordinates.

21. $\displaystyle\iint_R \sqrt{16x^2 + 9y^2}\, dA$, where R is the region enclosed by the ellipse $(x^2/9) + (y^2/16) = 1$.

22. $\displaystyle\iint_R e^{-(x^2+4y^2)}\, dA$, where R is the region enclosed by the ellipse $(x^2/4) + y^2 = 1$.

23. $\iint_R \sin(4x^2+9y^2)\,dA$, where R is the region in the first quadrant enclosed by the ellipse $4x^2+9y^2=1$ and the coordinate axes.

24. Show that the area of the ellipse

$$\frac{x^2}{a^2}+\frac{y^2}{b^2}=1$$

is πab.

25–26 If a, b, and c are positive constants, then the transformation $x=au$, $y=bv$, $z=cw$ can be rewritten as $x/a=u$, $y/b=v$, $z/c=w$, and hence it maps the spherical region

$$u^2+v^2+w^2\le 1$$

into the ellipsoidal region

$$\frac{x^2}{a^2}+\frac{y^2}{b^2}+\frac{z^2}{c^2}\le 1$$

In these exercises, perform the integration by transforming the ellipsoidal region of integration into a spherical region of integration and then evaluating the transformed integral in spherical coordinates.

25. $\iiint_G x^2\,dV$, where G is the region enclosed by the ellipsoid $9x^2+4y^2+z^2=36$.

26. Find the moment of inertia about the x-axis of the solid ellipsoid bounded by

$$\frac{x^2}{a^2}+\frac{y^2}{b^2}+\frac{z^2}{c^2}=1$$

given that $\delta(x,y,z)=1$. [See the definition preceding Exercise 43 of Section 15.7.]

FOCUS ON CONCEPTS

27–30 Find a transformation

$$u=f(x,y),\quad v=g(x,y)$$

that when applied to the region R in the xy-plane has as its image the region S in the uv-plane.

27.

28.

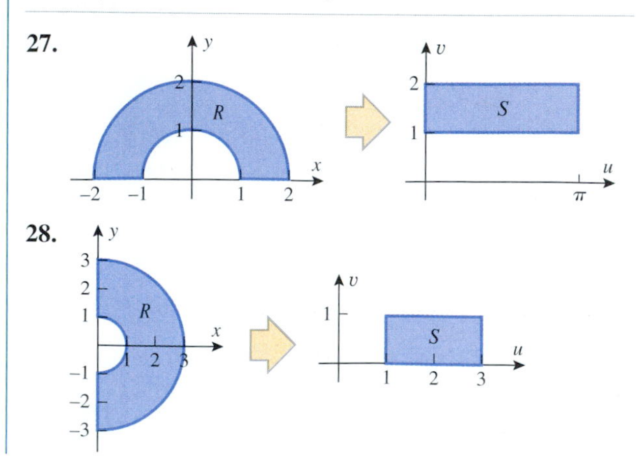

29.

30.

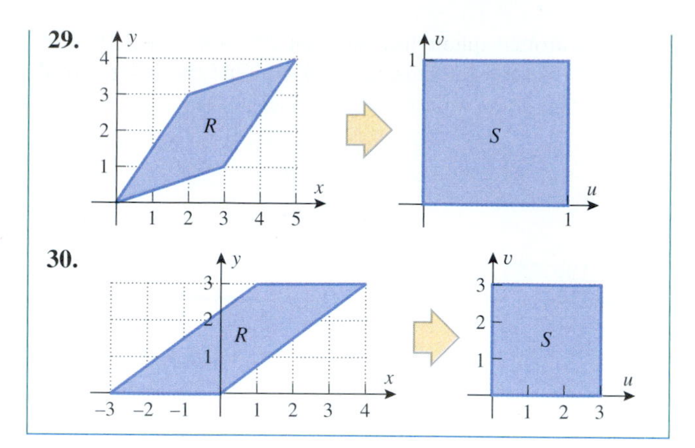

31–34 Evaluate the integral by making an appropriate change of variables.

31. $\iint_R \frac{y-4x}{y+4x}\,dA$, where R is the region enclosed by the lines $y=4x$, $y=4x+2$, $y=2-4x$, $y=5-4x$.

32. $\iint_R (x^2-y^2)\,dA$, where R is the rectangular region enclosed by the lines $y=-x$, $y=1-x$, $y=x$, $y=x+2$.

33. $\iint_R \frac{\sin(x-y)}{\cos(x+y)}\,dA$, where R is the triangular region enclosed by the lines $y=0$, $y=x$, $x+y=\pi/4$.

34. $\iint_R e^{(y-x)/(y+x)}\,dA$, where R is the region in the first quadrant enclosed by the trapezoid with vertices $(0,1)$, $(1,0)$, $(0,4)$, $(4,0)$.

35. Use an appropriate change of variables to find the area of the region in the first quadrant enclosed by the curves $y=x$, $y=2x$, $x=y^2$, $x=4y^2$.

36. Use an appropriate change of variables to find the volume of the solid bounded above by the plane $x+y+z=9$, below by the xy-plane, and laterally by the elliptic cylinder $4x^2+9y^2=36$. [*Hint:* Express the volume as a double integral in xy-coordinates, then use polar coordinates to evaluate the transformed integral.]

37. Use the transformation $u=x$, $v=z-y$, $w=xy$ to find

$$\iiint_G (z-y)^2 xy\,dV$$

where G is the region enclosed by the surfaces $x=1$, $x=3$, $z=y$, $z=y+1$, $xy=2$, $xy=4$.

38. Use the transformation $u=xy$, $v=yz$, $w=xz$ to find the volume of the region in the first octant that is enclosed by the hyperbolic cylinders $xy=1$, $xy=2$, $yz=1$, $yz=3$, $xz=1$, $xz=4$.

39. An astroidal sphere has equation $x^{2/3} + y^{2/3} + x^{2/3} = a^{2/3}$ (see Exercise 54 in Section 15.4). Find the volume of the astroidal sphere using a triple integral and the transformation

$$x = \rho(\sin\phi\cos\theta)^3$$
$$y = \rho(\sin\phi\sin\theta)^3$$
$$z = \rho(\cos\phi)^3$$

for which $0 \le \rho \le a, 0 \le \phi \le \pi, 0 \le \theta \le 2\pi$.

40. (a) Verify that

$$\begin{vmatrix} a_1 & b_1 \\ c_1 & d_1 \end{vmatrix}\begin{vmatrix} a_2 & b_2 \\ c_2 & d_2 \end{vmatrix} = \begin{vmatrix} a_1a_2 + b_1c_2 & a_1b_2 + b_1d_2 \\ c_1a_2 + d_1c_2 & c_1b_2 + d_1d_2 \end{vmatrix}$$

(b) If $x = x(u, v)$, $y = y(u, v)$ is a one-to-one transformation, then $u = u(x, y)$, $v = v(x, y)$. Assuming the necessary differentiability, use the result in part (a) and the chain rule to show that

$$\frac{\partial(x, y)}{\partial(u, v)} \cdot \frac{\partial(u, v)}{\partial(x, y)} = 1$$

41. In each part, confirm that the formula obtained in part (b) of Exercise 40 holds for the given transformation.

(a) $x = u - uv, \quad y = uv$

(b) $x = uv, \quad y = v^2 \quad (v > 0)$

(c) $x = \frac{1}{2}(u^2 + v^2), \quad y = \frac{1}{2}(u^2 - v^2) \quad (u > 0, v > 0)$

42–44 The formula obtained in part (b) of Exercise 40 is useful in integration problems where it is inconvenient or impossible to solve the transformation equations $u = f(x, y)$, $v = g(x, y)$ explicitly for x and y in terms of u and v. In these exercises, use the relationship

$$\frac{\partial(x, y)}{\partial(u, v)} = \frac{1}{\partial(u, v)/\partial(x, y)}$$

to avoid solving for x and y in terms of u and v.

42. Use the transformation $u = xy$, $v = xy^4$ to find

$$\iint_R \sin(xy)\, dA$$

where R is the region enclosed by the curves $xy = \pi$, $xy = 2\pi$, $xy^4 = 1$, $xy^4 = 2$.

43. Use the transformation $u = x^2 - y^2$, $v = x^2 + y^2$ to find

$$\iint_R xy\, dA$$

where R is the region in the first quadrant that is enclosed by the hyperbolas $x^2 - y^2 = 1$, $x^2 - y^2 = 4$ and the circles $x^2 + y^2 = 9$, $x^2 + y^2 = 16$.

44. Use the transformation $u = xy$, $v = x^2 - y^2$ to find

$$\iint_R (x^4 - y^4)e^{xy}\, dA$$

where R is the region in the first quadrant enclosed by the hyperbolas $xy = 1$, $xy = 3$, $x^2 - y^2 = 3$, $x^2 - y^2 = 4$.

45. The three-variable analog of the formula derived in part (b) of Exercise 40 is

$$\frac{\partial(x, y, z)}{\partial(u, v, w)} \cdot \frac{\partial(u, v, w)}{\partial(x, y, z)} = 1$$

Use this result to show that the volume V of the oblique parallelepiped that is bounded by the planes $x + y + 2z = \pm 3$, $x - 2y + z = \pm 2$, $4x + y + z = \pm 6$ is $V = 16$.

46. (a) Show that if R is the triangular region with vertices $(0, 0)$, $(1, 0)$, and $(0, 1)$, then

$$\iint_R f(x + y)\, dA = \int_0^1 uf(u)\, du$$

(b) Use the result in part (a) to evaluate the integral

$$\iint_R e^{x+y}\, dA$$

47. (a) Consider the transformation

$$x = r\cos\theta, \quad y = r\sin\theta, \quad z = z$$

from cylindrical to rectangular coordinates, where $r \ge 0$. Show that

$$\frac{\partial(x, y, z)}{\partial(r, \theta, z)} = r$$

(b) Consider the transformation

$$x = \rho\sin\phi\cos\theta, \quad y = \rho\sin\phi\sin\theta, \quad z = \rho\cos\phi$$

from spherical to rectangular coordinates, where $0 \le \phi \le \pi$. Show that

$$\frac{\partial(x, y, z)}{\partial(\rho, \phi, \theta)} = \rho^2\sin\phi$$

✔ QUICK CHECK ANSWERS 15.8

1. (a) The image is the region in the xy-plane enclosed by the parallelogram with vertices $(1, 3)$, $(-3, 5)$, $(-1, 11)$, and $(3, 9)$. (b) $u = \frac{1}{7}(x + 2y)$; $v = \frac{1}{7}(y - 3x)$ **2.** S is a region in the uv-plane and R is the image of S in the xy-plane under the transformation $x = x(u, v)$, $y = y(u, v)$ **3.** (a) 7 (b) $\int_0^2 \int_1^3 7e^{7u}\, du\, dv$ **4.** $2vw$

CHAPTER REVIEW EXERCISES

1. The double integral over a region R in the xy-plane is defined as

$$\iint_R f(x, y)\, dA = \lim_{n \to +\infty} \sum_{k=1}^{n} f(x_k^*, y_k^*)\, \Delta A_k$$

Describe the procedure on which this definition is based.

2. The triple integral over a solid G in an xyz-coordinate system is defined as

$$\iiint_G f(x, y, z)\, dV = \lim_{n \to +\infty} \sum_{k=1}^{n} f(x_k^*, y_k^*, z_k^*)\, \Delta V_k$$

Describe the procedure on which this definition is based.

3. (a) Express the area of a region R in the xy-plane as a double integral.
(b) Express the volume of a region G in an xyz-coordinate system as a triple integral.
(c) Express the area of the portion of the surface $z = f(x, y)$ that lies above the region R in the xy-plane as a double integral.

4. (a) Write down parametric equations for a sphere of radius a centered at the origin.
(b) Write down parametric equations for the right circular cylinder of radius a and height h that is centered on the z-axis, has its base in the xy-plane, and extends in the positive z-direction.

5. (a) In physical terms, what is meant by the center of gravity of a lamina?
(b) What is meant by the centroid of a lamina?
(c) Write down formulas for the coordinates of the center of gravity of a lamina in the xy-plane.
(d) Write down formulas for the coordinates of the centroid of a lamina in the xy-plane.

6. Suppose that you have a double integral over a region R in the xy-plane and you want to transform that integral into an equivalent double integral over a region S in the uv-plane. Describe the procedure you would use.

7. Let R be the region in the accompanying figure. Fill in the missing limits of integration in the iterated integral

$$\int_{\square}^{\square} \int_{\square}^{\square} f(x, y)\, dx\, dy$$

over R.

8. Let R be the region shown in the accompanying figure. Fill in the missing limits of integration in the sum of the iterated integrals

$$\int_0^2 \int_{\square}^{\square} f(x, y)\, dy\, dx + \int_2^3 \int_{\square}^{\square} f(x, y)\, dy\, dx$$

over R.

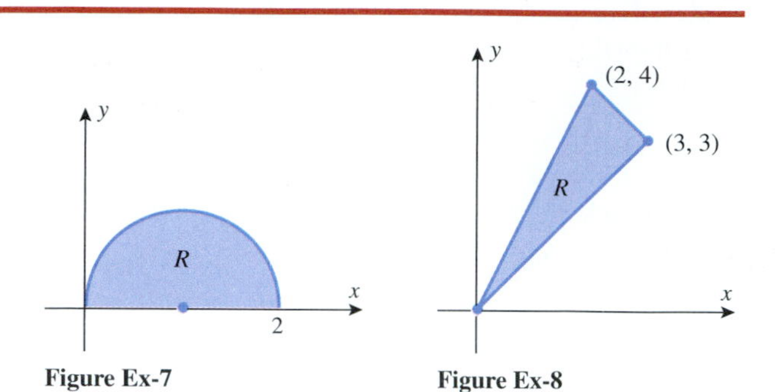

Figure Ex-7 **Figure Ex-8**

9. (a) Find constants a, b, c, and d such that the transformation $x = au + bv$, $y = cu + dv$ maps the region S in the accompanying figure into the region R.
(b) Find the area of the parallelogram R by integrating over the region S, and check your answer using a formula from geometry.

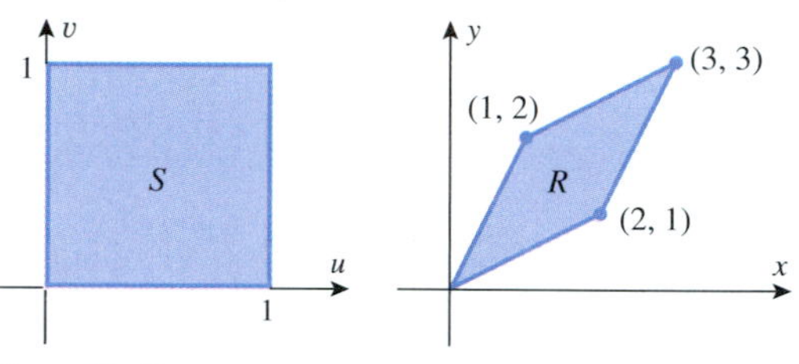

Figure Ex-9

10. Give a geometric argument to show that

$$0 < \int_0^{\pi} \int_0^{\pi} \sin \sqrt{xy}\, dy\, dx < \pi^2$$

11–12 Evaluate the iterated integral.

11. $\displaystyle\int_{1/2}^{1} \int_0^{2x} \cos(\pi x^2)\, dy\, dx$ **12.** $\displaystyle\int_0^2 \int_{-y}^{2y} xe^{y^3}\, dx\, dy$

13–14 Express the iterated integral as an equivalent integral with the order of integration reversed.

13. $\displaystyle\int_0^2 \int_0^{x/2} e^x e^y\, dy\, dx$ **14.** $\displaystyle\int_0^{\pi} \int_y^{\pi} \frac{\sin x}{x}\, dx\, dy$

15–16 Sketch the region whose area is represented by the iterated integral.

15. $\displaystyle\int_0^{\pi/2} \int_{\tan(x/2)}^{\sin x} dy\, dx$

16. $\displaystyle\int_{\pi/6}^{\pi/2} \int_a^{a(1+\cos\theta)} r\, dr\, d\theta \quad (a > 0)$

17–18 Evaluate the double integral.

17. $\displaystyle\iint_R x^2 \sin y^2\, dA$; R is the region that is bounded by $y = x^3$, $y = -x^3$, and $y = 8$.

18. $\iint_R (4 - x^2 - y^2)\, dA$; R is the sector in the first quadrant bounded by the circle $x^2 + y^2 = 4$ and the coordinate axes.

19. Convert to rectangular coordinates and evaluate:

$$\int_0^{\pi/2} \int_0^{2a \sin\theta} r \sin 2\theta \, dr \, d\theta$$

20. Convert to polar coordinates and evaluate:

$$\int_0^{\sqrt{2}} \int_x^{\sqrt{4-x^2}} 4xy \, dy \, dx$$

21–22 Find the area of the region using a double integral.

21. The region bounded by $y = 2x^3$, $2x + y = 4$, and the x-axis.

22. The region enclosed by the rose $r = \cos 3\theta$.

23. Convert to cylindrical coordinates and evaluate:

$$\int_{-2}^{2} \int_{-\sqrt{4-x^2}}^{\sqrt{4-x^2}} \int_{(x^2+y^2)^2}^{16} x^2 \, dz \, dy \, dx$$

24. Convert to spherical coordinates and evaluate:

$$\int_0^1 \int_0^{\sqrt{1-x^2}} \int_0^{\sqrt{1-x^2-y^2}} \frac{1}{1 + x^2 + y^2 + z^2} \, dz \, dy \, dx$$

25. Let G be the region bounded above by the sphere $\rho = a$ and below by the cone $\phi = \pi/3$. Express

$$\iiint_G (x^2 + y^2)\, dV$$

as an iterated integral in
(a) spherical coordinates (b) cylindrical coordinates
(c) rectangular coordinates.

26. Let $G = \{(x, y, z) : x^2 + y^2 \le z \le 4x\}$. Express the volume of G as an iterated integral in
(a) rectangular coordinates (b) cylindrical coordinates.

27–28 Find the volume of the solid using a triple integral.

27. The solid bounded below by the cone $\phi = \pi/6$ and above by the plane $z = a$.

28. The solid enclosed between the surfaces $x = y^2 + z^2$ and $x = 1 - y^2$.

29. Find the surface area of the portion of the hyperbolic paraboloid

$$\mathbf{r}(u, v) = (u + v)\mathbf{i} + (u - v)\mathbf{j} + uv\mathbf{k}$$

for which $u^2 + v^2 \le 4$.

30. Find the surface area of the portion of the spiral ramp

$$\mathbf{r}(u, v) = u \cos v\mathbf{i} + u \sin v\mathbf{j} + v\mathbf{k}$$

for which $0 \le u \le 2$, $0 \le v \le 3u$.

31–32 Find the equation of the tangent plane to the surface at the specified point.

31. $\mathbf{r} = u\mathbf{i} + v\mathbf{j} + (u^2 + v^2)\mathbf{k}$; $u = 1$, $v = 2$

32. $x = u \cosh v$, $y = u \sinh v$, $z = u^2$; $(-3, 0, 9)$

33–34 Find the centroid of the region.

33. The region bounded by $y^2 = 4x$ and $y^2 = 8(x - 2)$.

34. The upper half of the ellipse $(x/a)^2 + (y/b)^2 = 1$.

35–36 Find the centroid of the solid.

35. The solid cone with vertex $(0, 0, h)$ and with base the disk $x^2 + y^2 \le a^2$ in the xy-plane.

36. The solid bounded by $y = x^2$, $z = 0$, and $y + z = 4$.

37. Find the average distance from a point inside a sphere of radius a to the center. [See the definition preceding Exercise 27 of Section 15.5.]

38. Use the transformation $u = x - 3y$, $v = 3x + y$ to find

$$\iint_R \frac{x - 3y}{(3x + y)^2}\, dA$$

where R is the rectangular region enclosed by the lines $x - 3y = 0$, $x - 3y = 4$, $3x + y = 1$, and $3x + y = 3$.

39. Let G be the solid in 3-space defined by the inequalities

$$1 - e^x \le y \le 3 - e^x, \quad 1 - y \le 2z \le 2 - y, \quad y \le e^x \le y + 4$$

(a) Using the coordinate transformation

$$u = e^x + y, \quad v = y + 2z, \quad w = e^x - y$$

calculate the Jacobian $\partial(x, y, z)/\partial(u, v, w)$. Express your answer in terms of u, v, and w.

(b) Using a triple integral and the change of variables given in part (a), find the volume of G.

TOPICS IN VECTOR CALCULUS

I'm very good at integral and differential calculus, I know the scientific names of beings animalculous; In short, in matters vegetable, animal, and mineral, I am the very model of a modern Major-General.

—W. S. Gilbert
Librettist of the operetta The Mikado

The main theme of this chapter is the concept of a "flow." The body of mathematics that we will study here is concerned with analyzing flows of various types—the flow of a fluid or the flow of electricity, for example. Indeed, the early writings of Isaac Newton on calculus are replete with such nouns as "fluxion" and "fluent," which are rooted in the Latin *fluens* (to flow). We will begin this chapter by introducing the concept of a vector field, which is the mathematical description of a flow. In subsequent sections, we will introduce two new kinds of integrals that are used in a variety of applications to analyze properties of vector fields and flows. Finally, we conclude with three major theorems, Green's Theorem, the Divergence Theorem, and Stokes' Theorem. These theorems provide a deep insight into the nature of flows and are the basis for many of the most important principles in physics and engineering.

Photo: *Results in this chapter provide tools for analyzing and understanding the behavior of hurricanes and other fluid flows.*

16.1 VECTOR FIELDS

In this section we will consider functions that associate vectors with points in 2-space or 3-space. We will see that such functions play an important role in the study of fluid flow, gravitational force fields, electromagnetic force fields, and a wide range of other applied problems.

VECTOR FIELDS

To motivate the mathematical ideas in this section, consider a *unit* point mass located at any point in the universe. According to Newton's Law of Universal Gravitation, the Earth exerts an attractive force on the mass that is directed toward the center of the Earth and has a magnitude that is inversely proportional to the square of the distance from the mass to the Earth's center (Figure 16.1.1). This association of force vectors with points in space is called the Earth's *gravitational field*. A similar idea arises in fluid flow. Imagine a stream in which the water flows horizontally at every level, and consider the layer of water at a specific depth. At each point of the layer, the water has a certain velocity, which we can represent by a vector at that point (Figure 16.1.2). This association of velocity vectors with points in the two-dimensional layer is called the *velocity field* at that layer. These ideas are captured in the following definition.

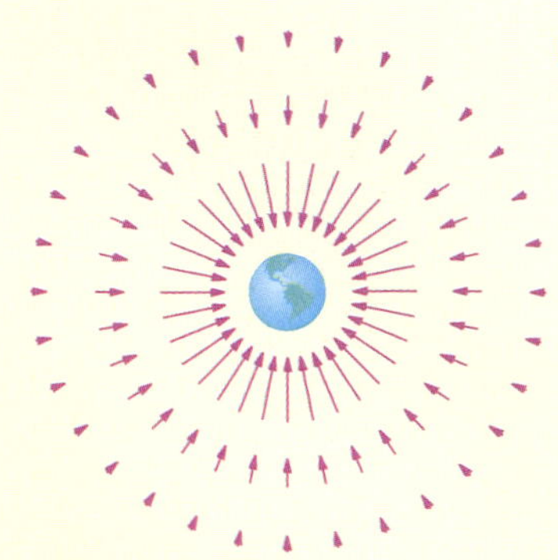

Figure 16.1.1

16.1.1 DEFINITION. A ***vector field*** in a plane is a function that associates with each point P in the plane a unique vector $\mathbf{F}(P)$ parallel to the plane. Similarly, a vector field in 3-space is a function that associates with each point P in 3-space a unique vector $\mathbf{F}(P)$ in 3-space.

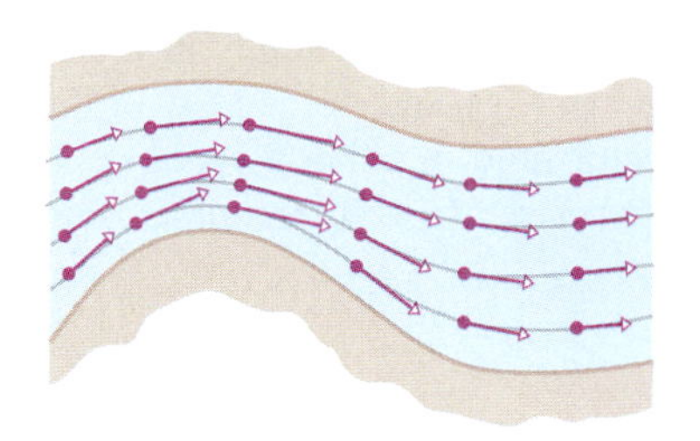

Figure 16.1.2

Observe that in this definition there is no reference to a coordinate system. However, for computational purposes it is usually desirable to introduce a coordinate system so that vectors can be assigned components. Specifically, if $\mathbf{F}(P)$ is a vector field in an xy-coordinate system, then the point P will have some coordinates (x, y) and the associated vector will have components that are functions of x and y. Thus, the vector field $\mathbf{F}(P)$ can be expressed as

$$\mathbf{F}(x, y) = f(x, y)\mathbf{i} + g(x, y)\mathbf{j}$$

Similarly, in 3-space with an xyz-coordinate system, a vector field $\mathbf{F}(P)$ can be expressed as

$$\mathbf{F}(x, y, z) = f(x, y, z)\mathbf{i} + g(x, y, z)\mathbf{j} + h(x, y, z)\mathbf{k}$$

GRAPHICAL REPRESENTATIONS OF VECTOR FIELDS

A vector field in 2-space can be pictured geometrically by drawing representative field vectors $\mathbf{F}(x, y)$ at some well-chosen points in the xy-plane. But, just as it is usually not possible to describe a plane curve completely by plotting finitely many points, so it is usually not possible to describe a vector field completely by drawing finitely many vectors. Nevertheless, such graphical representations can provide useful information about the general behavior of the field if the vectors are chosen appropriately. However, graphical representations of vector fields require a substantial amount of computation, so they are usually created using computers. Figure 16.1.3 shows four computer-generated vector fields. The vector field in part (a) might describe the velocity of the current in a stream at various depths. At the bottom of the stream the velocity is zero, but the speed of the current increases as the depth decreases. Points at the same depth have the same speed. The vector field in part (b) might describe the velocity at points on a rotating wheel. At the center of the wheel the velocity is zero, but the speed increases with the distance from the center. Points at the same distance from the center have the same speed. The vector field in part (c) might describe the repulsive force of an electrical charge—the closer to the charge, the greater the force of repulsion. Part (d) shows a vector field in 3-space. Such pictures tend to be cluttered and hence are of lesser value than graphical representations of vector fields in 2-space. Note also that the vectors in parts (b) and (c) are not to scale—their lengths have been compressed for clarity. We will follow this procedure throughout this chapter.

TECHNOLOGY MASTERY

If you have a graphing utility that can generate vector fields, read the relevant documentation and try to make reasonable duplicates of parts (a) and (b) of Figure 16.1.3.

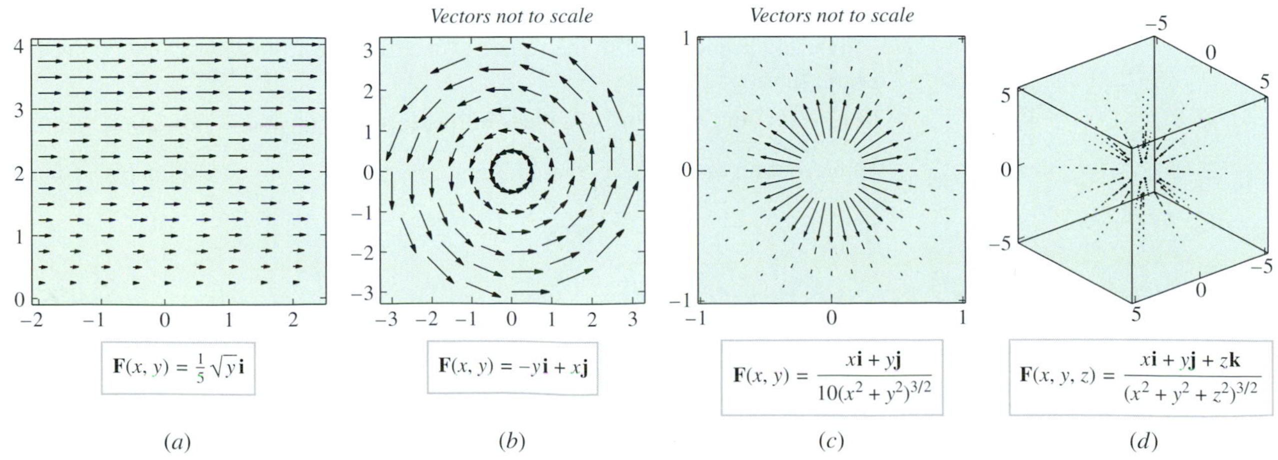

$\mathbf{F}(x, y) = \frac{1}{5}\sqrt{y}\,\mathbf{i}$

(a)

Vectors not to scale

$\mathbf{F}(x, y) = -y\mathbf{i} + x\mathbf{j}$

(b)

Vectors not to scale

$\mathbf{F}(x, y) = \dfrac{x\mathbf{i} + y\mathbf{j}}{10(x^2 + y^2)^{3/2}}$

(c)

$\mathbf{F}(x, y, z) = \dfrac{x\mathbf{i} + y\mathbf{j} + z\mathbf{k}}{(x^2 + y^2 + z^2)^{3/2}}$

(d)

Figure 16.1.3

A COMPACT NOTATION FOR VECTOR FIELDS

Sometimes it is helpful to denote the vector fields $\mathbf{F}(x, y)$ and $\mathbf{F}(x, y, z)$ entirely in vector notation by identifying (x, y) with the radius vector $\mathbf{r} = x\mathbf{i} + y\mathbf{j}$ and (x, y, z) with the radius vector $\mathbf{r} = x\mathbf{i} + y\mathbf{j} + z\mathbf{k}$. With this notation a vector field in either 2-space or 3-space can be written as $\mathbf{F}(\mathbf{r})$. When no confusion is likely to arise, we will sometimes omit the $\mathbf{r}$ altogether and denote the vector field as $\mathbf{F}$.

INVERSE-SQUARE FIELDS

According to Newton's Law of Universal Gravitation, particles with masses m and M attract each other with a force $\mathbf{F}$ of magnitude

$$\|\mathbf{F}\| = \frac{GmM}{r^2} \tag{1}$$

where r is the distance between the particles and G is a constant. If we assume that the particle of mass M is located at the origin of an xyz-coordinate system and $\mathbf{r}$ is the radius vector to the particle of mass m, then $r = \|\mathbf{r}\|$, and the force $\mathbf{F}(\mathbf{r})$ exerted by the particle of mass M on the particle of mass m is in the direction of the unit vector $-\mathbf{r}/\|\mathbf{r}\|$. Thus, it follows from (1) that

$$\mathbf{F}(\mathbf{r}) = -\frac{GmM}{\|\mathbf{r}\|^2}\frac{\mathbf{r}}{\|\mathbf{r}\|} = -\frac{GmM}{\|\mathbf{r}\|^3}\mathbf{r} \tag{2}$$

If m and M are constant, and we let $c = -GmM$, then this formula can be expressed as

$$\mathbf{F}(\mathbf{r}) = \frac{c}{\|\mathbf{r}\|^3}\mathbf{r}$$

Vector fields of this form arise in electromagnetic as well as gravitational problems. Such fields are so important that they have their own terminology.

16.1.2 DEFINITION. If $\mathbf{r}$ is a radius vector in 2-space or 3-space, and if c is a constant, then a vector field of the form

$$\mathbf{F}(\mathbf{r}) = \frac{c}{\|\mathbf{r}\|^3}\mathbf{r} \tag{3}$$

is called an ***inverse-square field***.

Observe that if $c > 0$ in (3), then $\mathbf{F}(\mathbf{r})$ has the same direction as $\mathbf{r}$, so each vector in the field is directed away from the origin; and if $c < 0$, then $\mathbf{F}(\mathbf{r})$ is oppositely directed to $\mathbf{r}$, so each vector in the field is directed toward the origin. In either case the magnitude of $\mathbf{F}(\mathbf{r})$ is inversely proportional to the square of the distance from the terminal point of $\mathbf{r}$ to the origin, since

$$\|\mathbf{F}(\mathbf{r})\| = \frac{|c|}{\|\mathbf{r}\|^3}\|\mathbf{r}\| = \frac{|c|}{\|\mathbf{r}\|^2}$$

We leave it for you to verify that in 2-space Formula (3) can be written in component form as

$$\mathbf{F}(x, y) = \frac{c}{(x^2 + y^2)^{3/2}}(x\mathbf{i} + y\mathbf{j}) \tag{4}$$

and in 3-space as

$$\mathbf{F}(x, y, z) = \frac{c}{(x^2 + y^2 + z^2)^{3/2}}(x\mathbf{i} + y\mathbf{j} + z\mathbf{k}) \tag{5}$$

[see parts (c) and (d) of Figure 16.1.3].

▶ **Example 1** ***Coulomb's law*** states that *the electrostatic force exerted by one charged particle on another is directly proportional to the product of the charges and inversely proportional to the square of the distance between them.* This has the same form as Newton's Law of Universal Gravitation, so the electrostatic force field exerted by a charged particle is an inverse-square field. Specifically, if a particle of charge Q is at the origin of a coordinate system, and if $\mathbf{r}$ is the radius vector to a particle of charge q, then the force $\mathbf{F}(\mathbf{r})$ that the particle of charge Q exerts on the particle of charge q is of the form

$$\mathbf{F}(\mathbf{r}) = \frac{qQ}{4\pi\epsilon_0\|\mathbf{r}\|^3}\mathbf{r}$$

where ϵ_0 is a positive constant (called the ***permittivity constant***). This formula is of form (3) with $c = qQ/4\pi\epsilon_0$. ◀

GRADIENT FIELDS

An important class of vector fields arises from the process of finding gradients. Recall that if ϕ is a function of three variables, then the gradient of ϕ is defined as

$$\nabla\phi = \frac{\partial\phi}{\partial x}\mathbf{i} + \frac{\partial\phi}{\partial y}\mathbf{j} + \frac{\partial\phi}{\partial z}\mathbf{k}$$

This formula defines a vector field in 3-space called the ***gradient field of*** $\boldsymbol{\phi}$. Similarly, the gradient of a function of two variables defines a gradient field in 2-space. At each point in a gradient field where the gradient is nonzero, the vector points in the direction in which the rate of increase of ϕ is maximum.

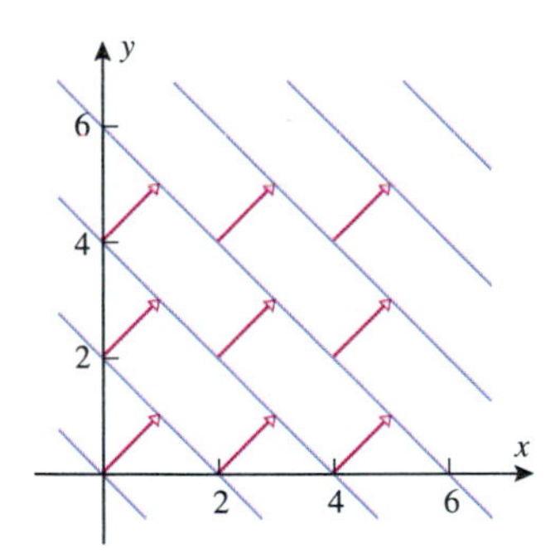

Figure 16.1.4

▶ **Example 2** Sketch the gradient field of $\phi(x, y) = x + y$.

Solution. The gradient of ϕ is

$$\nabla\phi = \frac{\partial\phi}{\partial x}\mathbf{i} + \frac{\partial\phi}{\partial y}\mathbf{j} = \mathbf{i} + \mathbf{j}$$

which is the same at each point. A portion of the vector field is sketched in Figure 16.1.4 together with some level curves of ϕ. Note that at each point, $\nabla\phi$ is normal to the level curve of ϕ through the point (Theorem 14.6.6). ◀

CONSERVATIVE FIELDS AND POTENTIAL FUNCTIONS

If $\mathbf{F}(\mathbf{r})$ is an arbitrary vector field in 2-space or 3-space, we can ask whether it is the gradient field of some function ϕ, and if so, how we can find ϕ. This is an important problem in various applications, and we will study it in more detail later. However, there is some terminology for such fields that we will introduce now.

16.1.3 DEFINITION. A vector field $\mathbf{F}$ in 2-space or 3-space is said to be ***conservative*** in a region if it is the gradient field for some function ϕ in that region; that is, if

$$\mathbf{F} = \nabla\phi$$

The function ϕ is called a ***potential function*** for $\mathbf{F}$ in the region.

▶ **Example 3** Inverse-square fields are conservative in any region that does not contain the origin. For example, in the two-dimensional case the function

$$\phi(x, y) = -\frac{c}{(x^2 + y^2)^{1/2}} \tag{6}$$

is a potential function for (4) in any region not containing the origin, since

$$\begin{aligned}\nabla\phi(x, y) &= \frac{\partial\phi}{\partial x}\mathbf{i} + \frac{\partial\phi}{\partial y}\mathbf{j} \\ &= \frac{cx}{(x^2 + y^2)^{3/2}}\mathbf{i} + \frac{cy}{(x^2 + y^2)^{3/2}}\mathbf{j} \\ &= \frac{c}{(x^2 + y^2)^{3/2}}(x\mathbf{i} + y\mathbf{j}) \\ &= \mathbf{F}(x, y)\end{aligned}$$

In a later section we will discuss methods for finding potential functions for conservative vector fields. ◀

DIVERGENCE AND CURL

We will now define two important operations on vector fields in 3-space—the *divergence* and the *curl* of the field. These names originate in the study of fluid flow, in which case the divergence relates to the way in which fluid flows toward or away from a point and the curl relates to the rotational properties of the fluid at a point. We will investigate the physical interpretations of these operations in more detail later, but for now we will focus only on their computation.

16.1.4 DEFINITION. If $\mathbf{F}(x, y, z) = f(x, y, z)\mathbf{i} + g(x, y, z)\mathbf{j} + h(x, y, z)\mathbf{k}$, then we define the ***divergence of* F**, written div **F**, to be the function given by

$$\operatorname{div} \mathbf{F} = \frac{\partial f}{\partial x} + \frac{\partial g}{\partial y} + \frac{\partial h}{\partial z} \tag{7}$$

16.1.5 DEFINITION. If $\mathbf{F}(x, y, z) = f(x, y, z)\mathbf{i} + g(x, y, z)\mathbf{j} + h(x, y, z)\mathbf{k}$, then we define the ***curl of* F**, written curl **F**, to be the vector field given by

$$\operatorname{curl} \mathbf{F} = \left(\frac{\partial h}{\partial y} - \frac{\partial g}{\partial z}\right)\mathbf{i} + \left(\frac{\partial f}{\partial z} - \frac{\partial h}{\partial x}\right)\mathbf{j} + \left(\frac{\partial g}{\partial x} - \frac{\partial f}{\partial y}\right)\mathbf{k} \tag{8}$$

Observe that div **F** and curl **F** depend on the point at which they are computed, and hence are more properly written as div $\mathbf{F}(x, y, z)$ and curl $\mathbf{F}(x, y, z)$. However, even though these functions are expressed in terms of x, y, and z, it can be proved that their values at a fixed point depend only on the point and not on the coordinate system selected. This is important in applications, since it allows physicists and engineers to compute the curl and divergence in any convenient coordinate system.

Before proceeding to some examples, we note that div **F** has scalar values, whereas curl **F** has vector values (i.e., curl **F** is itself a vector field). Moreover, for computational purposes it is useful to note that the formula for the curl can be expressed in the determinant form

$$\operatorname{curl} \mathbf{F} = \begin{vmatrix} \mathbf{i} & \mathbf{j} & \mathbf{k} \\ \dfrac{\partial}{\partial x} & \dfrac{\partial}{\partial y} & \dfrac{\partial}{\partial z} \\ f & g & h \end{vmatrix} \tag{9}$$

You should verify that Formula (8) results if the determinant is computed by interpreting a "product" such as $(\partial/\partial x)(g)$ to mean $\partial g/\partial x$. Keep in mind, however, that (9) is just a mnemonic device and not a true determinant, since the entries in a determinant must be numbers, not vectors and partial derivative symbols.

TECHNOLOGY MASTERY

Most computer algebra systems can compute gradient fields, divergence, and curl. If you have a CAS with these capabilities, read the relevant documentation and use your CAS to check the computations in Examples 2 and 4.

► Example 4 Find the divergence and the curl of the vector field

$$\mathbf{F}(x, y, z) = x^2y\mathbf{i} + 2y^3z\mathbf{j} + 3z\mathbf{k}$$

Solution. From (7)

$$\begin{aligned} \operatorname{div} \mathbf{F} &= \frac{\partial}{\partial x}(x^2y) + \frac{\partial}{\partial y}(2y^3z) + \frac{\partial}{\partial z}(3z) \\ &= 2xy + 6y^2z + 3 \end{aligned}$$

and from (9)

$$\begin{aligned} \operatorname{curl} \mathbf{F} &= \begin{vmatrix} \mathbf{i} & \mathbf{j} & \mathbf{k} \\ \dfrac{\partial}{\partial x} & \dfrac{\partial}{\partial y} & \dfrac{\partial}{\partial z} \\ x^2y & 2y^3z & 3z \end{vmatrix} \\ &= \left[\frac{\partial}{\partial y}(3z) - \frac{\partial}{\partial z}(2y^3z)\right]\mathbf{i} + \left[\frac{\partial}{\partial z}(x^2y) - \frac{\partial}{\partial x}(3z)\right]\mathbf{j} + \left[\frac{\partial}{\partial x}(2y^3z) - \frac{\partial}{\partial y}(x^2y)\right]\mathbf{k} \\ &= -2y^3\mathbf{i} - x^2\mathbf{k} \;\blacktriangleleft \end{aligned}$$

► Example 5 Show that the divergence of the inverse-square field

$$\mathbf{F}(x, y, z) = \frac{c}{(x^2 + y^2 + z^2)^{3/2}}(x\mathbf{i} + y\mathbf{j} + z\mathbf{k})$$

is zero.

Solution. The computations can be simplified by letting $r = (x^2 + y^2 + z^2)^{1/2}$, in which case **F** can be expressed as

$$\mathbf{F}(x, y, z) = \frac{cx\mathbf{i} + cy\mathbf{j} + cz\mathbf{k}}{r^3} = \frac{cx}{r^3}\mathbf{i} + \frac{cy}{r^3}\mathbf{j} + \frac{cz}{r^3}\mathbf{k}$$

We leave it for you to show that

$$\frac{\partial r}{\partial x} = \frac{x}{r}, \quad \frac{\partial r}{\partial y} = \frac{y}{r}, \quad \frac{\partial r}{\partial z} = \frac{z}{r}$$

Thus

$$\operatorname{div}\mathbf{F} = c\left[\frac{\partial}{\partial x}\left(\frac{x}{r^3}\right) + \frac{\partial}{\partial y}\left(\frac{y}{r^3}\right) + \frac{\partial}{\partial z}\left(\frac{z}{r^3}\right)\right] \tag{10}$$

But

$$\frac{\partial}{\partial x}\left(\frac{x}{r^3}\right) = \frac{r^3 - x(3r^2)(x/r)}{(r^3)^2} = \frac{1}{r^3} - \frac{3x^2}{r^5}$$

$$\frac{\partial}{\partial y}\left(\frac{y}{r^3}\right) = \frac{1}{r^3} - \frac{3y^2}{r^5}$$

$$\frac{\partial}{\partial z}\left(\frac{z}{r^3}\right) = \frac{1}{r^3} - \frac{3z^2}{r^5}$$

Substituting these expressions in (10) yields

$$\operatorname{div}\mathbf{F} = c\left[\frac{3}{r^3} - \frac{3x^2+3y^2+3z^2}{r^5}\right] = c\left[\frac{3}{r^3} - \frac{3r^2}{r^5}\right] = 0$$ ◀

THE ∇ OPERATOR

Thus far, the symbol ∇ that appears in the gradient expression $\nabla\phi$ has not been given a meaning of its own. However, it is often convenient to view ∇ as an operator

$$\nabla = \frac{\partial}{\partial x}\mathbf{i} + \frac{\partial}{\partial y}\mathbf{j} + \frac{\partial}{\partial z}\mathbf{k} \tag{11}$$

which when applied to $\phi(x, y, z)$ produces the gradient

$$\nabla\phi = \frac{\partial\phi}{\partial x}\mathbf{i} + \frac{\partial\phi}{\partial y}\mathbf{j} + \frac{\partial\phi}{\partial z}\mathbf{k}$$

We call (11) the ***del operator***. This is analogous to the derivative operator d/dx, which when applied to $f(x)$ produces the derivative $f'(x)$.

The del operator allows us to express the divergence of a vector field

$$\mathbf{F} = f(x, y, z)\mathbf{i} + g(x, y, z)\mathbf{j} + h(x, y, z)\mathbf{k}$$

in dot product notation as

$$\operatorname{div}\mathbf{F} = \nabla\cdot\mathbf{F} = \frac{\partial f}{\partial x} + \frac{\partial g}{\partial y} + \frac{\partial h}{\partial z} \tag{12}$$

and the curl of this field in cross-product notation as

$$\operatorname{curl}\mathbf{F} = \nabla\times\mathbf{F} = \begin{vmatrix} \mathbf{i} & \mathbf{j} & \mathbf{k} \\ \dfrac{\partial}{\partial x} & \dfrac{\partial}{\partial y} & \dfrac{\partial}{\partial z} \\ f & g & h \end{vmatrix} \tag{13}$$

THE LAPLACIAN ∇^2

The operator that results by taking the dot product of the del operator with itself is denoted by ∇^2 and is called the ***Laplacian operator***. This operator has the form

$$\nabla^2 = \nabla\cdot\nabla = \frac{\partial^2}{\partial x^2} + \frac{\partial^2}{\partial y^2} + \frac{\partial^2}{\partial z^2} \tag{14}$$

When applied to $\phi(x, y, z)$ the Laplacian operator produces the function

$$\nabla^2\phi = \frac{\partial^2\phi}{\partial x^2} + \frac{\partial^2\phi}{\partial y^2} + \frac{\partial^2\phi}{\partial z^2}$$

Note that $\nabla^2\phi$ can also be expressed as div $(\nabla\phi)$. The equation $\nabla^2\phi = 0$ or, equivalently,

$$\frac{\partial^2\phi}{\partial x^2} + \frac{\partial^2\phi}{\partial y^2} + \frac{\partial^2\phi}{\partial z^2} = 0$$

is known as ***Laplace's equation***. This partial differential equation plays an important role in a wide variety of applications, resulting from the fact that it is satisfied by the potential function for the inverse-square field.

✔ QUICK CHECK EXERCISES 16.1 *(See page 1111 for answers.)*

1. The function $\phi(x, y, z) = xy + yz + xz$ is a potential for the vector field $\mathbf{F} =$ ________.

2. The vector field $\mathbf{F}(x, y, z) =$ ________, defined for $(x, y, z) \neq (0, 0, 0)$, is always directed toward the origin and is of length equal to the distance from (x, y, z) to the origin.

3. An inverse-square field is one that can be written in the form $\mathbf{F}(\mathbf{r}) =$ ________.

4. The vector field

$$\mathbf{F}(x, y, z) = yz\mathbf{i} + xy^2\mathbf{j} + yz^2\mathbf{k}$$

has divergence ________ and curl ________.

Pierre-Simon de Laplace (1749–1827) French mathematician and physicist. Laplace is sometimes referred to as the French Isaac Newton because of his work in celestial mechanics. In a five-volume treatise entitled *Traité de Mécanique Céleste*, he solved extremely difficult problems involving gravitational interactions between the planets. In particular, he was able to show that our solar system is stable and not prone to catastrophic collapse as a result of these interactions. This was an issue of major concern at the time because Jupiter's orbit appeared to be shrinking and Saturn's expanding; Laplace showed that these were expected periodic anomalies. In addition to his work in celestial mechanics, he founded modern probability theory, showed with Lavoisier that respiration is a form of combustion, and developed methods that fostered many new branches of pure mathematics.

Laplace was born to moderately successful parents in Normandy, his father being a farmer and cider merchant. He matriculated in the theology program at the University of Caen at age 16 but left for Paris at age 18 with a letter of introduction to the influential mathematician d'Alembert, who eventually helped him undertake a career in mathematics. Laplace was a prolific writer, and after his election to the Academy of Sciences in 1773, the secretary wrote that the Academy had never received so many important research papers by so young a person in such a short time. Laplace had little interest in pure mathematics—he regarded mathematics merely as a tool for solving applied problems. In his impatience with mathematical detail, he frequently omitted complicated arguments with the statement, "It is easy to show that. . . ." He admitted, however, that as time passed he often had trouble reconstructing the omitted details himself!

At the height of his fame, Laplace served on many government committees and held the posts of Minister of the Interior and Chancellor of the Senate. He barely escaped imprisonment and execution during the period of the Revolution, probably because he was able to convince each opposing party that he sided with them. Napoleon described him as a great mathematician but a poor administrator who "sought subtleties everywhere, had only doubtful ideas, and . . . carried the spirit of the infinitely small into administration." In spite of his genius, Laplace was both egotistic and insecure, attempting to ensure his place in history by conveniently failing to credit mathematicians whose work he used—an unnecessary pettiness since his own work was so brilliant. However, on the positive side he was supportive of young mathematicians, often treating them as his own children. Laplace ranks as one of the most influential mathematicians in history.

EXERCISE SET 16.1 Graphing Utility C CAS

FOCUS ON CONCEPTS

1–2 Match the vector field $\mathbf{F}(x, y)$ with one of the plots, and explain your reasoning.

1. (a) $\mathbf{F}(x, y) = x\mathbf{i}$ (b) $\mathbf{F}(x, y) = \sin x\mathbf{i} + \mathbf{j}$

2. (a) $\mathbf{F}(x, y) = \mathbf{i} + \mathbf{j}$
   (b) $\mathbf{F}(x, y) = \dfrac{x}{\sqrt{x^2 + y^2}}\mathbf{i} + \dfrac{y}{\sqrt{x^2 + y^2}}\mathbf{j}$

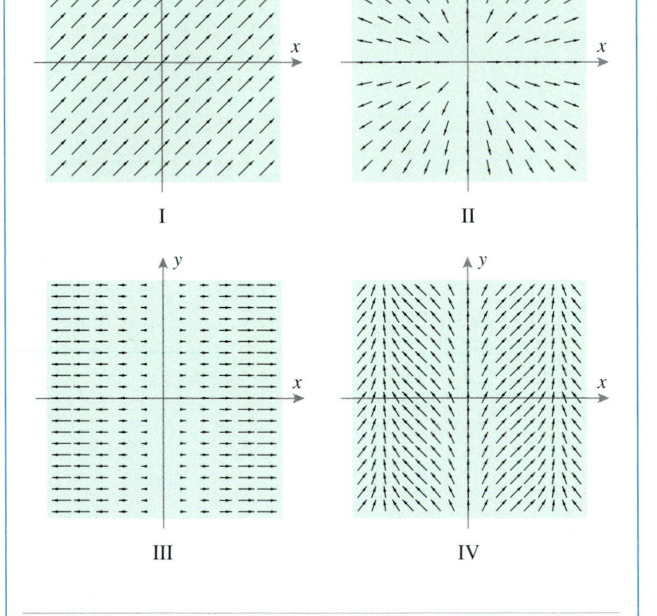

3–4 Determine whether the statement about the vector field $\mathbf{F}(x, y)$ is true or false. If false, explain why.

3. $\mathbf{F}(x, y) = x^2\mathbf{i} - y\mathbf{j}$.
   (a) $\|\mathbf{F}(x, y)\| \to 0$ as $(x, y) \to (0, 0)$.
   (b) If (x, y) is on the positive y-axis, then the vector points in the negative y-direction.
   (c) If (x, y) is in the first quadrant, then the vector points down and to the right.

4. $\mathbf{F}(x, y) = \dfrac{x}{\sqrt{x^2 + y^2}}\mathbf{i} - \dfrac{y}{\sqrt{x^2 + y^2}}\mathbf{j}$.
   (a) As (x, y) moves away from the origin, the lengths of the vectors decrease.
   (b) If (x, y) is a point on the positive x-axis, then the vector points up.
   (c) If (x, y) is a point on the positive y-axis, the vector points to the right.

5–8 Sketch the vector field by drawing some representative nonintersecting vectors. The vectors need not be drawn to scale, but they should be in reasonably correct proportion relative to each other.

5. $\mathbf{F}(x, y) = 2\mathbf{i} - \mathbf{j}$

6. $\mathbf{F}(x, y) = y\mathbf{j}, \quad y > 0$

7. $\mathbf{F}(x, y) = y\mathbf{i} - x\mathbf{j}$. [*Note:* Each vector in the field is perpendicular to the position vector $\mathbf{r} = x\mathbf{i} + y\mathbf{j}$.]

8. $\mathbf{F}(x, y) = \dfrac{x\mathbf{i} + y\mathbf{j}}{\sqrt{x^2 + y^2}}$. [*Note:* Each vector in the field is a unit vector in the same direction as the position vector $\mathbf{r} = x\mathbf{i} + y\mathbf{j}$.]

9–10 Use a graphing utility to generate a plot of the vector field.

9. $\mathbf{F}(x, y) = \mathbf{i} + \cos y\mathbf{j}$

10. $\mathbf{F}(x, y) = y\mathbf{i} - x\mathbf{j}$

11–12 Confirm that ϕ is a potential function for $\mathbf{F}(\mathbf{r})$ on some region, and state the region.

11. (a) $\phi(x, y) = \tan^{-1} xy$
    $\mathbf{F}(x, y) = \dfrac{y}{1 + x^2y^2}\mathbf{i} + \dfrac{x}{1 + x^2y^2}\mathbf{j}$
    (b) $\phi(x, y, z) = x^2 - 3y^2 + 4z^2$
    $\mathbf{F}(x, y, z) = 2x\mathbf{i} - 6y\mathbf{j} + 8z\mathbf{k}$

12. (a) $\phi(x, y) = 2y^2 + 3x^2y - xy^3$
    $\mathbf{F}(x, y) = (6xy - y^3)\mathbf{i} + (4y + 3x^2 - 3xy^2)\mathbf{j}$
    (b) $\phi(x, y, z) = x \sin z + y \sin x + z \sin y$
    $\mathbf{F}(x, y, z) = (\sin z + y \cos x)\mathbf{i} + (\sin x + z \cos y)\mathbf{j} + (\sin y + x \cos z)\mathbf{k}$

13–18 Find div $\mathbf{F}$ and curl $\mathbf{F}$.

13. $\mathbf{F}(x, y, z) = x^2\mathbf{i} - 2\mathbf{j} + yz\mathbf{k}$

14. $\mathbf{F}(x, y, z) = xz^3\mathbf{i} + 2y^4x^2\mathbf{j} + 5z^2y\mathbf{k}$

15. $\mathbf{F}(x, y, z) = 7y^3z^2\mathbf{i} - 8x^2z^5\mathbf{j} - 3xy^4\mathbf{k}$

16. $\mathbf{F}(x, y, z) = e^{xy}\mathbf{i} - \cos y\mathbf{j} + \sin^2 z\mathbf{k}$

17. $\mathbf{F}(x, y, z) = \dfrac{1}{\sqrt{x^2 + y^2 + z^2}}(x\mathbf{i} + y\mathbf{j} + z\mathbf{k})$

18. $\mathbf{F}(x, y, z) = \ln x\mathbf{i} + e^{xyz}\mathbf{j} + \tan^{-1}(z/x)\mathbf{k}$

19–20 Find $\nabla \cdot (\mathbf{F} \times \mathbf{G})$.

19. $\mathbf{F}(x, y, z) = 2x\mathbf{i} + \mathbf{j} + 4y\mathbf{k}$
    $\mathbf{G}(x, y, z) = x\mathbf{i} + y\mathbf{j} - z\mathbf{k}$

20. $\mathbf{F}(x, y, z) = yz\mathbf{i} + xz\mathbf{j} + xy\mathbf{k}$
    $\mathbf{G}(x, y, z) = xy\mathbf{j} + xyz\mathbf{k}$

21–22 Find $\nabla \cdot (\nabla \times \mathbf{F})$.

21. $\mathbf{F}(x, y, z) = \sin x\mathbf{i} + \cos(x - y)\mathbf{j} + z\mathbf{k}$

22. $\mathbf{F}(x, y, z) = e^{xz}\mathbf{i} + 3xe^{y}\mathbf{j} - e^{yz}\mathbf{k}$

23–24 Find $\nabla \times (\nabla \times \mathbf{F})$.

23. $\mathbf{F}(x, y, z) = xy\mathbf{j} + xyz\mathbf{k}$

24. $\mathbf{F}(x, y, z) = y^2x\mathbf{i} - 3yz\mathbf{j} + xy\mathbf{k}$

C **25.** Use a CAS to check the calculations in Exercises 19, 21, and 23.

C **26.** Use a CAS to check the calculations in Exercises 20, 22, and 24.

27–34 Let k be a constant, $\mathbf{F} = \mathbf{F}(x, y, z)$, $\mathbf{G} = \mathbf{G}(x, y, z)$, and $\phi = \phi(x, y, z)$. Prove the following identities, assuming that all derivatives involved exist and are continuous.

27. $\operatorname{div}(k\mathbf{F}) = k \operatorname{div} \mathbf{F}$

28. $\operatorname{curl}(k\mathbf{F}) = k \operatorname{curl} \mathbf{F}$

29. $\operatorname{div}(\mathbf{F} + \mathbf{G}) = \operatorname{div} \mathbf{F} + \operatorname{div} \mathbf{G}$

30. $\operatorname{curl}(\mathbf{F} + \mathbf{G}) = \operatorname{curl} \mathbf{F} + \operatorname{curl} \mathbf{G}$

31. $\operatorname{div}(\phi\mathbf{F}) = \phi \operatorname{div} \mathbf{F} + \nabla\phi \cdot \mathbf{F}$

32. $\operatorname{curl}(\phi\mathbf{F}) = \phi \operatorname{curl} \mathbf{F} + \nabla\phi \times \mathbf{F}$

33. $\operatorname{div}(\operatorname{curl} \mathbf{F}) = 0$

34. $\operatorname{curl}(\nabla\phi) = \mathbf{0}$

35. Rewrite the identities in Exercises 27, 29, 31, and 33 in an equivalent form using the notation $\nabla \cdot$ for divergence and $\nabla \times$ for curl.

36. Rewrite the identities in Exercises 28, 30, 32, and 34 in an equivalent form using the notation $\nabla \cdot$ for divergence and $\nabla \times$ for curl.

37–38 Verify that the radius vector $\mathbf{r} = x\mathbf{i} + y\mathbf{j} + z\mathbf{k}$ has the stated property.

37. (a) $\operatorname{curl} \mathbf{r} = \mathbf{0}$ (b) $\nabla\|\mathbf{r}\| = \dfrac{\mathbf{r}}{\|\mathbf{r}\|}$

38. (a) $\operatorname{div} \mathbf{r} = 3$ (b) $\nabla \dfrac{1}{\|\mathbf{r}\|} = -\dfrac{\mathbf{r}}{\|\mathbf{r}\|^3}$

39–40 Let $\mathbf{r} = x\mathbf{i} + y\mathbf{j} + z\mathbf{k}$, let $r = \|\mathbf{r}\|$, let f be a differentiable function of one variable, and let $\mathbf{F}(\mathbf{r}) = f(r)\mathbf{r}$.

39. (a) Use the chain rule and Exercise 37(b) to show that

$$\nabla f(r) = \frac{f'(r)}{r}\mathbf{r}$$

(b) Use the result in part (a) and Exercises 31 and 38(a) to show that $\operatorname{div} \mathbf{F} = 3f(r) + rf'(r)$.

40. (a) Use part (a) of Exercise 39, Exercise 32, and Exercise 37(a) to show that $\operatorname{curl} \mathbf{F} = \mathbf{0}$.

(b) Use the result in part (a) of Exercise 39 and Exercises 31 and 38(a) to show that

$$\nabla^2 f(r) = 2\frac{f'(r)}{r} + f''(r)$$

41. Use the result in Exercise 39(b) to show that the divergence of the inverse-square field $\mathbf{F} = \mathbf{r}/\|\mathbf{r}\|^3$ is zero.

42. Use the result of Exercise 39(b) to show that if $\mathbf{F}$ is a vector field of the form $\mathbf{F} = f(\|\mathbf{r}\|)\mathbf{r}$ and if $\operatorname{div} \mathbf{F} = 0$, then $\mathbf{F}$ is an inverse-square field. [*Suggestion:* Let $r = \|\mathbf{r}\|$ and multiply $3f(r) + rf'(r) = 0$ through by r^2. Then write the result as a derivative of a product.]

43. A curve C is called a ***flow line*** of a vector field $\mathbf{F}$ if $\mathbf{F}$ is a tangent vector to C at each point along C (see the accompanying figure).

(a) Let C be a flow line for $\mathbf{F}(x, y) = -y\mathbf{i} + x\mathbf{j}$, and let (x, y) be a point on C for which $y \neq 0$. Show that the flow lines satisfy the differential equation

$$\frac{dy}{dx} = -\frac{x}{y}$$

(b) Solve the differential equation in part (a) by separation of variables, and show that the flow lines are concentric circles centered at the origin.

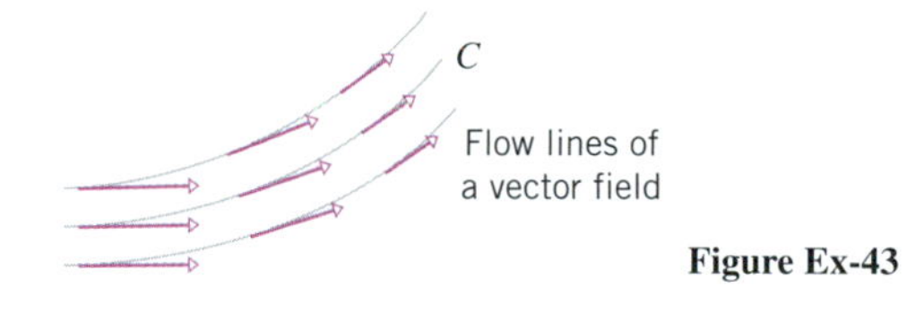

Figure Ex-43

44–46 Find a differential equation satisfied by the flow lines of $\mathbf{F}$ (see Exercise 43), and solve it to find equations for the flow lines of $\mathbf{F}$. Sketch some typical flow lines and tangent vectors.

44. $\mathbf{F}(x, y) = \mathbf{i} + x\mathbf{j}$

45. $\mathbf{F}(x, y) = x\mathbf{i} + \mathbf{j}$, $x > 0$

46. $\mathbf{F}(x, y) = x\mathbf{i} - y\mathbf{j}$, $x > 0$ and $y > 0$

✔ QUICK CHECK ANSWERS 16.1

1. $(y + z)\mathbf{i} + (x + z)\mathbf{j} + (x + y)\mathbf{k}$ **2.** $-\mathbf{r} = -x\mathbf{i} - y\mathbf{j} - z\mathbf{k}$ **3.** $\dfrac{c}{\|\mathbf{r}\|^3}\mathbf{r}$ **4.** $2xy + 2yz$; $z^2\mathbf{i} + y\mathbf{j} + (y^2 - z)\mathbf{k}$

16.2 LINE INTEGRALS

In earlier chapters we considered three kinds of integrals in rectangular coordinates: single integrals over intervals, double integrals over two-dimensional regions, and triple integrals over three-dimensional regions. In this section we will discuss integrals along curves in two- or three-dimensional space.

LINE INTEGRALS

The first goal of this section is to define what it means to integrate a function along a curve. To motivate the definition we will consider the problem of finding the mass of a very thin wire whose linear density function (mass per unit length) is known. We assume that we can model the wire by a smooth curve C between two points P and Q in 3-space (Figure 16.2.1). Given any point (x, y, z) on C, we let $f(x, y, z)$ denote the corresponding value of the density function. To compute the mass of the wire, we proceed as follows:

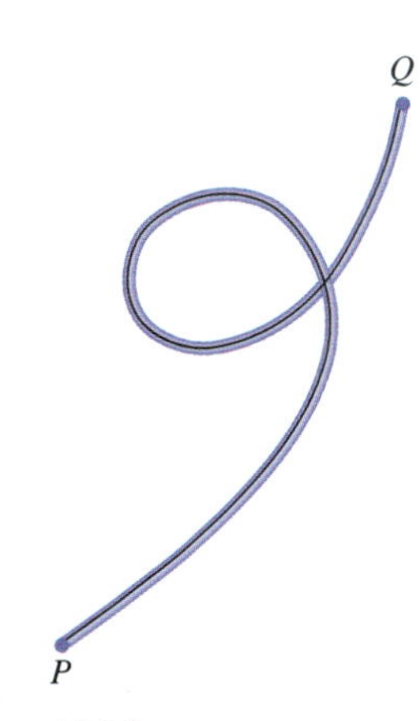

Figure 16.2.1

- Divide C into n very small sections using a succession of distinct partition points

$$P = P_0, P_1, P_2, \ldots, P_{n-1}, P_n = Q$$

as illustrated on the left side of Figure 16.2.2. Let ΔM_k be the mass of the kth section, and let Δs_k be the length of the arc between P_{k-1} and P_k.

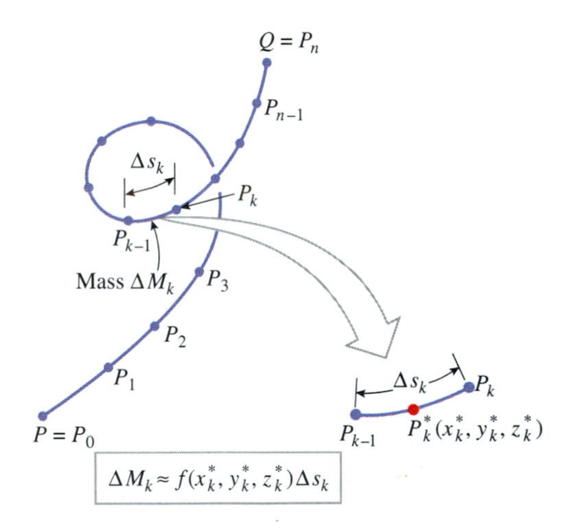

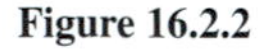
Figure 16.2.2

- Choose an arbitrary sampling point $P_k^*(x_k^*, y_k^*, z_k^*)$ on the kth arc, as illustrated on the right side of Figure 16.2.2. If Δs_k is very small, the value of f will not vary much along the kth section and we can approximate f along this section by the value $f(x_k^*, y_k^*, z_k^*)$. It follows that the mass of the kth section can be approximated by

$$\Delta M_k \approx f(x_k^*, y_k^*, z_k^*)\Delta s_k$$

- The mass M of the entire wire can then be approximated by

$$M = \sum_{k=1}^{n} \Delta M_k \approx \sum_{k=1}^{n} f(x_k^*, y_k^*, z_k^*)\Delta s_k \tag{1}$$

- We will use the expression $\max \Delta s_k \to 0$ to indicate the process of increasing n in such a way that the lengths of all the sections approach 0. It is plausible that the error in (1) will approach 0 as $\max \Delta s_k \to 0$ and the exact value of M will be given by

$$M = \lim_{\max \Delta s_k \to 0} \sum_{k=1}^{n} f(x_k^*, y_k^*, z_k^*)\Delta s_k \tag{2}$$

The limit in (2) is similar to the limit of Riemann sums used to define the definite integral of a function over an interval (Definition 6.5.1). With this similarity in mind, we make the following definition.

Although the term "curve integrals" is more descriptive, the integrals in Definition 16.2.1 are called "line integrals" for historical reasons.

16.2.1 DEFINITION. If C is a smooth curve in 2-space or 3-space, then the ***line integral of f with respect to s along C*** is

$$\int_C f(x, y)\, ds = \lim_{\max \Delta s_k \to 0} \sum_{k=1}^{n} f(x_k^*, y_k^*)\Delta s_k \quad \text{2-space} \tag{3}$$

or

$$\int_C f(x, y, z)\, ds = \lim_{\max \Delta s_k \to 0} \sum_{k=1}^{n} f(x_k^*, y_k^*, z_k^*)\Delta s_k \quad \text{3-space} \tag{4}$$

provided this limit exists and does not depend on the choice of partition or on the choice of sample points.

It is usually impractical to evaluate line integrals directly from Definition 16.2.1. However, the definition is important in the application and interpretation of line integrals. For example:

- If C is a curve in 3-space that models a thin wire, and if $f(x, y, z)$ is the linear density function of the wire, then it follows from (2) and Definition 16.2.1 that the mass M of the wire is given by

$$M = \int_C f(x, y, z)\, ds \tag{5}$$

  That is, to obtain the mass of a thin wire, we integrate the linear density function over the smooth curve that models the wire.

- If C is a smooth curve of arc length L, and f is identically 1, then it immediately follows from Definition 16.2.1 that

$$\int_C ds = \lim_{\max \Delta s_k \to 0} \sum_{k=1}^{n} \Delta s_k = \lim_{\max \Delta s_k \to 0} L = L \tag{6}$$

- If C is a curve in the xy-plane and $f(x, y)$ is a nonnegative continuous function defined on C, then $\int_C f(x, y)\, ds$ can be interpreted as the area A of the "sheet" that is swept out by a vertical line segment that extends upward from the point (x, y) to a height of $f(x, y)$ and moves along C from one endpoint to the other (Figure 16.2.3). To see why this is so, refer to Figure 16.2.4 and note the approximation

$$\Delta A_k \approx f(x_k^*, y_k^*)\Delta s_k$$

  It follows that

$$A = \sum_{k=1}^{n} \Delta A_k \approx \sum_{k=1}^{n} f(x_k^*, y_k^*)\Delta s_k$$

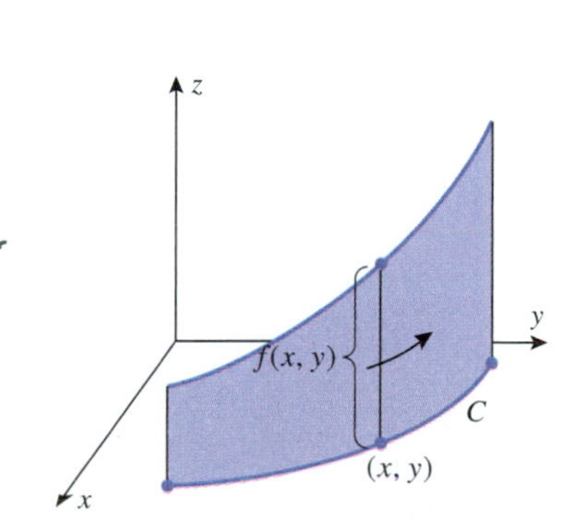

Figure 16.2.3

It is then plausible that

$$A = \lim_{\max \Delta s_k \to 0} \sum_{k=1}^{n} f(x_k^*, y_k^*)\Delta s_k = \int_C f(x, y)\, ds \tag{7}$$

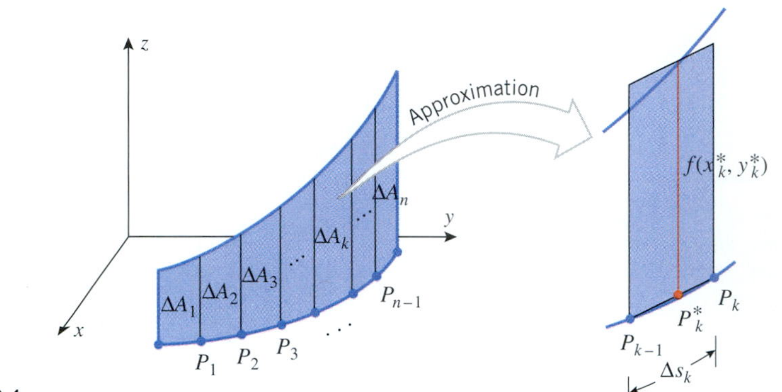

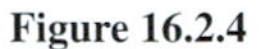
Figure 16.2.4

Since Definition 16.2.1 is closely modeled on Definition 6.5.1, it should come as no surprise that line integrals share many of the common properties of ordinary definite integrals. For example, we have

$$\int_C [f(x, y) + g(x, y)]\, ds = \int_C f(x, y)\, ds + \int_C g(x, y)\, ds$$

provided both line integrals on the right-hand side of this equation exist. Similarly, it can be shown that if f is continuous on C, then the line integral of f with respect to s along C exists.

EVALUATING LINE INTEGRALS

Except in simple cases, it will not be feasible to evaluate a line integral directly from (3) or (4). However, we will now show that it is possible to express a line integral as an ordinary definite integral, so that no special methods of evaluation are required. For example, suppose that C is a curve in the xy-plane that is smoothly parametrized by

$$\mathbf{r}(t) = x(t)\mathbf{i} + y(t)\mathbf{j} \qquad (a \le t \le b)$$

Moreover, suppose that each partition point P_k of C corresponds to a parameter value of t_k in $[a, b]$. The arc length of C between points P_{k-1} and P_k is then given by

$$\Delta s_k = \int_{t_{k-1}}^{t_k} \|\mathbf{r}'(t)\|\, dt \tag{8}$$

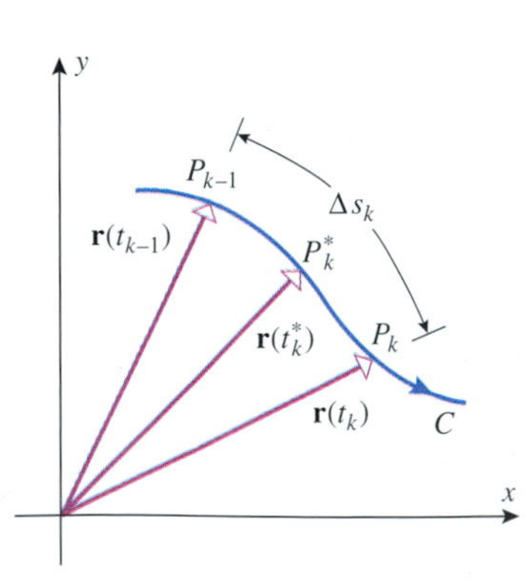

Figure 16.2.5

(Theorem 13.3.1). If we let $\Delta t_k = t_k - t_{k-1}$, then it follows from (8) and the Mean-Value Theorem for Integrals (Theorem 6.6.2) that there exists a point t_k^* in $[t_{k-1}, t_k]$ such that

$$\Delta s_k = \int_{t_{k-1}}^{t_k} \|\mathbf{r}'(t)\|\, dt = \|\mathbf{r}'(t_k^*)\|\Delta t_k$$

We let $P_k^*(x_k^*, y_k^*) = P_k^*(x(t_k^*), y(t_k^*))$ correspond to the parameter value t_k^* (Figure 16.2.5).

Since the parametrization of C is smooth, it can be shown that $\max \Delta s_k \to 0$ if and only if $\max \Delta t_k \to 0$ (Exercise 49). Furthermore, the composition $f(x(t), y(t))$ is a real-valued function defined on $[a, b]$ and we have

$$\begin{aligned}
\int_C f(x, y)\, ds &= \lim_{\max \Delta s_k \to 0} \sum_{k=1}^{n} f(x_k^*, y_k^*)\, \Delta s_k && \boxed{\text{Definition 16.2.1}} \\
&= \lim_{\max \Delta s_k \to 0} \sum_{k=1}^{n} f(x(t_k^*), y(t_k^*)) \|\mathbf{r}'(t_k^*)\| \Delta t_k && \boxed{\text{Substitution}} \\
&= \lim_{\max \Delta t_k \to 0} \sum_{k=1}^{n} f(x(t_k^*), y(t_k^*)) \|\mathbf{r}'(t_k^*)\| \Delta t_k \\
&= \int_a^b f(x(t), y(t)) \|\mathbf{r}'(t)\|\, dt && \boxed{\text{Definition 6.5.1}}
\end{aligned}$$

Therefore, if C is smoothly parametrized by

$$\mathbf{r}(t) = x(t)\mathbf{i} + y(t)\mathbf{j} \qquad (a \le t \le b)$$

then

$$\int_C f(x, y)\, ds = \int_a^b f(x(t), y(t)) \|\mathbf{r}'(t)\|\, dt \tag{9}$$

Similarly, if C is a curve in 3-space that is smoothly parametrized by

$$\mathbf{r}(t) = x(t)\mathbf{i} + y(t)\mathbf{j} + z(t)\mathbf{k} \qquad (a \le t \le b)$$

then

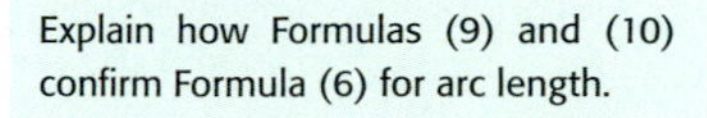
Explain how Formulas (9) and (10) confirm Formula (6) for arc length.

$$\int_C f(x, y, z)\, ds = \int_a^b f(x(t), y(t), z(t)) \|\mathbf{r}'(t)\|\, dt \tag{10}$$

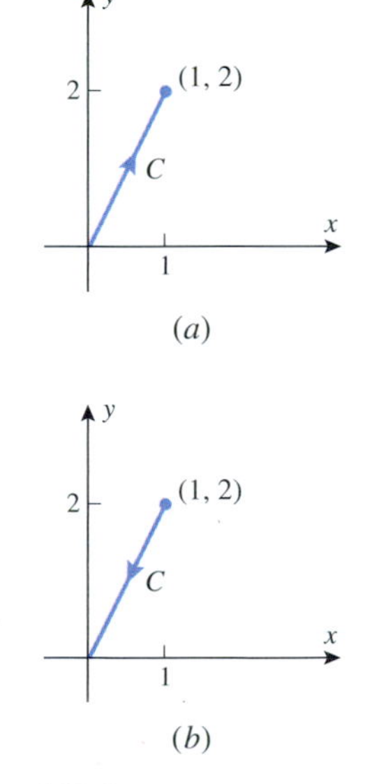

Figure 16.2.6

▶ **Example 1** Using the given parametrization, evaluate the line integral $\int_C (1 + xy^2)\, ds$.

(a) $C : \mathbf{r}(t) = t\mathbf{i} + 2t\mathbf{j}$ $(0 \le t \le 1)$ (see Figure 16.2.6*a*)

(b) $C : \mathbf{r}(t) = (1 - t)\mathbf{i} + (2 - 2t)\mathbf{j}$ $(0 \le t \le 1)$ (see Figure 16.2.6*b*)

Solution (a). Since $\mathbf{r}'(t) = \mathbf{i} + 2\mathbf{j}$, we have $\|\mathbf{r}'(t)\| = \sqrt{5}$ and it follows from Formula (9) that

$$\begin{aligned}
\int_C (1 + xy^2)\, ds &= \int_0^1 [1 + t(2t)^2]\sqrt{5}\, dt \\
&= \int_0^1 (1 + 4t^3)\sqrt{5}\, dt \\
&= \sqrt{5}\left[t + t^4\right]_0^1 = 2\sqrt{5}
\end{aligned}$$

Solution (b). Since $\mathbf{r}'(t) = -\mathbf{i} - 2\mathbf{j}$, we have $\|\mathbf{r}'(t)\| = \sqrt{5}$ and it follows from Formula (9) that

$$\begin{aligned}
\int_C (1 + xy^2)\, ds &= \int_0^1 [1 + (1 - t)(2 - 2t)^2]\sqrt{5}\, dt \\
&= \int_0^1 [1 + 4(1 - t)^3]\sqrt{5}\, dt \\
&= \sqrt{5}\left[t - (1 - t)^4\right]_0^1 = 2\sqrt{5} \quad \blacktriangleleft
\end{aligned}$$

Note that the integrals in parts (a) and (b) of Example 1 agree, even though the corresponding parametrizations of C have opposite orientations. This illustrates the important result that the value of a line integral of f with respect to s along C does not depend on an orientation of C. (This is because Δs_k is always positive; therefore, it does not matter in which *direction* along C we list the partition points of the curve in Definition 16.2.1.) Later in this section we will discuss line integrals that are defined only for oriented curves.

Formula (9) has an alternative expression for a curve C in the xy-plane that is given by parametric equations

$$x = x(t), \quad y = y(t) \qquad (a \le t \le b)$$

In this case, we write (9) in the expanded form

$$\int_C f(x, y)\, ds = \int_a^b f(x(t), y(t))\sqrt{\left(\frac{dx}{dt}\right)^2 + \left(\frac{dy}{dt}\right)^2}\, dt \tag{11}$$

Similarly, if C is a curve in 3-space that is parametrized by

$$x = x(t), \quad y = y(t), \quad z = z(t) \qquad (a \le t \le b)$$

then we write (10) in the form

$$\int_C f(x, y, z)\, ds = \int_a^b f(x(t), y(t), z(t))\sqrt{\left(\frac{dx}{dt}\right)^2 + \left(\frac{dy}{dt}\right)^2 + \left(\frac{dz}{dt}\right)^2}\, dt \tag{12}$$

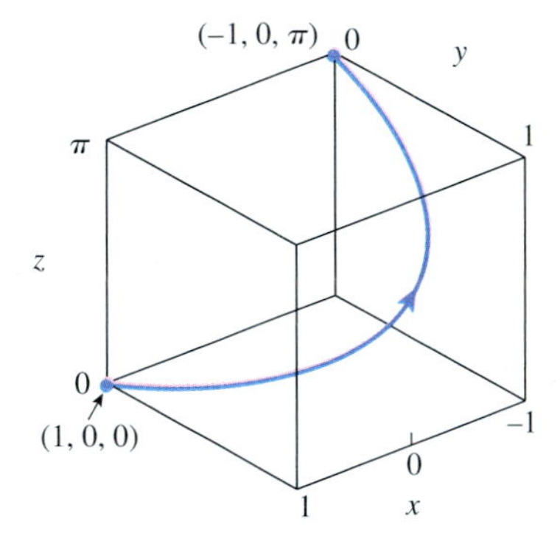

Figure 16.2.7

Example 2 Evaluate the line integral $\int_C (xy + z^3)\, ds$ from $(1, 0, 0)$ to $(-1, 0, \pi)$ along the helix C that is represented by the parametric equations

$$x = \cos t, \quad y = \sin t, \quad z = t \qquad (0 \le t \le \pi)$$

(Figure 16.2.7).

Solution. From (12)

$$\begin{aligned}
\int_C (xy + z^3)\, ds &= \int_0^\pi (\cos t \sin t + t^3)\sqrt{\left(\frac{dx}{dt}\right)^2 + \left(\frac{dy}{dt}\right)^2 + \left(\frac{dz}{dt}\right)^2}\, dt \\
&= \int_0^\pi (\cos t \sin t + t^3)\sqrt{(-\sin t)^2 + (\cos t)^2 + 1}\, dt \\
&= \sqrt{2}\int_0^\pi (\cos t \sin t + t^3)\, dt \\
&= \sqrt{2}\left[\frac{\sin^2 t}{2} + \frac{t^4}{4}\right]_0^\pi = \frac{\sqrt{2}\pi^4}{4}
\end{aligned}$$ ◀

If $\delta(x, y)$ is the linear density function of a wire that is modeled by a smooth curve C in the xy-plane, then an argument similar to the derivation of Formula (5) shows that the mass of the wire is given by $\int_C \delta(x, y)\, ds$.

Figure 16.2.8

Example 3 Suppose that a semicircular wire has the equation $y = \sqrt{25 - x^2}$ and that its mass density is $\delta(x, y) = 15 - y$ (Figure 16.2.8). Physically, this means the wire has a maximum density of 15 units at the base ($y = 0$) and that the density of the wire decreases linearly with respect to y to a value of 10 units at the top ($y = 5$). Find the mass of the wire.

Solution. The mass M of the wire can be expressed as the line integral

$$M = \int_C \delta(x, y)\, ds = \int_C (15 - y)\, ds$$

along the semicircle C. To evaluate this integral we will express C parametrically as

$$x = 5\cos t, \quad y = 5\sin t \qquad (0 \le t \le \pi)$$

Thus, it follows from (11) that

$$\begin{aligned}
M = \int_C (15 - y)\, ds &= \int_0^{\pi} (15 - 5\sin t)\sqrt{\left(\frac{dx}{dt}\right)^2 + \left(\frac{dy}{dt}\right)^2}\, dt \\
&= \int_0^{\pi} (15 - 5\sin t)\sqrt{(-5\sin t)^2 + (5\cos t)^2}\, dt \\
&= 5\int_0^{\pi} (15 - 5\sin t)\, dt \\
&= 5\big[15t + 5\cos t\big]_0^{\pi} \\
&= 75\pi - 50 \approx 185.6 \text{ units of mass}
\end{aligned}$$

In the special case where t is an arc length parameter, say $t = s$, it follows from Formulas (20) and (21) in Section 13.3 that the radicals in (11) and (12) reduce to 1 and the equations simplify to

$$\int_C f(x, y)\, ds = \int_a^b f(x(s), y(s))\, ds \tag{13}$$

and

$$\int_C f(x, y, z)\, ds = \int_a^b f(x(s), y(s), z(s))\, ds \tag{14}$$

respectively.

Example 4 Find the area of the surface extending upward from the circle $x^2 + y^2 = 1$ in the xy-plane to the parabolic cylinder $z = 1 - x^2$ (Figure 16.2.9).

Solution. It follows from (7) that the area A of the surface can be expressed as the line integral

$$A = \int_C (1 - x^2)\, ds \tag{15}$$

where C is the circle $x^2 + y^2 = 1$. This circle can be parametrized in terms of arc length as

$$x = \cos s, \quad y = \sin s \qquad (0 \le s \le 2\pi)$$

Thus, it follows from (13) and (15) that

$$\begin{aligned}
A = \int_C (1 - x^2)\, ds &= \int_0^{2\pi} (1 - \cos^2 s)\, ds \\
&= \int_0^{2\pi} \sin^2 s\, ds = \frac{1}{2}\int_0^{2\pi} (1 - \cos 2s)\, ds = \pi
\end{aligned}$$

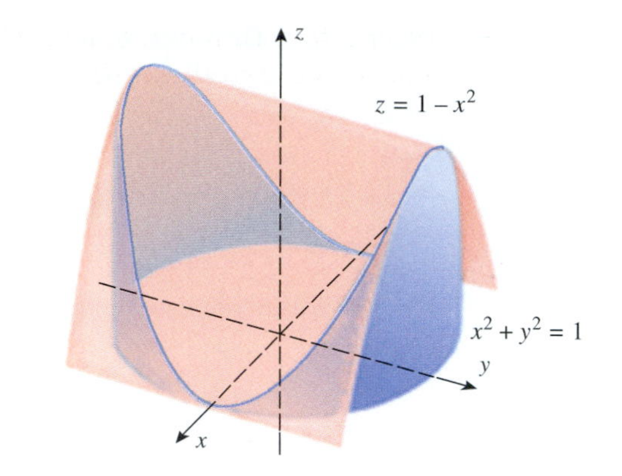

Figure 16.2.9

LINE INTEGRALS WITH RESPECT TO *x*, *y*, AND *z*

We now describe a second type of line integral in which we replace the "ds" in the integral by dx, dy, or dz. For example, suppose that f is a function defined on a smooth curve C in the xy-plane and that partition points of C are denoted by $P_k(x_k, y_k)$. Letting

$$\Delta x_k = x_k - x_{k-1} \quad \text{and} \quad \Delta y_k = y_k - y_{k-1}$$

we would like to define

$$\int_C f(x, y)\, dx = \lim_{\max \Delta s_k \to 0} \sum_{k=1}^{n} f(x_k^*, y_k^*)\Delta x_k \tag{16}$$

$$\int_C f(x, y)\, dy = \lim_{\max \Delta s_k \to 0} \sum_{k=1}^{n} f(x_k^*, y_k^*)\Delta y_k \tag{17}$$

However, unlike Δs_k, the values of Δx_k and Δy_k change sign if the order of the partition points along C is reversed. Therefore, in order to define the line integrals using Formulas (16) and (17), we must restrict ourselves to *oriented* curves C and to partitions of C in which the partition points are ordered in the direction of the curve. With this restriction, if the limit in (16) exists and does not depend on the choice of partition or sampling points, then we refer to (16) as the ***line integral of f with respect to x along C***. Similarly, (17) defines the ***line integral of f with respect to y along C***. If C is a smooth curve in 3-space, we can have ***line integrals of f with respect to x, y, and z along C***. For example,

Explain why Formula (16) implies that $\int_C dx = x_1 - x_0$, where x_1 and x_0 are the respective x-coordinates of the final and initial points of C. What about $\int_C dy$?

$$\int_C f(x, y, z)\, dx = \lim_{\max \Delta s_k \to 0} \sum_{k=1}^{n} f(x_k^*, y_k^*, z_k^*)\Delta x_k$$

and so forth. As was the case with line integrals with respect to s, line integrals of f with respect to x, y, and z exist if f is continuous on C.

Explain why Formula (16) implies that $\int_C f(x, y)\, dx = 0$ on any oriented segment parallel to the y-axis. What can you say about $\int_C f(x, y)\, dy$ on any oriented segment parallel to the x-axis?

The basic procedure for evaluating these line integrals is to find parametric equations for C, say

$$x = x(t), \quad y = y(t), \quad z = z(t) \qquad (a \le t \le b)$$

in which the orientation of C is in the direction of increasing t, and then express the integrand in terms of t. For example,

$$\int_C f(x, y, z)\, dz = \int_a^b f(x(t), y(t), z(t))z'(t)\, dt$$

[Such a formula is easy to remember—just substitute for x, y, and z using the parametric equations and recall that $dz = z'(t)\,dt$.]

▶ **Example 5** Evaluate $\int_C 3xy\,dx$, where C is the line segment joining $(0, 0)$ and $(1, 2)$ with the given orientation.

(a) Oriented from $(0, 0)$ to $(1, 2)$ as in Figure 16.2.6*a*.

(b) Oriented from $(1, 2)$ to $(0, 0)$ as in Figure 16.2.6*b*.

Solution (a). Using the parametrization

$$x = t, \quad y = 2t \qquad (0 \le t \le 1)$$

we have

$$\int_C 3xy\,dy = \int_0^1 3(t)(2t)(2)\,dt = \int_0^1 12t^2\,dt = 4t^3\Big]_0^1 = 4$$

Solution (b). Using the parametrization

$$x = 1 - t, \quad y = 2 - 2t \qquad (0 \le t \le 1)$$

we have

$$\int_C 3xy\,dy = \int_0^1 3(1-t)(2-2t)(-2)\,dt = \int_0^1 -12(1-t)^2\,dt = 4(1-t)^3\Big]_0^1 = -4$$ ◀

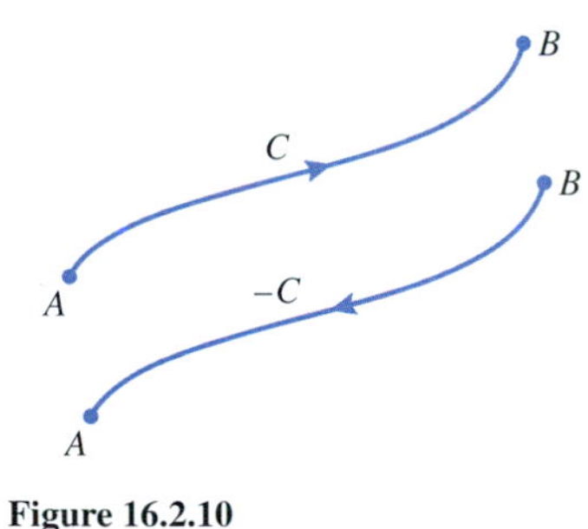

Figure 16.2.10

In Example 5, note that reversing the orientation of the curve changed the sign of the line integral. This is because reversing the orientation of a curve changes the sign of Δx_k in definition (16). Thus, unlike line integrals of functions with respect to s along C, reversing the orientation of C changes the sign of a line integral with respect to x, y, and z. If C is a smooth oriented curve, we will let $-C$ denote the oriented curve consisting of the same points as C but with the opposite orientation (Figure 16.2.10). We then have

$$\int_{-C} f(x, y)\,dx = -\int_C f(x, y)\,dx \quad \text{and} \quad \int_{-C} g(x, y)\,dy = -\int_C g(x, y)\,dy \tag{18–19}$$

while

$$\int_{-C} f(x, y)\,ds = \int_C f(x, y)\,ds \tag{20}$$

and similarly for line integrals in 3-space. Unless indicated otherwise, we will assume that parametric curves are oriented in the direction of increasing parameter.

Frequently, the line integrals with respect to x and y occur in combination, in which case we will dispense with one of the integral signs and write

$$\int_C f(x, y)\,dx + g(x, y)\,dy = \int_C f(x, y)\,dx + \int_C g(x, y)\,dy \tag{21}$$

We will use a similar convention for combinations of line integrals with respect to x, y, and z along curves in 3-space.

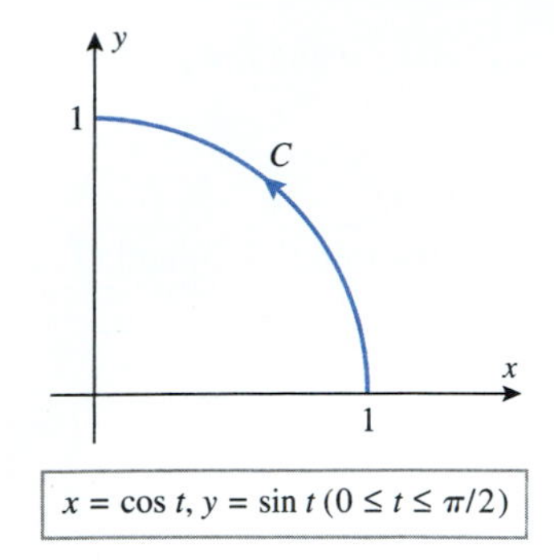

$x = \cos t,\ y = \sin t\ (0 \le t \le \pi/2)$

Figure 16.2.11

Example 6 Evaluate

$$\int_C 2xy\,dx + (x^2 + y^2)\,dy$$

along the circular arc C given by $x = \cos t$, $y = \sin t$ $(0 \le t \le \pi/2)$ (Figure 16.2.11).

Solution. We have

$$\int_C 2xy\,dx = \int_0^{\pi/2} (2\cos t \sin t)\left[\frac{d}{dt}(\cos t)\right] dt$$

$$= -2\int_0^{\pi/2} \sin^2 t \cos t\,dt = -\frac{2}{3}\sin^3 t\bigg]_0^{\pi/2} = -\frac{2}{3}$$

$$\int_C (x^2 + y^2)\,dy = \int_0^{\pi/2} (\cos^2 t + \sin^2 t)\left[\frac{d}{dt}(\sin t)\right] dt$$

$$= \int_0^{\pi/2} \cos t\,dt = \sin t\bigg]_0^{\pi/2} = 1$$

Thus, from (21)

$$\int_C 2xy\,dx + (x^2 + y^2)\,dy = \int_C 2xy\,dx + \int_C (x^2 + y^2)\,dy$$

$$= -\frac{2}{3} + 1 = \frac{1}{3}$$ ◄

It can be shown that if f and g are continuous functions on C, then combinations of line integrals with respect to x and y can be expressed in terms of a limit and can be evaluated together in a single step. For example, we have

$$\int_C f(x, y)\,dx + g(x, y)\,dy = \lim_{\max \Delta s_k \to 0} \sum_{k=1}^{n} [f(x_k^*, y_k^*)\Delta x_k + g(x_k^*, y_k^*)\,\Delta y_k] \tag{22}$$

and

$$\int_C f(x, y)\,dx + g(x, y)\,dy = \int_a^b [f(x(t), y(t))x'(t) + g(x(t), y(t))y'(t)]\,dt \tag{23}$$

Similar results hold for line integrals in 3-space. The evaluation of a line integral can sometimes be simplified by using Formula (23).

Example 7 Evaluate

$$\int_C (3x^2 + y^2)\,dx + 2xy\,dy$$

along the circular arc C given by $x = \cos t$, $y = \sin t$ $(0 \le t \le \pi/2)$ (Figure 16.2.11).

Solution. From (23) we have

$$\int_C (3x^2 + y^2)\,dx + 2xy\,dy = \int_0^{\pi/2} [(3\cos^2 t + \sin^2 t)(-\sin t) + 2(\cos t)(\sin t)(\cos t)]\,dt$$

$$= \int_0^{\pi/2} (-3\cos^2 t \sin t - \sin^3 t + 2\cos^2 t \sin t)\,dt$$

$$= \int_0^{\pi/2} (-\cos^2 t - \sin^2 t)(\sin t)\,dt = \int_0^{\pi/2} -\sin t\,dt$$

$$= \cos t\bigg]_0^{\pi/2} = -1$$ ◄

Compare the computations in Example 7 with those involved in computing

$$\int_C (3x^2 + y^2)\,dx + \int_C 2xy\,dy$$

It follows from (18) and (19) that

$$\int_{-C} f(x, y)\,dx + g(x, y)\,dy = -\int_{C} f(x, y)\,dx + g(x, y)\,dy \tag{24}$$

so that reversing the orientation of C changes the sign of a line integral in which x and y occur in combination. Similarly,

$$\begin{aligned}&\int_{-C} f(x, y, z)\,dx + g(x, y, z)\,dy + h(x, y, z)\,dz\\ &\quad = -\int_{C} f(x, y, z)\,dx + g(x, y, z)\,dy + h(x, y, z)\,dz\end{aligned} \tag{25}$$

INTEGRATING A VECTOR FIELD ALONG A CURVE

There is an alternative notation for line integrals with respect to x, y, and z that is particularly appropriate for dealing with problems involving vector fields. We will interpret $d\mathbf{r}$ as

$$d\mathbf{r} = dx\mathbf{i} + dy\mathbf{j} \quad \text{or} \quad d\mathbf{r} = dx\mathbf{i} + dy\mathbf{j} + dz\mathbf{k}$$

depending on whether C is in 2-space or 3-space. For an oriented curve C in 2-space and a vector field

$$\mathbf{F}(x, y) = f(x, y)\mathbf{i} + g(x, y)\mathbf{j}$$

we will write

$$\int_C \mathbf{F}\cdot d\mathbf{r} = \int_C (f(x, y)\mathbf{i} + g(x, y)\mathbf{j})\cdot(dx\mathbf{i} + dy\mathbf{j}) = \int_C f(x, y)\,dx + g(x, y)\,dy \tag{26}$$

Similarily, for a curve C in 3-space and vector field

$$\mathbf{F}(x, y, z) = f(x, y, z)\mathbf{i} + g(x, y, z)\mathbf{j} + h(x, y, z)\mathbf{k}$$

we will write

$$\begin{aligned}\int_C \mathbf{F}\cdot d\mathbf{r} &= \int_C (f(x, y, z)\mathbf{i} + g(x, y, z)\mathbf{j} + h(x, y, z)\mathbf{k})\cdot(dx\mathbf{i} + dy\mathbf{j} + dz\mathbf{k})\\ &= \int_C f(x, y, z)\,dx + g(x, y, z)\,dy + h(x, y, z)\,dz\end{aligned} \tag{27}$$

With these conventions, we are led to the following definition.

16.2.2 DEFINITION. If $\mathbf{F}$ is a continuous vector field and C is a smooth oriented curve, then the ***line integral of* F *along C*** is

$$\int_C \mathbf{F}\cdot d\mathbf{r} \tag{28}$$

The notation in Definition 16.2.2 makes it easy to remember the formula for evaluating the line integral of $\mathbf{F}$ along C. For example, suppose that C is an oriented curve in the plane given in vector form by

$$\mathbf{r} = \mathbf{r}(t) = x(t)\mathbf{i} + y(t)\mathbf{j} \qquad (a \le t \le b)$$

If we write

$$\mathbf{F}(\mathbf{r}(t)) = f(x(t), y(t))\mathbf{i} + g(x(t), y(t))\mathbf{j}$$

then

$$\int_C \mathbf{F} \cdot d\mathbf{r} = \int_a^b \mathbf{F}(\mathbf{r}(t)) \cdot \mathbf{r}'(t)\, dt \tag{29}$$

Formula (29) is also valid for oriented curves in 3-space.

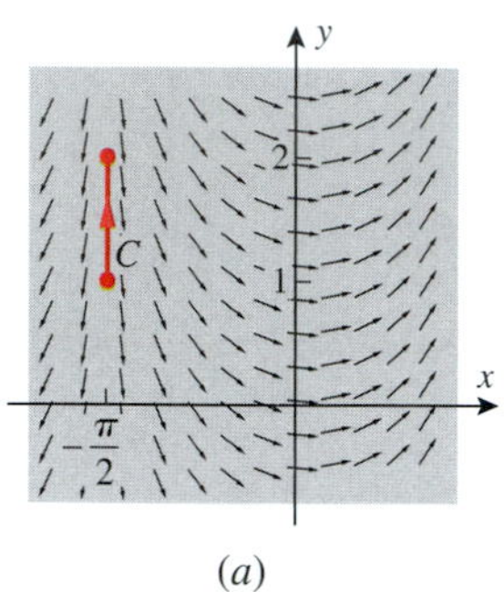

(a)

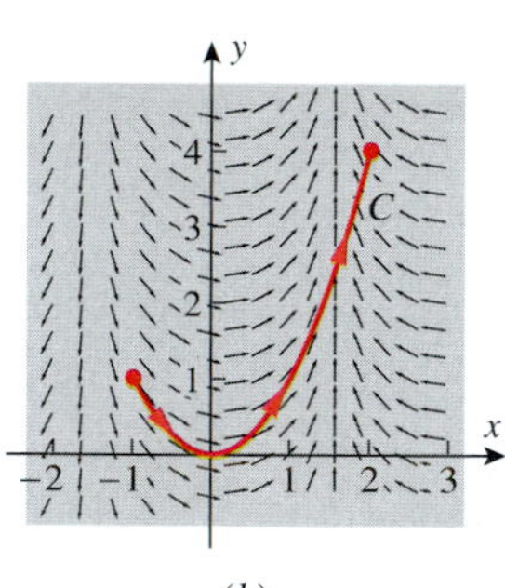

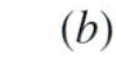

(b)

Figure 16.2.12

▸ **Example 8** Evaluate $\int_C \mathbf{F} \cdot d\mathbf{r}$ where $\mathbf{F}(x, y) = \cos x\mathbf{i} + \sin x\mathbf{j}$ and where C is the given oriented curve.

(a) $C : \mathbf{r}(t) = -\frac{\pi}{2}\mathbf{i} + t\mathbf{j} \quad (1 \le t \le 2)$ (see Figure 16.2.12*a*)

(b) $C : \mathbf{r}(t) = t\mathbf{i} + t^2\mathbf{j} \quad (-1 \le t \le 2)$ (see Figure 16.2.12*b*)

Solution (a). Using (27) we have

$$\int_C \mathbf{F} \cdot d\mathbf{r} = \int_1^2 \mathbf{F}(\mathbf{r}(t)) \cdot \mathbf{r}'(t)\, dt = \int_1^2 (-\mathbf{j}) \cdot \mathbf{j}\, dt = \int_1^2 (-1)\, dt = -1$$

Solution (b). Using (27) we have

$$\begin{aligned}\int_C \mathbf{F} \cdot d\mathbf{r} &= \int_{-1}^2 \mathbf{F}(\mathbf{r}(t)) \cdot \mathbf{r}'(t)\, dt = \int_{-1}^2 (\cos t\mathbf{i} + \sin t\mathbf{j}) \cdot (\mathbf{i} + 2t\mathbf{j})\, dt \\ &= \int_{-1}^2 (\cos t + 2t \sin t)\, dt = -2t\cos t + 3\sin t\Big]_{-1}^2 \\ &= -2\cos 1 - 4\cos 2 + 3(\sin 1 + \sin 2) \approx 5.83629\end{aligned}$$ ◂

If we let t denote an arc length parameter, say $t = s$, with $\mathbf{T} = \mathbf{r}'(s)$ the unit tangent vector field along C, then

$$\int_C \mathbf{F} \cdot d\mathbf{r} = \int_a^b \mathbf{F}(\mathbf{r}(s)) \cdot \mathbf{r}'(s)\, ds = \int_a^b \mathbf{F}(\mathbf{r}(s)) \cdot \mathbf{T}\, ds = \int_C \mathbf{F} \cdot \mathbf{T}\, ds$$

which shows that

$$\int_C \mathbf{F} \cdot d\mathbf{r} = \int_C \mathbf{F} \cdot \mathbf{T}\, ds \tag{30}$$

In words, the integral of a vector field along a curve has the same value as the integral of the tangential component of the vector field along the curve.

We can use (30) to interpret $\int_C \mathbf{F} \cdot d\mathbf{r}$ geometrically. If θ is the angle between $\mathbf{F}$ and $\mathbf{T}$ at a point on C, then at this point

$$\begin{aligned}\mathbf{F} \cdot \mathbf{T} &= \|\mathbf{F}\|\|\mathbf{T}\| \cos\theta && \text{Formula (4) in Section 12.3} \\ &= \|\mathbf{F}\| \cos\theta && \text{Since } \|\mathbf{T}\| = 1\end{aligned}$$

Thus,

$$-\|\mathbf{F}\| \le \mathbf{F} \cdot \mathbf{T} \le \|\mathbf{F}\|$$

and if $\mathbf{F} \ne \mathbf{0}$, then the sign of $\mathbf{F} \cdot \mathbf{T}$ will depend on the angle between the direction of $\mathbf{F}$ and the direction of C (Figure 16.2.13). That is, $\mathbf{F} \cdot \mathbf{T}$ will be positive where $\mathbf{F}$ has the same general direction as C, it will be 0 if $\mathbf{F}$ is normal to C, and it will be negative where $\mathbf{F}$ and C have more or less opposite directions. The line integral of $\mathbf{F}$ along C can be interpreted as the accumulated effect of the magnitude of **F** along C, the extent to which **F** and C have the same direction, and the arc length of C.

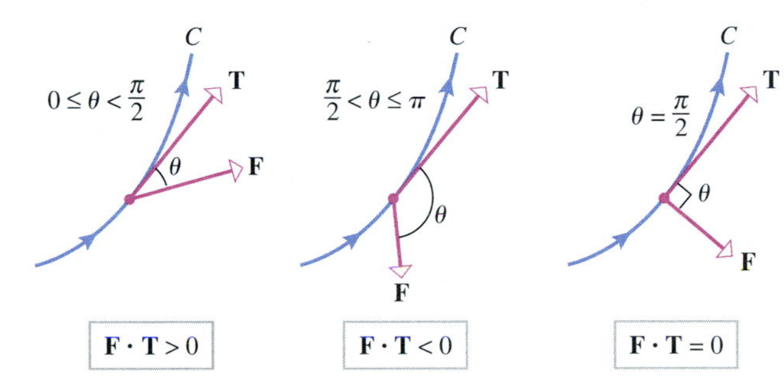

Figure 16.2.13

Vectors not to scale

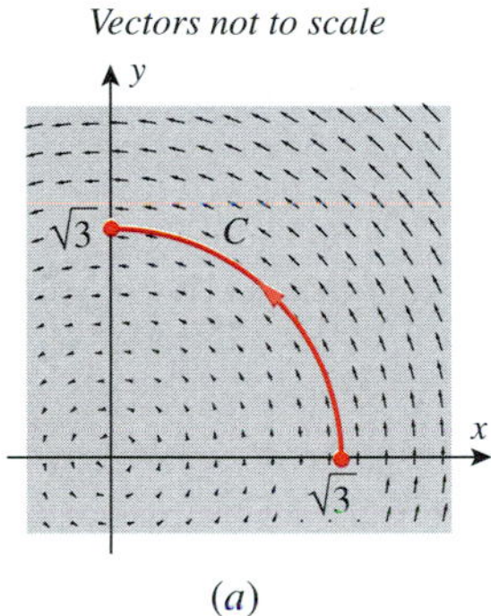

(a)

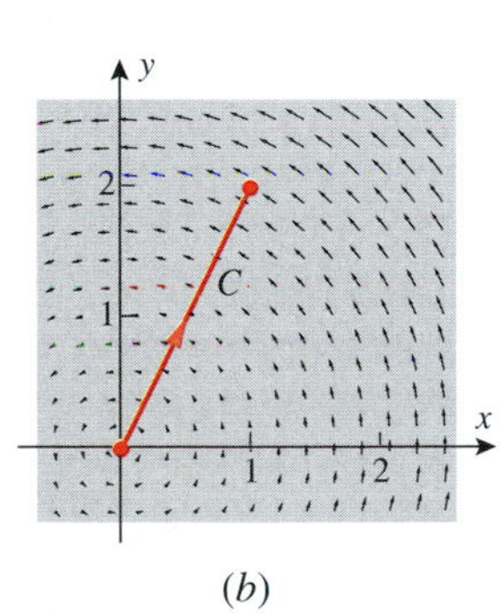

(b)

Figure 16.2.14

▶ **Example 9** Use (30) to evaluate $\int_C \mathbf{F}\cdot d\mathbf{r}$ where $\mathbf{F}(x, y) = -y\mathbf{i} + x\mathbf{j}$ and where C is the given oriented curve.

(a) $C : x^2 + y^2 = 3$ $\quad(0 \le x, y;$ oriented as in Figure 16.2.14$a)$

(b) $C : \mathbf{r}(t) = t\mathbf{i} + 2t\mathbf{j}$ $\quad(0 \le t \le 1;$ see Figure 16.2.14$b)$

Solution (a). At every point on C the direction of **F** and the direction of C are the same. (Why?) In addition, at every point on C

$$\|\mathbf{F}\| = \sqrt{(-y)^2 + x^2} = \sqrt{x^2 + y^2} = \sqrt{3}$$

Therefore, $\mathbf{F}\cdot\mathbf{T} = \|\mathbf{F}\|\cos(0) = \|\mathbf{F}\| = \sqrt{3}$, and

$$\int_C \mathbf{F}\cdot d\mathbf{r} = \int_C \mathbf{F}\cdot\mathbf{T}\,ds = \int_C \sqrt{3}\,ds = \sqrt{3}\int_C ds = \frac{3\pi}{2}$$

Solution (b). The vector field **F** is normal to C at every point. (Why?) Therefore,

$$\int_C \mathbf{F}\cdot d\mathbf{r} = \int_C \mathbf{F}\cdot\mathbf{T}\,ds = \int_C 0\,ds = 0$$ ◀

Refer to Figure 16.2.12 and explain the sign of each line integral in Example 8 geometrically. Exercises 5 and 6 take this geometric analysis further.

In light of (20) and (30), you might expect that reversing the orientation of C in $\int_C \mathbf{F}\cdot d\mathbf{r}$ would have no effect on the value of the line integral. However, reversing the orientation of C reverses the orientation of **T** in the integrand and hence reverses the sign of the integral; that is,

$$\int_{-C} \mathbf{F}\cdot\mathbf{T}\,ds = -\int_C \mathbf{F}\cdot\mathbf{T}\,ds \tag{31}$$

$$\int_{-C} \mathbf{F}\cdot d\mathbf{r} = -\int_C \mathbf{F}\cdot d\mathbf{r} \tag{32}$$

WORK AS A LINE INTEGRAL

An important application of line integrals with respect to x, y, and z is to the problem of defining the work performed by a variable force moving a particle along a curved path. In Section 7.7 we defined the work W performed by a force of constant magnitude acting on an object in the direction of motion (Definition 7.7.1), and later in that section we extended the definition to allow for a force of variable magnitude acting in the direction of motion (Definition 7.7.3). In Section 12.3 we took the concept of work a step further by defining the work W performed by a constant force $\mathbf{F}$ moving a particle in a straight line from point P to point Q. We defined the work to be

$$W = \mathbf{F} \cdot \overrightarrow{PQ} \tag{33}$$

[Formula (14) in Section 12.3]. Our next goal is to define a more general concept of work —the work performed by a variable force acting on a particle that moves along a curved path in 2-space or 3-space.

In many applications variable forces arise from force fields (gravitational fields, electromagnetic fields, and so forth), so we will consider the problem of work in that context. To motivate an appropriate definition for work performed by a force field, we will use a limit process, and since the procedure is the same for 2-space and 3-space, we will discuss it in detail for 2-space only. The idea is as follows:

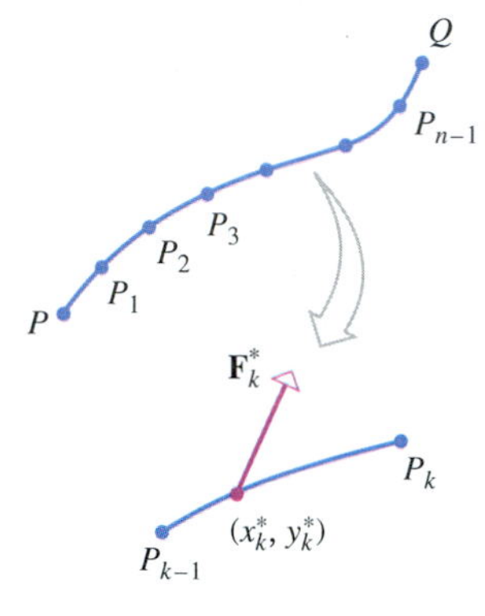

Figure 16.2.15

- Assume that a force field $\mathbf{F} = \mathbf{F}(x, y)$ moves a particle along a smooth curve C from a point P to a point Q. Divide C into n arcs using the partition points

$$P = P_0(x_0, y_0),\ P_1(x_1, y_1),\ P_2(x_2, y_2), \ldots,\ P_{n-1}(x_{n-1}, y_{n-1}),\ P_n(x_n, y_n) = Q$$

directed along C from P to Q, and denote the length of the kth arc by Δs_k. Let (x_k^*, y_k^*) be any point on the kth arc, and let

$$\mathbf{F}_k^* = \mathbf{F}(x_k^*, y_k^*) = f(x_k^*, y_k^*)\mathbf{i} + g(x_k^*, y_k^*)\mathbf{j}$$

be the force vector at this point (Figure 16.2.15).

- If the kth arc is small, then the force will not vary much, so we can approximate the force by the constant value $\mathbf{F}_k^*$ on this arc. Moreover, the direction of motion will not vary much over this small arc, so we can approximate the movement of the particle by the displacement vector

$$\overrightarrow{P_{k-1}P_k} = (\Delta x_k)\mathbf{i} + (\Delta y_k)\mathbf{j}$$

where $\Delta x_k = x_k - x_{k-1}$ and $\Delta y_k = y_k - y_{k-1}$.

- Since the work done by a constant force $\mathbf{F}_k^*$ moving a particle along a straight line from P_{k-1} to P_k is

$$\begin{aligned} \mathbf{F}_k^* \cdot \overrightarrow{P_{k-1}P_k} &= (f(x_k^*, y_k^*)\mathbf{i} + g(x_k^*, y_k^*)\mathbf{j}) \cdot ((\Delta x_k)\mathbf{i} + (\Delta y_k)\mathbf{j}) \\ &= f(x_k^*, y_k^*)\Delta x_k + g(x_k^*, y_k^*)\Delta y_k \end{aligned}$$

[Formula (33)], the work ΔW_k performed by the force field along the kth arc of C can be approximated by

$$\Delta W_k \approx f(x_k^*, y_k^*)\Delta x_k + g(x_k^*, y_k^*)\Delta y_k$$

The total work W performed by the force moving the particle over the entire curve C can then be approximated as

$$W = \sum_{k=1}^{n} \Delta W_k \approx \sum_{k=1}^{n} [f(x_k^*, y_k^*)\Delta x_k + g(x_k^*, y_k^*)\Delta y_k]$$

- As $\max \Delta s_k \to 0$, it is plausible that the error in this approximation approaches 0 and the exact work performed by the force field is

$$\begin{aligned} W &= \lim_{\max \Delta s_k \to 0} \sum_{k=1}^{n} [f(x_k^*, y_k^*)\Delta x_k + g(x_k^*, y_k^*)\Delta y_k] \\ &= \int_C f(x, y)\,dx + g(x, y)\,dy \qquad \text{Formula (22)} \\ &= \int_C \mathbf{F} \cdot d\mathbf{r} \qquad \text{Formula (26)} \end{aligned}$$

Thus, we are led to the following definition.

Note from Formula (30) that the work performed by a force field on a particle moving along a smooth curve C is obtained by integrating the scalar tangential component of force along C. This implies that the component of force orthogonal to the direction of motion of the particle has no effect on the work done.

16.2.3 DEFINITION. Suppose that under the influence of a continuous force field $\mathbf{F}$ a particle moves along a smooth curve C and that C is oriented in the direction of motion of the particle. Then the ***work performed by the force field*** on the particle is

$$\int_C \mathbf{F} \cdot d\mathbf{r} \tag{34}$$

For example suppose that force is measured in pounds and distance is measured in feet. It follows from part (a) of Example 9 that the work done by a force $\mathbf{F}(x, y) = -y\mathbf{i} + x\mathbf{j}$ acting on a particle moving along the circle $x^2 + y^2 = 3$ from $(\sqrt{3}, 0)$ to $(0, \sqrt{3})$ is $3\pi/2$ foot-pounds.

LINE INTEGRALS ALONG PIECEWISE SMOOTH CURVES

Thus far, we have only considered line integrals along smooth curves. However, the notion of a line integral can be extended to curves formed from finitely many smooth curves $C_1, C_2, \ldots, C_n$ joined end to end. Such a curve is called ***piecewise smooth*** (Figure 16.2.16). We define a line integral along a piecewise smooth curve C to be the sum of the integrals along the sections:

$$\int_C = \int_{C_1} + \int_{C_2} + \cdots + \int_{C_n}$$

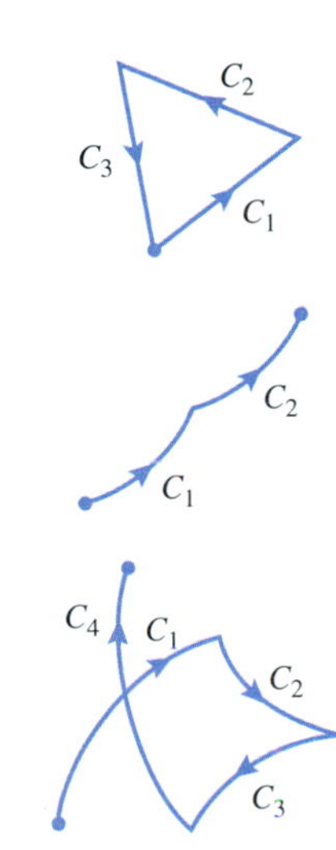

Figure 16.2.16

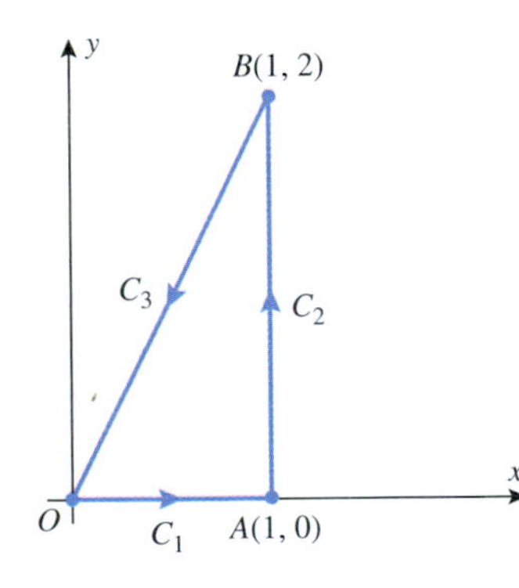

Figure 16.2.17

Example 10 Evaluate

$$\int_C x^2y\,dx + x\,dy$$

in a counterclockwise direction around the triangular path shown in Figure 16.2.17.

Solution. We will integrate over C_1, C_2, and C_3 separately and add the results. For each of the three integrals we must find parametric equations that trace the path of integration in the correct direction. For this purpose recall from Formula (7) of Section 13.1 that the graph of the vector-valued function

$$\mathbf{r}(t) = (1-t)\mathbf{r}_0 + t\mathbf{r}_1 \qquad (0 \le t \le 1)$$

is the line segment joining $\mathbf{r}_0$ and $\mathbf{r}_1$, oriented in the direction from $\mathbf{r}_0$ to $\mathbf{r}_1$. Thus, the line segments C_1, C_2, and C_3 can be represented in vector notation as

$$C_1\colon \mathbf{r}(t) = (1-t)\langle 0, 0\rangle + t\langle 1, 0\rangle = \langle t, 0\rangle$$
$$C_2\colon \mathbf{r}(t) = (1-t)\langle 1, 0\rangle + t\langle 1, 2\rangle = \langle 1, 2t\rangle$$
$$C_3\colon \mathbf{r}(t) = (1-t)\langle 1, 2\rangle + t\langle 0, 0\rangle = \langle 1-t, 2-2t\rangle$$

where t varies from 0 to 1 in each case. From these equations we obtain

$$\int_{C_1} x^2y\,dx + x\,dy = \int_{C_1} x^2y\,dx = \int_0^1 (t^2)(0)\frac{d}{dt}[t]\,dt = 0$$
$$\int_{C_2} x^2y\,dx + x\,dy = \int_{C_2} x\,dy = \int_0^1 (1)\frac{d}{dt}[2t]\,dt = 2$$
$$\int_{C_3} x^2y\,dx + x\,dy = \int_0^1 (1-t)^2(2-2t)\frac{d}{dt}[1-t]\,dt + \int_0^1 (1-t)\frac{d}{dt}[2-2t]\,dt$$
$$= 2\int_0^1 (t-1)^3\,dt + 2\int_0^1 (t-1)\,dt = -\tfrac{1}{2} - 1 = -\tfrac{3}{2}$$

Thus,

$$\int_C x^2y\,dx + x\,dy = 0 + 2 + (-\tfrac{3}{2}) = \tfrac{1}{2}$$

✔ QUICK CHECK EXERCISES 16.2

(See page 1128 for answers.)

1. The area of the surface extending upward from the line segment $y = x$ $(0 \leq x \leq 1)$ in the xy-plane to the plane $z = 2x + 1$ is ________.

2. Suppose that a wire has equation $y = 1 - x$ $(0 \leq x \leq 1)$ and that its mass density is $\delta(x, y) = 2 - x$. The mass of the wire is ________.

3. If C is the curve represented by the equations

$$x = \sin t, \quad y = \cos t, \quad z = t \qquad (0 \leq t \leq 2\pi)$$

then $\int_C y\,dx - x\,dy + dz =$ ________.

4. If C is the unit circle $x^2 + y^2 = 1$ oriented counterclockwise and $\mathbf{F}(x, y) = x\mathbf{i} + y\mathbf{j}$, then

$$\int_C \mathbf{F} \cdot d\mathbf{r} = \text{\_\_\_\_\_\_\_\_}$$

EXERCISE SET 16.2 Graphing Utility C CAS

FOCUS ON CONCEPTS

1. Let C be the line segment from $(0, 0)$ to $(0, 1)$. In each part, evaluate the line integral along C by inspection, and explain your reasoning.
   (a) $\int_C ds$ (b) $\int_C \sin xy\,dy$

2. Let C be the line segment from $(0, 2)$ to $(0, 4)$. In each part, evaluate the line integral along C by inspection, and explain your reasoning.
   (a) $\int_C ds$ (b) $\int_C e^{xy}\,dx$

3–4 Evaluate $\int_C \mathbf{F} \cdot d\mathbf{r}$ by inspection for the force field $\mathbf{F}(x, y) = \mathbf{i} + \mathbf{j}$ and the curve C shown in the figure. Explain your reasoning. [For clarity, the vectors in the force field are shown at less than true scale.]

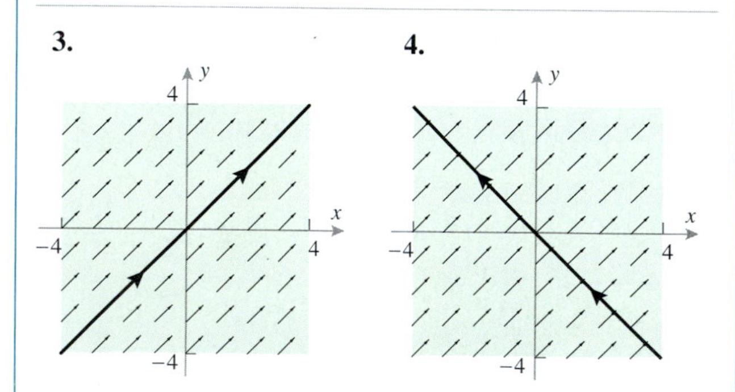

5. Use (30) to explain why the line integral in part (a) of Example 8 can be found by multiplying the length of the line segment C by -1.

6. (a) Use (30) to explain why the line integral in part (b) of Example 8 should be close to, but somewhat less than, the length of the parabolic curve C.
   (b) Verify the conclusion in part (a) of this exercise by computing the length of C and comparing the length with the value of the line integral.

7. Let C be the curve represented by the equations

$$x = 2t, \quad y = 3t^2 \qquad (0 \leq t \leq 1)$$

In each part, evaluate the line integral along C.
   (a) $\int_C (x - y)\,ds$ (b) $\int_C (x - y)\,dx$
   (c) $\int_C (x - y)\,dy$

8. Let C be the curve represented by the equations

$$x = t, \quad y = 3t^2, \quad z = 6t^3 \qquad (0 \leq t \leq 1)$$

In each part, evaluate the line integral along C.
   (a) $\int_C xyz^2\,ds$ (b) $\int_C xyz^2\,dx$
   (c) $\int_C xyz^2\,dy$ (d) $\int_C xyz^2\,dz$

9. In each part, evaluate the integral

$$\int_C (3x + 2y)\,dx + (2x - y)\,dy$$

along the stated curve.
   (a) The line segment from $(0, 0)$ to $(1, 1)$.
   (b) The parabolic arc $y = x^2$ from $(0, 0)$ to $(1, 1)$.
   (c) The curve $y = \sin(\pi x/2)$ from $(0, 0)$ to $(1, 1)$.
   (d) The curve $x = y^3$ from $(0, 0)$ to $(1, 1)$.

10. In each part, evaluate the integral

$$\int y\,dx + z\,dy - x\,dz$$

along the stated curve.
   (a) The line segment from $(0, 0, 0)$ to $(1, 1, 1)$.
   (b) The twisted cubic $x = t$, $y = t^2$, $z = t^3$ from $(0, 0, 0)$ to $(1, 1, 1)$.
   (c) The helix $x = \cos \pi t$, $y = \sin \pi t$, $z = t$ from $(1, 0, 0)$ to $(-1, 0, 1)$.

11–14 Evaluate the line integral with respect to s along the curve C.

11. $\displaystyle\int_C \frac{1}{1+x}\,ds$

$C: \mathbf{r}(t) = t\mathbf{i} + \frac{2}{3}t^{3/2}\mathbf{j} \quad (0 \le t \le 3)$

12. $\displaystyle\int_C \frac{x}{1+y^2}\,ds$

$C: x = 1 + 2t,\ y = t \quad (0 \le t \le 1)$

13. $\displaystyle\int_C 3x^2yz\,ds$

$C: x = t,\ y = t^2,\ z = \frac{2}{3}t^3 \quad (0 \le t \le 1)$

14. $\displaystyle\int_C \frac{e^{-z}}{x^2+y^2}\,ds$

$C: \mathbf{r}(t) = 2\cos t\,\mathbf{i} + 2\sin t\,\mathbf{j} + t\mathbf{k} \quad (0 \le t \le 2\pi)$

15–22 Evaluate the line integral along the curve C.

15. $\displaystyle\int_C (x+2y)\,dx + (x-y)\,dy$

$C: x = 2\cos t,\ y = 4\sin t \quad (0 \le t \le \pi/4)$

16. $\displaystyle\int_C (x^2-y^2)\,dx + x\,dy$

$C: x = t^{2/3},\ y = t \quad (-1 \le t \le 1)$

17. $\displaystyle\int_C -y\,dx + x\,dy$

$C: y^2 = 3x$ from $(3, 3)$ to $(0, 0)$

18. $\displaystyle\int_C (y-x)\,dx + x^2y\,dy$

$C: y^2 = x^3$ from $(1, -1)$ to $(1, 1)$

19. $\displaystyle\int_C (x^2+y^2)\,dx - x\,dy$

$C: x^2 + y^2 = 1$, counterclockwise from $(1, 0)$ to $(0, 1)$

20. $\displaystyle\int_C (y-x)\,dx + xy\,dy$

C: the line segment from $(3, 4)$ to $(2, 1)$

21. $\displaystyle\int_C yz\,dx - xz\,dy + xy\,dz$

$C: x = e^t,\ y = e^{3t},\ z = e^{-t} \quad (0 \le t \le 1)$

22. $\displaystyle\int_C x^2\,dx + xy\,dy + z^2\,dz$

$C: x = \sin t,\ y = \cos t,\ z = t^2 \quad (0 \le t \le \pi/2)$

23–24 Use a CAS to evaluate the line integrals along the given curves.

C **23.** (a) $\displaystyle\int_C (x^3+y^3)\,ds$

$C: \mathbf{r}(t) = e^t\mathbf{i} + e^{-t}\mathbf{j} \quad (0 \le t \le \ln 2)$

(b) $\displaystyle\int_C xe^z\,dx + (x-z)\,dy + (x^2+y^2+z^2)\,dz$

$C: x = \sin t,\ y = \cos t,\ z = t \quad (0 \le t \le \pi/2)$

C **24.** (a) $\displaystyle\int_C x^7y^3\,ds$

$C: x = \cos^3 t,\ y = \sin^3 t \quad (0 \le t \le \pi/2)$

(b) $\displaystyle\int_C x^5z\,dx + 7y\,dy + y^2z\,dz$

$C: \mathbf{r}(t) = t\mathbf{i} + t^2\mathbf{j} + \ln t\,\mathbf{k} \quad (1 \le t \le e)$

25–26 Evaluate $\int_C y\,dx - x\,dy$ along the curve C shown in the figure.

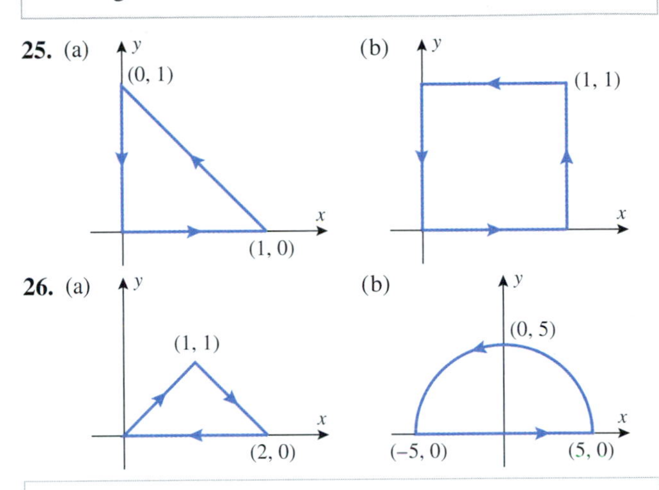

27–28 Evaluate $\int_C x^2z\,dx - yx^2\,dy + 3\,dz$ along the curve C shown in the figure.

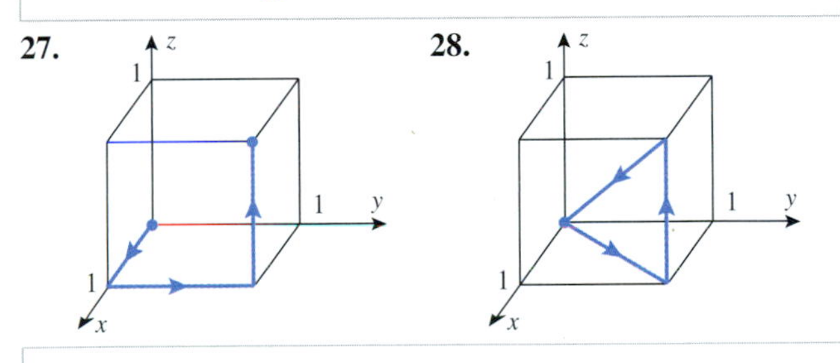

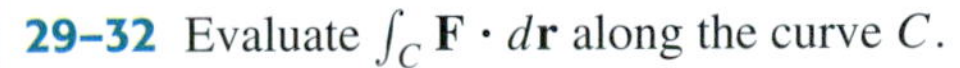

29–32 Evaluate $\int_C \mathbf{F} \cdot d\mathbf{r}$ along the curve C.

29. $\mathbf{F}(x, y) = x^2\mathbf{i} + xy\mathbf{j}$

$C: \mathbf{r}(t) = 2\cos t\,\mathbf{i} + 2\sin t\,\mathbf{j} \quad (0 \le t \le \pi)$

30. $\mathbf{F}(x, y) = x^2y\mathbf{i} + 4\mathbf{j}$

$C: \mathbf{r}(t) = e^t\mathbf{i} + e^{-t}\mathbf{j} \quad (0 \le t \le 1)$

31. $\mathbf{F}(x, y) = (x^2+y^2)^{-3/2}(x\mathbf{i} + y\mathbf{j})$

$C: \mathbf{r}(t) = e^t\sin t\,\mathbf{i} + e^t\cos t\,\mathbf{j} \quad (0 \le t \le 1)$

32. $\mathbf{F}(x, y, z) = z\mathbf{i} + x\mathbf{j} + y\mathbf{k}$

$C: \mathbf{r}(t) = \sin t\,\mathbf{i} + 3\sin t\,\mathbf{j} + \sin^2 t\,\mathbf{k} \quad (0 \le t \le \pi/2)$

33. Find the mass of a thin wire shaped in the form of the circular arc $y = \sqrt{9-x^2}$ $(0 \le x \le 3)$ if the density function is $\delta(x, y) = x\sqrt{y}$.

34. Find the mass of a thin wire shaped in the form of the curve $x = e^t\cos t$, $y = e^t\sin t$ $(0 \le t \le 1)$ if the density function δ is proportional to the distance from the origin.

35. Find the mass of a thin wire shaped in the form of the helix $x = 3\cos t$, $y = 3\sin t$, $z = 4t$ $(0 \le t \le \pi/2)$ if the density function is $\delta = kx/(1+y^2)$ $(k > 0)$.

36. Find the mass of a thin wire shaped in the form of the curve $x = 2t$, $y = \ln t$, $z = 4\sqrt{t}$ $(1 \le t \le 4)$ if the density function is proportional to the distance above the xy-plane.

37–40 Find the work done by the force field $\mathbf{F}$ on a particle that moves along the curve C.

37. $\mathbf{F}(x, y) = xy\mathbf{i} + x^2\mathbf{j}$
C: $x = y^2$ from $(0, 0)$ to $(1, 1)$

38. $\mathbf{F}(x, y) = (x^2 + xy)\mathbf{i} + (y - x^2y)\mathbf{j}$
C: $x = t,\ y = 1/t \quad (1 \le t \le 3)$

39. $\mathbf{F}(x, y, z) = xy\mathbf{i} + yz\mathbf{j} + xz\mathbf{k}$
C: $\mathbf{r}(t) = t\mathbf{i} + t^2\mathbf{j} + t^3\mathbf{k} \quad (0 \le t \le 1)$

40. $\mathbf{F}(x, y, z) = (x + y)\mathbf{i} + xy\mathbf{j} - z^2\mathbf{k}$
C: along line segments from $(0, 0, 0)$ to $(1, 3, 1)$ to $(2, -1, 4)$

41–42 Find the work done by the force field

$$\mathbf{F}(x, y) = \frac{1}{x^2 + y^2}\mathbf{i} + \frac{4}{x^2 + y^2}\mathbf{j}$$

on a particle that moves along the curve C shown in the figure.

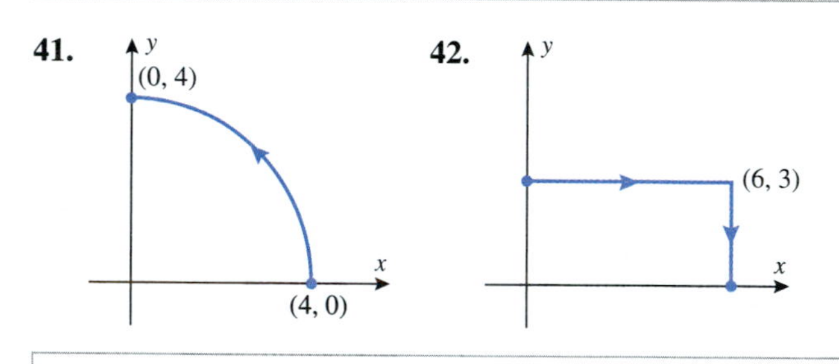

43–44 Use a line integral to find the area of the surface.

43. The surface that extends upward from the parabola $y = x^2$ $(0 \le x \le 2)$ in the xy-plane to the plane $z = 3x$.

44. The surface that extends upward from the semicircle $y = \sqrt{4 - x^2}$ in the xy-plane to the surface $z = x^2y$.

45. As illustrated in the accompanying figure, a sinusoidal cut is made in the top of a cylindrical tin can. Suppose that the base is modeled by the parametric equations $x = \cos t$, $y = \sin t$, $z = 0$ $(0 \le t \le 2\pi)$, and the height of the cut as a function of t is $z = 2 + 0.5\sin 3t$.

(a) Use a geometric argument to find the lateral surface area of the cut can.
(b) Write down a line integral for the surface area.
(c) Use the line integral to calculate the surface area.

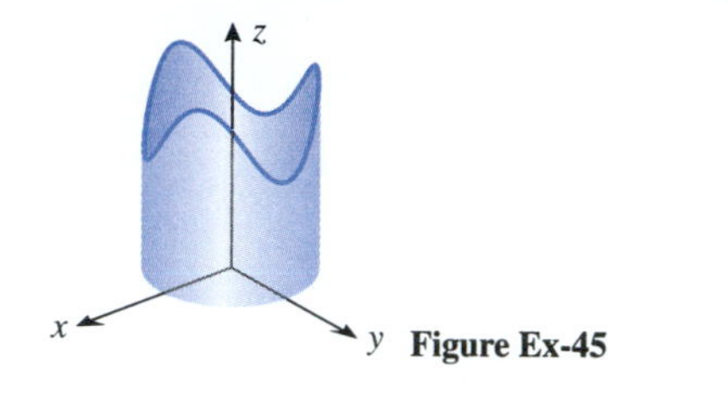

Figure Ex-45

46. Evaluate the integral $\displaystyle\int_{-C} \frac{x\,dy - y\,dx}{x^2 + y^2}$, where C is the circle $x^2 + y^2 = a^2$ traversed counterclockwise.

47. Suppose that a particle moves through the force field $\mathbf{F}(x, y) = xy\mathbf{i} + (x - y)\mathbf{j}$ from the point $(0, 0)$ to the point $(1, 0)$ along the curve $x = t$, $y = \lambda t(1 - t)$. For what value of λ will the work done by the force field be 1?

48. A farmer weighing 150 lb carries a sack of grain weighing 20 lb up a circular helical staircase around a silo of radius 25 ft. As the farmer climbs, grain leaks from the sack at a rate of 1 lb per 10 ft of ascent. How much work is performed by the farmer in climbing through a vertical distance of 60 ft in exactly four revolutions? [*Hint:* Find a vector field that represents the force exerted by the farmer in lifting his own weight plus the weight of the sack upward at each point along his path.]

49. Suppose that a curve C in the xy-plane is smoothly parametrized by

$$\mathbf{r}(t) = x(t)\mathbf{i} + y(t)\mathbf{j} \qquad (a \le t \le b)$$

In each part, refer to the notation used in the derivation of Formula (9).

(a) Let m and M denote the respective minimum and maximum values of $\|\mathbf{r}'(t)\|$ on $[a, b]$. Prove that

$$0 \le m(\max \Delta t_k) \le \max \Delta s_k \le M(\max \Delta t_k)$$

(b) Use part (a) to prove that $\max \Delta s_k \to 0$ if and only if $\max \Delta t_k \to 0$.

✔ QUICK CHECK ANSWERS 16.2

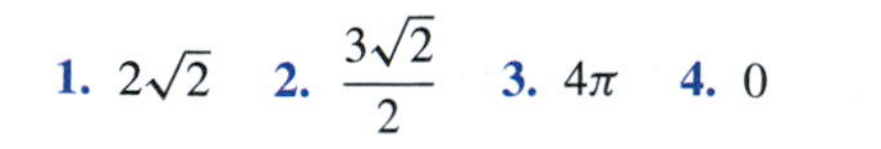

16.3 INDEPENDENCE OF PATH; CONSERVATIVE VECTOR FIELDS

In this section we will show that for certain kinds of vector fields the line integral of **F** *along a curve depends only on the endpoints of the curve and not on the curve itself. Such vector fields are of special importance in physics and engineering.*

WORK INTEGRALS

We saw in the last section that if **F** is a force field in 2-space or 3-space, then the work performed by the field on a particle moving along a parametric curve C from an initial point P to a final point Q is given by the integral

$$\int_C \mathbf{F}\cdot d\mathbf{r} \quad \text{or equivalently} \quad \int_C \mathbf{F}\cdot\mathbf{T}\,ds$$

Accordingly, we call an integral of this type a ***work integral***. Recall that a work integral can also be expressed in scalar form as

$$\int_C \mathbf{F}\cdot d\mathbf{r} = \int_C f(x,y)\,dx + g(x,y)\,dy \quad \boxed{\text{2-space}} \tag{1}$$

$$\int_C \mathbf{F}\cdot d\mathbf{r} = \int_C f(x,y,z)\,dx + g(x,y,z)\,dy + h(x,y,z)\,dz \quad \boxed{\text{3-space}} \tag{2}$$

where f, g, and h are the component functions of **F**.

INDEPENDENCE OF PATH

The parametric curve C in a work integral is called the ***path of integration***. One of the important problems in applications is to determine how the path of integration affects the work performed by a force field on a particle that moves from a fixed point P to a fixed point Q. We will show shortly that if the force field **F** is conservative (i.e., is the gradient of some potential function ϕ), then the work that the field performs on a particle that moves from P to Q does not depend on the particular path C that the particle follows. This is illustrated in the following example.

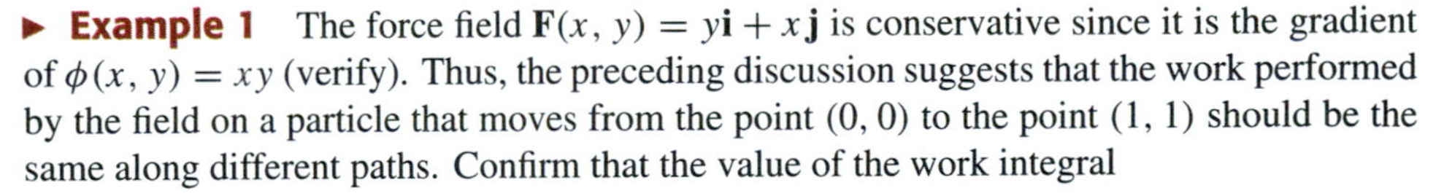

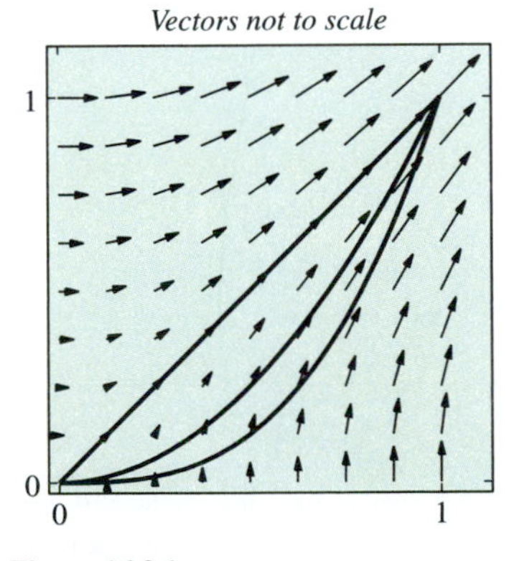

Figure 16.3.1

► **Example 1** The force field $\mathbf{F}(x,y) = y\mathbf{i} + x\mathbf{j}$ is conservative since it is the gradient of $\phi(x,y) = xy$ (verify). Thus, the preceding discussion suggests that the work performed by the field on a particle that moves from the point $(0,0)$ to the point $(1,1)$ should be the same along different paths. Confirm that the value of the work integral

$$\int_C \mathbf{F}\cdot d\mathbf{r}$$

is the same along the following paths (Figure 16.3.1):

(a) The line segment $y = x$ from $(0,0)$ to $(1,1)$.

(b) The parabola $y = x^2$ from $(0,0)$ to $(1,1)$.

(c) The cubic $y = x^3$ from $(0,0)$ to $(1,1)$.

Solution (a). With $x = t$ as the parameter, the path of integration is given by

$$x = t, \quad y = t \quad (0 \le t \le 1)$$

Thus,

$$\int_C \mathbf{F}\cdot d\mathbf{r} = \int_C (y\mathbf{i} + x\mathbf{j})\cdot(dx\mathbf{i} + dy\mathbf{j}) = \int_C y\,dx + x\,dy$$

$$= \int_0^1 2t\,dt = 1$$

Solution (b). With $x = t$ as the parameter, the path of integration is given by

$$x = t, \quad y = t^2 \qquad (0 \le t \le 1)$$

Thus,

$$\int_C \mathbf{F} \cdot d\mathbf{r} = \int_C y\,dx + x\,dy = \int_0^1 3t^2\,dt = 1$$

Solution (c). With $x = t$ as the parameter, the path of integration is given by

$$x = t, \quad y = t^3 \qquad (0 \le t \le 1)$$

Thus,

$$\int_C \mathbf{F} \cdot d\mathbf{r} = \int_C y\,dx + x\,dy = \int_0^1 4t^3\,dt = 1$$ ◀

THE FUNDAMENTAL THEOREM OF LINE INTEGRALS

Recall from the Fundamental Theorem of Calculus (Theorem 6.6.1) that if F is an antiderivative of f, then

$$\int_a^b f(x)\,dx = F(b) - F(a)$$

The following result is the analog of that theorem for line integrals in 2-space.

16.3.1 THEOREM (*The Fundamental Theorem of Line Integrals*). *Suppose that*

$$\mathbf{F}(x, y) = f(x, y)\mathbf{i} + g(x, y)\mathbf{j}$$

is a conservative vector field in some open region D containing the points (x_0, y_0) and (x_1, y_1) and that $f(x, y)$ and $g(x, y)$ are continuous in this region. If

$$\mathbf{F}(x, y) = \nabla\phi(x, y)$$

and if C is any piecewise smooth parametric curve that starts at (x_0, y_0), ends at (x_1, y_1), and lies in the region D, then

$$\int_C \mathbf{F}(x, y) \cdot d\mathbf{r} = \phi(x_1, y_1) - \phi(x_0, y_0) \tag{3}$$

or, equivalently,

$$\int_C \nabla\phi \cdot d\mathbf{r} = \phi(x_1, y_1) - \phi(x_0, y_0) \tag{4}$$

The value of

$$\int_C \mathbf{F} \cdot d\mathbf{r} = \int_C \mathbf{F} \cdot \mathbf{T}\,ds$$

depends on the magnitude of $\mathbf{F}$ along C, the alignment of $\mathbf{F}$ with the direction of C at each point, and the length of C. If $\mathbf{F}$ is conservative, these various factors always "balance out" so that the value of $\int_C \mathbf{F} \cdot d\mathbf{r}$ depends only on the initial and final points of C.

PROOF. We will give the proof for a smooth curve C. The proof for a piecewise smooth curve, which is left as an exercise, can be obtained by applying the theorem to each individual smooth piece and adding the results. Suppose that C is given parametrically by $x = x(t)$, $y = y(t)$ $(a \le t \le b)$, so that the initial and final points of the curve are

$$(x_0, y_0) = (x(a), y(a)) \quad \text{and} \quad (x_1, y_1) = (x(b), y(b))$$

Since $\mathbf{F}(x, y) = \nabla\phi$, it follows that

$$\mathbf{F}(x, y) = \frac{\partial\phi}{\partial x}\mathbf{i} + \frac{\partial\phi}{\partial y}\mathbf{j}$$

so

$$\int_C \mathbf{F}(x, y) \cdot d\mathbf{r} = \int_C \frac{\partial \phi}{\partial x}\,dx + \frac{\partial \phi}{\partial y}\,dy = \int_a^b \left[\frac{\partial \phi}{\partial x}\frac{dx}{dt} + \frac{\partial \phi}{\partial y}\frac{dy}{dt}\right] dt$$

$$= \int_a^b \frac{d}{dt}[\phi(x(t), y(t))]\,dt = \phi(x(t), y(t))\Big]_{t=a}^{b}$$

$$= \phi(x(b), y(b)) - \phi(x(a), y(a))$$

$$= \phi(x_1, y_1) - \phi(x_0, y_0)$$ ■

Stated informally, this theorem shows that *the value of a line integral along a piecewise smooth path in a conservative vector field is* ***independent of the path***; that is, the value of the integral depends on the endpoints and not on the actual path C. Accordingly, for line integrals along paths in conservative vector fields, it is common to express (3) and (4) as

$$\int_{(x_0, y_0)}^{(x_1, y_1)} \mathbf{F} \cdot d\mathbf{r} = \int_{(x_0, y_0)}^{(x_1, y_1)} \nabla\phi \cdot d\mathbf{r} = \phi(x_1, y_1) - \phi(x_0, y_0) \tag{5}$$

▶ **Example 2**

(a) Confirm that the force field $\mathbf{F}(x, y) = y\mathbf{i} + x\mathbf{j}$ in Example 1 is conservative by showing that $\mathbf{F}(x, y)$ is the gradient of $\phi(x, y) = xy$.

(b) Use the Fundamental Theorem of Line Integrals to evaluate $\displaystyle\int_{(0,0)}^{(1,1)} \mathbf{F} \cdot d\mathbf{r}$.

If $\mathbf{F}$ is conservative, then you have a choice of methods for evaluating $\int_C \mathbf{F} \cdot d\mathbf{r}$. You can work directly with the curve C, you can replace C with another curve that has the same endpoints as C, or you can apply (3).

Solution (a).

$$\nabla\phi = \frac{\partial \phi}{\partial x}\mathbf{i} + \frac{\partial \phi}{\partial y}\mathbf{j} = y\mathbf{i} + x\mathbf{j}$$

Solution (b). From (5) we obtain

$$\int_{(0,0)}^{(1,1)} \mathbf{F} \cdot d\mathbf{r} = \phi(1, 1) - \phi(0, 0) = 1 - 0 = 1$$

which agrees with the results obtained in Example 1 by integrating from $(0, 0)$ to $(1, 1)$ along specific paths. ◀

LINE INTEGRALS ALONG CLOSED PATHS

Parametric curves that begin and end at the same point play an important role in the study of vector fields, so there is some special terminology associated with them. A parametric curve C that is represented by the vector-valued function $\mathbf{r}(t)$ for $a \le t \le b$ is said to be ***closed*** if the initial point $\mathbf{r}(a)$ and the terminal point $\mathbf{r}(b)$ coincide; that is, $\mathbf{r}(a) = \mathbf{r}(b)$ (Figure 16.3.2).

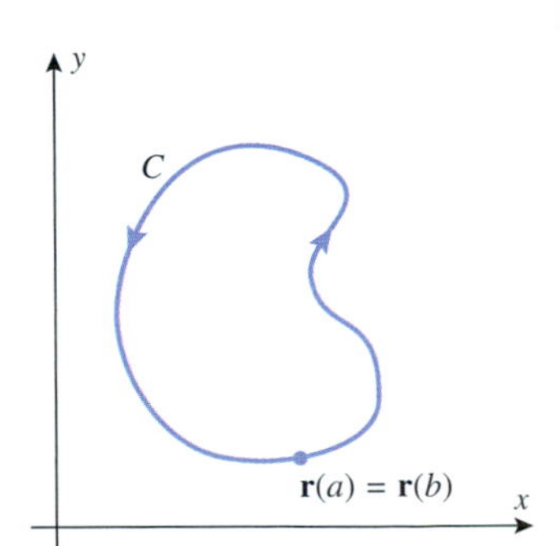

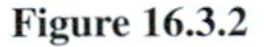

Figure 16.3.2

It follows from (5) that the line integral of a conservative vector field along a closed path C that begins and ends at (x_0, y_0) is zero. This is because the point (x_1, y_1) in (5) is the same as (x_0, y_0) and hence

$$\int_C \mathbf{F} \cdot d\mathbf{r} = \phi(x_1, y_1) - \phi(x_0, y_0) = 0$$

Our next objective is to show that the converse of this result is also true. That is, we want to show that under appropriate conditions a vector field whose line integral is zero along *all* closed paths must be conservative. For this to be true we will need to require that

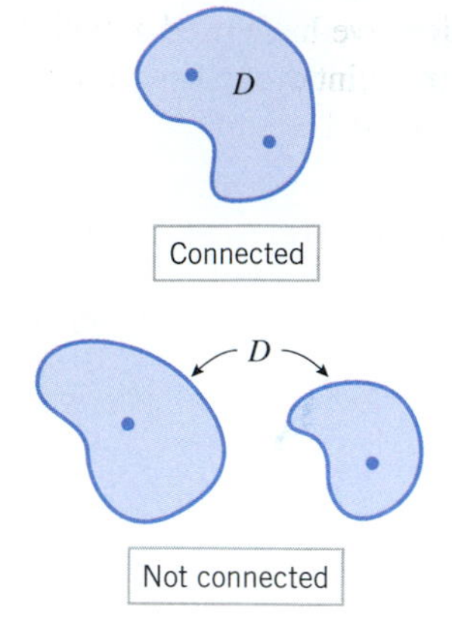

Figure 16.3.3

the domain D of the vector field be ***connected***, by which we mean that any two points in D can be joined by some piecewise smooth curve that lies entirely in D. Stated informally, D is connected if it does not consist of two or more separate pieces (Figure 16.3.3).

> **16.3.2 THEOREM.** *If $f(x, y)$ and $g(x, y)$ are continuous on some open connected region D, then the following statements are equivalent (all true or all false):*
>
> (a) $\mathbf{F}(x, y) = f(x, y)\mathbf{i} + g(x, y)\mathbf{j}$ *is a conservative vector field on the region D.*
>
> (b) $\displaystyle\int_C \mathbf{F} \cdot d\mathbf{r} = 0$ *for every piecewise smooth closed curve C in D.*
>
> (c) $\displaystyle\int_C \mathbf{F} \cdot d\mathbf{r}$ *is independent of the path from any point P in D to any point Q in D for every piecewise smooth curve C in D.*

This theorem can be established by proving three implications: $(a) \Rightarrow (b)$, $(b) \Rightarrow (c)$, and $(c) \Rightarrow (a)$. Since we showed above that $(a) \Rightarrow (b)$, we need only prove the last two implications. We will prove $(c) \Rightarrow (a)$ and leave the other implication as an exercise.

PROOF. $(c) \Rightarrow (a)$. We are assuming that $\int_C \mathbf{F} \cdot d\mathbf{r}$ is independent of the path for every piecewise smooth curve C in the region, and we want to show that there is a function $\phi = \phi(x, y)$ such that $\nabla\phi = \mathbf{F}(x, y)$ at each point of the region; that is,

$$\frac{\partial \phi}{\partial x} = f(x, y) \quad \text{and} \quad \frac{\partial \phi}{\partial y} = g(x, y) \tag{6}$$

Now choose a fixed point (a, b) in D, let (x, y) be any point in D, and define

$$\phi(x, y) = \int_{(a,b)}^{(x,y)} \mathbf{F} \cdot d\mathbf{r} \tag{7}$$

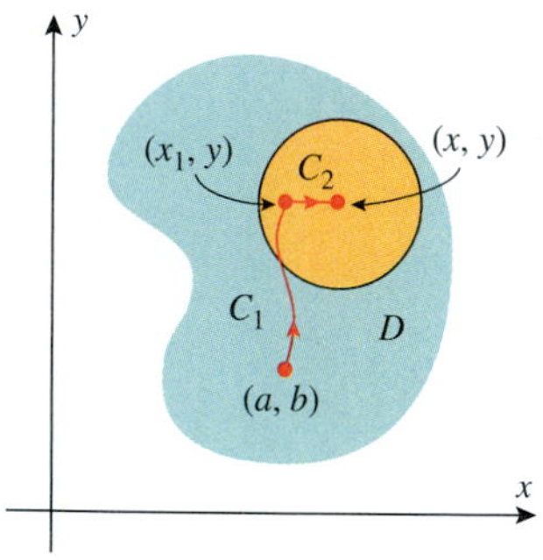

Figure 16.3.4

This is an unambiguous definition because we have assumed that the integral is independent of the path. We will show that $\nabla\phi = \mathbf{F}$. Since D is open, we can find a circular disk centered at (x, y) whose points lie entirely in D. As shown in Figure 16.3.4, choose any point (x_1, y) in this disk that lies on the same horizontal line as (x, y) such that $x_1 < x$. Because the integral in (7) is independent of path, we can evaluate it by first integrating from (a, b) to (x_1, y) along an arbitrary piecewise smooth curve C_1 in D, and then continuing along the horizontal line segment C_2 from (x_1, y) to (x, y). This yields

$$\phi(x, y) = \int_{C_1} \mathbf{F} \cdot d\mathbf{r} + \int_{C_2} \mathbf{F} \cdot d\mathbf{r} = \int_{(a,b)}^{(x_1,y)} \mathbf{F} \cdot d\mathbf{r} + \int_{C_2} \mathbf{F} \cdot d\mathbf{r}$$

Since the first term does not depend on x, its partial derivative with respect to x is zero and hence

$$\frac{\partial \phi}{\partial x} = \frac{\partial}{\partial x}\int_{C_2} \mathbf{F} \cdot d\mathbf{r} = \frac{\partial}{\partial x}\int_{C_2} f(x, y)\,dx + g(x, y)\,dy$$

However, the line integral with respect to y is zero along the horizontal line segment C_2, so this equation simplifies to

$$\frac{\partial \phi}{\partial x} = \frac{\partial}{\partial x}\int_{C_2} f(x, y)\,dx \tag{8}$$

To evaluate the integral in this expression, we treat y as a constant and express the line C_2 parametrically as

$$x = t, \quad y = y \qquad (x_1 \le t \le x)$$

At the risk of confusion, but to avoid complicating the notation, we have used x both as the dependent variable in the parametric equations and as the endpoint of the line segment. With the latter interpretation of x, it follows that (8) can be expressed as

$$\frac{\partial \phi}{\partial x} = \frac{\partial}{\partial x} \int_{x_1}^{x} f(t, y)\, dt$$

Now we apply Part 2 of the Fundamental Theorem of Calculus (Theorem 6.6.3), treating y as constant. This yields

$$\frac{\partial \phi}{\partial x} = f(x, y)$$

which proves the first part of (6). The proof that $\partial\phi/\partial y = g(x, y)$ can be obtained in a similar manner by joining (x, y) to a point (x, y_1) with a vertical line segment (Exercise 35). ■

A TEST FOR CONSERVATIVE VECTOR FIELDS

Although Theorem 16.3.2 is an important characterization of conservative vector fields, it is not an effective computational tool because it is usually not possible to evaluate the line integral over all possible piecewise smooth curves in D, as required in parts (b) and (c). To develop a method for determining whether a vector field is conservative, we will need to introduce some new concepts about parametric curves and connected sets. We will say that a parametric curve is ***simple*** if it does not intersect itself between its endpoints. A simple parametric curve may or may not be closed (Figure 16.3.5). In addition, we will say that a connected set D in 2-space is ***simply connected*** if no simple closed curve in D encloses points that are not in D. Stated informally, a connected set D is simply connected if it has no holes; a connected set with one or more holes is said to be ***multiply connected*** (Figure 16.3.6).

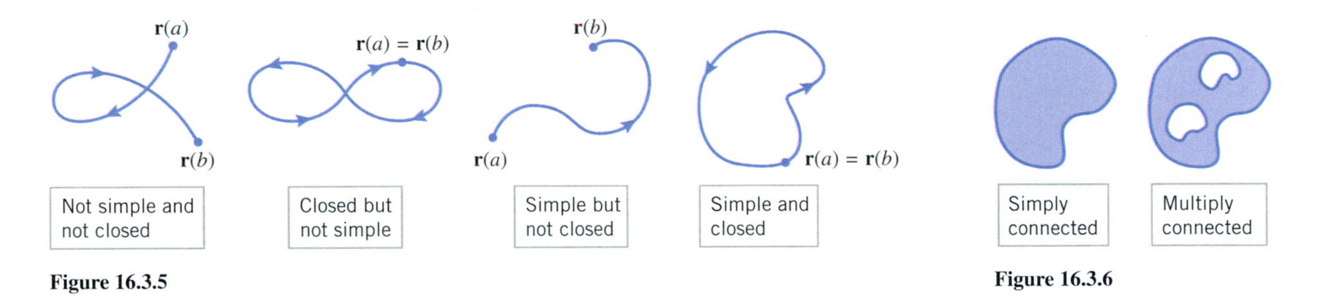

Figure 16.3.5

Figure 16.3.6

The following theorem is the primary tool for determining whether a vector field in 2-space is conservative.

16.3.3 THEOREM (*Conservative Field Test*). *If $f(x, y)$ and $g(x, y)$ are continuous and have continuous first partial derivatives on some open region D, and if the vector field $\mathbf{F}(x, y) = f(x, y)\mathbf{i} + g(x, y)\mathbf{j}$ is conservative on D, then*

$$\frac{\partial f}{\partial y} = \frac{\partial g}{\partial x} \tag{9}$$

at each point in D. Conversely, if D is simply connected and (9) holds at each point in D, then $\mathbf{F}(x, y) = f(x, y)\mathbf{i} + g(x, y)\mathbf{j}$ is conservative.

WARNING

In (9), the $\mathbf{i}$-component of $\mathbf{F}$ is differentiated with respect to y and the $\mathbf{j}$-component with respect to x. It is easy to get this backwards by mistake.

A complete proof of this theorem requires results from advanced calculus and will be omitted. However, it is not hard to see why (9) must hold if $\mathbf{F}$ is conservative. For this purpose suppose that $\mathbf{F} = \nabla\phi$, in which case we can express the functions f and g as

$$\frac{\partial\phi}{\partial x} = f \quad \text{and} \quad \frac{\partial\phi}{\partial y} = g \tag{10}$$

Thus,

$$\frac{\partial f}{\partial y} = \frac{\partial}{\partial y}\left(\frac{\partial\phi}{\partial x}\right) = \frac{\partial^2\phi}{\partial y\partial x} \quad \text{and} \quad \frac{\partial g}{\partial x} = \frac{\partial}{\partial x}\left(\frac{\partial\phi}{\partial y}\right) = \frac{\partial^2\phi}{\partial x\partial y}$$

But the mixed partial derivatives in these equations are equal (Theorem 14.3.2), so (9) follows.

▶ Example 3 Use Theorem 16.3.3 to determine whether the vector field

$$\mathbf{F}(x, y) = (y + x)\mathbf{i} + (y - x)\mathbf{j}$$

is conservative on some open set.

Solution. Let $f(x, y) = y + x$ and $g(x, y) = y - x$. Then

$$\frac{\partial f}{\partial y} = 1 \quad \text{and} \quad \frac{\partial g}{\partial x} = -1$$

Thus, there are no points in the xy-plane at which condition (9) holds, and hence $\mathbf{F}$ is not conservative on any open set. ◀

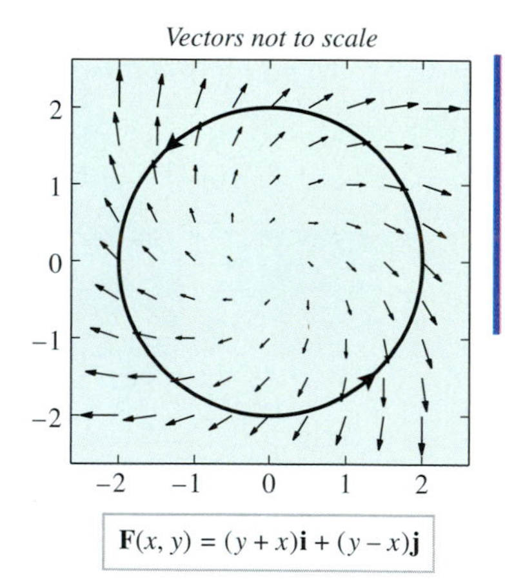

Figure 16.3.7

Since the vector field $\mathbf{F}$ in Example 3 is not conservative, it follows from Theorem 16.3.2 that there must exist piecewise smooth closed curves in every open connected set in the xy-plane on which

$$\int_C \mathbf{F}\cdot d\mathbf{r} = \int_C \mathbf{F}\cdot\mathbf{T}\,ds \neq 0$$

One such curve is the oriented circle shown in Figure 16.3.7. The figure suggests that $\mathbf{F}\cdot\mathbf{T} < 0$ at each point of C (why?), so $\int_C \mathbf{F}\cdot\mathbf{T}\,ds < 0$.

Once it is established that a vector field is conservative, a potential function for the field can be obtained by first integrating either of the equations in (10). This is illustrated in the following example.

▶ Example 4 Let $\mathbf{F}(x, y) = 2xy^3\mathbf{i} + (1 + 3x^2y^2)\mathbf{j}$.

(a) Show that $\mathbf{F}$ is a conservative vector field on the entire xy-plane.

(b) Find ϕ by first integrating $\partial\phi/\partial x$.

(c) Find ϕ by first integrating $\partial\phi/\partial y$.

Solution (a). Since $f(x, y) = 2xy^3$ and $g(x, y) = 1 + 3x^2y^2$, we have

$$\frac{\partial f}{\partial y} = 6xy^2 = \frac{\partial g}{\partial x}$$

so (9) holds for all (x, y).

Solution (b). Since the field $\mathbf{F}$ is conservative, there is a potential function ϕ such that

$$\frac{\partial\phi}{\partial x} = 2xy^3 \quad \text{and} \quad \frac{\partial\phi}{\partial y} = 1 + 3x^2y^2 \tag{11}$$

Integrating the first of these equations with respect to x (and treating y as a constant) yields

$$\phi = \int 2xy^3\,dx = x^2y^3 + k(y) \tag{12}$$

where $k(y)$ represents the "constant" of integration. We are justified in treating the constant of integration as a function of y, since y is held constant in the integration process. To find $k(y)$ we differentiate (12) with respect to y and use the second equation in (11) to obtain

$$\frac{\partial\phi}{\partial y} = 3x^2y^2 + k'(y) = 1 + 3x^2y^2$$

from which it follows that $k'(y) = 1$. Thus,

$$k(y) = \int k'(y)\,dy = \int 1\,dy = y + K$$

where K is a (numerical) constant of integration. Substituting in (12) we obtain

$$\phi = x^2y^3 + y + K$$

The appearance of the arbitrary constant K tells us that ϕ is not unique. As a check on the computations, you may want to verify that $\nabla\phi = \mathbf{F}$.

You can also use (7) to find a potential function for a conservative vector field. For example, find a potential function for the vector field in Example 4 by evaluating (7) on the line segment

$$\mathbf{r}(t) = t(x\mathbf{i}) + t(y\mathbf{j}) \quad (0 \le t \le 1)$$

from $(0, 0)$ to (x, y).

Solution (c). Integrating the second equation in (11) with respect to y (and treating x as a constant) yields

$$\phi = \int (1 + 3x^2y^2)\,dy = y + x^2y^3 + k(x) \tag{13}$$

where $k(x)$ is the "constant" of integration. Differentiating (13) with respect to x and using the first equation in (11) yields

$$\frac{\partial\phi}{\partial x} = 2xy^3 + k'(x) = 2xy^3$$

from which it follows that $k'(x) = 0$ and consequently that $k(x) = K$, where K is a numerical constant of integration. Substituting this in (13) yields

$$\phi = y + x^2y^3 + K$$

which agrees with the solution in part (b). ◀

▶ Example 5 Use the potential function obtained in Example 4 to evaluate the integral

$$\int_{(1,4)}^{(3,1)} 2xy^3\,dx + (1 + 3x^2y^2)\,dy$$

Solution. The integrand can be expressed as $\mathbf{F}\cdot d\mathbf{r}$, where $\mathbf{F}$ is the vector field in Example 4. Thus, using Formula (3) and the potential function $\phi = y + x^2y^3 + K$ for $\mathbf{F}$, we obtain

$$\begin{aligned}\int_{(1,4)}^{(3,1)} 2xy^3\,dx + (1 + 3x^2y^2)\,dy &= \int_{(1,4)}^{(3,1)} \mathbf{F}\cdot d\mathbf{r} = \phi(3,1) - \phi(1,4)\\ &= (10 + K) - (68 + K) = -58\end{aligned}$$ ◀

In the solution to Example 5, note that the constant K drops out. In future integration problems we will sometimes omit K from the computations.

▶ Example 6 Let $\mathbf{F}(x, y) = e^y\mathbf{i} + xe^y\mathbf{j}$ denote a force field in the xy-plane.

(a) Verify that the force field $\mathbf{F}$ is conservative on the entire xy-plane.

(b) Find the work done by the field on a particle that moves from $(1, 0)$ to $(-1, 0)$ along the semicircular path C shown in Figure 16.3.8.

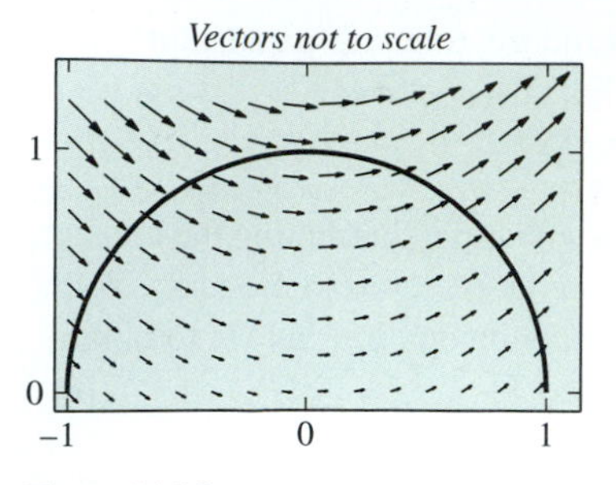

Figure 16.3.8

Solution (a). For the given field we have $f(x, y) = e^y$ and $g(x, y) = xe^y$. Thus,

$$\frac{\partial}{\partial y}(e^y) = e^y = \frac{\partial}{\partial x}(xe^y)$$

so (9) holds for all (x, y) and hence $\mathbf{F}$ is conservative on the entire xy-plane.

Solution (b). From Formula (34) of Section 16.2, the work done by the field is

$$W = \int_C \mathbf{F} \cdot d\mathbf{r} = \int_C e^y\, dx + xe^y\, dy \tag{14}$$

However, the calculations involved in integrating along C are tedious, so it is preferable to apply Theorem 16.3.1, taking advantage of the fact that the field is conservative and the integral is independent of path. Thus, we write (14) as

$$W = \int_{(1,0)}^{(-1,0)} e^y\, dx + xe^y\, dy = \phi(-1, 0) - \phi(1, 0) \tag{15}$$

As illustrated in Example 4, we can find ϕ by integrating either of the equations

$$\frac{\partial \phi}{\partial x} = e^y \quad \text{and} \quad \frac{\partial \phi}{\partial y} = xe^y \tag{16}$$

We will integrate the first. We obtain

$$\phi = \int e^y\, dx = xe^y + k(y) \tag{17}$$

Differentiating this equation with respect to y and using the second equation in (16) yields

$$\frac{\partial \phi}{\partial y} = xe^y + k'(y) = xe^y$$

from which it follows that $k'(y) = 0$ or $k(y) = K$. Thus, from (17)

$$\phi = xe^y + K$$

and hence from (15)

$$W = \phi(-1, 0) - \phi(1, 0) = (-1)e^0 - 1e^0 = -2 \blacktriangleleft$$

CONSERVATIVE VECTOR FIELDS IN 3-SPACE

All of the results in this section have analogs in 3-space: Theorems 16.3.1 and 16.3.2 can be extended to vector fields in 3-space simply by adding a third variable and modifying the hypotheses appropriately. For example, in 3-space, Formula (3) becomes

$$\int_C \mathbf{F}(x, y, z) \cdot d\mathbf{r} = \phi(x_1, y_1, z_1) - \phi(x_0, y_0, z_0) \tag{18}$$

Theorem 16.3.3 can also be extended to vector fields in 3-space. We leave it for the exercises to show that if $\mathbf{F}(x, y, z) = f(x, y, z)\mathbf{i} + g(x, y, z)\mathbf{j} + h(x, y, z)\mathbf{k}$ is a conservative field, then

$$\frac{\partial f}{\partial y} = \frac{\partial g}{\partial x}, \quad \frac{\partial f}{\partial z} = \frac{\partial h}{\partial x}, \quad \frac{\partial g}{\partial z} = \frac{\partial h}{\partial y} \tag{19}$$

that is, curl $\mathbf{F} = \mathbf{0}$. Conversely, a vector field satisfying these conditions on a suitably restricted region is conservative on that region if f, g, and h are continuous and have continuous first partial derivatives in the region. Some problems involving Formulas (18) and (19) are given in the review exercises at the end of this chapter.

CONSERVATION OF ENERGY

If $\mathbf{F}(x, y, z)$ is a conservative force field with a potential function $\phi(x, y, z)$, then we call $V(x, y, z) = -\phi(x, y, z)$ the ***potential energy*** of the field at the point (x, y, z). Thus, it follows from the 3-space version of Theorem 16.3.1 that the work W done by $\mathbf{F}$ on a particle that moves along any path C from a point (x_0, y_0, z_0) to a point (x_1, y_1, z_1) is related to the potential energy by the equation

$$W = \int_C \mathbf{F} \cdot d\mathbf{r} = \phi(x_1, y_1, z_1) - \phi(x_0, y_0, z_0) = -[V(x_1, y_1, z_1) - V(x_0, y_0, z_0)] \qquad (20)$$

That is, the work done by the field is the negative of the change in potential energy. In particular, it follows from the 3-space analog of Theorem 16.3.2 that if a particle traverses a piecewise smooth closed path in a conservative vector field, then the work done by the field is zero, and there is no change in potential energy. To take this a step further, suppose that a particle of mass m moves along any piecewise smooth curve (not necessarily closed) in a conservative force field $\mathbf{F}$, starting at (x_0, y_0, z_0) with velocity v_i and ending at (x_1, y_1, z_1) with velocity v_f. If $\mathbf{F}$ is the only force acting on the particle, then an argument similar to the derivation of Equation (5) in Section 7.7 shows that the work done on the particle by $\mathbf{F}$ is equal to the change in kinetic energy $\frac{1}{2}mv_f^2 - \frac{1}{2}mv_i^2$ of the particle. If we let V_i denote the potential energy at the starting point and V_f the potential energy at the final point, then it follows from (20)

$$\tfrac{1}{2}mv_f^2 - \tfrac{1}{2}mv_i^2 = -[V_f - V_i]$$

which we can rewrite as

$$\tfrac{1}{2}mv_f^2 + V_f = \tfrac{1}{2}mv_i^2 + V_i$$

This equation states that the total energy of the particle (kinetic energy + potential energy) does not change as the particle moves along a path in a conservative vector field. This result, called the ***conservation of energy principle***, explains the origin of the term "conservative vector field."

✔QUICK CHECK EXERCISES 16.3

(See page 1139 for answers.)

1. If C is a piecewise smooth curve from $(1, 2, 3)$ to $(4, 5, 6)$, then $$\int_C dx + 2\,dy + 3\,dz = \underline{\qquad}$$
2. If C is the portion of the circle $x^2 + y^2 = 1$ where $0 \le x$, oriented counterclockwise, and $f(x, y) = ye^x$ then $$\int_C \nabla f \cdot d\mathbf{r} = \underline{\qquad}$$
3. A potential function for the vector field $$\mathbf{F}(x, y, z) = yz\mathbf{i} + (xz + z)\mathbf{j} + (xy + y + 1)\mathbf{k}$$ is $\phi(x, y, z) = \underline{\qquad}$.
4. If a, b, and c are nonzero real numbers such that the vector field $x^5y^a\mathbf{i} + x^by^c\mathbf{j}$ is a conservative vector field, then $$a = \underline{\qquad}, \quad b = \underline{\qquad}, \quad c = \underline{\qquad}$$

EXERCISE SET 16.3 [C] CAS

1–6 Determine whether $\mathbf{F}$ is a conservative vector field. If so, find a potential function for it.

1. $\mathbf{F}(x, y) = x\mathbf{i} + y\mathbf{j}$
2. $\mathbf{F}(x, y) = 3y^2\mathbf{i} + 6xy\mathbf{j}$
3. $\mathbf{F}(x, y) = x^2y\mathbf{i} + 5xy^2\mathbf{j}$
4. $\mathbf{F}(x, y) = e^x \cos y\mathbf{i} - e^x \sin y\mathbf{j}$
5. $\mathbf{F}(x, y) = (\cos y + y \cos x)\mathbf{i} + (\sin x - x \sin y)\mathbf{j}$
6. $\mathbf{F}(x, y) = x \ln y\mathbf{i} + y \ln x\mathbf{j}$
7. (a) Show that the line integral $\int_C y^2\,dx + 2xy\,dy$ is independent of the path.
   (b) Evaluate the integral in part (a) along the line segment from $(-1, 2)$ to $(1, 3)$.
   (c) Evaluate the integral $\int_{(-1,2)}^{(1,3)} y^2\,dx + 2xy\,dy$ using Theorem 16.3.1, and confirm that the value is the same as that obtained in part (b).

8. (a) Show that the line integral $\int_C y \sin x\, dx - \cos x\, dy$ is independent of the path.
(b) Evaluate the integral in part (a) along the line segment from (0, 1) to $(\pi, -1)$.
(c) Evaluate the integral $\int_{(0,1)}^{(\pi,-1)} y \sin x\, dx - \cos x\, dy$ using Theorem 16.3.1, and confirm that the value is the same as that obtained in part (b).

9–14 Show that the integral is independent of the path, and use Theorem 16.3.1 to find its value.

9. $\displaystyle\int_{(1,2)}^{(4,0)} 3y\, dx + 3x\, dy$

10. $\displaystyle\int_{(0,0)}^{(1,\pi/2)} e^x \sin y\, dx + e^x \cos y\, dy$

11. $\displaystyle\int_{(0,0)}^{(3,2)} 2xe^y\, dx + x^2 e^y\, dy$

12. $\displaystyle\int_{(-1,2)}^{(0,1)} (3x - y + 1)\, dx - (x + 4y + 2)\, dy$

13. $\displaystyle\int_{(2,-2)}^{(-1,0)} 2xy^3\, dx + 3y^2x^2\, dy$

14. $\displaystyle\int_{(1,1)}^{(3,3)} \left(e^x \ln y - \frac{e^y}{x}\right) dx + \left(\frac{e^x}{y} - e^y \ln x\right) dy$, where x and y are positive.

15–18 Confirm that the force field **F** is conservative in some open connected region containing the points P and Q, and then find the work done by the force field on a particle moving along an arbitrary smooth curve in the region from P to Q.

15. $\mathbf{F}(x, y) = xy^2\mathbf{i} + x^2y\mathbf{j}$; $P(1, 1)$, $Q(0, 0)$

16. $\mathbf{F}(x, y) = 2xy^3\mathbf{i} + 3x^2y^2\mathbf{j}$; $P(-3, 0)$, $Q(4, 1)$

17. $\mathbf{F}(x, y) = ye^{xy}\mathbf{i} + xe^{xy}\mathbf{j}$; $P(-1, 1)$, $Q(2, 0)$

18. $\mathbf{F}(x, y) = e^{-y} \cos x\mathbf{i} - e^{-y} \sin x\mathbf{j}$; $P(\pi/2, 1)$, $Q(-\pi/2, 0)$

19–20 Find the exact value of $\int_C \mathbf{F} \cdot d\mathbf{r}$ using any method.

19. $\mathbf{F}(x, y) = (e^y + ye^x)\mathbf{i} + (xe^y + e^x)\mathbf{j}$
$C : \mathbf{r}(t) = \sin(\pi t/2)\mathbf{i} + \ln t\,\mathbf{j} \quad (1 \le t \le 2)$

20. $\mathbf{F}(x, y) = 2xy\mathbf{i} + (x^2 + \cos y)\mathbf{j}$
$C: \mathbf{r}(t) = t\mathbf{i} + t \cos(t/3)\mathbf{j} \quad (0 \le t \le \pi)$

C **21.** Use the numerical integration capability of a CAS or other calculating utility to approximate the value of the integral in Exercise 19 by direct integration. Confirm that the numerical approximation is consistent with the exact value.

C **22.** Use the numerical integration capability of a CAS or other calculating utility to approximate the value of the integral in Exercise 20 by direct integration. Confirm that the numerical approximation is consistent with the exact value.

FOCUS ON CONCEPTS

23–24 Is the vector field conservative? Explain.

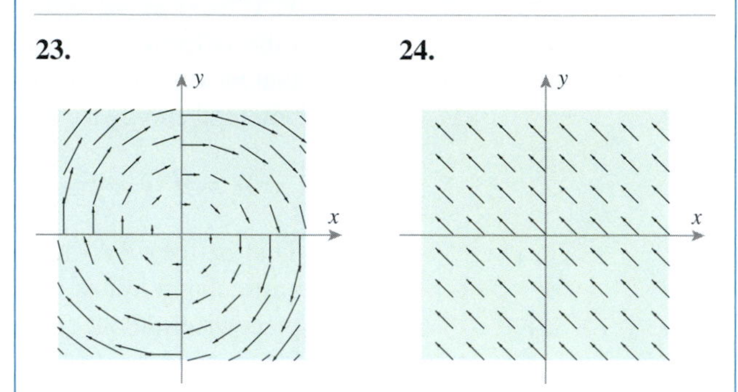

25. Suppose that C is a circle in the domain of a conservative vector field in the xy-plane whose component functions are continuous. Explain why there must be at least two points on C at which the vector field is normal to the circle.

26. Does the result in Exercise 25 remain true if the circle C is replaced by a square? Explain.

27. Prove: If

$$\mathbf{F}(x, y, z) = f(x, y, z)\mathbf{i} + g(x, y, z)\mathbf{j} + h(x, y, z)\mathbf{k}$$

is a conservative field and f, g, and h are continuous and have continuous first partial derivatives in a region, then

$$\frac{\partial f}{\partial y} = \frac{\partial g}{\partial x}, \quad \frac{\partial f}{\partial z} = \frac{\partial h}{\partial x}, \quad \frac{\partial g}{\partial z} = \frac{\partial h}{\partial y}$$

in the region.

28. Use the result in Exercise 27 to show that the integral

$$\int_C yz\, dx + xz\, dy + yx^2\, dz$$

is not independent of the path.

29. Find a nonzero function h for which

$$\mathbf{F}(x, y) = h(x)[x \sin y + y \cos y]\mathbf{i} + h(x)[x \cos y - y \sin y]\mathbf{j}$$

is conservative.

30. (a) In Example 3 of Section 16.1 we showed that

$$\phi(x, y) = -\frac{c}{(x^2 + y^2)^{1/2}}$$

is a potential function for the two-dimensional inverse-square field

$$\mathbf{F}(x, y) = \frac{c}{(x^2 + y^2)^{3/2}}(x\mathbf{i} + y\mathbf{j})$$

but we did not explain how the potential function $\phi(x, y)$ was obtained. Use Theorem 16.3.3 to show that the two-dimensional inverse-square field is conservative everywhere except at the origin, and then use the method of Example 4 to derive the formula for $\phi(x, y)$.

(b) Use an appropriate generalization of the method of Example 4 to derive the potential function

$$\phi(x, y, z) = -\frac{c}{(x^2 + y^2 + z^2)^{1/2}}$$

for the three-dimensional inverse-square field given by Formula (5) of Section 16.1.

31–32 Use the result in Exercise 30(b).

31. In each part, find the work done by the three-dimensional inverse-square field

$$\mathbf{F}(\mathbf{r}) = \frac{1}{\|\mathbf{r}\|^3}\mathbf{r}$$

on a particle that moves along the curve C.

(a) C is the line segment from $P(1, 1, 2)$ to $Q(3, 2, 1)$.

(b) C is the curve

$$\mathbf{r}(t) = (2t^2 + 1)\mathbf{i} + (t^3 + 1)\mathbf{j} + (2 - \sqrt{t})\mathbf{k}$$

where $0 \le t \le 1$.

(c) C is the circle in the xy-plane of radius 1 centered at $(2, 0, 0)$ traversed counterclockwise.

32. Let $\mathbf{F}(x, y) = \dfrac{y}{x^2 + y^2}\mathbf{i} - \dfrac{x}{x^2 + y^2}\mathbf{j}$.

(a) Show that

$$\int_{C_1} \mathbf{F} \cdot d\mathbf{r} \ne \int_{C_2} \mathbf{F} \cdot d\mathbf{r}$$

if C_1 and C_2 are the semicircular paths from $(1, 0)$ to $(-1, 0)$ given by

$$C_1\colon x = \cos t, \quad y = \sin t \qquad (0 \le t \le \pi)$$
$$C_2\colon x = \cos t, \quad y = -\sin t \qquad (0 \le t \le \pi)$$

(b) Show that the components of $\mathbf{F}$ satisfy Formula (9).

(c) Do the results in parts (a) and (b) violate Theorem 16.3.3? Explain.

33. Prove Theorem 16.3.1 if C is a piecewise smooth curve composed of smooth curves $C_1, C_2, \ldots, C_n$.

34. Prove that (*b*) implies (*c*) in Theorem 16.3.2. [*Hint:* Consider any two piecewise smooth oriented curves C_1 and C_2 in the region from a point P to a point Q, and integrate around the closed curve consisting of C_1 and $-C_2$.]

35. Complete the proof of Theorem 16.3.2 by showing that $\partial\phi/\partial y = g(x, y)$, where $\phi(x, y)$ is the function in (7).

✔ QUICK CHECK ANSWERS 16.3

1. 18 **2.** 2 **3.** $xyz + yz + z$ **4.** 6; 6; 5

16.4 GREEN'S THEOREM

In this section we will discuss a remarkable and beautiful theorem that expresses a double integral over a plane region in terms of a line integral around its boundary.

GREEN'S THEOREM

16.4.1 THEOREM (*Green's Theorem*). *Let R be a simply connected plane region whose boundary is a simple, closed, piecewise smooth curve C oriented counterclockwise. If $f(x, y)$ and $g(x, y)$ are continuous and have continuous first partial derivatives on some open set containing R, then*

$$\int_C f(x, y)\,dx + g(x, y)\,dy = \iint_R \left(\frac{\partial g}{\partial x} - \frac{\partial f}{\partial y}\right) dA \tag{1}$$

PROOF. For simplicity, we will prove the theorem for regions that are simultaneously type I and type II (see Definition 15.2.1). Such a region is shown in Figure 16.4.1. The crux of the proof is to show that

$$\int_C f(x, y)\,dx = -\iint_R \frac{\partial f}{\partial y}\,dA \quad \text{and} \quad \int_C g(x, y)\,dy = \iint_R \frac{\partial g}{\partial x}\,dA \tag{2–3}$$

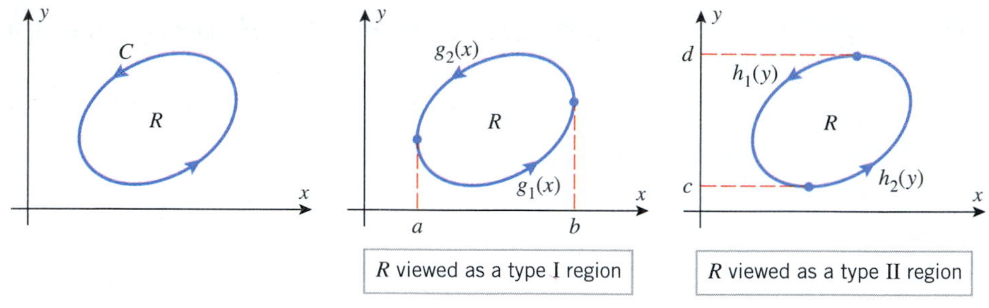

Figure 16.4.1

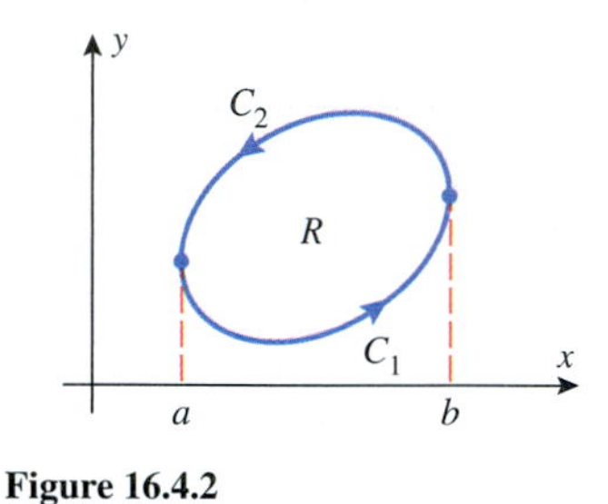

Figure 16.4.2

To prove (2), view R as a type I region and let C_1 and C_2 be the lower and upper boundary curves, oriented as in Figure 16.4.2. Then

$$\int_C f(x,y)\,dx = \int_{C_1} f(x,y)\,dx + \int_{C_2} f(x,y)\,dx$$

or, equivalently,

$$\int_C f(x,y)\,dx = \int_{C_1} f(x,y)\,dx - \int_{-C_2} f(x,y)\,dx \tag{4}$$

(This step will help simplify our calculations since C_1 and $-C_2$ are then both oriented left to right.) The curves C_1 and $-C_2$ can be expressed parametrically as

$$\begin{aligned} C_1&: x = t, \quad y = g_1(t) \quad (a \le t \le b) \\ -C_2&: x = t, \quad y = g_2(t) \quad (a \le t \le b) \end{aligned}$$

Thus, we can rewrite (4) as

$$\begin{aligned}
\int_C f(x,y)\,dx &= \int_a^b f(t, g_1(t))x'(t)\,dt - \int_a^b f(t, g_2(t))x'(t)\,dt \\
&= \int_a^b f(t, g_1(t))\,dt - \int_a^b f(t, g_2(t))\,dt \\
&= -\int_a^b [f(t, g_2(t)) - f(t, g_1(t))]\,dt \\
&= -\int_a^b \Big[f(t,y)\Big]_{y=g_1(t)}^{y=g_2(t)} dt = -\int_a^b \left[\int_{g_1(t)}^{g_2(t)} \frac{\partial f}{\partial y}\,dy\right] dt \\
&= -\int_a^b \int_{g_1(x)}^{g_2(x)} \frac{\partial f}{\partial y}\,dy\,dx = -\iint_R \frac{\partial f}{\partial y}\,dA
\end{aligned}$$

Since $x = t$

Supply the details for the proof of (3).

The proof of (3) is obtained similarly by treating R as a type II region. We omit the details. ■

▶ Example 1 Use Green's Theorem to evaluate

$$\int_C x^2y\,dx + x\,dy$$

along the triangular path shown in Figure 16.4.3.

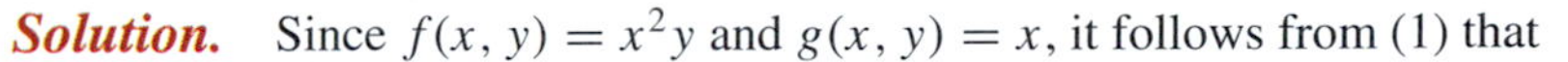

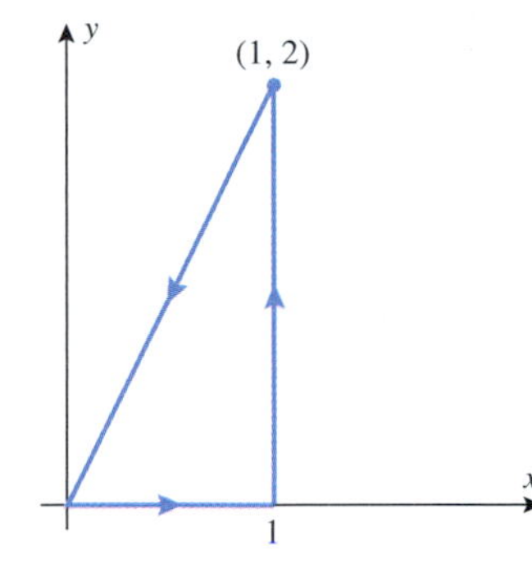

Figure 16.4.3

Solution. Since $f(x, y) = x^2y$ and $g(x, y) = x$, it follows from (1) that

$$\int_C x^2y\,dx + x\,dy = \iint_R \left[\frac{\partial}{\partial x}(x) - \frac{\partial}{\partial y}(x^2y)\right] dA = \int_0^1 \int_0^{2x} (1 - x^2)\,dy\,dx$$

$$= \int_0^1 (2x - 2x^3)\,dx = \left[x^2 - \frac{x^4}{2}\right]_0^1 = \frac{1}{2}$$

This agrees with the result obtained in Example 10 of Section 16.2, where we evaluated the line integral directly. Note how much simpler this solution is. ◀

A NOTATION FOR LINE INTEGRALS AROUND SIMPLE CLOSED CURVES

It is common practice to denote a line integral around a simple closed curve by an integral sign with a superimposed circle. With this notation Formula (1) would be written as

$$\oint_C f(x, y)\,dx + g(x, y)\,dy = \iint_R \left(\frac{\partial g}{\partial x} - \frac{\partial f}{\partial y}\right) dA$$

Sometimes a direction arrow is added to the circle to indicate whether the integration is clockwise or counterclockwise. Thus, if we wanted to emphasize the counterclockwise direction of integration required by Theorem 16.4.1, we could express (1) as

$$\oint_C f(x, y)\,dx + g(x, y)\,dy = \iint_R \left(\frac{\partial g}{\partial x} - \frac{\partial f}{\partial y}\right) dA \tag{5}$$

FINDING WORK USING GREEN'S THEOREM

It follows from Formula (26) of Section 16.2 that the integral on the left side of (5) is the work performed by the force field $\mathbf{F}(x, y) = f(x, y)\mathbf{i} + g(x, y)\mathbf{j}$ on a particle moving counterclockwise around the simple closed curve C. In the case where this vector field is conservative, it follows from Theorem 16.3.2 that the integrand in the double integral on the right side of (5) is zero, so the work performed by the field is zero, as expected. For vector fields that are not conservative, it is often more efficient to calculate the work around simple closed curves by using Green's Theorem than by parametrizing the curve.

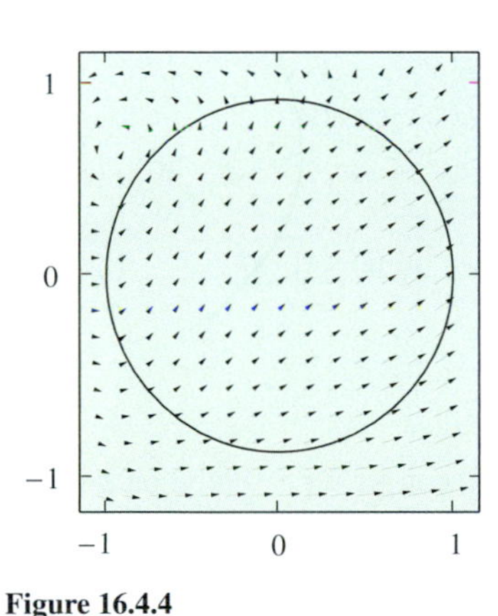

Figure 16.4.4

▶ **Example 2** Find the work done by the force field

$$\mathbf{F}(x, y) = (e^x - y^3)\mathbf{i} + (\cos y + x^3)\mathbf{j}$$

on a particle that travels once around the unit circle $x^2 + y^2 = 1$ in the counterclockwise direction (Figure 16.4.4).

George Green (1793–1841) English mathematician and physicist. Green left school at an early age to work in his father's bakery and consequently had little early formal education. When his father opened a mill, the boy used the top room as a study in which he taught himself physics and mathematics from library books. In 1828 Green published his most important work, *An Essay on the Application of Mathematical Analysis to the Theories of Electricity and Magnetism*. Although Green's Theorem appeared in that paper, the result went virtually unnoticed because of the small pressrun and local distribution. Following the death of his father in 1829, Green was urged by friends to seek a college education. In 1833, after four years of self-study to close the gaps in his elementary education, Green was admitted to Caius College, Cambridge. He graduated four years later, but with a disappointing performance on his final examinations—possibly because he was more interested in his own research. After a succession of works on light and sound, he was named to be Perse Fellow at Caius College. Two years later he died. In 1845, four years after his death, his paper of 1828 was published and the theories developed therein by this obscure, self-taught baker's son helped pave the way to the modern theories of electricity and magnetism.

Solution. The work W performed by the field is

$$W = \oint_C \mathbf{F} \cdot d\mathbf{r} = \oint_C (e^x - y^3)\,dx + (\cos y + x^3)\,dy$$

$$= \iint_R \left[\frac{\partial}{\partial x}(\cos y + x^3) - \frac{\partial}{\partial y}(e^x - y^3)\right] dA$$

Green's Theorem

$$= \iint_R (3x^2 + 3y^2)\,dA = 3\iint_R (x^2 + y^2)\,dA$$

$$= 3\int_0^{2\pi}\int_0^1 (r^2)r\,dr\,d\theta = \frac{3}{4}\int_0^{2\pi} d\theta = \frac{3\pi}{2}$$

We converted to polar coordinates.

FINDING AREAS USING GREEN'S THEOREM

Green's Theorem leads to some useful new formulas for the area A of a region R that satisfies the conditions of the theorem. Two such formulas can be obtained as follows:

$$A = \iint_R dA = \oint_C x\,dy \quad \text{and} \quad A = \iint_R dA = \oint_C (-y)\,dx$$

Set $f(x, y) = 0$ and $g(x, y) = x$ in (1).

Set $f(x, y) = -y$ and $g(x, y) = 0$ in (1).

A third formula can be obtained by adding these two equations together. Thus, we have the following three formulas that express the area A of a region R in terms of line integrals around the boundary:

Although the third formula in (6) looks more complicated than the other two, it often leads to simpler integrations. Each has advantages in certain situations.

$$A = \oint_C x\,dy = -\oint_C y\,dx = \frac{1}{2}\oint_C -y\,dx + x\,dy \tag{6}$$

Example 3 Use a line integral to find the area enclosed by the ellipse

$$\frac{x^2}{a^2} + \frac{y^2}{b^2} = 1$$

Solution. The ellipse, with counterclockwise orientation, can be represented parametrically by

$$x = a\cos t, \quad y = b\sin t \quad (0 \le t \le 2\pi)$$

If we denote this curve by C, then from the third formula in (6) the area A enclosed by the ellipse is

$$A = \frac{1}{2}\oint_C -y\,dx + x\,dy$$

$$= \frac{1}{2}\int_0^{2\pi} [(-b\sin t)(-a\sin t) + (a\cos t)(b\cos t)]\,dt$$

$$= \frac{1}{2}ab\int_0^{2\pi} (\sin^2 t + \cos^2 t)\,dt = \frac{1}{2}ab\int_0^{2\pi} dt = \pi ab$$

GREEN'S THEOREM FOR MULTIPLY CONNECTED REGIONS

Recall that a plane region is said to be simply connected if it has no holes and is said to be multiply connected if it has one or more holes (see Figure 16.3.6). At the beginning of this section we stated Green's Theorem for a counterclockwise integration around the

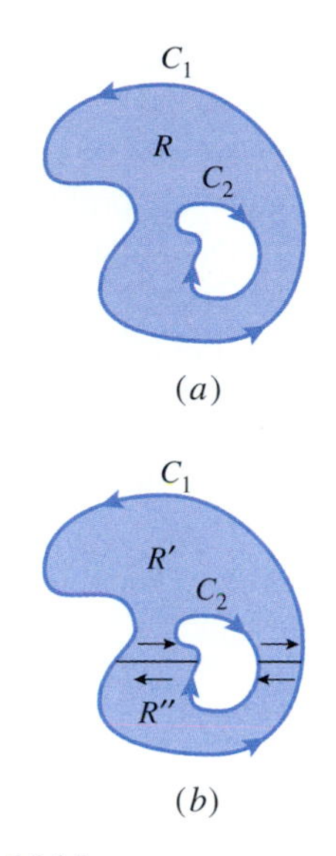

Figure 16.4.5

boundary of a simply connected region R (Theorem 16.4.1). Our next goal is to extend this theorem to multiply connected regions. To make this extension we will need to assume that *the region lies on the left when any portion of the boundary is traversed in the direction of its orientation.* This implies that the outer boundary curve of the region is oriented counterclockwise and the boundary curves that enclose holes have clockwise orientation (Figure 16.4.5a). If all portions of the boundary of a multiply connected region R are oriented in this way, then we say that the boundary of R has ***positive orientation***.

We will now derive a version of Green's Theorem that applies to multiply connected regions with positively oriented boundaries. For simplicity, we will consider a multiply connected region R with one hole, and we will assume that $f(x, y)$ and $g(x, y)$ have continuous first partial derivatives on some open set containing R. As shown in Figure 16.4.5b, let us divide R into two regions R' and R'' by introducing two "cuts" in R. The cuts are shown as line segments, but any piecewise smooth curves will suffice. If we assume that f and g satisfy the hypotheses of Green's Theorem on R (and hence on R' and R''), then we can apply this theorem to both R' and R'' to obtain

$$\iint_R \left(\frac{\partial g}{\partial x} - \frac{\partial f}{\partial y}\right) dA = \iint_{R'} \left(\frac{\partial g}{\partial x} - \frac{\partial f}{\partial y}\right) dA + \iint_{R''} \left(\frac{\partial g}{\partial x} - \frac{\partial f}{\partial y}\right) dA$$

$$= \oint_{\substack{\text{Boundary} \\ \text{of } R'}} f(x, y)\,dx + g(x, y)\,dy + \oint_{\substack{\text{Boundary} \\ \text{of } R''}} f(x, y)\,dx + g(x, y)\,dy$$

However, the two line integrals are taken in opposite directions along the cuts, and hence cancel there, leaving only the contributions along C_1 and C_2. Thus,

$$\iint_R \left(\frac{\partial g}{\partial x} - \frac{\partial f}{\partial y}\right) dA = \oint_{C_1} f(x, y)\,dx + g(x, y)\,dy + \oint_{C_2} f(x, y)\,dx + g(x, y)\,dy \tag{7}$$

which is an extension of Green's Theorem to a multiply connected region with one hole. Observe that the integral around the outer boundary is taken counterclockwise and the integral around the hole is taken clockwise. More generally, if R is a multiply connected region with n holes, then the analog of (7) involves a sum of $n + 1$ integrals, one taken counterclockwise around the outer boundary of R and the rest taken clockwise around the holes.

Sketch a proof of the version of Green's Theorem that applies to a multiply connected region with two holes.

▶ **Example 4** Evaluate the integral

$$\oint_C \frac{-y\,dx + x\,dy}{x^2 + y^2}$$

if C is a piecewise smooth simple closed curve oriented counterclockwise such that (a) C does not enclose the origin and (b) C encloses the origin.

Solution (a). Let

$$f(x, y) = -\frac{y}{x^2 + y^2}, \quad g(x, y) = \frac{x}{x^2 + y^2} \tag{8}$$

so that

$$\frac{\partial g}{\partial x} = \frac{y^2 - x^2}{(x^2 + y^2)^2} = \frac{\partial f}{\partial y}$$

if x and y are not both zero. Thus, if C does not enclose the origin, we have

$$\frac{\partial g}{\partial x} - \frac{\partial f}{\partial y} = 0 \tag{9}$$

on the simply connected region enclosed by C, and hence the given integral is zero by Green's Theorem.

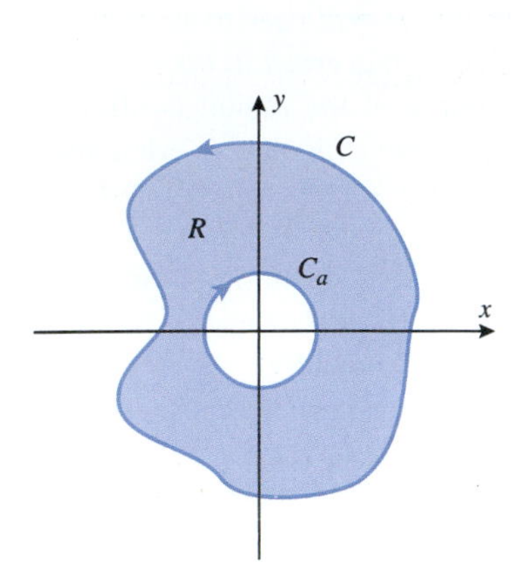

Figure 16.4.6

Solution (b). Unlike the situation in part (a), we cannot apply Green's Theorem directly because the functions $f(x, y)$ and $g(x, y)$ in (8) are discontinuous at the origin. Our problems are further compounded by the fact that we do not have a specific curve C that we can parametrize to evaluate the integral. Our strategy for circumventing these problems will be to replace C with a specific curve that produces the same value for the integral and then use that curve for the evaluation. To obtain such a curve, we will apply Green's Theorem for multiply connected regions to a region that does not contain the origin. For this purpose we construct a circle C_a with *clockwise* orientation, centered at the origin, and with sufficiently small radius a that it lies inside the region enclosed by C (Figure 16.4.6). This creates a multiply connected region R whose boundary curves C and C_a have the orientations required by Formula (7) and such that within R the functions $f(x, y)$ and $g(x, y)$ in (8) satisfy the hypotheses of Green's Theorem (the origin is outside of R). Thus, it follows from (7) and (9) that

$$\oint_C \frac{-y\,dx + x\,dy}{x^2 + y^2} + \oint_{C_a} \frac{-y\,dx + x\,dy}{x^2 + y^2} = \iint_R 0\,dA = 0$$

It follows from this equation that

$$\oint_C \frac{-y\,dx + x\,dy}{x^2 + y^2} = -\oint_{C_a} \frac{-y\,dx + x\,dy}{x^2 + y^2}$$

which we can rewrite as

$$\oint_C \frac{-y\,dx + x\,dy}{x^2 + y^2} = \oint_{-C_a} \frac{-y\,dx + x\,dy}{x^2 + y^2}$$

Reversing the orientation of C_a reverses the sign of the integral.

But C_a has clockwise orientation, so $-C_a$ has counterclockwise orientation. Thus, we have shown that the original integral can be evaluated by integrating counterclockwise around a circle of radius a that is centered at the origin and lies within the region enclosed by C. Such a circle can be expressed parametrically as $x = a\cos t$, $y = a\sin t$ $(0 \leq t \leq 2\pi)$; and hence

$$\begin{aligned}\oint_C \frac{-y\,dx + x\,dy}{x^2 + y^2} &= \int_0^{2\pi} \frac{(-a\sin t)(-a\sin t)\,dt + (a\cos t)(a\cos t)\,dt}{(a\cos t)^2 + (a\sin t)^2} \\ &= \int_0^{2\pi} 1\,dt = 2\pi\end{aligned}$$

✔ QUICK CHECK EXERCISES 16.4 (See page 1147 for answers.)

1. If C is the square with vertices $(\pm 1, \pm 1)$ oriented counterclockwise, then

$$\int_C -y\,dx + x\,dy = \underline{\qquad}$$

2. If C is the triangle with vertices $(0, 0)$, $(1, 0)$, and $(1, 1)$ oriented counterclockwise, then

$$\int_C 2xy\,dx + (x^2 + x)\,dy = \underline{\qquad}$$

3. Sometimes symmetry considerations can simplify an application of Green's Theorem. For example, if C is the unit circle centered at the origin and oriented counterclockwise, then

$$\int_C (y^3 - y - x)\,dx + (x^3 + x + y)\,dy = \underline{\qquad}$$

4. What region R and choice of functions $f(x, y)$ and $g(x, y)$ allow us to use Formula (1) of Theorem 16.4.1 to claim that

$$\int_0^1 \int_0^{\sqrt{1-x^2}} (2x + 2y)\,dy\,dx = \int_0^{\pi/2} (\sin^3 t + \cos^3 t)\,dt?$$

EXERCISE SET 16.4 [C] CAS

1–2 Evaluate the line integral using Green's Theorem and check the answer by evaluating it directly.

1. $\oint_C y^2\,dx + x^2\,dy$, where C is the square with vertices $(0, 0)$, $(1, 0)$, $(1, 1)$, and $(0, 1)$ oriented counterclockwise.

2. $\oint_C y\,dx + x\,dy$, where C is the unit circle oriented counterclockwise.

3–13 Use Green's Theorem to evaluate the integral. In each exercise, assume that the curve C is oriented counterclockwise.

3. $\oint_C 3xy\,dx + 2xy\,dy$, where C is the rectangle bounded by $x = -2$, $x = 4$, $y = 1$, and $y = 2$.

4. $\oint_C (x^2 - y^2)\,dx + x\,dy$, where C is the circle $x^2 + y^2 = 9$.

5. $\oint_C x\cos y\,dx - y\sin x\,dy$, where C is the square with vertices $(0, 0)$, $(\pi/2, 0)$, $(\pi/2, \pi/2)$, and $(0, \pi/2)$.

6. $\oint_C y\tan^2 x\,dx + \tan x\,dy$, where C is the circle $x^2 + (y+1)^2 = 1$.

7. $\oint_C (x^2 - y)\,dx + x\,dy$, where C is the circle $x^2 + y^2 = 4$.

8. $\oint_C (e^x + y^2)\,dx + (e^y + x^2)\,dy$, where C is the boundary of the region between $y = x^2$ and $y = x$.

9. $\oint_C \ln(1+y)\,dx - \dfrac{xy}{1+y}\,dy$, where C is the triangle with vertices $(0, 0)$, $(2, 0)$, and $(0, 4)$.

10. $\oint_C x^2 y\,dx - y^2 x\,dy$, where C is the boundary of the region in the first quadrant, enclosed between the coordinate axes and the circle $x^2 + y^2 = 16$.

11. $\oint_C \tan^{-1} y\,dx - \dfrac{y^2 x}{1+y^2}\,dy$, where C is the square with vertices $(0, 0)$, $(1, 0)$, $(1, 1)$, and $(0, 1)$.

12. $\oint_C \cos x\sin y\,dx + \sin x\cos y\,dy$, where C is the triangle with vertices $(0, 0)$, $(3, 3)$, and $(0, 3)$.

13. $\oint_C x^2 y\,dx + (y + xy^2)\,dy$, where C is the boundary of the region enclosed by $y = x^2$ and $x = y^2$.

14. Let C be the boundary of the region enclosed between $y = x^2$ and $y = 2x$. Assuming that C is oriented counterclockwise, evaluate the following integrals by Green's Theorem:

(a) $\oint_C (6xy - y^2)\,dx$ (b) $\oint_C (6xy - y^2)\,dy$

[C] **15.** Use a CAS to check Green's Theorem by evaluating both integrals in the equation

$$\oint_C e^y\,dx + ye^x\,dy = \iint_R \left[\frac{\partial}{\partial x}(ye^x) - \frac{\partial}{\partial y}(e^y)\right] dA$$

where
(a) C is the circle $x^2 + y^2 = 1$
(b) C is the boundary of the region enclosed by $y = x^2$ and $x = y^2$.

16. In Example 3, we used Green's Theorem to obtain the area of an ellipse. Obtain this area using the first and then the second formula in (6).

17. Use a line integral to find the area of the region enclosed by the astroid

$$x = a\cos^3\phi, \quad y = a\sin^3\phi \qquad (0 \le \phi \le 2\pi)$$

[See Exercise 29 of Section 7.4.]

18. Use a line integral to find the area of the triangle with vertices $(0, 0)$, $(a, 0)$, and $(0, b)$, where $a > 0$ and $b > 0$.

19. Use the formula

$$A = \frac{1}{2}\oint_C -y\,dx + x\,dy$$

to find the area of the region swept out by the line from the origin to the ellipse $x = a\cos t$, $y = b\sin t$ if t varies from $t = 0$ to $t = t_0$ $(0 \le t_0 \le 2\pi)$.

20. Use the formula

$$A = \frac{1}{2}\oint_C -y\,dx + x\,dy$$

to find the area of the region swept out by the line from the origin to the hyperbola $x = a\cosh t$, $y = b\sinh t$ if t varies from $t = 0$ to $t = t_0$ $(t_0 \ge 0)$.

FOCUS ON CONCEPTS

21. Suppose that $\mathbf{F}(x, y) = f(x, y)\mathbf{i} + g(x, y)\mathbf{j}$ is a vector field whose component functions f and g have continuous first partial derivatives. Let C denote a simple, closed, piecewise smooth curve oriented counterclockwise that bounds a region R contained in the domain of $\mathbf{F}$. We can think of $\mathbf{F}$ as a vector field in 3-space by writing it as

$$\mathbf{F}(x, y, z) = f(x, y)\mathbf{i} + g(x, y)\mathbf{j} + 0\mathbf{k}$$

With this convention, explain why

$$\int_C \mathbf{F}\cdot d\mathbf{r} = \iint_R \operatorname{curl}\mathbf{F}\cdot\mathbf{k}\,dA$$

22. Suppose that $\mathbf{F}(x, y) = f(x, y)\mathbf{i} + g(x, y)\mathbf{j}$ is a vector field on the xy-plane and that f and g have continuous first partial derivatives with $f_y = g_x$ everywhere. Use Green's Theorem to explain why

$$\int_{C_1} \mathbf{F}\cdot d\mathbf{r} = \int_{C_2} \mathbf{F}\cdot d\mathbf{r}$$

where C_1 and C_2 are the oriented curves in the accompanying figure. (Compare this result with Theorems 16.3.2 and 16.3.3.)

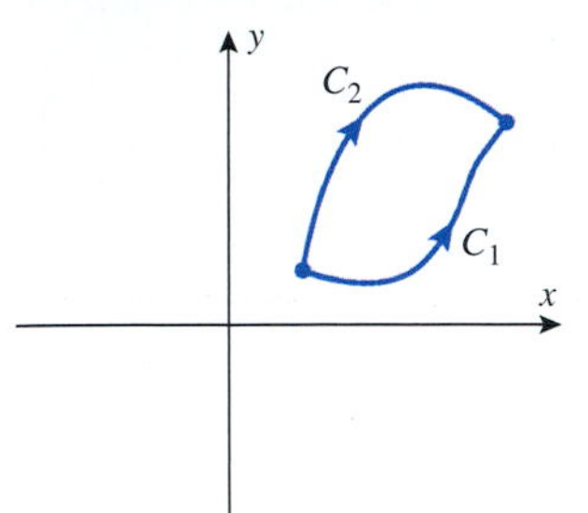

Figure Ex-22

23. Suppose that $f(x)$ and $g(x)$ are continuous functions with $g(x) \leq f(x)$. Let R denote the region bounded by the graph of f, the graph of g, and the vertical lines $x = a$ and $x = b$. Let C denote the boundary of R oriented counterclockwise. What familiar formula results from applying Green's Theorem to $\int_C (-y)\,dx$?

24. In the accompanying figure, C is a smooth oriented curve from $P(x_0, y_0)$ to $Q(x_1, y_1)$ that is contained inside the rectangle with corners at the origin and Q and outside the rectangle with corners at the origin and P.

(a) What region in the figure has area $\int_C x\,dy$?

(b) What region in the figure has area $\int_C y\,dx$?

(c) Express $\int_C x\,dy + \int_C y\,dx$ in terms of the coordinates of P and Q.

(d) Interpret the result of part (c) in terms of the Fundamental Theorem of Line Integrals.

(e) Interpret the result in part (c) in terms of integration by parts.

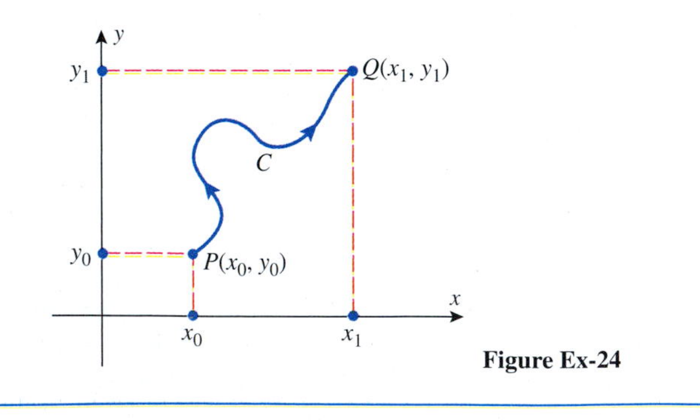

Figure Ex-24

25–26 Use Green's Theorem to find the work done by the force field $\mathbf{F}$ on a particle that moves along the stated path.

25. $\mathbf{F}(x, y) = xy\mathbf{i} + \left(\frac{1}{2}x^2 + xy\right)\mathbf{j}$; the particle starts at $(5, 0)$, traverses the upper semicircle $x^2 + y^2 = 25$, and returns to its starting point along the x-axis.

26. $\mathbf{F}(x, y) = \sqrt{y}\,\mathbf{i} + \sqrt{x}\,\mathbf{j}$; the particle moves counterclockwise one time around the closed curve given by the equations $y = 0$, $x = 2$, and $y = x^3/4$.

27. Evaluate $\oint_C y\,dx - x\,dy$, where C is the cardioid

$$r = a(1 + \cos\theta) \quad (0 \leq \theta \leq 2\pi)$$

28. Let R be a plane region with area A whose boundary is a piecewise smooth simple closed curve C. Use Green's Theorem to prove that the centroid $(\bar{x}, \bar{y})$ of R is given by

$$\bar{x} = \frac{1}{2A}\oint_C x^2\,dy, \quad \bar{y} = -\frac{1}{2A}\oint_C y^2\,dx$$

29–32 Use the result in Exercise 28 to find the centroid of the region.

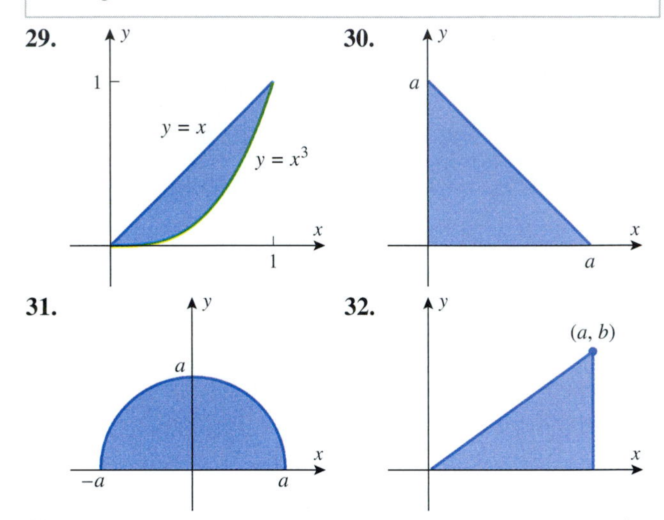

33. Find a simple closed curve C with counterclockwise orientation that maximizes the value of

$$\oint_C \tfrac{1}{3}y^3\,dx + \left(x - \tfrac{1}{3}x^3\right)dy$$

and explain your reasoning.

34. (a) Let C be the line segment from a point (a, b) to a point (c, d). Show that

$$\int_C -y\,dx + x\,dy = ad - bc$$

(b) Use the result in part (a) to show that the area A of a triangle with successive vertices (x_1, y_1), (x_2, y_2), and (x_3, y_3) going counterclockwise is

$$A = \tfrac{1}{2}[(x_1y_2 - x_2y_1) + (x_2y_3 - x_3y_2) + (x_3y_1 - x_1y_3)]$$

(c) Find a formula for the area of a polygon with successive vertices (x_1, y_1), (x_2, y_2), $\ldots$, (x_n, y_n) going counterclockwise.

(d) Use the result in part (c) to find the area of a quadrilateral with vertices $(0, 0)$, $(3, 4)$, $(-2, 2)$, $(-1, 0)$.

35–36 Evaluate the integral $\int_C \mathbf{F} \cdot d\mathbf{r}$, where C is the boundary of the region R and C is oriented so that the region is on the left when the boundary is traversed in the direction of its orientation.

35. $\mathbf{F}(x, y) = (x^2 + y)\mathbf{i} + (4x - \cos y)\mathbf{j}$; C is the boundary of the region R that is inside the square with vertices $(0, 0)$, $(5, 0)$, $(5, 5)$, $(0, 5)$ but is outside the rectangle with vertices $(1, 1)$, $(3, 1)$, $(3, 2)$, $(1, 2)$.

36. $\mathbf{F}(x, y) = (e^{-x} + 3y)\mathbf{i} + x\mathbf{j}$; C is the boundary of the region R inside the circle $x^2 + y^2 = 16$ and outside the circle $x^2 - 2x + y^2 = 3$.

✔QUICK CHECK ANSWERS 16.4

1. 8 **2.** $\frac{1}{2}$ **3.** 2π **4.** R is the region $x^2 + y^2 \le 1$ $(0 \le x, 0 \le y)$ and $f(x, y) = -y^2$, $g(x, y) = x^2$

16.5 SURFACE INTEGRALS

In previous sections we considered four kinds of integrals—integrals over intervals, double integrals over two-dimensional regions, triple integrals over three-dimensional solids, and line integrals along curves in two- or three-dimensional space. In this section we will discuss integrals over surfaces in three-dimensional space. Such integrals occur in problems involving fluid and heat flow, electricity, magnetism, mass, and center of gravity.

DEFINITION OF A SURFACE INTEGRAL

In this section we will define what it means to integrate a function $f(x, y, z)$ over a smooth parametric surface σ. To motivate the definition we will consider the problem of finding the mass of a curved lamina whose density function (mass per unit area) is known. Recall that in Section 15.6 we defined a *lamina* to be an idealized flat object that is thin enough to be viewed as a plane region. Analogously, a ***curved lamina*** is an idealized object that is thin enough to be viewed as a surface in 3-space. A curved lamina may look like a bent plate, as in Figure 16.5.1, or it may enclose a region in 3-space, like the shell of an egg. We will model the lamina by a smooth parametric surface σ. Given any point (x, y, z) on σ, we let $f(x, y, z)$ denote the corresponding value of the density function. To compute the mass of the lamina, we proceed as follows:

The thickness of a curved lamina is negligible.

Figure 16.5.1

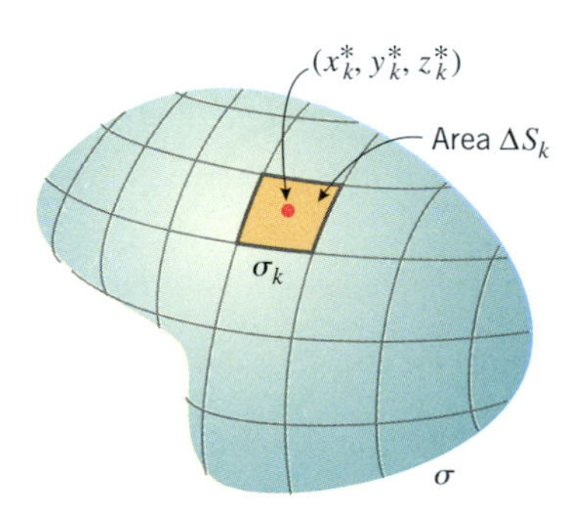

Figure 16.5.2

- As shown in Figure 16.5.2, we divide σ into n very small patches $\sigma_1, \sigma_2, \ldots, \sigma_n$ with areas $\Delta S_1, \Delta S_2, \ldots, \Delta S_n$, respectively. Let (x_k^*, y_k^*, z_k^*) be a sample point in the kth patch with ΔM_k the mass of the corresponding section.
- If the dimensions of σ_k are very small, the value of f will not vary much along the kth section and we can approximate f along this section by the value $f(x_k^*, y_k^*, z_k^*)$. It follows that the mass of the kth section can be approximated by

$$\Delta M_k \approx f(x_k^*, y_k^*, z_k^*)\Delta S_k$$

- The mass M of the entire lamina can then be approximated by

$$M = \sum_{k=1}^{n} \Delta M_k \approx \sum_{k=1}^{n} f(x_k^*, y_k^*, z_k^*)\Delta S_k \tag{1}$$

- We will use the expression $n \to \infty$ to indicate the process of increasing n in such a way that the maximum dimension of each patch approaches 0. It is plausible that the error in (1) will approach 0 as $n \to \infty$ and the exact value of M will be given by

$$M = \lim_{n \to \infty} \sum_{k=1}^{n} f(x_k^*, y_k^*, z_k^*)\Delta S_k \tag{2}$$

The limit in (2) is very similar to the limit used to find the mass of a thin wire [Formula (2) in Section 16.2]. By analogy to Definition 16.2.1, we make the following definition.

16.5.1 DEFINITION. If σ is a smooth parametric surface, then the ***surface integral*** of $f(x, y, z)$ over σ is

$$\iint_{\sigma} f(x, y, z)\, dS = \lim_{n \to \infty} \sum_{k=1}^{n} f(x_k^*, y_k^*, z_k^*) \Delta S_k \tag{3}$$

provided this limit exists and does not depend on the way the subdivisions of σ are made or how the sample points (x_k^*, y_k^*, z_k^*) are chosen.

It can be shown that the integral of f over σ exists if f is continuous on σ.

We see from (2) and Definition 16.5.1 that if σ models a lamina and if $f(x, y, z)$ is the density function of the lamina, then the mass M of the lamina is given by

$$M = \iint_{\sigma} f(x, y, z)\, dS \tag{4}$$

That is, to obtain the mass of a lamina, we integrate the density function over the smooth surface that models the lamina.

Note that if σ is a smooth surface of surface area S, and f is identically 1, then it immediately follows from Definition 16.5.1 that

$$\iint_{\sigma} dS = \lim_{n \to \infty} \sum_{k=1}^{n} \Delta S_k = \lim_{n \to \infty} S = S \tag{5}$$

EVALUATING SURFACE INTEGRALS

There are various procedures for evaluating surface integrals that depend on how the surface σ is represented. The following theorem provides a method for evaluating a surface integral when σ is represented parametrically.

16.5.2 THEOREM. *Let σ be a smooth parametric surface whose vector equation is*

$$\mathbf{r} = x(u, v)\mathbf{i} + y(u, v)\mathbf{j} + z(u, v)\mathbf{k}$$

where (u, v) varies over a region R in the uv-plane. If $f(x, y, z)$ is continuous on σ, then

$$\iint_{\sigma} f(x, y, z)\, dS = \iint_{R} f(x(u, v), y(u, v), z(u, v)) \left\| \frac{\partial \mathbf{r}}{\partial u} \times \frac{\partial \mathbf{r}}{\partial v} \right\| dA \tag{6}$$

Explain how to use Formula (6) to confirm Formula (5).

To motivate this result, suppose that the parameter domain R is subdivided as in Figure 15.4.10, and suppose that the point (x_k^*, y_k^*, z_k^*) in (3) corresponds to parameter values of u_k^* and v_k^*. If we use Formula (9) of Section 15.4 to approximate ΔS_k, and if we assume that the errors in the approximations approach zero as $n \to +\infty$, then it follows from (3) that

$$\iint_\sigma f(x, y, z)\, dS = \lim_{n \to +\infty} \sum_{k=1}^{n} f(x(u_k^*, v_k^*), y(u_k^*, v_k^*), z(u_k^*, v_k^*)) \left\| \frac{\partial \mathbf{r}}{\partial u} \times \frac{\partial \mathbf{r}}{\partial v} \right\| \Delta A_k$$

which suggests Formula (6).

Although Theorem 16.5.2 is stated for *smooth* parametric surfaces, Formula (6) remains valid even if $\partial \mathbf{r}/\partial u \times \partial \mathbf{r}/\partial v$ is allowed to equal $\mathbf{0}$ on the boundary of R.

Example 1 Evaluate the surface integral $\iint_\sigma x^2\, dS$ over the sphere $x^2 + y^2 + z^2 = 1$.

Solution. As in Example 9 of Section 15.4 (with $a = 1$), the sphere is the graph of the vector-valued function

$$\mathbf{r}(\phi, \theta) = \sin\phi\cos\theta\,\mathbf{i} + \sin\phi\sin\theta\,\mathbf{j} + \cos\phi\,\mathbf{k} \quad (0 \le \phi \le \pi,\ \ 0 \le \theta \le 2\pi) \tag{7}$$

and

$$\left\| \frac{\partial \mathbf{r}}{\partial \phi} \times \frac{\partial \mathbf{r}}{\partial \theta} \right\| = \sin\phi$$

Explain why the function $\mathbf{r}(\phi, \theta)$ given in (7) fails to be smooth on its domain.

From the $\mathbf{i}$-component of $\mathbf{r}$, the integrand in the surface integral can be expressed in terms of ϕ and θ as $x^2 = \sin^2\phi\cos^2\theta$. Thus, it follows from (6) with ϕ and θ in place of u and v and R as the rectangular region in the $\phi\theta$-plane determined by the inequalities in (7) that

$$\begin{aligned}
\iint_\sigma x^2\, dS &= \iint_R (\sin^2\phi\cos^2\theta) \left\| \frac{\partial \mathbf{r}}{\partial \phi} \times \frac{\partial \mathbf{r}}{\partial \theta} \right\| dA \\
&= \int_0^{2\pi} \int_0^{\pi} \sin^3\phi\cos^2\theta\, d\phi\, d\theta \\
&= \int_0^{2\pi} \left[\int_0^{\pi} \sin^3\phi\, d\phi \right] \cos^2\theta\, d\theta \\
&= \int_0^{2\pi} \left[\frac{1}{3}\cos^3\phi - \cos\phi \right]_0^{\pi} \cos^2\theta\, d\theta && \text{Formula (11), Section 8.3} \\
&= \frac{4}{3} \int_0^{2\pi} \cos^2\theta\, d\theta \\
&= \frac{4}{3} \left[\frac{1}{2}\theta + \frac{1}{4}\sin 2\theta \right]_0^{2\pi} = \frac{4\pi}{3} && \text{Formula (8), Section 8.3}
\end{aligned}$$

SURFACE INTEGRALS OVER $z = g(x, y)$, $y = g(x, z)$, AND $x = g(y, z)$

In the case where σ is a surface of the form $z = g(x, y)$, we can take $x = u$ and $y = v$ as parameters and express the equation of the surface as

$$\mathbf{r} = u\mathbf{i} + v\mathbf{j} + g(u, v)\mathbf{k}$$

in which case we obtain

$$\left\| \frac{\partial \mathbf{r}}{\partial u} \times \frac{\partial \mathbf{r}}{\partial v} \right\| = \sqrt{\left(\frac{\partial z}{\partial x}\right)^2 + \left(\frac{\partial z}{\partial y}\right)^2 + 1}$$

[see the derivation of Formula (11) in Section 15.4]. Thus, it follows from (6) that

$$\iint_{\sigma} f(x, y, z)\, dS = \iint_{R} f(x, y, g(x, y))\sqrt{\left(\frac{\partial z}{\partial x}\right)^2 + \left(\frac{\partial z}{\partial y}\right)^2 + 1}\, dA$$

Note that in this formula the region R lies in the xy-plane because the parameters are x and y. Geometrically, this region is the projection of σ on the xy-plane. The following theorem summarizes this result and gives analogous formulas for surface integrals over surfaces of the form $y = g(x, z)$ and $x = g(y, z)$.

16.5.3 THEOREM.

(*a*) *Let σ be a surface with equation $z = g(x, y)$ and let R be its projection on the xy-plane. If g has continuous first partial derivatives on R and $f(x, y, z)$ is continuous on σ, then*

$$\iint_{\sigma} f(x, y, z)\, dS = \iint_{R} f(x, y, g(x, y))\sqrt{\left(\frac{\partial z}{\partial x}\right)^2 + \left(\frac{\partial z}{\partial y}\right)^2 + 1}\, dA \tag{8}$$

(*b*) *Let σ be a surface with equation $y = g(x, z)$ and let R be its projection on the xz-plane. If g has continuous first partial derivatives on R and $f(x, y, z)$ is continuous on σ, then*

$$\iint_{\sigma} f(x, y, z)\, dS = \iint_{R} f(x, g(x, z), z)\sqrt{\left(\frac{\partial y}{\partial x}\right)^2 + \left(\frac{\partial y}{\partial z}\right)^2 + 1}\, dA \tag{9}$$

(*c*) *Let σ be a surface with equation $x = g(y, z)$ and let R be its projection on the yz-plane. If g has continuous first partial derivatives on R and $f(x, y, z)$ is continuous on σ, then*

$$\iint_{\sigma} f(x, y, z)\, dS = \iint_{R} f(g(y, z), y, z)\sqrt{\left(\frac{\partial x}{\partial y}\right)^2 + \left(\frac{\partial x}{\partial z}\right)^2 + 1}\, dA \tag{10}$$

Formulas (9) and (10) can be recovered from Formula (8). Explain how.

Example 2 Evaluate the surface integral

$$\iint_{\sigma} xz\, dS$$

where σ is the part of the plane $x + y + z = 1$ that lies in the first octant.

Solution. The equation of the plane can be written as

$$z = 1 - x - y$$

Consequently, we can apply Formula (8) with $z = g(x, y) = 1 - x - y$ and $f(x, y, z) = xz$. We have

$$\frac{\partial z}{\partial x} = -1 \quad \text{and} \quad \frac{\partial z}{\partial y} = -1$$

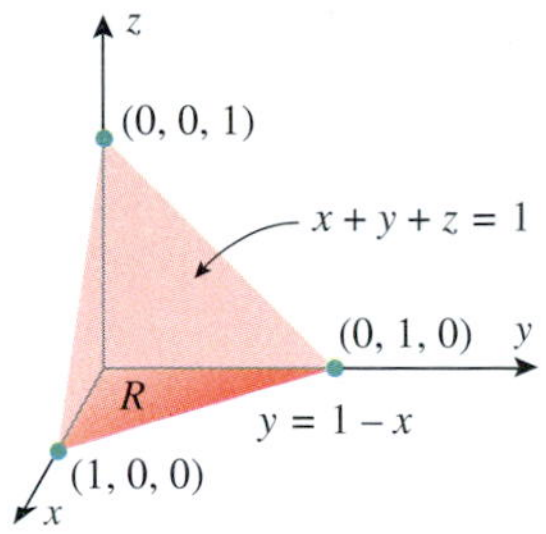

Figure 16.5.3

so (8) becomes

$$\iint_\sigma xz\,dS = \iint_R x(1-x-y)\sqrt{(-1)^2+(-1)^2+1}\,dA \tag{11}$$

where R is the projection of σ on the xy-plane (Figure 16.5.3). Rewriting the double integral in (11) as an iterated integral yields

$$\begin{aligned}\iint_\sigma xz\,dS &= \sqrt{3}\int_0^1\int_0^{1-x}(x-x^2-xy)\,dy\,dx\\ &= \sqrt{3}\int_0^1\left[xy-x^2y-\frac{xy^2}{2}\right]_{y=0}^{1-x}dx\\ &= \sqrt{3}\int_0^1\left(\frac{x}{2}-x^2+\frac{x^3}{2}\right)dx\\ &= \sqrt{3}\left[\frac{x^2}{4}-\frac{x^3}{3}+\frac{x^4}{8}\right]_0^1 = \frac{\sqrt{3}}{24}\end{aligned}$$

◀

▶ Example 3 Evaluate the surface integral

$$\iint_\sigma y^2z^2\,dS$$

where σ is the part of the cone $z=\sqrt{x^2+y^2}$ that lies between the planes $z=1$ and $z=2$ (Figure 16.5.4).

Solution. We will apply Formula (8) with

$$z = g(x,y) = \sqrt{x^2+y^2} \quad\text{and}\quad f(x,y,z)=y^2z^2$$

Thus,

$$\frac{\partial z}{\partial x} = \frac{x}{\sqrt{x^2+y^2}} \quad\text{and}\quad \frac{\partial z}{\partial y} = \frac{y}{\sqrt{x^2+y^2}}$$

so

$$\sqrt{\left(\frac{\partial z}{\partial x}\right)^2+\left(\frac{\partial z}{\partial y}\right)^2+1}=\sqrt{2}$$

(verify), and (8) yields

$$\iint_\sigma y^2z^2\,dS = \iint_R y^2\left(\sqrt{x^2+y^2}\right)^2\sqrt{2}\,dA = \sqrt{2}\iint_R y^2(x^2+y^2)\,dA$$

where R is the annulus enclosed between $x^2+y^2=1$ and $x^2+y^2=4$ (Figure 16.5.4). Using polar coordinates to evaluate this double integral over the annulus R yields

$$\begin{aligned}\iint_\sigma y^2z^2\,dS &= \sqrt{2}\int_0^{2\pi}\int_1^2(r\sin\theta)^2(r^2)r\,dr\,d\theta\\ &= \sqrt{2}\int_0^{2\pi}\int_1^2 r^5\sin^2\theta\,dr\,d\theta\\ &= \sqrt{2}\int_0^{2\pi}\left[\frac{r^6}{6}\sin^2\theta\right]_{r=1}^2 d\theta = \frac{21}{\sqrt{2}}\int_0^{2\pi}\sin^2\theta\,d\theta\\ &= \frac{21}{\sqrt{2}}\left[\frac{1}{2}\theta-\frac{1}{4}\sin 2\theta\right]_0^{2\pi} = \frac{21\pi}{\sqrt{2}}\end{aligned}$$

Formula (7), Section 8.3 ◀

Evaluate the integral in Example 3 with the help of Formula (6) and the parametrization

$\mathbf{r} = \langle r\cos\theta, r\sin\theta, r\rangle$

$(1\le r\le 2,\ 0\le\theta\le 2\pi)$

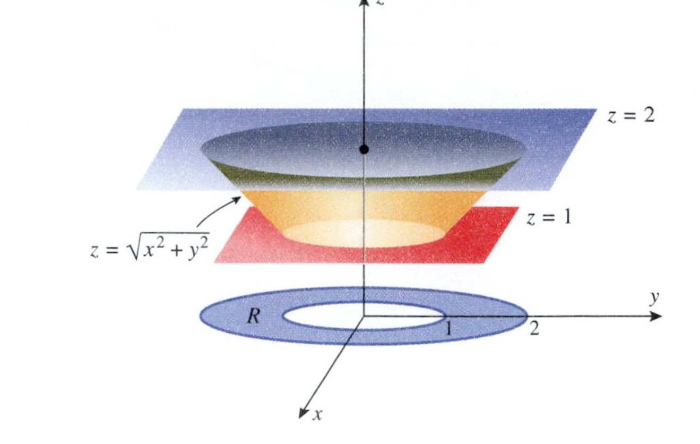

Figure 16.5.4

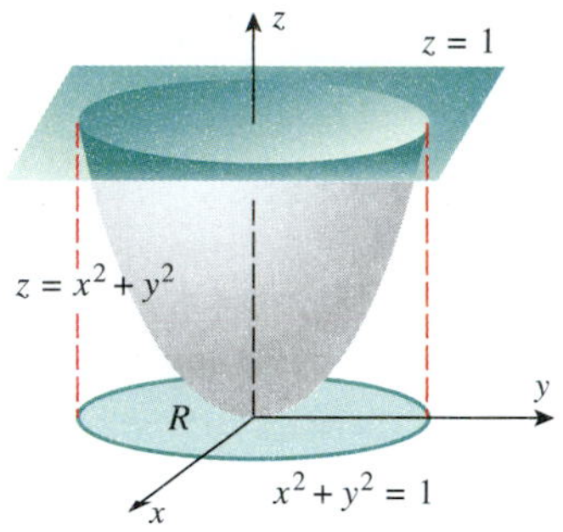

Figure 16.5.5

▶ **Example 4** Suppose that a curved lamina σ with constant density $\delta(x, y, z) = \delta_0$ is the portion of the paraboloid $z = x^2 + y^2$ below the plane $z = 1$ (Figure 16.5.5). Find the mass of the lamina.

Solution. Since $z = g(x, y) = x^2 + y^2$, it follows that

$$\frac{\partial z}{\partial x} = 2x \quad \text{and} \quad \frac{\partial z}{\partial y} = 2y$$

Therefore,

$$M = \iint_{\sigma} \delta_0 \, dS = \iint_{R} \delta_0 \sqrt{(2x)^2 + (2y)^2 + 1} \, dA = \delta_0 \iint_{R} \sqrt{4x^2 + 4y^2 + 1} \, dA \qquad (12)$$

where R is the circular region enclosed by $x^2 + y^2 = 1$. To evaluate (12) we use polar coordinates:

$$M = \delta_0 \int_0^{2\pi} \int_0^1 \sqrt{4r^2 + 1} \, r \, dr \, d\theta = \frac{\delta_0}{12} \int_0^{2\pi} (4r^2 + 1)^{3/2} \Big]_{r=0}^{1} d\theta$$

$$= \frac{\delta_0}{12} \int_0^{2\pi} (5^{3/2} - 1) \, d\theta = \frac{\pi \delta_0}{6} (5\sqrt{5} - 1) \blacktriangleleft$$

✔ QUICK CHECK EXERCISES 16.5 *(See page 1155 for answers.)*

1. Consider the surface integral $\iint_{\sigma} f(x, y, z) \, dS$.
   (a) If σ is a parametric surface whose vector equation is
   $$\mathbf{r} = x(u, v)\mathbf{i} + y(u, v)\mathbf{j} + z(u, v)\mathbf{k}$$
   to evaluate the integral replace dS by ________.
   (b) If σ is the graph of a function $z = g(x, y)$ with continuous first partial derivatives, to evaluate the integral replace dS by ________.
2. If σ is the triangular region with vertices $(1, 0, 0)$, $(0, 1, 0)$, and $(0, 0, 1)$, then
   $$\iint_{\sigma} (x + y + z) \, dS = \_\_\_\_\_\_\_\_$$
3. If σ is the sphere of radius 2 centered at the origin, then
   $$\iint_{\sigma} (x^2 + y^2 + z^2) \, dS = \_\_\_\_\_\_\_\_$$
4. If $f(x, y, z)$ is the mass density function of a curved lamina σ, then the mass of σ is given by the integral ________.

EXERCISE SET 16.5 C CAS

1–8 Evaluate the surface integral

$$\iint_\sigma f(x, y, z)\, dS$$

1. $f(x, y, z) = z^2$; σ is the portion of the cone $z = \sqrt{x^2 + y^2}$ between the planes $z = 1$ and $z = 2$.
2. $f(x, y, z) = xy$; σ is the portion of the plane $x + y + z = 1$ lying in the first octant.
3. $f(x, y, z) = x^2y$; σ is the portion of the cylinder $x^2 + z^2 = 1$ between the planes $y = 0$, $y = 1$, and above the xy-plane.
4. $f(x, y, z) = (x^2 + y^2)z$; σ is the portion of the sphere $x^2 + y^2 + z^2 = 4$ above the plane $z = 1$.
5. $f(x, y, z) = x - y - z$; σ is the portion of the plane $x + y = 1$ in the first octant between $z = 0$ and $z = 1$.
6. $f(x, y, z) = x + y$; σ is the portion of the plane $z = 6 - 2x - 3y$ in the first octant.
7. $f(x, y, z) = x + y + z$; σ is the surface of the cube defined by the inequalities $0 \le x \le 1$, $0 \le y \le 1$, $0 \le z \le 1$. [*Hint:* Integrate over each face separately.]
8. $f(x, y, z) = x^2 + y^2$; σ is the surface of the sphere $x^2 + y^2 + z^2 = a^2$.

9–10 Sometimes evaluating a surface integral results in an improper integral. When this happens, one can either attempt to determine the value of the integral using an appropriate limit or one can try another method. These exercises explore both approaches.

9. Consider the integral of $f(x, y, z) = z + 1$ over the upper hemisphere σ: $z = \sqrt{1 - x^2 - y^2}$ $(0 \le x^2 + y^2 \le 1)$.
   (a) Explain why evaluating this surface integral using (8) results in an improper integral.
   (b) Use (8) to evaluate the integral of f over the surface $\sigma_r : z = \sqrt{1 - x^2 - y^2}$ $(0 \le x^2 + y^2 \le r^2 < 1)$. Take the limit of this result as $r \to 1^-$ to determine the integral of f over σ.
   (c) Parametrize σ using spherical coordinates and evaluate the integral of f over σ using (6). Verify that your answer agrees with the result in part (b).
10. Consider the integral of $f(x, y, z) = \sqrt{x^2 + y^2 + z^2}$ over the cone $\sigma : z = \sqrt{x^2 + y^2}$ $(0 \le z \le 1)$.
   (a) Explain why evaluating this surface integral using (8) results in an improper integral.
   (b) Use (8) to evaluate the integral of f over the surface $\sigma_r : z = \sqrt{x^2 + y^2}$ $(0 < r^2 \le x^2 + y^2 \le 1)$. Take the limit of this result as $r \to 0^+$ to determine the integral of f over σ.
   (c) Parametrize σ using spherical coordinates and evaluate the integral of f over σ using (6). Verify that your answer agrees with the result in part (b).

FOCUS ON CONCEPTS

11–14 In some cases it is possible to use Definition 16.5.1 along with symmetry considerations to evaluate a surface integral without reference to a parametrization of the surface. In these exercises, σ denotes the unit sphere centered at the origin.

11. (a) Explain why it is possible to subdivide σ into patches and choose corresponding sample points (x_k^*, y_k^*, z_k^*) such that (i) the dimensions of each patch are as small as desired and (ii) for each sample point (x_k^*, y_k^*, z_k^*), there exists a sample point (x_j^*, y_j^*, z_j^*) with
$$x_k = -x_j, \quad y_k = y_j, \quad z_k = z_j$$
and with $\Delta S_k = \Delta S_j$.
   (b) Use Definition 16.5.1, the result in part (a), and the fact that surface integrals exist for continuous functions to prove that $\iint_\sigma x^n\, dS = 0$ for n an odd positive integer.
12. Use the argument in Exercise 11 to prove that if $f(x)$ is a continuous odd function of x and if $g(y, z)$ is a continuous function, then
$$\iint_\sigma f(x)g(y, z)\, dS = 0$$
13. (a) Explain why
$$\iint_\sigma x^2\, dS = \iint_\sigma y^2\, dS = \iint_\sigma z^2\, dS$$
   (b) Conclude from part (a) that
$$\iint_\sigma x^2\, dS = \frac{1}{3}\left[\iint_\sigma x^2\, dS + \iint_\sigma y^2\, dS + \iint_\sigma y^2\, dS\right]$$
   (c) Use part (b) to evaluate
$$\iint_\sigma x^2\, dS$$
   without performing an integration.
14. Use the results of Exercises 12 and 13 to evaluate
$$\iint_\sigma (x - y)^2\, dS$$
without performing an integration.

15–16 Set up, but do not evaluate, an iterated integral equal to the given surface integral by projecting σ on (a) the xy-plane, (b) the yz-plane, and (c) the xz-plane.

15. $\displaystyle\iint_\sigma xyz\, dS$, where σ is the portion of the plane $2x + 3y + 4z = 12$ in the first octant.

16. $\displaystyle\iint_{\sigma} xz\,dS$, where σ is the portion of the sphere $x^2 + y^2 + z^2 = a^2$ in the first octant.

C **17.** Use a CAS to confirm that the three integrals you obtained in Exercise 15 are equal, and find the exact value of the surface integral.

C **18.** Try to confirm with a CAS that the three integrals you obtained in Exercise 16 are equal. If you did not succeed, what was the difficulty?

19–20 Set up, but do not evaluate, two different iterated integrals equal to the given integral.

19. $\displaystyle\iint_{\sigma} xyz\,dS$, where σ is the portion of the surface $y^2 = x$ between the planes $z = 0$, $z = 4$, $y = 1$, and $y = 2$.

20. $\displaystyle\iint_{\sigma} x^2y\,dS$, where σ is the portion of the cylinder $y^2 + z^2 = a^2$ in the first octant between the planes $x = 0$, $x = 9$, $z = y$, and $z = 2y$.

C **21.** Use a CAS to confirm that the two integrals you obtained in Exercise 19 are equal, and find the exact value of the surface integral.

C **22.** Use a CAS to find the value of the surface integral

$$\iint_{\sigma} x^2yz\,dS$$

where the surface σ is the portion of the elliptic paraboloid $z = 5 - 3x^2 - 2y^2$ that lies above the xy-plane.

23–24 Find the mass of the lamina with constant density δ_0.

23. The lamina that is the portion of the circular cylinder $x^2 + z^2 = 4$ that lies directly above the rectangle $R = \{(x, y) : 0 \le x \le 1, 0 \le y \le 4\}$ in the xy-plane.

24. The lamina that is the portion of the paraboloid $2z = x^2 + y^2$ inside the cylinder $x^2 + y^2 = 8$.

25. Find the mass of the lamina that is the portion of the surface $y^2 = 4 - z$ between the planes $x = 0$, $x = 3$, $y = 0$, and $y = 3$ if the density is $\delta(x, y, z) = y$.

26. Find the mass of the lamina that is the portion of the cone $z = \sqrt{x^2 + y^2}$ between $z = 1$ and $z = 4$ if the density is $\delta(x, y, z) = x^2z$.

27. If a curved lamina has constant density δ_0, what relationship must exist between its mass and surface area? Explain your reasoning.

28. Show that if the density of the lamina $x^2 + y^2 + z^2 = a^2$ at each point is equal to the distance between that point and the xy-plane, then the mass of the lamina is $2\pi a^3$.

29–30 The centroid of a surface σ is defined by

$$\bar{x} = \frac{\displaystyle\iint_{\sigma} x\,dS}{\text{area of }\sigma}, \quad \bar{y} = \frac{\displaystyle\iint_{\sigma} y\,dS}{\text{area of }\sigma}, \quad \bar{z} = \frac{\displaystyle\iint_{\sigma} z\,dS}{\text{area of }\sigma}$$

Find the centroid of the surface.

29. The portion of the paraboloid $z = \frac{1}{2}(x^2 + y^2)$ below the plane $z = 4$.

30. The portion of the sphere $x^2 + y^2 + z^2 = 4$ above the plane $z = 1$.

31–34 Evaluate the integral $\iint_{\sigma} f(x, y, z)\,dS$ over the surface σ represented by the vector-valued function $\mathbf{r}(u, v)$.

31. $f(x, y, z) = xyz$; $\mathbf{r}(u, v) = u\cos v\,\mathbf{i} + u\sin v\,\mathbf{j} + 3u\mathbf{k}$ $(1 \le u \le 2,\ 0 \le v \le \pi/2)$

32. $f(x, y, z) = \dfrac{x^2 + z^2}{y}$; $\mathbf{r}(u, v) = 2\cos v\,\mathbf{i} + u\,\mathbf{j} + 2\sin v\mathbf{k}$ $(1 \le u \le 3,\ 0 \le v \le 2\pi)$

33. $f(x, y, z) = \dfrac{1}{\sqrt{1 + 4x^2 + 4y^2}}$;
$\mathbf{r}(u, v) = u\cos v\,\mathbf{i} + u\sin v\,\mathbf{j} + u^2\mathbf{k}$
$(0 \le u \le \sin v,\ 0 \le v \le \pi)$

34. $f(x, y, z) = e^{-z}$;
$\mathbf{r}(u, v) = 2\sin u\cos v\,\mathbf{i} + 2\sin u\sin v\,\mathbf{j} + 2\cos u\mathbf{k}$
$(0 \le u \le \pi/2, 0 \le v \le 2\pi)$

C **35.** Use a CAS to approximate the mass of the curved lamina $z = e^{-x^2-y^2}$ that lies above the region in the xy-plane enclosed by $x^2 + y^2 = 9$ given that the density function is $\delta(x, y, z) = \sqrt{x^2 + y^2}$.

C **36.** The surface σ shown in the accompanying figure, called a ***Möbius strip***, is represented by the parametric equations

$$x = (5 + u\cos(v/2))\cos v$$
$$y = (5 + u\cos(v/2))\sin v$$
$$z = u\sin(v/2)$$

where $-1 \le u \le 1$ and $0 \le v \le 2\pi$.

(a) Use a CAS to generate a reasonable facsimile of this surface.

(b) Use a CAS to approximate the location of the centroid of σ (see the definition preceding Exercise 29).

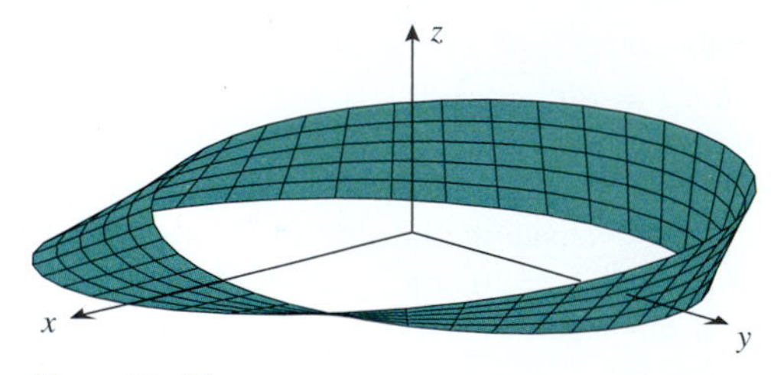

Figure Ex-36

✔ QUICK CHECK ANSWERS 16.5

1. (a) $\left\|\frac{\partial \mathbf{r}}{\partial u} \times \frac{\partial \mathbf{r}}{\partial v}\right\| dA$ (b) $\sqrt{\left(\frac{\partial z}{\partial x}\right)^2 + \left(\frac{\partial z}{\partial y}\right)^2 + 1}\, dA$ **2.** $\frac{\sqrt{3}}{2}$ **3.** 64π **4.** $\iint_{\sigma} f(x, y, z)\, dS$

16.6 APPLICATIONS OF SURFACE INTEGRALS; FLUX

In this section we will discuss applications of surface integrals in vector fields associated with fluid flow and electrostatic forces. However, the ideas that we will develop will be general in nature and applicable to other kinds of vector fields as well.

FLOW FIELDS

We will be concerned in this section with vector fields in 3-space that involve some type of "flow"—the flow of a fluid or the flow of charged particles in an electrostatic field, for example. In the case of fluid flow, the vector field $\mathbf{F}(x, y, z)$ represents the velocity of a fluid particle at the point (x, y, z), and the fluid particles flow along "streamlines" that are tangential to the velocity vectors (Figure 16.6.1a). In the case of an electrostatic field, $\mathbf{F}(x, y, z)$ is the force that the field exerts on a small unit of positive charge at the point (x, y, z), and such charges have acceleration in the directions of "electric lines" that are tangential to the force vectors (Figures 16.6.1b and 16.6.1c).

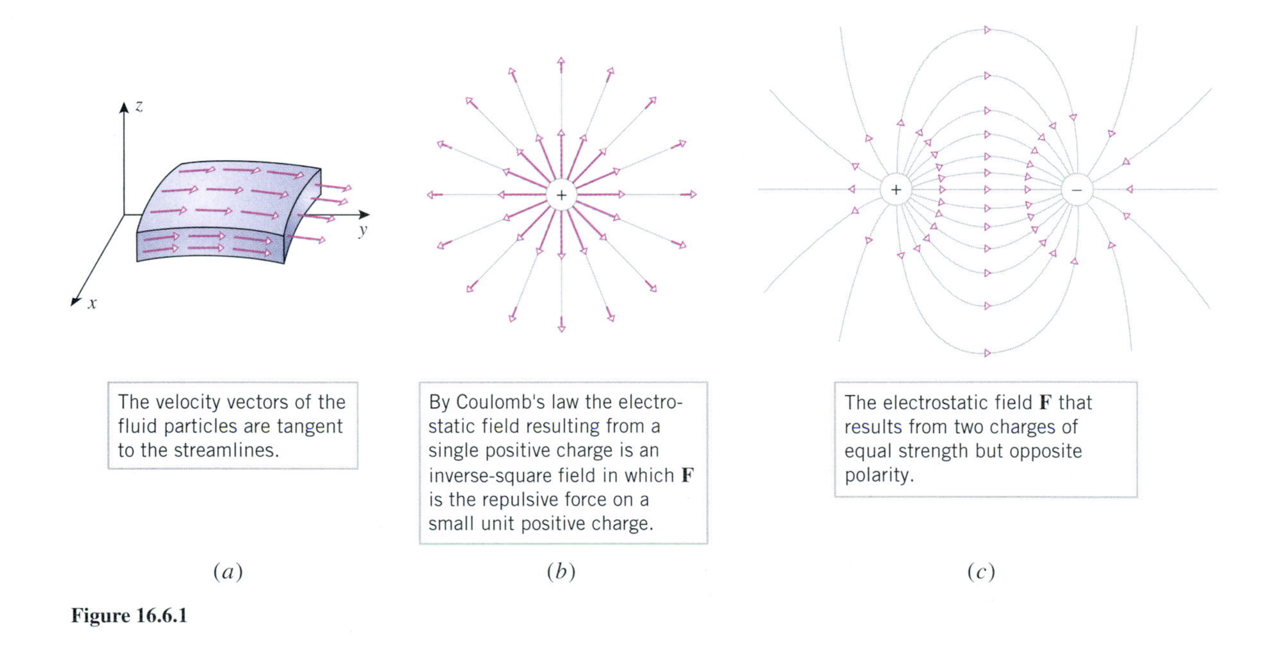

Figure 16.6.1

ORIENTED SURFACES

Our main goal in this section is to study flows of vector fields through permeable surfaces placed in the field. For this purpose we will need to consider some basic ideas about surfaces. Most surfaces that we encounter in applications have two sides—a sphere has an inside and an outside, and an infinite horizontal plane has a top side and a bottom side, for example. However, there exist mathematical surfaces with only one side. For example, Figure 16.6.2a

shows the construction of a surface called a ***Möbius strip*** [in honor of the German mathematician August Möbius (1790–1868)]. The Möbius strip has only one side in the sense that a bug can traverse the *entire* surface without crossing an edge (Figure 16.6.2*b*). In contrast, a sphere is two-sided in the sense that a bug walking on the sphere can traverse the inside surface or the outside surface but cannot traverse both without somehow passing through the sphere. A two-sided surface is said to be ***orientable***, and a one-sided surface is said to be ***nonorientable***. In the rest of this text we will only be concerned with orientable surfaces.

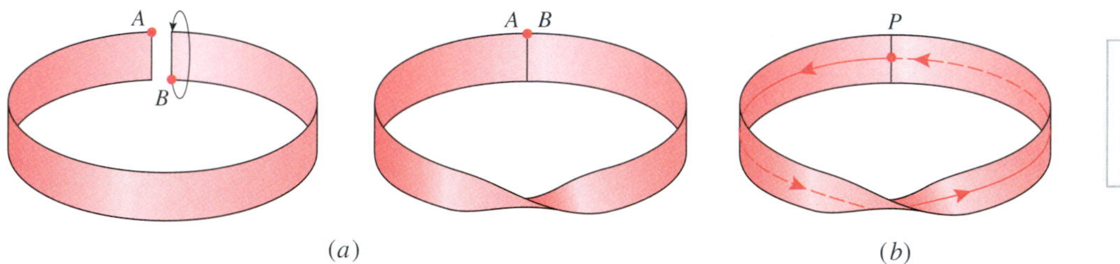

If an ant starts at P with its back facing you and makes one circuit around the strip, then its back will face away from you when it returns to P.

Figure 16.6.2

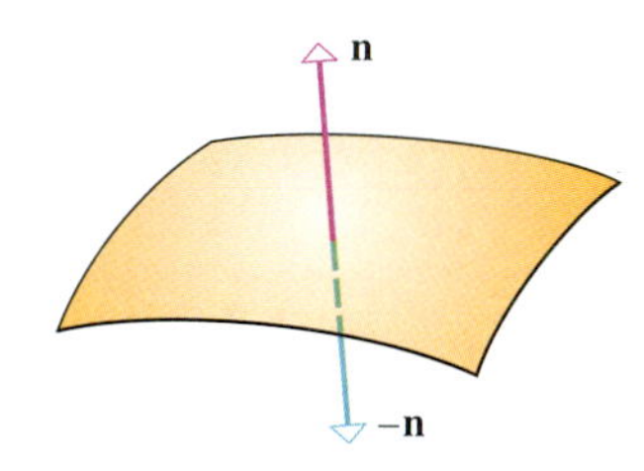

Figure 16.6.3

In applications, it is important to have some way of distinguishing between the two sides of an orientable surface. For this purpose let us suppose that σ is an orientable surface that has a unit normal vector $\mathbf{n}$ at each point. As illustrated in Figure 16.6.3, the vectors $\mathbf{n}$ and $-\mathbf{n}$ point to opposite sides of the surface and hence serve to distinguish between the two sides. It can be proved that if σ is a smooth orientable surface, then it is always possible to choose the direction of $\mathbf{n}$ at each point so that $\mathbf{n} = \mathbf{n}(x, y, z)$ varies continuously over the surface. These unit vectors are then said to form an ***orientation*** of the surface. It can also be proved that a smooth orientable surface has only two possible orientations. For example, the surface in Figure 16.6.4 is oriented up by the purple vectors and down by the green vectors. However, we cannot create a third orientation by mixing the two since this produces points on the surface at which there is an abrupt change in direction (across the black curve in the figure, for example).

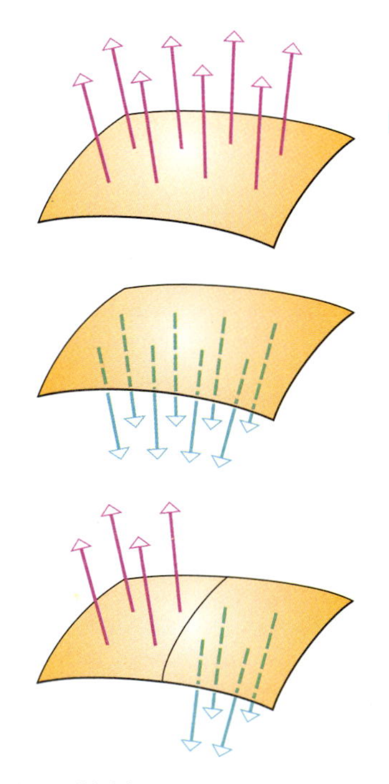

Figure 16.6.4

ORIENTATION OF A SMOOTH PARAMETRIC SURFACE

When a surface is expressed parametrically, the parametric equations create a natural orientation of the surface. To see why this is so, recall from Section 15.4 that if a smooth parametric surface σ is given by the vector equation

$$\mathbf{r} = x(u, v)\mathbf{i} + y(u, v)\mathbf{j} + z(u, v)\mathbf{k}$$

then the unit normal

$$\mathbf{n} = \mathbf{n}(u, v) = \frac{\dfrac{\partial \mathbf{r}}{\partial u} \times \dfrac{\partial \mathbf{r}}{\partial v}}{\left\| \dfrac{\partial \mathbf{r}}{\partial u} \times \dfrac{\partial \mathbf{r}}{\partial v} \right\|} \tag{1}$$

is a continuous vector-valued function of u and v. Thus, Formula (1) defines an orientation of the surface; we call this the ***positive orientation*** of the parametric surface and we say that $\mathbf{n}$ points in the ***positive direction*** from the surface. The orientation determined by $-\mathbf{n}$ is called the ***negative orientation*** of the surface and we say that $-\mathbf{n}$ points in the ***negative direction*** from the surface. For example, consider the cylinder that is represented parametrically by the vector equation

$$\mathbf{r}(u, v) = \cos u\mathbf{i} + v\mathbf{j} - \sin u\mathbf{k} \qquad (0 \le u \le 2\pi,\ 0 \le v \le 1)$$

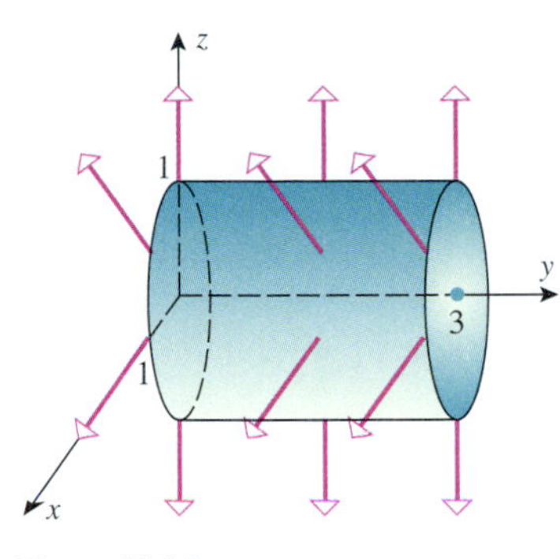

Figure 16.6.5

See if you can find a parametrization of the cylinder in which the positive direction is inward.

Then

$$\frac{\partial \mathbf{r}}{\partial u} \times \frac{\partial \mathbf{r}}{\partial v} = \cos u\mathbf{i} - \sin u\mathbf{k}$$

has unit length, so that Formula (1) becomes

$$\mathbf{n} = \frac{\partial \mathbf{r}}{\partial u} \times \frac{\partial \mathbf{r}}{\partial v} = \cos u\mathbf{i} - \sin u\mathbf{k}$$

Since $\mathbf{n}$ has the same $\mathbf{i}$- and $\mathbf{k}$-components as $\mathbf{r}$, the positive orientation of the cylinder is *outward* and the negative orientation is *inward* (Figure 16.6.5).

FLUX

In physics, the term *fluid* is used to describe both liquids and gases. Liquids are usually regarded to be ***incompressible***, meaning that the liquid has a uniform density (mass per unit volume) that cannot be altered by compressive forces. Gases are regarded to be ***compressible***, meaning that the density may vary from point to point and can be altered by compressive forces. In this text we will be concerned primarily with incompressible fluids. Moreover, we will assume that the velocity of the fluid at a fixed point does not vary with time. Fluid flows with this property are said to be in a ***steady state***.

Our next goal in this section is to define a fundamental concept of physics known as *flux* (from the Latin word *fluxus*, meaning "flow"). This concept is applicable in any vector field, but we will motivate it in the context of steady-state flow of an incompressible fluid. We consider the following problem.

16.6.1 PROBLEM. Suppose that an oriented surface σ is immersed in an incompressible, steady-state fluid flow and that the surface is permeable so that the fluid can flow through it freely in either direction. Find the net volume of fluid Φ that passes through the surface per unit of time, where the net volume is interpreted to mean the volume that passes through the surface in the positive direction minus the volume that passes through the surface in the negative direction.

To solve this problem, suppose that the velocity of the fluid at a point (x, y, z) on the surface σ is given by

$$\mathbf{F}(x, y, z) = f(x, y, z)\mathbf{i} + g(x, y, z)\mathbf{j} + h(x, y, z)\mathbf{k}$$

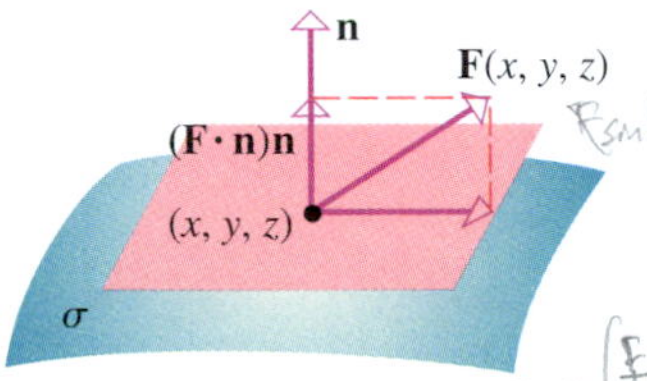

Let $\mathbf{n}$ be the unit normal toward the positive side of σ at the point (x, y, z). As illustrated in Figure 16.6.6, the velocity vector $\mathbf{F}$ can be resolved into two orthogonal components—a component $(\mathbf{F} \cdot \mathbf{n})\mathbf{n}$ that is perpendicular to the surface σ and a second component that is along the "face" of σ. The component of velocity along the face of the surface does not contribute to the flow through σ and hence can be ignored in our computations. Moreover, observe that the sign of $\mathbf{F} \cdot \mathbf{n}$ determines the direction of flow—a positive value means the flow is in the direction of $\mathbf{n}$ and a negative value means that it is opposite to $\mathbf{n}$.

Figure 16.6.6

To solve Problem 16.6.1, we subdivide σ into n patches $\sigma_1, \sigma_2, \ldots, \sigma_n$ with areas

$$\Delta S_1, \Delta S_2, \ldots, \Delta S_n$$

If the patches are small and the flow is not too erratic, it is reasonable to assume that the velocity does not vary much on each patch. Thus, if (x_k^*, y_k^*, z_k^*) is any point in the kth patch, we can assume that $\mathbf{F}(x, y, z)$ is constant and equal to $\mathbf{F}(x_k^*, y_k^*, z_k^*)$ throughout the patch and that the component of velocity across the surface σ_k is

$$\mathbf{F}(x_k^*, y_k^*, z_k^*) \cdot \mathbf{n}(x_k^*, y_k^*, z_k^*) \tag{2}$$

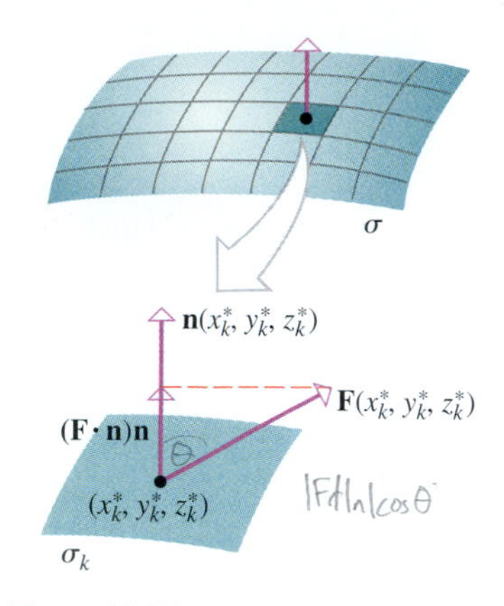

Figure 16.6.7

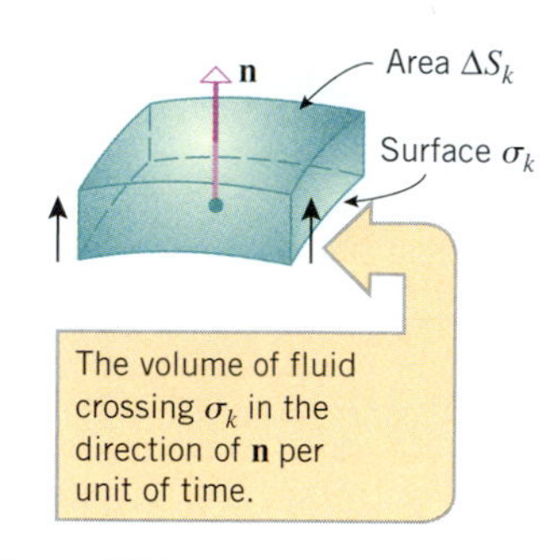

Figure 16.6.8

(Figure 16.6.7). Thus, we can interpret

$$\mathbf{F}(x_k^*, y_k^*, z_k^*) \cdot \mathbf{n}(x_k^*, y_k^*, z_k^*)\Delta S_k$$

as the approximate volume of fluid crossing the patch σ_k in the direction of $\mathbf{n}$ per unit of time (Figure 16.6.8). For example, if the component of velocity in the direction of $\mathbf{n}$ is $\mathbf{F}(x_k^*, y_k^*, z_k^*) \cdot \mathbf{n} = 25$ cm/s, and the area of the patch is $\Delta S_k = 2\text{ cm}^2$, then the volume of fluid ΔV_k crossing the patch in the direction of $\mathbf{n}$ per unit of time is approximately

$$\Delta V_k \approx \mathbf{F}(x_k^*, y_k^*, z_k^*) \cdot \mathbf{n}(x_k^*, y_k^*, z_k^*)\Delta S_k = 25\text{ cm/s} \cdot 2\text{ cm}^2 = 50\text{ cm}^3/\text{s}$$

In the case where the velocity component $\mathbf{F}(x_k^*, y_k^*, z_k^*) \cdot \mathbf{n}(x_k^*, y_k^*, z_k^*)$ is negative, the flow is in the direction opposite to $\mathbf{n}$, so that $-\Delta V_k$ is the approximate volume of fluid crossing the patch σ_k in the direction opposite to $\mathbf{n}$ per unit time. Thus, the sum

$$\sum_{k=1}^{n} \mathbf{F}(x_k^*, y_k^*, z_k^*) \cdot \mathbf{n}(x_k^*, y_k^*, z_k^*)\Delta S_k$$

measures the approximate net volume of fluid that crosses the surface σ in the direction of its orientation $\mathbf{n}$ per unit of time.

If we now increase n in such a way that the maximum dimension of each patch approaches zero, then it is plausible that the errors in the approximations approach zero, and the limit

$$\Phi = \lim_{n\to+\infty} \sum_{k=1}^{n} \mathbf{F}(x_k^*, y_k^*, z_k^*) \cdot \mathbf{n}(x_k^*, y_k^*, z_k^*)\Delta S_k \tag{3}$$

represents the exact net volume of fluid that crosses the surface σ in the direction of its orientation $\mathbf{n}$ per unit of time. The quantity Φ defined by Equation (3) is called the ***flux of F across σ***. The flux can also be expressed as the surface integral

$$\Phi = \iint_{\sigma} \mathbf{F}(x, y, z) \cdot \mathbf{n}(x, y, z)\, dS \tag{4}$$

If the fluid has mass density δ, then $\Phi\delta$ (volume $\times$ density) represents the net mass of fluid that passes through σ per unit of time.

A positive flux means that in one unit of time a greater volume of fluid passes through σ in the positive direction than in the negative direction, a negative flux means that a greater volume passes through the surface in the negative direction than in the positive direction, and a zero flux means that the same volume passes through the surface in each direction. Integrals of form (4) arise in other contexts as well and are called ***flux integrals***.

EVALUATING FLUX INTEGRALS

An effective formula for evaluating flux integrals can be obtained by applying Theorem 16.5.2 and using Formula (1) for $\mathbf{n}$. This yields

$$\begin{aligned}
\iint_{\sigma} \mathbf{F}\cdot\mathbf{n}\, dS &= \iint_{R} \mathbf{F}\cdot\mathbf{n} \left\| \frac{\partial \mathbf{r}}{\partial u} \times \frac{\partial \mathbf{r}}{\partial v} \right\| dA \\
&= \iint_{R} \mathbf{F}\cdot \frac{\dfrac{\partial \mathbf{r}}{\partial u} \times \dfrac{\partial \mathbf{r}}{\partial v}}{\left\| \dfrac{\partial \mathbf{r}}{\partial u} \times \dfrac{\partial \mathbf{r}}{\partial v} \right\|} \left\| \frac{\partial \mathbf{r}}{\partial u} \times \frac{\partial \mathbf{r}}{\partial v} \right\| dA \\
&= \iint_{R} \mathbf{F}\cdot \left(\frac{\partial \mathbf{r}}{\partial u} \times \frac{\partial \mathbf{r}}{\partial v} \right) dA
\end{aligned}$$

In summary, we have the following result.

16.6.2 THEOREM. *Let σ be a smooth parametric surface represented by the vector equation $\mathbf{r} = \mathbf{r}(u, v)$ in which (u, v) varies over a region R in the uv-plane. If the component functions of the vector field $\mathbf{F}$ are continuous on σ, and if $\mathbf{n}$ determines the positive orientation of σ, then*

$$\Phi = \iint_{\sigma} \mathbf{F} \cdot \mathbf{n}\, dS = \iint_{R} \mathbf{F} \cdot \left(\frac{\partial \mathbf{r}}{\partial u} \times \frac{\partial \mathbf{r}}{\partial v} \right) dA \tag{5}$$

where it is understood that the integrand on the right side of the equation is expressed in terms of u and v.

Although Theorem 16.6.2 was derived for smooth parametric surfaces, Formula (5) is valid more generally. For example, as long as σ has a continuous normal vector field $\mathbf{n}$ and the component functions of $\mathbf{r}(u, v)$ have continuous first partial derivatives, Formula (5) can be applied whenever $\partial \mathbf{r}/\partial u \times \partial \mathbf{r}/\partial v$ is a positive multiple of $\mathbf{n}$ in the *interior* of R. (That is, $\partial \mathbf{r}/\partial u \times \partial \mathbf{r}/\partial v$ is allowed to equal $\mathbf{0}$ on the boundary of R.)

Example 1 Find the flux of the vector field $\mathbf{F}(x, y, z) = z\mathbf{k}$ across the outward-oriented sphere $x^2 + y^2 + z^2 = a^2$.

Solution. The sphere with outward positive orientation can be represented by the vector-valued function

$$\mathbf{r}(\phi, \theta) = a \sin\phi \cos\theta \mathbf{i} + a \sin\phi \sin\theta \mathbf{j} + a \cos\phi \mathbf{k} \qquad (0 \le \phi \le \pi,\ \ 0 \le \theta \le 2\pi)$$

From this formula we obtain (see Example 9 of Section 15.4 for the computations)

$$\frac{\partial \mathbf{r}}{\partial \phi} \times \frac{\partial \mathbf{r}}{\partial \theta} = a^2 \sin^2\phi \cos\theta \mathbf{i} + a^2 \sin^2\phi \sin\theta \mathbf{j} + a^2 \sin\phi \cos\phi \mathbf{k}$$

Moreover, for points on the sphere we have $\mathbf{F} = z\mathbf{k} = a\cos\phi \mathbf{k}$; hence,

$$\mathbf{F} \cdot \left(\frac{\partial \mathbf{r}}{\partial \phi} \times \frac{\partial \mathbf{r}}{\partial \theta} \right) = a^3 \sin\phi \cos^2\phi$$

Thus, it follows from (5) with the parameters u and v replaced by ϕ and θ that

$$\begin{aligned}
\Phi &= \iint_{\sigma} \mathbf{F} \cdot \mathbf{n}\, dS \\
&= \iint_{R} \mathbf{F} \cdot \left(\frac{\partial \mathbf{r}}{\partial \phi} \times \frac{\partial \mathbf{r}}{\partial \theta} \right) dA \\
&= \int_0^{2\pi} \int_0^{\pi} a^3 \sin\phi \cos^2\phi \, d\phi \, d\theta \\
&= a^3 \int_0^{2\pi} \left[-\frac{\cos^3\phi}{3} \right]_0^{\pi} d\theta \\
&= \frac{2a^3}{3} \int_0^{2\pi} d\theta = \frac{4\pi a^3}{3}
\end{aligned}$$

Solve Example 1 using symmetry: First argue that the vector fields $x\mathbf{i}$, $y\mathbf{j}$, and $z\mathbf{k}$ will have the same flux across the sphere. Then define

$$\mathbf{H} = x\mathbf{i} + y\mathbf{j} + z\mathbf{k}$$

and explain why

$$\mathbf{H} \cdot \mathbf{n} = a$$

Use this to compute Φ.

> Reversing the orientation of the surface σ in (5) reverses the sign of $\mathbf{n}$, hence the sign of $\mathbf{F}\cdot\mathbf{n}$, and hence reverses the sign of Φ. This can also be seen physically by interpreting the flux integral as the volume of fluid per unit time that crosses σ in the positive direction minus the volume per unit time that crosses in the negative direction—reversing the orientation of σ changes the sign of the difference. Thus, in Example 1 an inward orientation of the sphere would produce a flux of $-4\pi a^3/3$.

ORIENTATION OF NONPARAMETRIC SURFACES

Nonparametric surfaces of the form $z = g(x, y)$, $y = g(z, x)$, and $x = g(y, z)$ can be expressed parametrically using the independent variables as parameters. More precisely, these surfaces can be represented by the vector equations

$$\mathbf{r} = u\mathbf{i} + v\mathbf{j} + g(u, v)\mathbf{k}, \quad \mathbf{r} = v\mathbf{i} + g(u, v)\mathbf{j} + u\mathbf{k}, \quad \mathbf{r} = g(u, v)\mathbf{i} + u\mathbf{j} + v\mathbf{k} \qquad (6\text{–}8)$$

$z = g(x, y)$ $\qquad$ $y = g(z, x)$ $\qquad$ $x = g(y, z)$

These representations impose positive and negative orientations on the surfaces in accordance with Formula (1). We leave it as an exercise to calculate $\mathbf{n}$ and $-\mathbf{n}$ in each case and to show that the positive and negative orientations are as shown in Table 16.6.1. (To assist with perspective, each graph is pictured as a portion of the surface of a small solid region.)

Table 16.6.1

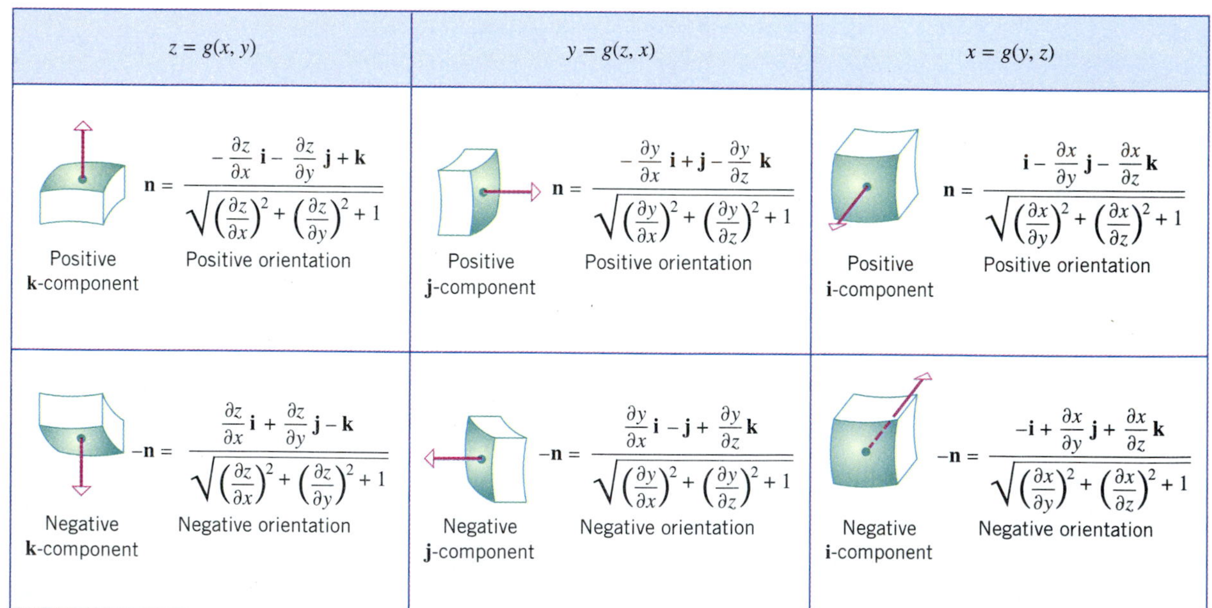

$z = g(x, y)$	$y = g(z, x)$	$x = g(y, z)$
$\mathbf{n} = \dfrac{-\frac{\partial z}{\partial x}\mathbf{i} - \frac{\partial z}{\partial y}\mathbf{j} + \mathbf{k}}{\sqrt{\left(\frac{\partial z}{\partial x}\right)^2 + \left(\frac{\partial z}{\partial y}\right)^2 + 1}}$ Positive **k**-component; Positive orientation	$\mathbf{n} = \dfrac{-\frac{\partial y}{\partial x}\mathbf{i} + \mathbf{j} - \frac{\partial y}{\partial z}\mathbf{k}}{\sqrt{\left(\frac{\partial y}{\partial x}\right)^2 + \left(\frac{\partial y}{\partial z}\right)^2 + 1}}$ Positive **j**-component; Positive orientation	$\mathbf{n} = \dfrac{\mathbf{i} - \frac{\partial x}{\partial y}\mathbf{j} - \frac{\partial x}{\partial z}\mathbf{k}}{\sqrt{\left(\frac{\partial x}{\partial y}\right)^2 + \left(\frac{\partial x}{\partial z}\right)^2 + 1}}$ Positive **i**-component; Positive orientation
$-\mathbf{n} = \dfrac{\frac{\partial z}{\partial x}\mathbf{i} + \frac{\partial z}{\partial y}\mathbf{j} - \mathbf{k}}{\sqrt{\left(\frac{\partial z}{\partial x}\right)^2 + \left(\frac{\partial z}{\partial y}\right)^2 + 1}}$ Negative **k**-component; Negative orientation	$-\mathbf{n} = \dfrac{\frac{\partial y}{\partial x}\mathbf{i} - \mathbf{j} + \frac{\partial y}{\partial z}\mathbf{k}}{\sqrt{\left(\frac{\partial y}{\partial x}\right)^2 + \left(\frac{\partial y}{\partial z}\right)^2 + 1}}$ Negative **j**-component; Negative orientation	$-\mathbf{n} = \dfrac{-\mathbf{i} + \frac{\partial x}{\partial y}\mathbf{j} + \frac{\partial x}{\partial z}\mathbf{k}}{\sqrt{\left(\frac{\partial x}{\partial y}\right)^2 + \left(\frac{\partial x}{\partial z}\right)^2 + 1}}$ Negative **i**-component; Negative orientation

The results in Table 16.6.1 can also be obtained using gradients. To see how this can be done, rewrite the equations of the surfaces as

$$z - g(x, y) = 0, \quad y - g(z, x) = 0, \quad x - g(y, z) = 0$$

> The dependent variable will increase as you move away from a surface
> $$z = g(x, y), \quad y = g(x, z)$$
> or
> $$x = g(y, z)$$
> in the direction of positive orientation.

Each of these equations has the form $G(x, y, z) = 0$ and hence can be viewed as a level surface of a function $G(x, y, z)$. Since the gradient of G is normal to the level surface, it follows that the unit normal $\mathbf{n}$ is either $\nabla G/\|\nabla G\|$ or $-\nabla G/\|\nabla G\|$. However, if $G(x, y, z) = z - g(x, y)$, then ∇G has a $\mathbf{k}$-component of 1; if $G(x, y, z) = y - g(z, x)$, then ∇G has a $\mathbf{j}$-component of 1; and if $G(x, y, z) = x - g(y, z)$, then ∇G has an $\mathbf{i}$-component of 1. Thus, it is evident from Table 16.6.1 that in all three cases we have

$$\mathbf{n} = \frac{\nabla G}{\|\nabla G\|} \tag{9}$$

Moreover, we leave it as an exercise to show that if the surfaces $z = g(x, y)$, $y = g(z, x)$, and $x = g(y, z)$ are expressed in vector forms (6), (7), and (8), then

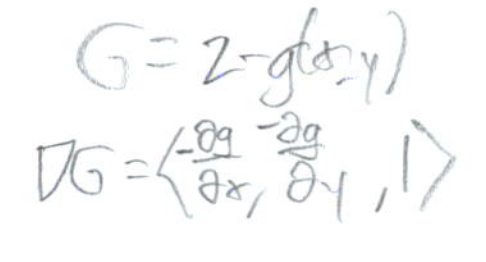

$$\nabla G = \frac{\partial \mathbf{r}}{\partial u} \times \frac{\partial \mathbf{r}}{\partial v} \tag{10}$$

[compare (1) and (9)]. Thus, we are led to the following version of Theorem 16.6.2 for non-parametric surfaces.

16.6.3 THEOREM. *Let σ be a smooth surface of the form $z = g(x, y)$, $y = g(z, x)$, or $x = g(y, z)$, and suppose that the component functions of the vector field $\mathbf{F}$ are continuous on σ. Suppose also that the equation for σ is rewritten as $G(x, y, z) = 0$ by taking g to the left side of the equation, and let R be the projection of σ on the coordinate plane determined by the independent variables of g. If σ has positive orientation, then*

$$\Phi = \iint_{\sigma} \mathbf{F} \cdot \mathbf{n}\, dS = \iint_{R} \mathbf{F} \cdot \nabla G\, dA \tag{11}$$

Formula (11) can either be used directly for computations or to derive some more specific formulas for each of the three surface types. For example, if $z = g(x, y)$, then we have $G(x, y, z) = z - g(x, y)$, so

$$\nabla G = -\frac{\partial g}{\partial x}\mathbf{i} - \frac{\partial g}{\partial y}\mathbf{j} + \mathbf{k} = -\frac{\partial z}{\partial x}\mathbf{i} - \frac{\partial z}{\partial y}\mathbf{j} + \mathbf{k}$$

Substituting this expression for ∇G in (11) and taking R to be the projection of the surface $z = g(x, y)$ on the xy-plane yields

$$\iint_{\sigma} \mathbf{F} \cdot \mathbf{n}\, dS = \iint_{R} \mathbf{F} \cdot \left(-\frac{\partial z}{\partial x}\mathbf{i} - \frac{\partial z}{\partial y}\mathbf{j} + \mathbf{k}\right) dA \tag{12}$$

σ of the form $z = g(x, y)$ and oriented up

$$\iint_{\sigma} \mathbf{F} \cdot \mathbf{n}\, dS = \iint_{R} \mathbf{F} \cdot \left(\frac{\partial z}{\partial x}\mathbf{i} + \frac{\partial z}{\partial y}\mathbf{j} - \mathbf{k}\right) dA \tag{13}$$

σ of the form $z = g(x, y)$ and oriented down

The derivations of the corresponding formulas when $y = g(z, x)$ and $x = g(y, z)$ are left as exercises.

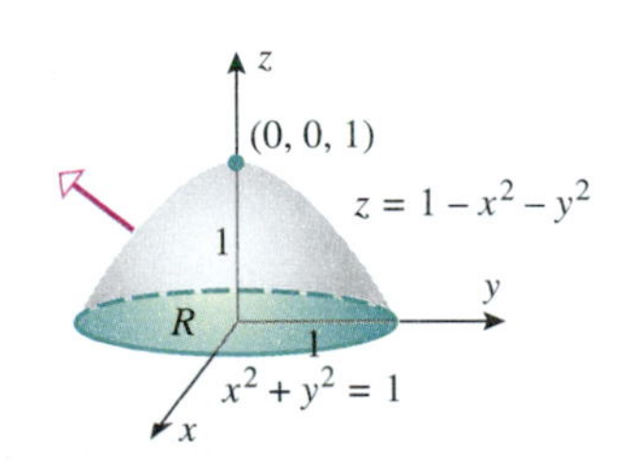

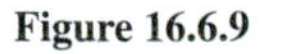

Figure 16.6.9

► **Example 2** Let σ be the portion of the surface $z = 1 - x^2 - y^2$ that lies above the xy-plane, and suppose that σ is oriented up, as shown in Figure 16.6.9. Find the flux of the vector field $\mathbf{F}(x, y, z) = x\mathbf{i} + y\mathbf{j} + z\mathbf{k}$ across σ.

Solution. From (12) the flux Φ is given by

$$\Phi = \iint_{\sigma} \mathbf{F} \cdot \mathbf{n}\, dS = \iint_{R} \mathbf{F} \cdot \left(-\frac{\partial z}{\partial x}\mathbf{i} - \frac{\partial z}{\partial y}\mathbf{j} + \mathbf{k} \right) dA$$

$$= \iint_{R} (x\mathbf{i} + y\mathbf{j} + z\mathbf{k}) \cdot (2x\mathbf{i} + 2y\mathbf{j} + \mathbf{k})\, dA$$

$$= \iint_{R} (x^2 + y^2 + 1)\, dA \qquad \text{Since } z = 1 - x^2 - y^2 \text{ on the surface}$$

$$= \int_0^{2\pi} \int_0^1 (r^2 + 1) r\, dr\, d\theta \qquad \text{Using polar coordinates to evaluate the integral}$$

$$= \int_0^{2\pi} \left(\frac{3}{4} \right) d\theta = \frac{3\pi}{2} \blacktriangleleft$$

✔QUICK CHECK EXERCISES 16.6 (See page 1164 for answers.)

In these exercises, let $\mathbf{F}(x, y, z)$ denote a vector field defined on a surface σ that is oriented by a unit normal vector field $\mathbf{n}(x, y, z)$, and let Φ denote the flux of $\mathbf{F}$ across σ.

1. (a) Φ is the value of the surface integral ________.
   (b) If σ is the unit sphere and $\mathbf{n}$ is the outward unit normal, then the flux of
   $$\mathbf{F}(x, y, z) = x\mathbf{i} + y\mathbf{j} + z\mathbf{k}$$
   across σ is $\Phi =$ ________.
2. (a) Assume that σ is parametrized by a vector-valued function $\mathbf{r}(u, v)$ whose domain is a region R in the uv-plane and that $\mathbf{n}$ is a positive multiple of
   $$\frac{\partial \mathbf{r}}{\partial u} \times \frac{\partial \mathbf{r}}{\partial v}$$
   Then the double integral over R whose value is Φ is ________.
   (b) Suppose that σ is the parametric surface
   $$\mathbf{r}(u, v) = u\mathbf{i} + v\mathbf{j} + (u + v)\mathbf{k} \qquad (0 \le u^2 + v^2 \le 1)$$
   and that $\mathbf{n}$ is a positive multiple of
   $$\frac{\partial \mathbf{r}}{\partial u} \times \frac{\partial \mathbf{r}}{\partial v}$$
   Then the flux of $\mathbf{F}(x, y, z) = x\mathbf{i} + y\mathbf{j} + z\mathbf{k}$ across σ is $\Phi =$ ________.
3. (a) Assume that σ is the graph of a function $z = g(x, y)$ over a region R in the xy-plane and that $\mathbf{n}$ has a positive $\mathbf{k}$-component for every point on σ. Then a double integral over R whose value is Φ is ________.
   (b) Suppose that σ is the triangular region with vertices $(1, 0, 0)$, $(0, 1, 0)$, and $(0, 0, 1)$ with upward orientation. Then the flux of
   $$\mathbf{F}(x, y, z) = x\mathbf{i} + y\mathbf{j} + z\mathbf{k}$$
   across σ is $\Phi =$ ________.
4. In the case of steady-state incompressible fluid flow, with $\mathbf{F}(x, y, z)$ the fluid velocity at (x, y, z) on σ, Φ can be interpreted as ________.

EXERCISE SET 16.6

FOCUS ON CONCEPTS

1. Suppose that the surface σ of the unit cube in the accompanying figure has an outward orientation. In each part, determine whether the flux of the vector field $\mathbf{F}(x, y, z) = z\mathbf{j}$ across the specified face is positive, negative, or zero.
   (a) The face $x = 1$ (b) The face $x = 0$
   (c) The face $y = 1$ (d) The face $y = 0$
   (e) The face $z = 1$ (f) The face $z = 0$

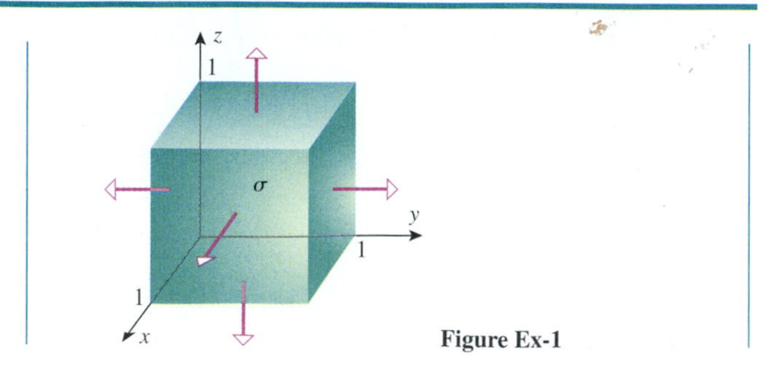

Figure Ex-1

2. Answer the questions posed in Exercise 1 for the vector field $\mathbf{F}(x, y, z) = x\mathbf{i} - z\mathbf{k}$.

3. Answer the questions posed in Exercise 1 for the vector field $\mathbf{F}(x, y, z) = x\mathbf{i} + y\mathbf{j} + z\mathbf{k}$.

4. Find the flux of the constant vector field $\mathbf{F}(x, y, z) = \mathbf{i}$ across the entire surface σ in Figure Ex-1. Explain your reasoning.

5. Let σ be the cylindrical surface that is represented by the vector-valued function $\mathbf{r}(u, v) = \cos v\mathbf{i} + \sin v\mathbf{j} + u\mathbf{k}$ with $0 \le u \le 1$ and $0 \le v \le 2\pi$.
   (a) Find the unit normal $\mathbf{n} = \mathbf{n}(u, v)$ that defines the positive orientation of σ.
   (b) Is the positive orientation inward or outward? Justify your answer.

6. Let σ be the conical surface that is represented by the parametric equations $x = r\cos\theta$, $y = r\sin\theta$, $z = r$ with $1 \le r \le 2$ and $0 \le \theta \le 2\pi$.
   (a) Find the unit normal $\mathbf{n} = \mathbf{n}(r, \theta)$ that defines the positive orientation of σ.
   (b) Is the positive orientation upward or downward? Justify your answer.

7–12 Find the flux of the vector field $\mathbf{F}$ across σ.

7. $\mathbf{F}(x, y, z) = x\mathbf{i} + y\mathbf{j} + 2z\mathbf{k}$; σ is the portion of the surface $z = 1 - x^2 - y^2$ above the xy-plane, oriented by upward normals.

8. $\mathbf{F}(x, y, z) = (x + y)\mathbf{i} + (y + z)\mathbf{j} + (z + x)\mathbf{k}$; σ is the portion of the plane $x + y + z = 1$ in the first octant, oriented by unit normals with positive components.

9. $\mathbf{F}(x, y, z) = x\mathbf{i} + y\mathbf{j} + 2z\mathbf{k}$; σ is the portion of the cone $z^2 = x^2 + y^2$ between the planes $z = 1$ and $z = 2$, oriented by upward unit normals.

10. $\mathbf{F}(x, y, z) = y\mathbf{j} + \mathbf{k}$; σ is the portion of the paraboloid $z = x^2 + y^2$ below the plane $z = 4$, oriented by downward unit normals.

11. $\mathbf{F}(x, y, z) = x\mathbf{k}$; the surface σ is the portion of the paraboloid $z = x^2 + y^2$ below the plane $z = y$, oriented by downward unit normals.

12. $\mathbf{F}(x, y, z) = x^2\mathbf{i} + yx\mathbf{j} + zx\mathbf{k}$; σ is the portion of the plane $6x + 3y + 2z = 6$ in the first octant, oriented by unit normals with positive components.

13–16 Find the flux of the vector field $\mathbf{F}$ across σ in the direction of positive orientation.

13. $\mathbf{F}(x, y, z) = x\mathbf{i} + y\mathbf{j} + \mathbf{k}$; σ is the portion of the paraboloid

$$\mathbf{r}(u, v) = u\cos v\mathbf{i} + u\sin v\mathbf{j} + (1 - u^2)\mathbf{k}$$

with $1 \le u \le 2$, $0 \le v \le 2\pi$.

14. $\mathbf{F}(x, y, z) = e^{-y}\mathbf{i} - y\mathbf{j} + x\sin z\mathbf{k}$; σ is the portion of the elliptic cylinder

$$\mathbf{r}(u, v) = 2\cos v\mathbf{i} + \sin v\mathbf{j} + u\mathbf{k}$$

with $0 \le u \le 5$, $0 \le v \le 2\pi$.

15. $\mathbf{F}(x, y, z) = \sqrt{x^2 + y^2}\,\mathbf{k}$; σ is the portion of the cone

$$\mathbf{r}(u, v) = u\cos v\mathbf{i} + u\sin v\mathbf{j} + 2u\mathbf{k}$$

with $0 \le u \le \sin v$, $0 \le v \le \pi$.

16. $\mathbf{F}(x, y, z) = x\mathbf{i} + y\mathbf{j} + z\mathbf{k}$; σ is the portion of the sphere

$$\mathbf{r}(u, v) = 2\sin u\cos v\mathbf{i} + 2\sin u\sin v\mathbf{j} + 2\cos u\mathbf{k}$$

with $0 \le u \le \pi/3$, $0 \le v \le 2\pi$.

17. Let σ be the surface of the cube bounded by the planes $x = \pm 1$, $y = \pm 1$, $z = \pm 1$, oriented by outward unit normals. In each part, find the flux of $\mathbf{F}$ across σ.
   (a) $\mathbf{F}(x, y, z) = x\mathbf{i}$
   (b) $\mathbf{F}(x, y, z) = x\mathbf{i} + y\mathbf{j} + z\mathbf{k}$
   (c) $\mathbf{F}(x, y, z) = x^2\mathbf{i} + y^2\mathbf{j} + z^2\mathbf{k}$

18. Let σ be the closed surface consisting of the portion of the paraboloid $z = x^2 + y^2$ for which $0 \le z \le 1$ and capped by the disk $x^2 + y^2 \le 1$ in the plane $z = 1$. Find the flux of the vector field $\mathbf{F}(x, y, z) = z\mathbf{j} - y\mathbf{k}$ in the outward direction across σ.

19–20 Find the flux of $\mathbf{F}$ across the surface σ by expressing σ parametrically.

19. $\mathbf{F}(x, y, z) = \mathbf{i} + \mathbf{j} + \mathbf{k}$; the surface σ is the portion of the cone $z = \sqrt{x^2 + y^2}$ between the planes $z = 1$ and $z = 2$, oriented by downward unit normals.

20. $\mathbf{F}(x, y, z) = x\mathbf{i} + y\mathbf{j} + z\mathbf{k}$; σ is the portion of the cylinder $x^2 + z^2 = 1$ between the planes $y = 1$ and $y = -2$, oriented by outward unit normals.

21. Let x, y, and z be measured in meters, and suppose that $\mathbf{F}(x, y, z) = 2x\mathbf{i} - 3y\mathbf{j} + z\mathbf{k}$ is the velocity vector (in m/s) of a fluid particle at the point (x, y, z) in a steady-state fluid flow.
   (a) Find the net volume of fluid that passes in the upward direction through the portion of the plane $x + y + z = 1$ in the first octant in 1 s.
   (b) Assuming that the fluid has a mass density of 806 kg/m^3, find the net mass of fluid that passes in the upward direction through the surface in part (a) in 1 s.

22. Let x, y, and z be measured in meters, and suppose that $\mathbf{F}(x, y, z) = -y\mathbf{i} + z\mathbf{j} + 3x\mathbf{k}$ is the velocity vector (in m/s) of a fluid particle at the point (x, y, z) in a steady-state incompressible fluid flow.
   (a) Find the net volume of fluid that passes in the upward direction through the hemisphere $z = \sqrt{9 - x^2 - y^2}$ in 1 s.
   (b) Assuming that the fluid has a mass density of 1060 kg/m^3, find the net mass of fluid that passes in the upward direction through the surface in part (a) in 1 s.

23. (a) Derive the analogs of Formulas (12) and (13) for surfaces of the form $x = g(y, z)$.
(b) Let σ be the portion of the paraboloid $x = y^2 + z^2$ for $x \leq 1$ and $z \geq 0$ oriented by unit normals with negative x-components. Use the result in part (a) to find the flux of
$$\mathbf{F}(x, y, z) = y\mathbf{i} - z\mathbf{j} + 8\mathbf{k}$$
across σ.

24. (a) Derive the analogs of Formulas (12) and (13) for surfaces of the form $y = g(z, x)$.
(b) Let σ be the portion of the paraboloid $y = z^2 + x^2$ for $y \leq 1$ and $z \geq 0$ oriented by unit normals with positive y-components. Use the result in part (a) to find the flux of
$$\mathbf{F}(x, y, z) = x\mathbf{i} + y\mathbf{j} + z\mathbf{k}$$
across σ.

25. Let $\mathbf{F} = \|\mathbf{r}\|^k\mathbf{r}$, where $\mathbf{r} = x\mathbf{i} + y\mathbf{j} + z\mathbf{k}$ and k is a constant. (Note that if $k = -3$, this is an inverse-square field.) Let σ be the sphere of radius a centered at the origin and oriented by the outward normal $\mathbf{n} = \mathbf{r}/\|\mathbf{r}\| = \mathbf{r}/a$.
(a) Find the flux of $\mathbf{F}$ across σ without performing any integrations. [*Hint:* The surface area of a sphere of radius a is $4\pi a^2$.]
(b) For what value of k is the flux independent of the radius of the sphere?

26. Let
$$\mathbf{F}(x, y, z) = a^2x\mathbf{i} + (y/a)\mathbf{j} + az^2\mathbf{k}$$
and let σ be the sphere of radius 1, centered at the origin and oriented outward. Approximate all values of a such that the flux of $\mathbf{F}$ across σ is 10.

✔ QUICK CHECK ANSWERS 16.6

1. (a) $\iint_\sigma \mathbf{F} \cdot \mathbf{n}\, dS$ (b) 4π **2.** (a) $\iint_R \mathbf{F} \cdot \left(\frac{\partial \mathbf{r}}{\partial u} \times \frac{\partial \mathbf{r}}{\partial v}\right) dA$ (b) 0 **3.** (a) $\iint_R \mathbf{F} \cdot \left(-\frac{\partial z}{\partial x}\mathbf{i} - \frac{\partial z}{\partial y}\mathbf{j} + \mathbf{k}\right) dA$ (b) $\frac{1}{2}$
4. the net volume of fluid crossing σ in the positive direction per unit time

16.7 THE DIVERGENCE THEOREM

In this section we will be concerned with flux across surfaces, such as spheres, that "enclose" a region of space. We will show that the flux across such surfaces can be expressed in terms of the divergence of the vector field, and we will use this result to give a physical interpretation of the concept of divergence.

ORIENTATION OF PIECEWISE SMOOTH CLOSED SURFACES

In the last section we studied flux across general surfaces. Here we will be concerned exclusively with surfaces that are boundaries of finite solids—the surface of a solid sphere, the surface of a solid box, or the surface of a solid cylinder, for example. Such surfaces are said to be ***closed***. A closed surface may or may not be smooth, but most of the surfaces that arise in applications are ***piecewise smooth***; that is, they consist of finitely many smooth surfaces joined together at the edges (a box, for example). We will limit our discussion to piecewise smooth surfaces that can be assigned an ***inward orientation*** (toward the interior of the solid) and an ***outward orientation*** (away from the interior). It is very difficult to make this concept mathematically precise, but the basic idea is that each piece of the surface is orientable, and oriented pieces fit together in such a way that the entire surface can be assigned an orientation (Figure 16.7.1).

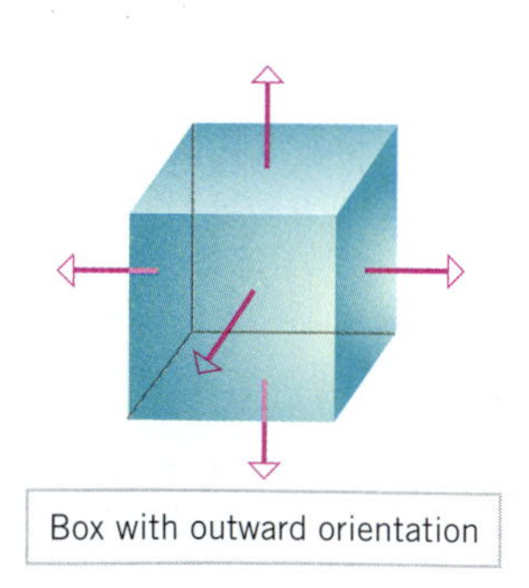

Box with outward orientation

Figure 16.7.1

THE DIVERGENCE THEOREM

In Section 16.1 we defined the divergence of a vector field
$$\mathbf{F}(x, y, z) = f(x, y, z)\mathbf{i} + g(x, y, z)\mathbf{j} + h(x, y, z)\mathbf{k}$$
as
$$\text{div}\,\mathbf{F} = \frac{\partial f}{\partial x} + \frac{\partial g}{\partial y} + \frac{\partial h}{\partial z}$$
but we did not attempt to give a physical explanation of its meaning at that time. The following result, known as the ***Divergence Theorem*** or ***Gauss's Theorem***, will provide us with a physical interpretation of divergence in the context of fluid flow.

16.7.1 THEOREM (*The Divergence Theorem*). *Let G be a solid whose surface σ is oriented outward. If*

$$\mathbf{F}(x, y, z) = f(x, y, z)\mathbf{i} + g(x, y, z)\mathbf{j} + h(x, y, z)\mathbf{k}$$

*where f, g, and h have continuous first partial derivatives on some open set containing G, and if **n** is the outward unit normal on σ, then*

$$\iint_{\sigma} \mathbf{F} \cdot \mathbf{n}\, dS = \iiint_{G} \operatorname{div} \mathbf{F}\, dV \tag{1}$$

Carl Friedrich Gauss (1777–1855) German mathematician and scientist. Sometimes called the "prince of mathematicians," Gauss ranks with Newton and Archimedes as one of the three greatest mathematicians who ever lived. His father, a laborer, was an uncouth but honest man who would have liked Gauss to take up a trade such as gardening or bricklaying; but the boy's genius for mathematics was not to be denied. In the entire history of mathematics there may never have been a child so precocious as Gauss—by his own account he worked out the rudiments of arithmetic before he could talk. One day, before he was even three years old, his genius became apparent to his parents in a very dramatic way. His father was preparing the weekly payroll for the laborers under his charge while the boy watched quietly from a corner. At the end of the long and tedious calculation, Gauss informed his father that there was an error in the result and stated the answer, which he had worked out in his head. To the astonishment of his parents, a check of the computations showed Gauss to be correct!

For his elementary education Gauss was enrolled in a squalid school run by a man named Büttner whose main teaching technique was thrashing. Büttner was in the habit of assigning long addition problems which, unknown to his students, were arithmetic progressions that he could sum up using formulas. On the first day that Gauss entered the arithmetic class, the students were asked to sum the numbers from 1 to 100. But no sooner had Büttner stated the problem than Gauss turned over his slate and exclaimed in his peasant dialect, "Ligget se'." (Here it lies.) For nearly an hour Büttner glared at Gauss, who sat with folded hands while his classmates toiled away. When Büttner examined the slates at the end of the period, Gauss's slate contained a single number, 5050—the only correct solution in the class. To his credit, Büttner recognized the genius of Gauss and with the help of his assistant, John Bartels, had him brought to the attention of Karl Wilhelm Ferdinand, Duke of Brunswick. The shy and awkward boy, who was then fourteen, so captivated the Duke that he subsidized him through preparatory school, college, and the early part of his career.

From 1795 to 1798 Gauss studied mathematics at the University of Göttingen, receiving his degree in absentia from the University of Helmstadt. For his dissertation, he gave the first complete proof of the fundamental theorem of algebra, which states that every polynomial equation has as many solutions as its degree. At age 19 he solved a problem that baffled Euclid, inscribing a regular polygon of 17 sides in a circle using straightedge and compass; and in 1801, at age 24, he published his first masterpiece, *Disquisitiones Arithmeticae*, considered by many to be one of the most brilliant achievements in mathematics. In that book Gauss systematized the study of number theory (properties of the integers) and formulated the basic concepts that form the foundation of that subject.

In the same year that the *Disquisitiones* was published, Gauss again applied his phenomenal computational skills in a dramatic way. The astronomer Giuseppi Piazzi had observed the asteroid Ceres for $\frac{1}{40}$ of its orbit, but lost it in the Sun. Using only three observations and the "method of least squares" that he had developed in 1795, Gauss computed the orbit with such accuracy that astronomers had no trouble relocating it the following year. This achievement brought him instant recognition as the premier mathematician in Europe, and in 1807 he was made Professor of Astronomy and head of the astronomical observatory at Göttingen.

In the years that followed, Gauss revolutionized mathematics by bringing to it standards of precision and rigor undreamed of by his predecessors. He had a passion for perfection that drove him to polish and rework his papers rather than publish less finished work in greater numbers—his favorite saying was "Pauca, sed matura" (Few, but ripe). As a result, many of his important discoveries were squirreled away in diaries that remained unpublished until years after his death.

Among his myriad achievements, Gauss discovered the Gaussian or "bell-shaped" error curve fundamental in probability, gave the first geometric interpretation of complex numbers and established their fundamental role in mathematics, developed methods of characterizing surfaces intrinsically by means of the curves that they contain, developed the theory of conformal (angle-preserving) maps, and discovered non-Euclidean geometry 30 years before the ideas were published by others. In physics he made major contributions to the theory of lenses and capillary action, and with Wilhelm Weber he did fundamental work in electromagnetism. Gauss invented the heliotrope, bifilar magnetometer, and an electrotelegraph.

Gauss was deeply religious and aristocratic in demeanor. He mastered foreign languages with ease, read extensively, and enjoyed mineralogy and botany as hobbies. He disliked teaching and was usually cool and discouraging to other mathematicians, possibly because he had already anticipated their work. It has been said that if Gauss had published all of his discoveries, the current state of mathematics would be advanced by 50 years. He was without a doubt the greatest mathematician of the modern era.

The proof of this theorem for a general solid G is too difficult to present here. However, we can give a proof for the special case where G is simultaneously a simple xy-solid, a simple yz-solid, and a simple zx-solid (see Figure 15.5.3 and the related discussion for terminology).

PROOF. Formula (1) can be expressed as

$$\iint_{\sigma} [f(x, y, z)\mathbf{i} + g(x, y, z)\mathbf{j} + h(x, y, z)\mathbf{k}] \cdot \mathbf{n}\, dS = \iiint_{G} \left(\frac{\partial f}{\partial x} + \frac{\partial g}{\partial y} + \frac{\partial h}{\partial z}\right) dV$$

so it suffices to prove the three equalities

$$\iint_{\sigma} f(x, y, z)\mathbf{i} \cdot \mathbf{n}\, dS = \iiint_{G} \frac{\partial f}{\partial x}\, dV \tag{2a}$$

$$\iint_{\sigma} g(x, y, z)\mathbf{j} \cdot \mathbf{n}\, dS = \iiint_{G} \frac{\partial g}{\partial y}\, dV \tag{2b}$$

$$\iint_{\sigma} h(x, y, z)\mathbf{k} \cdot \mathbf{n}\, dS = \iiint_{G} \frac{\partial h}{\partial z}\, dV \tag{2c}$$

Since the proofs of all three equalities are similar, we will prove only the third.

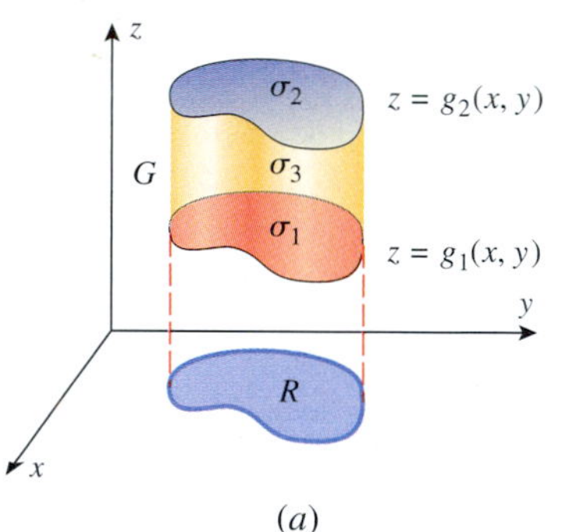

(a)

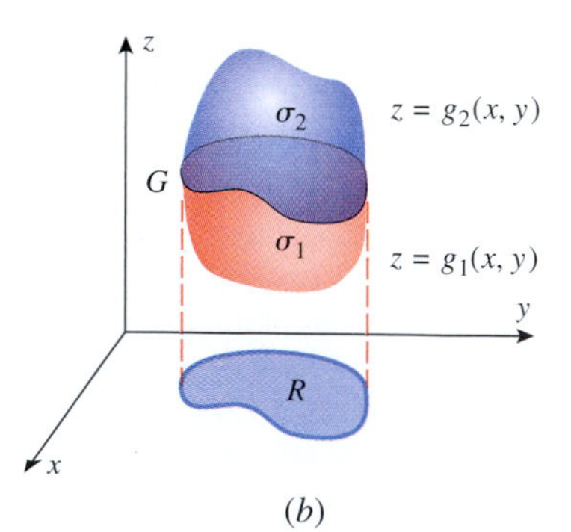

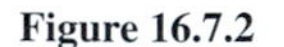
(b)

Figure 16.7.2

Suppose that G has upper surface $z = g_2(x, y)$, lower surface $z = g_1(x, y)$, and projection R on the xy-plane. Let σ_1 denote the lower surface, σ_2 the upper surface, and σ_3 the lateral surface (Figure 16.7.2a). If the upper surface and lower surface meet as in Figure 16.7.2b, then there is no lateral surface σ_3. Our proof will allow for both cases shown in those figures.

It follows from Theorem 15.5.2 that

$$\iiint_{G} \frac{\partial h}{\partial z}\, dV = \iint_{R} \left[\int_{g_1(x,y)}^{g_2(x,y)} \frac{\partial h}{\partial z}\, dz\right] dA = \iint_{R} \Big[h(x, y, z)\Big]_{z=g_1(x,y)}^{g_2(x,y)} dA$$

so

$$\iiint_{G} \frac{\partial h}{\partial z}\, dV = \iint_{R} [h(x, y, g_2(x, y)) - h(x, y, g_1(x, y))]\, dA \tag{3}$$

Next we will evaluate the surface integral in (2c) by integrating over each surface of G separately. If there is a lateral surface σ_3, then at each point of this surface $\mathbf{k} \cdot \mathbf{n} = 0$ since $\mathbf{n}$ is horizontal and $\mathbf{k}$ is vertical. Thus,

$$\iint_{\sigma_3} h(x, y, z)\mathbf{k} \cdot \mathbf{n}\, dS = 0$$

Therefore, regardless of whether G has a lateral surface, we can write

$$\iint_{\sigma} h(x, y, z)\mathbf{k} \cdot \mathbf{n}\, dS = \iint_{\sigma_1} h(x, y, z)\mathbf{k} \cdot \mathbf{n}\, dS + \iint_{\sigma_2} h(x, y, z)\mathbf{k} \cdot \mathbf{n}\, dS \tag{4}$$

On the upper surface σ_2, the outer normal is an upward normal, and on the lower surface σ_1, the outer normal is a downward normal. Thus, Formulas (12) and (13) of Section 16.6 imply that

$$\begin{aligned}\iint_{\sigma_2} h(x, y, z)\mathbf{k} \cdot \mathbf{n}\, dS &= \iint_{R} h(x, y, g_2(x, y))\mathbf{k} \cdot \left(-\frac{\partial z}{\partial x}\mathbf{i} - \frac{\partial z}{\partial y}\mathbf{j} + \mathbf{k}\right) dA \\ &= \iint_{R} h(x, y, g_2(x, y))\, dA\end{aligned} \tag{5}$$

and

$$\iint_{\sigma_1} h(x, y, z)\mathbf{k} \cdot \mathbf{n}\, dS = \iint_R h(x, y, g_1(x, y))\mathbf{k} \cdot \left(\frac{\partial z}{\partial x}\mathbf{i} + \frac{\partial z}{\partial y}\mathbf{j} - \mathbf{k}\right) dA$$
$$= -\iint_R h(x, y, g_1(x, y))\, dA \tag{6}$$

Substituting (5) and (6) into (4) and combining the terms into a single integral yields

$$\iint_{\sigma} h(x, y, z)\mathbf{k} \cdot \mathbf{n}\, dS = \iint_R [h(x, y, g_2(x, y)) - h(x, y, g_1(x, y))]\, dA \tag{7}$$

Explain how the derivation of (2c) should be modified to yield a proof of (2a) or (2b).

Equation (2c) now follows from (3) and (7). ■

In words, the Divergence Theorem states:

The flux of a vector field across a closed surface with outward orientation is equal to the triple integral of the divergence over the region enclosed by the surface.

This is sometimes called the ***outward flux*** across the surface.

USING THE DIVERGENCE THEOREM TO FIND FLUX

Sometimes it is easier to find the flux across a closed surface by using the Divergence Theorem than by evaluating the flux integral directly. This is illustrated in the following example.

▶ Example 1 Use the Divergence Theorem to find the outward flux of the vector field $\mathbf{F}(x, y, z) = z\mathbf{k}$ across the sphere $x^2 + y^2 + z^2 = a^2$.

Solution. Let σ denote the outward-oriented spherical surface and G the region that it encloses. The divergence of the vector field is

$$\operatorname{div} \mathbf{F} = \frac{\partial z}{\partial z} = 1$$

so from (1) the flux across σ is

$$\Phi = \iint_{\sigma} \mathbf{F} \cdot \mathbf{n}\, dS = \iiint_G dV = \text{volume of } G = \frac{4\pi a^3}{3}$$

Note how much simpler this calculation is than that in Example 1 of Section 16.6. ◀

The Divergence Theorem is usually the method of choice for finding the flux across closed piecewise smooth surfaces with multiple sections, since it eliminates the need for a separate integral evaluation over each section. This is illustrated in the next three examples.

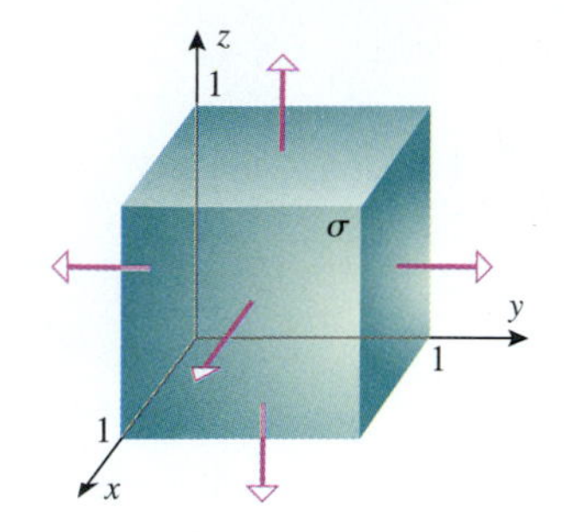

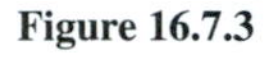
Figure 16.7.3

Let $\mathbf{F}(x, y, z)$ be the vector field in Example 2 and show that $\mathbf{F} \cdot \mathbf{n}$ is constant on each of the six faces of the cube in Figure 16.7.3. Use your computations to confirm the result in Example 2.

► **Example 2** Use the Divergence Theorem to find the outward flux of the vector field

$$\mathbf{F}(x, y, z) = 2x\mathbf{i} + 3y\mathbf{j} + z^2\mathbf{k}$$

across the unit cube in Figure 16.7.3.

Solution. Let σ denote the outward-oriented surface of the cube and G the region that it encloses. The divergence of the vector field is

$$\operatorname{div} \mathbf{F} = \frac{\partial}{\partial x}(2x) + \frac{\partial}{\partial y}(3y) + \frac{\partial}{\partial z}(z^2) = 5 + 2z$$

so from (1) the flux across σ is

$$\Phi = \iint_{\sigma} \mathbf{F} \cdot \mathbf{n}\, dS = \iiint_{G} (5 + 2z)\, dV = \int_0^1 \int_0^1 \int_0^1 (5 + 2z)\, dz\, dy\, dx$$

$$= \int_0^1 \int_0^1 \left[5z + z^2\right]_{z=0}^{1} dy\, dx = \int_0^1 \int_0^1 6\, dy\, dx = 6 \blacktriangleleft$$

► **Example 3** Use the Divergence Theorem to find the outward flux of the vector field

$$\mathbf{F}(x, y, z) = x^3\mathbf{i} + y^3\mathbf{j} + z^2\mathbf{k}$$

across the surface of the region that is enclosed by the circular cylinder $x^2 + y^2 = 9$ and the planes $z = 0$ and $z = 2$ (Figure 16.7.4).

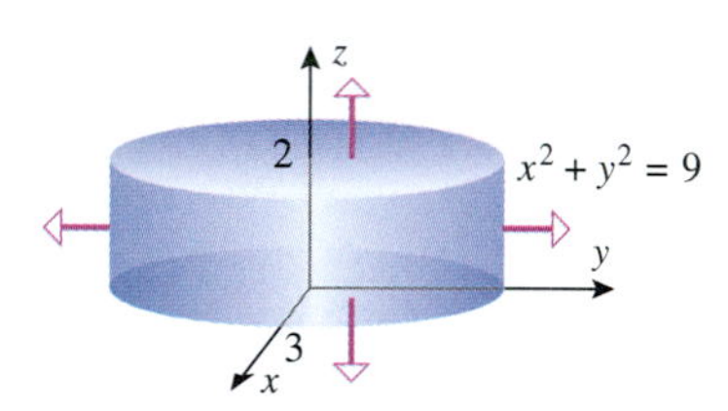

Figure 16.7.4

Solution. Let σ denote the outward-oriented surface and G the region that it encloses. The divergence of the vector field is

$$\operatorname{div} \mathbf{F} = \frac{\partial}{\partial x}(x^3) + \frac{\partial}{\partial y}(y^3) + \frac{\partial}{\partial z}(z^2) = 3x^2 + 3y^2 + 2z$$

so from (1) the flux across σ is

$$\Phi = \iint_{\sigma} \mathbf{F} \cdot \mathbf{n}\, dS = \iiint_{G} (3x^2 + 3y^2 + 2z)\, dV$$

$$= \int_0^{2\pi} \int_0^3 \int_0^2 (3r^2 + 2z) r\, dz\, dr\, d\theta$$

Using cylindrical coordinates

$$= \int_0^{2\pi} \int_0^3 \left[3r^3 z + z^2 r\right]_{z=0}^{2} dr\, d\theta$$

$$= \int_0^{2\pi} \int_0^3 (6r^3 + 4r)\, dr\, d\theta$$

$$= \int_0^{2\pi} \left[\frac{3r^4}{2} + 2r^2\right]_0^3 d\theta$$

$$= \int_0^{2\pi} \frac{279}{2}\, d\theta = 279\pi \blacktriangleleft$$

► **Example 4** Use the Divergence Theorem to find the outward flux of the vector field

$$\mathbf{F}(x, y, z) = x^3\mathbf{i} + y^3\mathbf{j} + z^3\mathbf{k}$$

across the surface of the region that is enclosed by the hemisphere $z = \sqrt{a^2 - x^2 - y^2}$ and the plane $z = 0$ (Figure 16.7.5).

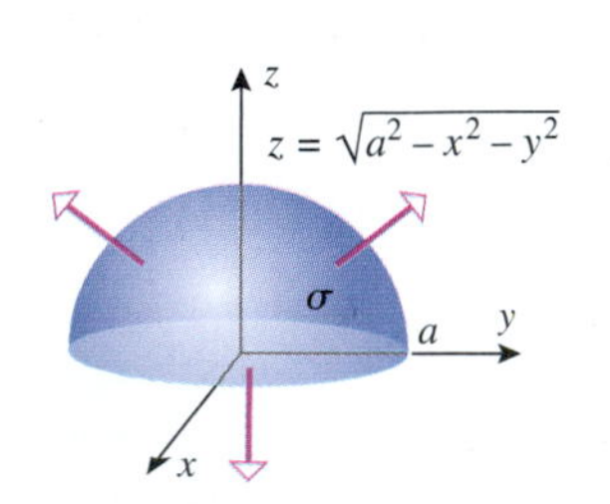

Figure 16.7.5

Solution. Let σ denote the outward-oriented surface and G the region that it encloses. The divergence of the vector field is

$$\operatorname{div} \mathbf{F} = \frac{\partial}{\partial x}(x^3) + \frac{\partial}{\partial y}(y^3) + \frac{\partial}{\partial z}(z^3) = 3x^2 + 3y^2 + 3z^2$$

so from (1) the flux across σ is

$$\begin{aligned}\Phi = \iint_{\sigma} \mathbf{F} \cdot \mathbf{n}\, dS &= \iiint_G (3x^2 + 3y^2 + 3z^2)\, dV \\ &= \int_0^{2\pi}\int_0^{\pi/2}\int_0^a (3\rho^2)\rho^2 \sin\phi \, d\rho\, d\phi\, d\theta \\ &= 3\int_0^{2\pi}\int_0^{\pi/2}\int_0^a \rho^4 \sin\phi\, d\rho\, d\phi\, d\theta \\ &= 3\int_0^{2\pi}\int_0^{\pi/2}\left[\frac{\rho^5}{5}\sin\phi\right]_{\rho=0}^{a} d\phi\, d\theta \\ &= \frac{3a^5}{5}\int_0^{2\pi}\int_0^{\pi/2} \sin\phi\, d\phi\, d\theta \\ &= \frac{3a^5}{5}\int_0^{2\pi} \big[-\cos\phi\big]_0^{\pi/2} d\theta \\ &= \frac{3a^5}{5}\int_0^{2\pi} d\theta = \frac{6\pi a^5}{5} \blacktriangleleft\end{aligned}$$

Using spherical coordinates

DIVERGENCE VIEWED AS FLUX DENSITY

The Divergence Theorem provides a way of interpreting the divergence of a vector field $\mathbf{F}$. Suppose that G is a *small* spherical region centered at the point P_0 and that its surface, denoted by $\sigma(G)$, is oriented outward. Denote the volume of the region by $\operatorname{vol}(G)$ and the flux of $\mathbf{F}$ across $\sigma(G)$ by $\Phi(G)$. If $\operatorname{div} \mathbf{F}$ is continuous on G, then across the small region G the value of $\operatorname{div} \mathbf{F}$ will not vary much from its value $\operatorname{div} \mathbf{F}(P_0)$ at the center, and we can reasonably approximate $\operatorname{div} \mathbf{F}$ by the constant $\operatorname{div} \mathbf{F}(P_0)$ on G. Thus, the Divergence Theorem implies that the flux $\Phi(G)$ of $\mathbf{F}$ across $\sigma(G)$ can be approximated by

$$\Phi(G) = \iint_{\sigma(G)} \mathbf{F} \cdot \mathbf{n}\, dS = \iiint_G \operatorname{div} \mathbf{F}\, dV \approx \operatorname{div} \mathbf{F}(P_0) \iiint_G dV = \operatorname{div} \mathbf{F}(P_0)\operatorname{vol}(G)$$

from which we obtain the approximation

$$\operatorname{div} \mathbf{F}(P_0) \approx \frac{\Phi(G)}{\operatorname{vol}(G)} \tag{8}$$

The expression on the right side of (8) is called the ***outward flux density of* F *across* G.** If we now let the radius of the sphere approach zero [so that $\operatorname{vol}(G)$ approaches zero], then it is plausible that the error in this approximation will approach zero, and the divergence of $\mathbf{F}$ at the point P_0 will be given exactly by

$$\operatorname{div} \mathbf{F}(P_0) = \lim_{\operatorname{vol}(G)\to 0} \frac{\Phi(G)}{\operatorname{vol}(G)}$$

which we can express as

$$\operatorname{div} \mathbf{F}(P_0) = \lim_{\operatorname{vol}(G)\to 0} \frac{1}{\operatorname{vol}(G)} \iint_{\sigma(G)} \mathbf{F} \cdot \mathbf{n}\, dS \tag{9}$$

Formula (9) is sometimes taken as the definition of divergence. This is a useful alternative to Definition 16.1.4 because it does not require a coordinate system.

This limit, which is called the ***outward flux density of* F *at* P_0**, tells us that *in a steady-state fluid flow*, div **F** *can be interpreted as the limiting flux per unit volume at a point*. Moreover, it follows from (8) that for a small spherical region G centered at a point P_0 in the flow, the outward flux across the surface of G can be approximated by

$$\Phi(G) \approx (\text{div}\,\mathbf{F}(P_0))(\text{vol}(G)) \tag{10}$$

SOURCES AND SINKS

If P_0 is a point in an incompressible fluid at which $\text{div}\,\mathbf{F}(P_0) > 0$, then it follows from (8) that $\Phi(G) > 0$ for a sufficiently small sphere G centered at P_0. Thus, there is a greater volume of fluid going out through the surface of G than coming in. But this can only happen if there is some point *inside* the sphere at which fluid is entering the flow (say by condensation, melting of a solid, or a chemical reaction); otherwise the net outward flow through the surface would result in a decrease in density within the sphere, contradicting the incompressibility assumption. Similarly, if $\text{div}\,\mathbf{F}(P_0) < 0$, there would have to be a point *inside* the sphere at which fluid is leaving the flow (say by evaporation); otherwise the net inward flow through the surface would result in an increase in density within the sphere. In an incompressible fluid, points at which $\text{div}\,\mathbf{F}(P_0) > 0$ are called ***sources*** and points at which $\text{div}\,\mathbf{F}(P_0) < 0$ are called ***sinks***. Fluid enters the flow at a source and drains out at a sink. In an incompressible fluid without sources or sinks we must have

$$\text{div}\,\mathbf{F}(P) = 0$$

at every point P. In hydrodynamics this is called the ***continuity equation for incompressible fluids*** and is sometimes taken as the defining characteristic of an incompressible fluid.

GAUSS'S LAW FOR INVERSE-SQUARE FIELDS

The Divergence Theorem applied to inverse-square fields (see Definition 16.1.2) produces a result called ***Gauss's Law for Inverse-Square Fields***. This result is the basis for many important principles in physics.

16.7.2 GAUSS'S LAW FOR INVERSE-SQUARE FIELDS. If

$$\mathbf{F}(\mathbf{r}) = \frac{c}{\|\mathbf{r}\|^3}\mathbf{r}$$

is an inverse-square field in 3-space, and if σ is a closed orientable surface that surrounds the origin, then the outward flux of **F** across σ is

$$\Phi = \iint_{\sigma} \mathbf{F}\cdot\mathbf{n}\,dS = 4\pi c \tag{11}$$

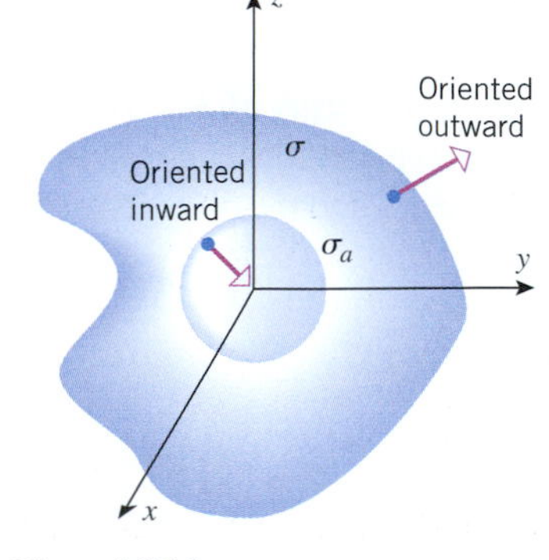

Figure 16.7.6

To derive this result, recall from Formula (5) of Section 16.1 that **F** can be expressed in component form as

$$\mathbf{F}(x, y, z) = \frac{c}{(x^2 + y^2 + z^2)^{3/2}}(x\mathbf{i} + y\mathbf{j} + z\mathbf{k}) \tag{12}$$

Since the components of **F** are not continuous at the origin, we cannot apply the Divergence Theorem across the solid enclosed by σ. However, we can circumvent this difficulty by constructing a sphere of radius a centered at the origin, where the radius is sufficiently small that the sphere lies entirely within the region enclosed by σ (Figure 16.7.6). We will denote the surface of this sphere by σ_a. The solid G enclosed between σ_a and σ is an example of a three-dimensional solid with an internal "cavity." Just as we were able to extend Green's Theorem to multiply connected regions in the plane (regions with holes), so it is possible to extend the Divergence Theorem to solids in 3-space with internal cavities, provided the surface integral in the theorem is taken over the *entire* boundary with the outside boundary of the solid oriented outward and the boundaries of the cavities oriented inward. Thus, if $\mathbf{F}$ is the inverse-square field in (12), and if σ_a is oriented inward, then the Divergence Theorem yields

$$\iiint_G \operatorname{div} \mathbf{F}\, dV = \iint_\sigma \mathbf{F}\cdot\mathbf{n}\, dS + \iint_{\sigma_a} \mathbf{F}\cdot\mathbf{n}\, dS \tag{13}$$

But we showed in Example 5 of Section 16.1 that $\operatorname{div} \mathbf{F} = 0$, so (13) yields

$$\iint_\sigma \mathbf{F}\cdot\mathbf{n}\, dS = -\iint_{\sigma_a} \mathbf{F}\cdot\mathbf{n}\, dS \tag{14}$$

sphere around origin

We can evaluate the surface integral over σ_a by expressing the integrand in terms of components; however, it is easier to leave it in vector form. At each point on the sphere the unit normal $\mathbf{n}$ points inward along a radius from the origin, and hence $\mathbf{n} = -\mathbf{r}/\|\mathbf{r}\|$. Thus, (14) yields

$$\begin{aligned}
\iint_\sigma \mathbf{F}\cdot\mathbf{n}\, dS &= -\iint_{\sigma_a} \frac{c}{\|\mathbf{r}\|^3}\mathbf{r}\cdot\left(-\frac{\mathbf{r}}{\|\mathbf{r}\|}\right) dS \\
&= \iint_{\sigma_a} \frac{c}{\|\mathbf{r}\|^4}(\mathbf{r}\cdot\mathbf{r})\, dS \\
&= \iint_{\sigma_a} \frac{c}{\|\mathbf{r}\|^2}\, dS \\
&= \frac{c}{a^2}\iint_{\sigma_a} dS && \|\mathbf{r}\| = a \text{ on } \sigma_a \\
&= \frac{c}{a^2}(4\pi a^2) && \text{The integral is the surface area of the sphere.} \\
&= 4\pi c
\end{aligned}$$

which establishes (11).

GAUSS'S LAW IN ELECTROSTATICS

It follows from Example 1 of Section 16.1 with $q = 1$ that a single charged particle of charge Q located at the origin creates an inverse-square field

$$\mathbf{F}(\mathbf{r}) = \frac{Q}{4\pi\epsilon_0\|\mathbf{r}\|^3}\mathbf{r}$$

in which $\mathbf{F}(\mathbf{r})$ is the electrical force exerted by Q on a unit positive charge ($q = 1$) located at the point with position vector $\mathbf{r}$. In this case Gauss's law (16.7.2) states that the outward flux Φ across any closed orientable surface σ that surrounds Q is

$$\Phi = \iint_\sigma \mathbf{F}\cdot\mathbf{n}\, dS = 4\pi\left(\frac{Q}{4\pi\epsilon_0}\right) = \frac{Q}{\epsilon_0}$$

This result, which is called ***Gauss's Law for Electric Fields***, can be extended to more than one charge. It is one of the fundamental laws in electricity and magnetism.

✔ QUICK CHECK EXERCISES 16.7

(See page 1173 for answers.)

1. Let G be a solid whose surface σ is oriented outward by the unit normal $\mathbf{n}$, and let $\mathbf{F}(x, y, z)$ denote a vector field whose component functions have continuous first partial derivatives on some open set containing G. The Divergence Theorem states that the surface integral ________ and the triple integral ________ have the same value.
2. The outward flux of $\mathbf{F}(x, y, z) = x\mathbf{i} + y\mathbf{j} + z\mathbf{k}$ across any unit cube is ________.
3. If $\mathbf{F}(x, y, z)$ is the velocity vector field for a steady-state incompressible fluid flow, then a point at which div $\mathbf{F}$ is positive is called a ________ and a point at which $\mathbf{F}$ is negative is called a ________. The continuity equation for an incompressible fluid states that ________.
4. If
$$\mathbf{F}(\mathbf{r}) = \frac{c}{\|\mathbf{r}\|^3}\mathbf{r}$$
is an inverse-square field, and if σ is a closed orientable surface that surrounds the origin, then Gauss's law states that the outward flux of $\mathbf{F}$ across σ is ________. On the other hand, if σ does not surround the origin, then it follows from the Divergence Theorem that the outward flux of $\mathbf{F}$ across σ is ________.

EXERCISE SET 16.7 C CAS

1–4 Verify Formula (1) in the Divergence Theorem by evaluating the surface integral and the triple integral.

1. $\mathbf{F}(x, y, z) = x\mathbf{i} + y\mathbf{j} + z\mathbf{k}$; σ is the surface of the cube bounded by the planes $x = 0$, $x = 1$, $y = 0$, $y = 1$, $z = 0$, $z = 1$.
2. $\mathbf{F}(x, y, z) = x\mathbf{i} + y\mathbf{j} + z\mathbf{k}$; σ is the spherical surface $x^2 + y^2 + z^2 = 1$.
3. $\mathbf{F}(x, y, z) = 2x\mathbf{i} - yz\mathbf{j} + z^2\mathbf{k}$; the surface σ is the paraboloid $z = x^2 + y^2$ capped by the disk $x^2 + y^2 \leq 1$ in the plane $z = 1$.
4. $\mathbf{F}(x, y, z) = xy\mathbf{i} + yz\mathbf{j} + xz\mathbf{k}$; σ is the surface of the cube bounded by the planes $x = 0$, $x = 2$, $y = 0$, $y = 2$, $z = 0$, $z = 2$.

5–15 Use the Divergence Theorem to find the flux of $\mathbf{F}$ across the surface σ with outward orientation.

5. $\mathbf{F}(x, y, z) = (x^2 + y)\mathbf{i} + z^2\mathbf{j} + (e^y - z)\mathbf{k}$; σ is the surface of the rectangular solid bounded by the coordinate planes and the planes $x = 3$, $y = 1$, and $z = 2$.
6. $\mathbf{F}(x, y, z) = z^3\mathbf{i} - x^3\mathbf{j} + y^3\mathbf{k}$, where σ is the sphere $x^2 + y^2 + z^2 = a^2$.
7. $\mathbf{F}(x, y, z) = (x - z)\mathbf{i} + (y - x)\mathbf{j} + (z - y)\mathbf{k}$; σ is the surface of the cylindrical solid bounded by $x^2 + y^2 = a^2$, $z = 0$, and $z = 1$.
8. $\mathbf{F}(x, y, z) = x\mathbf{i} + y\mathbf{j} + z\mathbf{k}$; σ is the surface of the solid bounded by the paraboloid $z = 1 - x^2 - y^2$ and the xy-plane.
9. $\mathbf{F}(x, y, z) = x^3\mathbf{i} + y^3\mathbf{j} + z^3\mathbf{k}$; σ is the surface of the cylindrical solid bounded by $x^2 + y^2 = 4$, $z = 0$, and $z = 3$.
10. $\mathbf{F}(x, y, z) = (x^2 + y)\mathbf{i} + xy\mathbf{j} - (2xz + y)\mathbf{k}$; σ is the surface of the tetrahedron in the first octant bounded by $x + y + z = 1$ and the coordinate planes.
11. $\mathbf{F}(x, y, z) = (x^3 - e^y)\mathbf{i} + (y^3 + \sin z)\mathbf{j} + (z^3 - xy)\mathbf{k}$, where σ is the surface of the solid bounded above by $z = \sqrt{4 - x^2 - y^2}$ and below by the xy-plane. [*Hint:* Use spherical coordinates.]
12. $\mathbf{F}(x, y, z) = 2xz\mathbf{i} + yz\mathbf{j} + z^2\mathbf{k}$, where σ is the surface of the solid bounded above by $z = \sqrt{a^2 - x^2 - y^2}$ and below by the xy-plane.
13. $\mathbf{F}(x, y, z) = x^2\mathbf{i} + y^2\mathbf{j} + z^2\mathbf{k}$; σ is the surface of the conical solid bounded by $z = \sqrt{x^2 + y^2}$ and $z = 1$.
14. $\mathbf{F}(x, y, z) = x^2y\mathbf{i} - xy^2\mathbf{j} + (z + 2)\mathbf{k}$; σ is the surface of the solid bounded above by the plane $z = 2x$ and below by the paraboloid $z = x^2 + y^2$.
15. $\mathbf{F}(x, y, z) = x^3\mathbf{i} + x^2y\mathbf{j} + xy\mathbf{k}$; σ is the surface of the solid bounded by $z = 4 - x^2$, $y + z = 5$, $z = 0$, and $y = 0$.
16. Prove that if $\mathbf{r} = x\mathbf{i} + y\mathbf{j} + z\mathbf{k}$ and σ is the surface of a solid G oriented by outward unit normals, then
$$\text{vol}(G) = \frac{1}{3}\iint_{\sigma} \mathbf{r} \cdot \mathbf{n}\, dS$$
where vol(G) is the volume of G.
17. Use the result in Exercise 16 to find the outward flux of the vector field $\mathbf{F}(x, y, z) = x\mathbf{i} + y\mathbf{j} + z\mathbf{k}$ across the surface σ of the cylindrical solid bounded by $x^2 + 4x + y^2 = 5$, $z = -1$, and $z = 4$.

FOCUS ON CONCEPTS

18. Let $\mathbf{F}(x, y, z) = a\mathbf{i} + b\mathbf{j} + c\mathbf{k}$ be a constant vector field and let σ be the surface of a solid G. Use the Divergence Theorem to show that the flux of $\mathbf{F}$ across σ is zero. Give an informal physical explanation of this result.
19. Find a vector field $\mathbf{F}(x, y, z)$ that has
(a) positive divergence everywhere
(b) negative divergence everywhere.

20. In each part, the figure shows a horizontal layer of the vector field of a fluid flow in which the flow is parallel to the xy-plane at every point and is identical in each layer (i.e., is independent of z). For each flow, what can you say about the sign of the divergence at the origin? Explain your reasoning.

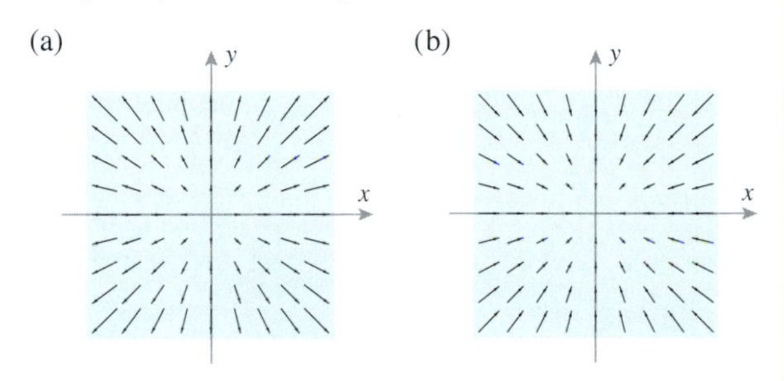

21. Let $\mathbf{F}(x, y, z)$ be a nonzero vector field in 3-space whose component functions have continuous first partial derivatives, and assume that $\operatorname{div} \mathbf{F} = 0$ everywhere. If σ is any sphere in 3-space, explain why there are infinitely many points on σ at which $\mathbf{F}$ is tangent to the sphere.

22. Does the result in Exercise 21 remain true if the sphere σ is replaced by a cube? Explain.

23–27 Prove the identity, assuming that $\mathbf{F}$, σ, and G satisfy the hypotheses of the Divergence Theorem and that all necessary differentiability requirements for the functions $f(x, y, z)$ and $g(x, y, z)$ are met.

23. $\displaystyle\iint_\sigma \operatorname{curl} \mathbf{F} \cdot \mathbf{n}\, dS = 0$ [*Hint:* See Exercise 33, Section 16.1.]

24. $\displaystyle\iint_\sigma \nabla f \cdot \mathbf{n}\, dS = \iiint_G \nabla^2 f\, dV$
$\left(\nabla^2 f = \dfrac{\partial^2 f}{\partial x^2} + \dfrac{\partial^2 f}{\partial y^2} + \dfrac{\partial^2 f}{\partial z^2}\right)$

25. $\displaystyle\iint_\sigma (f\nabla g) \cdot \mathbf{n}\, dS = \iiint_G (f\nabla^2 g + \nabla f \cdot \nabla g)\, dV$

26. $\displaystyle\iint_\sigma (f\nabla g - g\nabla f) \cdot \mathbf{n}\, dS = \iiint_G (f\nabla^2 g - g\nabla^2 f)\, dV$
[*Hint:* Interchange f and g in 25.]

27. $\displaystyle\iint_\sigma (f\mathbf{n}) \cdot \mathbf{v}\, dS = \iiint_G \nabla f \cdot \mathbf{v}\, dV$ ($\mathbf{v}$ a fixed vector)

28. Use the Divergence Theorem to find all positive values of k such that
$$\mathbf{F}(\mathbf{r}) = \frac{\mathbf{r}}{\|\mathbf{r}\|^k}$$
satisfies the condition $\operatorname{div} \mathbf{F} = 0$ when $\mathbf{r} \neq \mathbf{0}$.
[*Hint:* Modify the proof of (11).]

29–32 Determine whether the vector field $\mathbf{F}(x, y, z)$ is free of sources and sinks. If it is not, locate them.

29. $\mathbf{F}(x, y, z) = (y + z)\mathbf{i} - xz^3\mathbf{j} + (x^2 \sin y)\mathbf{k}$

30. $\mathbf{F}(x, y, z) = xy\mathbf{i} - xy\mathbf{j} + y^2\mathbf{k}$

31. $\mathbf{F}(x, y, z) = x^3\mathbf{i} + y^3\mathbf{j} + z^3\mathbf{k}$

32. $\mathbf{F}(x, y, z) = (x^3 - x)\mathbf{i} + (y^3 - y)\mathbf{j} + (z^3 - z)\mathbf{k}$

C **33.** Let σ be the surface of the solid G that is enclosed by the paraboloid $z = 1 - x^2 - y^2$ and the plane $z = 0$. Use a CAS to verify Formula (1) in the Divergence Theorem for the vector field
$$\mathbf{F} = (x^2y - z^2)\mathbf{i} + (y^3 - x)\mathbf{j} + (2x + 3z - 1)\mathbf{k}$$
by evaluating the surface integral and the triple integral.

✔ QUICK CHECK ANSWERS 16.7

1. $\displaystyle\iint_\sigma \mathbf{F} \cdot \mathbf{n}\, dS$; $\displaystyle\iiint_G \operatorname{div} \mathbf{F}\, dV$ **2.** 3 **3.** source; sink; $\operatorname{div} \mathbf{F} = 0$ **4.** $4\pi c$; 0

16.8 STOKES' THEOREM

In this section we will discuss a generalization of Green's Theorem to three dimensions that has important applications in the study of vector fields, particularly in the analysis of rotational motion of fluids. This theorem will also provide us with a physical interpretation of the curl of a vector field.

RELATIVE ORIENTATION OF CURVES AND SURFACES

We will be concerned in this section with oriented surfaces in 3-space that are bounded by simple closed parametric curves (Figure 16.8.1a). If σ is an oriented surface bounded by a simple closed parametric curve C, then there are two possible relationships between the orientations of σ and C, which can be described as follows. Imagine a person walking along the curve C with his or her head in the direction of the orientation of σ. The person is said to be walking in the ***positive direction*** of C relative to the orientation of σ if the surface is on the person's left (Figure 16.8.1*b*), and the person is said to be walking in the ***negative direction*** of C relative to the orientation of σ if the surface is on the person's right (Figure 16.8.1*c*). The positive direction of C establishes a right-hand relationship between the orientations of σ and C in the sense that if the fingers of the right hand are curled from the direction of C towards σ, then the thumb points (roughly) in the direction of the orientation of σ.

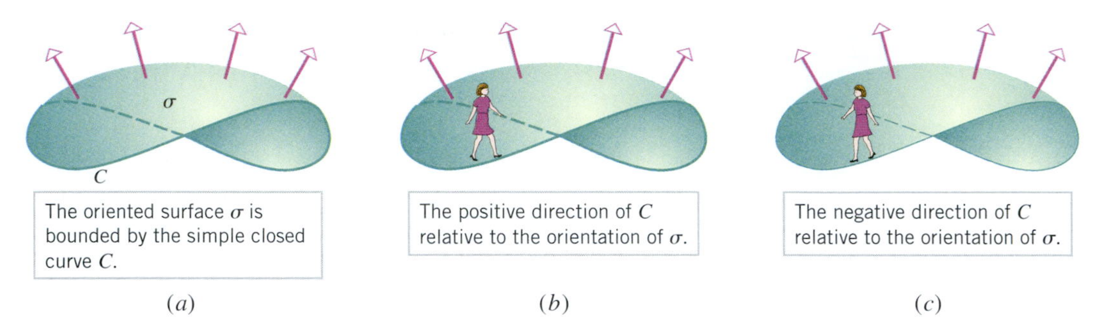

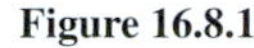
Figure 16.8.1

STOKES' THEOREM

In Section 16.1 we defined the curl of a vector field

$$\mathbf{F}(x, y, z) = f(x, y, z)\mathbf{i} + g(x, y, z)\mathbf{j} + h(x, y, z)\mathbf{k}$$

as

$$\operatorname{curl} \mathbf{F} = \left(\frac{\partial h}{\partial y} - \frac{\partial g}{\partial z}\right)\mathbf{i} + \left(\frac{\partial f}{\partial z} - \frac{\partial h}{\partial x}\right)\mathbf{j} + \left(\frac{\partial g}{\partial x} - \frac{\partial f}{\partial y}\right)\mathbf{k} = \begin{vmatrix} \mathbf{i} & \mathbf{j} & \mathbf{k} \\ \dfrac{\partial}{\partial x} & \dfrac{\partial}{\partial y} & \dfrac{\partial}{\partial z} \\ f & g & h \end{vmatrix} \tag{1}$$

but we did not attempt to give a physical explanation of its meaning at that time. The following result, known as ***Stokes' Theorem***, will provide us with a physical interpretation of the curl in the context of fluid flow.

16.8.1 THEOREM (*Stokes' Theorem*). *Let σ be a piecewise smooth oriented surface that is bounded by a simple, closed, piecewise smooth curve C with positive orientation. If the components of the vector field*

$$\mathbf{F}(x, y, z) = f(x, y, z)\mathbf{i} + g(x, y, z)\mathbf{j} + h(x, y, z)\mathbf{k}$$

are continuous and have continuous first partial derivatives on some open set containing σ, and if $\mathbf{T}$ is the unit tangent vector to C, then

$$\oint_C \mathbf{F} \cdot \mathbf{T}\, ds = \iint_\sigma (\operatorname{curl} \mathbf{F}) \cdot \mathbf{n}\, dS \tag{2}$$

The proof of this theorem is beyond the scope of this text, so we will focus on its applications.

Recall from Formulas (30) and (34) in Section 16.2 that if **F** is a force field, the integral on the left side of (2) represents the work performed by the force field on a particle that traverses the curve C. Thus, loosely phrased, Stokes' Theorem states:

> *The work performed by a force field on a particle that traverses a simple, closed, piecewise smooth curve C in the positive direction can be obtained by integrating the normal component of the curl over an oriented surface σ bounded by C.*

USING STOKES' THEOREM TO CALCULATE WORK

For computational purposes it is usually preferable to use Formula (30) in Section 16.2 to rewrite the formula in Stokes' Theorem as

$$\oint_C \mathbf{F} \cdot d\mathbf{r} = \iint_\sigma (\text{curl}\ \mathbf{F}) \cdot \mathbf{n}\, dS \qquad (3)$$

Stokes' Theorem is usually the method of choice for calculating work around piecewise smooth curves with multiple sections, since it eliminates the need for a separate integral evaluation over each section. This is illustrated in the following example.

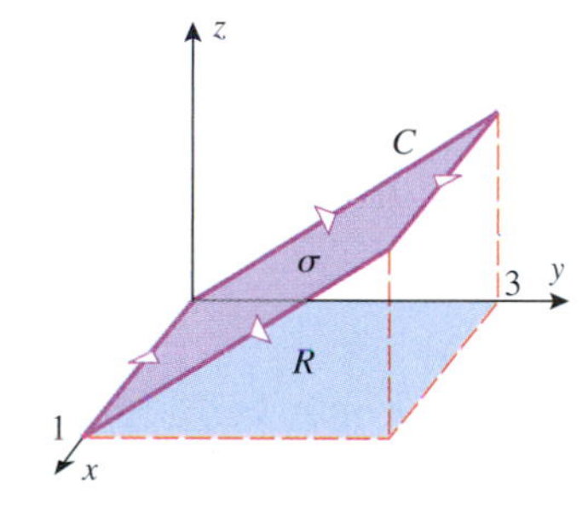

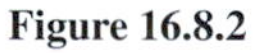
Figure 16.8.2

Example 1 Find the work performed by the force field

$$\mathbf{F}(x, y, z) = x^2\mathbf{i} + 4xy^3\mathbf{j} + y^2x\mathbf{k}$$

on a particle that traverses the rectangle C in the plane $z = y$ shown in Figure 16.8.2.

Solution. The work performed by the field is

$$W = \oint_C \mathbf{F} \cdot d\mathbf{r}$$

However, to evaluate this integral directly would require four separate integrations, one over each side of the rectangle. Instead, we will use Formula (3) to express the work as the surface integral

$$W = \iint_\sigma (\text{curl}\ \mathbf{F}) \cdot \mathbf{n}\, dS$$

George Gabriel Stokes (1819–1903) Irish mathematician and physicist. Born in Skreen, Ireland, Stokes came from a family deeply rooted in the Church of Ireland. His father was a rector, his mother the daughter of a rector, and three of his brothers took holy orders. He received his early education from his father and a local parish clerk. In 1837, he entered Pembroke College and after graduating with top honors accepted a fellowship at the college. In 1847 he was appointed Lucasian professor of mathematics at Cambridge, a position once held by Isaac Newton, but one that had lost its esteem through the years. By virtue of his accomplishments, Stokes ultimately restored the position to the eminence it once held. Unfortunately, the position paid very little and Stokes was forced to teach at the Government School of Mines during the 1850s to supplement his income.

Stokes was one of several outstanding nineteenth century scientists who helped turn the physical sciences in a more empirical direction. He systematically studied hydrodynamics, elasticity of solids, behavior of waves in elastic solids, and diffraction of light. For Stokes, mathematics was a tool for his physical studies. He wrote classic papers on the motion of viscous fluids that laid the foundation for modern hydrodynamics; he elaborated on the wave theory of light; and he wrote papers on gravitational variation that established him as a founder of the modern science of geodesy.

Stokes was honored in his later years with degrees, medals, and memberships in foreign societies. He was knighted in 1889. Throughout his life, Stokes gave generously of his time to learned societies and readily assisted those who sought his help in solving problems. He was deeply religious and vitally concerned with the relationship between science and religion.

in which the plane surface σ enclosed by C is assigned a *downward* orientation to make the orientation of C positive, as required by Stokes' Theorem.

Since the surface σ has equation $z = y$ and

$$\text{curl } \mathbf{F} = \begin{vmatrix} \mathbf{i} & \mathbf{j} & \mathbf{k} \\ \dfrac{\partial}{\partial x} & \dfrac{\partial}{\partial y} & \dfrac{\partial}{\partial z} \\ x^2 & 4xy^3 & xy^2 \end{vmatrix} = 2xy\mathbf{i} - y^2\mathbf{j} + 4y^3\mathbf{k}$$

it follows from Formula (13) of Section 16.6 with curl $\mathbf{F}$ replacing $\mathbf{F}$ that

$$\begin{aligned} W = \iint_{\sigma} (\text{curl } \mathbf{F}) \cdot \mathbf{n}\, dS &= \iint_{R} (\text{curl } \mathbf{F}) \cdot \left(\frac{\partial z}{\partial x}\mathbf{i} + \frac{\partial z}{\partial y}\mathbf{j} - \mathbf{k} \right) dA \\ &= \iint_{R} (2xy\mathbf{i} - y^2\mathbf{j} + 4y^3\mathbf{k}) \cdot (0\mathbf{i} + \mathbf{j} - \mathbf{k})\, dA \\ &= \int_0^1 \int_0^3 (-y^2 - 4y^3)\, dy\, dx \\ &= -\int_0^1 \left[\frac{y^3}{3} + y^4 \right]_{y=0}^{3} dx \\ &= -\int_0^1 90\, dx = -90 \blacktriangleleft \end{aligned}$$

Explain how the result in Example 1 shows that the given force field is not conservative.

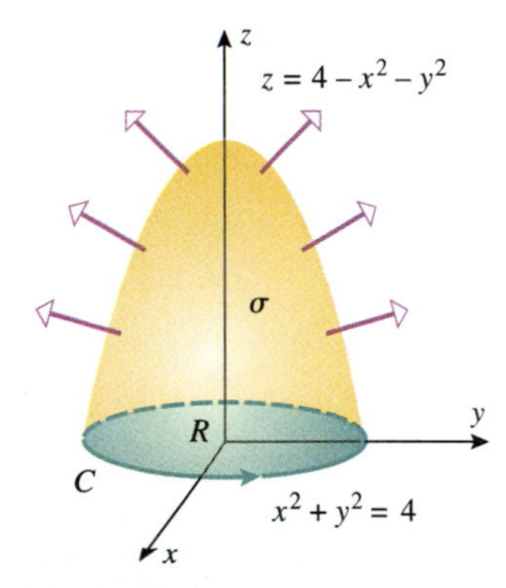

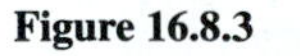

Figure 16.8.3

► **Example 2** Verify Stokes' Theorem for the vector field $\mathbf{F}(x, y, z) = 2z\mathbf{i} + 3x\mathbf{j} + 5y\mathbf{k}$, taking σ to be the portion of the paraboloid $z = 4 - x^2 - y^2$ for which $z \geq 0$ with upward orientation, and C to be the positively oriented circle $x^2 + y^2 = 4$ that forms the boundary of σ in the xy-plane (Figure 16.8.3).

Solution. We will verify Formula (3). Since σ is oriented up, the positive orientation of C is counterclockwise looking down the positive z-axis. Thus, C can be represented parametrically (with positive orientation) by

$$x = 2\cos t, \quad y = 2\sin t, \quad z = 0 \quad (0 \leq t \leq 2\pi) \tag{4}$$

Therefore,

$$\begin{aligned} \oint_C \mathbf{F} \cdot d\mathbf{r} &= \oint_C 2z\, dx + 3x\, dy + 5y\, dz \\ &= \int_0^{2\pi} [0 + (6\cos t)(2\cos t) + 0]\, dt \\ &= \int_0^{2\pi} 12\cos^2 t\, dt = 12\left[\frac{1}{2}t + \frac{1}{4}\sin 2t \right]_0^{2\pi} = 12\pi \end{aligned}$$

To evaluate the right side of (3), we start by finding curl $\mathbf{F}$. We obtain

$$\text{curl } \mathbf{F} = \begin{vmatrix} \mathbf{i} & \mathbf{j} & \mathbf{k} \\ \dfrac{\partial}{\partial x} & \dfrac{\partial}{\partial y} & \dfrac{\partial}{\partial z} \\ 2z & 3x & 5y \end{vmatrix} = 5\mathbf{i} + 2\mathbf{j} + 3\mathbf{k}$$

Since σ is oriented up and is expressed in the form $z = g(x, y) = 4 - x^2 - y^2$, it follows from Formula (12) of Section 16.6 with curl **F** replacing **F** that

$$\begin{aligned}
\iint_\sigma (\operatorname{curl} \mathbf{F}) \cdot \mathbf{n}\, dS &= \iint_R (\operatorname{curl} \mathbf{F}) \cdot \left(-\frac{\partial z}{\partial x}\mathbf{i} - \frac{\partial z}{\partial y}\mathbf{j} + \mathbf{k}\right) dA \\
&= \iint_R (5\mathbf{i} + 2\mathbf{j} + 3\mathbf{k}) \cdot (2x\mathbf{i} + 2y\mathbf{j} + \mathbf{k})\, dA \\
&= \iint_R (10x + 4y + 3)\, dA \\
&= \int_0^{2\pi} \int_0^2 (10r\cos\theta + 4r\sin\theta + 3) r\, dr\, d\theta \\
&= \int_0^{2\pi} \left[\frac{10r^3}{3}\cos\theta + \frac{4r^3}{3}\sin\theta + \frac{3r^2}{2}\right]_{r=0}^2 d\theta \\
&= \int_0^{2\pi} \left(\frac{80}{3}\cos\theta + \frac{32}{3}\sin\theta + 6\right) d\theta \\
&= \left[\frac{80}{3}\sin\theta - \frac{32}{3}\cos\theta + 6\theta\right]_0^{2\pi} = 12\pi
\end{aligned}$$

As guaranteed by Stokes' Theorem, the value of this surface integral is the same as the value of the line integral obtained above. Note, however, that the line integral was simpler to evaluate and hence would be the method of choice in this case. ◄

Observe that in Formula (3) the only relationships required between σ and C are that C be the boundary of σ and that C be positively oriented relative to the orientation of σ. Thus, if σ_1 and σ_2 are *different* oriented surfaces but have the *same* positively oriented boundary curve C, then it follows from (3) that

$$\iint_{\sigma_1} \operatorname{curl} \mathbf{F} \cdot \mathbf{n}\, dS = \iint_{\sigma_2} \operatorname{curl} \mathbf{F} \cdot \mathbf{n}\, dS$$

For example, the parabolic surface in Example 2 has the same positively oriented boundary C as the disk R in Figure 16.8.3 with upper orientation. Thus, the value of the surface integral in that example would not change if σ is replaced by R (or by any other oriented surface that has the positively oriented circle C as its boundary). This can be useful in computations because it is sometimes possible to circumvent a difficult integration by changing the surface of integration.

RELATIONSHIP BETWEEN GREEN'S THEOREM AND STOKES' THEOREM

It is sometimes convenient to regard a vector field

$$\mathbf{F}(x, y) = f(x, y)\mathbf{i} + g(x, y)\mathbf{j}$$

in 2-space as a vector field in 3-space by expressing it as

$$\mathbf{F}(x, y) = f(x, y)\mathbf{i} + g(x, y)\mathbf{j} + 0\mathbf{k} \tag{5}$$

If R is a region in the xy-plane enclosed by a simple, closed, piecewise smooth curve C, then we can treat R as a *flat* surface, and we can treat a surface integral over R as an ordinary double integral over R. Thus, if we orient R up and C counterclockwise looking down the positive z-axis, then Formula (3) applied to (5) yields

$$\oint_C \mathbf{F} \cdot d\mathbf{r} = \iint_R \text{curl } \mathbf{F} \cdot \mathbf{k}\, dA \tag{6}$$

But

$$\text{curl } \mathbf{F} = \begin{vmatrix} \mathbf{i} & \mathbf{j} & \mathbf{k} \\ \dfrac{\partial}{\partial x} & \dfrac{\partial}{\partial y} & \dfrac{\partial}{\partial z} \\ f & g & 0 \end{vmatrix} = -\frac{\partial g}{\partial z}\mathbf{i} + \frac{\partial f}{\partial z}\mathbf{j} + \left(\frac{\partial g}{\partial x} - \frac{\partial f}{\partial y}\right)\mathbf{k} = \left(\frac{\partial g}{\partial x} - \frac{\partial f}{\partial y}\right)\mathbf{k}$$

since $\partial g/\partial z = \partial f/\partial z = 0$. Substituting this expression in (6) and expressing the integrals in terms of components yields

$$\oint_C f\,dx + g\,dy = \iint_R \left(\frac{\partial g}{\partial x} - \frac{\partial f}{\partial y}\right) dA$$

which is Green's Theorem [Formula (1) of Section 16.4]. Thus, we have shown that Green's Theorem can be viewed as a special case of Stokes' Theorem.

CURL VIEWED AS CIRCULATION

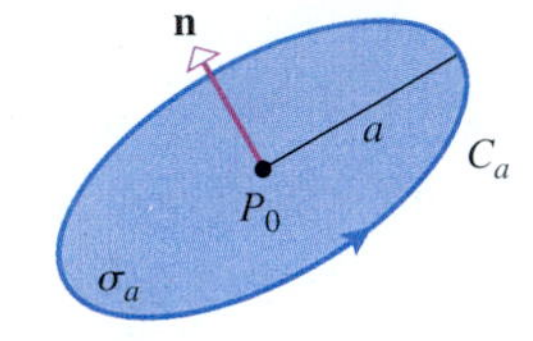

Figure 16.8.4

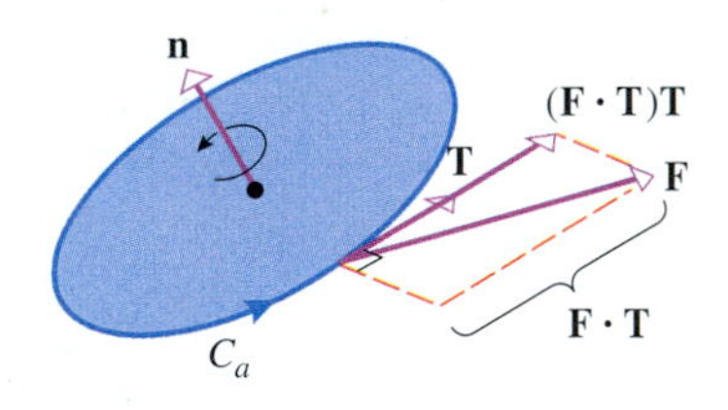

Figure 16.8.5

Stokes' Theorem provides a way of interpreting the curl of a vector field $\mathbf{F}$ in the context of fluid flow. For this purpose let σ_a be a small oriented disk of radius a centered at a point P_0 in a steady-state fluid flow, and let $\mathbf{n}$ be a unit normal vector at the center of the disk that points in the direction of orientation. Let us assume that the flow of liquid past the disk causes it to spin around the axis through $\mathbf{n}$, and let us try to find the direction of $\mathbf{n}$ that will produce the maximum rotation rate in the positive direction of the boundary curve C_a (Figure 16.8.4). For convenience, we will denote the area of the disk σ_a by $A(\sigma_a)$; that is, $A(\sigma_a) = \pi a^2$.

If the direction of $\mathbf{n}$ is fixed, then at each point of C_a the only component of $\mathbf{F}$ that contributes to the rotation of the disk about $\mathbf{n}$ is the component $\mathbf{F} \cdot \mathbf{T}$ tangent to C_a (Figure 16.8.5). Thus, for a fixed $\mathbf{n}$ the integral

$$\oint_{C_a} \mathbf{F} \cdot \mathbf{T}\, ds \tag{7}$$

can be viewed as a measure of the tendency for the fluid to flow in the positive direction around C_a. Accordingly, (7) is called the ***circulation of* F *around* C_a**. For example, in the extreme case where the flow is normal to the circle at each point, the circulation around C_a is zero, since $\mathbf{F} \cdot \mathbf{T} = 0$ at each point. The more closely that $\mathbf{F}$ aligns with $\mathbf{T}$ along the circle, the larger the value of $\mathbf{F} \cdot \mathbf{T}$ and the larger the value of the circulation.

To see the relationship between circulation and curl, suppose that curl $\mathbf{F}$ is continuous on σ_a, so that when σ_a is small the value of curl $\mathbf{F}$ at any point of σ_a will not vary much from the value of curl $\mathbf{F}(P_0)$ at the center. Thus, for a small disk σ_a we can reasonably approximate curl $\mathbf{F}$ by the constant value curl $\mathbf{F}(P_0)$ on σ_a. Moreover, because the surface σ_a is flat, the unit normal vectors that orient σ_a are all equal. Thus, the vector quantity $\mathbf{n}$ in Formula (3) can be treated as a constant, and we can write

$$\oint_{C_a} \mathbf{F} \cdot \mathbf{T}\, ds = \iint_{\sigma_a} (\text{curl } \mathbf{F}) \cdot \mathbf{n}\, dS \approx \text{curl } \mathbf{F}(P_0) \cdot \mathbf{n} \iint_{\sigma_a} dS$$

where the line integral is taken in the positive direction of C_a. But the last double integral in this equation represents the surface area of σ_a, so

$$\oint_{C_a} \mathbf{F} \cdot \mathbf{T}\, ds \approx [\text{curl } \mathbf{F}(P_0) \cdot \mathbf{n}]A(\sigma_a)$$

from which we obtain

$$\operatorname{curl} \mathbf{F}(P_0) \cdot \mathbf{n} \approx \frac{1}{A(\sigma_a)} \oint_{C_a} \mathbf{F} \cdot \mathbf{T}\, ds \tag{8}$$

Formula (9) is sometimes taken as a definition of curl. This is a useful alternative to Definition 16.1.5 because it does not require a coordinate system.

The quantity on the right side of (8) is called the ***circulation density of* F *around* C_a**. If we now let the radius a of the disk approach zero (with $\mathbf{n}$ fixed), then it is plausible that the error in this approximation will approach zero and the exact value of curl $\mathbf{F}(P_0) \cdot \mathbf{n}$ will be given by

$$\operatorname{curl} \mathbf{F}(P_0) \cdot \mathbf{n} = \lim_{a\to 0} \frac{1}{A(\sigma_a)} \oint_{C_a} \mathbf{F} \cdot \mathbf{T}\, ds \tag{9}$$

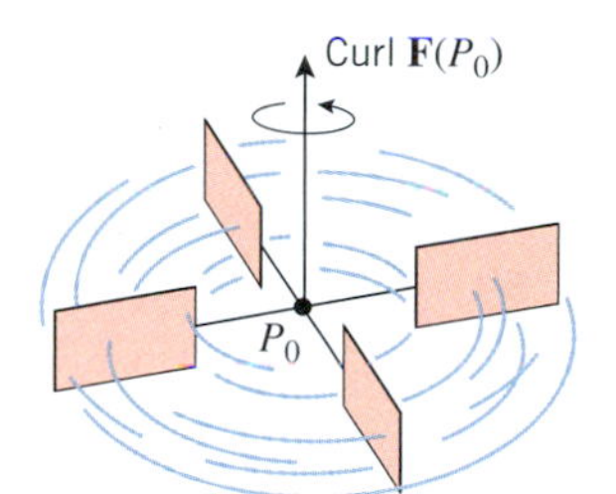

Figure 16.8.6

We call curl $\mathbf{F}(P_0) \cdot \mathbf{n}$ the ***circulation density of* F *at* P_0 *in the direction of* n**. This quantity has its maximum value when $\mathbf{n}$ is in the same direction as curl $\mathbf{F}(P_0)$; this tells us that *at each point in a steady-state fluid flow the maximum circulation density occurs in the direction of the curl*. Physically, this means that if a small paddle wheel is immersed in the fluid so that the pivot point is at P_0, then the paddles will turn most rapidly when the spindle is aligned with curl $\mathbf{F}(P_0)$ (Figure 16.8.6). If curl $\mathbf{F} = \mathbf{0}$ at each point of a region, then $\mathbf{F}$ is said to be ***irrotational*** in that region, since no circulation occurs about any point of the region.

✔QUICK CHECK EXERCISES 16.8

(See page 1181 for answers.)

1. Let σ be a piecewise smooth oriented surface that is bounded by a simple, closed, piecewise smooth curve C with positive orientation. If the component functions of the vector field $\mathbf{F}(x, y, z)$ have continuous first partial derivatives on some open set containing σ, and if $\mathbf{T}$ is the unit tangent vector to C, then Stokes' Theorem states that the line integral ________ and the surface integral ________ are equal.

2. We showed in Example 2 that the vector field

$$\mathbf{F}(x, y, z) = 2z\mathbf{i} + 3x\mathbf{j} + 5y\mathbf{k}$$

satisfies the equation curl $\mathbf{F} = 5\mathbf{i} + 2\mathbf{j} + 3\mathbf{k}$. It follows from Stokes' Theorem that if C is any circle of radius a in the xy-plane that is oriented counterclockwise when viewed from the positive z-axis, then

$$\int_C \mathbf{F} \cdot \mathbf{T}\, ds = \text{\_\_\_\_\_\_\_\_}$$

where $\mathbf{T}$ denotes the unit tangent vector to C.

3. (a) If σ_1 and σ_2 are two oriented surfaces that have the same positively oriented boundary curve C, and if the vector field $\mathbf{F}(x, y, z)$ has continuous first partial derivatives on some open set containing σ_1 and σ_2, then it follows from Stokes' Theorem that the surface integrals ________ and ________ are equal.

(b) Let $\mathbf{F}(x, y, z) = 2z\mathbf{i} + 3x\mathbf{j} + 5y\mathbf{k}$, let a be a positive number, and let σ be the portion of the paraboloid $z = a^2 - x^2 - y^2$ for which $z \geq 0$ with upward orientation. Using part (a) and Quick Check Exercise 2, it follows that

$$\iint_\sigma (\operatorname{curl} \mathbf{F}) \cdot \mathbf{n}\, dS = \text{\_\_\_\_\_\_\_\_}$$

4. For steady-state flow, the maximum circulation density occurs in the direction of the ________ of the velocity vector field for the flow.

EXERCISE SET 16.8 [C] CAS

1–4 Verify Formula (2) in Stokes' Theorem by evaluating the line integral and the double integral. Assume that the surface has an upward orientation.

1. $\mathbf{F}(x, y, z) = (x - y)\mathbf{i} + (y - z)\mathbf{j} + (z - x)\mathbf{k}$; σ is the portion of the plane $x + y + z = 1$ in the first octant.

2. $\mathbf{F}(x, y, z) = x^2\mathbf{i} + y^2\mathbf{j} + z^2\mathbf{k}$; σ is the portion of the cone $z = \sqrt{x^2 + y^2}$ below the plane $z = 1$.

3. $\mathbf{F}(x, y, z) = x\mathbf{i} + y\mathbf{j} + z\mathbf{k}$; σ is the upper hemisphere $z = \sqrt{a^2 - x^2 - y^2}$.

4. $\mathbf{F}(x, y, z) = (z - y)\mathbf{i} + (z + x)\mathbf{j} - (x + y)\mathbf{k}$; σ is the portion of the paraboloid $z = 9 - x^2 - y^2$ above the xy-plane.

5–12 Use Stokes' Theorem to evaluate $\oint_C \mathbf{F} \cdot d\mathbf{r}$.

5. $\mathbf{F}(x, y, z) = z^2\mathbf{i} + 2x\mathbf{j} - y^3\mathbf{k}$; C is the circle $x^2 + y^2 = 1$ in the xy-plane with counterclockwise orientation looking down the positive z-axis.

6. $\mathbf{F}(x, y, z) = xz\mathbf{i} + 3x^2y^2\mathbf{j} + yx\mathbf{k}$; C is the rectangle in the plane $z = y$ shown in Figure 16.8.2.

7. $\mathbf{F}(x, y, z) = 3z\mathbf{i} + 4x\mathbf{j} + 2y\mathbf{k}$; C is the boundary of the paraboloid shown in Figure 16.8.3.

8. $\mathbf{F}(x, y, z) = -3y^2\mathbf{i} + 4z\mathbf{j} + 6x\mathbf{k}$; C is the triangle in the plane $z = \frac{1}{2}y$ with vertices $(2, 0, 0)$, $(0, 2, 1)$, and $(0, 0, 0)$ with a counterclockwise orientation looking down the positive z-axis.

9. $\mathbf{F}(x, y, z) = xy\mathbf{i} + x^2\mathbf{j} + z^2\mathbf{k}$; C is the intersection of the paraboloid $z = x^2 + y^2$ and the plane $z = y$ with a counterclockwise orientation looking down the positive z-axis.

10. $\mathbf{F}(x, y, z) = xy\mathbf{i} + yz\mathbf{j} + zx\mathbf{k}$; C is the triangle in the plane $x + y + z = 1$ with vertices $(1, 0, 0)$, $(0, 1, 0)$, and $(0, 0, 1)$ with a counterclockwise orientation looking from the first octant toward the origin.

11. $\mathbf{F}(x, y, z) = (x - y)\mathbf{i} + (y - z)\mathbf{j} + (z - x)\mathbf{k}$; C is the circle $x^2 + y^2 = a^2$ in the xy-plane with counterclockwise orientation looking down the positive z-axis.

12. $\mathbf{F}(x, y, z) = (z + \sin x)\mathbf{i} + (x + y^2)\mathbf{j} + (y + e^z)\mathbf{k}$; C is the intersection of the sphere $x^2 + y^2 + z^2 = 1$ and the cone $z = \sqrt{x^2 + y^2}$ with counterclockwise orientation looking down the positive z-axis.

13. Consider the vector field given by the formula

$$\mathbf{F}(x, y, z) = (x - z)\mathbf{i} + (y - x)\mathbf{j} + (z - xy)\mathbf{k}$$

(a) Use Stokes' Theorem to find the circulation around the triangle with vertices $A(1, 0, 0)$, $B(0, 2, 0)$, and $C(0, 0, 1)$ oriented counterclockwise looking from the origin toward the first octant.

(b) Find the circulation density of $\mathbf{F}$ at the origin in the direction of $\mathbf{k}$.

(c) Find the unit vector $\mathbf{n}$ such that the circulation density of $\mathbf{F}$ at the origin is maximum in the direction of $\mathbf{n}$.

FOCUS ON CONCEPTS

14. (a) Let σ denote the surface of a solid G with $\mathbf{n}$ the outward unit normal vector field to σ. Assume that $\mathbf{F}$ is a vector field with continuous first-order partial derivatives on σ. Prove that

$$\iint_\sigma (\text{curl } \mathbf{F}) \cdot \mathbf{n}\, dS = 0$$

[*Hint:* Let C denote a simple closed curve on σ that separates the surface into two subsurfaces σ_1 and σ_2 that share C as their common boundary. Apply Stokes' Theorem to σ_1 and to σ_2 and add the results.]

(b) The vector field curl($\mathbf{F}$) is called the ***curl field*** of $\mathbf{F}$. In words, interpret the formula in part (a) as a statement about the flux of the curl field.

15–16 The figures in these exercises show a horizontal layer of the vector field of a fluid flow in which the flow is parallel to the xy-plane at every point and is identical in each layer (i.e., is independent of z). For each flow, state whether you believe that the curl is nonzero at the origin, and explain your reasoning. If you believe that it is nonzero, then state whether it points in the positive or negative z-direction.

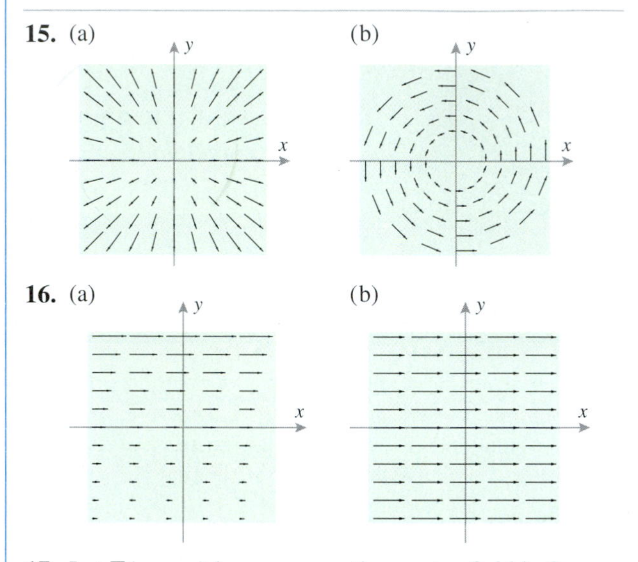

17. Let $\mathbf{F}(x, y, z)$ be a conservative vector field in 3-space whose component functions have continuous first partial derivatives. Explain how to use Formula (9) to prove that curl $\mathbf{F} = \mathbf{0}$.

18. In 1831 the physicist Michael Faraday discovered that an electric current can be produced by varying the magnetic flux through a conducting loop. His experiments showed that the electromotive force $\mathbf{E}$ is related to the magnetic induction $\mathbf{B}$ by the equation

$$\oint_C \mathbf{E} \cdot d\mathbf{r} = -\iint_\sigma \frac{\partial \mathbf{B}}{\partial t} \cdot \mathbf{n}\, dS$$

Use this result to make a conjecture about the relationship between curl $\mathbf{E}$ and $\mathbf{B}$, and explain your reasoning.

C **19.** Let σ be the portion of the paraboloid $z = 1 - x^2 - y^2$ for which $z \geq 0$, and let C be the circle $x^2 + y^2 = 1$ that forms the boundary of σ in the xy-plane. Assuming that σ is oriented up, use a CAS to verify Formula (2) in Stokes' Theorem for the vector field

$$\mathbf{F} = (x^2y - z^2)\mathbf{i} + (y^3 - x)\mathbf{j} + (2x + 3z - 1)\mathbf{k}$$

by evaluating the line integral and the surface integral.

✔QUICK CHECK ANSWERS 16.8

1. $\int_C \mathbf{F}\cdot\mathbf{T}\,ds$; $\iint_\sigma (\text{curl }\mathbf{F})\cdot\mathbf{n}\,dS$ **2.** $3\pi a^2$ **3.** (a) $\iint_{\sigma_1} (\text{curl }\mathbf{F})\cdot\mathbf{n}\,dS$; $\iint_{\sigma_2} (\text{curl }\mathbf{F})\cdot\mathbf{n}\,dS$ (b) $3\pi a^2$ **4.** curl

CHAPTER REVIEW EXERCISES

1. In words, what is a vector field? Give some physical examples of vector fields.

2. (a) Give a physical example of an inverse-square field $\mathbf{F}(\mathbf{r})$ in 3-space.
(b) Write a formula for a general inverse-square field $\mathbf{F}(\mathbf{r})$ in terms of the radius vector $\mathbf{r}$.
(c) Write a formula for a general inverse-square field $\mathbf{F}(x, y, z)$ in 3-space using rectangular coordinates.

3. Find an explicit coordinate expression for the vector field $\mathbf{F}(x, y)$ that at every point $(x, y) \neq (1, 2)$ is the unit vector directed from (x, y) to $(1, 2)$.

4. Find $\nabla\left(\dfrac{x+y}{x-y}\right)$.

5. Find $\text{curl}(z\mathbf{i} + x\mathbf{j} + y\mathbf{k})$.

6. Let

$$\mathbf{F}(x, y, z) = \frac{x}{x^2+y^2}\mathbf{i} + \frac{y}{x^2+y^2}\mathbf{j} + \frac{z}{x^2+y^2}\mathbf{k}$$

Sketch the level surface $\text{div }\mathbf{F} = 1$.

7. Assume that C is the parametric curve $x = x(t)$, $y = y(t)$, where t varies from a to b. In each part, express the line integral as a definite integral with variable of integration t.
(a) $\int_C f(x, y)\,dx + g(x, y)\,dy$ (b) $\int_C f(x, y)\,ds$

8. (a) Express the mass M of a thin wire in 3-space as a line integral.
(b) Express the length of a curve as a line integral.

9. Give a physical interpretation of $\int_C \mathbf{F}\cdot\mathbf{T}\,ds$.

10. State some alternative notations for $\int_C \mathbf{F}\cdot\mathbf{T}\,ds$.

11–13 Evaluate the line integral.

11. $\int_C (x-y)\,ds$; $C: x^2+y^2 = 1$

12. $\int_C x\,dx + z\,dy - 2y^2\,dz$;
$C: x = \cos t,\ y = \sin t,\ z = t \quad (0 \le t \le 2\pi)$

13. $\int_C \mathbf{F}\cdot d\mathbf{r}$ where $\mathbf{F}(x, y) = (x/y)\mathbf{i} - (y/x)\mathbf{j}$;
$\mathbf{r}(t) = t\mathbf{i} + 2t\mathbf{j} \quad (1 \le t \le 2)$

14. Find the work done by the force field

$$\mathbf{F}(x, y) = y^2\mathbf{i} + xy\mathbf{j}$$

moving a particle from $(0, 0)$ to $(1, 1)$ along the parabola $y = x^2$.

15. State the Fundamental Theorem of Line Integrals, including all required hypotheses.

16. Evaluate $\int_C \nabla f\cdot d\mathbf{r}$ where $f(x, y, z) = xy^2z^3$ and

$$\mathbf{r}(t) = t\mathbf{i} + (t^2+t)\mathbf{j} + \sin(3\pi t/2)\mathbf{k} \quad (0 \le t \le 1)$$

17. Let $\mathbf{F}(x, y) = y\mathbf{i} - 2x\mathbf{j}$.
(a) Find a nonzero function $h(x)$ such that $h(x)\mathbf{F}(x, y)$ is a conservative vector field.
(b) Find a nonzero function $g(y)$ such that $g(y)\mathbf{F}(x, y)$ is a conservative vector field.

18. Let $\mathbf{F}(x, y) = (ye^{xy} - 1)\mathbf{i} + xe^{xy}\mathbf{j}$.
(a) Show that $\mathbf{F}$ is a conservative vector field.
(b) Find a potential function for $\mathbf{F}$.
(c) Find the work performed by the force field on a particle that moves along the sawtooth curve represented by the parametric equations

$$x = t + \sin^{-1}(\sin t)$$
$$y = (2/\pi)\sin^{-1}(\sin t) \qquad (0 \le t \le 8\pi)$$

(see the accompanying figure).

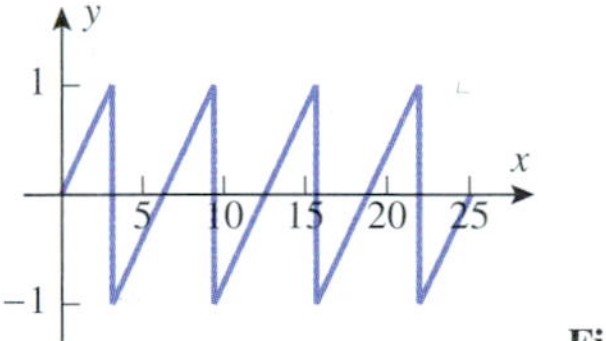

Figure Ex-18

19. State Green's Theorem, including all of the required hypotheses.

20. Express the area of a plane region as a line integral.

21. Let α and β denote angles that satisfy $0 < \beta - \alpha \le 2\pi$ and assume that $r = f(\theta)$ is a smooth polar curve with $f(\theta) > 0$ on the interval $[\alpha, \beta]$. Use the formula

$$A = \frac{1}{2}\int_C -y\,dx + x\,dy$$

to find the area of the region R enclosed by the curve $r = f(\theta)$ and the rays $\theta = \alpha$ and $\theta = \beta$.

22. (a) Use Green's Theorem to prove that

$$\int_C f(x)\,dx + g(y)\,dy = 0$$

if f and g are differentiable functions and C is a simple, closed, piecewise smooth curve.

(b) What does this tell you about the vector field $\mathbf{F}(x, y) = f(x)\mathbf{i} + g(y)\mathbf{j}$?

23. Assume that σ is the parametric surface

$$\mathbf{r} = x(u, v)\mathbf{i} + y(u, v)\mathbf{j} + z(u, v)\mathbf{k}$$

where (u, v) varies over a region R. Express the surface integral

$$\iint_{\sigma} f(x, y, z)\, dS$$

as a double integral with variables of integration u and v.

24. Evaluate $\iint_{\sigma} z\, dS$; $\sigma : x^2 + y^2 = 1 (0 \le z \le 1)$.

25. Do you think that the surface in the accompanying figure is orientable? Explain your reasoning.

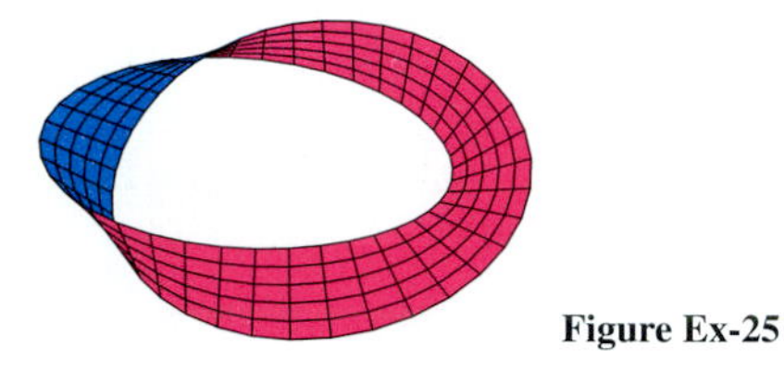

Figure Ex-25

26. Give a physical interpretation of $\iint_{\sigma} \mathbf{F} \cdot \mathbf{n}\, dS$.

27. Find the flux of $\mathbf{F}(x, y, z) = x\mathbf{i} + y\mathbf{j} + 2z\mathbf{k}$ through the portion of the paraboloid $z = 1 - x^2 - y^2$ that is on or above the xy-plane, with upward orientation.

28. Find the flux of $\mathbf{F}(x, y, z) = x\mathbf{i} + 2y\mathbf{j} + 3z\mathbf{k}$ through the unit sphere centered at the origin with outward orientation.

29. State the Divergence Theorem and Stokes' Theorem, including all required hypotheses.

30. Let G be a solid with the surface σ oriented by outward unit normals, suppose that ϕ has continuous first and second partial derivatives in some open set containing G, and let $D_{\mathbf{n}}\phi$ be the directional derivative of ϕ, where $\mathbf{n}$ is an outward unit normal to σ. Show that

$$\iint_{\sigma} D_{\mathbf{n}}\phi\, dS = \iiint_{G} \left[\frac{\partial^2\phi}{\partial x^2} + \frac{\partial^2\phi}{\partial y^2} + \frac{\partial^2\phi}{\partial z^2}\right] dV$$

31. Let σ be the sphere $x^2 + y^2 + z^2 = 1$, let $\mathbf{n}$ be an inward unit normal, and let $D_{\mathbf{n}}f$ be the directional derivative of $f(x, y, z) = x^2 + y^2 + z^2$. Use the result in Exercise 30 to evaluate the surface integral

$$\iint_{\sigma} D_{\mathbf{n}}f\, dS$$

32. Use Stokes' Theorem to evaluate $\iint_{\sigma} \operatorname{curl} \mathbf{F} \cdot \mathbf{n}\, dS$ where $\mathbf{F}(x, y, z) = (z - y)\mathbf{i} + (x + z)\mathbf{j} - (x + y)\mathbf{k}$ and σ is the portion of the paraboloid $z = 2 - x^2 - y^2$ on or above the plane $z = 1$, with upward orientation.

33. Let $\mathbf{F}(x, y, z) = f(x, y, z)\mathbf{i} + g(x, y, z)\mathbf{j} + h(x, y, z)\mathbf{k}$ and suppose that f, g, and h are continuous and have continuous first partial derivatives in a region. It was shown in Exercise 27 of Section 16.3 that if $\mathbf{F}$ is conservative in the region, then

$$\frac{\partial f}{\partial y} = \frac{\partial g}{\partial x}, \quad \frac{\partial f}{\partial z} = \frac{\partial h}{\partial x}, \quad \frac{\partial g}{\partial z} = \frac{\partial h}{\partial y}$$

there. Use this result to show that if $\mathbf{F}$ is conservative in an open spherical region, then $\operatorname{curl} \mathbf{F} = \mathbf{0}$ in that region.

34–35 With the aid of Exercise 33, determine whether $\mathbf{F}$ is conservative.

34. (a) $\mathbf{F}(x, y, z) = z^2\mathbf{i} + e^{-y}\mathbf{j} + 2xz\mathbf{k}$

(b) $\mathbf{F}(x, y, z) = xy\mathbf{i} + x^2\mathbf{j} + \sin z\mathbf{k}$

35. (a) $\mathbf{F}(x, y, z) = \sin x\mathbf{i} + z\mathbf{j} + y\mathbf{k}$

(b) $\mathbf{F}(x, y, z) = z\mathbf{i} + 2yz\mathbf{j} + y^2\mathbf{k}$

36. As discussed in Example 1 of Section 16.1, *Coulomb's law* states that the electrostatic force $\mathbf{F}(\mathbf{r})$ that a particle of charge Q exerts on a particle of charge q is given by the formula

$$\mathbf{F}(\mathbf{r}) = \frac{qQ}{4\pi\epsilon_0\|\mathbf{r}\|^3}\mathbf{r}$$

where $\mathbf{r}$ is the radius vector from Q to q and ϵ_0 is the permittivity constant.

(a) Express the vector field $\mathbf{F}(\mathbf{r})$ in coordinate form $\mathbf{F}(x, y, z)$ with Q at the origin.

(b) Find the work performed by the force field $\mathbf{F}$ on a charge q that moves along a straight line from $(3, 0, 0)$ to $(3, 1, 5)$.

EXPANDING THE CALCULUS HORIZON

Hurricane Modeling

Each year population centers throughout the world are ravaged by hurricanes, and it is the mission of the National Hurricane Center to minimize the damage and loss of life by issuing warnings and forecasts of hurricanes developing in the Caribbean, Atlantic, Gulf of Mexico, and Eastern Pacific regions. Your assignment as a trainee at the Center is to construct a simple mathematical model of a hurricane using basic principles of fluid flow and properties of vector fields.

Modeling Assumptions

You have been notified of a developing hurricane in the Bahamas (designated hurricane *Isaac*) and have been asked to construct a model of its velocity field. Because hurricanes are complicated three-dimensional fluid flows, you will have to make many simplifying assumptions about the structure of a hurricane and the properties of the fluid flow. Accordingly, you decide to model the moisture in Isaac as an ***ideal fluid***, meaning that it is ***incompressible*** and its ***viscosity*** can be ignored. An incompressible fluid is one in which the density of the fluid is the same at all points and cannot be altered by compressive forces. Experience has shown that water can be regarded as incompressible but water vapor cannot. However, incompressibility is a reasonable assumption for a basic hurricane model because a hurricane is not restricted to a closed container that would produce compressive forces.

All fluids have a certain amount of viscosity, which is a resistance to flow—oil and molasses have a high viscosity, whereas water has almost none at subsonic speeds. Thus, it is reasonable to ignore viscosity in a basic model. Next, you decide to assume that the flow is in a ***steady state***, meaning that the velocity of the fluid at any point does not vary with time. This is reasonable over very short time periods for hurricanes that move and change slowly. Finally, although hurricanes are three-dimensional flows, you decide to model a two-dimensional horizontal cross section, so you make the simplifying assumption that the fluid in the cross section flows horizontally.

The photograph of Isaac shown at the beginning of this module reveals a typical pattern of a Caribbean hurricane—a counterclockwise swirl of fluid around the ***eye*** through which the fluid exits the flow in the form of rain. The lower pressure in the eye causes an inward-rushing air mass, and circular winds around the eye contribute to the swirling effect.

Your first objective is to find an explicit formula for Isaac's velocity field $\mathbf{F}(x, y)$, so you begin by introducing a rectangular coordinate system with its origin at the eye and its y-axis pointing north. Moreover, based on the hurricane picture and your knowledge of meteorological theory, you decide to build up the velocity field for Isaac from the velocity fields of simpler flows—a counterclockwise "vortex flow" $\mathbf{F}_1(x, y)$ in which fluid flows counterclockwise in concentric circles around the eye and a "sink flow" $\mathbf{F}_2(x, y)$ in which the fluid flows in straight lines toward a sink at the eye. Once you find explicit formulas for $\mathbf{F}_1(x, y)$ and $\mathbf{F}_2(x, y)$, your plan is to use the ***superposition principle*** from fluid dynamics to express the velocity field for Isaac as $\mathbf{F}(x, y) = \mathbf{F}_1(x, y) + \mathbf{F}_2(x, y)$.

Modeling a Vortex Flow

A ***counterclockwise vortex flow*** of an ideal fluid around the origin has four defining characteristics (Figure 1a on the following page):

- The velocity vector at a point (x, y) is tangent to the circle that is centered at the origin and passes through the point (x, y).

- The direction of the velocity vector at a point (x, y) indicates a counterclockwise motion.
- The speed of the fluid is constant on circles centered at the origin.
- The speed of the fluid along a circle is inversely proportional to the radius of the circle (and hence the speed approaches $+\infty$ as the radius of the circle approaches 0).

In fluid dynamics, the ***strength*** k of a vortex flow is defined to be 2π times the speed of the fluid along the unit circle. If the strength of a vortex flow is known, then the speed of the fluid along any other circle can be found from the fact that speed is inversely proportional to the radius of the circle. Thus, your first objective is to find a formula for a vortex flow $\mathbf{F}_1(x, y)$ with a specified strength k.

Exercise 1 Show that

$$\mathbf{F}_1(x, y) = -\frac{k}{2\pi(x^2 + y^2)}(y\mathbf{i} - x\mathbf{j})$$

is a model for the velocity field of a counterclockwise vortex flow around the origin of strength k by confirming that

(a) $\mathbf{F}_1(x, y)$ has the four properties required of a counterclockwise vortex flow around the origin;

(b) k is 2π times the speed of the fluid along the unit circle.

Exercise 2 Use a graphing utility that can generate vector fields to generate a vortex flow of strength 2π.

Modeling a Sink Flow

A ***uniform sink flow*** of an ideal fluid toward the origin has four defining characteristics (Figure 1*b*):

- The velocity vector at every point (x, y) is directed toward the origin.
- The speed of the fluid is the same at all points on a circle centered at the origin.
- The speed of the fluid at a point is inversely proportional to its distance from the origin (from which it follows that the speed approaches $+\infty$ as the distance from the origin approaches 0).
- There is a sink at the origin at which fluid leaves the flow.

As with a vortex flow, the ***strength*** q of a uniform sink flow is defined to be 2π times the speed of the fluid at points on the unit circle. If the strength of a sink flow is known, then the speed of the fluid at any point in the flow can be found using the fact that the speed is inversely proportional to the distance from the origin. Thus, your next objective is to find a formula for a uniform sink flow $\mathbf{F}_2(x, y)$ with a specified strength q.

Exercise 3 Show that

$$\mathbf{F}_2(x, y) = -\frac{q}{2\pi(x^2 + y^2)}(x\mathbf{i} + y\mathbf{j})$$

is a model for the velocity field of a uniform sink flow toward the origin of strength q by confirming the following facts:

(a) $\mathbf{F}_2(x, y)$ has the four properties required of a uniform sink flow toward the origin. [A reasonable physical argument to confirm the existence of the sink will suffice.]

(b) q is 2π times the speed of the fluid at points on the unit circle.

Exercise 4 Use a graphing utility that can generate vector fields to generate a uniform sink flow of strength 2π.

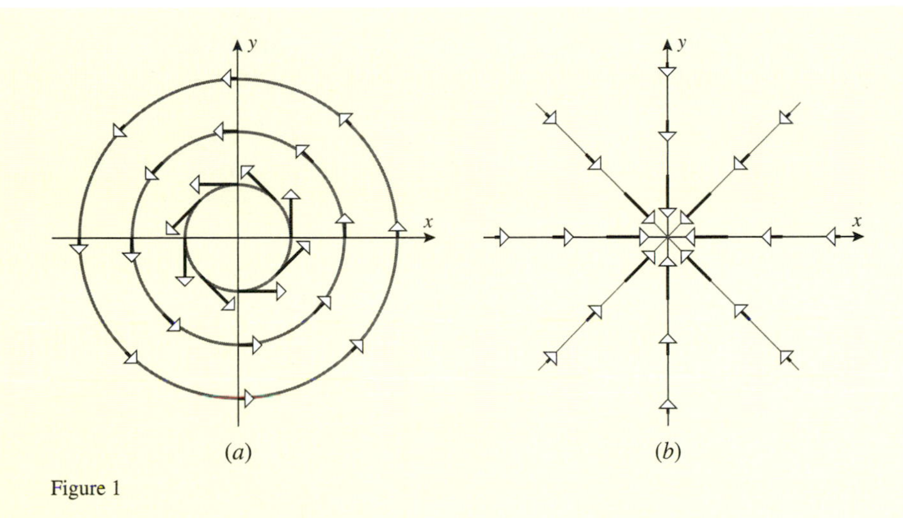

Figure 1

A Basic Hurricane Model

It now follows from Exercises 1 and 3 that the vector field $\mathbf{F}(x, y)$ for a hurricane model that combines a vortex flow around the origin of strength k and a uniform sink flow toward the origin of strength q is

$$\mathbf{F}(x, y) = -\frac{1}{2\pi(x^2 + y^2)}[(qx + ky)\mathbf{i} + (qy - kx)\mathbf{j}] \tag{1}$$

Exercise 5

(a) Figure 2 shows a vector field for a hurricane with vortex strength $k = 2\pi$ and sink strength $q = 2\pi$. Use a graphing utility that can generate vector fields to produce a reasonable facsimile of this figure.

(b) Make a conjecture about the effect of increasing k and keeping q fixed, and check your conjecture using a graphing utility.

(c) Make a conjecture about the effect of increasing q and keeping k fixed, and check your conjecture using a graphing utility.

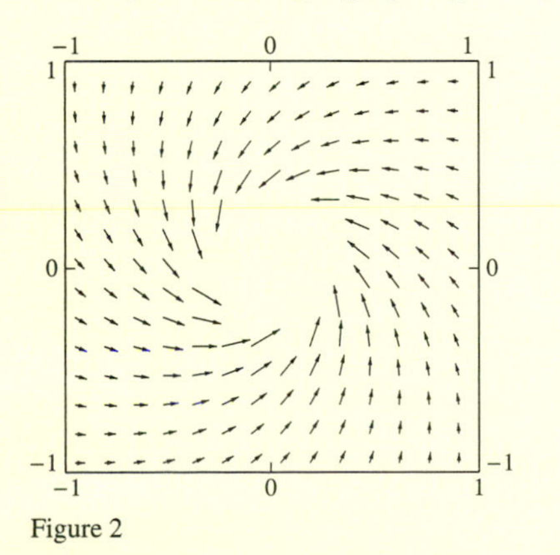

Figure 2

Modeling Hurricane Isaac

You are now ready to apply Formula (1) to obtain a model of the vector field $\mathbf{F}(x, y)$ of hurricane Isaac. You need some observational data to determine the constants k and q, so you call the Technical Support Branch of the Center for the latest information on hurricane Isaac. They report that 20 km from the eye the wind velocity has a component of 15 km/h toward the eye and a counterclockwise tangential component of 45 km/h.

Exercise 6

(a) Find the strengths k and q of the vortex and sink for hurricane Isaac.

(b) Find the vector field $\mathbf{F}(x, y)$ for hurricane Isaac.

(c) Estimate the size of hurricane Isaac by finding a radius beyond which the wind speed is less than 5 km/h.

Streamlines for the Basic Hurricane Model

The paths followed by the fluid particles in a fluid flow are called the ***streamlines*** of the flow. Thus, the vectors $\mathbf{F}(x, y)$ in the velocity field of a fluid flow are tangent to the streamlines. If the streamlines can be represented as the level curves of some function $\psi(x, y)$, then the function ψ is called a ***stream function*** for the flow. Since $\nabla\psi$ is normal to the level curves $\psi(x, y) = c$, it follows that $\nabla\psi$ is normal to the streamlines; and this in turn implies that

$$\nabla\psi \cdot \mathbf{F} = 0 \tag{2}$$

Your plan is to use this equation to find the stream function and then the streamlines of the basic hurricane model.

Since the vortex and sink flows that produce the basic hurricane model have a central symmetry, intuition suggests that polar coordinates may lead to simpler equations for the streamlines than rectangular coordinates. Thus, you decide to express the velocity vector $\mathbf{F}$ at a point (r, θ) in terms of the orthogonal unit vectors

$$\mathbf{u}_r = \cos\theta\,\mathbf{i} + \sin\theta\,\mathbf{j} \quad \text{and} \quad \mathbf{u}_\theta = -\sin\theta\,\mathbf{i} + \cos\theta\,\mathbf{j}$$

The vector $\mathbf{u}_r$, called the ***radial unit vector***, points away from the origin, and the vector $\mathbf{u}_\theta$, called the ***transverse unit vector***, is obtained by rotating $\mathbf{u}_r$ counterclockwise 90° (Figure 3).

Exercise 7 Show that the vector field for the basic hurricane model given in (1) can be expressed in terms of $\mathbf{u}_r$ and $\mathbf{u}_\theta$ as

$$\mathbf{F} = -\frac{1}{2\pi r}(q\mathbf{u}_r - k\mathbf{u}_\theta)$$

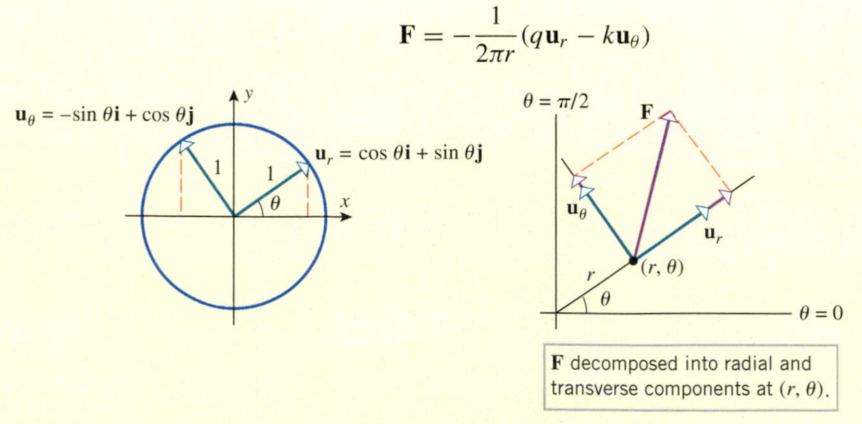

Figure 3

It follows from Exercise 75 of Section 14.6 that the gradient of the stream function can be expressed in terms of $\mathbf{u}_r$ and $\mathbf{u}_\theta$ as

$$\nabla\psi = \frac{\partial\psi}{\partial r}\mathbf{u}_r + \frac{1}{r}\frac{\partial\psi}{\partial\theta}\mathbf{u}_\theta$$

Exercise 8 Confirm that for the basic hurricane model the orthogonality condition in (2) is satisfied if

$$\frac{\partial \psi}{\partial r} = \frac{k}{r} \quad \text{and} \quad \frac{\partial \psi}{\partial \theta} = q$$

Exercise 9 By integrating the equations in Exercise 8, show that

$$\psi = k \ln r + q\theta$$

is a stream function for the basic hurricane model.

Exercise 10 Show that the streamlines for the basic hurricane model are logarithmic spirals of the form

$$r = Ke^{-q\theta/k} \quad (K > 0)$$

Exercise 11 Use a graphing utility to generate some typical streamlines for the basic hurricane model with vortex strength 2π and sink strength 2π.

Streamlines for Hurricane Isaac

Exercise 12 In Exercise 6 you found the strengths k and q of the vortex and sink for hurricane Isaac. Use that information to find a formula for the family of streamlines for Isaac; and then use a graphing utility to graph the streamline that passes through the point that is 20 km from the eye in the direction that is 45° NE from the eye.

Module by: *Josef S. Torok, Rochester Institute of Technology*
Howard Anton, Drexel University

ANSWERS TO ODD-NUMBERED EXERCISES

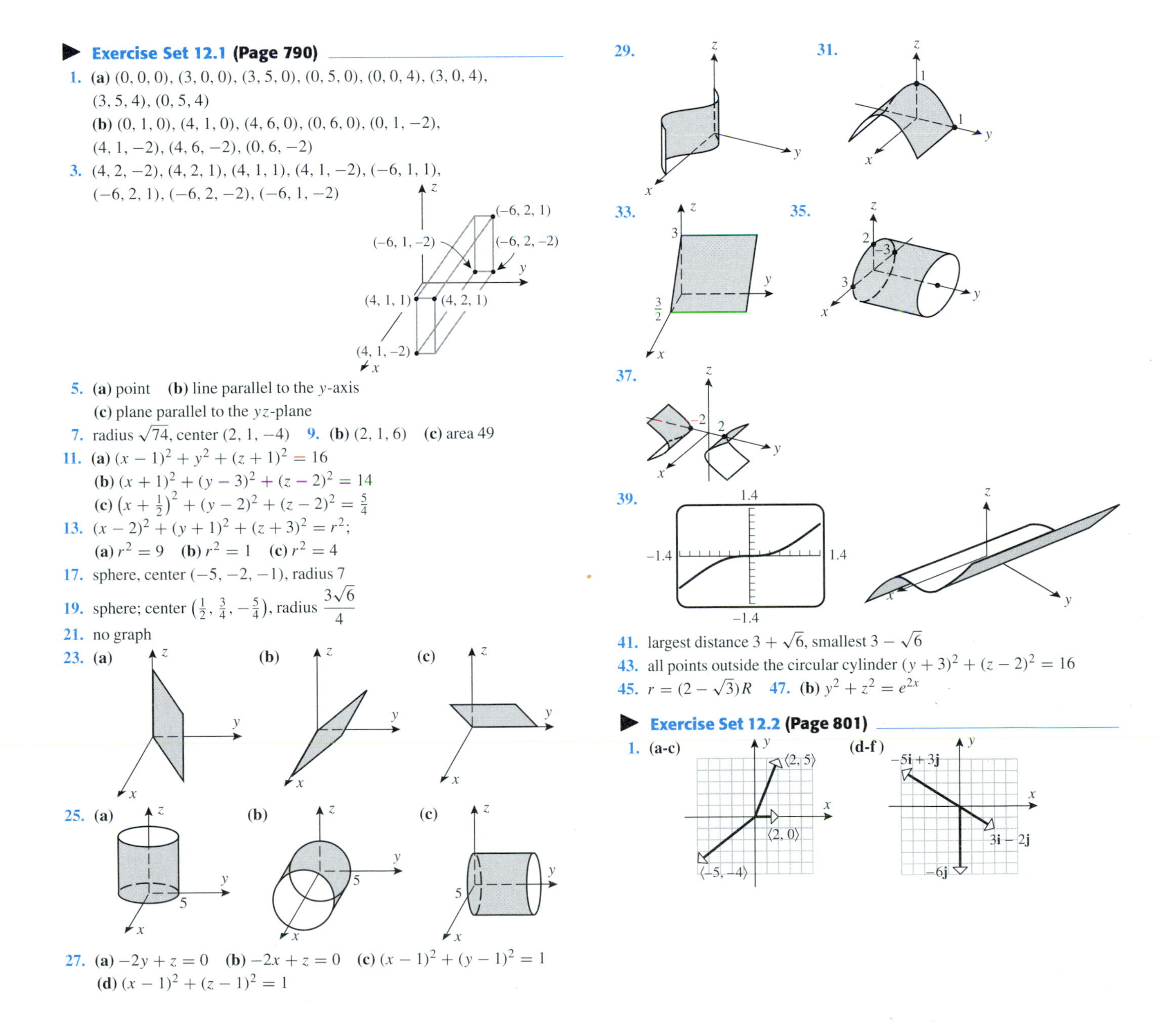

Exercise Set 12.1 (Page 790)

1. **(a)** (0, 0, 0), (3, 0, 0), (3, 5, 0), (0, 5, 0), (0, 0, 4), (3, 0, 4), (3, 5, 4), (0, 5, 4)
   (b) (0, 1, 0), (4, 1, 0), (4, 6, 0), (0, 6, 0), (0, 1, −2), (4, 1, −2), (4, 6, −2), (0, 6, −2)
3. (4, 2, −2), (4, 2, 1), (4, 1, 1), (4, 1, −2), (−6, 1, 1), (−6, 2, 1), (−6, 2, −2), (−6, 1, −2)
5. **(a)** point **(b)** line parallel to the y-axis **(c)** plane parallel to the yz-plane
7. radius $\sqrt{74}$, center (2, 1, −4) 9. **(b)** (2, 1, 6) **(c)** area 49
11. **(a)** $(x-1)^2+y^2+(z+1)^2=16$
    (b) $(x+1)^2+(y-3)^2+(z-2)^2=14$
    (c) $\left(x+\frac{1}{2}\right)^2+(y-2)^2+(z-2)^2=\frac{5}{4}$
13. $(x-2)^2+(y+1)^2+(z+3)^2=r^2$; **(a)** $r^2=9$ **(b)** $r^2=1$ **(c)** $r^2=4$
17. sphere, center (−5, −2, −1), radius 7
19. sphere; center $\left(\frac{1}{2}, \frac{3}{4}, -\frac{5}{4}\right)$, radius $\frac{3\sqrt{6}}{4}$
21. no graph
23. **(a)** **(b)** **(c)**
25. **(a)** **(b)** **(c)**
27. **(a)** $-2y+z=0$ **(b)** $-2x+z=0$ **(c)** $(x-1)^2+(y-1)^2=1$ **(d)** $(x-1)^2+(z-1)^2=1$
29. 31. 33. 35. 37. 39.
41. largest distance $3+\sqrt{6}$, smallest $3-\sqrt{6}$
43. all points outside the circular cylinder $(y+3)^2+(z-2)^2=16$
45. $r=(2-\sqrt{3})R$ 47. **(b)** $y^2+z^2=e^{2x}$

Exercise Set 12.2 (Page 801)

1. **(a-c)** **(d-f)**

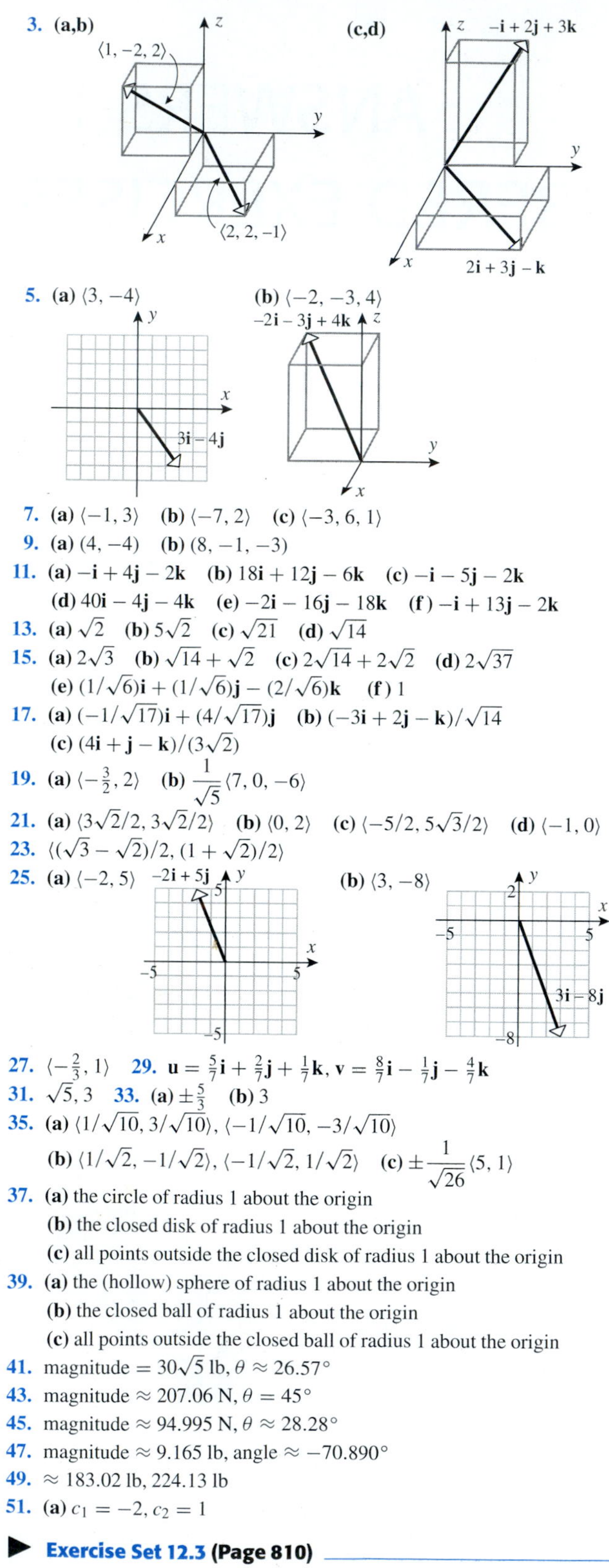

3. (a,b) (c,d)

5. (a) $\langle 3, -4\rangle$ (b) $\langle -2, -3, 4\rangle$

7. (a) $\langle -1, 3\rangle$ (b) $\langle -7, 2\rangle$ (c) $\langle -3, 6, 1\rangle$

9. (a) $(4, -4)$ (b) $(8, -1, -3)$

11. (a) $-\mathbf{i}+4\mathbf{j}-2\mathbf{k}$ (b) $18\mathbf{i}+12\mathbf{j}-6\mathbf{k}$ (c) $-\mathbf{i}-5\mathbf{j}-2\mathbf{k}$ (d) $40\mathbf{i}-4\mathbf{j}-4\mathbf{k}$ (e) $-2\mathbf{i}-16\mathbf{j}-18\mathbf{k}$ (f) $-\mathbf{i}+13\mathbf{j}-2\mathbf{k}$

13. (a) $\sqrt{2}$ (b) $5\sqrt{2}$ (c) $\sqrt{21}$ (d) $\sqrt{14}$

15. (a) $2\sqrt{3}$ (b) $\sqrt{14}+\sqrt{2}$ (c) $2\sqrt{14}+2\sqrt{2}$ (d) $2\sqrt{37}$ (e) $(1/\sqrt{6})\mathbf{i}+(1/\sqrt{6})\mathbf{j}-(2/\sqrt{6})\mathbf{k}$ (f) 1

17. (a) $(-1/\sqrt{17})\mathbf{i}+(4/\sqrt{17})\mathbf{j}$ (b) $(-3\mathbf{i}+2\mathbf{j}-\mathbf{k})/\sqrt{14}$ (c) $(4\mathbf{i}+\mathbf{j}-\mathbf{k})/(3\sqrt{2})$

19. (a) $\langle -\frac{3}{2}, 2\rangle$ (b) $\frac{1}{\sqrt{5}}\langle 7, 0, -6\rangle$

21. (a) $\langle 3\sqrt{2}/2, 3\sqrt{2}/2\rangle$ (b) $\langle 0, 2\rangle$ (c) $\langle -5/2, 5\sqrt{3}/2\rangle$ (d) $\langle -1, 0\rangle$

23. $\langle(\sqrt{3}-\sqrt{2})/2, (1+\sqrt{2})/2\rangle$

25. (a) $\langle -2, 5\rangle$ (b) $\langle 3, -8\rangle$

27. $\langle -\frac{2}{3}, 1\rangle$ **29.** $\mathbf{u}=\frac{5}{7}\mathbf{i}+\frac{2}{7}\mathbf{j}+\frac{1}{7}\mathbf{k}$, $\mathbf{v}=\frac{8}{7}\mathbf{i}-\frac{1}{7}\mathbf{j}-\frac{4}{7}\mathbf{k}$

31. $\sqrt{5}$, 3 **33.** (a) $\pm\frac{5}{3}$ (b) 3

35. (a) $\langle 1/\sqrt{10}, 3/\sqrt{10}\rangle$, $\langle -1/\sqrt{10}, -3/\sqrt{10}\rangle$ (b) $\langle 1/\sqrt{2}, -1/\sqrt{2}\rangle$, $\langle -1/\sqrt{2}, 1/\sqrt{2}\rangle$ (c) $\pm\frac{1}{\sqrt{26}}\langle 5, 1\rangle$

37. (a) the circle of radius 1 about the origin
(b) the closed disk of radius 1 about the origin
(c) all points outside the closed disk of radius 1 about the origin

39. (a) the (hollow) sphere of radius 1 about the origin
(b) the closed ball of radius 1 about the origin
(c) all points outside the closed ball of radius 1 about the origin

41. magnitude $= 30\sqrt{5}$ lb, $\theta \approx 26.57°$

43. magnitude ≈ 207.06 N, $\theta = 45°$

45. magnitude ≈ 94.995 N, $\theta \approx 28.28°$

47. magnitude ≈ 9.165 lb, angle $\approx -70.890°$

49. ≈ 183.02 lb, 224.13 lb

51. (a) $c_1 = -2$, $c_2 = 1$

▶ Exercise Set 12.3 (Page 810)

1. (a) -10; $\cos\theta = -1/\sqrt{5}$ (b) -3; $\cos\theta = -3/\sqrt{58}$ (c) 0; $\cos\theta = 0$ (d) -20; $\cos\theta = -20/(3\sqrt{70})$

3. (a) obtuse (b) acute (c) obtuse (d) orthogonal

5. $\sqrt{2}/2, 0, -\sqrt{2}/2, -1, -\sqrt{2}/2, 0, \sqrt{2}/2$

7. (a) vertex B (b) $82°, 60°, 38°$ **13.** $r = 7/5$

15. (a) $\alpha = \beta \approx 55°, \gamma \approx 125°$ (b) $\alpha \approx 48°, \beta \approx 132°, \gamma \approx 71°$

19. (a) $\approx 35°$ (b) $90°$

21. $64°, 41°, 60°$ **23.** $71°, 61°, 36°$

25. (a) $\left\langle \frac{2}{3}, \frac{4}{3}, \frac{4}{3}\right\rangle, \left\langle \frac{4}{3}, -\frac{7}{3}, \frac{5}{3}\right\rangle$
(b) $\left\langle -\frac{74}{49}, -\frac{111}{49}, \frac{222}{49}\right\rangle, \left\langle \frac{270}{49}, \frac{62}{49}, \frac{121}{49}\right\rangle$

27. (a) $\langle 1, 1\rangle + \langle -4, 4\rangle$ (b) $\left\langle 0, -\frac{8}{5}, \frac{4}{5}\right\rangle + \left\langle -2, \frac{13}{5}, \frac{26}{5}\right\rangle$
(c) $\mathbf{v} = \langle 1, 4, 1\rangle$ is orthogonal to $\mathbf{b}$.

29. $\sqrt{564/29}$ **31.** 98 N

33. (a)

500   500
−20   100   −20   100
−100   −100

(b) decrease (c) $40/\sqrt{65}$ ft

35. $-5\sqrt{3}$ J **37.** $W = 375$ ft·lb

45. (a) $40°$ (b) $x \approx -0.682328$

▶ Exercise Set 12.4 (Page 821)

1. (a) $-\mathbf{j}+\mathbf{k}$ **3.** $\langle 7, 10, 9\rangle$ **5.** $\langle -4, -6, -3\rangle$

7. (a) $\langle -20, -67, -9\rangle$ (b) $\langle -78, 52, -26\rangle$ (c) $\langle 0, -56, -392\rangle$ (d) $\langle 0, 56, 392\rangle$

9. $\frac{1}{\sqrt{2}}, -\frac{1}{\sqrt{2}}, 0$ **11.** $\pm\frac{1}{\sqrt{6}}\langle 2, 1, 1\rangle$ **13.** $\sqrt{59}$ **15.** $\sqrt{374}/2$

17. 80 **19.** -3 **21.** 16 **23.** (a) yes (b) yes (c) no

25. (a) 9 (b) $\sqrt{122}$ (c) $\sin^{-1}\left(\frac{9}{14}\right)$

27. (a) $2\sqrt{141/29}$ (b) $6/\sqrt{5}$ **29.** $\frac{2}{3}$ **33.** $\theta = \pi/4$

35. (a) $10\sqrt{2}$ lb·ft, direction of rotation about P is counterclockwise looking along $\overrightarrow{PQ}\times\mathbf{F} = -10\mathbf{i}+10\mathbf{k}$ toward its initial point
(b) 10 lb·ft, direction of rotation about P is counterclockwise looking along $-10\mathbf{i}$ toward its initial point
(c) 0 lb·ft, no rotation about P

37. ≈ 36.19 N·m **41.** $-8\mathbf{i}-20\mathbf{j}+2\mathbf{k}, -8\mathbf{i}-8\mathbf{k}$ **45.** 1.887850

▶ Exercise Set 12.5 (Page 828)

1. (a) $L_1: x = 1, y = t$, $L_2: x = t, y = 1$, $L_3: x = t, y = t$
(b) $L_1: x = 1, y = 1, z = t$, $L_2: x = t, y = 1, z = 1$, $L_3: x = 1, y = t, z = 1$, $L_4: x = t, y = t, z = t$

3. (a) $x = 3 + 2t, y = -2 + 3t$; line segment: $0 \le t \le 1$
(b) $x = 5 - 3t, y = -2 + 6t, z = 1 + t$; line segment: $0 \le t \le 1$

5. (a) $x = 2 + t, y = -3 - 4t$ (b) $x = t, y = -t, z = 1 + t$

7. (a) $P(2, -1)$, $\mathbf{v} = 4\mathbf{i} - \mathbf{j}$ (b) $P(-1, 2, 4)$, $\mathbf{v} = 5\mathbf{i} + 7\mathbf{j} - 8\mathbf{k}$

9. (a) $\langle -3, 4\rangle + t\langle 1, 5\rangle$; $-3\mathbf{i}+4\mathbf{j}+t(\mathbf{i}+5\mathbf{j})$
(b) $\langle 2, -3, 0\rangle + t\langle -1, 5, 1\rangle$; $2\mathbf{i}-3\mathbf{j}+t(-\mathbf{i}+5\mathbf{j}+\mathbf{k})$

11. $x = -5 + 2t, y = 2 - 3t$ **13.** $x = 3 + 4t, y = -4 + 3t$

15. $x = -1 + 3t, y = 2 - 4t, z = 4 + t$

17. $x = -2 + 2t, y = -t, z = 5 + 2t$

19. (a) $x = 7$ (b) $y = \frac{7}{3}$ (c) $x = \frac{-1 \pm \sqrt{85}}{6}, y = \frac{43 \mp \sqrt{85}}{18}$

21. $(-2, 10, 0)$; $(-2, 0, -5)$; The line does not intersect the yz-plane.

23. $(0, 4, -2), (4, 0, 6)$ **25.** $(1, -1, 2)$ **29.** The lines are parallel.

31. The points do not lie on the same line.

35. $\langle x, y\rangle = \langle -1, 2\rangle + t\langle 1, 1\rangle$

37. the point $1/n$ of the way from $(-2, 0)$ to $(1, 3)$
39. the line segment joining the points $(1, 0)$ and $(-3, 6)$
41. $(5, 2)$ **43.** $2\sqrt{5}$ **45.** distance $= \sqrt{35/6}$
47. **(a)** $x = x_0 + (x_1 - x_0)t,\ y = y_0 + (y_1 - y_0)t,\ z = z_0 + (z_1 - z_0)t$
(b) $x = x_1 + at,\ y = y_1 + bt,\ z = z_1 + ct$
49. **(b)** $\langle x, y, z\rangle = \langle 1 + 2t, -3 + 4t, 5 + t\rangle$
51. **(b)** $84°$ **(c)** $x = 7 + t,\ y = -1,\ z = -2 + t$
53. $x = t,\ y = 2 + t,\ z = 1 - t$
55. **(a)** $\sqrt{17}$ cm **(b)**

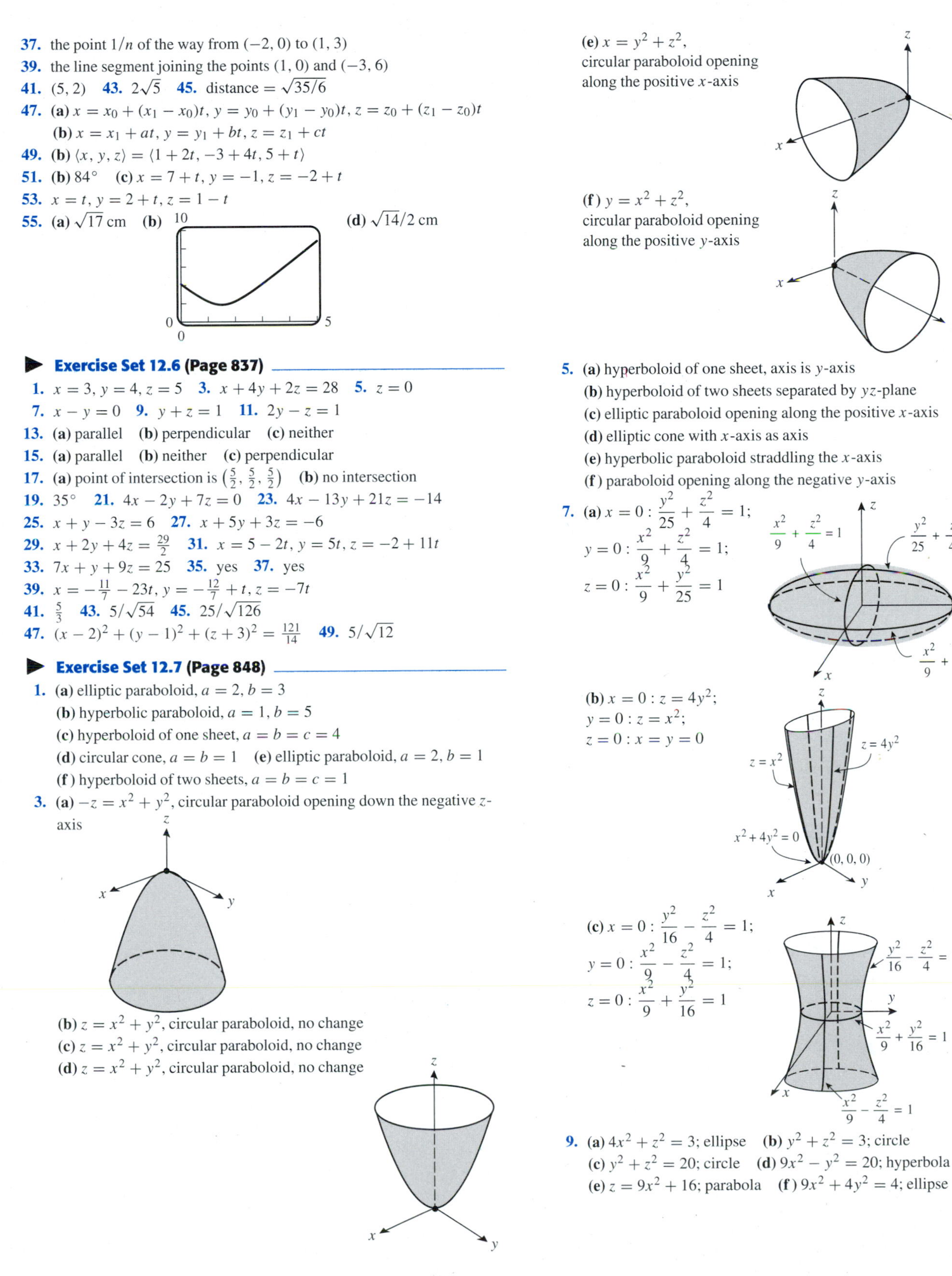

(d) $\sqrt{14}/2$ cm

Exercise Set 12.6 (Page 837)

1. $x = 3,\ y = 4,\ z = 5$ **3.** $x + 4y + 2z = 28$ **5.** $z = 0$
7. $x - y = 0$ **9.** $y + z = 1$ **11.** $2y - z = 1$
13. **(a)** parallel **(b)** perpendicular **(c)** neither
15. **(a)** parallel **(b)** neither **(c)** perpendicular
17. **(a)** point of intersection is $\left(\frac{5}{2}, \frac{5}{2}, \frac{5}{2}\right)$ **(b)** no intersection
19. $35°$ **21.** $4x - 2y + 7z = 0$ **23.** $4x - 13y + 21z = -14$
25. $x + y - 3z = 6$ **27.** $x + 5y + 3z = -6$
29. $x + 2y + 4z = \frac{29}{2}$ **31.** $x = 5 - 2t,\ y = 5t,\ z = -2 + 11t$
33. $7x + y + 9z = 25$ **35.** yes **37.** yes
39. $x = -\frac{11}{7} - 23t,\ y = -\frac{12}{7} + t,\ z = -7t$
41. $\frac{5}{3}$ **43.** $5/\sqrt{54}$ **45.** $25/\sqrt{126}$
47. $(x - 2)^2 + (y - 1)^2 + (z + 3)^2 = \frac{121}{14}$ **49.** $5/\sqrt{12}$

Exercise Set 12.7 (Page 848)

1. **(a)** elliptic paraboloid, $a = 2, b = 3$
(b) hyperbolic paraboloid, $a = 1, b = 5$
(c) hyperboloid of one sheet, $a = b = c = 4$
(d) circular cone, $a = b = 1$ **(e)** elliptic paraboloid, $a = 2, b = 1$
(f) hyperboloid of two sheets, $a = b = c = 1$
3. **(a)** $-z = x^2 + y^2$, circular paraboloid opening down the negative z-axis
(b) $z = x^2 + y^2$, circular paraboloid, no change
(c) $z = x^2 + y^2$, circular paraboloid, no change
(d) $z = x^2 + y^2$, circular paraboloid, no change
(e) $x = y^2 + z^2$, circular paraboloid opening along the positive x-axis
(f) $y = x^2 + z^2$, circular paraboloid opening along the positive y-axis

5. **(a)** hyperboloid of one sheet, axis is y-axis
(b) hyperboloid of two sheets separated by yz-plane
(c) elliptic paraboloid opening along the positive x-axis
(d) elliptic cone with x-axis as axis
(e) hyperbolic paraboloid straddling the x-axis
(f) paraboloid opening along the negative y-axis

7. **(a)** $x = 0: \dfrac{y^2}{25} + \dfrac{z^2}{4} = 1;$
$y = 0: \dfrac{x^2}{9} + \dfrac{z^2}{4} = 1;$
$z = 0: \dfrac{x^2}{9} + \dfrac{y^2}{25} = 1$

(b) $x = 0: z = 4y^2;$
$y = 0: z = x^2;$
$z = 0: x = y = 0$

(c) $x = 0: \dfrac{y^2}{16} - \dfrac{z^2}{4} = 1;$
$y = 0: \dfrac{x^2}{9} - \dfrac{z^2}{4} = 1;$
$z = 0: \dfrac{x^2}{9} + \dfrac{y^2}{16} = 1$

9. **(a)** $4x^2 + z^2 = 3$; ellipse **(b)** $y^2 + z^2 = 3$; circle
(c) $y^2 + z^2 = 20$; circle **(d)** $9x^2 - y^2 = 20$; hyperbola
(e) $z = 9x^2 + 16$; parabola **(f)** $9x^2 + 4y^2 = 4$; ellipse

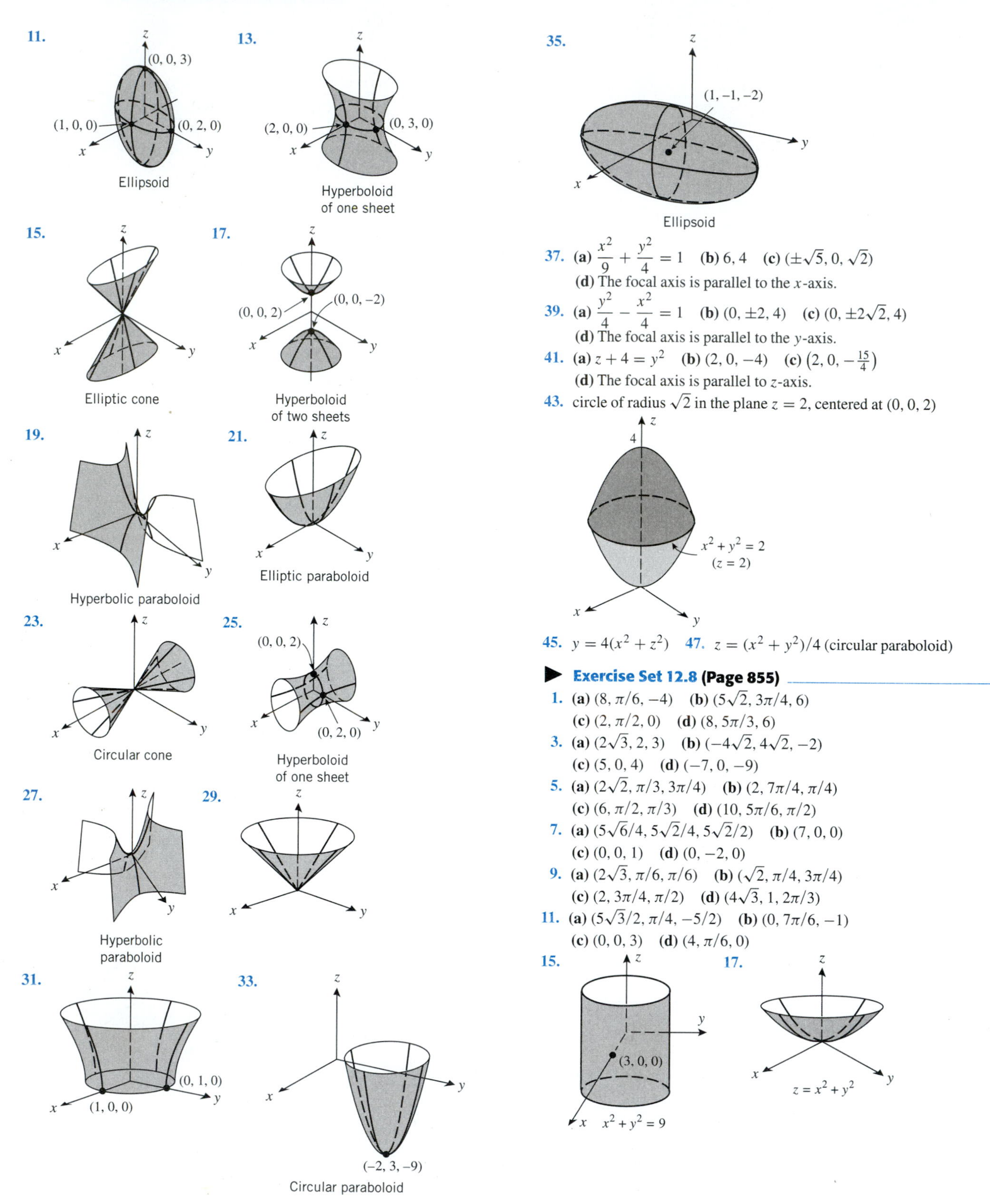

37. **(a)** $\dfrac{x^2}{9}+\dfrac{y^2}{4}=1$ **(b)** 6, 4 **(c)** $(\pm\sqrt{5}, 0, \sqrt{2})$
(d) The focal axis is parallel to the x-axis.

39. **(a)** $\dfrac{y^2}{4}-\dfrac{x^2}{4}=1$ **(b)** $(0, \pm 2, 4)$ **(c)** $(0, \pm 2\sqrt{2}, 4)$
(d) The focal axis is parallel to the y-axis.

41. **(a)** $z+4=y^2$ **(b)** $(2, 0, -4)$ **(c)** $\left(2, 0, -\frac{15}{4}\right)$
(d) The focal axis is parallel to z-axis.

43. circle of radius $\sqrt{2}$ in the plane $z=2$, centered at $(0, 0, 2)$

45. $y=4(x^2+z^2)$ 47. $z=(x^2+y^2)/4$ (circular paraboloid)

▶ Exercise Set 12.8 (Page 855)

1. **(a)** $(8, \pi/6, -4)$ **(b)** $(5\sqrt{2}, 3\pi/4, 6)$
(c) $(2, \pi/2, 0)$ **(d)** $(8, 5\pi/3, 6)$

3. **(a)** $(2\sqrt{3}, 2, 3)$ **(b)** $(-4\sqrt{2}, 4\sqrt{2}, -2)$
(c) $(5, 0, 4)$ **(d)** $(-7, 0, -9)$

5. **(a)** $(2\sqrt{2}, \pi/3, 3\pi/4)$ **(b)** $(2, 7\pi/4, \pi/4)$
(c) $(6, \pi/2, \pi/3)$ **(d)** $(10, 5\pi/6, \pi/2)$

7. **(a)** $(5\sqrt{6}/4, 5\sqrt{2}/4, 5\sqrt{2}/2)$ **(b)** $(7, 0, 0)$
(c) $(0, 0, 1)$ **(d)** $(0, -2, 0)$

9. **(a)** $(2\sqrt{3}, \pi/6, \pi/6)$ **(b)** $(\sqrt{2}, \pi/4, 3\pi/4)$
(c) $(2, 3\pi/4, \pi/2)$ **(d)** $(4\sqrt{3}, 1, 2\pi/3)$

11. **(a)** $(5\sqrt{3}/2, \pi/4, -5/2)$ **(b)** $(0, 7\pi/6, -1)$
(c) $(0, 0, 3)$ **(d)** $(4, \pi/6, 0)$

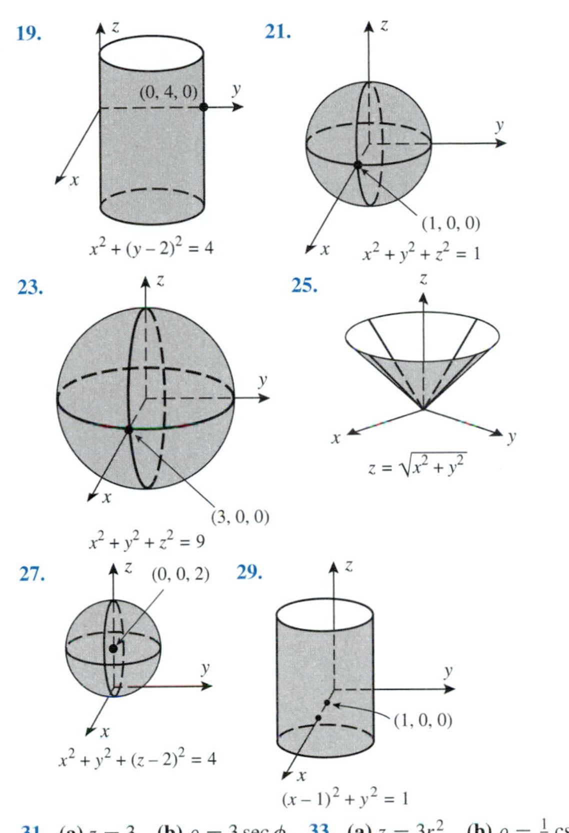

31. **(a)** $z = 3$ **(b)** $\rho = 3\sec\phi$ **33.** **(a)** $z = 3r^2$ **(b)** $\rho = \frac{1}{3}\csc\phi\cot\phi$
35. **(a)** $r = 2$ **(b)** $\rho = 2\csc\phi$ **37.** **(a)** $r^2 + z^2 = 9$ **(b)** $\rho = 3$
39. **(a)** $2r\cos\theta + 3r\sin\theta + 4z = 1$
(b) $2\rho\sin\phi\cos\theta + 3\rho\sin\phi\sin\theta + 4\rho\cos\phi = 1$
41. **(a)** $r^2\cos^2\theta = 16 - z^2$ **(b)** $\rho^2(1 - \sin^2\phi\sin^2\theta) = 16$
43. all points on or above the paraboloid $z = x^2 + y^2$ that are also on or below the plane $z = 4$
45. all points on or between concentric spheres of radii 1 and 3 centered at the origin
47. spherical $(4000, \pi/6, \pi/6)$, rectangular $(1000\sqrt{3}, 1000, 2000\sqrt{3})$
49. **(a)** $(10, \pi/2, 1)$ **(b)** $(0, 10, 1)$ **(c)** $(\sqrt{101}, \pi/2, \tan^{-1} 10)$
51. ≈ 2927 km

▶ Chapter 12. Review Exercises (Page 856)

3. **(b)** $-1/2, \pm\sqrt{3}/2$ **(d)** true
5. **(a)** $r^2 = 16$ **(b)** $r^2 = 25$ **(c)** $r^2 = 9$
7. $(7, 5)$
9. **(a)** $-\frac{3}{4}$ **(b)** $\frac{1}{7}$ **(c)** $(48 \pm 25\sqrt{3})/11$ **(d)** $c = \frac{4}{3}$
13. 13 ft·lb **15.** **(a)** $\sqrt{26}/2$ **(b)** $\sqrt{26}/3$
17. **(a)** 29 **(b)** $\dfrac{29}{\sqrt{65}}$ **19.** $x = 4 + t,\ y = 1 - t,\ z = 2$
21. $x + 5y - z - 2 = 0$ **23.** $a_1a_2 + b_1b_2 + c_1c_2 = 0$
25. **(a)** hyperboloid of one sheet **(b)** sphere **(c)** circular cone
27. **(a)** $z = x^2 - y^2$ **(b)** $xz = 1$

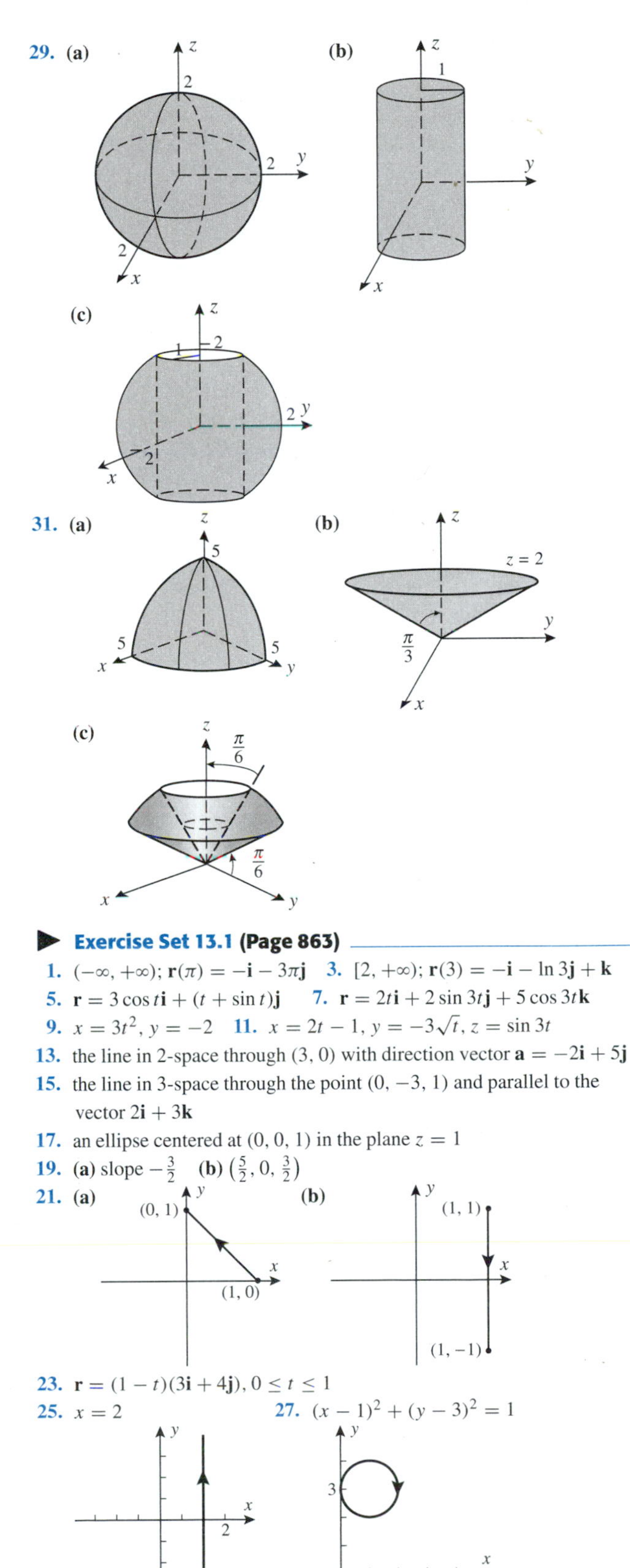

▶ Exercise Set 13.1 (Page 863)

1. $(-\infty, +\infty)$; $\mathbf{r}(\pi) = -\mathbf{i} - 3\pi\mathbf{j}$ **3.** $[2, +\infty)$; $\mathbf{r}(3) = -\mathbf{i} - \ln 3\mathbf{j} + \mathbf{k}$
5. $\mathbf{r} = 3\cos t\mathbf{i} + (t + \sin t)\mathbf{j}$ **7.** $\mathbf{r} = 2t\mathbf{i} + 2\sin 3t\mathbf{j} + 5\cos 3t\mathbf{k}$
9. $x = 3t^2,\ y = -2$ **11.** $x = 2t - 1,\ y = -3\sqrt{t},\ z = \sin 3t$
13. the line in 2-space through $(3, 0)$ with direction vector $\mathbf{a} = -2\mathbf{i} + 5\mathbf{j}$
15. the line in 3-space through the point $(0, -3, 1)$ and parallel to the vector $2\mathbf{i} + 3\mathbf{k}$
17. an ellipse centered at $(0, 0, 1)$ in the plane $z = 1$
19. **(a)** slope $-\frac{3}{2}$ **(b)** $\left(\frac{5}{2}, 0, \frac{3}{2}\right)$
21. **(a)** **(b)**
23. $\mathbf{r} = (1 - t)(3\mathbf{i} + 4\mathbf{j}),\ 0 \le t \le 1$
25. $x = 2$ **27.** $(x - 1)^2 + (y - 3)^2 = 1$

29. $x^2 - y^2 = 1, x \ge 1$ **31.** **33.** **35.** $x = t, y = t, z = 2t^2$

37. $\mathbf{r}(t) = t\mathbf{i} + t^2\mathbf{j} \pm \frac{1}{3}\sqrt{81 - 9t^2 - t^4}\,\mathbf{k}$ **43.** $c = 3/(2\pi)$

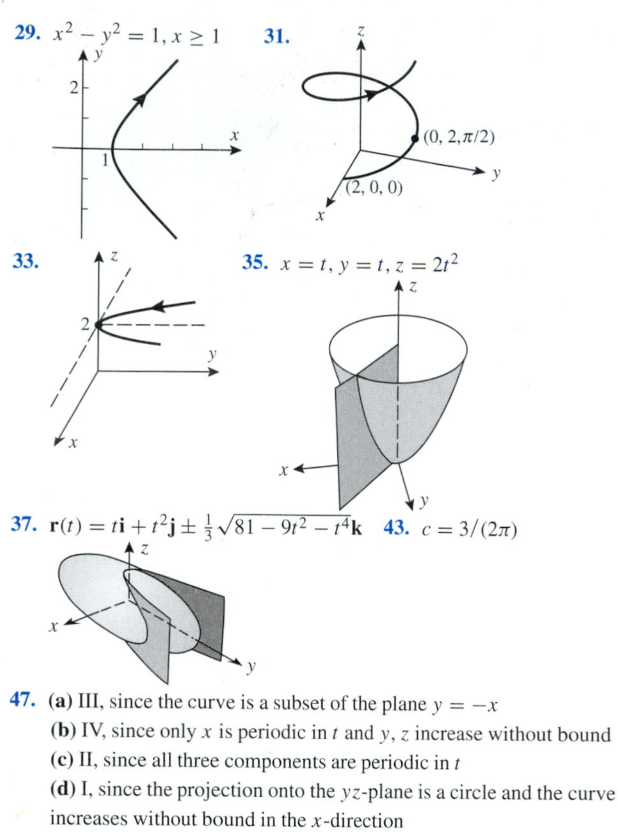

47. **(a)** III, since the curve is a subset of the plane $y = -x$
(b) IV, since only x is periodic in t and y, z increase without bound
(c) II, since all three components are periodic in t
(d) I, since the projection onto the yz-plane is a circle and the curve increases without bound in the x-direction

49. **(a)** $x = 3\cos t$ $y = 3\sin t, z = 9\cos^2 t$ **51.**
(b)

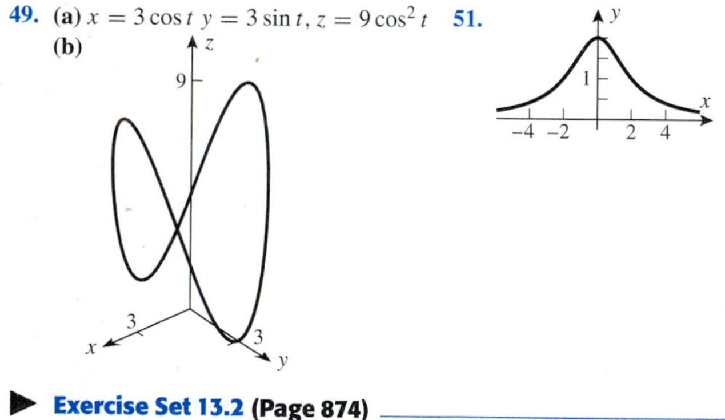

Exercise Set 13.2 (Page 874)

1. $\langle \frac{1}{3}, 0\rangle$ **3.** $2\mathbf{i} - 3\mathbf{j} + 4\mathbf{k}$ **5.** **(a)** continuous **(b)** not continuous

7. **9.** $(\sin t)\mathbf{j}$ **11.** $\mathbf{r}'(2) = \langle 1, 4\rangle$

13. $\mathbf{r}'(0) = \mathbf{j}$ **15.** $\mathbf{r}'(\pi/2) = -2\mathbf{k}$ **17.**

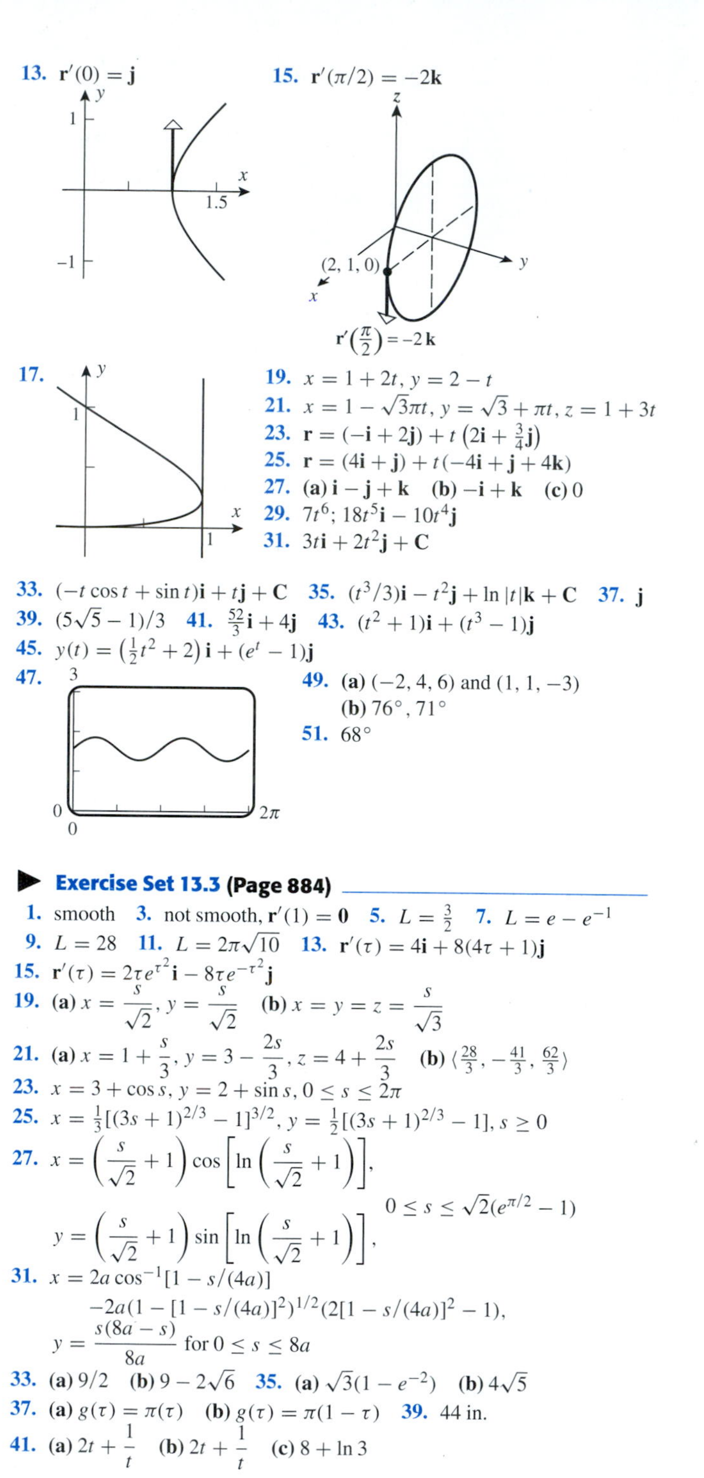

19. $x = 1 + 2t, y = 2 - t$
21. $x = 1 - \sqrt{3}\pi t, y = \sqrt{3} + \pi t, z = 1 + 3t$
23. $\mathbf{r} = (-\mathbf{i} + 2\mathbf{j}) + t\left(2\mathbf{i} + \frac{3}{4}\mathbf{j}\right)$
25. $\mathbf{r} = (4\mathbf{i} + \mathbf{j}) + t(-4\mathbf{i} + \mathbf{j} + 4\mathbf{k})$
27. **(a)** $\mathbf{i} - \mathbf{j} + \mathbf{k}$ **(b)** $-\mathbf{i} + \mathbf{k}$ **(c)** 0
29. $7t^6$; $18t^5\mathbf{i} - 10t^4\mathbf{j}$
31. $3t\mathbf{i} + 2t^2\mathbf{j} + \mathbf{C}$

33. $(-t\cos t + \sin t)\mathbf{i} + t\mathbf{j} + \mathbf{C}$ **35.** $(t^3/3)\mathbf{i} - t^2\mathbf{j} + \ln|t|\mathbf{k} + \mathbf{C}$ **37.** $\mathbf{j}$
39. $(5\sqrt{5} - 1)/3$ **41.** $\frac{52}{3}\mathbf{i} + 4\mathbf{j}$ **43.** $(t^2 + 1)\mathbf{i} + (t^3 - 1)\mathbf{j}$
45. $y(t) = \left(\frac{1}{2}t^2 + 2\right)\mathbf{i} + (e^t - 1)\mathbf{j}$
47.
49. **(a)** $(-2, 4, 6)$ and $(1, 1, -3)$
(b) $76°, 71°$
51. $68°$

Exercise Set 13.3 (Page 884)

1. smooth **3.** not smooth, $\mathbf{r}'(1) = \mathbf{0}$ **5.** $L = \frac{3}{2}$ **7.** $L = e - e^{-1}$
9. $L = 28$ **11.** $L = 2\pi\sqrt{10}$ **13.** $\mathbf{r}'(\tau) = 4\mathbf{i} + 8(4\tau + 1)\mathbf{j}$
15. $\mathbf{r}'(\tau) = 2\tau e^{\tau^2}\mathbf{i} - 8\tau e^{-\tau^2}\mathbf{j}$
19. **(a)** $x = \dfrac{s}{\sqrt{2}}, y = \dfrac{s}{\sqrt{2}}$ **(b)** $x = y = z = \dfrac{s}{\sqrt{3}}$
21. **(a)** $x = 1 + \dfrac{s}{3}, y = 3 - \dfrac{2s}{3}, z = 4 + \dfrac{2s}{3}$ **(b)** $\langle \frac{28}{3}, -\frac{41}{3}, \frac{62}{3}\rangle$
23. $x = 3 + \cos s, y = 2 + \sin s, 0 \le s \le 2\pi$
25. $x = \frac{1}{3}[(3s + 1)^{2/3} - 1]^{3/2}, y = \frac{1}{2}[(3s + 1)^{2/3} - 1], s \ge 0$
27. $x = \left(\dfrac{s}{\sqrt{2}} + 1\right)\cos\left[\ln\left(\dfrac{s}{\sqrt{2}} + 1\right)\right],$

$y = \left(\dfrac{s}{\sqrt{2}} + 1\right)\sin\left[\ln\left(\dfrac{s}{\sqrt{2}} + 1\right)\right],$ $0 \le s \le \sqrt{2}(e^{\pi/2} - 1)$

31. $x = 2a\cos^{-1}[1 - s/(4a)]$
$-2a(1 - [1 - s/(4a)]^2)^{1/2}(2[1 - s/(4a)]^2 - 1),$
$y = \dfrac{s(8a - s)}{8a}$ for $0 \le s \le 8a$
33. **(a)** $9/2$ **(b)** $9 - 2\sqrt{6}$ **35.** **(a)** $\sqrt{3}(1 - e^{-2})$ **(b)** $4\sqrt{5}$
37. **(a)** $g(\tau) = \pi(\tau)$ **(b)** $g(\tau) = \pi(1 - \tau)$ **39.** 44 in.
41. **(a)** $2t + \dfrac{1}{t}$ **(b)** $2t + \dfrac{1}{t}$ **(c)** $8 + \ln 3$

Exercise Set 13.4 (Page 891)

1. **(a)** **(b)**

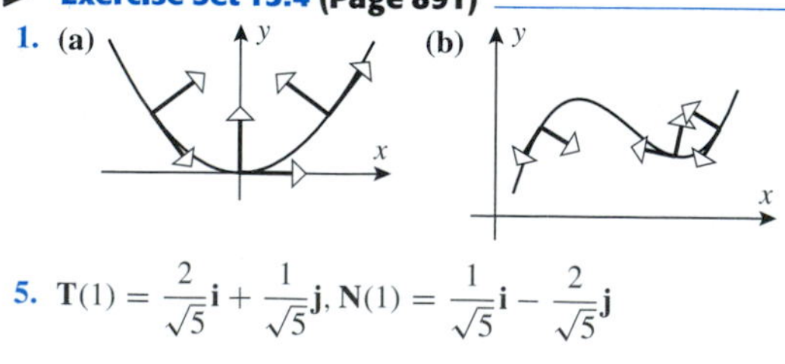

5. $\mathbf{T}(1) = \dfrac{2}{\sqrt{5}}\mathbf{i} + \dfrac{1}{\sqrt{5}}\mathbf{j}, \mathbf{N}(1) = \dfrac{1}{\sqrt{5}}\mathbf{i} - \dfrac{2}{\sqrt{5}}\mathbf{j}$

7. $\mathbf{T}\left(\frac{\pi}{3}\right)=-\frac{\sqrt{3}}{2}\mathbf{i}+\frac{1}{2}\mathbf{j},\ \mathbf{N}\left(\frac{\pi}{3}\right)=-\frac{1}{2}\mathbf{i}-\frac{\sqrt{3}}{2}\mathbf{j}$

9. $\mathbf{T}\left(\frac{\pi}{2}\right)=-\frac{4}{\sqrt{17}}\mathbf{i}+\frac{1}{\sqrt{17}}\mathbf{k},\ \mathbf{N}\left(\frac{\pi}{2}\right)=-\mathbf{j}$

11. $\mathbf{T}(0)=\frac{1}{\sqrt{3}}\mathbf{i}+\frac{1}{\sqrt{3}}\mathbf{j}+\frac{1}{\sqrt{3}}\mathbf{k},\ \mathbf{N}(0)=-\frac{1}{\sqrt{2}}\mathbf{i}+\frac{1}{\sqrt{2}}\mathbf{j}$

13. $x=s,\ y=1$ **15.** $\mathbf{B}=\frac{4}{5}\cos t\mathbf{i}-\frac{4}{5}\sin t\mathbf{j}-\frac{3}{5}\mathbf{k}$ **17.** $\mathbf{B}=-\mathbf{k}$

19. $\mathbf{T}\left(\frac{\pi}{4}\right)=\frac{\sqrt{2}}{2}(-\mathbf{i}+\mathbf{j}),\ \mathbf{N}\left(\frac{\pi}{4}\right)=-\frac{\sqrt{2}}{2}(\mathbf{i}+\mathbf{j}),$
$\mathbf{B}\left(\frac{\pi}{4}\right)=\mathbf{k}$ rectifying: $x+y=\sqrt{2}$; osculating: $z=1$;
normal: $-x+y=0$

23. $\mathbf{N}=-\sin t\mathbf{i}-\cos t\mathbf{j}$

Exercise Set 13.5 (Page 897)

1. $\kappa\approx 2$ **3.** **(a)** I is the curvature of II. **(b)** I is the curvature of II.

5. $\frac{6}{t(4+9t^2)^{3/2}}$ **7.** $\frac{12e^{2t}}{(9e^{6t}+e^{-2t})^{3/2}}$ **9.** $\frac{4}{17}$ **11.** $\frac{1}{2\cosh^2 t}$

13. $\kappa=\frac{2}{5},\ \rho=\frac{5}{2}$ **15.** $\kappa=\frac{\sqrt{2}}{3},\ \rho=\frac{3\sqrt{2}}{2}$ **17.** $\kappa=\frac{1}{4}$ **21.** 1

23. $\frac{1}{\sqrt{2}}$ **25.** $\frac{4}{5\sqrt{5}}$ **27.** $\frac{96}{125}$ **29.** $\frac{6}{5\sqrt{10}}$ **31.** $\frac{1}{\sqrt{2}}$

33. **(a)** $\rho(0)=\rho(\pi)=1$ **(b)** $\rho(0)=\frac{1}{2}$, $\rho\left(\frac{\pi}{2}\right)=4$

35.

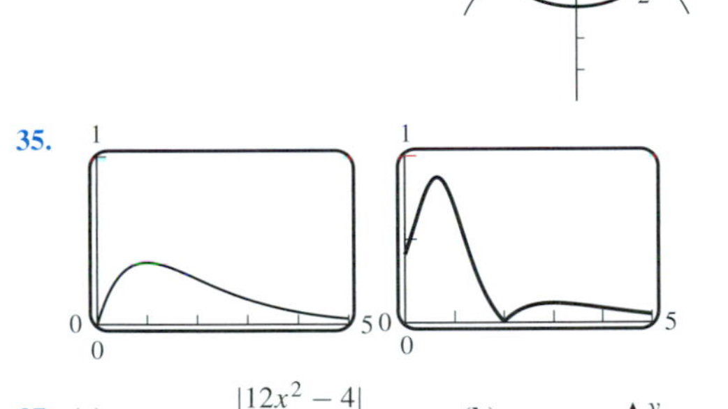

37. **(a)** $\kappa=\frac{|12x^2-4|}{[1+(4x^3-4x)^2]^{3/2}}$ **(b)**
(c) $\rho=\frac{1}{4}$ for $x=0$ and $\rho=\frac{1}{8}$ when $x=\pm 1$

41. $\frac{3}{2\sqrt{2}}$ **43.** $\frac{2}{3}$ **47.** $\rho=2|p|$ **49.** $(3,0),(-3,0)$

51. $\rho_{\min}=1/\sqrt{2};\ \rho_{\max}=2$

55. **(b)** $\rho=\sqrt{2}$ **(c)**

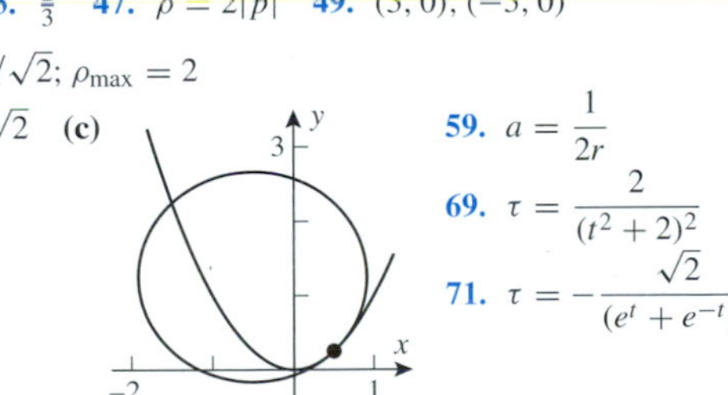

59. $a=\frac{1}{2r}$

69. $\tau=\frac{2}{(t^2+2)^2}$

71. $\tau=-\frac{\sqrt{2}}{(e^t+e^{-t})^2}$

Exercise Set 13.6 (Page 910)

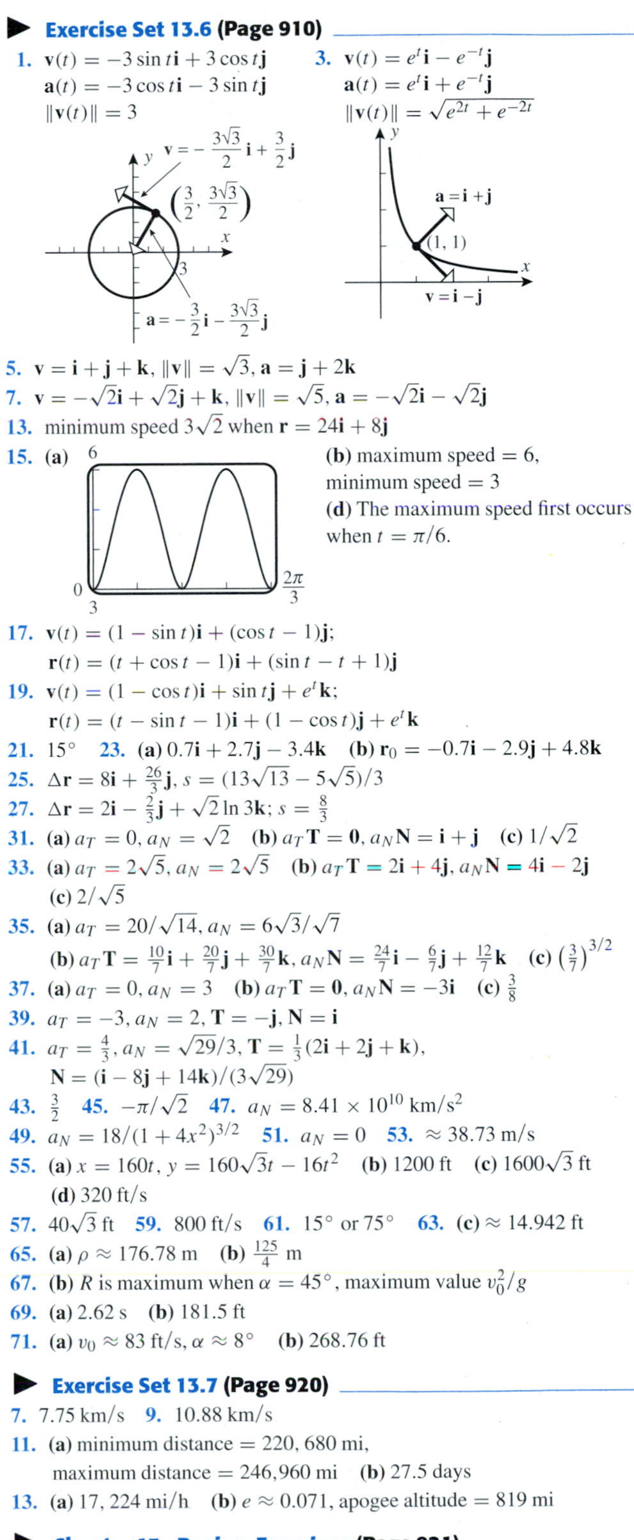

1. $\mathbf{v}(t)=-3\sin t\mathbf{i}+3\cos t\mathbf{j}$
$\mathbf{a}(t)=-3\cos t\mathbf{i}-3\sin t\mathbf{j}$
$\|\mathbf{v}(t)\|=3$

3. $\mathbf{v}(t)=e^t\mathbf{i}-e^{-t}\mathbf{j}$
$\mathbf{a}(t)=e^t\mathbf{i}+e^{-t}\mathbf{j}$
$\|\mathbf{v}(t)\|=\sqrt{e^{2t}+e^{-2t}}$

5. $\mathbf{v}=\mathbf{i}+\mathbf{j}+\mathbf{k},\ \|\mathbf{v}\|=\sqrt{3},\ \mathbf{a}=\mathbf{j}+2\mathbf{k}$

7. $\mathbf{v}=-\sqrt{2}\mathbf{i}+\sqrt{2}\mathbf{j}+\mathbf{k},\ \|\mathbf{v}\|=\sqrt{5},\ \mathbf{a}=-\sqrt{2}\mathbf{i}-\sqrt{2}\mathbf{j}$

13. minimum speed $3\sqrt{2}$ when $\mathbf{r}=24\mathbf{i}+8\mathbf{j}$

15. **(a)** **(b)** maximum speed $=6$, minimum speed $=3$
(d) The maximum speed first occurs when $t=\pi/6$.

17. $\mathbf{v}(t)=(1-\sin t)\mathbf{i}+(\cos t-1)\mathbf{j}$;
$\mathbf{r}(t)=(t+\cos t-1)\mathbf{i}+(\sin t-t+1)\mathbf{j}$

19. $\mathbf{v}(t)=(1-\cos t)\mathbf{i}+\sin t\mathbf{j}+e^t\mathbf{k}$;
$\mathbf{r}(t)=(t-\sin t-1)\mathbf{i}+(1-\cos t)\mathbf{j}+e^t\mathbf{k}$

21. $15°$ **23.** **(a)** $0.7\mathbf{i}+2.7\mathbf{j}-3.4\mathbf{k}$ **(b)** $\mathbf{r}_0=-0.7\mathbf{i}-2.9\mathbf{j}+4.8\mathbf{k}$

25. $\Delta\mathbf{r}=8\mathbf{i}+\frac{26}{3}\mathbf{j},\ s=(13\sqrt{13}-5\sqrt{5})/3$

27. $\Delta\mathbf{r}=2\mathbf{i}-\frac{2}{3}\mathbf{j}+\sqrt{2}\ln 3\mathbf{k};\ s=\frac{8}{3}$

31. **(a)** $a_T=0,\ a_N=\sqrt{2}$ **(b)** $a_T\mathbf{T}=\mathbf{0},\ a_N\mathbf{N}=\mathbf{i}+\mathbf{j}$ **(c)** $1/\sqrt{2}$

33. **(a)** $a_T=2\sqrt{5},\ a_N=2\sqrt{5}$ **(b)** $a_T\mathbf{T}=2\mathbf{i}+4\mathbf{j},\ a_N\mathbf{N}=4\mathbf{i}-2\mathbf{j}$
(c) $2/\sqrt{5}$

35. **(a)** $a_T=20/\sqrt{14},\ a_N=6\sqrt{3}/\sqrt{7}$
(b) $a_T\mathbf{T}=\frac{10}{7}\mathbf{i}+\frac{20}{7}\mathbf{j}+\frac{30}{7}\mathbf{k},\ a_N\mathbf{N}=\frac{24}{7}\mathbf{i}-\frac{6}{7}\mathbf{j}+\frac{12}{7}\mathbf{k}$ **(c)** $\left(\frac{3}{7}\right)^{3/2}$

37. **(a)** $a_T=0,\ a_N=3$ **(b)** $a_T\mathbf{T}=\mathbf{0},\ a_N\mathbf{N}=-3\mathbf{i}$ **(c)** $\frac{3}{8}$

39. $a_T=-3,\ a_N=2,\ \mathbf{T}=-\mathbf{j},\ \mathbf{N}=\mathbf{i}$

41. $a_T=\frac{4}{3},\ a_N=\sqrt{29}/3,\ \mathbf{T}=\frac{1}{3}(2\mathbf{i}+2\mathbf{j}+\mathbf{k}),$
$\mathbf{N}=(\mathbf{i}-8\mathbf{j}+14\mathbf{k})/(3\sqrt{29})$

43. $\frac{3}{2}$ **45.** $-\pi/\sqrt{2}$ **47.** $a_N=8.41\times 10^{10}\ \text{km/s}^2$

49. $a_N=18/(1+4x^2)^{3/2}$ **51.** $a_N=0$ **53.** ≈ 38.73 m/s

55. **(a)** $x=160t,\ y=160\sqrt{3}t-16t^2$ **(b)** 1200 ft **(c)** $1600\sqrt{3}$ ft
(d) 320 ft/s

57. $40\sqrt{3}$ ft **59.** 800 ft/s **61.** $15°$ or $75°$ **63.** **(c)** ≈ 14.942 ft

65. **(a)** $\rho\approx 176.78$ m **(b)** $\frac{125}{4}$ m

67. **(b)** R is maximum when $\alpha=45°$, maximum value v_0^2/g

69. **(a)** 2.62 s **(b)** 181.5 ft

71. **(a)** $v_0\approx 83$ ft/s, $\alpha\approx 8°$ **(b)** 268.76 ft

Exercise Set 13.7 (Page 920)

7. 7.75 km/s **9.** 10.88 km/s

11. **(a)** minimum distance $=220,680$ mi,
maximum distance $=246,960$ mi **(b)** 27.5 days

13. **(a)** 17,224 mi/h **(b)** $e\approx 0.071$, apogee altitude $=819$ mi

Chapter 13. Review Exercises (Page 921)

3. the circle of radius 3 in the xy-plane, with center at the origin

5. a parabola in the plane $x=-2$, vertex at $(-2,0,-1)$,
opening upward

11. $x=1+t,\ y=-t,\ z=t$ **13.** $(\sin t)\mathbf{i}-(\cos t)\mathbf{j}+\mathbf{C}$

15. $y(t) = \left(\frac{1}{3}t^3 + 1\right)\mathbf{i} + (t^2 + 1)\mathbf{j}$ **17.** 15/4

19. $\mathbf{r}(s) = \dfrac{s-3}{3}\mathbf{i} + \dfrac{12-2s}{3}\mathbf{j} + \dfrac{9+2s}{3}\mathbf{k}$ **25.** 3/5 **27.** 0

29. **(a)** speed **(b)** distance traveled
(c) distance of the particle from the origin

33. **(a)** $\mathbf{r}(t) = \left(\frac{1}{6}t^4 + t\right)\mathbf{i} + \left(\frac{1}{2}t^2 + 2t\right)\mathbf{j} - \left(\frac{1}{4}\cos 2t + t - \frac{1}{4}\right)\mathbf{k}$
(b) 3.475 **35.** 24.78 ft **37.** 36.50 km/s

Exercise Set 14.1 (Page 933)

1. **(a)** 5 **(b)** 3 **(c)** 1 **(d)** −2 **(e)** $9a^3 + 1$ **(f)** $a^3b^2 - a^2b^3 + 1$

3. **(a)** $x^2 - y^2 + 3$ **(b)** $3x^3y^4 + 3$ **5.** $x^3e^{x^3(3y+1)}$

7. **(a)** $t^2 + 3t^{10}$ **(b)** 0 **(c)** 3076

9. **(a)** WCI = 17.8°F **(b)** WCI = 22.6°F

11. **(a)** 66% **(b)** 73.5% **(c)** 60.6% **13.** **(a)** 19 **(b)** −9 **(c)** 3
(d) $a^6 + 3$ **(e)** $-t^8 + 3$ **(f)** $(a+b)(a-b)^2b^3 + 3$

15. $(y+1)e^{x^2(y+1)z^2}$ **17.** **(a)** $80\sqrt{\pi}$ **(b)** $n(n+1)/2$

19.

21.

23. **(a)** all points above or on the line $y = -2$ **(b)** all points on or within the sphere $x^2 + y^2 + z^2 = 25$ **(c)** all points in 3-space

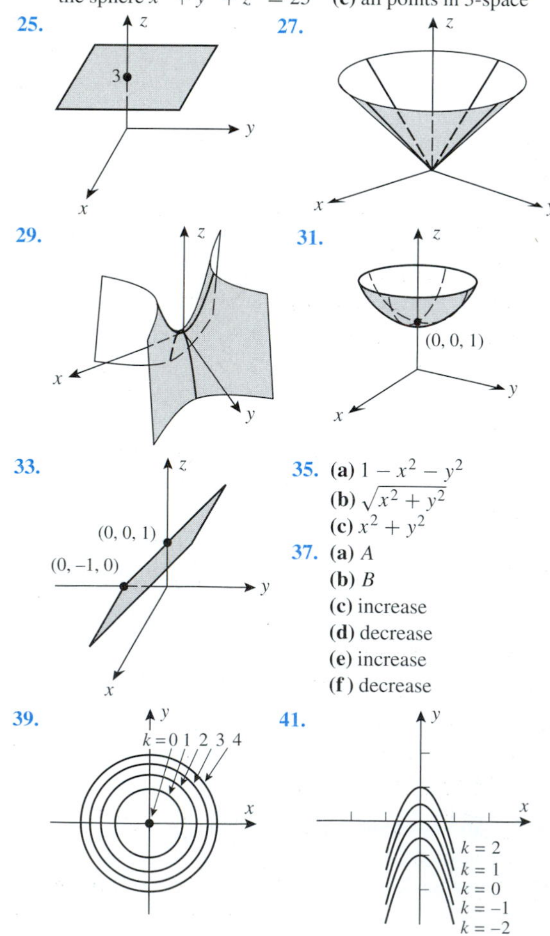

35. **(a)** $1 - x^2 - y^2$
(b) $\sqrt{x^2 + y^2}$
(c) $x^2 + y^2$

37. **(a)** A
(b) B
(c) increase
(d) decrease
(e) increase
(f) decrease

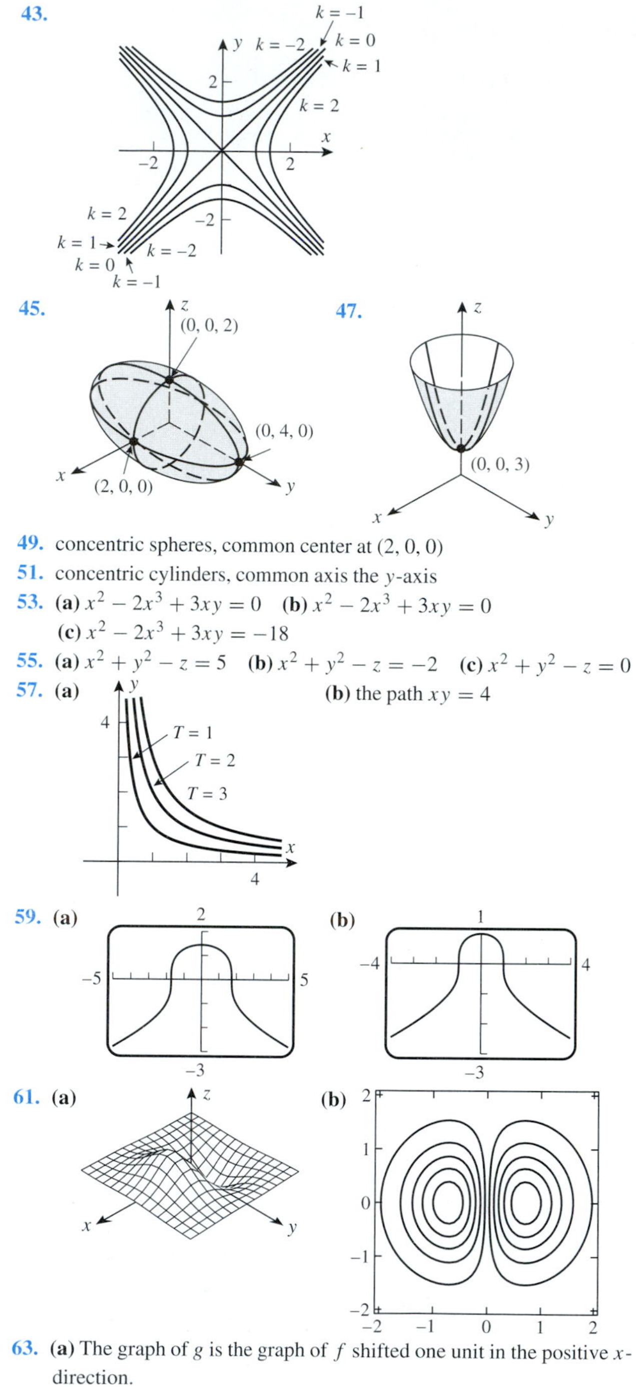

49. concentric spheres, common center at (2, 0, 0)

51. concentric cylinders, common axis the y-axis

53. **(a)** $x^2 - 2x^3 + 3xy = 0$ **(b)** $x^2 - 2x^3 + 3xy = 0$
(c) $x^2 - 2x^3 + 3xy = -18$

55. **(a)** $x^2 + y^2 - z = 5$ **(b)** $x^2 + y^2 - z = -2$ **(c)** $x^2 + y^2 - z = 0$

57. **(a)** **(b)** the path $xy = 4$

59. **(a)** **(b)**

61. **(a)** **(b)**

63. **(a)** The graph of g is the graph of f shifted one unit in the positive x-direction.
(b) The graph of g is the graph of f shifted one unit up the z-axis.
(c) The graph of g is the graph of f shifted one unit down the y-axis and then inverted with respect to the plane $z = 0$.

Exercise Set 14.2 (Page 944)

1. 35 **3.** −8 **5.** 0

7. **(a)** along $x = 0$ limit does not exist
(b) along $x = 0$ limit does not exist

9. 1 **11.** 0 **13.** 0 **15.** limit does not exist **17.** $\frac{8}{3}$ **19.** 0 **21.** 0

23. limit does not exist **25.** **(a)** no **(d)** no; yes **29.** $-\pi/2$ **31.** no

33. 35.

37. 39.

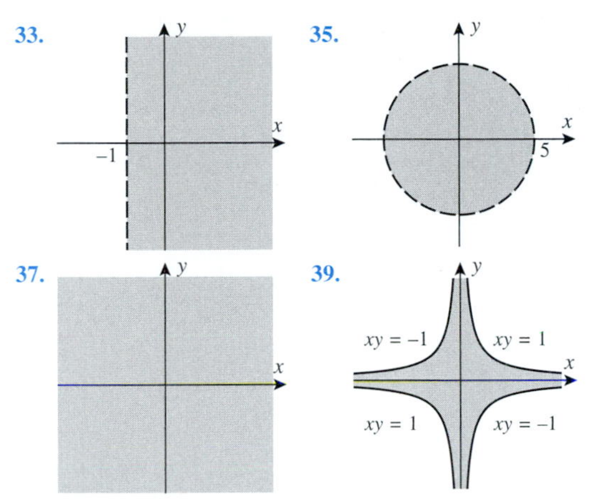

41. all of 3-space

43. all points not on the cylinder $x^2 + z^2 = 1$

▶ Exercise Set 14.3 (Page 955)

1. (a) $9x^2y^2$ (b) $6x^3y$ (c) $9y^2$ (d) $9x^2$ (e) $6y$ (f) $6x^3$ (g) 36 (h) 12

3. (a) $\frac{3}{8}$ (b) $\frac{1}{4}$ 5. (a) $-4\cos 7$ (b) $2\cos 7$

7. $\partial z/\partial x = -4$; $\partial z/\partial y = \frac{1}{2}$ 9. (a) 4.9 (b) 1.2

11. $z = f(x, y)$ has II as its graph, f_x has I as its graph, and f_y has III as its graph.

13. $8xy^3e^{x^2y^3}$, $12x^2y^2e^{x^2y^3}$

15. $x^3/(y^{3/5} + x) + 3x^2\ln(1 + xy^{-3/5})$, $-\frac{3}{5}x^4/(y^{8/5} + xy)$

17. $-\dfrac{y(x^2 - y^2)}{(x^2 + y^2)^2}, \dfrac{x(x^2 - y^2)}{(x^2 + y^2)^2}$

19. $(3/2)x^2y(5x^2 - 7)(3x^5y - 7x^3y)^{-1/2}$
$(1/2)x^3(3x^2 - 7)(3x^5y - 7x^3y)^{-1/2}$

21. $\dfrac{y^{-1/2}}{y^2 + x^2}, -\dfrac{xy^{-3/2}}{y^2 + x^2} - \dfrac{3}{2}y^{-5/2}\tan^{-1}\left(\dfrac{x}{y}\right)$

23. $-\frac{4}{3}y^2\sec^2 x(y^2\tan x)^{-7/3}, -\frac{8}{3}y\tan x(y^2\tan x)^{-7/3}$

25. $-6, -21$ 27. $1/\sqrt{17}, 8/\sqrt{17}$

29. (a) $2xy^4z^3 + y$ (b) $4x^2y^3z^3 + x$ (c) $3x^2y^4z^2 + 2z$ (d) $2y^4z^3 + y$ (e) $32z^3 + 1$ (f) 438

31. $2z/x, z/y, \ln(x^2y\cos z) - z\tan z$

33. $-y^2z^3/(1 + x^2y^4z^6), -2xyz^3/(1 + x^2y^4z^6), -3xy^2z^2/(1 + x^2y^4z^6)$

35. $yze^z\cos(xz), e^z\sin(xz), ye^z(\sin(xz) + x\cos(xz))$

37. $x/\sqrt{x^2 + y^2 + z^2}, y/\sqrt{x^2 + y^2 + z^2}, z/\sqrt{x^2 + y^2 + z^2}$

39. (a) e (b) $2e$ (c) e

41. (a) (b) 43. 4 45. −2

47. (a) $\partial V/\partial r = 2\pi rh$ (b) $\partial V/\partial h = \pi r^2$ (c) 48π (d) 64π

49. (a) $\dfrac{1}{5}\dfrac{\text{lb}}{\text{in}^2\cdot\text{K}}$ (b) $-\dfrac{25}{8}\dfrac{\text{in}^5}{\text{lb}}$

51. (a) $\dfrac{\partial V}{\partial \ell} = 6$ (b) $\dfrac{\partial V}{\partial w} = 15$ (c) $\dfrac{\partial V}{\partial h} = 10$

55. (a) $\pm\sqrt{6}/4$ 57. $-x/z, -y/z$

59. $-\dfrac{2x + yz^2\cos(xyz)}{xyz\cos(xyz) + \sin(xyz)}; -\dfrac{xz^2\cos(xyz)}{xyz\cos(xyz) + \sin(xyz)}$

61. $-x/w, -y/w, -z/w$

63. $-\dfrac{yzw\cos(xyz)}{2w + \sin(xyz)}, -\dfrac{xzw\cos(xyz)}{2w + \sin(xyz)}, -\dfrac{xyw\cos(xyz)}{2w + \cos(xyz)}$

65. $e^{x^2}, -e^{y^2}$

67. (a) $-\dfrac{\cos y}{4\sqrt{x^3}}$ (b) $-\sqrt{x}\cos y$ (c) $-\dfrac{1}{2\sqrt{x}}\sin y$ (d) $-\dfrac{1}{2\sqrt{x}}\sin y$

69. $-32y^3$ 71. $-e^x\sin y$ 73. $\dfrac{20}{(4x - 5y)^2}$ 75. $\dfrac{2(x - y)}{(x + y)^3}$

77. (a) $\dfrac{\partial^3 f}{\partial x^3}$ (b) $\dfrac{\partial^3 f}{\partial y^2\partial x}$ (c) $\dfrac{\partial^4 f}{\partial x^2\partial y^2}$ (d) $\dfrac{\partial^4 f}{\partial y^3\partial x}$

79. (a) $30xy^4 - 4$ (b) $60x^2y^3$ (c) $60x^3y^2$

81. (a) −30 (b) −125 (c) 150

83. (a) $15x^2y^4z^7 + 2y$ (b) $35x^3y^4z^6 + 3y^2$ (c) $21x^2y^5z^6$ (d) $42x^3y^5z^5$ (e) $140x^3y^3z^6 + 6y$ (f) $30xy^4z^7$ (g) $105x^2y^4z^6$ (h) $210xy^4z^6$

91. $\dfrac{\partial f}{\partial v} = 8vw^3x^4y^5, \dfrac{\partial f}{\partial w} = 12v^2w^2x^4y^5, \dfrac{\partial f}{\partial x} = 16v^2w^3x^3y^5,$
$\dfrac{\partial f}{\partial y} = 20v^2w^3x^4y^4$

93. $\dfrac{\partial f}{\partial v_1} = \dfrac{2v_1}{v_3^2 + v_4^2}, \dfrac{\partial f}{\partial v_2} = \dfrac{-2v_2}{v_3^2 + v_4^2}, \dfrac{\partial f}{\partial v_3} = \dfrac{-2v_3(v_1^2 - v_2^2)}{(v_3^2 + v_4^2)^2},$
$\dfrac{\partial f}{\partial v_4} = \dfrac{-2v_4(v_1^2 - v_2^2)}{(v_3^2 + v_4^2)^2}$

95. (a) 0 (b) 0 (c) 0 (d) 0 (e) $2(1 + yw)e^{yw}\sin z\cos z$ (f) $2xw(2 + yw)e^{yw}\sin z\cos z$

97. $-i\sin(x_1 + 2x_2 + \cdots + nx_n)$

99. (a) xy-plane, $12x^2 + 6x$ (b) $y \neq 0, -3x^2/y^2$

101. $f_x(2, -1) = 11, f_y(2, -1) = -8$

103. (b) does not exist if $y \neq 0$ and $x = -y$

▶ Exercise Set 14.4 (Page 965)

1. 5.04 3. 4.14 9. $dz = 7\,dx - 2\,dy$ 11. $dz = 3x^2y^2\,dx + 2x^3y\,dy$

13. $dz = \dfrac{y}{1 + x^2y^2}dx + \dfrac{x}{1 + x^2y^2}dy$ 15. $dw = 8\,dx - 3\,dy + 4\,dz$

17. $dw = 3x^2y^2z\,dx + 2x^3yz\,dy + x^3y^2\,dz$

19. $dw = \dfrac{yz}{1 + x^2y^2z^2}dx + \dfrac{xz}{1 + x^2y^2z^2}dy + \dfrac{xy}{1 + x^2y^2z^2}dz$

21. $df = 0.10, \Delta f = 0.1009$ 23. $df = 0.03, \Delta f \approx 0.029412$

25. $df = 0.96, \Delta f \approx 0.97929$

27. The increase in the area of the rectangle is given by the sum of the areas of the three small rectangles, and the total differential is given by the sum of the areas of the upper left and lower right rectangles.

29. (a) $L = \frac{1}{5} - \frac{4}{125}(x - 4) - \frac{3}{125}(y - 3)$ (b) 0.000176603

31. (a) $L = 0$ (b) 0.0024

33. (a) $L = 6 + 6(x - 1) + 3(y - 2) + 2(z - 3)$ (b) −0.000481

35. (a) $L = e + e(x - 1) - e(y + 1) - e(z + 1)$ (b) 0.01554

41. 0.5 43. 1, 1, −1, 2 45. (−1, 1) 47. (1, 0, 1) 49. 8%

51. r% 53. 0.3%

55. (a) $(r + s)$% (b) $(r + s)$% (c) $(2r + 3s)$% (d) $\left(3r + \frac{s}{2}\right)$%

57. ≈ 39 ft^2

▶ Exercise Set 14.5 (Page 975)

1. $42t^{13}$ 3. $3t^{-2}\sin(1/t)$ 5. $-\frac{10}{3}t^{7/3}e^{1-t^{10/3}}$ 7. $\dfrac{dw}{dt} = 165t^{32}$

9. $-2t\cos t^2$ 11. 3264 13. 0

17. $24u^2v^2 - 16uv^3 - 2v + 3, 16u^3v - 24u^2v^2 - 2u - 3$

19. $-\dfrac{2\sin u}{3\sin v}, -\dfrac{2\cos u\cos v}{3\sin^2 v}$ 21. $e^u, 0$

23. $3r^2\sin\theta\cos^2\theta - 4r^3\sin^3\theta\cos\theta, -2r^3\sin^2\theta\cos\theta + r^4\sin^4\theta + r^3\cos^3\theta - 3r^4\sin^2\theta\cos^2\theta$

25. $\dfrac{x^2 + y^2}{4x^2y^3}, \dfrac{y^2 - 3x^2}{4xy^4}$ 27. $\dfrac{\partial z}{\partial r} = \dfrac{2r\cos^2\theta}{r^2\cos^2\theta + 1}, \dfrac{\partial z}{\partial \theta} = \dfrac{-2r^2\cos\theta\sin\theta}{r^2\cos^2\theta + 1}$

29. $\dfrac{dw}{d\rho} = 2\rho(4\sin^2\phi + \cos^2\phi), \dfrac{\partial w}{\partial \phi} = 6\rho^2\sin\phi\cos\phi, \dfrac{dw}{d\theta} = 0$

31. $-\pi$ **33.** $\sqrt{3}e^{\sqrt{3}}, (2-4\sqrt{3})e^{\sqrt{3}}$ **35.** $-\dfrac{2xy^3}{3x^2y^2-\sin y}$

37. $-\dfrac{ye^{xy}}{xe^{xy}+ye^{y}+e^{y}}$ **41.** $\dfrac{2x+yz}{6yz-xy}, \dfrac{xz-3z^2}{6yz-xy}$

43. $\dfrac{ye^x}{15\cos 3z+3}, \dfrac{e^x}{15\cos 3z+3}$ **45.** -39 mi/h **47.** $-\frac{7}{36}\sqrt{3}$ rad/s

49. $16{,}200\pi$ in^3/year **51.** **(a)** 60 in^3/s **(b)** $\frac{26}{7}$ in/s

53. **(a)** 2 **(b)** 1 **(c)** 3 **(d)** -4

67. $\dfrac{\partial w}{\partial \rho} = (\sin\phi\cos\theta)\dfrac{\partial w}{\partial x} + (\sin\phi\sin\theta)\dfrac{\partial w}{\partial y} + (\cos\phi)\dfrac{\partial w}{\partial z}$,

$\dfrac{\partial w}{\partial \phi} = (\rho\cos\phi\cos\theta)\dfrac{\partial w}{\partial x} + (\rho\cos\phi\sin\theta)\dfrac{\partial w}{\partial y} - (\rho\sin\phi)\dfrac{\partial w}{\partial z}$,

$\dfrac{\partial w}{\partial \theta} = -(\rho\sin\phi\sin\theta)\dfrac{\partial w}{\partial x} + (\rho\sin\phi\cos\theta)\dfrac{\partial w}{\partial y}$

71. **(a)** $\dfrac{dw}{dt} = \displaystyle\sum_{i=1}^{4} \dfrac{\partial w}{\partial x_i}\dfrac{dx_i}{dt}$ **(b)** $\dfrac{\partial w}{\partial v_j} = \displaystyle\sum_{i=1}^{4} \dfrac{\partial w}{\partial x_i}\dfrac{\partial x_i}{\partial v_j}$, $j = 1, 2, 3$

Exercise Set 14.6 (Page 986)

1. $6\sqrt{2}$ **3.** $-3/\sqrt{10}$ **5.** -320 **7.** $-314/741$ **9.** 0 **11.** $-8\sqrt{2}$

13. $\sqrt{2}/4$ **15.** $72/\sqrt{14}$ **17.** $-8/63$ **19.** $1/2+\sqrt{3}/8$ **21.** $2\sqrt{2}$

23. $1/\sqrt{5}$ **25.** $-\frac{3}{2}e$ **27.** $3/\sqrt{11}$ **29.** **(a)** 5 **(b)** 10 **(c)** $-5\sqrt{5}$

31. III **33.** $4\mathbf{i}-8\mathbf{j}$

35. $\nabla w = \dfrac{x}{x^2+y^2+z^2}\mathbf{i} + \dfrac{y}{x^2+y^2+z^2}\mathbf{j} + \dfrac{z}{x^2+y^2+z^2}\mathbf{k}$

37. $-36\mathbf{i}-12\mathbf{j}$ **39.** $4(\mathbf{i}+\mathbf{j}+\mathbf{k})$

41.

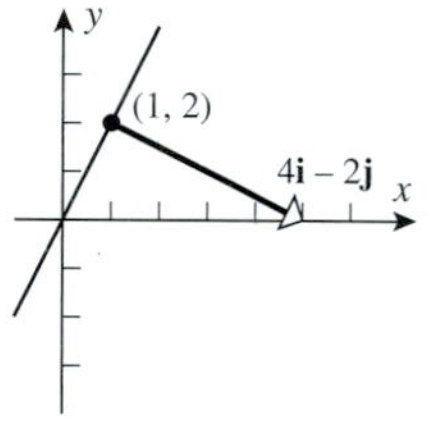

43.

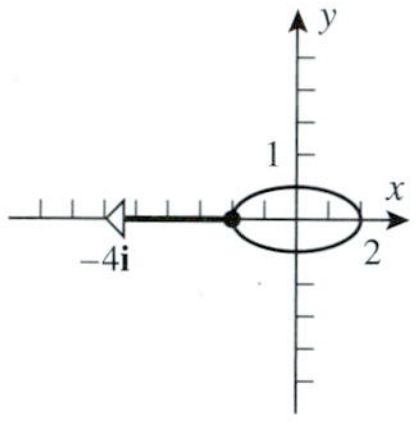

45. $\pm(-4\mathbf{i}+\mathbf{j})/\sqrt{17}$ **47.** $\mathbf{u} = (3\mathbf{i}-2\mathbf{j})/\sqrt{13}, \|\nabla f(-1,1)\| = 4\sqrt{13}$

49. $\mathbf{u} = (4\mathbf{i}-3\mathbf{j})/5, \|\nabla f(4,-3)\| = 1$ **51.** $\dfrac{1}{\sqrt{2}}(\mathbf{i}-\mathbf{j}), 3\sqrt{2}$

53. $\dfrac{1}{\sqrt{2}}(-\mathbf{i}+\mathbf{j}), \dfrac{1}{\sqrt{2}}$

55. $\mathbf{u} = -(\mathbf{i}+3\mathbf{j})/\sqrt{10}, -\|\nabla f(-1,-3)\| = -2\sqrt{10}$

57. $\mathbf{u} = (3\mathbf{i}-\mathbf{j})/\sqrt{10}, -\|\nabla f(\pi/6,\pi/4)\| = -\sqrt{5}$

59. $(\mathbf{i}-11\mathbf{j}+12\mathbf{k})/\sqrt{266}, -\sqrt{266}$ **61.** $8/\sqrt{29}$

63. **(a)** $\approx 1/\sqrt{2}$

(b)

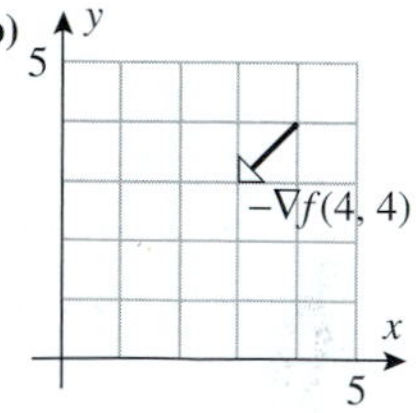

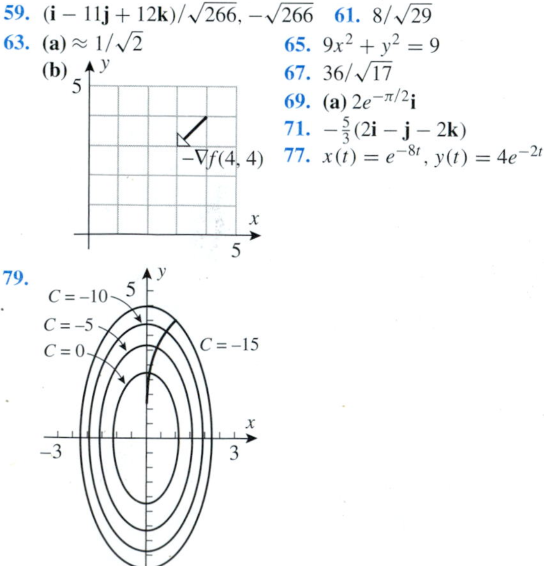

65. $9x^2+y^2=9$

67. $36/\sqrt{17}$

69. **(a)** $2e^{-\pi/2}\mathbf{i}$

71. $-\frac{5}{3}(2\mathbf{i}-\mathbf{j}-2\mathbf{k})$

77. $x(t) = e^{-8t}, y(t) = 4e^{-2t}$

79.

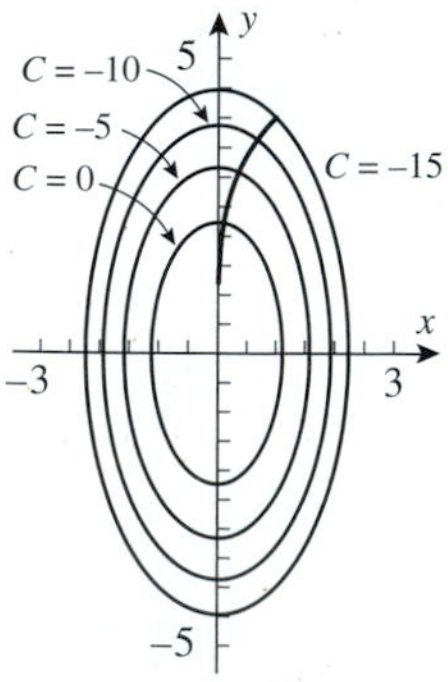

81. **(a)**

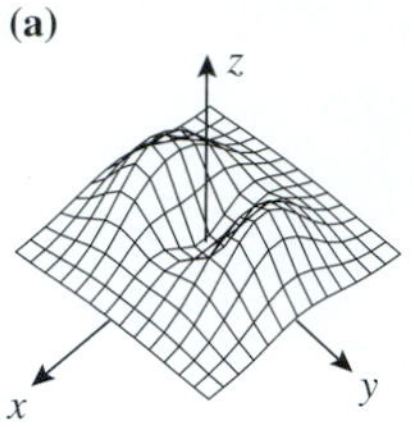

(c) $\nabla f = [2x - 2x(x^2+3y^2)]e^{-(x^2+y^2)}\mathbf{i} + [6y - 2y(x^2+3y^2)]e^{-(x^2+y^2)}\mathbf{j}$

(d) $x = y = 0$ or $x = 0, y = \pm 1$ or $x = \pm 1, y = 0$

Exercise Set 14.7 (Page 994)

1. tangent plane: $48x - 14y - z = 64$;
normal line: $x = 1+48t, y = -2-14t, z = 12-t$

3. tangent plane: $x - y - z = 0$;
normal line: $x = 1+t, y = -t, z = 1-t$

5. tangent plane: $3y - z = -1$;
normal line: $x = \pi/6, y = 3t, z = 1-t$

7. tangent plane: $3x - 4z = -25$;
normal line: $x = -3+(3t/4), y = 0, z = 4-t$

9. **(a)** all points on the x-axis or y-axis **(b)** $(0, -2, -4)$

11. $\left(\frac{1}{2}, -2, -\frac{3}{4}\right)$ **13.** **(a)** $(-2, 1, 5), (0, 3, 9)$ **(b)** $\dfrac{4}{3\sqrt{14}}, \dfrac{4}{\sqrt{222}}$

15. **(a)** $x+y+2z=6$ **(b)** $x = 2+t, y = 2+t, z = 1+2t$
(c) 35.26°

17. $\pm\dfrac{1}{\sqrt{365}}(\mathbf{i}-\mathbf{j}-19\mathbf{k})$ **21.** $(1, 2/3, 2/3), (-1, -2/3, -2/3)$

23. $x = 1+8t, y = -1+5t, z = 2+6t$

25. $x = 3+4t, y = -3-4t, z = 4-3t$

Exercise Set 14.8 (Page 1004)

1. **(a)** minimum at $(2, -1)$, no maxima
(b) maximum at $(0, 0)$, no minima **(c)** no maxima or minima

3. minimum at $(3, -2)$, no maxima **5.** relative minimum at $(0, 0)$

7. relative minimum at $(0, 0)$; saddle points at $(\pm 2, 1)$

9. saddle point at $(1, -2)$ **11.** relative minimum at $(2, -1)$

13. relative minima at $(-1, -1)$ and $(1, 1)$ **15.** saddle point at $(0, 0)$

17. no critical points **19.** relative maximum at $(-1, 0)$

21. saddle point at $(0, 0)$;
relative minima at $(1, 1)$
and $(-1, -1)$

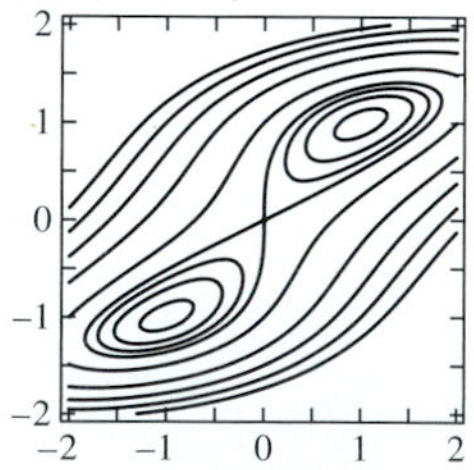

23. **(b)** relative minimum at $(0, 0)$

27. absolute maximum 0,
absolute minimum -12

29. absolute maximum 3,
absolute minimum -1

31. absolute maximum $\frac{33}{4}$,
absolute minimum $-\frac{1}{4}$

33. 16, 16, 16

35. maximum at $(1, 2, 2)$

37. $2a/\sqrt{3}, 2a/\sqrt{3}, 2a/\sqrt{3}$

39. length and width 2 ft, height 4 ft

41. **(a)** $x = 0$: minimum -3, maximum 0;
$x = 1$: minimum 3, maximum $13/3$;
$y = 0$: minimum 0, maximum 4;
$y = 1$: minimum -3, maximum 3
(b) $y = x$: minimum 0, maximum 3;
$y = 1-x$: maximum 4, minimum -3
(c) minimum -3, maximum $13/3$

43. length and width $\sqrt[3]{2V}$, height $\sqrt[3]{2V}/2$ **47.** $y = \frac{3}{4}x + \frac{19}{12}$

49. $y = 0.5x + 0.8$

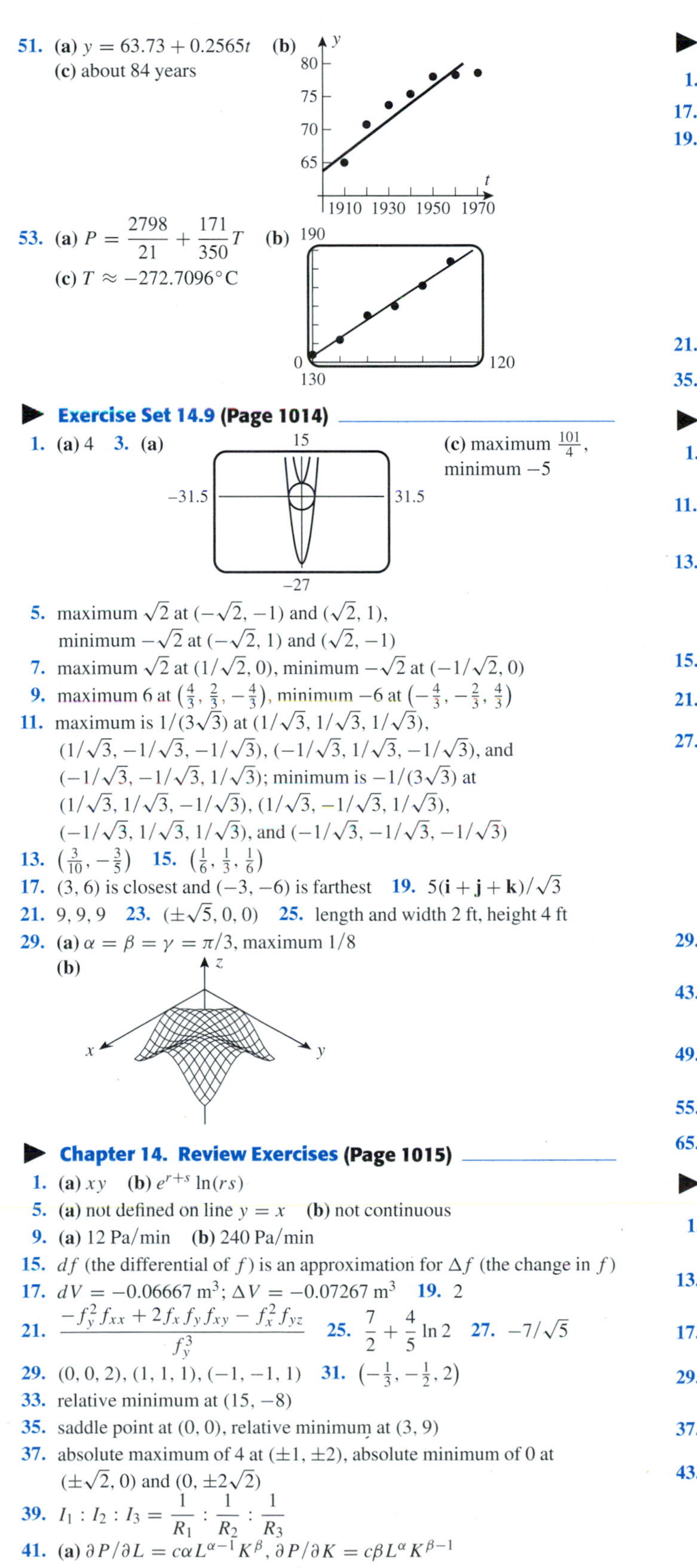

51. **(a)** $y = 63.73 + 0.2565t$ **(b)**
(c) about 84 years

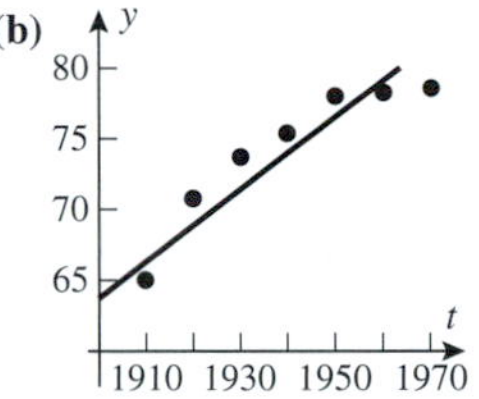

53. **(a)** $P = \dfrac{2798}{21} + \dfrac{171}{350}T$ **(b)**
(c) $T \approx -272.7096°\text{C}$

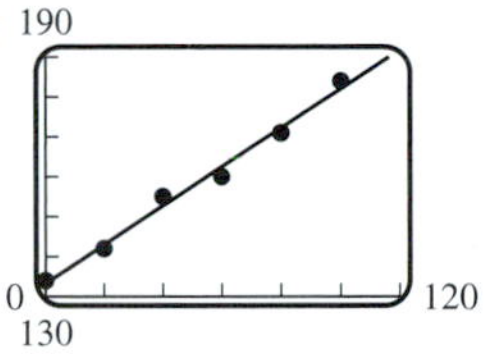

▶ **Exercise Set 14.9** (Page 1014)

1. **(a)** 4 **3.** **(a)**

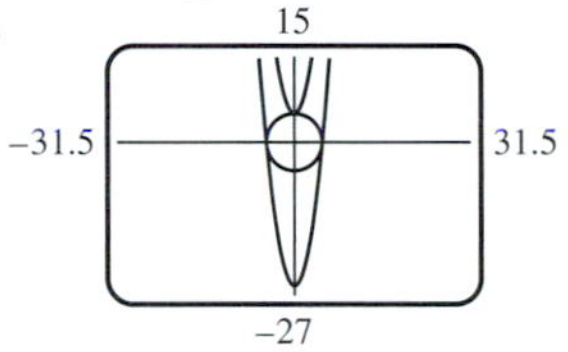

(c) maximum $\frac{101}{4}$, minimum -5

5. maximum $\sqrt{2}$ at $(-\sqrt{2}, -1)$ and $(\sqrt{2}, 1)$, minimum $-\sqrt{2}$ at $(-\sqrt{2}, 1)$ and $(\sqrt{2}, -1)$
7. maximum $\sqrt{2}$ at $(1/\sqrt{2}, 0)$, minimum $-\sqrt{2}$ at $(-1/\sqrt{2}, 0)$
9. maximum 6 at $\left(\frac{4}{3}, \frac{2}{3}, -\frac{4}{3}\right)$, minimum -6 at $\left(-\frac{4}{3}, -\frac{2}{3}, \frac{4}{3}\right)$
11. maximum is $1/(3\sqrt{3})$ at $(1/\sqrt{3}, 1/\sqrt{3}, 1/\sqrt{3})$, $(1/\sqrt{3}, -1/\sqrt{3}, -1/\sqrt{3})$, $(-1/\sqrt{3}, 1/\sqrt{3}, -1/\sqrt{3})$, and $(-1/\sqrt{3}, -1/\sqrt{3}, 1/\sqrt{3})$; minimum is $-1/(3\sqrt{3})$ at $(1/\sqrt{3}, 1/\sqrt{3}, -1/\sqrt{3})$, $(1/\sqrt{3}, -1/\sqrt{3}, 1/\sqrt{3})$, $(-1/\sqrt{3}, 1/\sqrt{3}, 1/\sqrt{3})$, and $(-1/\sqrt{3}, -1/\sqrt{3}, -1/\sqrt{3})$
13. $\left(\frac{3}{10}, -\frac{3}{5}\right)$ **15.** $\left(\frac{1}{6}, \frac{1}{3}, \frac{1}{6}\right)$
17. (3, 6) is closest and (−3, −6) is farthest **19.** $5(\mathbf{i} + \mathbf{j} + \mathbf{k})/\sqrt{3}$
21. 9, 9, 9 **23.** $(\pm\sqrt{5}, 0, 0)$ **25.** length and width 2 ft, height 4 ft
29. **(a)** $\alpha = \beta = \gamma = \pi/3$, maximum 1/8
(b)

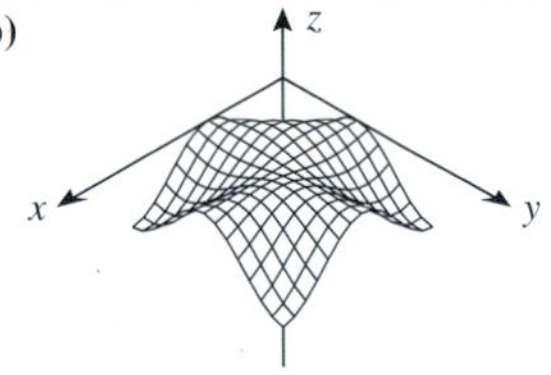

▶ **Chapter 14. Review Exercises** (Page 1015)

1. **(a)** xy **(b)** $e^{r+s}\ln(rs)$
5. **(a)** not defined on line $y = x$ **(b)** not continuous
9. **(a)** 12 Pa/min **(b)** 240 Pa/min
15. df (the differential of f) is an approximation for Δf (the change in f)
17. $dV = -0.06667\ \text{m}^3$; $\Delta V = -0.07267\ \text{m}^3$ **19.** 2
21. $\dfrac{-f_y^2 f_{xx} + 2f_x f_y f_{xy} - f_x^2 f_{yz}}{f_y^3}$ **25.** $\dfrac{7}{2} + \dfrac{4}{5}\ln 2$ **27.** $-7/\sqrt{5}$
29. (0, 0, 2), (1, 1, 1), (−1, −1, 1) **31.** $\left(-\frac{1}{3}, -\frac{1}{2}, 2\right)$
33. relative minimum at (15, −8)
35. saddle point at (0, 0), relative minimum at (3, 9)
37. absolute maximum of 4 at (±1, ±2), absolute minimum of 0 at $(\pm\sqrt{2}, 0)$ and $(0, \pm 2\sqrt{2})$
39. $I_1 : I_2 : I_3 = \dfrac{1}{R_1} : \dfrac{1}{R_2} : \dfrac{1}{R_3}$
41. **(a)** $\partial P/\partial L = c\alpha L^{\alpha-1}K^{\beta}$, $\partial P/\partial K = c\beta L^{\alpha}K^{\beta-1}$

▶ **Exercise Set 15.1** (Page 1024)

1. 7 **3.** 2 **5.** 2 **7.** 3 **9.** $1 - \ln 2$ **11.** $\dfrac{1 - \ln 2}{2}$ **13.** 0 **15.** $\frac{1}{3}$
17. **(a)** 37/4 **(b)** exact value = 28/3; differ by 1/12
19. **(a)** **(b)**

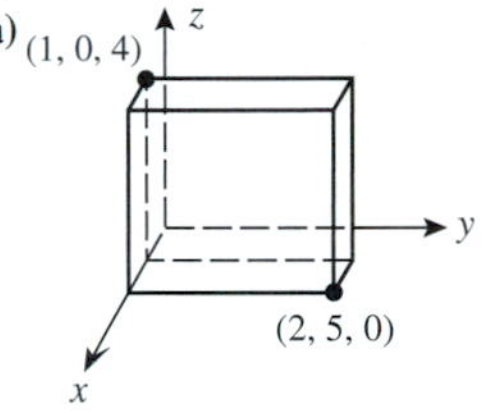

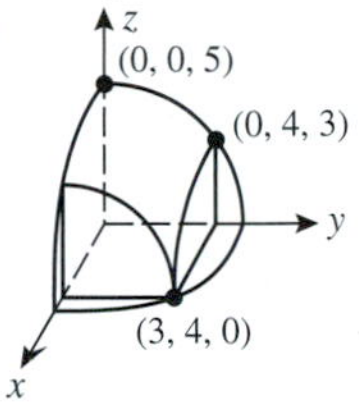

21. 19 **23.** 8 **25.** $\dfrac{1}{3\pi}$ **27.** $1 - \dfrac{2}{\pi}$ **29.** $\frac{14}{3}$ °C **31.** 1.381737122
35. first integral equals $\frac{1}{2}$, second equals $-\frac{1}{2}$; no

▶ **Exercise Set 15.2** (Page 1033)

1. $\frac{1}{40}$ **3.** 9 **5.** $\dfrac{\pi}{2}$ **7.** 1 **9.** $\frac{1}{12}$
11. **(a)** $\displaystyle\int_0^2\int_0^{x^2} f(x, y)\,dy\,dx$ **(b)** $\displaystyle\int_0^4\int_{\sqrt{y}}^2 f(x, y)\,dx\,dy$
13. **(a)** $\displaystyle\int_1^2\int_{-2x+5}^3 f(x, y)\,dy\,dx + \int_2^4\int_1^3 f(x, y)\,dy\,dx + \int_4^5\int_{2x-7}^3 f(x, y)\,dy\,dx$ **(b)** $\displaystyle\int_1^3\int_{(5-y)/2}^{(y+7)/2} f(x, y)\,dx\,dy$
15. **(a)** 16/3 **(b)** 38 **17.** 576 **19.** 0
21. $\dfrac{\sqrt{17} - 1}{2}$ **23.** $\frac{50}{3}$ **25.** $-\frac{7}{60}$
27. **(a)**

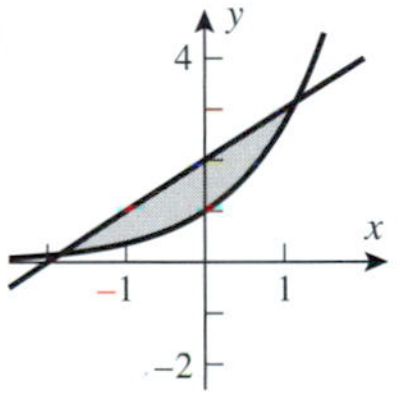

(b) (−1.8414, 0.1586), (1.1462, 3.1462)
(c) −0.4044
(d) −0.4044

29. $\sqrt{2} - 1$ **31.** 32 **33.** 12 **35.** 27π **37.** 170 **39.** $\dfrac{27\pi}{2}$ **41.** $\frac{2000}{3}$
43. $\dfrac{\pi}{2}$ **45.** $\displaystyle\int_0^{\sqrt{2}}\int_{y^2}^2 f(x, y)\,dx\,dy$ **47.** $\displaystyle\int_1^{e^2}\int_{\ln x}^2 f(x, y)\,dy\,dx$
49. $\displaystyle\int_0^{\pi/2}\int_0^{\sin x} f(x, y)\,dy\,dx$ **51.** $\dfrac{1 - e^{-16}}{8}$ **53.** $\dfrac{e^8 - 1}{3}$
55. $\dfrac{1 - \cos 8}{3}$ **57.** **(a)** 0 **(b)** tan 1 **59.** 0 **61.** $\dfrac{\pi}{2} - \ln 2$ **63.** $\frac{2}{3}$ °C
65. 0.676089

▶ **Exercise Set 15.3** (Page 1041)

1. $\frac{1}{6}$ **3.** $\frac{2}{9}a^3$ **5.** 0 **7.** $\dfrac{3\pi}{2}$ **9.** $\dfrac{\pi}{16}$ **11.** $\displaystyle\int_{\pi/6}^{5\pi/6}\int_2^{4\sin\theta} f(r, \theta)r\,dr\,d\theta$
13. $8\displaystyle\int_0^{\pi/2}\int_1^3 r\sqrt{9 - r^2}\,dr\,d\theta$ **15.** $2\displaystyle\int_0^{\pi/2}\int_0^{\cos\theta}(1 - r^2)r\,dr\,d\theta$
17. $\dfrac{64\sqrt{2}}{3}\pi$ **19.** $\dfrac{5\pi}{32}$ **21.** $\dfrac{27\pi}{16}$ **23.** $(1 - e^{-1})\pi$ **25.** $\dfrac{\pi}{8}\ln 5$ **27.** $\dfrac{\pi}{8}$
29. $\frac{16}{9}$ **31.** $\dfrac{\pi}{2}\left(1 - \dfrac{1}{\sqrt{1 + a^2}}\right)$ **33.** $\dfrac{\pi}{4}(\sqrt{5} - 1)$ **35.** $\pi a^2 h$
37. $\dfrac{(3\pi - 4)a^2 c}{9}$ **39.** $\dfrac{4\pi}{3} + 2\sqrt{3} - 2$ **41.** **(b)** $\dfrac{\pi}{4}$
43. **(a)** 1.173108605 **(b)** 1.173108605 **45.** $\dfrac{1}{5} + \dfrac{\pi}{2}$

Exercise Set 15.4 (Page 1053)

1. (a) (b) (c)

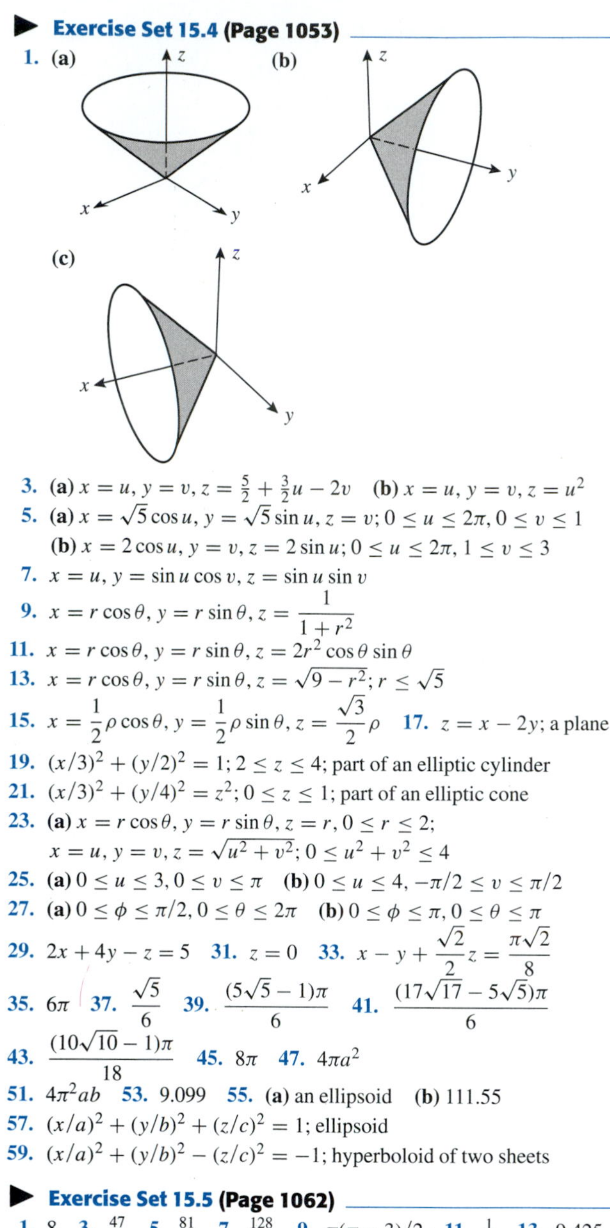

3. (a) $x = u, y = v, z = \frac{5}{2} + \frac{3}{2}u - 2v$ (b) $x = u, y = v, z = u^2$
5. (a) $x = \sqrt{5}\cos u, y = \sqrt{5}\sin u, z = v; 0 \le u \le 2\pi, 0 \le v \le 1$
(b) $x = 2\cos u, y = v, z = 2\sin u; 0 \le u \le 2\pi, 1 \le v \le 3$
7. $x = u, y = \sin u \cos v, z = \sin u \sin v$
9. $x = r\cos\theta, y = r\sin\theta, z = \dfrac{1}{1+r^2}$
11. $x = r\cos\theta, y = r\sin\theta, z = 2r^2\cos\theta\sin\theta$
13. $x = r\cos\theta, y = r\sin\theta, z = \sqrt{9-r^2}; r \le \sqrt{5}$
15. $x = \dfrac{1}{2}\rho\cos\theta, y = \dfrac{1}{2}\rho\sin\theta, z = \dfrac{\sqrt{3}}{2}\rho$ 17. $z = x - 2y$; a plane
19. $(x/3)^2 + (y/2)^2 = 1; 2 \le z \le 4$; part of an elliptic cylinder
21. $(x/3)^2 + (y/4)^2 = z^2; 0 \le z \le 1$; part of an elliptic cone
23. (a) $x = r\cos\theta, y = r\sin\theta, z = r, 0 \le r \le 2$;
$x = u, y = v, z = \sqrt{u^2+v^2}; 0 \le u^2 + v^2 \le 4$
25. (a) $0 \le u \le 3, 0 \le v \le \pi$ (b) $0 \le u \le 4, -\pi/2 \le v \le \pi/2$
27. (a) $0 \le \phi \le \pi/2, 0 \le \theta \le 2\pi$ (b) $0 \le \phi \le \pi, 0 \le \theta \le \pi$
29. $2x + 4y - z = 5$ 31. $z = 0$ 33. $x - y + \dfrac{\sqrt{2}}{2}z = \dfrac{\pi\sqrt{2}}{8}$
35. 6π 37. $\dfrac{\sqrt{5}}{6}$ 39. $\dfrac{(5\sqrt{5}-1)\pi}{6}$ 41. $\dfrac{(17\sqrt{17}-5\sqrt{5})\pi}{6}$
43. $\dfrac{(10\sqrt{10}-1)\pi}{18}$ 45. 8π 47. $4\pi a^2$
51. $4\pi^2 ab$ 53. 9.099 55. (a) an ellipsoid (b) 111.55
57. $(x/a)^2 + (y/b)^2 + (z/c)^2 = 1$; ellipsoid
59. $(x/a)^2 + (y/b)^2 - (z/c)^2 = -1$; hyperboloid of two sheets

Exercise Set 15.5 (Page 1062)

1. 8 3. $\frac{47}{3}$ 5. $\frac{81}{5}$ 7. $\frac{128}{15}$ 9. $\pi(\pi-3)/2$ 11. $\frac{1}{6}$ 13. 9.425
15. 4 17. $\frac{256}{15}$

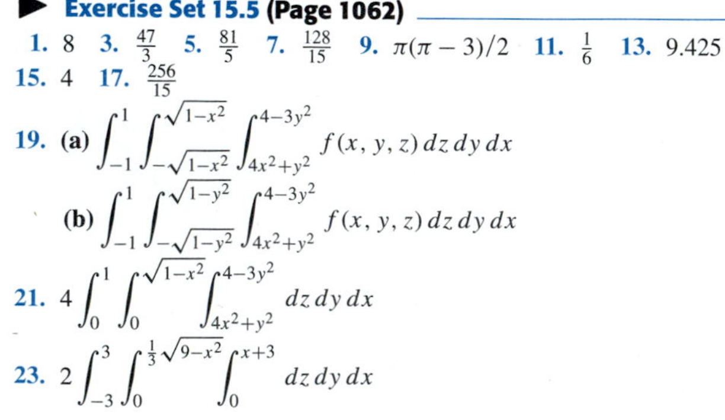

19. (a) $\displaystyle\int_{-1}^{1}\int_{-\sqrt{1-x^2}}^{\sqrt{1-x^2}}\int_{4x^2+y^2}^{4-3y^2} f(x,y,z)\,dz\,dy\,dx$
(b) $\displaystyle\int_{-1}^{1}\int_{-\sqrt{1-y^2}}^{\sqrt{1-y^2}}\int_{4x^2+y^2}^{4-3y^2} f(x,y,z)\,dz\,dy\,dx$
21. $\displaystyle 4\int_{0}^{1}\int_{0}^{\sqrt{1-x^2}}\int_{4x^2+y^2}^{4-3y^2} dz\,dy\,dx$
23. $\displaystyle 2\int_{-3}^{3}\int_{0}^{\frac{1}{3}\sqrt{9-x^2}}\int_{0}^{x+3} dz\,dy\,dx$
25. (a) (b) (c)

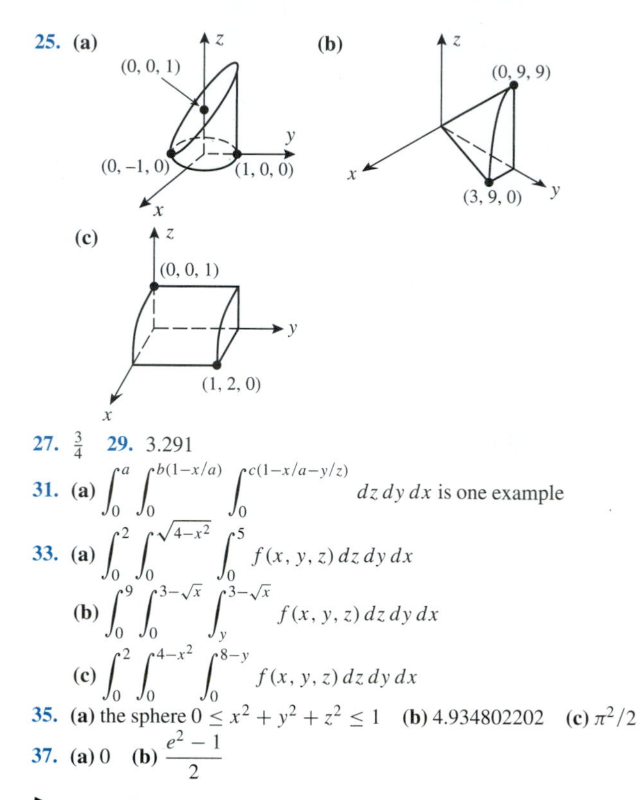

27. $\frac{3}{4}$ 29. 3.291
31. (a) $\displaystyle\int_{0}^{a}\int_{0}^{b(1-x/a)}\int_{0}^{c(1-x/a-y/z)} dz\,dy\,dx$ is one example
33. (a) $\displaystyle\int_{0}^{2}\int_{0}^{\sqrt{4-x^2}}\int_{0}^{5} f(x,y,z)\,dz\,dy\,dx$
(b) $\displaystyle\int_{0}^{9}\int_{0}^{3-\sqrt{x}}\int_{y}^{3-\sqrt{x}} f(x,y,z)\,dz\,dy\,dx$
(c) $\displaystyle\int_{0}^{2}\int_{0}^{4-x^2}\int_{0}^{8-y} f(x,y,z)\,dz\,dy\,dx$
35. (a) the sphere $0 \le x^2 + y^2 + z^2 \le 1$ (b) 4.934802202 (c) $\pi^2/2$
37. (a) 0 (b) $\dfrac{e^2-1}{2}$

Exercise Set 15.6 (Page 1073)

1. (a) positive: m_2 is at the fulcrum, so it can be ignored; masses m_1 and m_3 are equidistant from position 5, but $m_1 < m_3$, so the beam will rotate clockwise. (b) The fulcrum should be placed $\frac{50}{7}$ units to the right of m_1
3. $\left(\frac{1}{2}, \frac{1}{2}\right)$ 5. $\left(\frac{2}{3}, \frac{1}{3}\right)$ 7. $\left(\frac{5}{14}, \frac{38}{35}\right)$ 9. $\left(0, \dfrac{4(b^3-a^3)}{3\pi(b^2-a^2)}\right)$ 11. $\left(\frac{1}{2}, \frac{1}{2}\right)$
13. $M = \frac{13}{20}$, center of gravity $\left(\frac{190}{273}, \frac{6}{13}\right)$
15. $M = a^4/8$, center of gravity $(8a/15, 8a/15)$
17. $\left(\frac{1}{2}, \frac{1}{2}, \frac{1}{2}\right)$ 19. $\left(\frac{1}{4}, \frac{1}{4}, \frac{1}{4}\right)$ 21. $\left(\frac{1}{2}, 0, \frac{3}{5}\right)$ 23. $(3a/8, 3a/8, 3a/8)$
25. $M = a^4/2$, center of gravity $(a/3, a/2, a/2)$
27. $M = \frac{1}{6}$, center of gravity $\left(0, \frac{16}{35}, \frac{1}{2}\right)$ 29. (a) $\left(\frac{5}{8}, \frac{5}{8}\right)$ (b) $\left(\frac{2}{3}, \frac{1}{2}\right)$
31. (1.177406, 0.353554, 0.231557) 35. $\left(\dfrac{128}{105\pi}, \dfrac{128}{105\pi}\right)$
39. $2\pi^2 abk$ 41. $(a/3, b/3)$

Exercise Set 15.7 (Page 1084)

1. $\dfrac{\pi}{4}$ 3. $\dfrac{\pi}{16}$
5. the region is bounded by the xy-plane and the upper half of a sphere of radius 1 centered at the origin; $f(r,\theta,z) = z$
7. the region is the portion of the first octant inside a sphere of radius 1 centered at the origin; $f(\rho,\theta,\phi) = \rho\cos\phi$
9. $\dfrac{81\pi}{2}$ 11. $\frac{152}{3}\pi + \frac{80}{3}\pi\sqrt{5}$ 13. $\dfrac{64\pi}{3}$ 15. $\dfrac{11\pi a^3}{3}$ 17. $\dfrac{\pi a^6}{48}$
19. $\dfrac{32(2\sqrt{2}-1)\pi}{15}$
21. (a) $\frac{5}{2}(-8 + 3\ln 3)\ln(\sqrt{5}-2)$ (b) $f(x,y,z) = \dfrac{y^3}{x^3\sqrt{1+z^2}}$;
G is the cylindrical wedge $1 \le r \le 4, \dfrac{\pi}{6} \le \theta \le \dfrac{\pi}{3}, -2 \le z \le 2$
23. $\dfrac{4\pi a^3}{3}$ 25. $\dfrac{27\pi}{4}$ 27. $\pi k a^4$

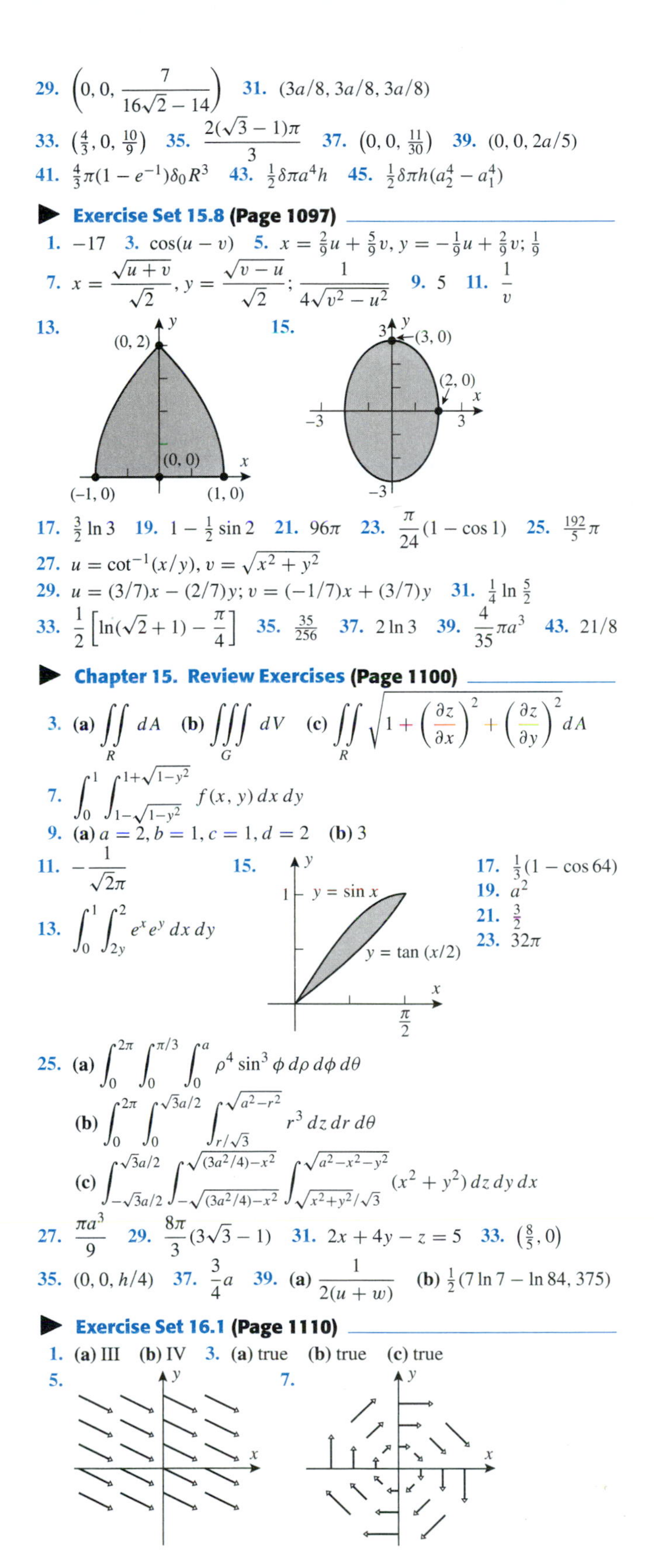

29. $\left(0, 0, \dfrac{7}{16\sqrt{2}-14}\right)$ **31.** $(3a/8, 3a/8, 3a/8)$

33. $\left(\frac{4}{3}, 0, \frac{10}{9}\right)$ **35.** $\dfrac{2(\sqrt{3}-1)\pi}{3}$ **37.** $\left(0, 0, \frac{11}{30}\right)$ **39.** $(0, 0, 2a/5)$

41. $\frac{4}{3}\pi(1-e^{-1})\delta_0 R^3$ **43.** $\frac{1}{2}\delta\pi a^4 h$ **45.** $\frac{1}{2}\delta\pi h(a_2^4 - a_1^4)$

Exercise Set 15.8 (Page 1097)

1. -17 **3.** $\cos(u-v)$ **5.** $x = \frac{2}{9}u + \frac{5}{9}v$, $y = -\frac{1}{9}u + \frac{2}{9}v$; $\frac{1}{9}$

7. $x = \dfrac{\sqrt{u+v}}{\sqrt{2}}$, $y = \dfrac{\sqrt{v-u}}{\sqrt{2}}$; $\dfrac{1}{4\sqrt{v^2-u^2}}$ **9.** 5 **11.** $\dfrac{1}{v}$

13. **15.**

17. $\frac{3}{2}\ln 3$ **19.** $1 - \frac{1}{2}\sin 2$ **21.** 96π **23.** $\dfrac{\pi}{24}(1-\cos 1)$ **25.** $\frac{192}{5}\pi$

27. $u = \cot^{-1}(x/y)$, $v = \sqrt{x^2+y^2}$

29. $u = (3/7)x - (2/7)y$; $v = (-1/7)x + (3/7)y$ **31.** $\frac{1}{4}\ln\frac{5}{2}$

33. $\dfrac{1}{2}\left[\ln(\sqrt{2}+1) - \dfrac{\pi}{4}\right]$ **35.** $\frac{35}{256}$ **37.** $2\ln 3$ **39.** $\dfrac{4}{35}\pi a^3$ **43.** $21/8$

Chapter 15. Review Exercises (Page 1100)

3. **(a)** $\displaystyle\iint_R dA$ **(b)** $\displaystyle\iiint_G dV$ **(c)** $\displaystyle\iint_R \sqrt{1+\left(\frac{\partial z}{\partial x}\right)^2+\left(\frac{\partial z}{\partial y}\right)^2}\, dA$

7. $\displaystyle\int_0^1 \int_{1-\sqrt{1-y^2}}^{1+\sqrt{1-y^2}} f(x,y)\,dx\,dy$

9. **(a)** $a = 2, b = 1, c = 1, d = 2$ **(b)** 3

11. $-\dfrac{1}{\sqrt{2\pi}}$ **15.**

17. $\frac{1}{3}(1-\cos 64)$ **19.** a^2 **21.** $\frac{3}{2}$ **23.** 32π

13. $\displaystyle\int_0^1 \int_{2y}^{2} e^x e^y\,dx\,dy$

25. **(a)** $\displaystyle\int_0^{2\pi}\int_0^{\pi/3}\int_0^a \rho^4 \sin^3\phi\, d\rho\, d\phi\, d\theta$

(b) $\displaystyle\int_0^{2\pi}\int_0^{\sqrt{3}a/2}\int_{r/\sqrt{3}}^{\sqrt{a^2-r^2}} r^3\, dz\, dr\, d\theta$

(c) $\displaystyle\int_{-\sqrt{3}a/2}^{\sqrt{3}a/2}\int_{-\sqrt{(3a^2/4)-x^2}}^{\sqrt{(3a^2/4)-x^2}}\int_{\sqrt{x^2+y^2}/\sqrt{3}}^{\sqrt{a^2-x^2-y^2}} (x^2+y^2)\, dz\, dy\, dx$

27. $\dfrac{\pi a^3}{9}$ **29.** $\dfrac{8\pi}{3}(3\sqrt{3}-1)$ **31.** $2x + 4y - z = 5$ **33.** $\left(\frac{8}{5}, 0\right)$

35. $(0, 0, h/4)$ **37.** $\dfrac{3}{4}a$ **39.** **(a)** $\dfrac{1}{2(u+w)}$ **(b)** $\frac{1}{2}(7\ln 7 - \ln 84, 375)$

Exercise Set 16.1 (Page 1110)

1. **(a)** III **(b)** IV **3.** **(a)** true **(b)** true **(c)** true

5. **7.**

9. **11.** **(a)** all x, y **(b)** all x, y

13. $\operatorname{div}\mathbf{F} = 2x + y$, $\operatorname{curl}\mathbf{F} = z\mathbf{i}$

15. $\operatorname{div}\mathbf{F} = 0$, $\operatorname{curl}\mathbf{F} = (40x^2z^4 - 12xy^3)\mathbf{i} + (14y^3z + 3y^4)\mathbf{j} - (16xz^5 + 21y^2z^2)\mathbf{k}$

17. $\operatorname{div}\mathbf{F} = \dfrac{2}{\sqrt{x^2+y^2+z^2}}$, $\operatorname{curl}\mathbf{F} = 0$ **19.** $4x$ **21.** 0

23. $(1+y)\mathbf{i} + x\mathbf{j}$

35. $\nabla\cdot(k\mathbf{F}) = k\nabla\cdot\mathbf{F}$, $\nabla\cdot(\mathbf{F}+\mathbf{G}) = \nabla\cdot\mathbf{F} + \nabla\cdot\mathbf{G}$, $\nabla\cdot(\phi\mathbf{F}) = \phi\nabla\cdot\mathbf{F} + \nabla\phi\cdot\mathbf{F}$, $\nabla\cdot(\nabla\times\mathbf{F}) = 0$ **43.** **(b)** $x^2 + y^2 = K$

45. $\dfrac{dy}{dx} = \dfrac{1}{x}$, $y = \ln x + K$

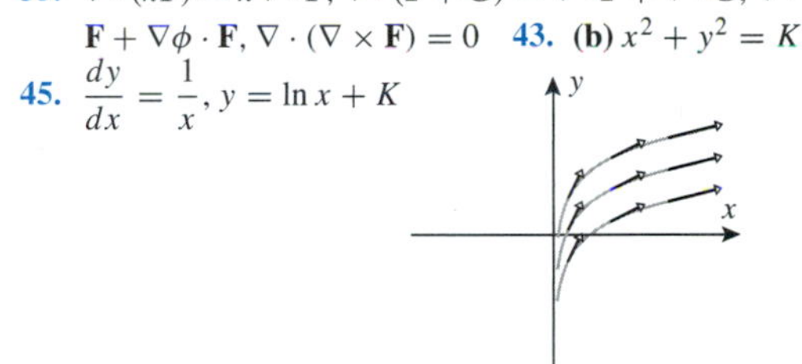

Exercise Set 16.2 (Page 1126)

1. **(a)** 1 **(b)** 0 **3.** 16

7. **(a)** $-\frac{11}{108}\sqrt{10} - \frac{1}{36}\ln(\sqrt{10}-3) - \frac{4}{27}$ **(b)** 0 **(c)** $-\frac{1}{2}$

9. **(a)** 3 **(b)** 3 **(c)** 3 **(d)** 3 **11.** 2 **13.** $\frac{13}{20}$ **15.** $1-\pi$ **17.** 3

19. $-1-(\pi/4)$ **21.** $1-e^3$

23. **(a)** $63\sqrt{17}/64 + \dfrac{1}{4}\ln(4+\sqrt{17}) - \dfrac{1}{8}\ln\dfrac{\sqrt{17}+1}{\sqrt{17}-1} - \dfrac{1}{4}\ln(\sqrt{2}+1) + \dfrac{1}{8}\ln\dfrac{\sqrt{2}+1}{\sqrt{2}-1}$ **(b)** $1/2 - \pi/4$

25. **(a)** -1 **(b)** -2 **27.** $\frac{5}{2}$ **29.** 0 **31.** $1-e^{-1}$ **33.** $6\sqrt{3}$

35. $5k\tan^{-1}3$ **37.** $\frac{3}{5}$ **39.** $\frac{27}{28}$ **41.** $\frac{3}{4}$ **43.** $\dfrac{17\sqrt{17}-1}{4}$

45. **(b)** $S = \displaystyle\int_C z(t)\,dt$ **(c)** 4π **47.** $\lambda = -12$

Exercise Set 16.3 (Page 1137)

1. conservative, $\phi = \dfrac{x^2}{2} + \dfrac{y^2}{2} + K$ **3.** not conservative

5. conservative, $\phi = x\cos y + y\sin x + K$

7. **(b)** 13 **9.** -6 **11.** $9e^2$ **13.** 32 **15.** $W = -\frac{1}{2}$

17. $W = 1 - e^{-1}$ **19.** $\ln 2 - 1$ **21.** ≈ -0.307 **23.** no

29. $h(x) = Ce^x$

31. **(a)** $W = -\dfrac{1}{\sqrt{14}} + \dfrac{1}{\sqrt{16}}$ **(b)** $W = -\dfrac{1}{\sqrt{14}} + \dfrac{1}{\sqrt{6}}$ **(c)** $W = 0$

Exercise Set 16.4 (Page 1145)

1. 0 **3.** 0 **5.** 0 **7.** 8π **9.** -4 **11.** -1 **13.** 0

15. **(a)** ≈ -3.550999378 **(b)** ≈ -0.269616482 **17.** $\frac{3}{8}a^2\pi$ **19.** $\frac{1}{2}abt_0$

23. Formula (1) of Section 7.1 **25.** $\frac{250}{3}$ **27.** $-3\pi a^2$ **29.** $\left(\frac{8}{15}, \frac{8}{21}\right)$

31. $\left(0, \dfrac{4a}{3\pi}\right)$ **33.** the circle $x^2 + y^2 = 1$ **35.** 69

Exercise Set 16.5 (Page 1153)

1. $\dfrac{15}{2}\pi\sqrt{2}$ **3.** $\dfrac{\pi}{4}$ **5.** $-\dfrac{\sqrt{2}}{2}$ **7.** 9

9. **(b)** $2\pi\left[1-\sqrt{1-r^2}+\dfrac{r^2}{2}\right]\to 3\pi$ as $r\to 1^-$
(c) $\mathbf{r}(\phi,\theta)=\sin\phi\cos\theta\mathbf{i}+\sin\phi\sin\theta\mathbf{j}+\cos\phi\mathbf{k}$, $0\le\theta\le 2\pi, 0\le\phi\le\pi/2$;
$$\iint (1+z)\,dS=\int_0^{2\pi}\int_0^{\pi/2}(1+\cos\phi)\sin\phi\,d\phi\,d\theta=3\pi$$
13. **(c)** $4\pi/3$
15. **(a)** $\dfrac{\sqrt{29}}{16}\displaystyle\int_0^6\int_0^{(12-2x)/3}xy(12-2x-3y)\,dy\,dx$
(b) $\dfrac{\sqrt{29}}{4}\displaystyle\int_0^3\int_0^{(12-4z)/3}yz(12-3y-4z)\,dy\,dz$
(c) $\dfrac{\sqrt{29}}{9}\displaystyle\int_0^3\int_0^{6-2z}xz(12-2x-4z)\,dx\,dz$
17. $\dfrac{18\sqrt{29}}{5}$
19. $\displaystyle\int_0^4\int_1^2 y^3z\sqrt{4y^2+1}\,dy\,dz;\ \frac{1}{2}\int_0^4\int_1^4 xz\sqrt{1+4x}\,dx\,dz$
21. $\dfrac{391\sqrt{17}}{15}-\dfrac{5\sqrt{5}}{3}$ **23.** $\frac{4}{3}\pi\delta_0$ **25.** $\frac{1}{4}(37\sqrt{37}-1)$ **27.** $M=\delta_0 S$
29. $(0,0,149/65)$ **31.** $\dfrac{93}{\sqrt{10}}$ **33.** $\dfrac{\pi}{4}$ **35.** 57.895751

▶ Exercise Set 16.6 (Page 1162)

1. **(a)** zero **(b)** zero **(c)** positive **(d)** negative **(e)** zero **(f)** zero
3. **(a)** positive **(b)** zero **(c)** positive **(d)** zero **(e)** positive **(f)** zero
5. **(a)** $n=-\cos v\mathbf{i}-\sin v\mathbf{j}$ **(b)** inward **7.** 2π **9.** $\dfrac{14\pi}{3}$ **11.** 0
13. 18π **15.** $\frac{4}{9}$ **17.** **(a)** 8 **(b)** 24 **(c)** 0 **19.** 3π
21. **(a)** $0\ \text{m}^3/\text{s}$ **(b)** $0\ \text{kg/s}$ **23.** **(b)** $32/3$
25. **(a)** $4\pi a^{k+3}$ **(b)** $k=-3$

▶ Exercise Set 16.7 (Page 1172)

1. 3 **3.** $\dfrac{4\pi}{3}$ **5.** 12 **7.** $3\pi a^2$ **9.** 180π **11.** $\dfrac{192\pi}{5}$ **13.** $\dfrac{\pi}{2}$
15. $\dfrac{4608}{35}$ **17.** 135π **29.** no sources or sinks
31. sources at all points except the origin, no sinks **33.** $\dfrac{7\pi}{4}$

▶ Exercise Set 16.8 (Page 1179)

1. $\frac{3}{2}$ **3.** 0 **5.** 2π **7.** 16π **9.** 0 **11.** πa^2
13. **(a)** $\frac{3}{2}$ **(b)** -1 **(c)** $-\dfrac{1}{\sqrt{2}}\mathbf{j}-\dfrac{1}{\sqrt{2}}\mathbf{k}$ **19.** $-\dfrac{5\pi}{4}$

▶ Chapter 16. Review Exercises (Page 1181)

3. $\dfrac{1-x}{\sqrt{(1-x)^2+(2-y)^2}}\mathbf{i}+\dfrac{2-y}{\sqrt{(1-x)^2+(2-y)^2}}\mathbf{j}$ **5.** $\mathbf{i}+\mathbf{j}+\mathbf{k}$
7. **(a)** $\displaystyle\int_a^b\left[f(x(t),y(t))\frac{dx}{dt}+g(x(t),y(t))\frac{dy}{dt}\right]dt$
(b) $\displaystyle\int_a^b f(x(t),y(t))\sqrt{x'(t)^2+y'(t)^2}\,dt$
11. 0 **13.** $-7/2$ **17.** **(a)** $h(x)=Cx^{-3/2}$ **(b)** $g(y)=C/y^3$
23. $\displaystyle\iint_R f(x(u,v),y(u,v),z(u,v))\|r_u\times r_v\|\,du\,dv$ **25.** yes **27.** 2π
31. -8π **35.** **(a)** conservative **(b)** not conservative

PHOTO CREDITS

Chapter 12
Page 786: Craig Aurness/Corbis Images. Page 820: World Perspectives/Stone/Getty Images.

Chapter 13
Page 859: Courtesy Cedar Point. Page 860: Ken Eward/Biografx/Photo Researchers.

Chapter 14
Page 924: Stone/Getty Images. Page 954: Leverett Bradley/Stone/Getty Images.

Chapter 15
Page 1018: Stone/Getty Images.

Chapter 16
Page 1102: Images and animation produced by Hal Pierce, Laboratory for Atmospheres, NASA Goddard Space and Flight Center.

INDEX

RATIONAL FUNCTIONS CONTAINING POWERS OF $a + bu$ IN THE DENOMINATOR

60. $\displaystyle\int \frac{u\,du}{a+bu} = \frac{1}{b^2}[bu - a\ln|a+bu|] + C$

61. $\displaystyle\int \frac{u^2\,du}{a+bu} = \frac{1}{b^3}\left[\frac{1}{2}(a+bu)^2 - 2a(a+bu) + a^2\ln|a+bu|\right] + C$

62. $\displaystyle\int \frac{u\,du}{(a+bu)^2} = \frac{1}{b^2}\left[\frac{a}{a+bu} + \ln|a+bu|\right] + C$

63. $\displaystyle\int \frac{u^2\,du}{(a+bu)^2} = \frac{1}{b^3}\left[bu - \frac{a^2}{a+bu} - 2a\ln|a+bu|\right] + C$

64. $\displaystyle\int \frac{u\,du}{(a+bu)^3} = \frac{1}{b^2}\left[\frac{a}{2(a+bu)^2} - \frac{1}{a+bu}\right] + C$

65. $\displaystyle\int \frac{du}{u(a+bu)} = \frac{1}{a}\ln\left|\frac{u}{a+bu}\right| + C$

66. $\displaystyle\int \frac{du}{u^2(a+bu)} = -\frac{1}{au} + \frac{b}{a^2}\ln\left|\frac{a+bu}{u}\right| + C$

67. $\displaystyle\int \frac{du}{u(a+bu)^2} = \frac{1}{a(a+bu)} + \frac{1}{a^2}\ln\left|\frac{u}{a+bu}\right| + C$

RATIONAL FUNCTIONS CONTAINING $a^2 \pm u^2$ IN THE DENOMINATOR ($a > 0$)

68. $\displaystyle\int \frac{du}{a^2+u^2} = \frac{1}{a}\tan^{-1}\frac{u}{a} + C$

69. $\displaystyle\int \frac{du}{a^2-u^2} = \frac{1}{2a}\ln\left|\frac{u+a}{u-a}\right| + C$

70. $\displaystyle\int \frac{du}{u^2-a^2} = \frac{1}{2a}\ln\left|\frac{u-a}{u+a}\right| + C$

71. $\displaystyle\int \frac{bu+c}{a^2+u^2}\,du = \frac{b}{2}\ln(a^2+u^2) + \frac{c}{a}\tan^{-1}\frac{u}{a} + C$

INTEGRALS OF $\sqrt{a^2+u^2}$, $\sqrt{a^2-u^2}$, $\sqrt{u^2-a^2}$ AND THEIR RECIPROCALS ($a > 0$)

72. $\displaystyle\int \sqrt{u^2+a^2}\,du = \frac{u}{2}\sqrt{u^2+a^2} + \frac{a^2}{2}\ln(u+\sqrt{u^2+a^2}) + C$

73. $\displaystyle\int \sqrt{u^2-a^2}\,du = \frac{u}{2}\sqrt{u^2-a^2} - \frac{a^2}{2}\ln|u+\sqrt{u^2-a^2}| + C$

74. $\displaystyle\int \sqrt{a^2-u^2}\,du = \frac{u}{2}\sqrt{a^2-u^2} + \frac{a^2}{2}\sin^{-1}\frac{u}{a} + C$

75. $\displaystyle\int \frac{du}{\sqrt{u^2+a^2}} = \ln(u+\sqrt{u^2+a^2}) + C$

76. $\displaystyle\int \frac{du}{\sqrt{u^2-a^2}} = \ln|u+\sqrt{u^2-a^2}| + C$

77. $\displaystyle\int \frac{du}{\sqrt{a^2-u^2}} = \sin^{-1}\frac{u}{a} + C$

POWERS OF u MULTIPLYING OR DIVIDING $\sqrt{a^2-u^2}$ OR ITS RECIPROCAL

78. $\displaystyle\int u^2\sqrt{a^2-u^2}\,du = \frac{u}{8}(2u^2-a^2)\sqrt{a^2-u^2} + \frac{a^4}{8}\sin^{-1}\frac{u}{a} + C$

79. $\displaystyle\int \frac{\sqrt{a^2-u^2}\,du}{u} = \sqrt{a^2-u^2} - a\ln\left|\frac{a+\sqrt{a^2-u^2}}{u}\right| + C$

80. $\displaystyle\int \frac{\sqrt{a^2-u^2}\,du}{u^2} = -\frac{\sqrt{a^2-u^2}}{u} - \sin^{-1}\frac{u}{a} + C$

81. $\displaystyle\int \frac{u^2\,du}{\sqrt{a^2-u^2}} = -\frac{u}{2}\sqrt{a^2-u^2} + \frac{a^2}{2}\sin^{-1}\frac{u}{a} + C$

82. $\displaystyle\int \frac{du}{u\sqrt{a^2-u^2}} = -\frac{1}{a}\ln\left|\frac{a+\sqrt{a^2-u^2}}{u}\right| + C$

83. $\displaystyle\int \frac{du}{u^2\sqrt{a^2-u^2}} = -\frac{\sqrt{a^2-u^2}}{a^2u} + C$

POWERS OF u MULTIPLYING OR DIVIDING $\sqrt{u^2 \pm a^2}$ OR THEIR RECIPROCALS

84. $\displaystyle\int u\sqrt{u^2+a^2}\,du = \frac{1}{3}(u^2+a^2)^{3/2} + C$

85. $\displaystyle\int u\sqrt{u^2-a^2}\,du = \frac{1}{3}(u^2-a^2)^{3/2} + C$

86. $\displaystyle\int \frac{du}{u\sqrt{u^2+a^2}} = -\frac{1}{a}\ln\left|\frac{a+\sqrt{u^2+a^2}}{u}\right| + C$

87. $\displaystyle\int \frac{du}{u\sqrt{u^2-a^2}} = \frac{1}{a}\sec^{-1}\left|\frac{u}{a}\right| + C$

88. $\displaystyle\int \frac{\sqrt{u^2-a^2}\,du}{u} = \sqrt{u^2-a^2} - a\sec^{-1}\left|\frac{u}{a}\right| + C$

89. $\displaystyle\int \frac{\sqrt{u^2+a^2}\,du}{u} = \sqrt{u^2+a^2} - a\ln\left|\frac{a+\sqrt{u^2+a^2}}{u}\right| + C$

90. $\displaystyle\int \frac{du}{u^2\sqrt{u^2\pm a^2}} = \mp\frac{\sqrt{u^2\pm a^2}}{a^2u} + C$

91. $\displaystyle\int u^2\sqrt{u^2+a^2}\,du = \frac{u}{8}(2u^2+a^2)\sqrt{u^2+a^2} - \frac{a^4}{8}\ln(u+\sqrt{u^2+a^2}) + C$

92. $\displaystyle\int u^2\sqrt{u^2-a^2}\,du = \frac{u}{8}(2u^2-a^2)\sqrt{u^2-a^2} - \frac{a^4}{8}\ln|u+\sqrt{u^2-a^2}| + C$

93. $\displaystyle\int \frac{\sqrt{u^2+a^2}}{u^2}\,du = -\frac{\sqrt{u^2+a^2}}{u} + \ln(u+\sqrt{u^2+a^2}) + C$

94. $\displaystyle\int \frac{\sqrt{u^2-a^2}}{u^2}\,du = -\frac{\sqrt{u^2-a^2}}{u} + \ln|u+\sqrt{u^2-a^2}| + C$

95. $\displaystyle\int \frac{u^2}{\sqrt{u^2+a^2}}\,du = \frac{u}{2}\sqrt{u^2+a^2} - \frac{a^2}{2}\ln(u+\sqrt{u^2+a^2}) + C$

96. $\displaystyle\int \frac{u^2}{\sqrt{u^2-a^2}}\,du = \frac{u}{2}\sqrt{u^2-a^2} + \frac{a^2}{2}\ln|u+\sqrt{u^2-a^2}| + C$

INTEGRALS CONTAINING $(a^2+u^2)^{3/2}$, $(a^2-u^2)^{3/2}$, $(u^2-a^2)^{3/2}$ ($a > 0$)

97. $\displaystyle\int \frac{du}{(a^2-u^2)^{3/2}} = \frac{u}{a^2\sqrt{a^2-u^2}} + C$

98. $\displaystyle\int \frac{du}{(u^2\pm a^2)^{3/2}} = \pm\frac{u}{a^2\sqrt{u^2\pm a^2}} + C$

99. $\displaystyle\int (a^2-u^2)^{3/2}\,du = -\frac{u}{8}(2u^2-5a^2)\sqrt{a^2-u^2} + \frac{3a^4}{8}\sin^{-1}\frac{u}{a} + C$

100. $\displaystyle\int (u^2+a^2)^{3/2}\,du = \frac{u}{8}(2u^2+5a^2)\sqrt{u^2+a^2} + \frac{3a^4}{8}\ln(u+\sqrt{u^2+a^2}) + C$

101. $\displaystyle\int (u^2-a^2)^{3/2}\,du = \frac{u}{8}(2u^2-5a^2)\sqrt{u^2-a^2} + \frac{3a^4}{8}\ln|u+\sqrt{u^2-a^2}| + C$